Handbook of Proteases in Cancer

The therapeutic features of protease-induced cancer are covered in detail in this book. Proteases make up to two percent of the human genome and play a critical role in the tumour microenvironment. The book delves into the applications of natural, synthetic, and non-coding RNAs in cancer therapy. It highlights how effective targeting of relevant proteases can help in cancer diagnosis and treatment. It covers the systems biology and bioinformatics approach in cancer drug development. This book:

- Provides insights on how proteases can be used as potential drug targets or as diagnostic and prognostic biomarkers.
- Covers the role of antiproteases and natural and synthetic compounds to ameliorate protease-induced cancer progression and metastasis.
- Discusses the systems biology and bioinformatics-based approaches in cancer drug development.
- Provides insights on the use of non-coding RNAs in protease-induced cancer therapy.
- Covers stem-cell therapy and nanotechnology-based therapies in protease-induced cancer.

The book is meant for researchers and professionals in cancer research, biochemistry, physiology, and pharmacology.

Dr. Sajal Chakraborti is associated with the Department of Biochemistry and Biophysics, University of Kalyani, Kalyani, West Bengal, India. He conducted postdoctoral research at The Johns Hopkins University School of Medicine, New York Medical College, The University of Utah, and the University of Florida. He has more than 35 years of teaching and independent research experience in Biochemistry and Cell Biology. He has published more than 150 original research papers, 25 review articles, and edited 17 books.

Handbook of Proteases in Cancer

Therapeutic Aspects

Edited by

Sajal Chakraborti

CRC Press
Taylor & Francis Group
Boca Raton London New York

CRC Press is an imprint of the
Taylor & Francis Group, an **informa** business

First edition published 2025
by CRC Press
2385 NW Executive Center Drive, Suite 320, Boca Raton FL 33431

and by CRC Press
4 Park Square, Milton Park, Abingdon, Oxon, OX14 4RN

CRC Press is an imprint of Taylor & Francis Group, LLC

ISBN: 978-1-032-48952-0 (hbk)
ISBN: 978-1-032-49627-6 (pbk)
ISBN: 978-1-003-39469-3 (ebk)

DOI: 10.1201/9781003394693

Typeset in Times
by KnowledgeWorks Global Ltd.

Dedication

This book is dedicated to Dr. Carlos López-Otín in recognition of his exceptional research on the role of proteolytic systems in pathological processes, particularly in cancer. Dr. Carlos López-Otín is a Professor of Biochemistry and Molecular Biology at the University of Oviedo, Spain. He was born in Sabiñánigo, Spain, in 1958. His academic training took place at the Universities of Zaragoza and Complutense of Madrid, where he was awarded PhD in 1984. His scientific work has been developed at the Centro Ramón y Cajal in Madrid, the Severo Ochoa Molecular Biology Center in Madrid, and the Universities of Lund, Sweden; New York and Harvard, USA. Currently, Dr. Carlos combines his teaching responsibilities at the University of Oviedo with his research lines on Cancer and Aging Biology as well as on the Functional Analysis of Genomes. Soon after joining the University of Oviedo, he focused on the study of the involvement of proteolytic systems in pathological processes, especially in cancer. His research in this field has yielded the identification of more than 60 novel human proteases dysregulated in different malignant tumours, which have delineated several important biological information in regard to the functional role of many of these genes, especially in cancer.

Dr. Carlos has also proposed a global approach for the study of proteases in health and disease through novel concepts such as Degradomics and Degradome, together with the development of high-throughput methods that tackle these concepts. Since 2009, Dr. Carlos co-directs the Spanish contribution to the International Cancer Genome Consortium (ICGC-CLL). This scientific work has led to the unveiling of the complete tumour genome sequence of hundreds of patients with chronic lymphocytic leukaemia and to the identification of recurrent mutations in several genes, which have become preferential targets for therapeutic intervention in this type of neoplasia. Dr. Carlos's laboratory in molecular oncology has also led to the discovery of several proteases with paradoxical anti-tumour and anti-metastatic properties, as well as two new genes implicated in the progression of head and neck carcinomas. The research of his group at the University of Oviedo has also contributed to clarifying the molecular mechanisms associated with physiological ageing and their links with cancer. They have also reported the first integrative analysis of the molecular and cellular hallmarks of ageing and proposed different strategies for the metabolic control of longevity. Moreover, they have developed a method to facilitate the cell reprogramming of aged human cells to pluripotent cells with embryonic properties. D. Carlos's group has also identified a new loss-of-proteostasis mechanism of myeloid transformation, connecting age-associated alterations in haematopoietic stem cells with the pathogenesis of haematological malignancies. His research has been collected in about 400 articles in international journals and have been cited more than 38,000 times, with an aggregate Hirsch index of h = 95.

Dr. Carlos López-Otín is a member of numerous scientific societies, committees, and journal editorial boards. He is a member of the Spanish Royal Academy of Sciences and the European Academy, among others. Throughout his scientific career, he has received several national and international awards and distinctions, such as the Doctorate Honoris Causa from the International University Menendez Pelayo and the University of Zaragoza (Spain) and the Autonomous University of Chile, the Rey Jaime I Award in Research, the European 25th FEBS Jubilee Award in Biochemistry, the Mexico Award in Science and Technology, and the Santiago Ramón y Cajal National Research Award. In 2017, he was awarded an ERC Advanced Grant from the European Union.

Dr. Carlos is undoubtedly a legendary figure in the research field on different aspects of protease in cancer. He has the excellent ability to motivate young researchers. We feel honoured to dedicate this book to Dr. Carlos López-Otín and wish him good health and success in his long, fruitful activities.

Contents

SECTION III Epigenetic Tools and Stem Cell Therapy

SECTION IV Nanotechnology-Based Therapies

SECTION V Bioinformatics and System Biology

Preface

Life is 10% what happens to us and 90% how we react to it.

Leslie Medley-Russell **(Ovarian cancer survivor)**

Proteases, for a long time, have received the attention of biochemists as degrading enzymes. One of the most significant biological processes is proteolysis. Proteases are a class of enzymes known for their proteolytic action. These enzymes are widely distributed and carry out important biological reactions. Recent research has demonstrated that proteases have a role in tumour growth and progression at both the primary and metastatic sites. There is a favourable link between the release of several proteases and the aggressiveness of the tumour. While transformed tumour cells are the primary source of significant damage, proteases play a critical role in many biological processes carried out by normal cells.

Proteases and associated anti-proteases balance each other out in biological systems, and imbalances cause a variety of illnesses, including cancer. All five classes of proteases – serine, cysteine, aspartate, threonine, and matrix metalloproteases – are involved in the processes leading up to the formation of a tumour, its growth, metastasis, and invasion into another site. Numerous studies have demonstrated a link between tumour growth and lysosomal cysteine protease activity. Trypsin is a common and well-known digestive serine protease that has been linked to a number of malignancies by promoting invasion, metastasis, and proliferation. Trypsin-expressing colorectal cancer has a worse prognosis and a shorter disease-free survival time.

Proteases possess outstanding diversity in their functions. Because of their unique properties, proteases are a major focus of attention as potential drug targets or as diagnostic and prognostic biomarkers. Protease inhibitors have been used as anticancer medications due to their role in regulating the initiation and progression of cancer. This book is divided into multiple parts that explain the role of proteases in different stages of carcinogenesis as well as some medications that are based on protease inhibitors.

The role of proteases as pharmacological targets and how they have been used to create anticancer medications are covered in detail in this book. This book discusses the structure and function of proteases as well as the chemical and biological justification for drug design that relates to the exploration of proteases to discover effective chemotherapeutics to combat cancer. Furthermore, the book covers the therapeutic aspects of drug design while placing sufficient attention on the underlying mechanisms. For those studying therapeutic aspects of protease-induced cancer, it is a valuable resource.

We present 33 state-of-the-art chapters in this book, which are divided into the following sections: Section I, General Aspects – where three chapters focus on the general mechanisms associated with the therapeutics of some types of proteases induced cancer; Section II, Proteases in Cancer Therapy – where 12 chapters demonstrate the role of protease inhibitors in therapy against several types of cancer; Section III, Epigenetic Tools and Stem Cell Therapy, where 8 chapters discuss the clinical potentials of epigenetic targeting and stem cell-mediated interventions in cancers in relation to different types of proteases; Section IV, Nanotechnology-Based Therapies, where 5 chapters are presented that deal with the nanoparticles and nano-delivery systems for therapy in different types of cancers induced by proteases; and Section V, Bioinformatics and System Biology, where 5 chapters explore system biology and the omics approaches in the therapy of protease-induced cancer.

With their research, passions, and issues, the contributors for the book shed light on our comprehension of the functions of various proteases. It is our belief that the chapters included in this book will prove beneficial to researchers working on therapeutics of protease-related cancer research. While some of these chapters highlight more recent notions, others focus on more established ones. We have a sliver of optimism for these innovative ideas.

Sajal Chakraborti
Kalyani, West Bengal, India

Contributors

Noha A. Ahmed
Zoology Department
Beni-Suef University
Beni-Suef, Egypt

Osama M. Ahmed
Physiology Division
Zoology Department
Faculty of Science
Beni-Suef University
Beni-Suef, Egypt

Aubaidah Akhtar
Department of Cancer Pharmacology
CSIR-Indian Institute of Integrative Medicine
Jammu, India

Kuan Ping Ang
School of Biosciences, Faculty of Health and Medical Sciences
Taylor's University
Subang Jaya, Selangor, Malaysia

Pushparaj Annadurai
C.P.R. Environmental Education Center
Chennai, India

Sherlin Rosita Arockiaraj
Department of Biotechnology and Bioinformatics
Bishop Heber College
Tiruchirappalli, India

Sivakumar Arumugam
School of Biosciences and Technology
Vellore Institute of Technology
Vellore, India

Dhanu A. S.
School of Life Sciences
JSS Academy of Higher Education & Research
Mysuru, India

Samina Aslam
Institute of Chemistry
University of Sargodha
Sargodha, Pakistan

Mir Owais Ayaz
Academy of Scientific & Innovative Research
Ghaziabad, India

and

CSIR-Indian Institute of Integrative Medicine
Jammu, India

Manal N. Abdel Azeem
Zoology Department
Beni-Suef University
Beni-Suef, Egypt

Abdul Basit Baba
Advanced Centre for Human Genetics
Sher-i-Kashmir
Institute of Medical Sciences (SKIMS)
Srinagar, India

Neetu Badesra
Department of Cancer Pharmacology
CSIR-Indian Institute of Integrative Medicine
Jammu, India

Angshuman Bagchi
Department of Biochemistry and Biophysics
University of Kalyani
Kalyani, India

Karnati Balaji
Vignan's Foundation for Science Technology and Research
Guntur, India

Purushothaman Balakrishnan
TanBio R & D Solution
Tiruvarur, India

Sourav Banerjee
Division of Cellular and Systems Medicine
University of Dundee
Dundee, United Kingdom

Gowrav Baradwaj
School of Life Sciences
JSS Academy of Higher Education & Research
Mysuru, India

Kanthesh M. Basalingappa
Division of Molecular Biology
School of Life Sciences
JSS Academy of Higher Education & Research (Deemed to be University)
Mysuru, India

Gaurav Kumar Bhagat
Department of Biochemistry and Biophysics
University of Kalyani
Kalyani, India

Mahima Bhardwaj
Vignan's Foundation for Science Technology and Research
Guntur, India

Rajabrata Bhuyan
Department of Bioscience and Biotechnology
Banasthali Vidyapith
Tonk, India

Sima Biswas
Department of Biochemistry and Biophysics
University of Kalyani
Kalyani, India

Anuj Kumar Borah
Department of Biochemistry
Royal School of Bio-Sciences
The Assam Royal Global University
Guwahati, India

Subhendu Chakrabarty
Department of Microbiology
M.U.C. Women's College
Burdwan, India

Sajal Chakraborti
Department of Biochemistry and Biophysics
University of Kalyani
Kalyani, India

Tapati Chakraborti
Department of Biochemistry and Biophysics
University of Kalyani
Kalyani, India

Sahiti Chamarthy
Department of Biotechnology
Koneru Lakshmaiah Education Foundation
Vaddeswaram, India

Anupam Chatterjee
Royal School of Bio-Sciences
The Assam Royal Global University
Guwahati, India

Shubhra Chaturvedi
Radiopharmaceutical Chemistry Lab
Radiological Nuclear and Imaging Sciences
Institute of Nuclear Medicine and Allied Sciences (INMAS)-DRDO
Delhi, India

Vishakha Chaudhary
Radiopharmaceutical Chemistry Lab
Radiological Nuclear and Imaging Sciences
Institute of Nuclear Medicine and Allied Sciences (INMAS)-DRDO
Delhi, India

Sudipta Choudhury
Division of Molecular Biology
JSS Academy of Higher Education & Research
Mysuru, India

Dibyapriya Roy Chowdhury
Department of Biochemistry and Biophysics
University of Kalyani
Kalyani, India

Somasri Dam
Department of Microbiology
The University of Burdwan
Burdwan, India

Mohd Jamal Dar
Academy of Scientific & Innovative Research
Ghaziabad, India

and

CSIR-Indian Institute of Integrative Medicine
Jammu, India

Mohmmad Saleem Dar
Academy of Scientific & Innovative Research
Ghaziabad, India

and

CSIR-Indian Institute of Integrative Medicine
Jammu, India

Amlan Das
Department of Biochemistry
Royal School of Bio-Sciences
The Assam Royal Global University
Guwahati, India

Janumala Mary Edna
Department of Biotechnology
Koneru Lakshmaiah Education Foundation
Vaddeswaram, India

Abinaya Elango
Department of Pharmacology
Chettinad Hospital and Research Institute
Chettinad Academy of Research and Education
Chengalpattu, India

Ayman El-Meghawry El-Kenawy
Department of Pathology
College of Medicine
Taif University
Al Hawiyah, Saudi Arabia

and

Genetic Engineering and Biotechnology Institute
University of Sadat City
Sadat, Egypt

Mahasweta Ghosh
Department of Biochemistry and Biophysics
University of Kalyani
Kalyani, India

Priya Ghosh
School of Medical Science and Technology
Indian Institute of Technology Kharagpur
Kharagpur, India

Shazia Akram Ghumman
Institute of Chemistry
University of Sargodha
Sargodha, Pakistan

Daniel A. Gideon
Department of Biochemistry
St. Joseph's University
Bengaluru, India

Dipanjan Guha
Bioinformatics Infrastructure Facility Center
University of Kalyani
Kalyani, India

Soundharya H.
Department of Biotechnology
School of Bio Sciences and Technology
Vellore Institute of Technology
Vellore, India

Sara Hasan
Institute of Chemistry
University of Sargodha
Sargodha, Pakistan

Avipsa Hazra
Division of Molecular Biology
JSS Academy of Higher Education & Research
Mysuru, India

Md Mehedi Hossain
Department of Cancer Pharmacology
CSIR-Indian Institute of Integrative Medicine
Jammu, India

Uju Dorathy Iliemene
Biochemistry Department
Science & Technology Faculty
Bingham University
Karu, Nigeria

Mehwish Ishtiaq
Institute of Chemistry
University of Sargodha
Sargodha, Pakistan

Shivani Jambulingam
Department of Biotechnology and Bioinformatics
Bishop Heber College
Tiruchirappalli, India

Thomas Jebastin
Department of Biotechnology and Bioinformatics
Bishop Heber College
Tiruchirappalli, India

Girish Kanavi K.
Division of Molecular Biology
JSS Academy of Higher Education & Research
Mysuru, India

Anmol Kapoor
Department of Biochemistry
Central University of Rajasthan
Ajmer, India

Deepmala Karmakar
Department of Biochemistry
University of Calcutta
Kolkata, India

Aqsa Khan
Amity Institute of Molecular Medicine and Stem Cell Research
Amity University
Noida, India

Ragini Khanna
Department of Microbiology
JSS Academy of Higher Education & Research
Mysuru, India

Babiangshisha Kharbuli
Department of Biochemistry
Royal School of Bio-Sciences
The Assam Royal Global University
Guwahati, India

Ammu V. V. V. Ravi Kiran
Department of Pharmacology
JSS College of Pharmacy
JSS Academy of Higher Education & Research
Ooty, India

Praveen T. Krishnamurthy
Department of Pharmacology
JSS College of Pharmacy
JSS Academy of Higher Education & Research
Ooty, India

Garikapati Kusuma Kumari
Department of Pharmacology
JSS College of Pharmacy
JSS Academy of Higher Education & Research
Ooty, India

Reena Kumari
Institute of Bioresources and Sustainable Development
Tadong-Gangtok, India

and

Manipur University
Canchipur-Imphal, India

Rajlakshmi Laha
Department of Microbiology
The University of Burdwan
Burdwan, India

Andreja Leskovac
Department of Physical Chemistry
National Institute of the Republic of Serbia
University of Belgrade
Belgrade, Serbia

Mubarak Labaran Liman
Biochemistry Department
Science and Technology Faculty
Bingham University
Karu, Nigeria

Tan Bee Ling
Department of Healthcare Professional
Management and Science University
Selangor, Malaysia

Chung Yeng Looi
School of Biosciences, Faculty of Health and Medical Sciences

and

Digital Health and Medical Advancement Impact Lab
Taylor's University
Subang Jaya, Selangor, Malaysia

Chandi C. Mandal
Department of Biochemistry
Central University of Rajasthan
Ajmer, India

Mahitosh Mandal
School of Medical Science and Technology
Indian Institute of Technology Kharagpur
Kharagpur, India

Jay Shree Mathivanan
Department of Biotechnology and Bioinformatics
Bishop Heber College
Tiruchirappalli, India

Janaki Ramaiah Mekala
Department of Biotechnology
Koneru Lakshmaiah Education Foundation
Guntur, India

Amy Dunne Miller
Division of Cellular and Systems Medicine
University of Dundee
Dundee, United Kingdom

A. K. Mishra
Radiological Nuclear and Imaging Sciences
Institute of Nuclear Medicine and Allied Sciences
Timarpur, India

Anamika Mishra
Department of Biotechnology
School of Bio Sciences and Technology
Vellore Institute of Technology
Vellore, India

Rajakishore Mishra
Centre for Life Sciences
University of Jharkhand
Ranchi, India

Siddavaram Nagini
Department of Biochemistry & Biotechnology
Annamalai University
Annamalainagar, India

Aniket Naha
Medical Biotechnology and Computational Drug Designing Laboratory
Pushpagiri Research Centre
Tiruvalla, India

Suvendu Nandi
School of Medical Science and Technology
Indian Institute of Technology Kharagpur
Kharagpur, India

Anirban Nandy
Department of Biochemistry and Biophysics
University of Kalyani
Kalyani, India

Soumitra Nath
Department of Biotechnology
Gurucharan College
Silchar, India

Sobia Noreen
Institute of Chemistry
University of Sargodha
Sargodha, Pakistan

Paul T. Olonishuwa
Biochemistry Department
Science and Technology Faculty
Bingham University
Karu, Nigeria

Amit Pal
ICMR-National Institute of Cholera and Enteric Diseases
Kolkata, India

Suchetana Pal
Department of Microbiology
The University of Burdwan
Burdwan, India

Abhinav Parashar
Department of Biological Sciences
PDPIAS, CHARUSAT
Changa, India

Sabra Parveen
Department of Cancer Pharmacology
CSIR-Indian Institute of Integrative Medicine
Jammu, India

Sandra Petrovic
Department of Physical Chemistry
Vinca Institute of Nuclear Sciences
National Institute of the Republic of Serbia
University of Belgrade
Belgrade, Serbia

Presenjit
Radiological Nuclear and Imaging Sciences
Institute of Nuclear Medicine and Allied Sciences
Defence Research and Development Organization (DRDO)
Delhi, India

Arjun K. R.
Division of Molecular Biology
JSS Academy of Higher Education & Research
Mysuru, India

Mythreyi R.
Division of Molecular Biology
JSS Academy of Higher Education & Research
Mysuru, India

Parthiban R.
Department of Translational Research
Malabar Cancer Centre
Thalassery, India

Arunkumar Radhakrishnan
Department of Pharmacology
Chettinad Hospital and Research Institute
Chettinad Academy of Research and Education
Chengalpattu, India

Amit Kumar Rai
Institute of Bioresources and Sustainable Development
Gangtok, India

Yetirajam Rajesh
School of Medical Science and Technology
Indian Institute of Technology Kharagpur
Kharagpur, India

Mekala Janaki Ramaiah
Department of Biotechnology
Koneru Lakshmaiah Education Foundation
Vaddeswaram, India

Prasanna Srinivasan Ramalingam
School of Biosciences and Technology
Vellore Institute of Technology
Vellore, India

Asma Mohin Rojewale
Department of Biological Research
Glenmark Pharmaceuticals Limited
Glenmark Research Centre
Mumbai, India

Soumya R. S.
Pushpagiri Research Centre
Tiruvalla, India

Subhasish Sarkar
Department of Microbiology
The University of Burdwan
Burdwan, India

Praveen Kumar Chandra Sekar
Human Cytogenetics and Genomics Laboratory
Faculty of Allied Health Sciences
Chettinad Academy of Research and Education
Kelambakkam, India

Chandrabose Selvaraj
Center for Global Health Research
Saveetha Medical College
Chennai, India

Sanghamitra Sengupta
Department of Biochemistry
University of Calcutta
Kolkata, India

Shinjinee Sengupta
Amity Institute of Molecular Medicine and Stem Cell Research (AIMMSCR)
Amity University
Noida, India

Farheen Showket
Department of Cancer Pharmacology
CSIR-Indian Institute of Integrative Medicine
Jammu, India

Nilabja Sikdar
Human Genetics Unit
Indian Statistical Institute
Kolkata, India

Chongtham Sovachandra Singh
Department of Biochemistry
Royal School of Bio-Sciences
The Assam Royal Global University
Guwahati, India

Kaman Singh
Department of Chemistry
Bhimrao Ambedkar University
Lucknow, India

Kshetrimayum Birla Singh
Department of Life Sciences
Manipur University
Imphal, India

Madhu Tanya Singh
Department of Chemistry
JSS College of Pharmacy
JSS Academy of Higher Education and Research
Ooty, India

Namrata Singh
Department of Biological Research
Glenmark Pharmaceuticals Limited
Glenmark Research Centre
Mumbai, India

Nanda Singh
ICMR-National Institute of Cholera and Enteric Diseases
Royal School of Bio-Sciences
Kolkata, India

Sanjeev Kumar Singh
Department of Bioinformatics
Alagappa University
Karaikudi, India

V. P. Singh
Department of Pharmacology
Banaras Hindu University
Varanasi, India

Nallapaty Srilakshmi
Department of Pharmacy
Koneru Lakshmaiah Education Foundation
Vaddeswaram, India

Jayakumar Subramaniyan
Department of Biochemistry
SRM Arts and Science College
Kattankulathur, India

Katari Sukeerthi
Department of Pharmacy
Koneru Lakshmaiah Education Foundation
Vaddeswaram, India

Vasudha Tandon
Division of Cellular and Systems Medicine
University of Dundee
Dundee, United Kingdom

Rima Tapader
ICMR-National Institute of Cholera and Enteric Diseases
Kolkata, India

Yin Sim Tor
School of Biosciences
Taylor's University
Selangor, Malaysia

Gopenath T. S.
Department of Biotechnology & Bioinformatics
JSS Academy of Higher Education & Research
Mysuru, India

Nibu Varghese
Department of Biosciences
Mar Athanasios College for Advanced Studies
Tiruvalla, India

Ramakrishnan Veerabathiran
Human Cytogenetics and Genomics Laboratory
Faculty of Allied Health Sciences
Chettinad Academy of Research and Education
Kelambakkam, India

Devadasan Velmurugan
Department of Biotechnology
SRM Institute of Science & Technology
Chennai, India

Sivakumar Vijayaraghavalu
Department of Life Sciences
Manipur University
Imphal, India

and

Department of Medical Sciences & Technology
IIT Madras
Madras, Tamil Nadu

Won Fen Wong
Department of Medical Microbiology
Faculty of Medicine
Universiti Malaya
Kuala Lumpur, Malaysia

Nour Y. S. Yassin
Zoology Department
Beni-Suef University
Beni-Suef, Egypt

Section I

General Aspects

1 Influence of Serine Proteases on Bone Morphogenetic Proteins

Insights for Advancing Treatment Paradigms in Breast Cancer

Anmol Kapoor and Chandi C. Mandal
Central University of Rajasthan, Ajmer, India

1.1 INTRODUCTION

Breast cancer stands as one of the deadliest pathophysiological disorders, and accounts for high mortality rates among women globally. Its prevalence is projected to increase significantly in the near future [1]. In recent years, numerous efforts have been made toward improving the diagnosis and treatment of breast cancer cases. However, a knowledge gap still exists when it comes to developing effective clinical strategies for proper prevention and management. Moreover, due to diverse responses and complex characteristics exhibited by breast cancer, it is quite challenging to mitigate complications and improve patient survival. Studies have reported that in advanced stages, breast cancer cells break away from the primary site and tend to spread through the bloodstream, indicating metastatic behavior [2]. This has revealed valuable insights into the underlying molecular mechanisms driving this more aggressive spread. Therefore, a comprehensive study is needed to unravel the possible factors contributing to the diverse breast cancer responses, which can be further assessed as the potential targets for the development of novel therapeutics.

In this context, bone morphogenetic proteins (BMPs), belonging to the TGF-β family of proteins along with their associated signaling pathways, have emerged as an unexplored area of research. Initially identified as the growth factors secreted from bones, BMPs are involved in various developmental processes including bone and cartilage formation [3]. However, there has been increased attention on their implications in governing cancer development and dissemination. Among various cancer types, breast cancer comes into consideration first, as it is known to have a propensity to metastasize to the bone environment. Earlier studies have suggested that bone matrix proteins are the factors responsible for promoting mineralization and osteoblast-like characteristics in the breast cancer cells, which ultimately lead to their dissemination and colonization to the bone microenvironment [4]. Thus, this led researchers to focus on BMPs as the key factors to uncovering novel insights into the molecular mechanisms that drive the aggressive behavior of breast cancer.

Similarly, members of the serine protease family are known to modulate the functionality of BMPs. These molecules act as secreted enzymes and have been identified as regulators of matrix mineralization by inhibiting BMP signaling [5]. Furthermore, studies have highlighted their

DOI: 10.1201/9781003394693-2

significant role in influencing the tumor microenvironment in diverse aspects, thereby leading to tumor progression and dissemination [6, 7]. However, the precise role of serine proteases in determining the functions of BMPs and their complexity concerning cancer development and progression requires detailed investigation. Due to highly context-dependent effects varying across different cellular and tissue types, BMPs have emerged to induce multifaceted responses in cancer including breast cancer [8]. These molecules are sometimes involved in the transition of normal cells to cancerous cells, promoting proliferation, migration, and invasion in different tissues and organs of the body. Yet, in certain scenarios, BMPs have demonstrated suppressive effects on cancer growth through activating apoptosis, inhibiting proliferation, and promoting differentiation [9]. These conflicting behaviors of BMPs reflect the intricacies in their involvement in maintaining tissue homeostasis. Concerning this complexity and the uncertain role of BMPs in cancer, researchers have encountered challenges in dissecting the spatiotemporal dynamics of BMPs. Although earlier research works focused on developing BMPs as potential therapeutic agents to mitigate various health-related disorders, among the diverse players, only BMP2 stands out as a potent clinical agent where its osteoinductive properties are utilized in promoting bone regeneration and improving bone as well as spine-related surgical outcomes [10]. However, despite extensive research on BMPs, including BMP2 and involved signaling pathways, researchers have faced limited success and huge challenges in utilizing BMPs as the treatment approach against cancer-related responses [11]. Hence, understanding the precise mechanistic roles of BMPs across diverse cancers, with a specific focus on breast cancer, remains imperative for advancing clinical investigations.

This chapter discusses the BMPs and their associated signaling pathways involved in various disorders, including cancer. It aims to provide comprehensive knowledge about the connection of extensively studied bone morphogenetic protein BMP2 and its effects on breast cancer responses. This will give us further insight into how various BMPs, including BMP2, can alter different associated signaling pathways and lead to cancer-like characteristics in different tissues. In addition to this, it sheds light on the influence of serine proteases within the tumor microenvironment, thus modulating tumor responses and developing metastatic properties. Overall, this chapter suggests that BMPs hold important potential for the scientific community and clinicians to delve deeper into the underlying molecular mechanisms behind their functioning and to discover these bone-inducing factors as promising medication strategies for preventing complications, primarily concerning metastasis of breast cancer.

1.2 BMPs: RECEPTOR-MEDIATED SIGNALING

1.2.1 Classification of BMPs and Signaling Receptors

BMPs are derived from the matrix of de-mineralized bones and are responsible for inducing bone formation. They are members of the transforming growth factor beta (TGF-β) family of proteins and are involved in governing immense functionalities based on cell and tissue types, such as organ formation, cell proliferation, angiogenesis, apoptosis, and maintaining homeostasis in the body. These proteins are identified as large inactive pre-proteins formed of a signal peptide, a pro-domain, and a highly conserved carboxylic domain containing cysteine residues [12]. Further, dissociation of the signal peptide leads to dimerization of the pro-BMPs that are cleaved by certain proteolytic enzymes in the trans-Golgi network and finally secretes as a mature BMP ligand in the extracellular environment [13]. These mature ligands are ready to interact with their respective receptors, initiating signaling processes.

Based on their structural and functional characteristics, BMPs are classified into several subgroups such as BMP2/4, BMP5/6/7/8, BMP9/10, and GDF5/6/7. Moreover, these ligands possess the ability to form homo- and heterodimers with each other and govern signaling cascade with different potencies [14]. The specific combination of these ligands involved, can trigger diverse cellular responses and exert their influence through distinct receptors located on the cellular membrane. The receptors involved, described as type I and type II receptors, assemble as heteromeric complexes and enhance the binding affinity of these ligands to them. Out of these, there are four transmembrane type I receptors (ACVR1L/ALK1, ACVR1/ALK2, BMPRIA/ALK3, and BMPRIB/ALK6) and three type II receptors (BMPRII, ActRIIA, and ActRIIB) that are known to be utilized by ligands of the BMP family [15]. Upon binding of the BMP ligands to the respective receptors, a series of intracellular downstream signaling cascades begins.

1.2.2 BMP Signaling Pathways

BMPs are known to initiate signaling through both canonical (SMAD-dependent) and non-canonical (SMAD-independent) pathways. The receptors involved in these signaling processes exhibit different modes of pathway activation. Studies have reported two modes of activation, one through pre-formed heteromeric complexes and in the other, receptors are recruited after ligand binding. In the first mode, BMP receptors exist as pre-assembled transmembrane complexes, which serve as constitutive active responses. In the other mode, the receptors recruited after ligand binding undergo a sequential process of activation. These two modes of receptor activation influence downstream signaling cascades in different ways [16].

1.2.2.1 Canonical Pathway (SMAD-dependent Pathway)

The canonical pathway involves type II receptors, known to possess constitutive activation, which further transphosphorylates type I receptors at specific conserved cytoplasmic domains. This domain exhibits serine/threonine kinase activity that further phosphorylates intracellular proteins of the SMAD family (Smad1/5/8) in the cytoplasm. The phosphorylated Smad1/5/8 binds to Smad4, the central mediator of the canonical BMP signaling pathway, to form transfer complexes for entering the nucleus and regulating target gene expression upon binding to BMP response-binding elements [15].

1.2.2.2 Non-canonical Pathways (SMAD-independent Pathways)

Non-canonical BMP signaling pathways deviate from the canonical paradigm or are derived from alternative biogenesis pathways. These pathways were described to be executed in a SMAD-independent manner and utilized distinct branches including mitogen-activated protein kinase (MAPK), RAS, TGF-β-activated kinase 1 (Tak1), Phosphoinositide 3-kinase (PI3K)/protein kinase B (AKT), and Rho-GTPases pathways. The exact pathway activated is dependent on both the BMP ligand and receptors involved [17].

These pathways are involved in cross-talks with other different signaling pathways, such as Notch, Sonic-hedgehog (Shh), Janus kinase (JAK)-signal transducer and activator of transcription (STAT), and Wnt/β-catenin pathways [18]. These diverse pathways and interactions with BMP signaling allow for fine-tuned modulation of downstream cellular responses in a temporal and context-dependent manner [19, 20].

1.2.3 Signaling Regulation

BMP signaling plays a crucial role in tissue homeostasis and disease prevention. Concerning balance functioning, BMP signaling is tightly regulated at multiple levels. The availability of BMP ligands is a critical factor in BMP signaling, and their activity is defined to be modulated by several factors, including proteases, extracellular matrix (ECM) proteins, secreted inhibitors, and membrane-bound modulators [21]. Moreover, BMP receptor activity is described to be regulated by various mechanisms, including ligand-induced receptor assembling, receptor internalization, and phosphorylation by other kinases. The downstream functioning of the SMADs effector proteins, which are activated by BMP receptor-kinase activity, is considered to be regulated by several factors, such as protein phosphatases and co-repressors. In addition, this signaling is also governed by membrane-associated co-receptors, such as repulsive guidance molecules classified as RGMa, RGMb, and RGMc with Endoglin, which may enhance or inhibit BMP signaling response. Intracellular inhibitors (I-Smads), such as Smad6 and Smad7, are described as regulating BMP signaling by inhibiting Smad phosphorylation and promoting Smad degradation as well as extracellular inhibitors, such as Noggin, Chordin, and Gremlin, which are described to prevent BMP ligand-receptor interaction [15]. Other than these layers of controls ensuring precise regulation of BMP signaling, the interactions with other signaling pathways are required to be explored in detail to understand the hidden mysteries behind disruption in tissue homeostasis and the exhibition of several health disorders.

1.3 COMPLEX BEHAVIOR OF BMPs CAUSING DIVERSE PATHOLOGIES: CANCER IN FOCUS

Abnormal BMP signaling has been associated with a wide range of pathologies, including cancer, cardiovascular disease, and other developmental disorders. BMPs such as BMP2 and BMP4 have been found to promote the calcification of heart valves, which can lead to valve stenosis and heart failure [22]. BMP2 is shown to induce the development of atherosclerosis and vascular calcification [23]. BMPs play a critical role in the development and maintenance of healthy joints. However, their dysregulation has been implicated in the development of osteoarthritis, a degenerative joint disease characterized by the loss of cartilage homeostasis [24]. BMPs are described as inducing the development of the skull, and their dysregulation can lead to craniosynostosis, a condition in which the bones of the skull fuse prematurely. This can lead to abnormal skull shape, increased intracranial pressure, and developmental delays. BMP2 and BMP4 are found to have differential spatial regulation in patients with craniosynostosis [25]. Moreover, certain genetic ablations are known to be involved in the BMP-associated pathways causing bone-related deformities. Fibrodysplasia ossificans progressiva (FOP) is a rare genetic disorder characterized by the formation of bone in soft tissues, such as muscles and tendons. Studies have reported that mutations in the gene encoding the activin receptor A type I (ACVR1), a type I BMP receptor, are responsible for the development of FOP [26]. Similarly, another hereditary disorder, hereditary hemorrhagic telangiectasia (HHT), is known to be characterized by the formation of abnormal blood vessels that are prone to bleeding [27]. Dysregulated BMP signaling has been implicated in the development of HHT and mutations in the genes encoding BMP9 and BMP10 have been found in some patients with this disease [28].

Concerning these disorders, BMPs have gained attention in recent years for their role in cancer due to their complex and multifaceted functions in cancer development and progression. Aberrant expression of various BMPs has been reported in several cancer tissues, and biological function studies have elicited the dual role of BMPs referring to both cancer development and suppression [9]. The expression of specific BMPs remains controversial, and there is a knowledge gap in the behavior of different types of BMPs. Earlier studies have identified several BMPs causing various cancer-related responses. BMP2 is found to induce motility and invasiveness in gastric cancer cells, promoting tumor growth, migration, and invasiveness through the MAPK, PI3K/AKT, or NF-κB pathway [29]. Moreover, thorough upregulation of STAT3 levels leads to stimulating migration and invasion in human colon cancer cells. This suggests that BMP2-induced STAT3-mediated induction of colon cancer cell metastasis requires alterations in cancer stem-cell markers [30]. However, other studies found that BMP2 can also induce growth suppression and enhance the chemosensitivity of human colon cancer cells, suggesting a potential therapeutic strategy for colon cancer [31]. BMPs modulate the tumor microenvironment and can produce cancer progression. BMP4 promotes breast cancer cell invasion and migration, suggesting a correlation with the overexpression of ECM metalloproteinase-1 (MMP-1) producing and C-X-C chemokine receptor type 4 (CXCR4) genes in the breast cancer cell lines [32]. In contrast, BMP4 acts as an autocrine modulator for SMAD7 expression, which further activates the canonical pathway to cause suppression of breast cancer metastasis [33]. In gastric cancer, the role of BMP4 is to promote epithelial-mesenchymal transition (EMT) by upregulating inhibitors of DNA-binding 1 (ID1) expression [34]. However, BMP4 suppression induces glioblastoma progression through canonical SMAD1/5/8 dependent pathway inhibition [35]. Moreover, BMP6 suppression is also described as promoting breast cancer cell proliferation and chemo-resistance through ERK signaling pathway activation [36]. In contrast, it has been found to inhibit MDA-MB-231 breast cancer cell metastasis by downregulating MMP-1 expressions [37]. BMP7 suppression leads to the inhibition of EMT-related genes and cell invasion in cervical cancer through p-SMAD1/5/9 downregulation [38]. Similarly, BMP7 downregulation is also suggested to inhibit A2780 ovarian cancer cells through p-SMAD1/5/9, ID2, and cyclin D1 protein suppression [39]. However, rhBMP7 expression is found to inhibit the proliferation and invasion of SBC-3 and SBC-5 cells of small cell lung cancer (SCLC) through SMAD2, SMAD4, and P21 downregulation in the cells [40]. Furthermore, BMPs are also described as promoting cancer response through vascularization. BMP9 expression is correlated with angiogenesis-related genes and promotes angiogenesis in hepatocellular carcinoma through ID1 transcription factor activation [41]. However, BMP9 can promote its anti-proliferative effects in hepatocellular carcinoma through the p38MAPK/ID3/p21 pathway, which leads to cell cycle arrest [42]. BMP10 suppresses hepatocellular carcinoma growth through STAT3 signaling inhibition [43]. Similarly, BMP10 over-expression is correlated with the suppression of proliferative and invasive properties of ovarian cancer cells [44]. Thus, this complexity in the action of BMPs suggests their participation in causing diverse cancer-related responses by influencing numerous cellular signaling pathways and, therefore, it has become a challenging task for scientists to find a way to target their context-dependent functioning.

1.4 SERINE PROTEASES: POSSIBLE AVENUES IN BREAST CANCER PROGRESSION AND DISTINCT LINKS WITH BMPs

1.4.1 Multifaceted Role of Serine Proteases in Breast Cancer Development and Progression

Serine proteases are the key enzymes involved in regulating various biological processes. However, dysregulation in their activity represents a key event in breast cancer. Studies have mentioned their potential to modulate the ECM in diverse aspects and promote breast cancer invasion and metastasis [45]. To prevent the metastatic behavior of breast cancer, these proteins have emerged out as promising therapeutic targets [46]. Out of numerous serine proteases, a novel sub-family called type II transmembrane serine proteases (TTSPs), described as cell surface-anchored proteases, contributes to cancer progression through various molecular mechanisms, garnering attention toward anticancer research [47, 48]. High expression levels of transmembrane serine protease 2 (TMPRSS2) indicated a poor prognosis in the case of breast invasive carcinoma and showed a significant correlation with tumor immune infiltration [49]. However, other families of serine proteases have been described as preventing cancer-related responses. The high-temperature requirement A (HtrA) family is another key family that has been extensively studied in the context of cancer responses [50]. HtrA1 is defined as a tumor suppressor in various cancers. Its downregulation is suggested to promote epithelial-mesenchymal transition and induce breast cancer cell invasion [51]. The studies primarily highlight a knowledge gap, indicating a need for in-depth investigations into the molecular mechanisms to uncover their precise roles in breast cancer responses.

1.4.2 Possible Link between Serine Proteases and BMPs during Signaling Regulation

Several serine proteases, as peptide cleaving enzymes, are known to have a crucial function in BMP activation. The inactive precursor form of BMP called proBMP possesses a pro-domain that acts as a cleaving site for these proteases. Furin, as one of the members of serine proteases, is known to cleave the pro-domain of BMP4 sequentially to generate an active BMP4 signaling molecule, thus regulating its function [52]. TMPRSS6 plays a role in downregulating BMP/SMAD signaling by cleaving a signaling co-receptor molecule; hemojuvelin (HJV), thus, was discovered to maintain systemic iron homeostasis [53]. Similarly, HtrA1 as a secretory enzyme inhibits TGF-β signaling, acting as an inhibitor for binding of several ligands, including BMP2 and BMP4, to their receptors [54]. Moreover, it has been reported that serine proteases possess the ability to potentiate BMP4/7 heterodimer activity, which in turn induces de-differentiation and cell cycle re-entry of muscle cells during limb generation in newts [55]. A study revealed a cross-talk between BMP2 and a serine protease called urokinase-type plasminogen activator (uPA) during myogenic differentiation. BMP2 is demonstrated to inhibit uPA activity through the p38-MAPK signaling pathway, which in turn hinders myogenic differentiation of C2C12 cells. Notably, the exogenous implication of uPA is shown to counteract BMP2's inhibitory effect, thereby promoting myogenesis [56]. While these studies suggest a potential association of serine proteases with BMP-mediated regulation of cellular events, the connection with cancer progression, including breast cancer responses, remains unexplored.

1.5 INFLUENCE OF BMP2 IN BREAST CARCINOMA DEVELOPMENT, PROGRESSION, AND REGRESSION

Out of numerous BMPs, BMP2 is the first characterized BMP that is well studied. It received U.S. Food and Drug Administration (FDA) approval for its clinical use as an osteoinductive factor in its recombinant form during spinal fusion surgeries and bone regeneration [57]. Unfortunately, to date, none of the BMPs have been clinically approved as a treatment strategy for cancer. Moreover, the clinically active therapeutic use of recombinant form (rhBMP2) is known to produce other side effects including inflammation, swelling, and uncontrolled bone formation [58, 59]. Breast cancer, as the most prevalent type of cancer, is often metastasizes to the bones [4, 60]. The entire process behind its establishment in different tissue environments is complex, starting from cancer cell invasion in the surrounding stroma, intravasation into circulation through blood vessels, and further extravasation in distant tissues followed by establishing secondary tumor growth there [61]. To reiterate the in-depth reasons behind the whole process in response to BMP2, several *in vitro* and xenograft-based studies have been done. However, the treatment options that have been suggested are limited and still meant to mitigate breast cancer development rather than focusing on breast cancer-induced deformities in other tissues.

1.5.1 BMP2 Functioning to Induce Diverse Breast Cancer-related Responses

BMP2 has emerged with diverse possibilities in producing multiple breast cancer-like characteristics, which is likely to be dependent on several factors, including the expression levels of BMP2 and its receptors, the cellular context, and the presence of other signaling partners (Figure 1.1). Few studies have demonstrated that BMP2 can inhibit breast cancer cell proliferation, migration, and invasion. BMP2 induces the expression of the cyclin-dependent kinase inhibitor p21, which leads to G1 cell cycle arrest and the inhibition of breast cancer cell growth [62]. It also regulates PTEN protein levels by decreasing its conjugation with the ubiquitin proteins involved in the degradative pathway. Exposure of BMP2 to the MCF-7 breast cancer cells prevents PTEN degradation and leads to control cell proliferation [63]. Moreover, BMP2 is suggested to induce apoptotic effects in MCF-7 breast cancer cells by regulating expressions of apoptosis-related gene protein kinase R (PKR) and its activated substrate eIF2α [64]. However, previous studies have also suggested that BMP2 promoted breast cancer cell growth through the activation of the genes involved in cell migration and invasion. It is mentioned that BMP2 can promote EMT and stemness of breast cancer cells by regulating Rb and CD44 signaling. Rb expression is shown to be suppressed by BMP2, whereas CD44 expression is positively correlated with the BMP2 expression, suggesting a combined role of both pathways in causing breast cancer metastasis [65]. *In vitro* and *in vivo* studies conducted previously suggested that BMP2 induces invasiveness and migratory characteristics in MCF-7 breast cancer cells through the induction of metastatic-related gene BCSG1 [66]. Moreover, when BMP2

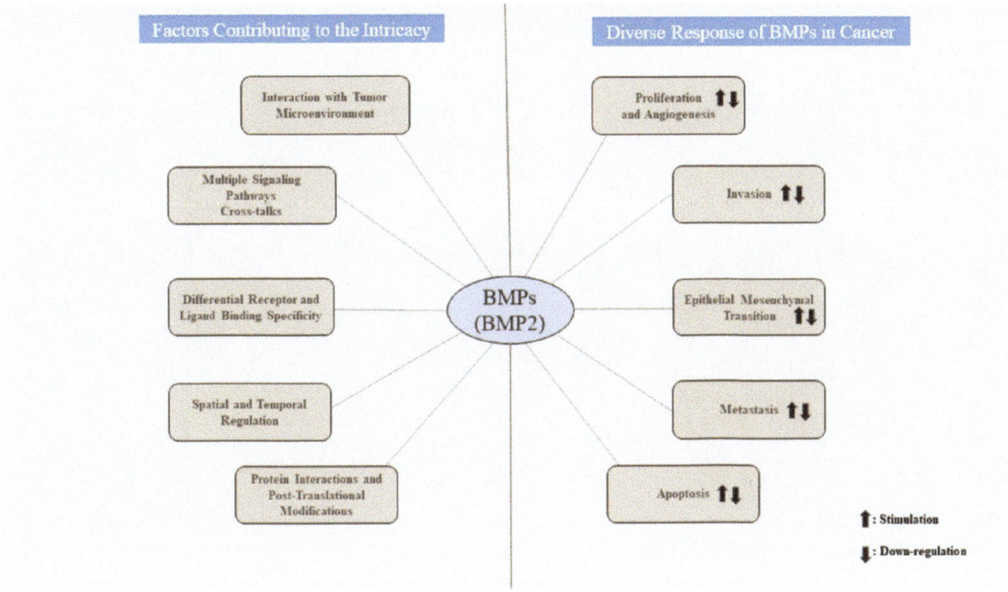

FIGURE 1.1 Schematic diagram showing factors influencing the multifaceted responses of BMPs in cancer.

is transfected into MCF-7 breast cancer cells, it reflects the same responses through suppression of hypoxia-generated cell death via activation of MAPK, ERK1/2, and the ID1 axis [67]. Similarly, the influence of recombinant BMP2 expression modulates cytoskeletal machinery in the MCF-7 cells and promote invasiveness and migration of these cells, suggested by the downregulation of epithelial marker E-cadherin [68]. Regarding the breast cancer sub-class-specific response, BMP2 overexpression is reported to induce the transformation of normal mammary phenotype to luminal-like characteristics of breast cancer [69]. Overall, these studies highlight a disparity in the functioning of BMP2 in terms of breast cancer responses, depicting a high demand for advanced research to dissect the roles of BMP2 in a more specific way.

1.5.2 Breast Tumor Microenvironment and Dissemination: BMP2-mediated Responses

The breast tumor microenvironment is responsible for offering a dynamic niche for the interaction of tumor cells with various components including stromal cells, immune cells, ECM, and secreted factors including serine proteases. These interactions sometimes serve an important role in regulating the aggressive nature of the tumor cells and further mediate tumor dissemination to distant tissues, most often in bones (Figure 1.2). In the breast tumor stroma region, BMP2 is suggested to promote Tenascin-W expression as an ECM glycoprotein responsible for inducing metastatic behavior of the mammary tissues through interaction with $\alpha 8$ integrin. This response is regulated via the p38MAPK and JNK signaling pathways [70]. BMP2 serves as the factor stimulating the expression of bone-related genes in breast cancer cells, which leads to the induction of osteoblast-like features in them. The transcription factor RUNX2 is known to regulate the transcription of genes involved in the osteomimetic transformation of breast cancer cells [71]. BMP2 also induces microcalcification in the case of breast cancer through activation of the Akt/mTOR pathway, which ultimately correlates with poor prognosis [72]. Moreover, BMP2 is suggested to increase osteoblast-like characteristics and calcification in breast cancer cells through a mechanism of BMP2-mediated Nox4 signaling, promoting osteoblast differentiation [73, 74].

In a similar context, serine proteases also regulate ECM regions through several unknown mechanisms and are defined to have a link with breast cancer progression [7]. On the other hand, these

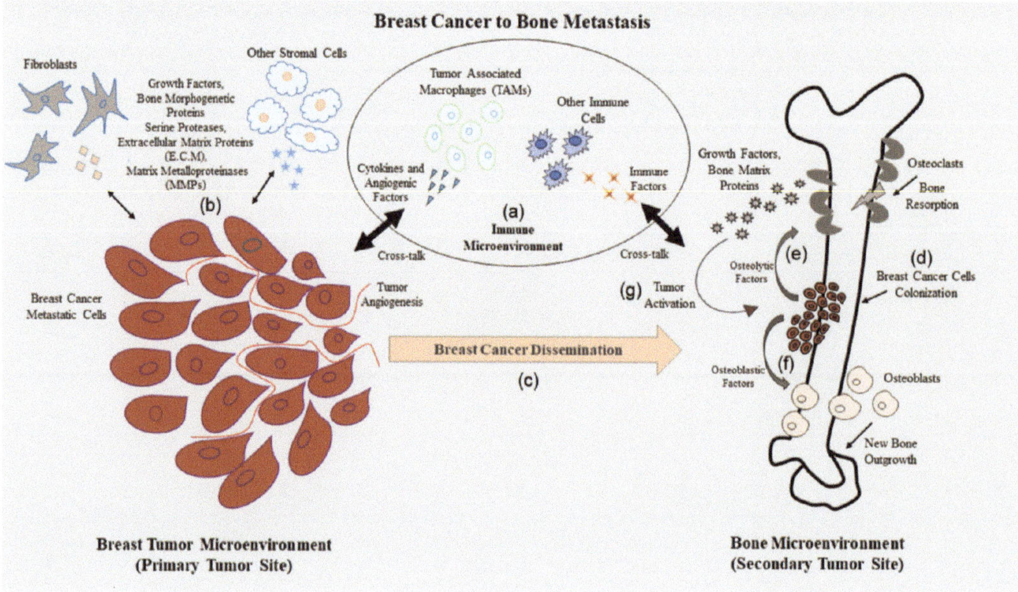

FIGURE 1.2 Interplay of cellular events and microenvironmental factors driving breast cancer to bone metastasis. Breast cancer progression is influenced by several factors secreted in the surrounding microenvironments that lead to its dissemination in the bones: (a) Secretion of immune regulators including cytokines and angiogenic factors by the immune microenvironment interact with the primary breast tissue site; (b) Dynamic interplay among these factors combined with the secretion of different stromal cells appear to play a critical role in activating metastatic response in breast cancer cells; (c) The metastatic breast cancer cells undergo dissemination in the blood circulation; (d) Subsequently, these cells extravasate into the bone microenvironment, leading to colonization and the establishment of secondary tumors. The interplay between established breast cancer cells, bone-resident cells, and the immune microenvironment impacts bone remodeling dynamics, which initiates secretion of (e) osteoblastic factors that trigger osteoblasts functions, promoting mineralization and new bone formation. Simultaneously, osteolytic factors (f) promote osteoclasts activity, leading to the degradation of the bone matrix and involved in the secretion of bone matrix proteins with several growth factors (g). The secreted bone matrix proteins influence immune cells to secrete immune factors, which further triggers the progression of colonized breast cancer cells within the bone microenvironment.

proteases play a crucial role in regulating other similar characteristics that suggest their cross-talk with the BMP2 functionality. Serine protease HtrA1 is reported to have a protective role in preventing microcalcification. Over-expression of HtrA1 is reported to cleave matrix proteins that lead to modulating osteoblast gene expressions via inhibition of BMP2 signaling [75]. Additionally, a study presented the relation of circulating serine protease known as Factor VII-activating protease (FSAP) in activating pro-BMP2, which further indicates an interplay of BMP2 and FSAP in regulating cell differentiation and calcification in blood vessels [76]. However, none of the studies depicted the direct relation of serine proteases with the BMP2-mediated breast cancer progression. Future studies targeting the interaction of BMP signaling and serine proteases in terms of breast tumor development and progression will require full comprehension of the specific mechanisms that could be involved in tumor dissemination.

1.6 CONCLUSION

An increasing number of breast cancer cases worldwide highlights an urgent need for innovative ways to address this challenging disease. The investigations on BMPs and their related signaling pathways shed light on their connection with breast cancer responses. Numerous scientific studies have drawn our attention to BMPs and their partners as key regulators of diverse processes in cancer. However, despite extensive research, there is still a lack of breakthroughs in understanding precise spatiotemporal dynamics of BMPs in terms of specific cancer responses. Due to the conflicting status of BMPs, including BMP2 functioning, researchers have identified no effective clinically proven therapeutics targeting BMP-induced cancers.

This chapter serves as a call to action for future research and showcases an opportunity for scientists and clinicians to explore the molecular pathways associated with BMPs more deeply. By focusing, on the interaction between tumor cells and the surrounding extracellular environment, further ideas could unravel the possible factors contributing to tumor dissemination. The functionalities of various signaling effectors including serine proteases in the context of distinct cancer responses require in-depth analysis. Hence, exploring an undefined cross-talk between serine proteases and different BMPs holds the promise to unveil the possible biomarkers for various cancer types including breast cancer. This knowledge can drive advancements in offering successful non-invasive diagnostic strategies for cancer mitigation. By harnessing the potential of BMP functionality in response to serine proteases, future research can expand to produce effective treatment therapies to combat cancer-related disorders.

REFERENCES

1. Arnold M, Morgan E, Rumgay H, et al. Current and future burden of breast cancer: Global statistics for 2020 and 2040. Breast. 2022;66:15–23. DOI: 10.1016/j.breast.2022.08.010
2. Wang R, Zhu Y, Liu X, et al. The Clinicopathological features and survival outcomes of patients with different metastatic sites in stage IV breast cancer. BMC Cancer. 2019;19(1):1091. DOI: 10.1186/s12885-019-6311-z
3. Urist MR. Bone: formation by autoinduction. Science. 1965;150(3698):893–9. DOI: 10.1126/science.150.3698.893
4. Cox RF, Jenkinson A, Pohl K, et al. Osteomimicry of mammary adenocarcinoma cells *in vitro*; increased expression of bone matrix proteins and proliferation within a 3D collagen environment. PLoS One. 2012;7(7):e41679. DOI: 10.1371/journal.pone.0041679
5. Canfield AE, Hadfield KD, Rock CF, et al. HtrA1: a novel regulator of physiological and pathological matrix mineralization? Biochem Soc Trans. 2007;35(Pt 4):669–71. DOI: 10.1042/BST0350669
6. Tagirasa R, Yoo E. Role of serine proteases at the tumor-stroma interface. Front Immunol. 2022;13:832418. https://doi.org/10.3389/fimmu.2022.832418
7. Bergum C, Zoratti G, Boerner J, et al. Strong expression association between matriptase and its substrate prostasin in breast cancer. J Cell Physiol. 2012;227(4):1604–9. DOI: 10.1002/jcp.22877
8. Sharma T, Kapoor A, Mandal CC. Duality of bone morphogenetic proteins in cancer: A comprehensive analysis. J Cell Physiol. 2022;237(8):3127–63. DOI: 10.1002/jcp.30785
9. Bach DH, Park HJ, Lee SK. The dual role of bone morphogenetic proteins in cancer. Mol Ther Oncolytics. 2018;8:1–13. DOI: 10.1016/j.omto.2017.10.002
10. McKay WF, Peckham SM, Badura JM. A comprehensive clinical review of recombinant human bone morphogenetic protein-2 (INFUSE Bone Graft). Int Orthop. 2007;31(6):729–34. DOI: 10.1007/s00264-007-0418-6
11. Kapoor A, Mandal CC. A perspective on bone morphogenetic proteins: dilemma behind cancer- related responses. Curr Drug Targets. 2023;24(5):382–7. DOI: 10.2174/1389450124666230201144605
12. Horbelt D, Denkis A, Knaus P. A portrait of transforming growth factor β superfamily signalling: background matters. Int J Biochem Cell Biol. 2012;44(3):469–74. DOI: 10.1016/j.biocel.2011.12.013
13. Jain AP, Pundir S, Sharma A. Bone morphogenetic proteins: the anomalous molecules. J Indian Soc Periodontol. 2013;17(5):583–6. doi: 10.4103/0972-124X.119275
14. Mueller TD, Nickel J. Promiscuity and specificity in BMP receptor activation. FEBS Lett. 2012;586(14):1846–59. DOI: 10.1016/j.febslet.2012.02.043
15. Gomez-Puerto MC, Iyengar PV, García de Vinuesa A, et al. Bone morphogenetic protein receptor signal transduction in human disease. J Pathol. 2019;247(1):9–20. DOI: 10.1002/path.5170
16. Nohe A, Hassel S, Ehrlich M, et al. The mode of bone morphogenetic protein (BMP) receptor oligomerization determines different BMP-2 signaling pathways. J Biol Chem. 2002;277(7):5330–8. DOI: 10.1074/jbc.M102750200

17. Zhang YE. Non-Smad signaling pathways of the TGF-β family. Cold Spring Harb Perspect Biol. 2017;9(2). DOI: 10.1101/cshperspect.a022129
18. Manzari-Tavakoli A, Babajani A, Farjoo MH, et al. The cross-talks among bone morphogenetic protein (BMP) signaling and other prominent pathways involved in neural differentiation. Front Mol Neurosci. 2022;15:827275. DOI: 10.3389/fnmol.2022.827275
19. Nohe A, Keating E, Knaus P, et al. Signal transduction of bone morphogenetic protein receptors. Cell Signal. 2004;16(3):291–9. DOI: 10.1016/j.cellsig.2003.08.011
20. Broege A, Pham L, Jensen ED, et al. Bone morphogenetic proteins signal via SMAD and mitogen-activated protein (MAP) kinase pathways at distinct times during osteoclastogenesis. J Biol Chem. 2013;288(52):37230–40. DOI: 10.1074/jbc.M113.496950
21. Bragdon B, Moseychuk O, Saldanha S, et al. Bone morphogenetic proteins: a critical review. Cell Signal. 2011;23(4):609–20. DOI: 10.1016/j.cellsig.2010.10.003
22. Yang P, Troncone L, Augur ZM, et al. The role of bone morphogenetic protein signaling in vascular calcification. Bone. 2020;141:115542. DOI: 10.1016/j.bone.2020.115542
23. Li X, Yang HY, Giachelli CM. BMP-2 promotes phosphate uptake, phenotypic modulation, and calcification of human vascular smooth muscle cells. Atherosclerosis. 2008;199(2):271–7. DOI: 10.1016/j.atherosclerosis.2007.11.031
24. Nakase T, Miyaji T, Tomita T, et al. Localization of bone morphogenetic protein-2 in human osteoarthritic cartilage and osteophyte. Osteoarthritis Cartilage. 2003;11(4):278–84. DOI: 10.1016/s1063-4584(03)00004-9
25. Whitton A, Hyzy SL, Britt C, et al. Differential spatial regulation of BMP molecules is associated with single-suture craniosynostosis. J Neurosurg Pediatr. 2016;18(1):83–91. DOI: 10.3171/2015.12.PEDS15414
26. Kaplan FS, Al Mukaddam M, Stanley A, et al. Fibrodysplasia ossificans progressiva (FOP): A disorder of osteochondrogenesis. Bone. 2020;140:115539. DOI: 10.1016/j.bone.2020.115539
27. Kritharis A, Al-Samkari H, Kuter DJ. Hereditary hemorrhagic telangiectasia: diagnosis and management from the hematologist's perspective. Haematologica. 2018;103(9):1433–43. DOI: 10.3324/haematol.2018.193003
28. Tillet E, Bailly S. Emerging roles of BMP9 and BMP10 in hereditary hemorrhagic telangiectasia. Front Genet. 2014;5:456. DOI: 10.3389/fgene.2014.00456
29. Kang MH, Oh SC, Lee HJ, et al. Metastatic function of BMP-2 in gastric cancer cells: the role of PI3K/AKT, MAPK, the NF-κB pathway, and MMP-9 expression. Exp Cell Res. 2011;317(12):1746–62. DOI: 10.1016/j.yexcr.2011.04.006
30. Kim BR, Oh SC, Lee DH, et al. BMP-2 induces motility and invasiveness by promoting colon cancer stemness through STAT3 activation. Tumour Biol. 2015;36(12):9475–86. DOI: 10.1007/s13277-015-3681-y
31. Vishnubalaji R, Yue S, Alfayez M, et al. Bone morphogenetic protein 2 (BMP2) induces growth suppression and enhances chemosensitivity of human colon cancer cells. Cancer Cell Int. 2016;16:77. DOI: 10.1186/s12935-016-0355-9
32. Guo D, Huang J, Gong J. Bone morphogenetic protein 4 (BMP4) is required for migration and invasion of breast cancer. Mol Cell Biochem. 2012;363(1-2):179–90. DOI: 10.1007/s11010-011-1170-1
33. Eckhardt BL, Cao Y, Redfern AD, et al. Activation of canonical BMP4-SMAD7 signaling suppresses breast cancer metastasis. Cancer Res. 2020;80(6):1304–15. DOI: 10.1158/0008-5472.CAN-19-0743
34. Deng G, Chen Y, Guo C, et al. BMP4 promotes the metastasis of gastric cancer by inducing epithelial-mesenchymal transition via ID1. J Cell Sci. 2020;133(11). DOI: 10.1158/0008-5472.CAN-19-0743
35. Zhao X, Sun Q, Dou C, et al. BMP4 inhibits glioblastoma invasion by promoting E-cadherin and claudin expression. Front Biosci (Landmark Ed). 2019;24(6):1060–70. DOI: 10.2741/4768
36. Lian WJ, Liu G, Liu YJ, et al. Downregulation of BMP6 enhances cell proliferation and chemoresistance via activation of the ERK signaling pathway in breast cancer. Oncol Rep. 2013;30(1):193–200. DOI: 10.3892/or.2013.2462
37. Hu F, Zhang Y, Li M, et al. BMP-6 inhibits the metastasis of MDA-MB-231 breast cancer cells by regulating MMP-1 expression. Oncol Rep. 2016;35(3):1823–30. DOI: 10.3892/or.2015.4540
38. Sun R, Guan H, Liu W, et al. Expression of BMP7 in cervical cancer and inhibition of epithelial-mesenchymal transition by BMP7 knockdown in HeLa cells. Int J Mol Med. 2020;45(5):1417–24. DOI: 10.3892/ijmm.2020.4519
39. Guan H, Li J, Sun R, et al. Expression of BMP7 in ovarian cancer and biological effect of BMP7 knockdown on ovarian cancer cells. Onco Targets Ther. 2019;12:7897–909. DOI: 10.2147/OTT.S217975
40. Shen W, Pang H, Xin B, et al. Biological effects of BMP7 on small-cell lung cancer cells and its bone metastasis. Int J Oncol. 2018;53(3):1354–62. DOI: 10.3892/ijo.2018.4469

41. Chen H, Nio K, Tang H, et al. BMP9-ID1 signaling activates HIF-1α and VEGFA expression to promote tumor angiogenesis in hepatocellular carcinoma. Int J Mol Sci. 2022;23(3). DOI: 10.3390/ijms23031475
42. Jung JW, Yoon SM, Kim S, et al. Bone morphogenetic protein-9 is a potent growth inhibitor of hepatocellular carcinoma and reduces the liver cancer stem cells population. Oncotarget. 2016;7(45): 73754–68. DOI: 10.18632/oncotarget.12062
43. Yuan YM, Ma N, Zhang EB, et al. BMP10 suppresses hepatocellular carcinoma progression via PTPRS-STAT3 axis. Oncogene. 2019;38(48):7281–93. DOI: 10.1038/s41388-019-0943-y
44. Jin Y, Zheng W, Li L, et al. Loss of BMP-10 is correlated with poor survival in ovarian cancer. Pathol Res Pract. 2019;215(1):121–6. DOI: 10.1016/j.prp.2018.10.025
45. Tang L, Han X. The urokinase plasminogen activator system in breast cancer invasion and metastasis. Biomed Pharmacother. 2013;67(2):179–82. DOI: 10.1016/j.biopha.2012.10.003
46. Sananes A, Cohen I, Allon I, et al. Serine protease inhibitors decrease metastasis in prostate, breast, and ovarian cancers. Mol Oncol. 2023. DOI: 10.1002/1878-0261.13513
47. Martin CE, List K. Cell surface-anchored serine proteases in cancer progression and metastasis. Cancer Metastasis Rev. 2019;38(3):357–87. DOI: 10.1007/s10555-019-09811-7
48. Kim S. TMPRSS4, a type II transmembrane serine protease, as a potential therapeutic target in cancer. Exp Mol Med. 2023;55(4):716–24. DOI: 10.1038/s12276-023-00975-5
49. Xiao X, Shan H, Niu Y, et al. TMPRSS2 serves as a prognostic biomarker and correlated with immune infiltrates in breast invasive cancer and lung adenocarcinoma. Front Mol Biosci. 2022;9:647826. DOI: 10.3389/fmolb.2022.647826
50. Wu L, Li X, Li Z, et al. HtrA serine proteases in cancers: A target of interest for cancer therapy. Biomed Pharmacother. 2021;139:111603. DOI: 10.1016/j.biopha.2021.111603
51. Lehner A, Magdolen V, Schuster T, et al. Downregulation of serine protease HTRA1 is associated with poor survival in breast cancer. PLoS One. 2013;8(4):e60359. DOI: 10.1371/journal.pone.0060359
52. Sopory S, Nelsen SM, Degnin C, et al. Regulation of bone morphogenetic protein-4 activity by sequence elements within the prodomain. J Biol Chem. 2006;281(45):34021–31. DOI: 10.1074/jbc.M605330200
53. Finberg KE, Whittlesey RL, Fleming MD, et al. Down-regulation of Bmp/Smad signaling by Tmprss6 is required for maintenance of systemic iron homeostasis. Blood. 2010;115(18):3817–26. DOI: 10.1182/blood-2009-05-224808
54. Oka C, Tsujimoto R, Kajikawa M, et al. HtrA1 serine protease inhibits signaling mediated by Tgfbeta family proteins. Development. 2004;131(5):1041–53. DOI: 10.1242/dev.00999
55. Wagner I, Wang H, Weissert PM, et al. Serum proteases potentiate BMP-induced cell cycle re-entry of dedifferentiating muscle cells during newt limb regeneration. Dev Cell. 2017;40(6):608–17.e6. DOI: 10.1016/j.devcel.2017.03.002
56. Santibanez JF, Obradović H, Krstić J. BMP2 downregulates urokinase-type plasminogen activator via p38 MAPK: Implications in C2C12 cells myogenic differentiation. Acta Histochem. 2021;123(6):151774. DOI: 10.1016/j.acthis.2021.151774
57. Halloran D, Durbano HW, Nohe A. Bone morphogenetic protein-2 in development and bone homeostasis. J Dev Biol. 2020;8(3). DOI: 10.3390/jdb8030019
58. Haidar ZS, Hamdy RC, Tabrizian M. Delivery of recombinant bone morphogenetic proteins for bone regeneration and repair. Part A: current challenges in BMP delivery. Biotechnol Lett. 2009;31(12): 1817–24. DOI: 10.1007/s10529-009-0099-x
59. Tan Y, Montgomery SR, Aghdasi BG, et al. The effect of corticosteroid administration on soft-tissue inflammation associated with rhBMP-2 use in a rodent model of inflammation. Spine (Phila Pa 1976). 2013;38(10):806–13. DOI: 10.1097/BRS.0b013e31827eca09
60. Mandal CC. Osteolytic metastasis in breast cancer: effective prevention strategies. Expert Rev Anticancer Ther. 2020;20(9):797–811. DOI: 10.1080/14737140.2020.1807950
61. Fidler IJ. The pathogenesis of cancer metastasis: the 'seed and soil' hypothesis revisited. Nat Rev Cancer. 2003;3(6):453–8. DOI: 10.1038/nrc1098
62. Chen A, Wang D, Liu X, et al. Inhibitory effect of BMP-2 on the proliferation of breast cancer cells. Mol Med Rep. 2012;6(3):615–20. DOI: 10.3892/mmr.2012.962
63. Waite KA, Eng C. BMP2 exposure results in decreased PTEN protein degradation and increased PTEN levels. Hum Mol Genet. 2003;12(6):679–84. DOI: 10.1093/hmg/ddg069
64. Steinert S, Kroll TC, Taubert I, et al. Differential expression of cancer-related genes by single and permanent exposure to bone morphogenetic protein 2. J Cancer Res Clin Oncol. 2008;134(11):1237–45. DOI: 10.1007/s00432-008-0396-0
65. Huang P, Chen A, He W, et al. BMP-2 induces EMT and breast cancer stemness through Rb and CD44. Cell Death Discov. 2017;3:17039. DOI: 10.1038/cddiscovery.2017.39

66. Clement JH, Raida M, Sänger J, et al. Bone morphogenetic protein 2 (BMP-2) induces *in vitro* invasion and *in vivo* hormone independent growth of breast carcinoma cells. Int J Oncol. 2005;27(2):401–7. DOI: 10.3892/ijo.27.2.401
67. Raida M, Clement JH, Ameri K, et al. Expression of bone morphogenetic protein 2 in breast cancer cells inhibits hypoxic cell death. Int J Oncol. 2005;26(6):1465–70. DOI:10.3892/ijo.26.6.1465
68. Jin H, Pi J, Huang X, et al. BMP2 promotes migration and invasion of breast cancer cells via cytoskeletal reorganization and adhesion decrease: an AFM investigation. Appl Microbiol Biotechnol. 2012;93(4):1715–23. DOI: 10.1007/s00253-011-3865-3
69. Chapellier M, Bachelard-Cascales E, Schmidt X, et al. Disequilibrium of BMP2 levels in the breast stem cell niche launches epithelial transformation by overamplifying BMPR1B cell response. Stem Cell Reports. 2015;4(2):239–54. DOI: 10.1016/j.stemcr.2014.12.007
70. Scherberich A, Tucker RP, Degen M, et al. Tenascin-W is found in malignant mammary tumors, promotes alpha8 integrin-dependent motility and requires p38MAPK activity for BMP-2 and TNF-alpha induced expression *in vitro*. Oncogene. 2005;24(9):1525–32. DOI: 10.1038/sj.onc.1208342
71. Tan CC, Li GX, Tan LD, et al. Breast cancer cells obtain an osteomimetic feature via epithelial-mesenchymal transition that have undergone BMP2/RUNX2 signaling pathway induction. Oncotarget. 2016;7(48):79688–705. DOI: 10.18632/oncotarget.12939
72. Wang S, Gu M, Jiang H, et al. BMP-2 upregulates the AKT/mTOR pathway in breast cancer with microcalcification and indicates a poor prognosis. Clin Transl Oncol. 2020;22(8):1263–71. DOI: 10.1007/s12094-019-02248-y
73. Sharma T, Sharma A, Maheshwari R, et al. Docosahexaenoic Acid (DHA) Inhibits Bone Morphogenetic Protein-2 (BMP-2) Elevated Osteoblast Potential of Metastatic Breast Cancer (MDA-MB-231) Cells in Mammary Microcalcification. Nutr Cancer. 2020;72(5):873–83. DOI: 10.1080/01635581.2019.1651879
74. Mandal CC, Ganapathy S, Gorin Y, et al. Reactive oxygen species derived from Nox4 mediate BMP2 gene transcription and osteoblast differentiation. Biochem J. 2011;433(2):393–402. DOI: 10.1042/BJ20100357
75. Hadfield KD, Rock CF, Inkson CA, et al. HtrA1 inhibits mineral deposition by osteoblasts: requirement for the protease and PDZ domains. J Biol Chem. 2008;283(9):5928–38. DOI: 10.1074/jbc.M709299200
76. Roedel EK, Schwarz E, Kanse SM. The factor VII-activating protease (FSAP) enhances the activity of bone morphogenetic protein-2 (BMP-2). J Biol Chem. 2013;288(10):7193–203. DOI: 10.1074/jbc.M112.433029

2 Proteases in Cancer Theranostics

Deepmala Karmakar and Sanghamitra Sengupta
University of Calcutta, Kolkata, India

2.1 INTRODUCTION

Our knowledge about proteases has significantly changed since the seminal study by P. A. Levene entitled "The Cleavage Products of Proteoses" in the *Journal of Biological Chemistry* (Issue I) published October 1, 1905 (Levene, 1905). Proteases have occupied the leading edge of biological research after a centennial period, and more than 350,000 articles on these degradative enzymes are in the scientific literature. For many years, protease studies remained restricted to their original roles associated with protein degradation. The recognition that proteases catalyze precise proteolytic reactions beyond their degradative functions, producing diverse protein products, has inaugurated a new era in protease research. Present knowledge thrives on an extensive assortment of findings demonstrating the relevance of proteases in the control of multiple biological processes such as regulating the fate, localization, and activity of several proteins, modifying protein-protein interactions, generating new bioactive products that contribute to the exchange of cellular information, and transducing and amplifying molecular signals. As a direct result of its multi-faceted roles, proteases influence a vast repertoire of processes such as DNA replication and transcription, cell proliferation and differentiation, tissue morphogenesis and remodeling, heat shock and unfolded protein responses, neurogenesis, ovulation, angiogenesis, wound repair, blood coagulation, inflammation, immunity, stem cell mobilization, autophagy, necrosis, senescence, and apoptosis. Any perturbations in the proteolytic process may thus lead to pathological conditions such as cancer, inflammatory and cardiovascular diseases, and neurodegenerative disorders, suggesting their suitability as valuable drug targets (López-Otín & Bond, 2008). This synthesis will provide fresh insights into cancer biology and opportunities to develop new diagnostic and treatment strategies.

2.2 CANCER HALLMARKS AND PROTEASE SIGNALING

Tumor progression is a multi-step process in which transformed cells achieve their ability to invade and metastasize. Despite being quite diverse, the molecular mechanisms pertaining to proteolytic activities are mostly tumor-suppressive. These affect almost each stage of tumor progression including proliferation, inflammation, invasion, angiogenesis, evading growth suppressor signals, and survival by resisting apoptosis (López-Otín & Matrisian, 2007) (Figure 2.1). However, the network of individual proteases, with their wide-ranging targets, activation mechanisms, and influence on tumor progression and metastasis, are context-specific and can only be explained concerning tumor type and stage-specific. To gain a better perspective of the therapeutic strategies targeting protease, a discussion regarding protease and its roles in association with cancer hallmarks is necessary.

DOI: 10.1201/9781003394693-3

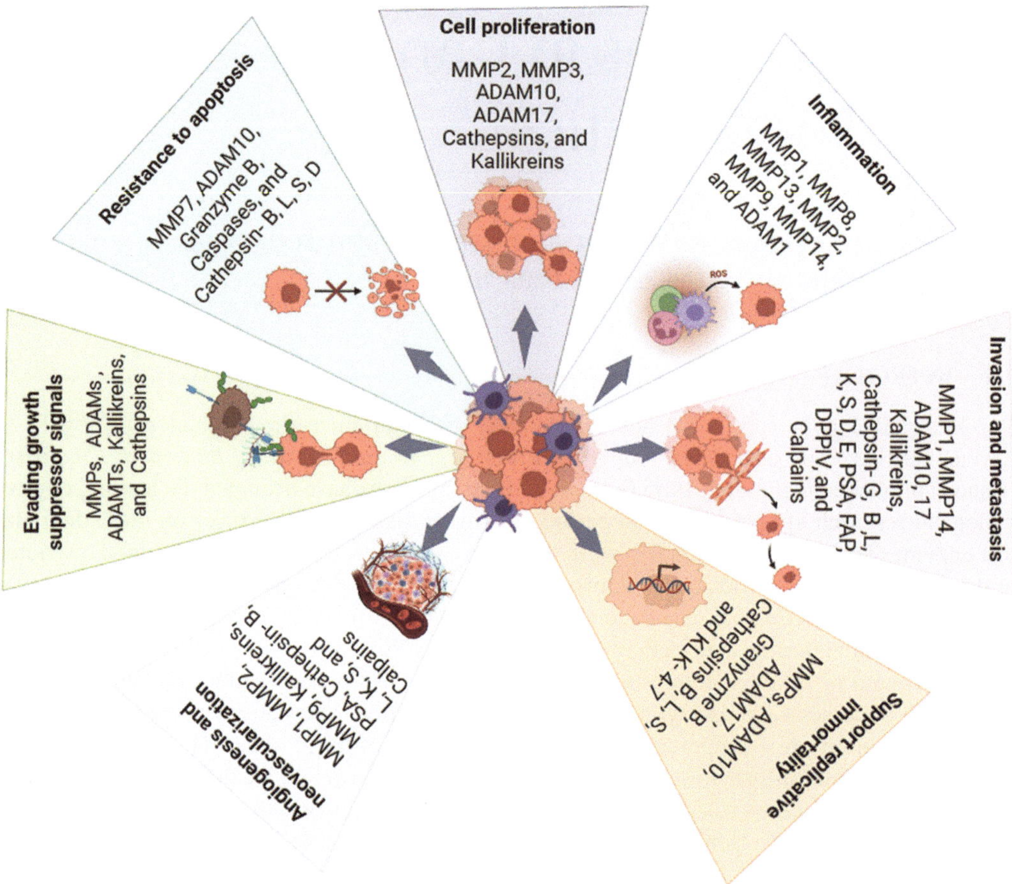

FIGURE 2.1 Examples of major proteases involved in different stages of cancer development and progression.

2.2.1 Proliferation

Protease activity often regulates multiple signaling pathways and growth factors in controlling cell proliferation. Numerous matrix metalloproteinases (MMPs), such as MMP2, MMP9, and MMP14, affect the activity of transforming growth factor beta (TGF-β), which could augment tumor growth (Kessenbrock et al., 2010). A disintegrin and metalloproteinase—ADAM10 and ADAM17—can modify the epidermal growth factor receptor (EGFR) functionality, resulting in tumor cell proliferation (Kessenbrock et al., 2010). Kallikrein (KLK)-2, also regulates the availability of growth factors (Borgoño & Diamandis, 2004). Additionally, cathepsin B, L, and S play critical roles in lysosomal catabolism necessary for cellular proliferation (Olson & Joyce, 2015).

2.2.2 Inflammation

Sustained inflammation is another emerging hallmark of cancer. Proteases such as ADAM17 play active roles in the immune evasion of tumor cells and promote the activity of TNF-α, a pro-inflammatory cytokine, while MMP8 is responsible for generating the tripeptide (N-acetyl Pro-Gly-Pro) that attracts neutrophils to sites of inflammation (Kessenbrock et al., 2010). Cathepsins are also implicated in tissue inflammation when expressed by tumor-associated macrophages (Olson & Joyce, 2015). Suppression of natural killer (NK) cells ADAM17 and processing of cytokines such as CCL8 and CCL11 by MMPs are a few notable examples (Kessenbrock et al., 2010).

2.2.3 Invasion and Metastasis

Cancer metastasis to distant organs is enabled by the extracellular matrix (ECM) degradation, which is mediated by numerous proteases, including cathepsins (e.g., cathepsin B, L, and, S), MMPs (e.g., MMP1 and MMP14), kallikreins (e.g., KLK3 and KLK6) and certain serine proteases such as hepsin, matriptase, and urokinase plasminogen activator (Dudani et al., 2018).

2.2.4 Angiogenesis and Vascularization

The angiogenic switch that regulates the supply of metabolites and nutrients for tumor growth also involves numerous proteases. MMPs, cathepsins, and KLKs are among the significant players implicated in both malignant angiogenesis and vasculogenesis (Deryugina & Quigley, 2010).

2.2.5 Evading Growth Suppressor Signals

Proteases influence cytokine expression by divergent signaling pathways and proteolytically affect their activity in post-translational processing. For instance, cathepsin S acts as a transcriptional mediator of the pro-inflammatory chemokine, CCL2/MCP-1, via type I transmembrane glycoprotein CD74 in cancer cells. Cathepsin S downregulation expression leads to a decrement in the recruitment of the macrophages to the tumor site and consequently reduces tumor growth in the mice (Wilkinson et al., 2015). The activation of the urokinase-plasminogen activator, uPA-uPAR axis, augments interleukin 4 (IL-4) expression via ERK1/2 signaling and TGF-β in cancer cells by a yet unknown mechanism (Hu, Jo, et al., 2014). Furthermore, MMP2 and MMP14 that are secreted extracellularly, bind to latent TGF-β-binding protein and activate TGF-β, which is an important component of ECM (Kessenbrock et al., 2010).

2.2.6 Resistance to Apoptosis

Cancer cells evade mechanisms often involving proteases to prevent cell death and sustain cell survival. Examples include Fas ligand being modified by ADAM10 and MMP7 toward suppressing the apoptotic processes mediated by the caspases (Kessenbrock et al., 2010); upregulation of cathepsin S following ionizing radiation, led to improved cell survival and reduced therapeutic efficacy (Olson & Joyce, 2015). In glioblastoma, elevated levels of cathepsin L causes an anti-apoptotic effect by inducing caspase 3 synthesis and ultimately leading to drug resistance (Kenig et al., 2011). Caspases 2, 8, 9, and 10 together initiate a cascade, which in turn directly or indirectly activates various effector caspases (3, 6, and 7), inducing the apoptotic pathway (Mason & Joyce, 2011). Caspase downregulation can lead to reduced apoptosis, improved cancer cell survival, and therapy resistance.

The reports mentioned previously provide concrete evidence of the profound involvement of protease in the development and progression of cancer. Therefore, efforts to integrate the proteases in the diagnosis and therapeutic processes have increased. The next section confers the vast implications of cancer proteases that can be harnessed for functional diagnostic imaging to translate their applications as protease-based drug delivery systems or activators of prodrugs to pave the way toward the next generation of clinical modalities.

2.3 PROTEASES AND CANCER DIAGNOSTICS

Considering the widespread dysregulation of protease expression and activity, atypical proteolysis may be considered as an emerging important hallmark of cancer. Currently, more than 100 proteases are associated with different stages of cancer development. Thus, there is a strong inclination to exploit their potential in the context of cancer diagnosis. Current technological advancement has enabled sensitive and robust detection of proteases in various settings, paving the way toward profiling of protease activity patterns in cancer. These include activity-based probes, substrate-based probes, and several nanosystems that generate strong reporter signals for techniques such as positron-emission tomography (PET) or magnetic resonance imaging (MRI) following their interaction with the target protease.

2.3.1 Activity-based Probes

Activity-based probes (ABPs) are designed to carry a reactive group known as a warhead, reporter, and linker. The warhead group specifically reacts with the protease's active site, while the reporter generates a signal upon interaction with the protease. A cocktail of active site probes in combination with mass spectrometry was used for comprehensive profiling of the metalloprotease family (Sieber et al., 2006). Activity-based fluorescent probe profiling has been useful for metalloproteases (Saghatelian et al., 2004), cysteine proteases such as cathepsins (Edgington-Mitchell et al., 2017), serine proteases (Pan et al., 2006), aspartic proteases (Shi et al., 2011), caspases (Poreba et al., 2019), and, recently, also for multi-domain proteasomes (Hewings et al., 2017). ABPs have substantially improved our understanding of protease function in cancer and the tumor microenvironment, as revealed by a cathepsin S activity-based probe applied to image tumor-associated macrophages (Verdoes et al., 2012). A major disadvantage of activity-based probes is that they block the protease upon binding, which causes unwanted activity in the experimental system that may dampen signal amplification. Designing a selective product with sufficient specificity toward the target enzyme without interference from other closely related proteases in the system poses another challenge (Mahmood & Weissleder, 2003).

2.3.2 Substrate-based Probes

The substrate-based probes use protease cleavage on the target site in their natural environment to produce a fluorescent signal (Yhee et al., 2012). These probes are designed either as Förster resonance energy transfer (FRET) probes where the emission spectra are shifted or as quenched substrates that fluoresce more intensely after cleavage, resulting in a quantifiable protease activity measurement in successful cancer detection (Hu, Gehrig, et al., 2014). For instance, the PerkinElmer's ProSense probe can generate signal upon cathepsin cleavage resolved millimeter-sized tumors in a mouse lung cancer model using 3D fluorescence-molecular tomography (Grimm et al., 2005).

Substrate-based probes have been successfully applied to detect cathepsins and caspases (Poreba et al., 2018), neutrophil proteases (Kasperkiewicz et al., 2017) and kallikreins (Gruba & Bielecka, 2019). A hybrid combinatorial substrate library-based assay has been used in biopsied specimens for the screening of protease activity (Tholen et al., 2020). Substrate probes provide an added benefit of not inactivating the target protease and, hence, do not immediately influence on-site protease activity. Consequently, the protease can hydrolyze more substrates, generating an amplified signal, which is counteracted by on-site diffusion away from the target site. One approach to troubleshoot this issue is to improve the reporter target site retention after protease cleavage (Hu, Vats, et al., 2014).

Magnetic resonance tuning has been developed that separates a paramagnetic enhancer molecule from a superparamagnetic quencher using a protease-cleavable substrate. This new technology could overcome many limitations associated with *in vivo* optical imaging protease activity (Dudani et al., 2018).

A third approach involves changes in probes that can be imaged upon substrate proteolysis. One technique releases a cell-penetrating peptide, termed activatable cell-penetrating peptide (ACPP), to image cancer cells after proteolysis by MMPs (Jiang et al., 2004). This sensitivity has been improved by using ratiometric ACPPs, where two dyes are placed on the same ACPP (Savariar et al., 2013). Alternatively, nanoparticles may be used for dual fluorescence and MRI (Olson et al., 2010). A hydrophobic moiety containing a PET agent can be released from a hydrophilic domain after proteolysis to enhance tumor-specific PET signals (Chuang et al., 2012).

There are several examples of substrate-based imaging probes for protease detection. For example, antibodies developed to detect tumors expressing active and upregulated matriptase (Darragh et al., 2010). Additionally, quenched ABPs have been used to image inflammatory status of tumors *in vivo* as demonstrated by a probe specific for legumain (Edgington et al., 2013), which is highly expressed in tumor-associated macrophages (Luo et al., 2006). An advantage of ABP over antibodies is their capacity to enter cells to image intracellular proteases. These probes have also shown a potential in imaging proteolytic activity in the excised tissues. It may prove valuable for a rapid analysis of surgical samples (Withana et al., 2016). However, one disadvantage is that imaging often limits the scope of multiplexing.

2.3.3 Protease Responsive Nanoparticles

The idea behind developing protease-responsive nanoparticles is that the proteolytic cleavage releases small reporters, which can be readily voided by the kidneys and detected in urine samples. Synthetic biomarkers offer greater sensitivity in tumor detection due to the renal concentration of reporter fragments. These fragments multiplexed with heavy isotope-encoded peptide barcodes can be easily detected by mass spectrometry (Kwong et al., 2013) or ligand-encoded barcodes detected by enzyme-linked immunoassay (ELISA) (Warren, Gaylord, et al., 2014), and paper-based point-of-care assays (Warren, Kwong, et al., 2014). A ten-plex synthetic biomarker is broadly responsive to MMPs and underscored its potential to monitor fibrogenesis with distinct etiologies in a mouse model of liver fibrosis and colorectal cancer (Kwong et al., 2013).

2.3.4 Molecular Imaging in *In Vivo* Systems

Molecular imaging of proteases in cancer has been explored with a wide variety of agents such as polymers, peptides, proteins, and nanoparticles labeled with multiple types of imaging tags such as fluorescent dyes, quantum dots (QDs), magnetic nanoparticles, radioisotopes, luciferins, and Gd-chelates (Table 2.1). The application of different protease-sensitive nanomaterials suitable for optical imaging approaches has raised hope in protease-based diagnosis (Smith & Gambhir, 2017). Above all, fluorescence imaging is the most thoroughly validated, and many protease imaging probes are already commercially available. QDs consisting of CdSe or ZnS have the quality to exhibit FRET effects and, thus, can be used as fluorescent reporters for *in vivo* imaging of protease activity (Larson et al., 2003). The FRET effect between the QDs and a suitable fluorescent labeled protease-cleavable peptide results in the emission of fluorescence

TABLE 2.1
A Summary of Protease-Based Molecular Imaging Modalities in Cancer

Modalities	Probes	Target	Sensitivity	Tissue Penetration	Clinical Significance
Fluorescence	Activity-based	MMPs, uPA, cathepsins	++	Low limit	Low
PET	Inhibitor-based	MMPs	+++	No limit	High
SPECT	Inhibitor-based	MMPs	+++	No limit	High
CT	Activity-based	Cathepsin	+	Low limit	Low
MRI	Activity-based	MMPs	+	No limit	Low

spectra upon protease cleavage. This idea has been employed to design protease reporters for MMPs (Li et al., 2014), caspase-1, collagenase, chymotrypsin and thrombin (Medintz et al., 2006), uPAR (Lowe et al., 2012), caspase-3 (Boeneman et al., 2009) and kallikreins (Breger et al., 2014). Recently, the concept was also employed to monitor the MMP2/9 activities in the tumor microenvironment (Jeong et al., 2017). Translating the above modality, which may apply to clinical settings, is mainly hindered due to the toxicity and high costs of materials for QD preparation.

Despite many advances in protease-based disease state detection, the techniques developed are yet to find broad applications. Nevertheless, near-infrared (NIR) fluorescent protease reporters have great potential in image-guided surgery. Prosense® 680 (PerkinElmer), a cathepsin-sensitive poly-L-glutamic acid-based quenched fluorescent probe, has successfully detected margins around the tumor in image-guided surgery (Blau et al., 2018). Additionally, another cathepsin-sensitive quenched fluorescence activity-based probe that can demarcate surgical margins of the tumor has been designed for intravenous application (Suurs et al., 2020). LUM015, a NIR FRET-based probe that is sensitive to cysteine cathepsins, has been tested in a mouse-human Phase I co-clinical trial, offering the first promising human data leveraging the utility of this technique (Whitley et al., 2016).

Since the scope of fluorescent labels for the in-depth tissue imaging is weakened by the limitation of fluorescence; PET, single-photon emission computed tomography (SPECT), MRI, and computed tomography (CT) may be considered as potential tools for non-invasive protease activity tracing approaches. Given the dependence of tracers and contrast agents in CT imaging, *in vivo* imaging of proteases has been attempted with MRI. Paramagnetic $Gd^{3}+$ compounds are conventionally preferred as contrast agents in MRI. A Tetraazacyclododecane tetraacetic acid (Gd-DOTA) probe coupled with MMP2 substrate has led to solubility changes that allowed possible imaging of MMP2 activity in tumors (Jastrzebska et al., 2009). MRI protease-sensitive imaging is endowed with higher promises for further expansion.

In most of the cases, the following concepts may be combined to understand the *in vivo* observations such as signal intensity vs. specific protease activity, permeability vs. cleavage, and affinity vs. specificity. *In vivo* protease activity imaging coupled with non-optical techniques is a field in its infancy but with substantial promise. The next decade will likely witness dramatic technological advances required to bring novel protease imaging agents to clinical trials (Yang et al., 2009).

2.4 PROTEASES AND CANCER THERAPY

The diverse tumor-promoting roles of proteases have underscored their importance as promising therapeutic targets. For example, inhibiting MMPs and several cathepsins has been recognized as one potential anti-cancer therapeutic approach. Early studies targeted MMPs with inhibitors based on the hypothesis that broad inhibition would hamper cancer progression. However, these treatments mostly failed due to the broad-spectrum nature of these therapeutics (Coussens et al., 2002). Improved understanding of protease biology paved the way for more selective therapeutic strategies targeting proteases. The following section systematically details recent advances in cancer therapeutics centered on proteases and their inhibitors.

2.4.1 Naturally Occurring Protease Inhibitors

Deubiquitinating enzymes are proteases that remove ubiquitin conjugates from proteins, thereby regulating protein turnover. These enzymes are involved in various cellular functions and, thus, are emerging as promising therapeutic targets for cancer and other diseases. As an important member of the ubiquitin-specific proteolytic enzyme family, ubiquitin-specific peptidase 20 (USP20) removes ubiquitin moieties from ubiquitin-labeled proteins. The involvement of USP20 in the progression of various cancer types, such as adult T-cell leukemia, lung, breast, colon, and gastric cancers, is well documented. Recently, GlaxoSmithKline (GSK) screened one inhibitor of USP20 sharing approximately 59% identity, which affects USP33 expression and has strong homology at the amino and carboxy termini. This novel USP20 inhibitor is expected to provide effective treatment options for cancer types with overexpressed USP20 (Li, Ye, et al., 2022).

Poly-(ADP-ribose) polymerase inhibitors (PARPi) selectively kill breast and ovarian cancer cells with homologous recombination defects due to BRCA1 or BRCA2 mutations. USP15 regulates homologous recombination and cancer cell response to PARPi. Mechanistically, USP15 is recruited at the DNA double-strand breaks (DSBs) and is phosphorylated. The phosphorylated USP15 deubiquitinates the BARD1-BRCT domain, which binds poly-ADP-ribose to regulate BARD1-BRCA1 accumulation at the DSBs and promote BARD1-HP1γ interaction, resulting in BRCA1/BARD1 retention at DSBs. USP15 knockout mice present genomic instability *in vivo*. USP15, thus, emerges as a novel regulator of homologous recombination and a potential biomarker for treatment using PARP inhibitors in cancers (Peng et al., 2019). Analysis of the Dependency and Connectivity Maps, Cancer Cell Line Encyclopedia, and protein-protein interaction databases developed a portal for deubiquitinating enzymes improved our understanding of this protease for the incompletely characterized deubiquitinating enzyme inhibitor for example USPL1 and USP32 and those already targeted with investigational cancer therapeutics such as USP14 and USP7 (Doherty & Mills, 2022).

An induction of STING-mediated type I interferons can augment the resultant anti-tumor activity. However, the STING pathway is generally suppressed in cancer. A deubiquitinase USP35, commonly upregulated in ovarian cancer (OCa), can deubiquitinate and inactivate STING and is responsible for diminished infiltration of CD8+ T-cell and poor prognosis in OCa patients. USP35 knockdown induces OCa cells to the DNA-damage chemotherapeutic drug cisplatin, highlighting its significance as a potential therapeutic target for OCa (Zhang et al., 2021). Programmed death-1 (PD-1) and its ligand PD-L1 are regarded as important targets of anti-tumor immunotherapy and are involved in tumor cell invasion during immune surveillance. Post-translational modification of PD-L1 has potential involvement in immunosuppression. Pancreatic cancer tissues have been shown to exhibit elevated USP8 levels, which deubiquitinate PD-L1. Clinically, the expression of USP8 is significantly correlated with the tumor-node-metastasis stage. The USP8 inhibitor and anti-PD-L1 combination therapy effectively suppresses pancreatic tumor growth by activating cytotoxic T-cells. This evidence highlights the importance of targeting USP8, which represents a novel therapeutic strategy by sensitizing PD-L1-targeted pancreatic cancer to immunotherapy in pancreatic cancer (Yang et al., 2023).

Our current understanding regarding the regulatory functions of Yes-associated protein (YAP) and transcriptional co-activator with PDZ-binding (TAZ) motif can be disrupted in cancer. YAP and TAZ are the important downstream effectors that belong to the Hippo pathway. They function and regulate tissue regeneration, tissue, homeostasis, organ size, and tumorigenesis (Moroishi et al., 2015). USP7 potently antagonizes ubiquitin-proteasomal degradation of TAZ and enhances its nuclear retention and transcriptional output. Remarkably, overexpression of USP7 is correlated with TAZ upregulation, tumor aggressiveness, and unfavorable prognosis in patients with head and neck squamous cell carcinomas (HNSCCs). Pharmacological inhibition of USP7 has significantly suppressed tumor growth in both xenograft and PDX models. This evidence elaborates a novel role of USP7 as an essential regulator of TAZ and defines the USP7-TAZ signaling axis as a novel biomarker and effective therapeutic target for HNSCC (Li, Dai, et al., 2022).

Neutrophil elastase is a major inflammatory serine proteinase released by neutrophils and is aberrantly regulated in chronic lung diseases. Recent research shows that neutrophil elastase (ELANE) can kill multiple cancer cell types while leaving the non-malignant cells. ELANE proteolytically cleaves and releases the CD95 death domain, which interacts with the isoforms of histone H1 to selectively eradicate cancer cells. Further, it attenuates the growth of primary tumors and generates a CD8+ T-cell-mediated abscopal effect to attack distant metastatic cells. In summary, ELANE is raising hopes for developing it as a broad anti-cancer therapy that effectively kills diverse cancer cells with minimal toxicity to non-cancer cells (Cui et al., 2021). Caseinolytic protease P is a mitochondrial serine protease. In mammalian cells, the heterodimerization of protease P with its chaperone results in a complex that maintains mitochondrial integrity by degrading mitochondrial misfolded or damaged proteins. The fact that caseinolytic protease P is upregulated in different primary and metastatic human tumors and its overexpression desensitizes cells to cisplatin motivates researchers to discover small modulators of protease P activity. Both activators and inhibitors are shown to impair oxidative phosphorylation (OXPHOS) in cancer cells and to induce apoptosis, underscoring the importance of this protease as a novel theranostic marker for the treatment of cancer (Cormio et al., 2021).

Recently, the term "PANoptosis" has been proposed, which includes three programmed cell death modes, namely pyroptosis, apoptosis, and necroptosis. Caspase-8 contributes to pyroptosis by the cleaving the gastrin family of proteins. Furthermore, enzymatically inactive caspase-8 (C362S) can induce necroptosis and pyroptosis by degradation of mice embryonic lethality and inflammatory tissue. Thus, modulation of caspase-8 enzyme activity is closely related to "PANoptosis." Together, the caspase-8-mediated "PANoptosis" provides new strategies and targets for cancer (Jiang et al., 2021).

2.4.2 Synthetic Protease Inhibitors

The earliest synthetic inhibitors were pseudo-peptides mimicking the cleavage sites of MMP substrates. A broad-spectrum hydroxamate-based inhibitor Batimastat (BB-94) was amongst the first MMPI trial in humans that was administered intraperitoneally. However, clinical trials with Batimastat were not encouraging, and later it was replaced by another orally available peptidomimetic MMPI called Marimastat (BB-2516). It inhibits the activity of many MMPs, such as MMP1, 2, 3, 7, 9, 12, and 13. As Marimastat has multiple targets, it causes musculoskeletal pain in patients after a sustained treatment (Rasmussen & McCann, 1997). Irrespective of this constraint, Marimastat is as effective as conventional therapy with gemcitabine in treating pancreatic carcinoma patients (Bramhall et al., 2002). Moreover, Marimastat alone could improve the survival rate and the time to disease progression in advanced gastric cancer patients and, while in combination with temozolomide, could improve survival in patients with glioblastoma multiforme (Mohanty et al., 2017).

Recently, an array of non-peptidomimetics MMPIs with higher specificity and oral bioavailability have been synthesized. They are based on the 3D structure of MMP zinc-binding sites. For instance, BMS-275291 has particular interest because of its minimum musculoskeletal side effects. Its effectiveness has been assessed in advanced prostate cancer, lung cancer, and AIDS-related Kaposi sarcoma. Moreover, effectiveness of the non-peptide MMPIs, such as tetracycline derivatives and bisphosphonates, are also tested in clinical trials.

An additional MMP inhibitor class was produced to fit the variable S1′ pocket of MMPs (Devel et al., 2012). MMPs contain an S1′ pocket near the catalytic Zn^{2+} ion, essential for recognizing the preferred substrate (Fischer & Riedl, 2019). Modification of the S1′ pocket of MMP12 and MMP13 facilitated the design of efficient MMP inhibitors (Nara et al., 2017). MMPIs based on protein therapeutics such as specific antibodies and TIMPs are preferred for advanced selectivity attributable to the large surface of protein-protein interaction and, thus, reduce toxicity. Moreover, employing protein engineering techniques such as surface display and directed evolution has also improved this class of MMPI drugs' selectivity, affinity, and stability (Levin et al., 2017).

Chimeric antigen receptor (CAR) T-cell therapy has shown a remarkable therapeutic activity in relapsed and refractory B-cell malignancies (Wang et al., 2020). However, this therapy leads to severe adverse events such as cytokine release syndrome (CRS) neurotoxicity and possesses a relatively limited success of CAR T-cell therapy in solid tumors that cannot be overlooked. The self-amplifying CAR T-cells exhibit high patient-to-patient variability in expansion as well as activity. As a result, constitutive CAR T-cell platforms provide inadequate prospect to control the activity and dosing of these potent immune effectors following the onset of toxicity. Very recently, a high-performance, US Food and Drug Administration (FDA)-approved drug-regulatable system called Signal Neutralization by an Inhibitable Protease (SNIP) has been advanced with favorable pharmacokinetics in humans. Besides providing solutions to safety and efficacy barriers, the platform has no leaky activity and outdoes constitutive CARs in multiple models of orthotopic solid tumor. Additionally, in an *in vivo* on-target off-tumor toxicity mouse model, reduced drug dosing in SNIP CAR T-cells has been considered as a therapeutic window. It flushes out tumor cells that express high levels of antigen while spare healthy tissue with lower antigen levels (Labanieh et al., 2022).

2.5 PROTEASE INHIBITORS IN CLINICAL TRIALS

Several preclinical and clinical data suggested that MMPs play a pivotal role in cancer progression. In the past 50 years, pharmaceutical industries have focused on developing novel drugs targeting the development of different malignant cells. MMPIs are used to treat cancers such as leukemias, lymphomas, testicular cancer, lung, gastrointestinal, and oropharyngeal cancer. Unfortunately, all the therapies had side effects. MMPIs may inhibit malignant growth by enhancing fibrosis around malignant lesions, preventing tumor invasion, apoptosis, and angiogenesis (Chaudhary et al., 2013).

MMPIs have also been designed based on its interaction with zinc ion and substrate-binding pockets. The zinc-binding group MMPIs can be categorized into hydroxamates, amino carboxylates, sulfhydryls, carboxylates, and phosphoric acid-based derivatives. Hydroxamates, due to their significant zinc-binding abilities, are preferred for their clinical development as MMPIs. The NH and OH groups of the hydroxamates interact with Glu and Ala residues, respectively, through hydrogen bonds (Baidya et al., 2021). Inhibitors of MMPs falling into peptidomimetics, non-peptidomimetics, naturally occurring MMPIs, and tetracycline-like derivatives are listed in Table 2.2.

Furthermore, the FDA has approved a variety of proteases for therapeutic application. In 1978, primary neonatal kidney cell derived uPA, was approved for clinical application and favored for its ability to dissolve clots in blood vessels and intravenous catheters. uPA is also involved in the degradation of ECM proteins, cell invasion, and metastasis; thus, potentiating the repurposing of this protease as a target in cancer treatment and diagnostics (Dass et al., 2008).

Oral administration of proteases and digestive enzymes cocktails for treating pancreatic insufficiency applies to treating cancer, cystic fibrosis, and several other diseases. Trizytek™ (Eli Lilly or Altus-135), comprising α-amylase, lipase, and an alkaline elastolytic protease from the fungus *Aspergillus melleus*, has been developed (Luisetti et al., 1991). A Phase III clinical trial of Trizytek™ has been completed. The patients treated with Trizytek™ displayed a statistically significant improvement in the coefficient of absorbed nitrogen, signifying that the protease component of the mixture was improving protein digestion. Despite the intense research in the field, certain drugs, namely MMP inhibitors (doxycycline) and proteasome (multi-catalytic protease complex) inhibitors (carfilzomib, ixazomib, and bortezomib) have been approved by the FDA. Proteasome inhibitors combined with DNA-damaging drugs such as melphalan and doxorubicin have shown a promising effect in the treatment of multiple myeloma and also exhibited their potential use in developing an anti-tumor therapy (Hideshima et al., 2005).

TABLE 2.2
Protease Inhibitors Reported in Advanced Stages of Clinical Development and Their Corresponding Therapeutic Targets

Protease Inhibitors (Company)	Chemical Nature	Specificity	Cancer	Clinical Status	References
COL-3, MetaStat (Collagenex)	Modified tetracyclines	MMP1, MMP2, MMP8, MMP9, and MMP13	Soft tissue sarcoma, prostate	Phase II	Chaudhary et al., 2013 Hidalgo & Eckhardt, 2001
WX-UK1 (Wilex)	3-amidinophenylalanine-based inhibitor	uPA	Pancreatic, hypopharynx, head, and neck, gastric, ovarian	Phase I/II	Ertongur et al., 2004 Heinemann et al., 2013
AE-941 (AEterna), (Neovastat, a naturally occurring inhibitor)	Shark cartilage extract	MMPs 2, 9, 12	Multiple myeloma and plasma cell neoplasm, bone and breast	Phase III	Weber et al., 2002
Bowman-Birk inhibitors (BBI) concentrate, (naturally occurring inhibitor)	MEROPS inhibitor family I12, clan IF	Broad	Colon	Phase II	Weed et al., 1985
AG3340, Prinomastat (Agouron)	Non-peptidomimetic inhibitor	Broad (MMP)	Non-small cell lung, breast, colon, and prostate	Phase III	Baidya et al., 2021 Hidalgo & Eckhardt, 2001
CGS-27023A, MMI 270 B (Novartis)	Non-peptidomimetic inhibitor	Broad	Melanoma	Phase I/II	Hidalgo & Eckhardt, 2001 Nakamura et al., 2003
BMS-275291 (Bristol-Myers)	Contains a thiol zinc- binding group	Broad (MMP)	Prostate, Kaposi sarcoma	Phase III	Brinker et al., 2008
Batimastat (BB-94)	Hydroxamate-based inhibitor	MMPs 1, 2, 7, and 9	Ovarian	Phase I/II	Erba et al., 1999
Marimastat (BB-2516)	Collagen-mimicking hydroxamate	MMPs 1, 2, 3, 7, 9, and 12	Gastric, colorectal, pancreatic	Phase I/II/III	Steward, 1999
Prinomastat (AG 3340)	Non-peptidic collagen-mimicking MMP inhibitor	MMPs 2, 3, 9, 13, and 14	Advanced non-small-cell lung cancer	Phase III	Hidalgo & Eckhardt, 2001
MMI 270 B (CGS27023A)	Broad nonpeptidic inhibitor	MMPs 1, 2, and 3	Breast, ovarian, prostate, bladder, colon, glioblastoma, and lung	Phase I	Levitt et al., 2001

2.6 CONCLUSION AND FUTURE PERSPECTIVES

In short, the complexity of proteolytic signaling in cancer has gone through significant conceptual and technological revolutions. As far as the therapy is concerned, it is important to contemplate whether direct protease targeting is a better strategy than one that uses them in combination with other therapies. New approaches, including protease-responsive nanomaterials and protease-activatable drugs, are raising hopes. Although multi-functional platforms can improve clinical outcomes and pave the way for proteases toward personalized medicine, a few notable aspects require further consideration. First, selecting the appropriate protease target is vital in theranostic system design, as some proteases may prove more practical for imaging or drug delivery based on their location in, on, or around cancer cells. Signal reception and efficient delivery are susceptible to protease location, quantity, and activity, which can vary and are not uniform among all cancer types. Second, profiling patient-to-patient variance in protease activity for a given cancer type remains to be addressed. Global assessment of protease activity in human samples in association with other omics data is another unmet area of research. These issues can be addressed by using molecules that possess both imaging and therapeutic properties, and work should be done to develop these types of theranostic probes. Addressing these challenges in the future will unleash the full potential of proteases for cancer diagnosis, disease staging, therapy stratification, and evaluation of therapeutic response.

DECLARATIONS

Conflict of interest: The authors declare no conflicts of interest.

Consent for publishing: All authors have approved the final version of the manuscript and given consent for publication.

Funding: Not Applicable.

REFERENCES

Baidya, S. K., Amin, S. A., & Jha, T. (2021). Outline of gelatinase inhibitors as anti-cancer agents: a patent mini-review for 2010-present. *Eur J Med Chem*, *213*, 113044. https://doi.org/10.1016/j.ejmech.2020.113044

Blau, R., Epshtein, Y., Pisarevsky, E., Tiram, G., Israeli Dangoor, S., Yeini, E., … Satchi-Fainaro, R. (2018). Image-guided surgery using near-infrared Turn-ON fluorescent nanoprobes for precise detection of tumor margins. *Theranostics*, *8*(13), 3437–3460. https://doi.org/10.7150/thno.23853

Boeneman, K., Mei, B. C., Dennis, A. M., Bao, G., Deschamps, J. R., Mattoussi, H., & Medintz, I. L. (2009). Sensing caspase 3 activity with quantum dot-fluorescent protein assemblies. *J Am Chem Soc*, *131*(11), 3828–3829. https://doi.org/10.1021/ja809721j

Borgoño, C. A., & Diamandis, E. P. (2004). The emerging roles of human tissue kallikreins in cancer. *Nat Rev Cancer*, *4*(11), 876–890. https://doi.org/10.1038/nrc1474

Bramhall, S., Schulz, J., Nemunaitis, J., Brown, P., Baillet, M., & Buckels, J. (2002). A double-blind placebo-controlled, randomised study comparing gemcitabine and marimastat with gemcitabine and placebo as first line therapy in patients with advanced pancreatic cancer. *Br J Cancer*, *87*(2), 161–167.

Breger, J. C., Sapsford, K. E., Ganek, J., Susumu, K., Stewart, M. H., & Medintz, I. L. (2014). Detecting kallikrein proteolytic activity with peptide-quantum dot nanosensors. *ACS Appl Mater Interfaces*, *6*(14), 11529–11535. https://doi.org/10.1021/am502135h

Brinker, B. T., Krown, S. E., Lee, J. Y., Cesarman, E., Chadburn, A., Kaplan, L. D., … Von Roenn, J. H. (2008). Phase 1/2 trial of BMS-275291 in patients with human immunodeficiency virus-related Kaposi sarcoma: a multicenter trial of the AIDS Malignancy Consortium. *Cancer*, *112*(5), 1083–1088. https://doi.org/10.1002/cncr.23108

Chaudhary, A. K., Pandya, S., Ghosh, K., & Nadkarni, A. (2013). Matrix metalloproteinase and its drug targets therapy in solid and hematological malignancies: an overview. *Mutat Res*, *753*(1), 7–23. https://doi.org/10.1016/j.mrrev.2013.01.002

Chuang, C. H., Chuang, K. H., Wang, H. E., Roffler, S. R., Shiea, J. T., Tzou, S. C., … Cheng, T. L. (2012). *In vivo* positron emission tomography imaging of protease activity by generation of a hydrophobic product from a noninhibitory protease substrate. *Clin Cancer Res*, *18*(1), 238–247. https://doi.org/10.1158/1078-0432.ccr-11-0608

Cormio, A., Sanguedolce, F., & Pesce, V. (2021). Mitochondrial caseinolytic protease P: a possible novel prognostic marker and therapeutic target in cancer. *Int J Mol Sci*, *22*(12). https://doi.org/10.3390/ijms22126228

Coussens, L. M., Fingleton, B., & Matrisian, L. M. (2002). Matrix metalloproteinase inhibitors and cancer: trials and tribulations. *Science*, *295*(5564), 2387–2392. https://doi.org/10.1126/science.1067100

Cui, C., Chakraborty, K., Tang, X. A., Zhou, G., Schoenfelt, K. Q., Becker, K. M., … Becker, L. (2021). Neutrophil elastase selectively kills cancer cells and attenuates tumorigenesis. *Cell, 184*(12), 3163–3177. e3121. https://doi.org/10.1016/j.cell.2021.04.016

Darragh, M. R., Schneider, E. L., Lou, J., Phojanakong, P. J., Farady, C. J., Marks, J. D., … Craik, C. S. (2010). Tumor detection by imaging proteolytic activity. *Cancer Res, 70*(4), 1505–1512. https://doi.org/10.1158/0008-5472.can-09-1640

Dass, K., Ahmad, A., Azmi, A. S., Sarkar, S. H., & Sarkar, F. H. (2008). Evolving role of uPA/uPAR system in human cancers. *Cancer Treat Rev, 34*(2), 122–136. https://doi.org/10.1016/j.ctrv.2007.10.005

Deryugina, E. I., & Quigley, J. P. (2010). Pleiotropic roles of matrix metalloproteinases in tumor angiogenesis: contrasting, overlapping and compensatory functions. *Biochim Biophys Acta, 1803*(1), 103–120. https://doi.org/10.1016/j.bbamcr.2009.09.017

Devel, L., Beau, F., Amoura, M., Vera, L., Cassar-Lajeunesse, E., Garcia, S., … Dive, V. (2012). Simple pseudo-dipeptides with a P2' glutamate: a novel inhibitor family of matrix metalloproteases and other metzincins. *J Biol Chem, 287*(32), 26647–26656. https://doi.org/10.1074/jbc.M112.380782

Doherty, L. M., & Mills, C. E. (2022). Integrating multi-omics data reveals function and therapeutic potential of deubiquitinating enzymes. *eLife, 11*. https://doi.org/10.7554/eLife.72879

Dudani, J. S., Warren, A. D., & Bhatia, S. N. (2018). Harnessing protease activity to improve cancer care. *Annu Rev Cancer Biol, 2*(1), 353–376. https://doi.org/10.1146/annurev-cancerbio-030617-050549

Edgington, L. E., Verdoes, M., Ortega, A., Withana, N. P., Lee, J., Syed, S., … Bogyo, M. (2013). Functional imaging of legumain in cancer using a new quenched activity-based probe. *J Am Chem Soc, 135*(1), 174–182. https://doi.org/10.1021/ja307083b

Edgington-Mitchell, L. E., Bogyo, M., & Verdoes, M. (2017). Live cell imaging and profiling of cysteine cathepsin activity using a quenched activity-based probe. *Methods Mol Biol, 1491*, 145–159. https://doi.org/10.1007/978-1-4939-6439-0_11

Erba, E., Ronzoni, S., Bassano, L., Giavazzi, R., & D'Incalci, M. (1999). The metalloproteinase inhibitor batimastat (BB-94) causes cell cycle phase perturbations in ovarian cancer cells. *Ann Oncol, 10*(5), 589–591. https://doi.org/10.1023/a:1026438801080

Ertongur, S., Lang, S., Mack, B., Wosikowski, K., Muehlenweg, B., & Gires, O. (2004). Inhibition of the invasion capacity of carcinoma cells by WX-UK1, a novel synthetic inhibitor of the urokinase-type plasminogen activator system. *Int J Cancer, 110*(6), 815–824. https://doi.org/10.1002/ijc.20192

Fischer, T., & Riedl, R. (2019). Inhibitory antibodies designed for matrix metalloproteinase modulation. *Molecules, 24*(12). https://doi.org/10.3390/molecules24122265

Grimm, J., Kirsch, D. G., Windsor, S. D., Kim, C. F., Santiago, P. M., Ntziachristos, V., … Weissleder, R. (2005). Use of gene expression profiling to direct *in vivo* molecular imaging of lung cancer. *Proc Natl Acad Sci USA, 102*(40), 14404–14409. https://doi.org/10.1073/pnas.0503920102

Gruba, N., & Bielecka, E. (2019). Development of chemical tools to monitor human Kallikrein 13 (KLK13) activity. *Int. J. Mol. Sci., 20*(7), 1557. https://doi.org/10.3390/ijms20071557

Heinemann, V., Ebert, M. P., Laubender, R. P., Bevan, P., Mala, C., & Boeck, S. (2013). Phase II randomised proof-of-concept study of the urokinase inhibitor upamostat (WX-671) in combination with gemcitabine compared with gemcitabine alone in patients with non-resectable, locally advanced pancreatic cancer. *Br J Cancer, 108*(4), 766–770. https://doi.org/10.1038/bjc.2013.62

Hewings, D. S., Flygare, J. A., Wertz, I. E., & Bogyo, M. (2017). Activity-based probes for the multicatalytic proteasome. *FEBS J, 284*(10), 1540–1554. https://doi.org/10.1111/febs.14016

Hidalgo, M., & Eckhardt, S. G. (2001). Development of matrix metalloproteinase inhibitors in cancer therapy. *J Natl Cancer Inst, 93*(3), 178–193. https://doi.org/10.1093/jnci/93.3.178

Hideshima, T., Bradner, J. E., Wong, J., Chauhan, D., Richardson, P., Schreiber, S. L., & Anderson, K. C. (2005). Small-molecule inhibition of proteasome and aggresome function induces synergistic antitumor activity in multiple myeloma. *Proc Natl Acad Sci USA, 102*(24), 8567–8572. https://doi.org/10.1073/pnas.0503221102

Hu, H. Y., Gehrig, S., Reither, G., Subramanian, D., Mall, M. A., Plettenburg, O., & Schultz, C. (2014). FRET-based and other fluorescent proteinase probes. *Biotechnol J, 9*(2), 266–281. https://doi.org/10.1002/biot.201300201

Hu, H. Y., Vats, D., Vizovisek, M., Kramer, L., Germanier, C., Wendt, K. U., … Schultz, C. (2014). *In vivo* imaging of mouse tumors by a lipidated cathepsin S substrate. *Angew Chem Int Ed Engl, 53*(29), 7669–7673. https://doi.org/10.1002/anie.201310979

Hu, J., Jo, M., Eastman, B. M., Gilder, A. S., Bui, J. D., & Gonias, S. L. (2014). uPAR induces expression of transforming growth factor β and interleukin-4 in cancer cells to promote tumor-permissive conditioning of macrophages. *Am J Pathol, 184*(12), 3384–3393. https://doi.org/10.1016/j.ajpath.2014.08.003

Jastrzebska, B., Lebel, R., Therriault, H., McIntyre, J. O., Escher, E., Guérin, B., ... Lepage, M. (2009). New enzyme-activated solubility-switchable contrast agent for magnetic resonance imaging: from synthesis to *in vivo* imaging. *J Med Chem*, *52*(6), 1576–1581. https://doi.org/10.1021/jm801411h

Jeong, S., Song, J., Lee, W., Ryu, Y. M., Jung, Y., Kim, S. Y., ... Kim, S. (2017). Cancer-microenvironment-sensitive activatable quantum dot probe in the second near-infrared window. *Nano Lett*, *17*(3), 1378–1386. https://doi.org/10.1021/acs.nanolett.6b04261

Jiang, M., Qi, L., Li, L., Wu, Y., Song, D., & Li, Y. (2021). Caspase-8: a key protein of cross-talk signal way in "PANoptosis" in cancer. *Int J Cancer*, *149*(7), 1408–1420.

Jiang, T., Olson, E. S., Nguyen, Q. T., Roy, M., Jennings, P. A., & Tsien, R. Y. (2004). Tumor imaging by means of proteolytic activation of cell-penetrating peptides. *Proc Natl Acad Sci USA*, *101*(51), 17867–17872. https://doi.org/10.1073/pnas.0408191101

Kasperkiewicz, P., Altman, Y., D'Angelo, M., Salvesen, G. S., & Drag, M. (2017). Toolbox of fluorescent probes for parallel imaging reveals uneven location of serine proteases in neutrophils. *J Am Chem Soc*, *139*(29), 10115–10125. https://doi.org/10.1021/jacs.7b04394

Kenig, S., Frangež, R., Pucer, A., & Lah, T. (2011). Inhibition of cathepsin L lowers the apoptotic threshold of glioblastoma cells by up-regulating p53 and transcription of caspases 3 and 7. *Apoptosis*, *16*(7), 671–682. https://doi.org/10.1007/s10495-011-0600-6

Kessenbrock, K., Plaks, V., & Werb, Z. (2010). Matrix metalloproteinases: regulators of the tumor microenvironment. *Cell*, *141*(1), 52–67. https://doi.org/10.1016/j.cell.2010.03.015

Kwong, G. A., von Maltzahn, G., Murugappan, G., Abudayyeh, O., Mo, S., Papayannopoulos, I. A., ... Bhatia, S. N. (2013). Mass-encoded synthetic biomarkers for multiplexed urinary monitoring of disease. *Nat Biotechnol*, *31*(1), 63–70. https://doi.org/10.1038/nbt.2464

Labanieh, L., Majzner, R. G., Klysz, D., Sotillo, E., Fisher, C. J., Vilches-Moure, J. G., ... Mackall, C. L. (2022). Enhanced safety and efficacy of protease-regulated CAR-T cell receptors. *Cell*, *185*(10), 1745–1763.e1722. https://doi.org/10.1016/j.cell.2022.03.041

Larson, D. R., Zipfel, W. R., Williams, R. M., Clark, S. W., Bruchez, M. P., Wise, F. W., & Webb, W. W. (2003). Water-soluble quantum dots for multiphoton fluorescence imaging *in vivo*. *Science*, *300*(5624), 1434–1436. https://doi.org/10.1126/science.1083780

Levene, P. (1905). The cleavage products of proteoses. *J Biol Chem*, *1*(1), 45–58.

Levin, M., Udi, Y., Solomonov, I., & Sagi, I. (2017). Next generation matrix metalloproteinase inhibitors — Novel strategies bring new prospects. *Biochimica et Biophysica Acta (BBA) - Molecular Cell Research*, *1864*(11, Part A), 1927–1939. https://doi.org/https://doi.org/10.1016/j.bbamcr.2017.06.009

Levitt, N. C., Eskens, F. A., O'Byrne, K. J., Propper, D. J., Denis, L. J., Owen, S. J., ... Verweij, J. (2001). Phase I and pharmacological study of the oral matrix metalloproteinase inhibitor, MMI270 (CGS27023A), in patients with advanced solid cancer. *Clin Cancer Res*, *7*(7), 1912–1922.

Li, J., Dai, Y., Ge, H., Guo, S., Zhang, W., Wang, Y., ... Jiang, H. (2022). The deubiquitinase USP7 promotes HNSCC progression via deubiquitinating and stabilizing TAZ. *Cell Death Dis*, *13*(8), 677. https://doi.org/10.1038/s41419-022-05113-z

Li, Q., Ye, C., Tian, T., Jiang, Q., Zhao, P., Wang, X., ... Ruan, J. (2022). The emerging role of ubiquitin-specific protease 20 in tumorigenesis and cancer therapeutics. *Cell Death Dis*, *13*(5), 434. https://doi.org/10.1038/s41419-022-04853-2

Li, X., Deng, D., Xue, J., Qu, L., Achilefu, S., & Gu, Y. (2014). Quantum dots based molecular beacons for *in vitro* and *in vivo* detection of MMP-2 on tumor. *Biosens Bioelectron*, *61*, 512–518. https://doi.org/10.1016/j.bios.2014.05.035

López-Otín, C., & Bond, J. S. (2008). Proteases: multifunctional enzymes in life and disease. *J Biol Chem*, *283*(45), 30433–30437. https://doi.org/10.1074/jbc.R800035200

López-Otín, C., & Matrisian, L. M. (2007). Emerging roles of proteases in tumour suppression. *Nat Rev Cancer*, *7*(10), 800–808. https://doi.org/10.1038/nrc2228

Lowe, S. B., Dick, J. A., Cohen, B. E., & Stevens, M. M. (2012). Multiplex sensing of protease and kinase enzyme activity via orthogonal coupling of quantum dot-peptide conjugates. *ACS Nano*, *6*(1), 851–857. https://doi.org/10.1021/nn204361s

Luisetti, M., Piccioni, P. D., Dyne, K., Donnini, M., Bulgheroni, A., Pasturenzi, L., ... Peona, V. (1991). Some properties of the alkaline proteinase from Aspergillus melleus. *Int J Tissue React*, *13*(4), 187–192.

Luo, Y., Zhou, H., Krueger, J., Kaplan, C., Lee, S. H., Dolman, C., ... Xiang, R. (2006). Targeting tumor-associated macrophages as a novel strategy against breast cancer. *J Clin Invest*, *116*(8), 2132–2141. https://doi.org/10.1172/jci27648

Mahmood, U., & Weissleder, R. (2003). Near-infrared optical imaging of proteases in cancer. *Mol Cancer Ther*, *2*(5), 489–496.

Mason, S. D., & Joyce, J. A. (2011). Proteolytic networks in cancer. *Trends Cell Biol*, *21*(4), 228–237. https://doi.org/10.1016/j.tcb.2010.12.002
Medintz, I. L., Clapp, A. R., Brunel, F. M., Tiefenbrunn, T., Uyeda, H. T., Chang, E. L., ... Mattoussi, H. (2006). Proteolytic activity monitored by fluorescence resonance energy transfer through quantum-dot-peptide conjugates. *Nat Mater*, *5*(7), 581–589. https://doi.org/10.1038/nmat1676
Mohanty, S., Chen, Z., Li, K., Morais, G. R., Klockow, J., Yerneni, K., ... Cheshier, S. (2017). A novel theranostic strategy for MMP-14–Expressing glioblastomas impacts survival. *Mol Cancer Ther*, *16*(9), 1909–1921.
Nakamura, E. S., Koizumi, K., Yamaura, T., & Saiki, I. (2003). Anti-tumor angiogenic effect of a matrix metalloproteinase inhibitor MMI270. *Anticancer Res*, *23*(1a), 411–417.
Nara, H., Kaieda, A., Sato, K., Naito, T., Mototani, H., Oki, H., ... Kori, M. (2017). Discovery of novel, highly potent, and selective matrix metalloproteinase (MMP)-13 inhibitors with a 1,2,4-triazol-3-yl moiety as a zinc binding group using a structure-based design approach. *J Med Chem*, *60*(2), 608–626. https://doi.org/10.1021/acs.jmedchem.6b01007
Olson, E. S., Jiang, T., Aguilera, T. A., Nguyen, Q. T., Ellies, L. G., Scadeng, M., & Tsien, R. Y. (2010). Activatable cell penetrating peptides linked to nanoparticles as dual probes for *in vivo* fluorescence and MR imaging of proteases. *Proc Natl Acad Sci USA*, *107*(9), 4311–4316. https://doi.org/10.1073/pnas.0910283107
Olson, O. C., & Joyce, J. A. (2015). Cysteine cathepsin proteases: regulators of cancer progression and therapeutic response. *Nat Rev Cancer*, *15*(12), 712–729. https://doi.org/10.1038/nrc4027
Pan, Z., Jeffery, D. A., Chehade, K., Beltman, J., Clark, J. M., Grothaus, P., ... Baruch, A. (2006). Development of activity-based probes for trypsin-family serine proteases. *Bioorg Med Chem Lett*, *16*(11), 2882–2885. https://doi.org/10.1016/j.bmcl.2006.03.012
Peng, Y., Liao, Q., Tan, W., Peng, C., Hu, Z., Chen, Y., ... Pei, H. (2019). The deubiquitylating enzyme USP15 regulates homologous recombination repair and cancer cell response to PARP inhibitors. *Nature Communications*, *10*(1), 1224. https://doi.org/10.1038/s41467-019-09232-8
Poreba, M., Groborz, K., Navarro, M., Snipas, S. J., Drag, M., & Salvesen, G. S. (2019). Caspase selective reagents for diagnosing apoptotic mechanisms. *Cell Death Differ*, *26*(2), 229–244. https://doi.org/10.1038/s41418-018-0110-y
Poreba, M., Rut, W., Vizovisek, M., Groborz, K., Kasperkiewicz, P., & Finlay, D. (2018). Selective imaging of cathepsin L in breast cancer by fluorescent activity-based probes. *Chemical Science*, *9*(8), 2113–2129. https://doi.org/10.1039/c7sc04303a
Rasmussen, H. S., & McCann, P. P. (1997). Matrix metalloproteinase inhibition as a novel anticancer strategy: a review with special focus on batimastat and marimastat. *Pharmacol Ther*, *75*(1), 69–75.
Saghatelian, A., Jessani, N., Joseph, A., Humphrey, M., & Cravatt, B. F. (2004). Activity-based probes for the proteomic profiling of metalloproteases. *Proc Natl Acad Sci USA*, *101*(27), 10000–10005. https://doi.org/10.1073/pnas.0402784101
Savariar, E. N., Felsen, C. N., Nashi, N., Jiang, T., Ellies, L. G., Steinbach, P., ... Nguyen, Q. T. (2013). Real-time *in vivo* molecular detection of primary tumors and metastases with ratiometric activatable cell-penetrating peptides. *Cancer Res*, *73*(2), 855–864. https://doi.org/10.1158/0008-5472.can-12-2969
Shi, H., Uttamchandani, M., & Yao, S. Q. (2011). Applying small molecule microarrays and resulting affinity probe cocktails for proteome profiling of mammalian cell lysates. *Chem Asian J*, *6*(10), 2803–2815. https://doi.org/10.1002/asia.201100523
Sieber, S. A., Niessen, S., Hoover, H. S., & Cravatt, B. F. (2006). Proteomic profiling of metalloprotease activities with cocktails of active-site probes. *Nat Chem Biol*, *2*(5), 274–281. https://doi.org/10.1038/nchembio781
Smith, B. R., & Gambhir, S. S. (2017). Nanomaterials for *in vivo* imaging. *Chem Rev*, *117*(3), 901–986. https://doi.org/10.1021/acs.chemrev.6b00073
Steward, W. P. (1999). Marimastat (BB2516): current status of development. *Cancer Chemother Pharmacol*, *43*(Suppl), S56–S60. https://doi.org/10.1007/s002800051099
Suurs, F. V., Qiu, S. Q., Yim, J. J., Schröder, C. P., Timmer-Bosscha, H., Bensen, E. S., ... van Dam, G. M. (2020). Fluorescent image-guided surgery in breast cancer by intravenous application of a quenched fluorescence activity-based probe for cysteine cathepsins in a syngeneic mouse model. *EJNMMI Res*, *10*(1), 111. https://doi.org/10.1186/s13550-020-00688-0
Tholen, M., Yim, J. J., Groborz, K., Yoo, E., Martin, B. A., van den Berg, N. S., ... Bogyo, M. (2020). Design of optical-imaging probes by screening of diverse substrate libraries directly in disease-tissue extracts. *Angew Chem Int Ed*, *59*(43), 19143–19152. https://doi.org/10.1002/anie.202006719

Verdoes, M., Edgington, L. E., Scheeren, F. A., Leyva, M., Blum, G., Weiskopf, K., ... Bogyo, M. (2012). A nonpeptidic cathepsin S activity-based probe for noninvasive optical imaging of tumor-associated macrophages. *Chem Biol*, *19*(5), 619–628. https://doi.org/10.1016/j.chembiol.2012.03.012

Wang, M., Munoz, J., Goy, A., Locke, F. L., Jacobson, C. A., Hill, B. T., ... Topp, M. S. (2020). KTE-X19 CAR T-cell therapy in relapsed or refractory mantle-cell lymphoma. *N Engl J Med*, *382*(14), 1331–1342. https://doi.org/10.1056/NEJMoa1914347

Warren, A. D., Gaylord, S. T., Ngan, K. C., Dumont Milutinovic, M., Kwong, G. A., Bhatia, S. N., & Walt, D. R. (2014). Disease detection by ultrasensitive quantification of microdosed synthetic urinary biomarkers. *J Am Chem Soc*, *136*(39), 13709–13714. https://doi.org/10.1021/ja505676h

Warren, A. D., Kwong, G. A., Wood, D. K., Lin, K. Y., & Bhatia, S. N. (2014). Point-of-care diagnostics for noncommunicable diseases using synthetic urinary biomarkers and paper microfluidics. *Proc Natl Acad Sci USA*, *111*(10), 3671–3676. https://doi.org/10.1073/pnas.1314651111

Weber, M. H., Lee, J., & Orr, F. W. (2002). The effect of Neovastat (AE-941) on an experimental metastatic bone tumor model. *Int J Oncol*, *20*(2), 299–303.

Weed, H. G., McGandy, R. B., & Kennedy, A. R. (1985). Protection against dimethylhydrazine-induced adenomatous tumors of the mouse colon by the dietary addition of an extract of soybeans containing the Bowman-Birk protease inhibitor. *Carcinogenesis*, *6*(8), 1239–1241. https://doi.org/10.1093/carcin/6.8.1239

Whitley, M. J., Cardona, D. M., Lazarides, A. L., Spasojevic, I., Ferrer, J. M., Cahill, J., ... Brigman, B. E. (2016). A mouse-human phase 1 co-clinical trial of a protease-activated fluorescent probe for imaging cancer. *Sci Transl Med*, *8*(320), 320ra324. https://doi.org/10.1126/scitranslmed.aad0293

Wilkinson, R. D., Williams, R., Scott, C. J., & Burden, R. E. (2015). Cathepsin S: therapeutic, diagnostic, and prognostic potential. *Biol Chem*, *396*(8), 867–882. https://doi.org/10.1515/hsz-2015-0114

Withana, N. P., Ma, X., McGuire, H. M., Verdoes, M., van der Linden, W. A., Ofori, L. O., ... Bogyo, M. (2016). Non-invasive imaging of idiopathic pulmonary fibrosis using cathepsin protease probes. *Sci Rep*, *6*(1), 19755. https://doi.org/10.1038/srep19755

Wysocka, M., Wojtysiak, A., Brzezińska-Bodal, M., Sychowska, K., Kalińska, M., Magoch, M., ... Falkowski, K.

Yang, H., Zhang, X., Lao, M., Sun, K., He, L., Xu, J., ... Bai, X. (2023). Targeting ubiquitin-specific protease 8 sensitizes anti-programmed death-ligand 1 immunotherapy of pancreatic cancer. *Cell Death Differ*, *30*(2), 560–575. https://doi.org/10.1038/s41418-022-01102-z

Yang, Y., Hong, H., Zhang, Y., & Cai, W. (2009). Molecular imaging of proteases in cancer. *Cancer Growth Metastasis*, *2*, 13–27. https://doi.org/10.4137/cgm.s2814

Yhee, J. Y., Kim, S. A., Koo, H., Son, S., Ryu, J. H., Youn, I. C., ... Kim, K. (2012). Optical imaging of cancer-related proteases using near-infrared fluorescence matrix metalloproteinase-sensitive and cathepsin B-sensitive probes. *Theranostics*, *2*(2), 179–189. https://doi.org/10.7150/thno.3716

Zhang, J., Chen, Y., Chen, X., Zhang, W., Zhao, L., Weng, L., ... Ge, X. (2021). Deubiquitinase USP35 restrains STING-mediated interferon signaling in ovarian cancer. *Cell Death Differ*, *28*(1), 139–155. https://doi.org/10.1038/s41418-020-0588-y

3 Structural Aspects of Proteases in Cancer Therapeutics

A Perspective

Anirban Nandy, Sima Biswas, and Angshuman Bagchi
University of Kalyani, Kalyani, Nadia, India

3.1 INTRODUCTION

Cancer, with its ability to infiltrate virtually any type of tissue in the human body, is a formidable adversary of human health and still poses a challenging threat that continues to baffle scientists, clinicians and patients alike (Hausman, 2019). In this chapter, we embark on a journey to unfold the complexity of cancer, shedding light on its diverse manifestations, the milestones made in terms of cancer research and finally on the cutting-edge advancements in its diagnosis and treatment.

3.2 THE LANDSCAPE OF CANCER

The term "cancer" is used as an umbrella term for a group of diseases characterized by uncontrolled cell proliferation, growth and potential metastasis (Torre et al., 2016). One of the most formidable aspects of comprehending the cancer landscape is to understand the genetic mutations that underlie its development. The discovery and study of oncogenes and tumour suppressor genes (TSG) has revolutionized our understanding that how normal cellular processes can be affected, paving the therapeutic ways in terms of targeting the specific genetic alterations without affecting the neighbouring tissues (Wang et al., 2018).

3.3 ADVANCEMENTS IN DIAGNOSIS AND EARLY DETECTION

A quite formidable challenge lies in the early diagnosis and detection of cancer. As this disease is known to creep silently in its early stages, the detection used to occur generally at advanced and less treatable stages. However, some remarkable strides in diagnostic techniques have been witnessed by the medical community. High-resolution imaging techniques, such as magnetic resonance imaging (MRI) and positron emission tomography (PET), now offer extremely fine visualization of tumours and metastatic sites. Moreover, techniques like liquid biopsies involving the analysis of circulating tumour DNA and other biomarkers in the blood, exhibit great promise in detecting cancer in its earliest and treatable stages, enabling timely intervention and patient outcomes (Schiffman et al., 2015).

Traditional approaches like radiation therapy and chemotherapy, although effective, often come with severe side effects due to their non-specificity and indiscriminate behaviours. Here enters precision medicine, which solely focuses on the identification of specific genetic alterations in a patient's tumour and minimizing its effects on the neighbouring healthy cells. One of the most remarkable breakthroughs in cancer treatment is the development of immunotherapy. This type of diagnostic

DOI: 10.1201/9781003394693-4

capitalizes on the body's immune system to recognize and nullify cancer cells. Unprecedented success has been shown by immune checkpoint inhibitors. They function by blocking the mechanisms tumours use to evade the body's immune detection (Bhinder et al., 2021). Furthermore, CAR-T cell therapy, which is a type of adoptive cell transfer, genetically involves modifying a patient's T cells to target the particular cancer antigens. This ground-breaking therapy has exhibited remarkable responses in patients with haematologic malignancies, thus opening new paths for personalized cancer treatment (Sterner & Sterner, 2021).

Significant challenges still persist in the field of personalized cancer treatment, and this is quite prominent in various fields as follows:

1. *Tumour Heterogeneity*: Tumours can vary greatly in terms of their cellular characteristics and mutations at the genetic level, leading to the high heterogeneity of cancer. Identifying and administering the right medicines for the right mutations or biomarkers for each patient can be quite challenging.
2. *Biomarker Discovery*: Reliable biomarker identification is one of the most crucial steps in understanding the treatment as well as the progression of the disease. Identifying them remains a challenging task.
3. *Resistance to Targeted Therapies*: The cancer cells do have the unique property in terms of developing resistance to the specific targeted therapies over time. Overcoming this challenge can lead to highly effective treatments.
4. *Regulatory Approval and Drug Development*: The development of new targeted therapies and receiving approval from the regulatory agencies is a quite lengthy and complex process. Streamlining this process without compromising safety is a challenge for researchers.

Although a lot of advanced technologies have emerged to fight this lethal disease, still these previously mentioned issues need to be addressed. In the complex scenario of cancer biology, the role of extracellular proteases has emerged to be a pivotal field of research (Lippman & Hawk, 2009). These enzymes, generally responsible for the breakdown of proteins within the extracellular matrix (ECM), have stolen the spotlight due to their massive involvement in various stages of cancer.

3.4 CANCER AND PLANT PROTEASES

Plant proteases are a subset of proteolytic enzymes that are not only involved in the hydrolysis of peptide bonds but are also responsible for the treatment of cancer. Proteases derived from plants have been studied extensively and have exhibited anti-cancer properties through induction of apoptosis, inhibition of angiogenesis and modulation of the microenvironment components of several types of tumours. By targeting the major signalling pathways and by promoting cellular deaths in cancerous cells, these proteases have shown promising results by promoting death in cancer cells.

This chapter presents a comprehensive analysis of the pre-clinical studies highlighting the potential of plant proteases in the treatment of cancer. For example, we have bromelain from pineapple (Hikisz & Bernasinska-Slomczewska, 2021), papain from papaya (Jedeszko & Sloane, 2004) or nattokinase from soybeans. Certain factors have put these proteases in the limelight of cancer research. These proteases are known to exhibit anti-proliferative effects, inhibition of metastasis and are quite sensitive towards conventional chemotherapeutic agents. The intricate interplay between cancer cells and plant proteases underscores their promise as means of alternative therapies.

Plants and their medicinal products serve as the raw materials for multiple drugs. Plant extracts that contain proteolytic enzymes have been in use since time immemorial (De Feo, 1992). Protease inhibitors are positioned in the group of low-molecular-weight proteins, which need to be studied further. Studies have found that the onset of breast, prostate and colon cancers are quite low in the population that consumes a high amount of seeds like rice, maize and beans (Correa, 1981). The most extensively studied protease inhibitors found in legumes are the serine protease Bowman-Birk inhibitors (BBI) and Kunitz-type inhibitors (KTI) (Ryan, 1990). But amongst all, BBI is considered to be one of the most preferable candidates to work with as it is resistant to relatively high temperatures as well as acidic environment, and therefore, is considered to be one of the most suitable candidates in inhibiting several disease processes like cancer (Fields et al., 2012).

3.5 MECHANISM OF ACTION OF BOWMAN-BIRK INHIBITORS

BBIs are responsible for the breakdown of dietary proteins and inhibiting serine proteases like trypsin and chymotrypsin in both humans as well as in herbivorous animals (Kennedy, 1993).

Binding to Active Sites: BBIs possess a 3D structure that allows them to bind tightly to the active site of serine proteases. This strong and irreversible binding leads to the inhibition of protease activity.

Competitive Inhibition: Occupying the active site and functioning as competitive inhibitors, BBIs prevent the protease from cleaving the target protein, thus slowing the digestion of proteins.

Multiple Protease Targets: BBIs are known to inhibit both trypsin and chymotrypsin, thus, are responsible for targeting multiple protease targets. In humans, BBIs are known to inhibit proteases involved in cancer metastasis and tumour growth. By inhibiting these proteases, BBIs may help in slowing the progression of certain cancer types. Also, BBIs are involved in the modulation of inflammation by inhibiting proteases involved in the inflammatory responses (Chen et al., 2020).

Many plant protease inhibitors (PPI) are under clinical trial and further approvals need to be made. The most well-studied PPIs are from plant families as follows: *Solanaceae*, *Fabaceae* and *Poaceae* (Srikanth & Chen, 2016). Plant biotechnology is emerging as a revolutionary field and is quite useful in inducing the production and accumulation of desirable products and ultimately leading to the formation of therapeutic products (Constabel, 1990).

An increment in proteolytic activity is being shown by malignant cells that aids in the digestion of ECM (Lu et al., 2012). ECM-degrading proteases are known to play crucial roles in angiogenesis as they can act both as positive and negative regulators of endothelial cell proliferation as well as vascular morphogenesis. Tumour progression and metastases are heavily dependent on oxygen and nutrient supply and this is highly upregulated by various types of proteases present in the tumour and in the surrounding tissues (Yang et al., 2009). The major cause of death in cancer patients appears to occur due to metastases. Thus, reducing the metastatic progression is considered to be one of the greatest challenges in developing drugs for cancer therapies (Yang et al., 2020a,b).

3.6 ROLE OF PROTEASE INHIBITORS IN CANCER

These are the molecules that are designed to block the activity of certain proteases, thus interfering with vital processes in cancer development. The following section will provide an overview of the mechanisms by which protease inhibitors show their anti-cancer effects.

1. *Matrix metalloproteinase (MMP) Inhibitors*: Synthetic MMP inhibitors, such as Batimastat and Marimastat function by binding onto the active site of MMPs, thus preventing them from degrading the extracellular components. The latest developments have mainly focused on specific or selective inhibitors, for example, GM6001 is known to target specific MMP subtypes, thus avoiding the side effects associated with broad-spectrum inhibitors. Moreover, nanotechnology-based drug delivery systems are known to enhance the target specific delivery of MMP inhibitors to the tumour sites, therefore minimizing systemic

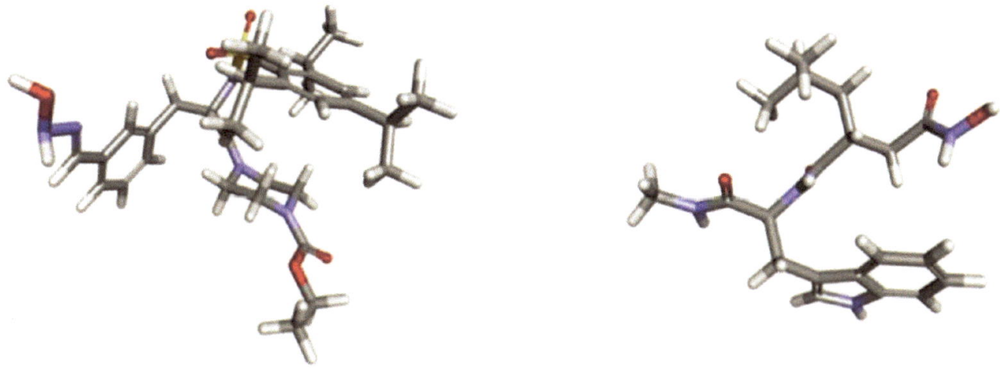

WX 671 GM6001

FIGURE 3.1 Stick representations of protease inhibitors WX671 and GM600.

toxicity. The latest developments in MMP inhibition focus on selective inhibitors like GM6001, which target specific MMP subtypes, avoiding the side effects associated with broad-spectrum inhibitors. Additionally, nanotechnology-based delivery systems enhance the targeted delivery of MMP inhibitors to tumour sites, minimizing systemic toxicity (Mondal et al., 2020).

2. *Serine Protease Inhibitors* (*SERPIN*): It is known that uPA plays a pivotal role in cancer cell invasion by activating plasmin, which is responsible for degrading ECM components (Harish & Uppuluri, 2018). Serine protease inhibitors such as serpins and aprotinin are known to suppress the activity of uPA and other serine proteases (Figure 3.1). Aprotinin has shown promising results in experimental models. These inhibitors have raised concerns due to their side effects and limited bioavailability (Soleyman-Jahi et al., 2019). Recent studies have identified candidates like WX-671, which is a potent uPA inhibitor, and SERPINB3, an endogenous uPA inhibitor.
3. *SERPINB3*: It is also called squamous cell carcinoma antigen 1 (SCCA1), SERPINB3 is a member of the serine protease inhibitor (SERPIN) super family. This is the protein which plays a role in the regulation of protease activity, particularly serine proteases. Here, we will be discussing some key points about SERPINB3.
   a. *Protease Inhibition*: SERPINB3 is involved in the inhibition of serine proteases. SERPINB3 achieves this by forming a covalent complex with the target protease, irreversibly inhibiting its activity (Sun et al., 2017).
   b. *Tissue Expression*: In certain types of cancers like squamous cell carcinoma (SCC) and hepato-cellular carcinoma (HCC), SERPINB3 is being expressed and mostly elevated levels are witnessed. In these cancers, increased expression of SERPINB3 has been associated with disease progression and aggressiveness of the disease (Pontisso, 2014).
   c. *Clinical Significance*: High levels of SERPINB3 in blood or tissue samples may serve as an indicator of cancer presence and severity, thus making it a useful tool for both diagnostic and prognostic purposes. SERPINB3 also functions in immune responses and regulation of inflammation and plays a role in modulating immune cell activity and cytokine production (Figure 3.2, Table 3.1). Targeting SERPINB3 or studying its interaction with specific proteases may have potential applications in the treatment of cancer by disrupting pathways that play a pivotal role in the progression of tumour growth and metastasis (Lauko et al., 2022).

These molecules are known for their capability for potential targeted therapy with reduced adverse effects (Soman & Nair, 2022).

4. *Cysteine Protease Inhibitors*: Cathepsins, a family of cysteine proteases, are known for their action towards tumour invasion and metastasis. Small molecules like E-64 and its derivatives can inhibit cathepsin activity, leading to the reduction of cancer cell invasiveness in pre-clinical studies. However, clinical translation has been quite challenging due to issues related to their specificity and toxicity (Calkins & Sloane, 1995).

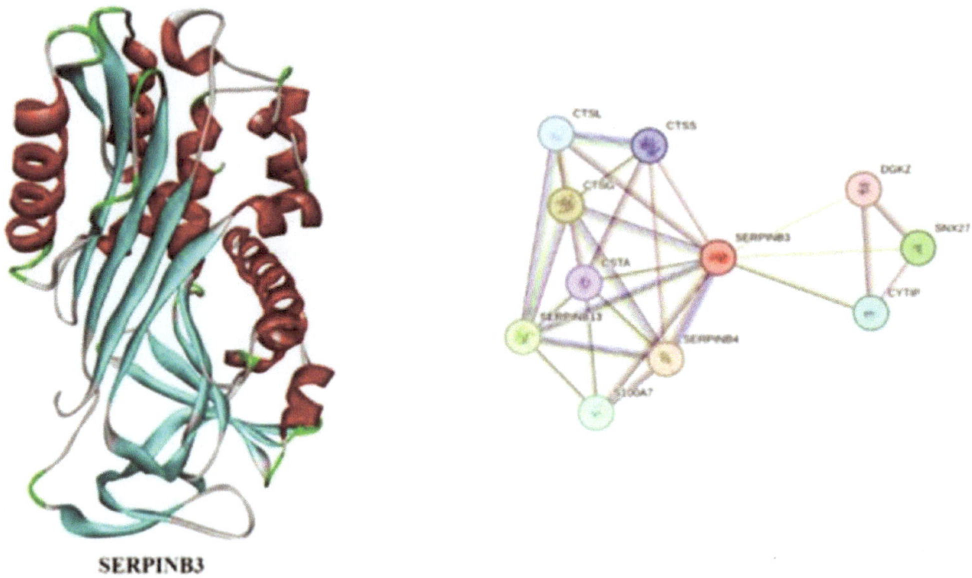

FIGURE 3.2 Cartoon representation of SERPINB3 (PDB ID: 2ZV6) and its interaction with its partners as obtained from network analysis. Red: Helices; Cyan: Strand; Green: Coils. *Abbreviations*: CTSL, Cathepsin L; CTSS, Cathepsin S mRNA; CTSG, Cathepsin G; CTSA, Cathepsin A.

TABLE 3.1
Details of the Binding Partners of SERPINB3

Name of the Gene	Function
SERPINB4	It helps in modulating the immune response against tumour cells by the hosts.
CTSG	It exerts protease by having trypsin and chymotrypsin like specificity. It can cleave off the complement C3 and possesses antibacterial activity against the deadly Gram-negative bacterium *Pseudomonas aeruginosa*.
SERPINB13	It participates in proliferation or differentiation of keratinocytes. It belongs to the Serpin family
SNX27	Helps in the retrograde transport from endosome to the plasma membrane. It subsequently promotes the recycling of the internalized transmembrane proteins.
S100A7	It exhibits the ability to bind to Calcium ions.
CYTIP	Also known as Cytohesin-interacting protein. It binds to Cytohesin-1 (CYTH-1) and then modifies the activation of ARFs by CYTH1.
CTSL	This is a thiol protease. Probably it has its role in the overall degradation of proteins in lysosomes
CTSS	This is also a thiol protease and plays a key role in the removal of invariant chain from class II MHC molecules.
CSTA	In the lower levels of the epidermis, CSTA helps in desmosome-mediated cell-cell adhesion.
DGKZ	This binds to 1,2-diacylglycerols; however, it lacks substrate specificity.

Source: From the STRING database, https://string-db.org/

3.7 CLINICAL TRIALS AND CURRENT STATUS

The following sheds light on the potential of protease inhibitors in cancer therapy and the latest clinical trials with their outcomes.

1. *MMP Inhibitors*: MMP inhibitors have faced a quite few setbacks due to off-target effects and limited efficacy. However, newer selective MMP inhibitors such as like Prinomastat and SB-3CT have shown promise in pre-clinical studies. Certain clinical trials are underway to assess their efficacy and safety in various cancer types, including glioblastoma and metastatic melanoma (Pavlaki & Zucker, 2003).
2. *Serine Protease Inhibitors*: A uPA inhibitor called WX-671 has entered Phase I clinical trials for the treatment of solid tumours. SERPINB3, another uPA inhibitor, is also under investigation in the early phase trials for ovarian as well as head and neck cancers (Danalev, 2012).
3. *Cysteine Protease Inhibitors*: Cathepsin inhibitors used in clinical trials have faced quite a few challenges due to issues in their specificity and toxicity. Ongoing research and studies have largely focused on how to reduce off-target effects to improve their pharmacokinetics (Liu et al., 2018).

Previously, protease inhibitors were generally considered as protein-degrading enzymes. But this point of view has changed dramatically, and protease inhibitors have been put in the spotlight as studies have found out their importance ranging from signalling molecules to performing various functions in many biological activities like inflammation, blood clotting, hormone processing and apoptosis. Recently, protease inhibitors have been studied extensively as therapeutic agents, focusing on treating various types of human cancers. Many PPIs are under further evaluation and BBIs, as mentioned earlier, have exhibited some of the most promising results in cancer prevention. For designing and development of anti-SARS-Cov-2 drug discovery, the SARS-Cov-2 papain-like protease has been considered a vital potential target. *In silico* molecular docking has been performed to identify the possible inhibitors of SARS-Cov-2 papain-like protease by studying anti-fungal, anti-viral and anti-cancer compounds. Out of all the drugs, Neobavaisoflavone's binding energy has been found to be the highest for SARS-Cov-2 papain-like protease. Blocking the papain-like protease is an important strategy in fighting viruses and the above-mentioned drug can be a prominent therapeutic intervention against COVID-19 (Li et al., 2021).

3.8 CONCLUSION AND FUTURE PERSPECTIVES

A vast variety of natural as well as synthetic products have been studied, out of which protease inhibitors like Primomastat, Marimastst, Tanomasta and Rebimastat have shown to be promising candidates for anti-cancerous agents. When it comes to diagnosis, most of them have failed to cross even the earliest of stages due to their adverse pharmacological effects and toxicity, which is non-specific in nature and, thus, is responsible for creating side-effects. To overcome this battle, researchers have primarily shifted their attention towards a nano-sized drug delivery system, since it is highly specific in nature and a leap in the results of clinical trials have been reported (Rudzińska et al., 2021). In this type of diagnosis, materials such as dendrimers, polymers, liposomes, carbon or silicon materials and magnetic nanoparticles that lie in the nanoscale range (10 nm to <1000 nm) are being used (Rizvi & Saleh, 2018). Gold nanoclusters (GNC) have been recently developed for drug delivery. A reduction-sensitive cis-platin prodrug has been conjugated with GNC and finally been activated with a folic acid target ligand, targeting breast tumours (Zhou et al., 2016).

Combination therapies using nanoparticles, including a combination of MMP inhibitors (Marimastat), Urokinase type plasminogen activator inhibitors and cysteine cathepsins inhibitors, have tremendously improved the therapeutic properties of unknown medicines. Thus, we can conclude by saying that by combining protease inhibitors with nanoparticle-based drug delivery systems can be exploited as a powerful medical tool that might help to overcome the battle against chemo resistance with minimal or no side effects.

ACKNOWLEDGEMENTS

The funding and infrastructural support from the DBT-sponsored BIF Centre (BT/PR40162/BTIS/137/48/2022 & SAN No. 102/IFD/SAN/653/2022-2023), University of Kalyani are duly acknowledged. Furthermore, Prof. Angshuman Bagchi would like to acknowledge the funding support from DBT, Government of India, for the extramural National Network Project (BT/PR40192/BTIS/137/69/2023). The authors acknowledge the help from University of Kalyani.

DECLARATIONS

Conflict of interest: The authors declare no conflicts of interest.

Consent for publishing: All authors have approved the final version of the manuscript and given consent for publication.

REFERENCES

Bhinder, B., Gilvary, C., Madhukar, N. S., & Elemento, O. (2021). Artificial intelligence in cancer research and precision medicine. Cancer Discovery, 11(4), 900–915.

Calkins, C. C., & Sloane, B. F. (1995). Mammalian cysteine protease inhibitors: biochemical properties and possible roles in tumor progression. Biological Chemistry Hoppe-Seyler, 376(2), 71–80.

Chen, X., Chen, D., Huang, L., Chen, X., Zhou, M., Xi, X., … Wang, L. (2020). Identification and target-modification of SL-BBI: a novel Bowman–Birk type trypsin inhibitor from sylviranalatouchii. Biomolecules, 10(9), 1254.

Constabel, F. (1990). Medicinal plant biotechnology1. Planta Medica, 56(05), 421–425.

Correa, P. (1981). Epidemiological correlations between diet and cancer frequency. Cancer research, 41(9_Part_2), 3685–3689.

Danalev, D. (2012). Inhibitors of serine proteinases from blood coagulation cascade-view on current developments. Mini Reviews in Medicinal Chemistry, 12(8), 721–730.

De Feo, V. (1992). Medicinal and magical plants in the northern Peruvian Andes. FITOTERAPIA-MILANO, 63, 417–440.

Fields, C., Mallee, P., Muzard, J., & Lee, G. U. (2012). Isolation of Bowman-Birk-Inhibitor from soybean extracts using novel peptide probes and high gradient magnetic separation. Food Chemistry, 134(4), 1831–1838.

Harish, B. S., & Uppuluri, K. B. (2018). Microbial serine protease inhibitors and their therapeutic applications. International Journal of Biological Macromolecules, 107, 1373–1387.

Hausman, D. M. (2019). What is cancer? Perspectives in Biology and Medicine, 62(4), 778–784.

Hikisz, P., & Bernasinska-Slomczewska, J. (2021). Beneficial properties of bromelain. Nutrients, 13(12), 4313.

Jedeszko, C., & Sloane, B. F. (2004). Cysteine cathepsins in human cancer.

Kennedy, A. R. (1993). Cancer prevention by protease inhibitors. Preventive Medicine, 22(5), 796–811.

Lauko, A., Volovetz, J., Turaga, S. M., Bayik, D., Silver, D. J., Mitchell, K., … Lathia, J. D. (2022). SerpinB3 drives cancer stem cell survival in glioblastoma. Cell Reports, 40(11), 111348.

Li, D., Luan, J., & Zhang, L. (2021). Molecular docking of potential SARS-CoV-2 papain-like protease inhibitors. Biochemical and Biophysical Research Communications, 538, 72–79.

Lippman, S. M., & Hawk, E. T. (2009). Cancer prevention: from 1727 to milestones of the past 100 years. Cancer Research, 69(13), 5269–5284.

Liu, C. L., Guo, J., Zhang, X., Sukhova, G. K., Libby, P., & Shi, G. P. (2018). Cysteine protease cathepsins in cardiovascular disease: from basic research to clinical trials. Nature Reviews Cardiology, 15(6), 351–370.

Lu, S. H. J., Jeon, A. H. W., Schmitt-Ulms, G., Qamar, S., Dodd, R., McDonald, B., … George-Hyslop, P. S. (2012). Vigilin interacts with signal peptide peptidase. Proteome Science, 10(1), 1–13.

Mondal, S., Adhikari, N., Banerjee, S., Amin, S. A., & Jha, T. (2020). Matrix metalloproteinase-9 (MMP-9) and its inhibitors in cancer: a minireview. European Journal of Medicinal Chemistry, 194, 112260.

Pavlaki, M., & Zucker, S. (2003). Matrix metalloproteinase inhibitors (MMPIs): the beginning of phase I or the termination of phase III clinical trials. Cancer and Metastasis Reviews, 22, 177–203.

Pontisso, P. (2014). Role of SERPINB3 in hepatocellular carcinoma. Annals of Hepatology, 13(6), 722–727.

Rizvi, S. A., & Saleh, A. M. (2018). Applications of nanoparticle systems in drug delivery technology. Saudi Pharmaceutical Journal, 26(1), 64–70.

Rudzińska, M., Daglioglu, C., Savvateeva, L. V., Kaci, F. N., Antoine, R., & Zamyatnin, A. A. Jr. (2021). Current status and perspectives of protease inhibitors and their combination with nanosized drug delivery systems for targeted cancer therapy. Drug Design, Development and Therapy, 15, 9–20.

Ryan, C. A. (1990). Protease inhibitors in plants: genes for improving defenses against insects and pathogens. Annual Review of Phytopathology, 28(1), 425–449.

Schiffman, J. D., Fisher, P. G., & Gibbs, P. (2015). Early detection of cancer: past, present, and future. American Society of Clinical Oncology Educational Book, 35(1), 57–65.

Soleyman-Jahi, S., Sadeghi, F., Afshari, Z., Barati, T., Ghasemi, S., Muhammadnejad, S., … Zendehdel, K. (2019). Anti-neoplastic effects of aprotinin on human breast cancer cell lines: in vitro study. Advances in Clinical and Experimental Medicine, 28(2), 151–157.

Soman, A., & Nair, S. A. (2022). Unfolding the cascade of SERPINA3: inflammation to cancer. Biochimica et Biophysica Acta (BBA)-Reviews on Cancer, 1877, 188760.

Srikanth, S., & Chen, Z. (2016). Plant protease inhibitors in therapeutics-focus on cancer therapy. Frontiers in Pharmacology, 7, 470.

Sterner, R. C., & Sterner, R. M. (2021). CAR-T cell therapy: current limitations and potential strategies. Blood Cancer Journal, 11(4), 69.

Sun, Y., Sheshadri, N., & Zong, W. X. (2017). SERPINB3 and B4: from biochemistry to biology. Seminars in Cell & Developmental Biology, 62, 170–177.

Torre, L. A., Siegel, R. L., Ward, E. M., & Jemal, A. (2016). Global cancer incidence and mortality rates and trends—an update. Cancer Epidemiology, Biomarkers & Prevention, 25(1), 16–27.

Wang, L. H., Wu, C. F., Rajasekaran, N., & Shin, Y. K. (2018). Loss of tumor suppressor gene function in human cancer: an overview. Cellular Physiology and Biochemistry, 51(6), 2647–2693.

Yang, L., Yong, L., Zhu, X., Feng, Y., Fu, Y., Kong, D., … Zhou, T. Y. (2020a). Disease progression model of 4T1 metastatic breast cancer. Journal of Pharmacokinetics and Pharmacodynamics, 47, 105–116.

Yang, P., Yuan, W., He, J., Wang, J., Yu, L., Jin, X., … Zhang, Y. (2009). Overexpression of EphA2, MMP-9, and MVD-CD34 in hepatocellular carcinoma: implications for tumor progression and prognosis. Hepatology Research, 39(12), 1169–1177.

Yang, P., Zhang, S., Chen, X., Liu, X., Wang, Z., & Li, Y. (2020b). Recent developments in polydopamine fluorescent nanomaterials. Materials Horizons, 7(3), 746–761.

Zhou, F., Feng, B., Yu, H., Wang, D., Wang, T., Liu, J., … Li, Y. (2016). Cisplatin prodrug-conjugated gold nanocluster for fluorescence imaging and targeted therapy of the breast cancer. Theranostics, 6(5), 679.

Section II

Proteases in Cancer Therapy

4 Ubiquitin-Specific Proteases

The Role and Potential in Cancer Immunotherapy

Tan Bee Ling
Management and Science University, Shah Alam, Selangor, Malaysia

Ayman El-Meghawry El-Kenawy
College of Medicine, Taif University,
Al Hawiyah, Ta'if, Saudi Arabia
Genetic Engineering and Biotechnology Institute,
University of Sadat City, Sadat, Egypt

Yin Sim Tor
Taylor's University, Subang Jaya, Selangor, Malaysia

4.1 INTRODUCTION

Tumors exert metastatic, invasion, and growth properties. In recent decades, cancers' capacity to avoid immune detection and evade the immune system has drawn an interest among researchers [1]. One of the vital modes of action of immune evasion is the production of immunosuppressive cytokines and mediators. These immunosuppressive factors can suppress the proliferation and stimulation of immune cells, hence decreasing the immune response toward tumors [2]. Indeed, tumors may downregulate major histocompatibility complex (MHC) molecule levels, which may facilitate the promotion of tumor antigens to T cells, thus decreasing their ability to transmit antigens to T lymphocytes [3]. In addition, tumors can recruit special immunoregulatory cells, such as myeloid-derived suppressor cells (MDSCs) and regulatory T cells (Tregs), to a specific microenvironment. These cells contribute to the immunosuppressive milieu by releasing factors that suppress the activity of immune cells. Tregs are capable of suppressing activity exhibited by effector T cells; while MDSCs can prevent neutrophils, dendritic cells (DCs), and natural killer (NK) cells from being activated. The recruitment of immunoregulatory cells can stimulate metastasis and tumor angiogenesis, thereby facilitating tumor progression and proliferation. In this regard, the tumor exerts a different mode of action to evade the immune system. Increasing the immune response against tumors may represent a promising approach for cancer treatment.

The regulation of ubiquitination is linked to numerous processes of tumorigenesis, such as regulation of the tumor immune microenvironment. The ubiquitin-proteasome system plays a vital role in many cellular proteins for cancer genesis [4]. The ubiquitin-proteasome system consists of 26S proteasome, ubiquitin-ligase (E3), ubiquitin-conjugating enzyme (E2), and ubiquitin-activating enzyme (E1). Ubiquitin chains can be removed from the substrates with the facilitation of deubiquitinating enzymes (DUBs), thus leading to the remodeling and alteration of the biological processes and stabilization of target proteins.

DOI: 10.1201/9781003394693-6

The DUBs consist of a relatively small group of proteins, in which the cysteine protease activity targets some predominant proteins involved in autophagy, senescence, apoptosis, and tumorigenesis. DUBs facilitate an untapped means of mediating key regulatory proteins that support oncogenic progression and transformation via multiple contacts with ubiquitinated protein substrates. Indeed, deubiquitination-altered biological processes are associated with the immune microenvironment, stemness, and tumor metabolism [5]. In recent decades, tumor microenvironment modulation approaches have drawn great interest in cancer immunotherapy [6].

Immunotherapy has emerged as a potential treatment approach specific to various types of cancer by utilizing multifarious biological mechanisms to improve the immune response toward tumors. Among the mechanisms, the ubiquitin–proteasome system (UPS) is a predominant non-lysosomal pathway specific to protein degradation, mediated by DUBs that counterbalance ubiquitin signaling. An emerging study indicates that the UPS is a potential therapeutic target for the treatment of cancer, as numerous UPSs are differentially activated or expressed in microenvironments or tumors, suggesting this is an ideal candidate for drug development [7]. The tumor has evolved in different modes of action to evade the immune system, facilitating tumor progression and hindering the anti-tumor immune response. Unlike conventional chemotherapy, which targets both healthy and cancerous cells, immunotherapy aims to harness the immune system of the patients to selectively remove the tumor cells [8]. For instance, immune checkpoint inhibitors (ICIs) block the connection between ligands and immune checkpoint molecules, thereby enhancing anti-tumor activity and preventing the exhaustion of T cells. Furthermore, immunotherapy drugs such as nivolumab and pembrolizumab have been demonstrated to effectively treat several solid tumors, such as renal cell carcinoma, melanoma, bladder cancer, and non-small cell lung cancer. Nonetheless, not all cancers respond equally to immunotherapy. Resistance against treatment can develop in more than 60% of patients who initially respond to immunotherapy, as well as advanced therapies such as ICIs, which displayed limited efficacy [9]. Therefore, developing new therapeutic modalities capable of strengthening immunotherapy efficacy is the most significant goal. Intriguingly, combining immunotherapy with chemotherapy, radiation therapy, or targeted therapy may represent a promising approach to enhance therapeutic rates and overcome resistance.

The controlled modulation of protein turnover can maintain the function and stability of cell structure. Nearly 30% of proteins that are newly synthesized in mammalian cells have a short half-life and degrade rapidly [10]. To achieve a high protein turnover rate, the cells have evolved a specialized system for the selective and controlled unwanted degradation of protein. The specialized system includes autophagy, lysosomal degradation, and the ubiquitin-proteasome pathway. Among these specialized systems, the ubiquitin-proteasome pathway is a typical non-lysosomal protein degradation pathway. Ubiquitination acts as a vital post-translational modification process in the modulation of DNA repair, gene regulation, signal transduction, and protein inactivation/activation. This process involves ubiquitin molecules being covalently connected to substrate protein through iso-peptide bonds under the catalysis of the E_1-E_2-E_3 ligase cascade. Deubiquitination is a reversible ubiquitination process. In deubiquitination, DUBs play a vital role in modulating the covalently interacted ubiquitin moieties from protein substrates. DUBs reverse ubiquitin attachment for counterbalancing ubiquitination signaling and stimulating ubiquitin molecule recycling [11]. Modulating DUBs can facilitate cell physiology and biology, whereas DUB turnover defects can lead to the development of diseases, for instance, cancers, infections, inflammatory and autoimmune disorders, and neurodegenerative disease. Nearly 100 DUBs are identified in humans, which can be divided into six families based on the function and structure, for instance, Zinc Finger USP, the motif interacting with the DUB family consisting of ubiquitin, Machado-Joseph disease protein proteases, ovarian tumor proteases, ubiquitin C-terminal hydrolases, and ubiquitin-specific proteases (USPs). Several studies revealed that USPs can modulate the efficacy of immunotherapy via mediating immune response and immune cell function in the tumor microenvironment [12]. Some of the USP expression in tumor cells is associated with tumor immune escape and the modulation of drug resistance.

USPs, the largest DUB family, exerts the strongest variety, which plays a crucial role in the processing of antigen, regulating immune response, and modulating immune cell function during tumor progression. The previous study found that the expression of several USP family members in tumor cells are involved in the immune microenvironment and tumor immune escape. This chapter explores the potential of targeting USPs in cancer immunotherapy. We also provided insight into the prospects of USP inhibitors in the treatment of various cancers. This chapter will provide insight into the function and structure of USPs in cancer immunity.

4.2 UBIQUITIN-PROTEASOME SYSTEM

Emerging research evidence indicates that USPs are potential therapeutic targets in various diseases. USPs contain more than 50 members, contributing to the largest DUBs family. DUBs modulate ubiquitin processing and removal, which may be functionally crucial, but the underlying molecular mechanisms are poorly understood. Nearly 100 human DUBs have been identified, in which most of them are cysteine-proteases, containing conserved histidine (H) and cysteine (C) in catalytic sites. DUBs are classified into five superfamilies, namely zinc-containing metalloproteases, ubiquitin C-terminal hydrolase (UCH), Machado-Josephin domain superfamily, USPs, and ovarian tumor protease. Like kinases, dysregulation of the ubiquitination system's components can lead to various diseases including tumorigenesis [13].

Ubiquitin can be removed by USPs from specific protein substrates, allowing protein localization and protein salvage. Importantly, all USPs feature highly conserved USP domains, which are made up of three domains that resemble the right hand's fingers, thumb, and palm. The finger domain manages the interactions with distal ubiquitin; the catalytic site is located between the thumb and palm domains [14]. Although their relative structural diversity includes terminal extensions and additional domains, many USPs shared a common characteristic of a typical conformational change. USPs cause the transition to a catalytically active state from an inactive form upon ubiquitin binding. The first exhibited X-ray structure of USP protein was the catalytic core of HAUSP/USP7. In 2005, the crystal structure of a 45-kDa catalytic domain of USP14 was reported [15]. A study by Ward et al. [16] demonstrated that the structure of the deubiquitinase USP15 is an open ubiquitin-binding channel with a misaligned catalytic triad.

4.3 USP-MEDIATED SIGNALING PATHWAYS IMPLICATED IN CANCERS

Recurrent USP mutation is rare except for the CYLD enzyme. However, familial cylindromatosis, a genetic condition that can contribute to the predisposition to develop multiple skin tumors, exerts a germline mutation of tumor-suppressor gene *CYLD*. Chromosomal translocation, which involves a USP, is the fusion of the promoter of CDH11 to the full-length USP6 gene that contributes to the upregulation of the transcript USP6 expression. USP16 and USP43 are fusion gene partners with RUNX, which exerts high-frequency gene alterations in hematologic diseases, for instance, acute myeloid leukemia and chronic myelomonocytic leukemia [17]. The transcriptional activity of USP18 is upregulated in mice expressing AML fusion protein AML1–ETO. Intriguingly, a study by De Pitta et al. [18] found overexpression of USP33/VDU1 in B-cell acute lymphoblastic leukemia. Upregulation of USP50 transcriptional activity has also been demonstrated in AML [19]. Increased mRNA levels of USP9x have been associated with poor prognosis in several myelomas and are linked to the accumulation of Mcl-1 in patients with follicular lymphoma [20]. Hence, USPs are implicated in numerous hematologic malignancies.

USPs play a crucial role in cancer-associated pathways. The nuclear factor kappa B (NF-κB) pathway is constitutively stimulated in many cancers, subsequently leading to metastasis and tumorigenesis. Indeed, some USPs are demonstrated as negative regulators in NF-κB signaling. CYLD interacts with the NEMO component of the IκB kinase complex and mediates the activity via TRAF2 deubiquitination. Suppression of CYLD promotes resistance toward apoptosis, in which this effect can be modulated by suppressing NF-κB signaling using aspirin derivatives [21]. In addition, USP31 has also been reported as a TRAF2-interacting protein that modulates NF-κB stimulation by tumor necrosis factor (TNF) receptor superfamily members [22]. Furthermore, USP4 also interacts and possesses deubiquitinating activity on several NF-κB pathway-associated molecules, for instance, TAK1 kinase, TRAF6, and TRAF2, and negatively modulates interleukin-1 beta (IL-1β)- and TNFα-induced cancer cell migration. Moreover, USP4 negatively mediates TNFα-induced apoptosis and receptor-interacting protein 1 (RIP1)-modulated NF-κB stimulation in neck and head squamous cell carcinomas via the deubiquitination of K63-associated ubiquitin from RIP1 and binding to RIP1 [23]. A study by Sun et al. [24] found that USP11 decreases TNFα-mediated NF-κB stimulation via mediation of IκBα stability by its deubiquitination. USP14 removes the ubiquitin chain of I-κB and thus triggers the I-κB phosphorylation and stimulates the release of cytokine in lung epithelial cells. COP9 signalosome (CSN) modulates the activity and assembly of cullin-RING ubiquitin ligases (CRL), which are involved in the Iκ-α ubiquitination. A study by Schweitzer et al. [25] found that USP15 is a CSN-associated DUB that reverses the CSN ubiquitinating activity on IκBα, thereby leading to a negatively regulating NF-κB signaling.

4.4 THE OVEREXPRESSION OF USPs IN CANCERS

USPs belong to cysteine proteases, which are aberrantly expressed in microenvironments or tumors, indicating they are promising candidates as targets for drug development. Several USPs were overexpressed in breast cancer including USP3, USP37, USP25, USP22, USP20, USP18, USP14, USP9X, USP7, USP4, and USP1 [26].

Of all 100 DUBs, USP7 is the most commonly studied. USP7 is a 135kDa cysteine protease containing about 1102 amino acids, for instance, 1 N-terminal TNF receptor-associated factor (TRAF)-like domain, 5 C-terminal ubiquitin-like domains (UBL1-5), and 1 catalytic core domain [27]. In general, the catalytic domain of USP7 (residues 208-560) is located between the UBL domain and TRAF-like domain and is flanked by an N-terminal domain. The catalytic core domain contains amino acid residues including Asp481, His464, and Cys223, which play a crucial role in cleaving the iso-peptide bonds between substrate protein and ubiquitin, to cooperatively involved in deubiquitination of substrate. UBL domains 1, 2, and 3 bind with a wide range of proteins, whereas UBL domains 5 and 4 may lead to the completion of deubiquitinated activity. Interestingly, only USP7 showed the special TRAF-like domain (residue 53-205), which is required for substrate recognition. The catalytic activity may partially rearrange through conformational change after interacting and recognizing a ubiquitinated substrate by a TRAF-like domain. Researchers have contributed a significant advance in understanding how USP7 mediates the immune response of cancer patients in recent decades. The highly immunosuppressive Forkhead box protein P3 (Foxp3)+Tregs can hinder the anti-tumor responses found in effective T cells. Several studies have revealed that higher USP7 expression facilitates the proliferation of tumors by altering the immunosuppressive properties of Foxp3+Treg cells [28]. Figure 4.1 shows the effect by targeting suppression of USP7 on anti-tumor immune response.

Tumor-associated macrophages (TAMs) are infiltrating inflammatory cells differentiating into M2 and M1 phenotypes based on the signals within the tumor microenvironment. The M1 phenotype is characterized by its ability to trigger an anti-tumor immune response and inhibit tumor proliferation, whereas the M2 phenotype is linked to immune inhibition and tumor promotion. USP7 is

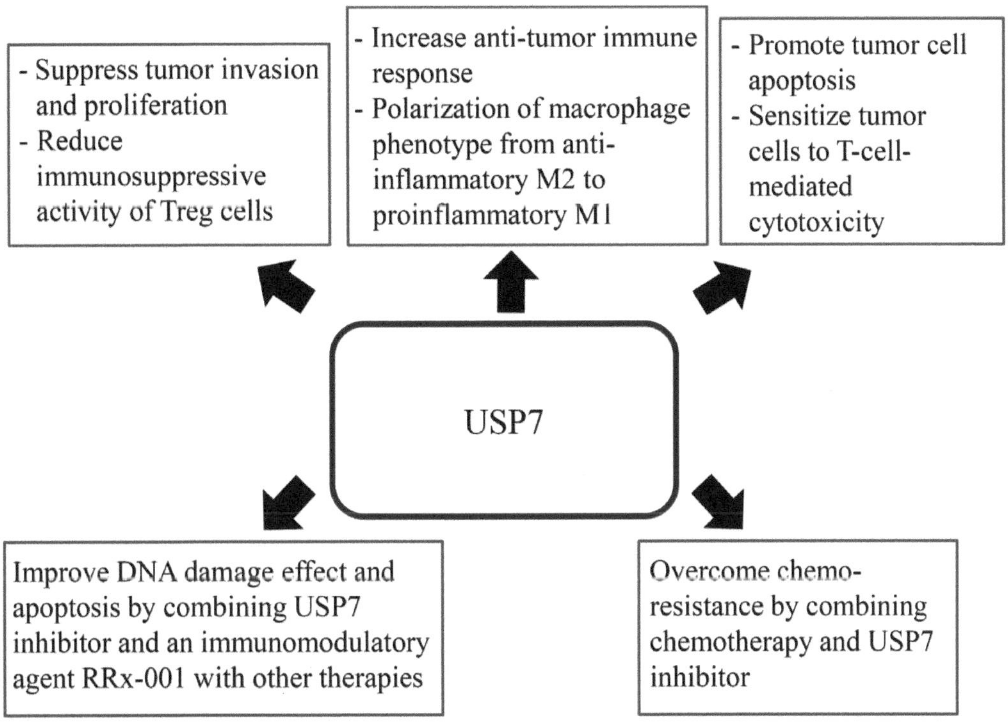

FIGURE 4.1 Effect by targeting suppression of ubiquitin-specific protease (USP7) on anti-tumor immune response. USP suppresses tumor invasion and proliferation. USP 7 decreases immunosuppressive activity of Treg cells. It stimulates anti-tumor immune response and assists in the polarization of macrophage phenotype to proinflammatory M1 from anti-inflammatory M2. USP7 also promotes tumor cell apoptosis and sensitizes tumor cells to T-cell-mediated cytotoxicity. USP7 improves DNA damage effect and apoptosis via the combination of an immunomodulatory agent RRx-001 and USP7 inhibitor with other therapies. It also overcomes chemoresistance by combining USP7 inhibitor and chemotherapy.

an indispensable gene that influences M2 and M1 macrophage homeostasis. During the inhibition of USP7, M2 macrophages alter the functions and phenotype, thereby enhancing the differentiation and growth of CD8+ T cell clusters *in vitro* [29]. The study showed that treatment of USP7 inhibitor enhanced the infiltration of M1 macrophages and CD8+T cells expressing interferon-gamma (IFN-γ) and decreased the tumor proliferation via modulation of the p38 mitogen-activated protein kinase (MAPK) pathway. Intriguingly, the study found that the therapeutic effect was weakened by TAM depletion [29].

Emerging studies showed that overexpression of USP7 in several cancers, including gastric tumors, which was found to be associated with PD-L1 expression. Thus, USP7 may facilitate the stabilization of PD-L1 protein expression, which plays a crucial role in tumor immune evasion. Nonetheless, Dai et al. [29] found that suppression of USP7 increased the expression of PD-L1 in lung cancer cells, suggesting that the regulatory effect of USP7 on PD-L1 levels is complex and may be context-dependent. Furthermore, USP7 inhibitors combined with PD-L1 inhibition and PD-1 suppression have been demonstrated as promising in enhancing anti-tumor responses. In addition, USP7 also mediates the stimulation of essential transcriptional factors to modulate the immune process, for instance, the p53 tumor suppressor gene. USP7 was also found to stabilize and deubiquitinate p53 to induce apoptosis and suppress tumor cell proliferation [30]. By contrast, USP7 can also stabilize and interact with E3 ubiquitin ligase MDM2, a negative regulator of p53 [31]. During the USP7 deubiquitination process toward these two proteins, overexpression of USP7 leads MDM2 to be deubiquitinated compared to p53, thereby accelerating p53 phosphorylation. Hence, suppression of USP7 leads to the phosphorylation of MDM2, when the MDM2 levels are insufficient for ubiquitination, and subsequently enhances the stability of p53 to increase the efficacy of therapeutic both *in vivo* and *in vitro* models. APG-115 is a type of MDM2 antagonist that functions as a p53 activator. A Phase 1b clinical trial study by Fang et al. [32] evaluated the effectiveness and safety of APG-115 with anti-PD-1 pembrolizumab therapy in mutant p53 solid tumor patients. USP7 inhibitors have been demonstrated to improve the PD-L1 expression in tumor cells, thereby enhancing the anti-PD-1 therapy efficacy. Hence, it may be highly effective to enhance cancer immunotherapy with the combination of MDM2 blockage and anti-PD-1 therapy plus USP7 suppression [32].

USP22 contains an open reading frame, which takes charge of encoding a peptide with 525 amino acids (Figure 4.2). The structure of USP22 is predominantly described based on the ubiquitin carboxyl-terminal hydrolase 8 (Ubp8) and yeast homologue. The C-terminal catalytic domain with an N-terminal zinc finger produces USP22/Ubp8. By contrast, the zinc-finger ubiquitin-binding domain (ZnF UBP) of USP22/Ubp8 fails to attach with ubiquitin directly. Thus, it interacts with some proteins to form a tetrameric deubiquitinase module (DUBm). Indeed, the DUBm deubiquitinates target proteins to modify expression. Of all the substrate proteins, far-upstream element–binding protein 1 (FBP1), nuclear factor of activated T cells 2 (NFATC2), B cell-specific Moloney murine leukemia virus integration site 1 (BMI-1), sirtuin 1 (SIRT1), c-Myc, and cyclin D1 have been identified. These proteins markedly impact the cellular processes that are essential for tumor survival and proliferation, DNA repair, and cell-cycle progression [33].

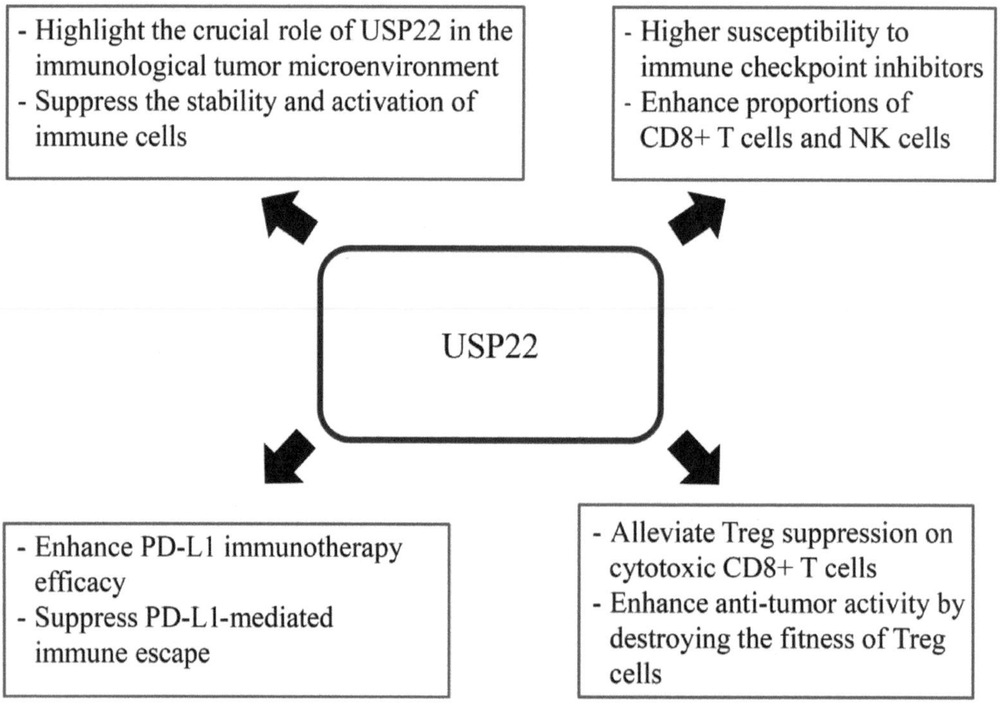

FIGURE 4.2 The effect of targeting suppression of ubiquitin-specific protease 22 (USP22) on anti-tumor immune response. USP22 inhibits the activation and stability of immune cells. It also highlighted the crucial nature of USP22 in the immunological tumor microenvironment. USP22 improves fractions of natural killer and CD8+ T cells. USP22 shows greater susceptibility to immune checkpoint inhibitors. USP22 inhibits PD-L1-mediated immune escape and improves PD-L1 immunotherapy efficacy. USP22 improves anti-tumor response by destroying the fitness of Treg cells and alleviates Treg inhibition on cytotoxic CD8+ T cells.

USP22 plays a crucial role in the immune system by modulating the phenotypic switching of B cells and T cells, development, and growth. USP22 can stabilize and deubiquitinate NFATC2 to stimulate T cells and increase the release of IL-2 levels. USP22 is also required for the development of invariant NK T cells in the early stage. Moreover, USP22 triggers the activation of T-box transcription factor (T-bet) genes and IL-2 receptor beta (IL-2Rβ) through the deubiquitinating H2A [34]. USP22 can also repair the breakage of programmed DNA by H2BK120 deubiquitination *in vivo*. USP22 ablation could lead to damage to classic non-homologous end joining caused by the primary B cells and the lack of phosphorylated histone H2AX. Collectively, USP22 plays a crucial role in the regulation of immune cells and may serve as a potential immunotherapeutic target. USP22 is possibly affected by the response to immunotherapy as well as being capable of reprogramming the TME. In general, tumors can be categorized as "cold" and "hot" based on the infiltration by lymphocytes. The "cold" tumor is resistant to immunotherapy and lacks immune cell infiltration, whereas the "hot" tumor is capable of responding to immunotherapy and shows immunogenicity. Targeting USP22 has been demonstrated to improve the response to immunotherapy in pancreatic ductal adenocarcinoma and linked to the increase in NK cells and CD8+T cells, thereby leading to the "cold" tumor changing into "hot" [2]. A similar finding was also observed in liver tumors when inhibiting USP22 enhanced tumor immunogenicity, improved the sensitivity toward anti-PD-L1 immunotherapy, and promoted T-cell infiltration as well as cisplatin-based chemotherapy [35]. Knockdown USP22 may enhance the efficacy of anti-PD-1/PD-L1, improve NK cell activity, and decrease tumor metastasis [2].

4.5 SMALL MOLECULE INHIBITOR-MODULATED TARGETING OF USPs

Emerging evidence indicates that molecular signaling pathways and USPs are closely connected; many efforts have been made to develop USPs inhibitors. Targeting the USP therapy approach has drawn a great deal of attention. In recent decades, great breakthroughs have been made in the development of small-molecule USPs inhibitors. Among all USP families, USP14 is the most studied USP family member. USP14 consists of 494 amino acids attached with a catalytic USP domain at its C-terminus and UBL domain at its N-terminus. In general, proteasome activity is regulated by the UBL domain; whereas deubiquitinase activity is modulated by the C-terminal USP domain. The previous study indicated that the two surface loops, BL2 and BL1, on free USP14 prevent active site binding to the C-terminus of ubiquitin [36]. The attachment of 26S proteasome by USP14 in an autoinhibited state stimulates the deubiquitination function. More than 60 USP inhibitors were discovered and two of them, namely, VLX1570 and b-AP15, are under clinical trial research for multiple myeloma treatment. Table 4.1 summarizes several USPs inhibitors in breast cancer.

A study reported by Wang et al. [44] indicates that IU1 is a highly selective inhibitor of proteasome-bound USP14. IU1 has been utilized in *in vitro* models. Since DUBs are highly conserved, an earlier study has predominantly focused on covalent inhibitors, in which the compounds are attached to active site cysteines via covalent bonds. Nonetheless, this compound commonly exerts poor selectivity for the DUBs family and cannot be utilized clinically [45]. The IU1 blocks the binding to USP14 from the C-terminus of ubiquitin by binding to the thumb-palm cleft part of the catalytic domain of USP14. This study suggested that allosteric mediation via steric retardation may serve as a viable strategy for the USP inhibitors discovery. Another class of USP14 inhibitors is the IU2 series, which belongs to the tricyclic thiophene derivatives. The suppressive effect of IU2 may prevent the entry of ubiquitinated substrates by inhibiting the ubiquitin-binding pocket.

Nonetheless, the low suppressive efficiency of IU1 may prevent the development of drugs targeting USP14. USP14 not only plays a crucial role in inflammatory and immune responses but also affects tumor development. Therefore, suppression of USP14 activity hinders caspase-1 stimulation and IL-1β release, suggesting the therapeutic potential in autoinflammatory diseases [46]. Suppression of USP14 may demonstrate a diverse biological effect and thereby lead to unpredictable toxicity. For instance, dysregulation of USP14 affects synaptic activity and thus contributes to

TABLE 4.1
Ubiquitin-specific Peptidase Inhibitors in Breast Cancer

Target	USP Inhibitors	Breast Cancer Subtype	Models	Pathways	References
USP9x	WP1130	ER-negative breast cancer	*In vitro*	Mcl-1	[37]
USP14	IU1	AR-positive breast cancer	*In vitro*, *In vivo*	Wnt/β-catenin, PI3K/Akt pathways	[38]
	Auranofin	ER-positive breast cancer	*In vitro*	Mitochondrial apoptosis	[39]
USP7/47	P5091	Breast cancer	*In vitro*	Epithelial-mesenchymal transition	[40, 41]
USP7	Costunolide	Metastatic TNBC, breast cancer	*In vitro*	p38MAPK pathways, Akt/14-3-3 pathway, c-Myc/p53, cell cycle regulation, NF-κB signaling	[41]
USP2	6-TG	BRCA2-defective PARP inhibitor-resistant breast cancer, BRCA1-mutant breast cancer, TNBC	*In vitro*, *In vivo*	DNA repair, PI3K-Akt, apoptosis pathway, lncRNA-miRNA-mRNA ceRNA network, DNMT1	[42]
USP1	Pimozide	Estrogen receptor-negative breast cancer, TNBC	*In vitro*, *In vivo*	STAT3, vimentin, MMP-9, EMT, Akt signaling pathway, cell cycle	[43]

Abbreviations: Akt, Protein kinase B; ceRNA, Competing endogenous RNA; DNMT1, DNA (cytosine-5-)-methyltransferase 1; EMT, Epithelial mesenchymal transition; lncRNA, Long non-coding RNA; MAPK, Mitogen-activated protein kinase; Mcl-1, Myeloid cell leukemia-1; miRNA, MicroRNA; MMP-9, Matrix metallopeptidase 9; NFκB, Nuclear factor kappa-B; PI3K, Phosphoinositide 3-kinase; STAT3, Signal transducer and activator of transcription 3; TNBC, Triple-negative breast cancer.

neuronal dysfunction. In this regard, it is vitally important to identify a specific inhibitor targeting the connection of USP14 with several of its substrates.

USP22 possesses tumor-promoting activity in several types of tumors and inhibits anti-tumor immunity by modulating PD-L1 in tumors [47]. Emerging research evidence revealed that prescribing USP22 inhibitors is desirable. The USP22 has been evaluated in cancer in recent decades, yet USP22 inhibitors have not been reported. A study by Morgan et al. [48] evaluated the cyclic peptides in a high-throughput manner using RaPID, which contains 1012 structurally unique cyclic peptides, and ultimately evaluated the ability to suppress DUB activity based on binding affinity using an *in vitro* model to create highly specific and effective DUB inhibitor. The ubiquitin variants (UbVs) targeting specific USPs may provide a new direction for designing small molecule inhibitors of USP22. A study by Tang et al. [49] created a new library of combinatorial UbVs with high specificity and affinity for their cognate target domains found in USP. UbV probes can act as a potential target for suppression of USPs. This inhibitory mode of action can be extended to other USPs

4.6 CONCLUSION AND FUTURE PERSPECTIVE

The development of immunotherapy is a potent approach to cancer treatment. Nonetheless, this therapy does not benefit all cancer patients although it has great potential. Sipuleucel-T, an autologous cellular immunotherapy made from antigen-presenting cells, is the first US Food and Drug Administration-approved cancer vaccine used for metastatic castration-resistant prostate cancer, suggesting the prospect of immunotherapy in prostate cancer. USPs are one of the specialized category of DUBs that emerged as a potential therapeutic target in tumorigenesis. USPs are involved in several crucial signaling pathways. USPs have gained attention as a potential target, and USPs inhibitors have begun to progressively appear. Emerging evidence indicates a significant insight into how USP members influence cancer immunotherapy. Indeed, USP7, USP22, and USP14 are widely studied USP members involved in the immune process. Moreover, USP18, USP9X, USP15, and USP8 are also in the spotlight from the number of studies investigating their crucial immunomodulatory functions in cancer progression. Several other USPs have also demonstrated to be involved in the immune process of cancer, including USP5, USP6, USP10, USP12, USP20, USP25, USP35, USP43, USP44, USP48, and USP51. Of all these USP members, some have been shown to participate in the immunotherapy effect and modulate the critical PD1/PD-L1 signaling pathway. USP5, USP9X, USP18, USP8, USP22, and USP7 have been reported to bind with PD-L1 directly to trigger the stabilization and deubiquitination [47, 50]. Suppressing these USP members sensitizes tumor cells to immunosurveillance and improves the efficacy of anti-PD-1/anti-PD-L1 therapy. These USP inhibitors, including WP1130 and P5091, have been demonstrated to enhance the therapeutic efficacy of PD-L1/anti-PD-1 via suppression of the PD-L1 deubiquitination. Nonetheless, these therapeutic and regulatory effects are not unique, and inconsistent findings have been found. For instance, suppression of both USP8 and UPS7 has also been demonstrated to contribute to increased PD-L1 expression. Inhibition of USP48 and USP12 expression reduced the therapeutic efficacy to PD-1 inhibitors and promoted the resistance to anti-PD-1 therapy [51]. It is hypothesized that immune-cell infiltration conditions and different TME, different tumor cell states, and various cancer types contribute to the complexity of the regulatory effect on PD-L1/PD-1. Moreover, PD-L1/PD-1 shows a wide range of protein translational modifications (PTMs), for instance, acetylation, SUMOylation, palmitoylation, phosphorylation, and glycosylation. The study of USP deubiquitinases in cancer immunotherapy has demonstrated a promising result, particularly in preclinical and basic research. Nevertheless, there was limited evidence on the role of USPs in modulating the immune responses and mechanisms of immune and tumor cells. Overall, further studies are warranted on the PD-1/PD-L1 deubiquitination expression on USPs and PTMs. Data available on the USP members involved in PD-1/PD-L1 regulation are limited, and further investigation on the mechanism may provide a new insight into the design of a therapeutic approach to medicate PD-1/PD-L1 pathway by targeting related USPs in cancer immunotherapy.

The specificity and development of USP inhibitors are still huge challenges that need to be fully elucidated. The most crucial part is the potential for off-target effects. This could be due to the USP involvement in different biological processes, and thereby the suppression of USPs could potentially contribute to off-target effects and affect several pathways. Indeed, most of the USP inhibitors lack accurate targeting specificity. The efficacy of targeting USPs is limited by the heterogeneity of tumors because different tumors may have different modes of action of immune invasion. Therefore, targeting a single USP may not be sufficient to overcome the mode of action possibly hindering the efficacy of targeting USPs alone. There is still the need of a more precise mode of action of interaction between USPs and small molecules as well as screening methods to advance the development of USP inhibitors. One approach is to conduct high-screening assays to identify the compounds selectively inhibiting specific USP family members related to cancer immunity. Taken together, the synergistic USP suppression mechanisms and the creation of combinatorial therapeutic strategies for cancer immunotherapy should be the primary goals of future studies.

DECLARATIONS

Conflict of interest: The authors declare no conflicts of interest.

Consent for publishing: All authors have approved the final version of the manuscript and given consent for publication.

Funding: Not Applicable.

REFERENCES

1. Guo, S., Feng, J., Li, Z., Yang, S., Qiu, X., Xu, Y., & Shen, Z. (2023). Improved cancer immunotherapy strategies by nanomedicine. *Nanomed NanoBiotechnol*, 15(3), e1873. https://doi.org/10.1002/wnan.1873
2. Li, J., Yuan, S., Norgard, R. J., Yan, F., Yamazoe, T., Blanco, A., & Stanger, B. Z. (2020). Tumor cell-intrinsic USP22 suppresses antitumor immunity in pancreatic cancer. *Cancer Immunol Res*, 8(3), 282–291. https://doi.org/10.1158/2326-6066.CIR-19-0661
3. Janelle, V., Rulleau, C., Testa, S. D., Carli, C. & Delisle, J.-S. (2020). T-cell immunotherapies targeting histocompatibility and tumor antigens in hematological malignancies. *Front Immunol*, 11, 276. https://doi.org/10.3389/fimmu.2020.00276
4. Sharma, A., Khan, H., Singh, T. G., Grewal, A. K., Najda, A., Kawecka-Radomska, M., Kamel, M., Altyar, A. A. E. & Abdel-Daim, M. M. (2021). Pharmacological modulation of ubiquitin-proteasome pathways in oncogenic signaling. *Int J Mol Sci*, 22(21), 11971. https://doi.org/10.3390/ijms222111971
5. Guo, J., Zhao, J., Sun, L., & Yang, C. (2022). Role of ubiquitin specific proteases in the immune microenvironment of prostate cancer: A new direction. *Front Oncol*, 12, 955718. https://doi.org/10.3389/fonc.2022.955718
6. Aghanejad, A., Bonab, S. F., Sepehri, M., Haghighi, F. S., Tarighatnia, A., Kreiter, C., Nader, N. D., & Tohidkia, M. R. (2022). A review on targeting tumor microenvironment: The main paradigm shift in the mAb-based immunotherapy of solid tumors. *Int J Biol Macromol*, 207, 592–610. https://doi.org/10.1016/j.ijbiomac.2022.03.057
7. Huang, M.-L., Shen, G.-T. & Li, N. L. (2022). Emerging potential of ubiquitin-specific proteases and ubiquitin-specific proteases inhibitors in breast cancer treatment. *World J Clin Cases*, 10(32), 11690–11701. https://doi.org/10.12998%2Fwjcc.v10.i32.11690
8. Bailly, C., Thuru, X., & Quesnel, B. (2020). Combined cytotoxic chemotherapy and immunotherapy of cancer: modern times. *Nar Cancer*, 2(1), zcaa002. https://doi.org/10.1093/narcan/zcaa002
9. Pérez-Ruiz, E., Melero, I., Kopecka, J., Sarmento-Ribeiro, A. B., García-Aranda, M., & De Las Rivas, J. (2020). Cancer immunotherapy resistance based on immune checkpoints inhibitors: Targets, biomarkers, and remedies. *Drug Resist Updat*, 53, 100718. https://doi.org/10.1016/j.drup.2020.100718
10. Tong, M., Smeekens, J. M., Xiao, H., & Wu, R. (2020). Systematic quantification of the dynamics of newly synthesized proteins unveiling their degradation pathways in human cells. *Chem Sci*, 11(13), 3557–3568. https://doi.org/10.1039/c9sc06479f
11. Dagar, G., Kumar, R., Yadav, K. K., Dingh, M., & Pandita, T. K. (2023). Ubiquitination and deubiquitination: Implications on cancer therapy. *Biochimica et Biophysica Acta (BBA) – Gene Regulatory Mechanisms*, 1866(4), 194979. https://doi.org/10.1016/j.bbagrm.2023.194979
12. Zhou, X., Fu, C., & Chen X. (2023). The role of ubiquitin pathway-mediated regulation of immune checkpoints in cancer immunotherapy. *Cancer*, 129(11), 1649–1661. https://doi.org/10.1002/cncr.34729
13. Fhu, C. W., & Ali, Z. (2021). Dysregulation of the ubiquitin proteasome system in human malignancies: A window for therapeutic intervention. *Cancers*, 13(7), 1513. https://doi.org/10.3390/cancers13071513
14. Young, M. J., Hsu, K. C., Lin, T. E., Chang, W. C., & Hung, J. J. (2019). The role of ubiquitin-specific peptidases in cancer progression. *J Biomed Sci*, 26, 42.
15. Hu, M., Li, P., Song, L., Jeffrey, P. D., Chenova, T. A., Wilkinson, K. D., Cohen, R. E., & Shi, Y. (2005). Structure and mechanisms of the proteasome-associated deubiquitinating enzyme USP14. *EMBO J*, 24(21), 3747–3756. https://doi.org/10.1038/sj.emboj.7600832
16. Ward, S. J., Gratton, H. E., Indrayudha, P., Michavila, C., Mukhopadhyay, R., Maurer, S. K., Caulton, S. G., Emsley, J., & Dreveny, I. (2018). The structure of the deubiquitinase USP15 reveals a misaligned catalytic triad and an open ubiquitin-binding channel. *J Biol Chem*, 293(45), 17362–17374. https://doi.org/10.1074/jbc.RA118.003857
17. Masetti, R., Togni, M., Astolfi, A., Pigazzi, M., Indio, V., Rivalta, B., Manara, E., Rutella, S., Basso, G., Pession, A., & Locatelli, F. (2014). Whole transcriptome sequencing of a paediatric case of de novo acute myeloid leukaemia with del(5q) reveals RUNX1-USP42 and PRDM16-SKI fusion transcripts. *Br J Haematol*, 166(3), 449–452. https://doi.org/10.1111/bjh.12855
18. De Pitta, C., Tombolan, L., Campo Dell'Orto, M., Accordi, B., te Kronnie, G., Romualdi, C., Vitulo, N., Basso, G., & Lanfranchi, G. (2005). A leukemia-enriched cDNA microarray platform identifies new transcripts with relevance to the biology of pediatric acute lymphoblastic leukemia. *Haematologica*, 90(7), 890–898
19. Figueroa, M. E., Lugthart, S., Li, Y., Erpelinck-Verschueren, C., Deng, X., Christos, P. J., Schifano, E., Booth, J., van Putten, W., Skrabanek, L., Campagne, F., Mazumdar, M., Greally, J.M., Valk, P. J. M., Löwenberg, B., Delwel, R., & Melnick, A. (2010). DNA methylation signatures identify biologically distinct subtypes in acute myeloid leukemia. *Cancer Cell*, 17(1), 13–27. https://doi.org/10.1016/j.ccr.2009.11.020

20. Schwickart, M., Huang, X., Lill, J. R., Liu, J., Ferrando, R., French, D. M., Maecker, H., O'Rourke, K., Bazan, F., Eastham-Anderson, J., Yue, P., Dournan, D., Huang, D. C. S., & Dixit, V. M. (2010). Deubiquitinase USP9X stabilizes MCL1 and promotes tumour cell survival. *Nature*, 463(7277), 103–107. https://doi.org/10.1038/nature08646
21. Harhaj, E. W., & Dixit, V. M. (2012). Regulation of NF-kappaB by deubiquitinases. *Immunol Rev*, 246(1), 107–224. https://doi.org/10.1111/j.1600-065X.2012.01100.x
22. Tzimas, C., Michailidou, G., Arsenakis, M., Kieff, E., Mosialos, G., & Hatzivassiliou, E. G. (2006). Human ubiquitin specific protease 31 is a deubiquitinating enzyme implicated in activation of nuclear factor-kappaB. *Cell Signal*, 18(1), 83–92. https://doi.org/10.1016/j.cellsig.2005.03.017
23. Hou, X., Wang, L., Zhang, L., Pan, X., & Zhao, W. (2013). Ubiquitin-specific protease 4 promotes TNF-alpha–induced apoptosis by deubiquitination of RIP1 in head and neck squamous cell carcinoma. *FEBS Lett*, 587, 311–316. https://doi.org/10.1016/j.febslet.2012.12.016
24. Sun, W., Tan, X., Shi, Y., Xu, G., Mao, R., Gu, X., Fan, Y., Yu, Y., Burlingame, S., Zhang, H., Rednam, S. P., Lu, X., Zhang, T., Fu, S., Cao, G., Qin, J. & Yang, J. (2010). USP11 negatively regulates TNFalpha-induced NF-kappaB activation by targeting on IkappaBalpha. *Cell Signal*, 22(3), 386–394. https://doi.org/10.1016/j.cellsig.2009.10.008
25. Schweitzer, K., Bozko, P. M., Dubiel, W., & Naumann, M. (2007). CSN controls NFkappaB by deubiquitinylation of IkappaBalpha. *EMBO J*, 26(6), 1532–1541. https://doi.org/10.1038/sj.emboj.7601600
26. Nijman, S. M., Luna-Vargas, M. P., Velds, A., Brummelkamp, T. R., Dirac, A. M., Sixma, T. K., & Bernards, R. (2005). A genomic and functional inventory of deubiquitinating enzymes. *Cell*, 123, 773–786. https://doi.org/10.1016/j.cell.2005.11.007
27. Kim, R. Q., & Sixma, T. K. (2017). Regulation of USP7: a High incidence of E3 complexes. *J Mol Biol*, 429(22), 3395–3408. https://doi.org/10.1016/j.jmb.2017.05.028
28. Wang, L., Kumar, S., Dahiya, S., Wang, F., Wu, J., Newick, K., Han, R., Samanta, A., Beier, U. H., Akimova, T., Bhatti, T. R., Nicholson, B., Kodrasov, M. P., Agarwal, S., Sterner, D. R., Gu, W., Weinstock, J., Butt, T. R., Albelda, S. M., & Hancock, W. W. (2016). Ubiquitin-specific protease-7 inhibition impairs Tip60-dependent Foxp3+ T-regulatory cell function and promotes antitumor immunity. *EBioMedicine*, 13, 99–112. https://doi.org/10.1016/j.ebiom.2016.10.018
29. Dai, X., Lu, L., Deng, S., Meng, J., Wan, C., Huang, J., Sun, Y., Hu, Y., Wu, B., Wu, G., Lovell, J. F., Jin, H., & Yang, K (2020). USP7 targeting modulates anti-tumor immune response by reprogramming tumor-associated macrophages in lung cancer. *Theranostics*, 10(20), 9332–9347. https://doi.org/10.7150/thno.47137
30. Li, M., Chen, D., Shiloh, A., Luo, J., Nikolaev, A. Y., Qin, J., & Gu, W. (2002). Deubiquitination of P53 by HAUSP is an important pathway for P53 stabilization. *Nature*, 416(6881), 648–653. https://doi.org/10.1038/nature737
31. Qi, S. M., Cheng, G., Cheng, X. D., Xu, Z., Xu, B., Zhang, W. D., & Qin, J.-J. (2020). Targeting USP7-mediated deubiquitination of MDM2/MDMX-P53 pathway for cancer therapy: are we there yet? *Front Cell Dev Biol*, 8, 233. https://doi.org/10.3389/fcell.2020.00233
32. Fang, D. D., Tang, Q., Kong, Y., Wang, Q., Gu, J., Fang, X., Zou, P., Rong, T., Wang, J., Yang, D., & Zhai, Y. (2019). MDM2 Inhibitor APG-115 Synergizes with PD-1 Blockade through enhancing antitumor immunity in the tumor microenvironment. *J Immunother Cancer*, 7(1), 327. https://doi.org/10.1186/s40425-019-0750-6
33. Jeusset, L. M., & McManus, K. J. (2017). Ubiquitin specific peptidase 22 regulates histone H2B mono-ubiquitination and exhibits both oncogenic and tumor suppressor roles in cancer. *Cancers*, 9(12), 167. https://doi.org/10.3390/cancers9120167
34. Zhang, Y., Wang, Y., Gao, B., Sun, Y., Cao, L., Genardi, S. M., Wang, C.-R., Li, H. B., Sun, Z., Yang, Y., & Fang, D. (2020). USP22 controls iNKT immunity through MED1 suppression of histone H2A monoubiquitination. *J Exp Med*, 217(5), e20182218.
35. Huang, X., Zhang, Q., Lou, Y., Wang, J., Zhao, X., Wang, L., Zhang, X., Li, S., Zhao, Y., Chen, Q., Liang, T., & Bai, X. (2019). USP22 Deubiquitinates CD274 to suppress anticancer immunity. *Cancer Immunol Res*, 7(10), 1580–1590. https://doi.org/10.1158/2326-6066.CIR-18-0910
36. Wang, F., Ning, S., Yu, B., & Wang, Y. (2022). USP14: Structure, function, and target inhibition. *Front Pharmacol*, 12, 801328. https://doi.org/10.3389/fphar.2021.801328
37. Fu, P., Du, F., Liu, Y., Yao, M., Zhang, S., Zheng, X., & Zheng, S. (2017). WP1130 increases cisplatin sensitivity through inhibition of usp9x in estrogen receptor-negative breast cancer cells. *Am J Transl Res*, 9(4), 1783–1791.
38. Xia, X., Huang, C., Liao, Y., Liu, Y., He, J., Guo, Z., Jiang, L., Wang, X., Liu, J., & Huang, H. (2019). Inhibition of USP14 enhances the sensitivity of breast cancer to enzalutamide. *J Exp Clin Cancer Res*, 38(1), 220. https://doi.org/10.1186/s13046-019-1227-7

39. Joo, M. K., Shin, S., Ye, D. J., An, H. G., Kwon, T. U., Baek, H. S., Kwon, Y. J., & Chun, Y. J. (2021). Combined treatment with auranofin and trametinib induces synergistic apoptosis in breast cancer cells. *J Toxicol Environ Health A*, 84, 84–94. https://doi.org/10.1080/15287394.2020.1835762
40. Silvestrini, V. C., Thomé, C. H., Albuquerque, D., de Souza Palma, C., Ferreira, G. A., Lanfredi, G. P., Masson, A. P., Delsin, L. E. A., Ferreira, F. U., de Souza, F. C., de Godoy, L. M. F., Aquino, A., Carrilho, E., Panepucci, R. A., Covas, D. T., & Faça, V. M. (2020). Proteomics analysis reveals the role of ubiquitin specific protease (USP47) in Epithelial to Mesenchymal Transition (EMT) induced by TGFβ2 in breast cells. *J Proteomics*, 219, 103734. https://doi.org/10.1016/j.jprot.2020.103734
41. Liu, D., Zeng, M., Pi, J. W., Liu, M. J., Ding, W. Z., Mei, X. Y., Liu, J. L., & Cao, X. Y. (2021). Exploring the potential mechanism of Costunolide-induced MCF-7 cells apoptosis by multi-spectroscopy, molecular docking and cell experiments. *Chem Biodivers*, 18, e2001069. https://doi.org/10.1002/cbdv.202001069
42. Zhang, D., An, X., Yu, H., & Li, Z. (2021). The regulatory effect of 6-TG on lncRNA-miRNA-mRNA ceRNA network in triple-negative breast cancer cell line. *Biosci Rep*, 41(2), BSR20203890. https://doi.org/10.1042/BSR20203890
43. Dees, S., Pontiggia, L., Jasmin, J. F., & Mercier, I. (2020). Phosphorylated STAT3 (Tyr705) as a biomarker of response to pimozide treatment in triple-negative breast cancer. *Cancer Biol Ther*, 21(6), 506–521. https://doi.org/10.1080/15384047.2020.1726718
44. Wang, Z., Kang, W., Li, O., Qi, F., Wan, J., You, Y., He, P., Suo, Z., Zheng, Y., & Liu, H.-M. (2021). Abrogation of USP7 is an alternative strategy to downregulate PD-L1 and sensitize gastric cancer cells to T cells killing. *Acta Pharm Sin B*, 11(3), 694–707. https://doi.org/10.1016/j.apsb.2020.11.005
45. D'Arcy, P., Wang, X., & Linder, S. (2015). Deubiquitinase inhibition as a cancer therapeutic strategy. *Pharmacol Ther*, 147, 32–54. https://doi.org/10.1016/j.pharmthera.2014.11.002
46. Lopez-Castejon, G., Luheshi, N. M., Compan, V., High, S., Whitehead, R. C., Flitsch, S., Kirov, A., Prudovsky, I., Swanton, E., & Brough, D. (2013). Deubiquitinases regulate the activity of caspase-1 and interleukin-1β secretion via assembly of the inflammasome. *J Biol Chem*, 288(4), 2721–2733. https://doi.org/10.1074/jbc.M112.422238
47. Li, L., Yu, R., Cai, T., Chen, Z., Lan, M., Zou, T., Wang, B., Wang, Q., Zhao, Y. & Cai, Y. (2020). Effects of immune cells and cytokines on inflammation and immunosuppression in the tumor microenvironment. *Int Immunopharmacol*, 88, 106939. https://doi.org/10.1016/j.intimp.2020.106939
48. Morgan, M., Ikenoue, T., Suga, H., & Wolberger, C. (2022). Potent macrocycle inhibitors of the human SAGA deubiquitinating module. *Cell Chem Biol*, 29(4), 544–554. https://doi.org/10.1016/j.chembiol.2021.12.004
49. Tang, J. Q., Veggiani, G., Singer, A., Teyra, J., Chung, J., & Sidhu, S. S. (2021). A panel of engineered ubiquitin variants targeting the family of domains found in ubiquitin specific proteases (DUSPs). *J Mol Biol*, 433(24), 167300. https://doi.org/10.1016/j.jmb.2021.167300
50. Wang, Y., Sun, Q., Mu, N., Sun, X., Wang, Y., Fan, S., Su, L., & Liu, X. (2020). The deubiquitinase USP22 regulates PD-L1 degradation in human cancer cells. *Cell Commun Signal*, 18(1), 112. https://doi.org/10.1186/s12964-020-00612-y
51. Ren, Y., Feng, M., Hao, X., Liu, X., Li, J., Li, P., Gao, J., Qi, Q., Du, L., Wang, C., Wang, Q. & Wang, Y. (2023). USP48 stabilizes gasdermin E to promote pyroptosis in cancer. *Cancer Res*, 83(7), 1074–1093. https://doi.org/10.1158/0008-5472.CAN-22-1812

5 Proteasome, the Recycle Bin of the Cancer Cell

Amy Dunne Miller, Vasudha Tandon, and Sourav Banerjee
University of Dundee, Dundee, UK

Nilabja Sikdar
Indian Statistical Institute, Kolkata, India

5.1 INTRODUCTION

The correct function of a cellular organism requires a fine balance between the synthesis and degradation of proteins. In a healthy environment, cells are continuously synthesising proteins to grow and to maintain basic functions. Additionally, protein requirements change in response to a stimulus or with changes in cellular needs. Of the amino acids required for such a large demand only a small proportion is available from diet (Young, Steffee, Pencharz, Winterer, & Scrimshaw, 1975). Therefore, most amino acids are obtained through the degradation and recycling of proteins that are not needed.

Protein degradation mainly occurs through two processes: autophagy and the ubiquitination proteasome pathway (UPP). Autophagy, coming from the Greek words '*phagy*' meaning eat and '*auto*' meaning self, is a term used to describe lysosomal-driven degradation of both complex cellular structures and proteins (Dikic & Elazar, 2018). While autophagy can in some instances be selective towards specific targets, keeping the dynamic protein turnover in perfect harmony requires a system with the capacity for much greater protein specificity and tighter regulation. The bulk of substrate specific protein turnover is in fact carried out by the UPP.

The multi-step approach to the ubiquitin proteasome system allows for high substrate specificity and various regulation points. Ubiquitination involves covalent binding of ubiquitin to lysine residues on the surface of target substrates. This is mediated by three enzymes in a stepwise manner (Figure 5.1). Firstly, the ubiquitin-activating enzyme E (E1) activates and binds to ubiquitin in the presence of adenosine triphosphate (ATP). Next, ubiquitin is transferred from E1 to the ubiquitin conjugating enzyme (E2). Lastly, ubiquitin ligase (E3) recruits a substrate and the ubiquitin-bound E2, enabling the transfer of ubiquitin onto the substrate (Hershko, Heller, Elias, & Ciechanover, 1983). Once the substrate is ubiquitinated, it is recognised by the 26S proteasome for degradation. There are two E1 ligases, about 40 E2 ligases and more than 600 E3 ligases, the latter conveys the specificity of the UPP system in combination with E2 ligases. Ubiquitination by E3 ligases also varies. The most potent signals for degradation are polyubiquitin chains formed by the formation of isopeptide bonds between a C-terminal Gly residue and the ε-NH2 group of (linear) or through Lys48-Lys11 links (branched). However, proteins may also be monoubiquitinated and multi-monoubiquitinated, both of which can be sufficient for degradation (Braten et al., 2016; Jevtić, Haakonsen, & Rapé, 2021).

The proteasome digests polypeptides into short peptides between two and ten residues in length. Ensuring the proteolysis of the right substrates take place, this process also comprises multiple tightly regulated steps. First, the ubiquitinated substrate must bind to the proteasome and commit to proteolysis (Peth, Uchiki, & Goldberg, 2010). Then, it must be de-ubiquitinated by one of the proteasomes multiple de-ubiquitinating enzymes (DUBs) (Verma et al., 2002). Lastly, the protein is

DOI: 10.1201/9781003394693-7

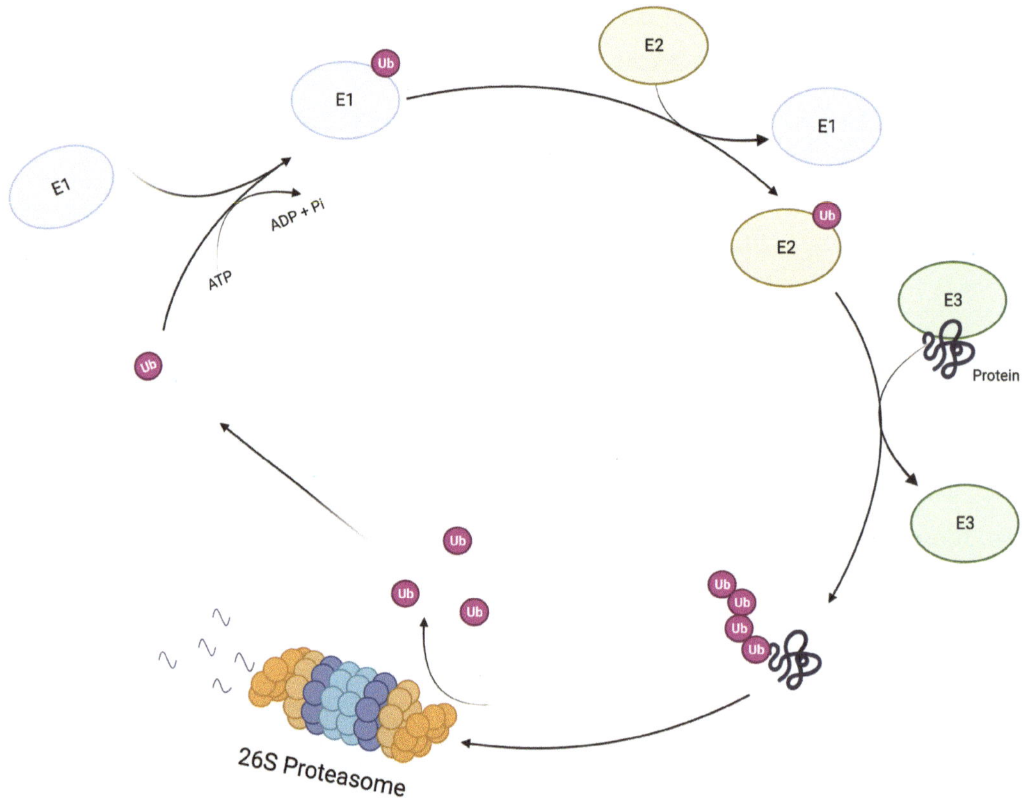

FIGURE 5.1 The ubiquitin proteasome pathway. Protein degradation through the multi-step process of the UPP. (Created with Biorender.com.)

unfolded and translocated into the core in an energy dependent manner where it is finally digested (Smith et al., 2005).

It is estimated that about 80% of protein degradation occurs through the proteasome, consequently the proteasome influences most cellular processes. It is no wonder that there has been extensive research on proteasomal structure, function, and regulation in an effort to understand its role in disease and most importantly to build therapeutic tools targeting the 26S proteasome.

5.1.1 Brief History of the Proteasome: Discovery

Schoenheimer was the first to describe protein turnover with the use of ^{15}N-labelled amino acids. Until then, it was believed that proteins were stably expressed in the tissue while dietary protein was necessary only for fuel (Schoenheimer, Ratner, & Ritten-berg, 1939; Schoenheimer, 1946). In 1953, Melvin Simpson used rat liver slices to study the release of radioactively labelled methionine and leucine. In anaerobic conditions in which energy supplies are low, their release was impeded, implying that proteolysis requires energy (Simpson, 1953). Nevertheless, the notion of protein turnover was still not widely accepted.

Reports emerged of cytosolic granules with a membrane-like barrier containing several enzymes such as β-glucuronidase and acid phosphatase are described between 1952 and 1955, thus the lysosome was discovered (Duve, Gianetto, Appelmans, & Wattiaux, 1953; Gianetto & De Duve, 1955). At the time, the lysosome seemed to provide logical mechanism for the observed continuous protein synthesis and degradation. However, it would soon become obvious that protein specific degradation could not be accounted for by lysosomal degradation.

It wasn't until 1974 that the components of the UPP started to be identified. First, ubiquitin was isolated from the bovine thymus and named thymine. Due to its perceived effect in promoting T cell differentiation following the injection of its purified form in mice, it was believed to be a thymic hormone (Goldstein, 1974). Further investigation showed its expression was conserved in most cell types and several other organisms such as yeast and plants and, therefore, was later re-named ubiquitin (Goldstein et al., 1975). At this time, it was believed to act through the stimulation of adenylate cyclase, and it wasn't until 1979 that ubiquitin's involvement in the immune response was disproven (Figure 5.2) (Low & Goldstein, 1979).

Goldberg and colleagues in 1977 described for the first time protein degradation in a lysosome independent manner, and requiring ATP, using cell-free fractions obtained from reticulocytes (Etlinger & Goldberg, 1977). At the same time, more reports came out of two distinct proteolytic systems which were dependent on cellular conditions. One in which both abnormal and normal proteins were degraded during growth and one which degrades normal proteins in the absence of nutrients (Mandelstam, 1958; Poole, Ohkuma, & Warburton, 1977). Shortly after, two cell-free fractions are isolated from reticulocytes, both of which are needed for globulin degradation. Furthermore, fraction one appeared to contain a heat-stable polypeptide that improved the proteolysis capacity of fraction II (Ciehanover, Hod, & Hershko, 1978). At this point in time, research in this area began to accelerate with great strides being made in the understanding of the proteolytic systems. In 1980, the heat-stable polypeptide, called ATP-dependent proteolysis factor 1 (APF1), was shown to form a high molecular conjugate when combined with fraction II in an ATP dependent manner. Additionally, APF1 was found to bind covalently to other reticulocyte protein substrates. A chain reaction in which APF1 recognises and binds to substrates, subsequently leading to their degradation through proteolysis, was proposed (Ciechanover, Heller, Elias, Haas, & Hershko, 1980; Hershko, Ciechanover, Heller, Haas, & Rose, 1980). Shortly after, APF1 is identified as ubiquitin (Ciechanover, Elias, Heller, Ferber, & Hershko, 1980; Wilkinson, Urban, & Haas, 1980).

It was in 1983, when the three enzymes required for ubiquitination were first isolated using chromatography of reticulocyte extracts on ubiquitin-Sepharose. E1 ubiquitin activating enzyme was found and described the necessity for two other enzymes, E2 and E3, for the final conjugation of ubiquitin on proteins in a stepwise manner (Hershko et al., 1983). Next, using a thermolabile E1 enzyme, it was demonstrated that heat-inactivation of E1 led to the inhibition of degradation of both abnormal proteins and normal short-lived proteins (Ciechanover, Finley, & Varshavsky, 1984). Thus, a solid link between ubiquitination and degradation was established. Using chromatography and a glycerol gradient it was possible to partially purify a large enzyme that was able to degrade ubiquitin-lysozyme substrates in the presence of ATP, but not alone or with added free ubiquitin (Hough, Pratt, & Rechsteiner, 1986). This was later named the 26S proteasome and, simultaneously, the 20S complex is identified with similar catalytic activity to the 26S and likened to the multi-catalytic proteinase complex described by Wilk and Orlowski (Figure 5.2) (Hough, Pratt, & Rechsteiner, 1987). The 20S complex is suggested to be part of the 26S proteasome, although it was thought unlikely.

The structure of the 26S proteasome was mapped in the following decades, with the first electron microscopy images of the mammalian 26S show the appearance of the 20S CP with two asymmetric masses attached to either end (Ventii et al., 2008). This was closely followed by the first crystal structures of both archaeal and yeast 20S CP (Groll et al., 1997; Löwe et al., 1995). Subnanomolar resolution of the 26S yeast proteasome structure was achieved in 2010 which gave useful insight on how subunits of the 19S RP interacted with the 20S CP (Bohn et al., 2010). In subsequent years as cryo-EM technology has improved, it has become possible to obtain atomic structures of substrate-engaged human and yeast 26S proteasome, leading to a better understanding of

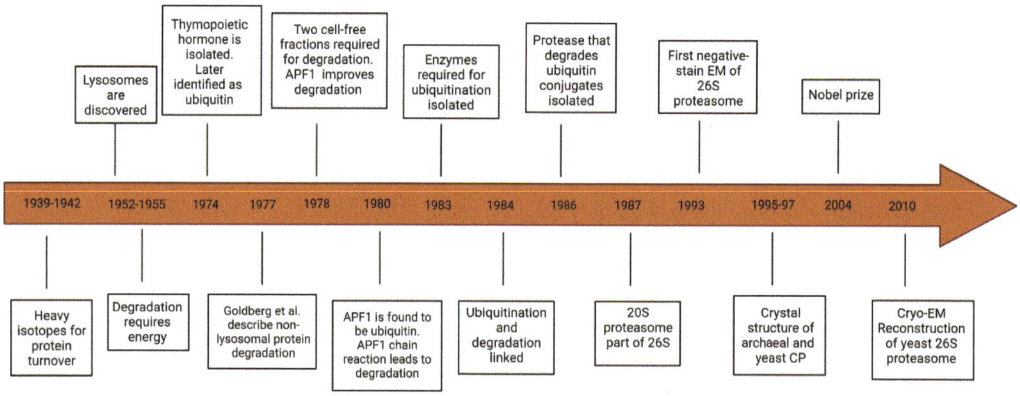

FIGURE 5.2 Discovery of the proteasome timeline. Notable discoveries in the establishment of the 26S proteasome in the UPP. (Adapted from Ciechanover, 2005; created with biorender.com.)

how proteolysis is coordinated (Figure 5.2) (de la Peña, Goodall, Gates, Lander, & Martin, 2018; Dong et al., 2019).

In 2004, the Nobel Prize in Chemistry was awarded to Aaron Ciechanover, Avram Hershko and Irwin Rose for their contributions towards defining the ubiquitin-proteasome system. These initial discoveries have been pivotal in opening research in protein turnover leading to incredible findings in the context of cell regulation and disease.

5.2 STRUCTURE OF THE PROTEASOME

Electron microscopy demonstrated that the 26S proteasome is a large multi-catalytic protease composed of two complexes: a 20S core particle (CP) and a 19S regulatory particle (RP) (Lupas, Koster, & Baumeister, 1993; Peters, Cejka, Harris, Kleinschmidt, & Baumeister, 1993). The 20S CP performs the catalytic activity of the proteasome while the 19S RP recognises ubiquitin chains and governs substrate entry to the proteolytic chamber (Figure 5.3).

The 20S has a barrel-shaped structure of four stacked rings, two outer α-rings and two inner β-rings (Figure 5.3) (Baumeister et al., 1988; Groll et al., 1997; Unno et al., 2002). Each of these are composed of seven distinct subunits (Löwe et al., 1995; Pühler et al., 1992). The two outer alpha-rings act as a gate to prevent uncontrolled substrate entry through their tightly interwoven (N)-terminal tails (Groll et al., 2000; Finley, 2009). The gate opens upon binding of the C-terminal hydrophobic-tyrosine-X motif (HbYX), found in 19S ATPases, to pockets found between the 20S subunits (Smith et al., 2007). Of the seven β-subunits, four are inactive because they either contain unprocessed (β3 and β4) or intermediately processed (β6 and β7) pro-peptides. The other three β-subunits are active and have distinct catalytic capacities. β1 has caspase-like/PGPH (peptidyl glutamyl-peptide hydrolysing) activity, β2 shows trypsin-like activity and β5 has chymotrypsin-like activity (Arendt & Hochstrasser, 1997; Groll et al., 1997; Unno et al., 2002).

The 19S RP can cap one or both ends of the 20s barrel. It is formed of two subcomplexes: the lid and the base with a total of 17 distinct subunits: nine in the base subcomplex and eight in the lid subcomplex (Figure 5.3). The lid is composed of non-ATPase subunits (Rpn): Rpn3, Rpn5-9, Rpn11-12 and Sem1. Rpn11 is the intrinsic DUB of the mammalian proteasome which cleaves ubiquitin once a ubiquitinated substrate has bound to ubiquitin receptors found in the RP base (Verma et al., 2002). Additionally, auxiliary DUBs can be found bound to the lid such as ubiquitin C-terminal hydrolase 14 (USP14) and ubiquitin C-terminal hydrolase 37 (UCH37) (Lee et al., 2016; Vander Linden et al., 2015). As previously mentioned, both the channel opening and the protein unfolding require energy. The base has six AAA+-ATPase subunits known as RP triple-A proteins (Rpt) 1–6. These

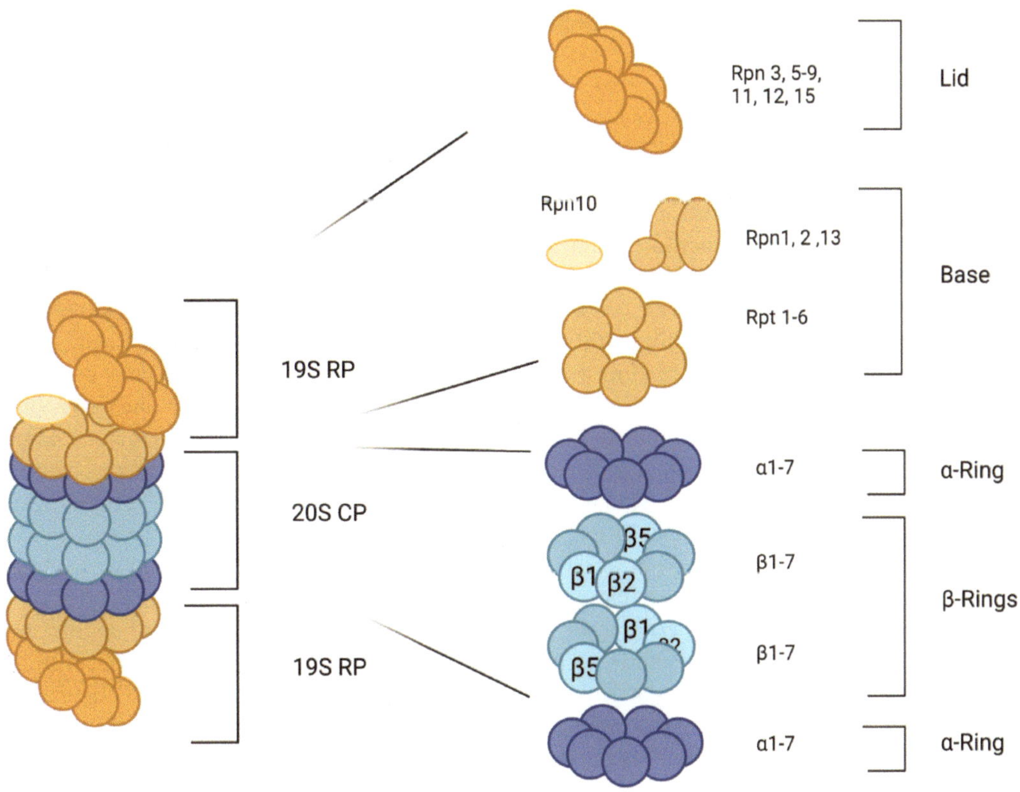

FIGURE 5.3 26S proteasome structure. (Adapted from Murata, Yashiroda, & Tanaka, 2009; created with biorender.com.)

form a heterohexameric ring which is in direct contact with the outer alpha-ring of the CP (Tomko, Funakoshi, Schneider, Wang, & Hochstrasser, 2010). Additionally, the base has four more RP non-ATPase subunits (rpn): Rpn 1, Rpn2, Rpn10 and Rpn13. Rpn1 and Rpn2 act mainly as scaffolds, mediating binding of proteasomal subunits and proteasome-associated proteins (Elsasser et al., 2002; Hamazaki et al., 2006). Rpn10 and 13 are intrinsic ubiquitin receptors (Husnjak et al., 2008; Mayor, Graumann, Bryan, MacCoss, & Deshaies, 2007; Schreiner et al., 2008). The interactions between base and lid occur at Rpn3, Rpn7, Rpn8 and Rpn11 of the lid and Rpt 3, Rpt 6 and Rpn2 of the base (Lander et al., 2012).

5.3 TISSUE-SPECIFIC PROTEASOME COMPLEXES

There are three known tissue-specific proteasome complexes: the immunoproteasome, the thymoproteasome and the spermatoproteasome. Within these three tissues, the proteasome has clear tissue-specific functions.

5.3.1 The Immunoproteasome

The immunoproteasome is named such as it is constitutively expressed in cells of the immune system (Barton, Cruz, Rangwala, Deepe, & Monaco, 2002). It differs from the canonical proteasome in the catalytic β1, β2 and β5 subunits of the 20S which are replaced by β1i, β2i and β5i, respectively, which change the proteasome's peptidase activity: enhancing trypsin and chymotrypsin-like activity and reducing caspase-like activity (Akiyama et al., 1994; Driscoll, Brown, Finley, & Monaco,

1993; Gaczynska, Goldberg, Tanaka, Hendil, & Rock, 1996). This switch to immunoproteasome subunits can be activated in other cell types and upregulated in the immune system by interferon-γ (IFN-γ) (Table 5.1) (Akiyama et al., 1994; Groettrup et al., 1996).

The immunoproteasome is involved in antigen processing as well as T helper (Th) cell differentiation. Two of the subunits, β1i and β5i, are encoded in the major histocompatibility complex (MHC) gene locus (Martinez & Monaco, 1991). Using a screen of proteasome inhibitors, we now know that the immunoproteasome is necessary for the MHC I but not the MHC II pathway (Craiu et al., 1997; Rock et al., 1994). The immunoproteasome cleaves protein antigens into peptide fragments which then can bind to the peptide ligand cleft found in class I molecules. Only when this has occurred can the molecules fold into their trimeric complex conformation and be released from the endoplasmic reticulum (ER) to be presented at the cell surface for antigen recognition by cytotoxic T cells. There are two requirements for peptide binding to class I molecules: the peptides must be 8 to 9 aa (amino acid) long and the C terminus of the peptide must contain basic or hydrophobic anchor residues (Madden, 1995; Rammensee, Bachmann, Emmerich, Bachor, & Stevanović, 1999). Only the immunoproteasome and not the canonical proteasome can achieve this due to the differences between the β1i subunit and the β1 subunit. This is because the β1 subunit accepts acidic P1 residues while the β1i subunit binds to hydrophobic P1 residues (Huber et al., 2012). Furthermore, the replacement of β5 with β5i benefits the antigen-processing pathway by providing larger hydrophobic aa chains which are preferred by some types of class I molecules (Table 5.1) (Huber et al., 2012).

The immunoproteasome may have other functions, although they are not yet well defined. For example, it appears that T cells require an immunoproteasome to survive inflammation, with β5i and β2i subunits appearing to be particularly important (Moebius, van den Broek, Groettrup, & Basler, 2010). This weakness can be exploited in the context of autoimmune diseases (Muchamuel et al., 2009; Schmidt et al., 2010). This immunoproteasome function may be a consequence of promoting the differentiation Th1- and Th17-type proinflammatory cells and suppressing regulatory T cells (Treg) (Tabel 5.1) (Kalim, Basler, Kirk, & Groettrup, 2012).

5.3.2 The Thymoproteasome

The discovery of the thymoproteasome occurred with the finding of an open reading frame homologous to β5 and β5i expressed only in the epithelial cells of the thymus. The β5t appears to be involved in the selection of CD8+ cytotoxic T cells (Murata et al., 2007). β5t proteasomes have a reduced chymotrypsin-like activity compared to β5 canonical proteasomes caused by a change in the S1 pockets of a few hydrophobic amino acids to more hydrophilic amino acids. This translates to an increased preference for cleavage of hydrophilic amino acids by the β5t, leading to weakened MHC class I ligands. In the context of the thymus, these are important for positive selection, as the affinity model suggests that T cell receptors with low affinity binding to the MHC/peptide complex leads to positive selection, while high affinity binding leads to negative selection (Table 5.1) (Murata et al., 2007; Ohigashi & Takahama, 2021).

TABLE 5.1
Tissue-specific Proteasomes

Tissue Specific Proteasome	Subunits	Peptidase Activity	Functions
Immunoproteasome	β1i, β2i and β5i	Increased trypsin + Chymotrypsin activity Decreased caspase activity	Antigen presentation T helper differentiation
Thymoproteasome	β5t	Decreased chymotrypsin activity	T cell selection
Spermatoproteasome	Multiple subunits	N/A	Sperm individualisation

5.3.3 The Spermatoproteasome

The testis-specific proteasome was discovered in the 1990s, using *Drosophila melanogaster*. First, two alternative α4 subunits were identified, α4T1 and α4T2 (Yuan, Miller, & Belote, 1996). However, since then, 12 more subunits of the 26S have been shown to have two or even three isoforms which are exclusively expressed in the testis (Ma, Katz, & Belote, 2002). Specifically, these are expressed only in mid to late stages of spermatogenesis. In elongated spermatids, proteasomes containing alternate subunits were reported along the individualisation complex (IC), which mediates sperm individualisation by enveloping single spermatids a plasma membrane (Zhong & Belote, 2007). The function of the proteasome interaction with the IC is not yet understood, although knock-out of the α6T alternative subunit in *drosophila* led to infertility due to defective sperm individualisation (Zhong & Belote, 2007). In mammals, an alternative α4 subunit of the 20S, α4s/PSMA8, was found to be essential for correct meiotic exit (Gómez et al., 2019). Furthermore, in the mammalian testis, the PA200 regulator, a 200 kDa monomeric non-ATPase proteasome activator, was highly expressed and can form a single or double cap complex. PA200-deficient male mice suffer from low fertility. It is postulated that α4s subunits allows docking of PA200 more easily (Table 5.1) (Khor et al., 2006).

5.4 REGULATION AND FURTHER DETAILED FUNCTION

Initially, it was believed the proteasome didn't require much regulation as it is considered part of the housekeeping machinery. This idea has now been thoroughly disproven. The proteasome's involvement in complex and diverse pathways requires tight regulation both at the transcriptional and the post-translational level.

5.4.1 Regulation of Subunit Abundance

The abundance of proteasome subunits is tightly controlled. In yeast one transcription factor, Rpn4, regulates the transcription of all subunits by binding to a nonameric box (5′-GGTGGCAA-3′) which can be found in most proteasome subunits named proteasome-associated controlled elements (PACE) (Mannhaupt, Schnall, Karpov, Vetter, & Feldmann, 1999). In mammals, there are two known transcription factors involved in regulating proteasome subunit expression: nuclear factor erythroid 2-related factor 1 (NRF1) and NRF2. Each plays a role in regulating proteasomal expression under different conditions. NRF1 is an ER membrane protein that is continuously translocated to the cytosol where it ubiquitinated and degraded under normal conditions. In circumstances where the proteasome is compromised, NRF1 evades proteasomal degradation and is cleaved by aspartyl protease protein DDI1 homologue 2 (DDI2). Cleavage of NRF1 allows for its translocation to the nucleus where it induces expression of proteasomal subunits (Lehrbach & Ruvkun, 2016; Radhakrishnan et al., 2010). On the other hand, NRF2 promotes proteasome subunit expression upon its stabilisation during redox stress (Kwak, Wakabayashi, Greenlaw, Yamamoto, & Kensler, 2003). Additionally, NRF2 induction of the proteasome has been reported in a p53-dependent manner in several cancer types (Walerych et al., 2016).

5.4.2 Regulation of Subunit Assembly

5.4.2.1 Regulation of RP Assembly

The RP assembly chaperone (RAC) Adc17 was the first to be identified as a regulator of RP assembly in yeast. Adc17 interacts with Rpt6 assisting in the pairing of Rpt6-Rpt3 ATPase, an important step for RP assembly (Hanssum et al., 2014). Adc17, and other RACs, are inducible under stress condition such as heat shock and ER stress upon TOR complex 1 inhibition and subsequent map kinase 1 (MAPK1) activation. An increase in proteasomal capacity under these challenging conditions is essential for maintaining cell viability (Rousseau & Bertolotti, 2016).

5.4.2.2 Regulation of CP Assembly

Molecular mechanisms of CP assembly are poorly understood. There is some evidence to indicate the Rpn4 may regulate CP chaperone expression (Shirozu, Yashiroda, & Murata, 2015). Additionally, ER stress in yeast induces the expression of Pba1 and Pba2 followed by increased CP assembly. On the other hand, accumulation of mitochondrial precursors in the cytosol and disturbed mitochondrial protein import led to an increase of Pba3 and Pba4 (Rousseau & Bertolotti, 2016). Similarly, the stabilisation and dimerisation of PAC1 and PAC2 (Pba1–2 human homologues) was reported upon ER stressed and mediated proteasomal activity upregulation through inactive rhomboid protein 1 (iRhom1) (Lee et al., 2015). Furthermore, the transmembrane domain recognition complex (TRC) is known to be involved in CP assembly as loss of TRC components leads to a decrease of CP assembly. It has been suggested that TRCs may mediate incorporation of the β-subunit rings into the α-subunits (Akahane, Sahara, Yashiroda, Tanaka, & Murata, 2013). Lastly, the miR-101 tumour suppressor also regulates CP assembly by targeting the mRNA of chaperone POMP, decreasing POMP protein levels which in turn decreases proteasome assembly (Zhang et al., 2015).

5.4.2.3 Regulation of RP-CP Association

The association of the RP-CP complex is crucial for degradation as without the RP, the N-terminal tails of the α-rings of the CP impede substrate entry into the catalytic core by forming a ring. Association between both components can occur spontaneously *in vitro* (Liu et al., 2006). However, some evidence suggests that this process does require further regulation *in vivo*. Several factors have been implicated in RP-CP assembly such as heat shock protein 90 (Hsp90) and RP component Rpn6 (Imai, Maruya, Yashiroda, Yahara, & Tanaka, 2003; Vilchez et al., 2012). Another mechanism by which proteasome assembly is regulated is through HEAT-like repeat protein Ecm9. Ecm9 detects faulty RP-CP association and impedes 26S activity by inhibiting both CP gate opening and ATPase activity (De La Mota-Peynado et al., 2013; Lee, De la Mota-Peynado, & Roelofs, 2011). Additionally, Ecm9 can also cause dissociation of RP-CP by binding between the RP and the CP under oxidative stress conditions (Wang et al., 2017).

5.4.3 Regulation of Proteasomal Activity through Post-Translational Modifications

A large-scale analysis in yeast has revealed that the 26S proteasome can be modified with more than 345 post-translational modifications. These include: phosphorylation, ubiquitylation, poly(ADP-ribosyl)ation, succinylation, N-myristoylation, O-glycosylation, N-methylation, oxidation, N-acetylation, sumoylation and truncation (Kikuchi et al., 2010). The most extensively studied have been phosphorylation events and the kinases/phosphatases responsible.

5.4.3.1 Kinases Regulating Proteasomal Activity

Many kinases have been identified as proteasomal regulators with the abilities to modulate assembly and activity of the proteasome components.

5.4.3.1.1 Dual-specificity Tyrosine-Regulated Kinase

The Dixon lab demonstrated in 2016 the importance of 19S subunit Rpt3-Thr25 phosphorylation in regulating proteasomal activity. Their studies demonstrated Rpt3-Thr25 phosphorylation is upregulated during the G2/M phase of the cell cycle and was downregulated during serum starvation when cell proliferation is reduced (Guo et al., 2016). Furthermore, using CRISPR-Cas9 to introduce a threonine to alanine (T25A) mutation impeding its phosphorylation and a fluorogenic peptide, they demonstrated a reduction in proteasomal activity in the absence of Rpt3-Thr25 phosphorylation. Furthermore, using a kinome-wide screen, they discovered dual-specificity tyrosine-regulated kinase 2 (DYRK2) as the main kinase responsible for phosphorylating Rpt3-Thr25. Said phosphorylation appears to improve substrate stimulated ATP hydrolysis and subsequently improve substrate unfolding and translocation (Guo et al., 2016).

5.4.3.1.2 Protein Kinase A

cAMP dependent protein kinase A (PKA) phosphorylates 19S subunit Rpn6 which promotes proteasomal activity and degradation of short-lived proteins (Zhang et al., 2007). This was discovered using a combination of pharmacological and chemical assays. For example, stimulation of adenylate cyclase led to an increase in both PKA and proteasome activity as did the activation of cAMP (Lokireddy, Kukushkin, & Goldberg, 2015; Zhang et al., 2007). Most importantly, the effects observed on proteasomal activity could be reversed with the use a PKA inhibitor as with the use of a protein phosphatase. Reinforcing these findings, the overexpression of a phospho-mimetic mutant of Rpn6 led to enhanced proteolytic activity. Therefore, PKA phosphorylation of Rpn6 clearly regulates proteasome activity (Lokireddy et al., 2015).

5.4.3.1.3 Protein Kinase G

Like the discovery of PKA involvement in proteasome regulation, cGMP-dependent protein kinase G (PKG) was shown to be involved using a wide array of pharmacological and chemical assays. Firstly, the inhibition of phosphodiesterase 5, which hydrolyses cGMP, led to increased peptidase activity by increasing the speed of both hydrolysis of peptides and ATP consumption, indicating regulation of the 19S RP (VerPlank, Tyrkalska, Fleming, Rubinsztein, & Goldberg, 2020). When PKG is inhibited, or a phosphatase is stimulated, these effects are reversed. However, the phosphorylation site is yet to be determined (VerPlank et al., 2020).

5.4.3.1.4 Proviral Insertion in Murine (PIM) Kinases

Phosphorylation of Rpn1-ser361 was found to be within the ten most common S/T phospho-sites of the human proteasome. Phospho-proteomic analysis of wild type HaCaT and S361A cells demonstrated the importance of this phosphorylation site as a change in protein levels was observed in more than 300 proteins, many of which were involved in cell cycle regulation, apoptosis as well as cellular components of the mitochondria and ER. Several kinases were found to phosphorylate Rpn1-ser361 including PIM1-3, MAP4K1/2, PKA and NEK6; with PIM1-3 knock-out leading to decreased proteasomal activity (Liu et al., 2020).

5.4.3.1.5 Src Kinase

About one-third of phosphorylation sites of the 26S proteasome are phospho-tyrosine (pTyr) sites. Among these, the Rpt2-y439 is the most phosphorylated site of the 19S subunit (Chen et al., 2021). Phosphorylation at this site appeared to impede binding of 19S RP to the 20S CP, leading to a decrease in proteasomal activity. It was also determined that phosphorylation of Y439 was dependent on N-terminal myristoylation and occurred at the cell membrane. Src is also myristoylated on the N-terminal and colocalised with Rpt2-Y439 at the membrane. Inhibition of Src led to a decrease in Y439 phosphorylation as did the introduction of a kinase-deficient mutant Src. Many of the proteins regulated by this phospho-site were found to be pro-oncogenic proteins (Chen et al., 2021).

5.4.3.2 Phosphatases Regulating Proteasomal Activity

5.4.3.2.1 Ubiquitin Like Domain Containing CTD Phosphatase 1 (UBLCP1)

Using co-immunoprecipitation techniques, it was demonstrated that UBLCP1 interacted with components of the 26S proteasome, specifically binding to Rpn1 of the 19S RP. Fluorescent microscopy showed strong localisation of UBLCP1 in the nucleus, with its nuclear localisation being dependant on proteasomal binding. Additionally, UBLCP1 was shown to dephosphorylate the nuclear proteasome, reducing its activity using a combination of *in vitro* and *in vivo* methods (Guo et al., 2011). Dephosphorylation of the proteasome by UBLCP1 regulated the assembly of the 26S proteasome. When UBLCP1 was depleted, there was an increase in the size of the proteasome and increased RP and CP interaction, both of which correlate with an increase in proteasomal activity (Guo et al., 2011). Furthermore, UBLCP1 is also involved in dephosphorylating the Rpn1-S361 site described earlier (Liu et al., 2020).

5.4.3.2.2 Protein Tyrosine Phosphatase Non-Receptor Type 2 (PTPN2)

PTPN2 was found to associate with the 26S proteasome in Rpn11 pull down and interactome studies (Kumar et al., 2017). Knock-out of PTPN2 led to an increase in RPT-Y439 phosphorylation. PTPN2 can act by targeting and dephosphorylating SRC, thus reducing Y439 phosphorylation indirectly or by directly dephosphorylating the RPT2-Y439 (Chen et al., 2021).

5.5 THE PROTEASOME IN CANCER

Dysregulation of the proteasome in cancer can occur at the genetic, epigenetic, transcriptional, or post-translational level which can lead to changes in expression, assembly, localisation and function. Genomic instability is a prominent hallmark of cancer, with most cancer cells being aneuploid. Proteasomal genes are no exception. The consequences of such dysregulation range from uncontrolled cell growth, improved survival, immune avoidance, drug resistance and to promoting epithelial to mesenchymal transition and stemness.

5.5.1 Breast Cancer

Triple-negative breast (TNB) cancers, named after their lack of oestrogen, progesterone and Her2 receptors, are some of the most aggressive forms of breast cancer. A genome-wide siRNA screen of two TNB lines, one basal-like and one myoepithelial, identified proteasomal genes to correlate with poor-prognosis. Using proteasomal inhibitors, it was also demonstrated that basal-like TNB, but no other TNB lines, were dependent on proteasomal activity or showed proteasome addiction (Petrocca et al., 2013). Conversely, expression of the immunoproteasome in TNBs has been proposed to improve immune-mediated clearance. High immunoproteasome expression in TNBs appears to improve prognosis, whereas loss of immunoproteasome expression may lead to immune evasion (Adwal et al., 2020).

5.5.2 Multiple Myeloma

A feature of multiple myeloma (MM) cells is that they secrete abnormally high amounts of immunoglobulins. This makes MM highly reliant on proteasomal activity to avoid proteotoxic and ER stress (Meister et al., 2007). Additionally, upregulation of the proteasome in MM also leads to degradation of tumour suppressor genes such as p53 and the inhibitor of nuclear factor-κB (NF-κB), IκB. IκB degradation allows NF-κB translocation to the nucleus where it promotes transcription of growth factors, cell adhesion molecules, pro-angiogenic factors and anti-apoptotic factors (Palombella, Rando, Goldberg, & Maniatis, 1994; Wang, Guttridge, Mayo, & Baldwin, 1999). Therefore, the proteasome is heavily involved in promoting multiple myeloma carcinogenesis.

5.5.3 Lymphoma

Many of the pathways regulating the cell cycle and replication are regulated partly by the proteasome. Dysregulation through increased proteolysis of cell cycle inhibitors can lead to uncontrolled cell division. Degradation of p27 cyclin dependent kinase inhibitor (CDKi) leads to early entry into S phase. Increased degradation of p27 in mantel cell lymphoma (MCL) was directly linked to poor survival (Chiarle et al., 2000). Another example is Geminin, a DNA replication negative regulator. Unbalanced levels of Geminin, caused by the 26S proteasome, have been associated with the progression of peripheral B cell lymphoma and in mantle cell lymphoma with genetic instabilities (Obermann et al., 2005; Pinyol et al., 2006). Cyclin D is involved in regulating the transition from G1 to S phase and is also regulated by the 26S proteasome. Overexpression of cyclin D can lead to shortened G1 phase and increased proliferation which has been observed in MCL (Hachem & Gartenhaus, 2005; Resnitzky, Gossen, Bujard, & Reed, 1994). Furthermore, as seen in multiple myeloma, the proteasomal regulation of transcription factors such as c-myc and NF-κB can also play a role in various lymphomas which necessitates the constitutive activation of NF-κB for their survival (Davis, Brown, Siebenlist, & Staudt, 2001).

5.5.4 Non-small-cell Lung Carcinoma (NSCLC)

PSME4/PA200 proteasome regulator is upregulated in NSCLC tumours. PSME4 is a well-characterised proteasomal activator as described previously. Upregulation in NSCLC patients correlates with poor prognosis, larger tumour size and, importantly, correlates with a decreased response to immunotherapy. These effects appear to be caused by an enhanced ability for immune evasion (Javitt et al., 2023). PSME4 was found to change proteasomal cleavage reducing antigen diversity in MHC class I molecule presentation. Additionally, knock-down of PSM4 led to increased infiltration of cytotoxic CD8+ T cells, suggesting that targeting PSM4 could improve responsiveness to immunotherapy (Javitt et al., 2023).

5.5.5 Immunoproteasome in Cancer

The expression of immunoproteasome subunits were reported in various solid and haematopoietic tumours. However, their role in malignancies is difficult to pinpoint. On the one hand, downregulation of the immunoproteasome may lead to immune evasion, while upregulation of the immunoproteasome could improve resistance to proteotoxicity. Inducible expression of the immunoproteasome subunits has been observed in many cancer types, but one appears to be implicated in most, β5i. Overexpression of β5i was correlated to poor prognosis in many cancer types: renal cell carcinoma, glioma, colorectal cancer and adenocarcinomas (Kwon et al., 2016; Piotrowska, Niezgoda, Młynarczyk, Acewicz, & Kasacka, 2020; Yang et al., 2018). However, in some types of malignancies, the reverse is true. For example, in NSCLC and TNB cancer (as mentioned earlier), increased β5i correlated with improved overall survival (Kiuchi et al., 2021).

5.6 CONCLUSION

The 26S proteasome function, structure and regulation has been well characterised since the initial discovery of the ubiquitin-proteasome pathway for protein degradation. This has given invaluable insight into one of the most integral components of the cell, playing a role in a multitude of cellular functions. The consequences of uncontrolled proteasomal activity are clear, as observed by proteasomal reliance of multiple cancer types to propagate and survive. Nevertheless, it is also apparent that the function of the proteasome in cancer is much more nuanced and could be both context and cell type dependent.

REFERENCES

Adwal, A., Kalita-de Croft, P., Shakya, R., Lim, M., Kalaw, E., Taege, L. D., … Saunus, J. M. (2020). Tradeoff between metabolic i-proteasome addiction and immune evasion in triple-negative breast cancer. *Life Sci Alliance*, *3*(7). doi:10.26508/lsa.201900562

Akahane, T., Sahara, K., Yashiroda, H., Tanaka, K., & Murata, S. (2013). Involvement of Bag6 and the TRC pathway in proteasome assembly. *Nat Commun*, *4*, 2234. doi:10.1038/ncomms3234

Akiyama, K., Kagawa, S., Tamura, T., Shimbara, N., Takashina, M., Kristensen, P., … Ichihara, A. (1994). Replacement of proteasome subunits X and Y by LMP7 and LMP2 induced by interferon-gamma for acquirement of the functional diversity responsible for antigen processing. *FEBS Lett*, *343*(1), 85–88. doi:10.1016/0014-5793(94)80612-8

Arendt, C. S., & Hochstrasser, M. (1997). Identification of the yeast 20S proteasome catalytic centers and subunit interactions required for active-site formation. *Proc Natl Acad Sci USA*, *94*(14), 7156–7161. doi:10.1073/pnas.94.14.7156

Barton, L. F., Cruz, M., Rangwala, R., Deepe, G. S., Jr., & Monaco, J. J. (2002). Regulation of immunoproteasome subunit expression *in vivo* following pathogenic fungal infection1. *J Immun*, *169*(6), 3046–3052. doi:10.4049/jimmunol.169.6.3046

Baumeister, W., Dahlmann, B., Hegerl, R., Kopp, F., Kuehn, L., & Pfeifer, G. (1988). Electron microscopy and image analysis of the multicatalytic proteinase. *FEBS Lett*, *241*(1–2), 239–245. doi:10.1016/0014-5793(88)81069-x

Bohn, S., Beck, F., Sakata, E., Walzthoeni, T., Beck, M., Aebersold, R., … Nickell, S. (2010). Structure of the 26S proteasome from Schizosaccharomyces pombe at subnanometer resolution. *Proc Natl Acad Sci USA*, *107*(49), 20992–20997. doi:10.1073/pnas.1015530107

Braten, O., Livneh, I., Ziv, T., Admon, A., Kehat, I., Caspi, L. H., … Ciechanover, A. (2016). Numerous proteins with unique characteristics are degraded by the 26S proteasome following monoubiquitination. *Proc Natl Acad Sci USA*, *113*(32), E4639–E4647. doi:10.1073/pnas.1608644113

Chen, L., Zhang, Y., Shu, X., Chen, Q., Wei, T., Wang, H., … Guo, X. (2021). Proteasome regulation by reversible tyrosine phosphorylation at the membrane. *Oncogene*, *40*(11), 1942–1956. doi:10.1038/s41388-021-01674-z

Chiarle, R., Budel, L. M., Skolnik, J., Frizzera, G., Chilosi, M., Corato, A., … Inghirami, G. (2000). Increased proteasome degradation of cyclin-dependent kinase inhibitor p27 is associated with a decreased overall survival in mantle cell lymphoma. *Blood*, *95*(2), 619–626.

Ciechanover, A. (2005). Proteolysis: from the lysosome to ubiquitin and the proteasome. *Nat Rev Mol Cell Biol*, *6*, 79–87.

Ciechanover, A., Elias, S., Heller, H., Ferber, S., & Hershko, A. (1980). Characterization of the heat-stable polypeptide of the ATP-dependent proteolytic system from reticulocytes. *J Biol Chem*, *255*(16), 7525–7528.

Ciechanover, A., Finley, D., & Varshavsky, A. (1984). Ubiquitin dependence of selective protein degradation demonstrated in the mammalian cell cycle mutant ts85. *Cell*, *37*(1), 57–66. doi:10.1016/0092-8674(84)90300-3

Ciechanover, A., Heller, H., Elias, S., Haas, A. L., & Hershko, A. (1980). ATP-dependent conjugation of reticulocyte proteins with the polypeptide required for protein degradation. *Proc Natl Acad Sci USA*, *77*(3), 1365–1368. doi:10.1073/pnas.77.3.1365

Ciehanover, A., Hod, Y., & Hershko, A. (1978). A heat-stable polypeptide component of an ATP-dependent proteolytic system from reticulocytes. *Biochem Biophys Res Commun*, *81*(4), 1100–1105. doi:10.1016/0006-291x(78)91249-4

Craiu, A., Gaczynska, M., Akopian, T., Gramm, C. F., Fenteany, G., Goldberg, A. L., & Rock, K. L. (1997). Lactacystin and clasto-lactacystin β-lactone modify multiple proteasome β-subunits and inhibit intracellular protein degradation and major histocompatibility complex class I antigen presentation. *J Biol Chem*, *272*(20), 13437–13445. doi: 10.1074/jbc.272.20.13437

Davis, R. E., Brown, K. D., Siebenlist, U., & Staudt, L. M. (2001). Constitutive nuclear factor κB activity is required for survival of activated B cell-like diffuse large B cell lymphoma cells. *J Exp Med*, *194*(12), 1861–1874. doi:10.1084/jem.194.12.1861

De La Mota-Peynado, A., Lee, S. Y., Pierce, B. M., Wani, P., Singh, C. R., & Roelofs, J. (2013). The proteasome-associated protein Ecm29 inhibits proteasomal ATPase activity and *in vivo* protein degradation by the proteasome. *J Biol Chem*, *288*(41), 29467–29481. doi:10.1074/jbc.M113.491662

de la Peña, A. H., Goodall, E. A., Gates, S. N., Lander, G. C., & Martin, A. (2018). Substrate-engaged 26*S* proteasome structures reveal mechanisms for ATP-hydrolysis–driven translocation. *Science*, *362*(6418), eaav0725. doi:10.1126/science.aav0725

Dikic, I., & Elazar, Z. (2018). Mechanism and medical implications of mammalian autophagy. *Nat Rev Mol Cell Biol*, *19*(6), 349–364. doi:10.1038/s41580-018-0003-4

Dong, Y., Zhang, S., Wu, Z., Li, X., Wang, W. L., Zhu, Y., … Mao, Y. (2019). Cryo-EM structures and dynamics of substrate-engaged human 26S proteasome. *Nature*, *565*(7737), 49–55. doi:10.1038/s41586-018-0736-4

Driscoll, J., Brown, M. G., Finley, D., & Monaco, J. J. (1993). MHC-linked LMP gene products specifically alter peptidase activities of the proteasome. *Nature*, *365*(6443), 262–264. doi:10.1038/365262a0

Duve, C. D., Gianetto, R., Appelmans, F., & Wattiaux, R. (1953). Enzymic content of the mitochondria fraction. *Nature*, *172*(4390), 1143–1144. doi:10.1038/1721143a0

Elsasser, S., Gali, R. R., Schwickart, M., Larsen, C. N., Leggett, D. S., Müller, B., ... Finley, D. (2002). Proteasome subunit Rpn1 binds ubiquitin-like protein domains. *Nat Cell Biol*, *4*(9), 725–730. doi:10.1038/ncb845

Etlinger, J. D., & Goldberg, A. L. (1977). A soluble ATP-dependent proteolytic system responsible for the degradation of abnormal proteins in reticulocytes. *Proc Natl Acad Sci USA*, *74*(1), 54–58. https://www.jstor.org/stable/66512

Finley, D. (2009). Recognition and processing of ubiquitin-protein conjugates by the proteasome. *Annu Rev Biochem*, *78*, 477–513. doi:10.1146/annurev.biochem.78.081507.101607

Gaczynska, M., Goldberg, A. L., Tanaka, K., Hendil, K. B., & Rock, K. L. (1996). Proteasome subunits X and Y alter peptidase activities in opposite ways to the interferon-gamma-induced subunits LMP2 and LMP7. *J Biol Chem*, *271*(29), 17275–17280. doi:10.1074/jbc.271.29.17275

Gianetto, R., & De Duve, C. (1955). Tissue fractionation studies. 4. Comparative study of the binding of acid phosphatase, beta-glucuronidase and cathepsin by rat-liver particles. *Biochem J*, *59*(3), 433–438. doi:10.1042/bj0590433

Goldstein, G. (1974). Isolation of bovine thymin: a polypeptide hormone of the thymus. *Nature*, *247*(5435), 11–14. doi:10.1038/247011a0

Goldstein, G., Scheid, M., Hammerling, U., Schlesinger, D. H., Niall, H. D., & Boyse, E. A. (1975). Isolation of a polypeptide that has lymphocyte-differentiating properties and is probably represented universally in living cells. *Proc Natl Acad Sci USA*, *72*(1), 11–15. doi:10.1073/pnas.72.1.11

Gómez, H. L., Felipe-Medina, N., Condezo, Y. B., Garcia-Valiente, R., Ramos, I., Suja, J. A., ... Pendas, A. M. (2019). The PSMA8 subunit of the spermatoproteasome is essential for proper meiotic exit and mouse fertility. *PLoS Genet*, *15*(8), e1008316. doi:10.1371/journal.pgen.1008316

Groettrup, M., Kraft, R., Kostka, S., Standera, S., Stohwasser, R., & Kloetzel, P. M. (1996). A third interferon-gamma-induced subunit exchange in the 20S proteasome. *Eur J Immunol*, *26*(4), 863–869. doi:10.1002/eji.1830260421

Groll, M., Bajorek, M., Köhler, A., Moroder, L., Rubin, D. M., Huber, R., ... Finley, D. (2000). A gated channel into the proteasome core particle. *Nature Structural Biology*, *7*(11), 1062–1067. doi:10.1038/80992

Groll, M., Ditzel, L., Löwe, J., Stock, D., Bochtler, M., Bartunik, H. D., & Huber, R. (1997). Structure of 20S proteasome from yeast at 2.4 A resolution. *Nature*, *386*(6624), 463–471. doi:10.1038/386463a0

Guo, X., Engel, J. L., Xiao, J., Tagliabracci, V. S., Wang, X., Huang, L., & Dixon, J. E. (2011). UBLCP1 is a 26S proteasome phosphatase that regulates nuclear proteasome activity. *Proc Natl Acad Sci USA*, *108*(46), 18649–18654. doi:10.1073/pnas.1113170108

Guo, X., Wang, X., Wang, Z., Banerjee, S., Yang, J., Huang, L., & Dixon, J. E. (2016). Site-specific proteasome phosphorylation controls cell proliferation and tumorigenesis. *Nat Cell Biol*, *18*(2), 202–212. doi:10.1038/ncb3289

Hachem, A., & Gartenhaus, R. B. (2005). Oncogenes as molecular targets in lymphoma. *Blood*, *106*(6), 1911–1923. doi:10.1182/blood-2004-12-4621

Hamazaki, J., Iemura, S., Natsume, T., Yashiroda, H., Tanaka, K., & Murata, S. (2006). A novel proteasome interacting protein recruits the deubiquitinating enzyme UCH37 to 26S proteasomes. *EMBO J*, *25*(19), 4524–4536. doi:10.1038/sj.emboj.7601338

Hanssum, A., Zhong, Z., Rousseau, A., Krzyzosiak, A., Sigurdardottir, A., & Bertolotti, A. (2014). An inducible chaperone adapts proteasome assembly to stress. *Mol Cell*, *55*(4), 566–577. doi:10.1016/j.molcel.2014.06.017

Hershko, A., Ciechanover, A., Heller, H., Haas, A. L., & Rose, I. A. (1980). Proposed role of ATP in protein breakdown: conjugation of protein with multiple chains of the polypeptide of ATP-dependent proteolysis. *Proc Natl Acad Sci USA*, *77*(4), 1783–1786. doi:10.1073/pnas.77.4.1783

Hershko, A., Heller, H., Elias, S., & Ciechanover, A. (1983). Components of ubiquitin-protein ligase system. Resolution, affinity purification, and role in protein breakdown. *J Biol Chem*, *258*(13), 8206–8214.

Hough, R., Pratt, G., & Rechsteiner, M. (1986). Ubiquitin-lysozyme conjugates. Identification and characterization of an ATP-dependent protease from rabbit reticulocyte lysates. *J Biol Chem*, *261*(5), 2400–2408.

Hough, R., Pratt, G., & Rechsteiner, M. (1987). Purification of two high molecular weight proteases from rabbit reticulocyte lysate. *J Biol Chem*, *262*(17), 8303–8313.

Huber, E. M., Basler, M., Schwab, R., Heinemeyer, W., Kirk, C. J., Groettrup, M., & Groll, M. (2012). Immuno- and constitutive proteasome crystal structures reveal differences in substrate and inhibitor specificity. *Cell*, *148*(4), 727–738. doi:10.1016/j.cell.2011.12.030

Husnjak, K., Elsasser, S., Zhang, N., Chen, X., Randles, L., Shi, Y., … Dikic, I. (2008). Proteasome subunit Rpn13 is a novel ubiquitin receptor. *Nature*, *453*(7194), 481–488. doi:10.1038/nature06926

Imai, J., Maruya, M., Yashiroda, H., Yahara, I., & Tanaka, K. (2003). The molecular chaperone Hsp90 plays a role in the assembly and maintenance of the 26S proteasome. *EMBO J*, *22*(14), 3557–3567. doi:10.1093/emboj/cdg349

Javitt, A., Shmueli, M. D., Kramer, M. P., Kolodziejczyk, A. A., Cohen, I. J., Radomir, L., … Merbl, Y. (2023). The proteasome regulator PSME4 modulates proteasome activity and antigen diversity to abrogate anti-tumor immunity in NSCLC. *Nat Cancer*, *4*(5), 629–647. doi:10.1038/s43018-023-00557-4

Jevtić, P., Haakonsen, D. L., & Rapé, M. (2021). An E3 ligase guide to the galaxy of small-molecule-induced protein degradation. *Cell Chem Biol*, *28*(7), 1000–1013. doi:10.1016/j.chembiol.2021.04.002

Kalim, K. W., Basler, M., Kirk, C. J., & Groettrup, M. (2012). Immunoproteasome subunit LMP7 deficiency and inhibition suppresses Th1 and Th17 but enhances regulatory T cell differentiation. *J Immun*, *189*(8), 4182–4193. doi:10.4049/jimmunol.1201183

Khor, B., Bredemeyer, A. L., Huang, C. Y., Turnbull, I. R., Evans, R., Maggi Jr, L. B., … Sleckman, B. P. (2006). Proteasome activator PA200 is required for normal spermatogenesis. *Mol Cell Biol*, *26*(8), 2999–3007. doi:10.1128/MCB.26.8.2999-3007.2006

Kikuchi, J., Iwafune, Y., Akiyama, T., Okayama, A., Nakamura, H., Arakawa, N., … Hirano, H. (2010). Co- and post-translational modifications of the 26S proteasome in yeast. *Proteomics*, *10*(15), 2769–2779. doi:10.1002/pmic.200900283

Kiuchi, T., Tomaru, U., Ishizu, A., Imagawa, M., Iwasaki, S., Suzuki, A., … Kasahara, M. (2021). Expression of the immunoproteasome subunit β5i in non-small cell lung carcinomas. *J Clin Pathol*, *74*(5), 300–306. doi:10.1136/jclinpath-2020-206618

Kumar, P., Munnangi, P., Chowdary, K. R., Shah, V. J., Shinde, S. R., Kolli, N. R., … Maddika, S. (2017). A human tyrosine phosphatase interactome mapped by proteomic profiling. *J Proteome Res*, *16*(8), 2789–2801. doi:10.1021/acs.jproteome.7b00065

Kwak, M. K., Wakabayashi, N., Greenlaw, J. L., Yamamoto, M., & Kensler, T. W. (2003). Antioxidants enhance mammalian proteasome expression through the Keap1-Nrf2 signaling pathway. *Mol Cell Biol*, *23*(23), 8786–8794. doi:10.1128/mcb.23.23.8786-8794.2003

Kwon, C. H., Park, H. J., Choi, Y. R., Kim, A., Kim, H. W., Choi, J. H., … Park do, Y. (2016). PSMB8 and PBK as potential gastric cancer subtype-specific biomarkers associated with prognosis. *Oncotarget*, *7*(16), 21454–21468. doi:10.18632/oncotarget.7411

Lander, G. C., Estrin, E., Matyskiela, M. E., Bashore, C., Nogales, E., & Martin, A. (2012). Complete subunit architecture of the proteasome regulatory particle. *Nature*, *482*(7384), 186–191. doi:10.1038/nature10774

Lee, B. H., Lu, Y., Prado, M. A., Shi, Y., Tian, G., Sun, S., … Finley, D. (2016). USP14 deubiquitinates proteasome-bound substrates that are ubiquitinated at multiple sites. *Nature*, *532*(7599), 398–401. doi:10.1038/nature17433

Lee, S. Y., De la Mota-Peynado, A., & Roelofs, J. (2011). Loss of Rpt5 protein interactions with the core particle and Nas2 protein causes the formation of faulty proteasomes that are inhibited by Ecm29 protein. *J Biol Chem*, *286*(42), 36641–36651. doi:10.1074/jbc.M111.280875

Lee, W., Kim, Y., Park, J., Shim, S., Lee, J., Hong, S. H., … Jung, Y. K. (2015). iRhom1 regulates proteasome activity via PAC1/2 under ER stress. *Sci Rep*, *5*, 11559. doi:10.1038/srep11559

Lehrbach, N. J., & Ruvkun, G. (2016). Proteasome dysfunction triggers activation of SKN-1A/Nrf1 by the aspartic protease DDI-1. *eLife*, *5*. doi:10.7554/eLife.17721

Liu, C.-W., Li, X., Thompson, D., Wooding, K., Chang, T.-l., Tang, Z., … DeMartino, G. N. (2006). ATP binding and ATP hydrolysis play distinct roles in the function of 26S proteasome. *Molecular Cell*, *24*(1), 39–50. doi:10.1016/j.molcel.2006.08.025

Liu, X., Xiao, W., Zhang, Y., Wiley, S. E., Zuo, T., Zheng, Y., … Guo, X. (2020). Reversible phosphorylation of Rpn1 regulates 26S proteasome assembly and function. *Proc Natl Acad Sci USA*, *117*(1), 328–336. doi:10.1073/pnas.1912531117

Lokireddy, S., Kukushkin, N. V., & Goldberg, A. L. (2015). cAMP-induced phosphorylation of 26S proteasomes on Rpn6/PSMD11 enhances their activity and the degradation of misfolded proteins. *Proc Natl Acad Sci USA*, *112*(52), E7176–E7185. doi:10.1073/pnas.1522332112

Low, T. L., & Goldstein, A. L. (1979). The chemistry and biology of thymosin. II. Amino acid sequence analysis of thymosin alpha1 and polypeptide beta1. *J Biol Chem*, *254*(3), 987–995.

Löwe, J., Stock, D., Jap, B., Zwickl, P., Baumeister, W., & Huber, R. (1995). Crystal structure of the 20S proteasome from the archaeon *T. acidophilum* at 3.4 A resolution. *Science*, *268*(5210), 533–539. doi:10.1126/science.7725097

Lupas, A., Koster, A. J., & Baumeister, W. (1993). Structural features of 26S and 20S proteasomes. *Enzyme Protein*, *47*(4-6), 252–273. doi:10.1159/000468684

Ma, J., Katz, E., & Belote, J. M. (2002). Expression of proteasome subunit isoforms during spermatogenesis in Drosophila melanogaster. *Insect Mol Biol*, *11*(6), 627–639. doi:10.1046/j.1365-2583.2002.00374.x

Madden, D. R. (1995). The three-dimensional structure of peptide-MHC complexes. *Annu Rev Immunol*, *13*, 587–622. doi:10.1146/annurev.iy.13.040195.003103

Mandelstam, J. (1958). Turnover of protein in growing and non-growing populations of *Escherichia coli*. *Biochem J*, *69*(1), 110–119. doi:10.1042/bj0690110

Mannhaupt, G., Schnall, R., Karpov, V., Vetter, I., & Feldmann, H. (1999). Rpn4p acts as a transcription factor by binding to PACE, a nonamer box found upstream of 26S proteasomal and other genes in yeast. *FEBS Lett*, *450*(1-2), 27–34. doi:10.1016/s0014-5793(99)00467-6

Martinez, C. K., & Monaco, J. J. (1991). Homology of proteasome subunits to a major histocompatibility complex-linked LMP gene. *Nature*, *353*(6345), 664–667. doi:10.1038/353664a0

Mayor, T., Graumann, J., Bryan, J., MacCoss, M. J., & Deshaies, R. J. (2007). Quantitative profiling of ubiquitylated proteins reveals proteasome substrates and the substrate repertoire influenced by the Rpn10 receptor pathway. *Mol Cell Proteomics*, *6*(11), 1885–1895. doi:10.1074/mcp.M700264-MCP200

Meister, S., Schubert, U., Neubert, K., Herrmann, K., Burger, R., Gramatzki, M., ... Voll, R. E. (2007). Extensive immunoglobulin production sensitizes myeloma cells for proteasome inhibition. *Cancer Res*, *67*(4), 1783–1792. doi:10.1158/0008-5472.Can-06-2258

Moebius, J., van den Broek, M., Groettrup, M., & Basler, M. (2010). Immunoproteasomes are essential for survival and expansion of T cells in virus-infected mice. *Eur J Immunol*, *40*(12), 3439–3449. doi:10.1002/eji.201040620

Muchamuel, T., Basler, M., Aujay, M. A., Suzuki, E., Kalim, K. W., Lauer, C., ... Groettrup, M. (2009). A selective inhibitor of the immunoproteasome subunit LMP7 blocks cytokine production and attenuates progression of experimental arthritis. *Nat Med*, *15*(7), 781–787. doi:10.1038/nm.1978

Murata, S., Sasaki, K., Kishimoto, T., Niwa, S.-I., Hayashi, H., Takahama, Y., & Tanaka, K. (2007). Regulation of CD8+ T cell development by thymus-specific proteasomes. *Science*, *316*(5829), 1349–1353. doi:10.1126/science.1141915

Murata, S., Yashiroda, H., & Tanaka, K. (2009). Molecular mechanisms of proteasome assembly. *Nat Rev Mol Cell Biol*, *10*(2), 104–115. doi:10.1038/nrm2630

Obermann, E. C., Eward, K. L., Dogan, A., Paul, E. A., Loddo, M., Munson, P., ... Stoeber, K. (2005). DNA replication licensing in peripheral B-cell lymphoma. *J Pathol*, *205*(3), 318–328. doi:10.1002/path.1695

Ohigashi, I., & Takahama, Y. (2021). Thymoproteasome optimizes positive selection of CD8(+) T cells without contribution of negative selection. *Adv Immunol*, *149*, 1–23. doi:10.1016/bs.ai.2021.03.001

Palombella, V. J., Rando, O. J., Goldberg, A. L., & Maniatis, T. (1994). The ubiquitin-proteasome pathway is required for processing the NF-κB1 precursor protein and the activation of NF-κB. *Cell*, *78*(5), 773–785. doi:10.1016/s0092-8674(94)90482-0

Peters, J. M., Cejka, Z., Harris, J. R., Kleinschmidt, J. A., & Baumeister, W. (1993). Structural features of the 26 S proteasome complex. *J Mol Biol*, *234*(4), 932–937. doi:10.1006/jmbi.1993.1646

Peth, A., Uchiki, T., & Goldberg, A. L. (2010). ATP-dependent steps in the binding of ubiquitin conjugates to the 26S proteasome that commit to degradation. *Mol Cell*, *40*(4), 671–681. doi:10.1016/j.molcel.2010.11.002

Petrocca, F., Altschuler, G., Tan, S. M., Mendillo, M. L., Yan, H., Jerry, D. J., ... Lieberman, J. (2013). A genome-wide siRNA screen identifies proteasome addiction as a vulnerability of basal-like triple-negative breast cancer cells. *Cancer Cell*, *24*(2), 182–196. doi:10.1016/j.ccr.2013.07.008

Pinyol, M., Salaverria, I., Bea, S., Fernández, V., Colomo, L., Campo, E., & Jares, P. (2006). Unbalanced expression of licensing DNA replication factors occurs in a subset of mantle cell lymphomas with genomic instability. *Int J Cancer*, *119*(12), 2768–2774. doi:10.1002/ijc.22146

Piotrowska, Ż., Niezgoda, M., Młynarczyk, G., Acewicz, M., & Kasacka, I. (2020). Comparative assessment of the WNT/β-catenin pathway, CacyBP/SIP, and the immunoproteasome subunit LMP7 in various histological types of renal cell carcinoma. *Front Oncol*, *10*, 566637. doi:10.3389/fonc.2020.566637

Poole, B., Ohkuma, S., & Warburton, M. J. (1977). The accumulation of weakly basic substances in lysosomes and the inhibition of intracellular protein degradation. *Acta Biol Med Ger*, *36*(11–12), 1777–1788.

Pühler, G., Weinkauf, S., Bachmann, L., Müller, S., Engel, A., Hegerl, R., & Baumeister, W. (1992). Subunit stoichiometry and three-dimensional arrangement in proteasomes from *Thermoplasma acidophilum*. *EMBO J*, *11*(4), 1607–1616. doi:10.1002/j.1460-2075.1992.tb05206.x

Radhakrishnan, S. K., Lee, C. S., Young, P., Beskow, A., Chan, J. Y., & Deshaies, R. J. (2010). Transcription factor Nrf1 mediates the proteasome recovery pathway after proteasome inhibition in mammalian cells. *Mol Cell*, *38*(1), 17–28. doi:10.1016/j.molcel.2010.02.029

Rammensee, H., Bachmann, J., Emmerich, N. P., Bachor, O. A., & Stevanović, S. (1999). SYFPEITHI: database for MHC ligands and peptide motifs. *Immunogenetics, 50*(3–4), 213–219. doi:10.1007/s002510050595

Resnitzky, D., Gossen, M., Bujard, H., & Reed, S. I. (1994). Acceleration of the G1/S phase transition by expression of cyclins D1 and E with an inducible system. *Mol Cell Biol, 14*(3), 1669–1679. doi:10.1128/mcb.14.3.1669-1679.1994

Rock, K. L., Gramm, C., Rothstein, L., Clark, K., Stein, R., Dick, L., ... Goldberg, A. L. (1994). Inhibitors of the proteasome block the degradation of most cell proteins and the generation of peptides presented on MHC class I molecules. *Cell, 78*(5), 761–771. doi:10.1016/S0092-8674(94)90462-6

Rousseau, A., & Bertolotti, A. (2016). An evolutionarily conserved pathway controls proteasome homeostasis. *Nature, 536*(7615), 184–189. doi:10.1038/nature18943

Schmidt, N., Gonzalez, E., Visekruna, A., Kühl, A. A., Loddenkemper, C., Mollenkopf, H., ... Joeris, T. (2010). Targeting the proteasome: partial inhibition of the proteasome by bortezomib or deletion of the immunosubunit LMP7 attenuates experimental colitis. *Gut, 59*(7), 896–906. doi:10.1136/gut.2009.203554

Schoenheimer, R. (1946). The dynamic state of body constituents. *The dynamic state of body constituents.*

Schoenheimer, R., Ratner, S., & Ritten-berg, D. (1939). Studies in protein metabolism. 10. The metabolic activity of body proteins investigated with 1 (-)-leucine containing two isotopes. *J Biol Chem, 130*, 703–732.

Schreiner, P., Chen, X., Husnjak, K., Randles, L., Zhang, N., Elsasser, S., ... Groll, M. (2008). Ubiquitin docking at the proteasome through a novel pleckstrin-homology domain interaction. *Nature, 453*(7194), 548–552. doi:10.1038/nature06924

Shirozu, R., Yashiroda, H., & Murata, S. (2015). Identification of minimum Rpn4-responsive elements in genes related to proteasome functions. *FEBS Lett, 589*(8), 933–940. doi:10.1016/j.febslet.2015.02.025

Simpson, M. V. (1953). The release of labeled amino acids from the proteins of rat liver slices. *J Biol Chem, 201*(1), 143–154.

Smith, D. M., Chang, S. C., Park, S., Finley, D., Cheng, Y., & Goldberg, A. L. (2007). Docking of the proteasomal ATPases' carboxyl termini in the 20S proteasome's alpha ring opens the gate for substrate entry. *Mol Cell, 27*(5), 731–744. doi:10.1016/j.molcel.2007.06.033

Smith, D. M., Kafri, G., Cheng, Y., Ng, D., Walz, T., & Goldberg, A. L. (2005). ATP binding to PAN or the 26S ATPases causes association with the 20S proteasome, gate opening, and translocation of unfolded proteins. *Mol Cell, 20*(5), 687–698. doi:10.1016/j.molcel.2005.10.019

Tomko, R. J., Jr., Funakoshi, M., Schneider, K., Wang, J., & Hochstrasser, M. (2010). Heterohexameric ring arrangement of the eukaryotic proteasomal ATPases: Implications for proteasome structure and assembly. *Molecular Cell, 38*(3), 393–403. doi:10.1016/j.molcel.2010.02.035

Unno, M., Mizushima, T., Morimoto, Y., Tomisugi, Y., Tanaka, K., Yasuoka, N., & Tsukihara, T. (2002). The structure of the mammalian 20S proteasome at 2.75 A resolution. *Structure, 10*(5), 609–618. doi:10.1016/s0969-2126(02)00748-7

Vander Linden, R. T., Hemmis, C. W., Schmitt, B., Ndoja, A., Whitby, F. G., Robinson, H., ... Hill, C. P. (2015). Structural basis for the activation and inhibition of the UCH37 deubiquitylase. *Mol Cell, 57*(5), 901–911. doi:10.1016/j.molcel.2015.01.016

Ventii, K. H., Devi, N. S., Friedrich, K. L., Chernova, T. A., Tighiouart, M., Van Meir, E. G., & Wilkinson, K. D. (2008). BRCA1-associated protein-1 is a tumor suppressor that requires deubiquitinating activity and nuclear localization. *Cancer Res, 68*(17), 6953–6962. doi:10.1158/0008-5472.Can-08-0365

Verma, R., Aravind, L., Oania, R., McDonald, W. H., Yates, J. R., 3rd, Koonin, E. V., & Deshaies, R. J. (2002). Role of Rpn11 metalloprotease in deubiquitination and degradation by the 26S proteasome. *Science, 298*(5593), 611–615. doi:10.1126/science.1075898

VerPlank, J. J. S., Tyrkalska, S. D., Fleming, A., Rubinsztein, D. C., & Goldberg, A. L. (2020). cGMP via PKG activates 26S proteasomes and enhances degradation of proteins, including ones that cause neurodegenerative diseases. *Proc Natl Acad Sci USA, 117*(25), 14220–14230. doi:10.1073/pnas.2003277117

Vilchez, D., Morantte, I., Liu, Z., Douglas, P. M., Merkwirth, C., Rodrigues, A. P., ... Dillin, A. (2012). RPN-6 determines C. elegans longevity under proteotoxic stress conditions. *Nature, 489*(7415), 263–268. doi:10.1038/nature11315

Walerych, D., Lisek, K., Sommaggio, R., Piazza, S., Ciani, Y., Dalla, E., ... Del Sal, G. (2016). Proteasome machinery is instrumental in a common gain-of-function program of the p53 missense mutants in cancer. *Nat Cell Biol, 18*(8), 897–909. doi:10.1038/ncb3380

Wang, C. Y., Guttridge, D. C., Mayo, M. W., & Baldwin, A. S., Jr. (1999). NF-κB induces expression of the Bcl-2 homologue A1/Bfl-1 to preferentially suppress chemotherapy-induced apoptosis. *Mol Cell Biol, 19*(9), 5923–5929. doi:10.1128/mcb.19.9.5923

Wang, X., Chemmama, I. E., Yu, C., Huszagh, A., Xu, Y., Viner, R., … Huang, L. (2017). The proteasome-interacting Ecm29 protein disassembles the 26S proteasome in response to oxidative stress. *J Biol Chem*, *292*(39), 16310–16320. doi:10.1074/jbc.M117.803619

Wilkinson, K. D., Urban, M. K., & Haas, A. L. (1980). Ubiquitin is the ATP-dependent proteolysis factor I of rabbit reticulocytes. *J Biol Chem*, *255*(16), 7529–7532.

Yang, B. Y., Song, J. W., Sun, H. Z., Xing, J. C., Yang, Z. H., Wei, C. Y., … Zhang, Y. S. (2018). PSMB8 regulates glioma cell migration, proliferation, and apoptosis through modulating ERK1/2 and PI3K/AKT signaling pathways. *Biomed Pharmacother*, *100*, 205–212. doi:10.1016/j.biopha.2018.01.170

Young, V. R., Steffee, W. P., Pencharz, P. B., Winterer, J. C., & Scrimshaw, N. S. (1975). Total human body protein synthesis in relation to protein requirements at various ages. *Nature*, *253*(5488), 192–194. doi:10.1038/253192a0

Yuan, X., Miller, M., & Belote, J. M. (1996). Duplicated proteasome subunit genes in drosophila melanogaster encoding testes-specific isoforms. *Genetics*, *144*(1), 147–157. doi:10.1093/genetics/144.1.147

Zhang, F., Hu, Y., Huang, P., Toleman, C. A., Paterson, A. J., & Kudlow, J. E. (2007). Proteasome function is regulated by cyclic AMP-dependent protein kinase through phosphorylation of Rpt6. *J Biol Chem*, *282*(31), 22460–22471. doi:10.1074/jbc.M702439200

Zhang, X., Schulz, R., Edmunds, S., Krüger, E., Markert, E., Gaedcke, J., … Dobbelstein, M. (2015). MicroRNA-101 suppresses tumor cell proliferation by acting as an endogenous proteasome inhibitor via targeting the proteasome assembly factor POMP. *Mol Cell*, *59*(2), 243–257. doi:10.1016/j.molcel.2015.05.036

Zhong, L., & Belote, J. M. (2007). The testis specific proteasome subunit Prosα6T of D. melanogaster is required for individualization and nuclear maturation during spermatogenesis. *Development*, *134*(19), 3517–3525. doi:10.1242/dev.004770

6 The Role of Caspase-3 in Post-Chemotherapy Breast and Colorectal Cancer Relapse and Metastasis

Andreja Leskovac and Sandra Petrovic
National Institute of the Republic of Serbia,
University of Belgrade, Belgrade, Serbia

6.1 INTRODUCTION

Despite significant advancements in cancer prevention, detection, therapy, and monitoring over the years, cancer poses a serious threat to human health. Its morbidity and mortality rates are increasing annually worldwide [1]. Chemotherapeutic interventions are considered the most effective in treating cancer, used alone or combined with surgery or radiotherapy. However, their clinical efficacy and application are hindered by challenges such as their detrimental impact on normal tissues, lack of precise targeting [2], and the occurrence of drug resistance [1, 3]. Targeting enzymes that control crucial cellular processes holds promise for treating cancer effectively [4, 5]. By inhibiting or enhancing their function, we could potentially impede the spread of tumors, metastasis, and angiogenesis and improve the efficacy of various cancer therapies [6].

A fundamental characteristic of human cancer is its ability to avoid apoptosis. As a result, numerous current cancer therapies focus on addressing apoptosis evasion by targeting different caspase pathways [7]. Caspases are the evolutionary conserved family of cysteine proteases playing central roles in programmed cell death and inflammation [8]. Based on structural characteristics, functional roles, and substrate preferences, they have been classified into two main categories: inflammatory caspases (caspase-1, -4, -5, -11) and apoptotic caspases [9]. Apoptotic caspases are subdivided into initiator/apical and effector/executioner caspases. In human cells, when apoptosis is triggered, initiator caspases (caspase-8, -9, and -10) become activated, subsequently initiating the cleavage and activation of executioner caspases (caspase-3, -6, and -7) [10]. Activated executioner caspases, in response, enhance mitochondrial outer membrane permeabilization (MOMP) and initiator caspase activation, creating a positive feedback loop that ensures efficient caspase activation and cell elimination [11]. Caspase-3 is considered a canonical executor of apoptosis. When initiated by caspase-8 or caspase-9, it cleaves multiple essential proteins in the cell, ultimately triggering apoptosis.

The activity of caspases is controlled through various mechanisms, encompassing gene transcription and post-translational changes. The process of alternative splicing in caspase genes results in the production of both complete and truncated proteins, and these variations can have antagonistic functions [12]. The alternative splicing of the caspase-3 gene generates two transcripts, caspase-3 and caspase-3s, which may have different functions [13]. For instance, the expression of the short isoform caspase-3s in 293T cells was reported to prevent DNA fragmentation and poly(ADP-ribose) polymerase 1 (PARP1) cleavage in response to apoptotic stimuli, thus antagonizing caspase-3 apoptotic activity [13].

DOI: 10.1201/9781003394693-8

Executioner caspases also play a role in pyroptosis, a pro-inflammatory programmed cell death involving inflammatory caspases and the gasdermin D (GSDMD) protein family [14]. GSDMD consists of a pore-forming N-terminal part, an inhibitory C-terminal domain, and a linker region between them. During pyroptosis, activated inflammatory caspases cleave GSDMD's linker region, releasing the N-terminal fragment to perforate the cell membrane, causing cell lysis [14]. Caspase-3 was also identified to cleave another gasdermin family protein, gasdermin E (GSDME), releasing the N-terminal fragment, which can switch cell death from apoptosis to pyroptosis [15]. This process is usually observed under specific conditions, such as the presence of some chemotherapeutic drugs, TNF-α, or at high GSDME expression [16]. Namely, elevated expression of GSDME enables cytotoxic drugs to trigger cancer cell death *via* caspase-3-mediated pyroptosis [17]. When GSDME is expressed at high levels, caspase-3 can cleave it, leading to the initiation of pyroptosis; otherwise, the apoptosis is activated [17]. Chemotherapy can induce both apoptosis and pyroptosis, both of which have a significant impact on the effectiveness of cancer treatment. However, pyroptosis represents a double-sided phenomenon, as it can induce adverse effects in normal tissue while concurrently facilitating anti-tumor responses by modulating the anti-tumor immune system [18]. Consequently, it is critical to carefully control the type of cell death triggered by chemotherapy and strike a balance between cell apoptosis and pyroptosis when employing caspase-3-activating chemotherapy to enhance the effectiveness of treatments while minimizing toxicity [18].

While various physiological and pathological factors can induce apoptosis, not every cell will necessarily die. Dysregulation in the pathway controlling cell death can give rise to various irregularities and disorders, with cancer being the primary concern [19]. An extensive body of evidence has revealed the intricate roles of caspase-3 in cancer initiation and treatment. It was found that even in normal cells, executioner caspase activation might induce genomic instability and oncogenic transformation; activated executioner caspases can initiate the activation of caspase-activated DNase (CAD) and impair the functioning of DNA repair proteins such as PARP1 and the ataxia-telangiectasia mutated (ATM) protein [20].

When considering cancer cells, numerous anti-cancer treatments, such as cytotoxic drugs, radiation therapy, and immunotherapy, induce tumor cell death by activating caspase-3. As such, caspase-3 activation is usually considered an indicator of the effectiveness of cancer therapy [21]. However, recent research shows that cancer cells can survive executioner caspase activation as a response to stress, leading to molecular and phenotypic changes among the survivors [11]. Several studies have demonstrated that following periodic anti-cancer chemotherapy, cancer cells that survive executioner caspase activation attained enhanced malignant behaviors [11, 22]. Caspase-3 has a role in facilitating cancer development after cytotoxic drug treatment, promoting tumor regeneration after radiotherapy *via* paracrine signaling, fostering tumor growth by establishing a pro-angiogenic microenvironment, and many other non-apoptotic effects [1, 21]. Therefore, the activation of executioner caspases can initiate pro-survival and pro-proliferation signals. Caspase-3 directly affects the activation of several pro-angiogenic genes, notably vascular endothelial growth factor A (VEGFA), promoting oncogenic angiogenesis and tumorigenesis through transcriptional regulation. At the same time, the activation of caspase-3 indirectly diminished the expression of a set of pro-apoptotic genes, resulting in increased resistance to cytotoxic treatments [1].

It has been shown that neither the rate, peak level, nor the extent of caspase activity could reliably forecast whether cells would undergo cell death or survive [23]. Consequently, cells can survive direct activation of executioner caspases, and alterations in their state can affect the result of caspase activity, which could potentially be lethal. These differences in cellular responses might contribute to the incomplete eradication of tumor cells when exposed to cancer treatments [23]. In this regard, a growing body of evidence has shown that cells can survive apoptotic stress, even following the activation of executioner caspases, through a phenomenon referred to as anastasis [24].

This chapter provides the latest insights into the functions of caspase-3 and its potential relevance in cancer promotion and relapse after chemotherapy, proposing future research directions toward efficient anti-cancer therapy.

6.2 THE ROLE OF CASPASE CASCADE IN THE CELL DEATH RECOVERY

As discussed, recent findings reveal that cells can undergo anastasis after activating the effector caspase cascade [25]. For example, previous studies on cervical HeLa cancer cells have shown caspase-3 activation induced by ethanol treatment, but cellular function and morphology returned to normal upon ethanol removal [25]. Similar effects have been observed in ovarian and breast cancer cells treated with paclitaxel, melanoma cells after dacarbazine treatment, and HeLa cells exposed to TNFα plus cycloheximide or etoposide [11]. Various factors can impact the decision between cellular survival and death, encompassing inherent variations in the pro-apoptotic and anti-apoptotic protein expression, such as those within the Bcl-2 family, inhibitors of apoptosis proteins (IAPs), and caspases [11]. This decision is also influenced by stress-induced signals that can either promote cell death or survival.

Activating apoptotic caspases and triggering cell death can simultaneously initiate pro-survival signals. For instance, activated executioner caspases can cleave proteins like Ras GTPase activating protein (p120 RasGAP), the cyclin-dependent kinase inhibitor p27Kipl, the inositol-requiring enzyme 1 (IRE1), and AMP-activated protein kinase α1 (AMPK-α1), generating products with anti-apoptotic functions [26]. For example, it was shown that even in slightly stressed cells, low levels of caspase-3 activity can activate the anti-apoptotic Akt kinase by partially cleaving RasGAP [27]. Akt activation promotes cell survival by stimulating the mechanistic target of rapamycin (mTOR) or nuclear factor kappa B (NF-κB) pathway and inhibits the amplification of caspase activity by inactivating pro-apoptotic proteins [28].

Moreover, both initiator caspases and executioner caspases, when activated, can stimulate the release of wingless-related integration site (Wnt), prostaglandin E2 (PGE2), epidermal growth factor receptor (EGFR) ligands, and fractalkine (FKN), which support the survival and proliferation of neighboring cells [29]. The caspase-activated cells can recover even after mitochondrial fragmentation, cytosolic and nuclear components condensation, and cell shrinkage, thus indicating that apoptosis may be reversible during the cellular execution phase [30]. While it remains unclear whether cells employ identical or distinct mechanisms to regulate activated caspases in the context of anastasis and non-apoptotic caspase activity, several genes that are selectively upregulated during the reversal of apoptosis might function as inhibitors of the caspases [31]. For instance, a group of heat shock proteins (HSPs) is considered to play a notable role. HSP27 has been demonstrated to interact with procaspase-3, thereby hindering caspase-9-mediated proteolysis, a crucial step for the activation of caspase-3 [25].

Additionally, HSP27, HSP70, and HSP90 can inhibit apoptosome formation and the subsequent activation of caspase-9, effectively impeding further activation of the caspase cascade [25]. Furthermore, anastatic cells are characterized by increased expression of the mouse double minute 2 homolog (MDM2) proto-oncogene, which acts as an inhibitor of p53 and suppresses the p53-mediated pro-apoptotic response to DNA damage during the reversal of apoptosis [31]. Furthermore, the expression of MDM2 can activate the X-linked inhibitor of the apoptosis protein (XIAP), which stops the progression of the apoptotic cascade by inhibiting the activated initiator and executioner caspases and facilitating the degradation of pro-apoptogenic mitochondrial protein Smac/DIABLO [32]. It has been shown that transcriptional activity remains active during anastasis, enabling cells to repair damage and restore their normal morphology. In addition, anastatic cells also exhibit an increase in the expression of angiogenic factors, such as angiopoietin-like 4 (ANGPTL4) and VEGFA, which promote angiogenesis and enhance vascular permeability [31]. Still, further research is needed to comprehensively understand the factors involved in anastasis-related processes.

Therapeutic approaches to treat cancer include surgery, radiotherapy, and chemotherapy, which are based on the initiation of apoptosis and cell-cycle arrest [3]. However, the response of cancer cells to treatments can be influenced by several factors, such as changes in epigenetic markers, genetic background, and various environmental stimuli. Cancer therapy faces a significant challenge in the form of drug resistance, which is primarily linked to various proteins involved in DNA repair, as well as mechanisms related to cell survival and apoptosis. Notably, proteins such as NF-κB, PARP-1, MDM2, cellular FLICE inhibitory protein (c-FLIP), survivin, myeloid cell leukemia sequence 1 (MCL-1), B-cell lymphoma 2 (BCL-2), and B-cell lymphoma extra-large (BCL-XL) play a crucial role in drug resistance and cancer recurrence [22]. These factors contribute to drug resistance by enhancing the aggressive characteristics of cancer cells, including their ability to evade the immune system, enter a state of dormancy, exhibit stem-cell-like properties, metastasize, and migrate [22]. Several reports suggested a link between anastasis and drug resistance in cancer treatment, indicating that anastasis-related genes and pathways facilitate resistance to chemotherapy, promote metastasis, and ensure cell survival [22, 33].

6.3 THE ROLE OF CASPASE-3 IN POST-CHEMOTHERAPY CANCER RELAPSE AND METASTASIS

In recent years, the ability of cancer cells to survive after the activation of executioner caspases following exposure to chemotherapy has been observed in several cancer cell lines, including breast cancer, cervical cancer, ovarian cancer, colorectal cancer, and melanoma [34, 35]. Anastasis has been found to induce different characteristics in surviving cancer cells, with variations noted among various cancer types. For example, surviving cervical cancer cells become more migratory and resistant to drugs, while melanoma cells primarily exhibit increased migration without a significant change in drug resistance [34, 35]. However, the mechanisms responsible for these anastasis-induced changes in cell behavior remain largely unclear. Here, we discuss recently reported mechanisms of the anastasis process and the relevance of caspase-3 in breast and colorectal cancer promotion and relapse after chemotherapy.

6.3.1 Breast Cancer Relapse and Metastasis

Breast cancer is the most frequently diagnosed cancer type and the leading cause of cancer-related fatalities worldwide [36]. Breast cancer treatment is often determined by the molecular subtype of cancer, which is defined by the presence or absence of hormone receptors, comprising progesterone receptor (PR) and estrogen receptor (ER), as well as the presence of human epidermal growth factor receptor 2 (HER2) [37]. Despite significant advancements in diagnostic and treatment methods, patient mortality rates remain elevated due to chemotherapy resistance, relapse, and the development of distant metastases.

Numerous investigations have indicated that breast cancers exhibiting a high apoptosis index tend to have a more favorable prognosis than those displaying lower or absent levels of apoptosis. However, in some cases, advanced breast cancer has shown an overexpression of apoptosis factors. Several studies have reported that caspase-3 expression decreases the likelihood of breast cancer development [38], while others found the opposite [39].

The loss of caspase-3 expression and the failure to release cytochrome c from the mitochondria are usually linked to resistance to multiple drugs [40]. To address this issue, introducing caspase-3 expression into the human MCF-7 breast tumor cell line (which lacks caspase-3) was performed. It led to the re-establishment of the apoptotic response of cells when treated with the topoisomerase II inhibitor, etoposide [41]. In contrast, opposite results were obtained in experiments involving breast cancer cells treated with staurosporine [42]. The authors used a fluorescent dye to mark active caspase-3/7, then separated and cultured cells with and without caspase-3/7 activation following drug treatment. Notably, breast cancer cells that recovered after staurosporine-induced executioner caspase activation displayed enhanced *in vivo* tumor growth and metastasis, along with a transformation from non-stem cells into cancer stem cells [42].

Therefore, whether high levels of caspase-3 expression or its reduction can predict unfavorable outcomes in individuals with breast cancer is still a matter of debate. Some previous reports indicated that caspase-3s variants affect the activation of caspase-3 and the apoptotic process in breast cancer, adjusting the response to chemotherapy treatments [40]. The authors performed experiments in MCF-7 and HBL100 breast cancer cells using chemotherapeutic treatments such as etoposide and methotrexate. They suggested that caspase-3s, produced through the alternative splicing of caspase-3 pre-mRNA, have an inhibitory effect on caspase-3 activation and the apoptosis process in breast cancer cells exposed to cytotoxic drugs. This impact seems to result from caspase-3s interacting with procaspase-3, preventing its recruitment into the apoptosome and subsequent activation [40]. Indeed, it has been shown that dysregulation of microRNA (miRNA) and splicing events contribute to the development of cancer and drug resistance in various types of cancers such as breast, cervical, prostate, colorectal, ovarian, and leukemia [43]. Alternative splicing can modify the binding site in the 3′ UTR of target mRNA, which, in turn, can impact the protein produced and potentially alter the drug-binding affinity of target proteins, ultimately leading to drug resistance [43]. Nonetheless, the relationship between miRNA-mRNA splicing in the context of drug resistance in cancer still requires comprehensive elucidation.

Despite numerous studies addressing this issue, the limited availability of substantial sample sizes has hindered deeper investigation. In this regard, to comprehensively assess how caspase-3 expression impacts the clinical characteristics and survival of patients with breast cancer, Yang and colleagues conducted a meta-analysis of 3091 cases [44]. The analysis showed that increased caspase-3 expression was significantly associated with poor overall survival. In addition, the authors suggested that caspase-3 might serve as a risk factor for the PR and HER2 subtypes of breast cancer [44].

A recent study has demonstrated that anastasis, induced by chemotherapeutic drugs Adriamycin and cisplatin, makes breast cancer cells more proliferative and migratory, resulting in enhanced cancer growth and metastasis [24]. It was shown that chemotherapy-induced anastasis triggers the epigenetic de-repression of the gene responsible for encoding cadherin 12 (CDH12). This process, in turn, promotes the proliferation and migration of breast cancer by activating the extracellular signal-regulated kinase/cyclic-AMP response element binding protein (ERK/CREB) signaling pathway. Furthermore, it was demonstrated that the activation of caspases led to a reduction in DNA methylation and histone modifications in the CDH12 promoter, resulting in enhanced CDH12 expression [24]. Thus, CDH12 can be considered a positive regulator for post-chemotherapy recurrence and metastasis of breast cancer.

The investigations described here suggest the mechanisms behind how anastasis contributes to the increased aggressiveness of breast cancer and provide novel targets for therapeutic interventions aimed at preventing post-chemotherapy cancer relapse and metastasis.

6.3.2 Colorectal Cancer Relapse and Metastasis

Colorectal cancer ranks third in global cancer prevalence and stands as the second leading cause of cancer-related deaths [36]. Chemotherapy is a frequently employed method for treating colorectal cancer. Nevertheless, certain cancer cells can survive chemotherapy, persist, and give rise to cancer relapse and metastasis. While several factors, including variations in drug metabolism, abnormal gene regulation, the presence of cancer stem cells, and modifications in cell death pathways, have been suggested as potential contributors to chemotherapy resistance, the precise mechanisms involved remain elusive. At this point, we have briefly discussed studies that shed light on a previously unrecognized aspect of caspase-3's role in colorectal cancer and its potential implications for treatment.

In their earlier study, Brown and colleagues reported an unexpected role of caspase-3 in preventing necrosis induced by DNA-damaging agents in colon cancer cells [45]. Namely, caspase-3 knockout or knockdown colon cancer cells maintained their ability to undergo apoptosis but exhibited increased sensitivity to DNA-damaging agents, resulting in receptor-interacting serine/threonine-protein kinase 1 (RIPK1)- and caspase-8-dependent necrosis, both *in vitro* and *in vivo*. These findings suggested a potential new strategy to enhance the susceptibility of cancer cells to chemotherapy. The research indicated that caspase-3 deficiency heightens the sensitivity of colon cancer cells to genotoxic agents by promoting necrosis through RIPK1, pro-caspase-8, and ROS mechanisms without affecting apoptosis [45]. Caspase-3 appears to control the interaction between RIPK1 and pro-caspase-8 to inhibit necrosis in cancer cells lacking RIPK3 expression, indicating variations in cell death regulation between normal and cancer cells. Furthermore, caspase-3 may play a role in connecting cell death and innate immunity, potentially involving the release of an extracellular inflammatory cytokine protein, high mobility group box 1 (HMGB1) [46].

A recent study that used CRISPR technology to create colon cancer cell lines with caspase-3 knockout (CASP3KO) compared their behavior to control cells in both *in vivo* and *in vitro* settings [21]. The results revealed that caspase-3 contributes to drug and radiation resistance and regulates colon cancer cells' migration, invasion, and metastasis. The most significant finding was that caspase-3 ablation completely prevented HCT116 cancer cells from metastasizing to the lung, highlighting its previously unknown role in colon cancer [21].

The study also demonstrated that caspase-3 regulates epithelial-mesenchymal transition (EMT), a process crucial for cell migration and invasion. CASP3KO cells exhibited reduced EMT characteristics compared to parental HCT116 cells, explaining why the loss of caspase-3 impairs their metastatic abilities [21]. Given that caspase-3 promotes resistance to radiotherapy and chemotherapy, as well as cancer cell invasion and metastasis, combining caspase-3-targeted agents with these treatments may offer a promising approach for colon cancer therapy.

Furthermore, it has been demonstrated that exposure of colorectal cancer cells to chemotherapeutic drugs activates NF-κB and increases expression of a cIAP2, a protein that belongs to the IAP protein family, promoting anastasis [47]. The activated NF-κB and cIAP2 form a positive feedback loop that enhances the migration and metastasis of these anastatic cells. The study, using lineage tracing, has shown that anastatic colorectal cancer cells became more mobile, metastatic, and resistant to chemotherapy due to elevated cIAP2 expression and increased NF-κB activity [47]. The study also suggests that cIAP2 and NF-κB can both positively and negatively affect cell migration through different downstream molecules. It was reported that XIAP and cIAPs impede cell migration by facilitating the ubiquitination and subsequent degradation of c-Raf and Rac1 [48]. Conversely, XIAP can promote the migration of cancer cells by blocking the action of the RhoGDP dissociation inhibitor (RhoGDI) or activating NF-κB [49]. Likewise, it was demonstrated that reducing the expression of cIAP2 or inhibiting NF-κB activity suppresses anastasis in colorectal cancer cells exposed to chemotherapeutic drugs [47]. Nevertheless, the precise molecular mechanisms involving cIAP2, NF-κB, and other regulators identified in previous studies require further investigation.

6.4 CONCLUSION AND FUTURE PERSPECTIVES

Cancer is a substantial global health concern, and chemotherapy is a common approach for treatment. However, resistance to chemotherapy in some cancer cells leads to relapse and metastasis, and despite various proposed mechanisms, the exact reasons for this resistance are not yet fully understood. Caspase-3 is a key mediator of apoptosis, typically activated when cells are exposed to cytotoxic drugs, radiation therapy, or immunotherapy. It is commonly used as a marker to assess the effectiveness of cancer treatments. However, recent findings suggest that caspase-3 also has non-apoptotic functions. Growing evidence suggests that caspase-3 promotes stress-induced cancer cell growth, migration, invasiveness, and angiogenesis. Our knowledge about the involvement of caspase-3 in cancer is still incomplete, so the key task is further exploring the intricate molecular aspects of caspase-3-mediated cell death and how these insights can improve cancer therapy. It appears that the regulation of caspase-3 activity and its interaction with other cellular factors are critical in determining whether a cell proceeds with apoptosis or undergoes anastasis. Understanding these processes could have implications for the therapy of cancer and other diseases involving cell death pathways.

An in-depth investigation of caspase pathways has discovered complex interactions between apoptosis, necroptosis, and pyroptosis, necessitating additional research to comprehend any possible clinical effects. Therefore, manipulating caspase-3 activity could provide a novel approach to increase cancer cell sensitivity to chemotherapy and radiotherapy while also inhibiting cancer cell invasion and metastasis. In this regard, it is essential to have a deeper understanding of the anti-cancer mechanisms of caspase-3 and its role in tumor growth and proliferation, particularly concerning apoptosis and pyroptosis. This knowledge can contribute to the development of more effective anti-cancer drugs. Additionally, a critical aspect for future applications involves harnessing the control over GSDME's function to switch between these two cell death modes selectively; this would involve inducing pyroptosis to reduce drug resistance in tumors while simultaneously mitigating pyroptosis in normal tissues to prevent harm. As shown, GSDME-triggered pyroptosis can release inflammatory substances and danger-associated molecular patterns, leading to a potent inflammatory reaction that helps modulate the tumor-immune microenvironment and activates an anti-cancer immune response [18]. As caspase-3 plays a crucial role in both GSDME-mediated pyroptosis and apoptosis, promoting the shift from apoptosis to pyroptosis appears to hold promise as a potential strategy for cancer therapy.

ACKNOWLEDGMENTS

The authors acknowledge the support provided by the Ministry of Science, Technological Development and Innovation of the Republic of Serbia (No. 451-03-66/2024-03/200017).

DECLARATIONS

Conflict of interest: The authors declare no conflicts of interest.

Consent for publishing: All authors have approved the final version of the manuscript and given consent for publication.

REFERENCES

1. Li, Y., Zhang, C., Li, G., et al. (2021). Protease-triggered bioresponsive drug delivery for the targeted theranostics of malignancy. *Acta Pharmaceutica Sinica B*, 11, 2220–2242.
2. Nešić, M., Popović, I., Leskovac, A., et al. (2016). Testing the photo-sensitive nanocomposite system for potential controlled metallo-drug delivery. *Optical and Quantum Electronics*, 48, 119.
3. Petrovic, S., Vasic, V., Mitrovic, T., et al. (2017). The impact of concentration and administration time on the radiomodulating properties of undecylprodigiosin in vitro. *Archives of Industrial Hygiene and Toxicology*, 68, 1–8.
4. Petrovic, V., Petrovic, S., Joksic, G., et al. (2014). Inhibition of Na(+)/K(+)-ATPase and cytotoxicity of a few selected gold(III) complexes. *Journal of Inorganic Biochemistry*, 140, 228–235.
5. Bondzic, A.M., Colovic, M.B., Janjic, G.V., et al. (2017). The influence of oxo-bridged binuclear gold(III) complexes on Na/K-ATPase activity: a joint experimental and theoretical approach. *JBIC Journal of Biological Inorganic Chemistry*, 22, 819–832.
6. Leskovac, A., Petrovic, S., Lazarevic-Pasti, T., et al. (2018). Ruthenium(II)-N-alkyl phenothiazine complexes as potential anticancer agents. *JBIC Journal of Biological Inorganic Chemistry*, 23, 689–704.
7. Boice, A., Bouchier-Hayes, L. (2020). Targeting apoptotic caspases in cancer. *Biochimica et Biophysica Acta (BBA) - Molecular Cell Research*, 1867, 118688.
8. Eckhart, L., Ballaun, C., Uthman, A., et al. (2005). Identification and Characterization of a Novel Mammalian Caspase with Proapoptotic Activity. *Journal of Biological Chemistry*, 280, 35077–35080.

9. Scott, A.M., Saleh, M. (2007). The inflammatory caspases: guardians against infections and sepsis. *Cell Death & Differentiation*, 14, 23–31.
10. Eckhart, L., Ballaun, C., Hermann, M., et al. (2008). Identification of novel mammalian caspases reveals an important role of gene loss in shaping the human caspase repertoire. *Molecular Biology and Evolution*, 25, 831–841.
11. Sun, G. (2023). Death and survival from executioner caspase activation. *Seminars in Cell & Developmental Biology*, 156, 66–73.
12. Mercatante, D., Kole, R. (2000). Modification of alternative splicing pathways as a potential approach to chemotherapy. *Pharmacology & Therapeutics*, 85, 237–243.
13. Huang, Y., Shin, N.-H., Sun, Y., et al. (2001). Molecular cloning and characterization of a novel caspase-3 variant that attenuates apoptosis induced by proteasome inhibition. *Biochemical and Biophysical Research Communications*, 283, 762–769.
14. Kesavardhana, S., Malireddi, R.K.S., Kanneganti, T.D. (2020). Caspases in cell death, inflammation, and pyroptosis. *Annual Review of Immunology*, 38, 567–595.
15. Wang, Y., Gao, W., Shi, X., et al. (2017). Chemotherapy drugs induce pyroptosis through caspase-3 cleavage of a gasdermin. *Nature*, 547, 99–103.
16. Zhang, Z., Zhang, Y., Xia, S., et al. (2020). Gasdermin E suppresses tumour growth by activating antitumour immunity. *Nature*, 579, 415–420.
17. Jiang, M., Qi, L., Li, L., et al. (2020). The caspase-3/GSDME signal pathway as a switch between apoptosis and pyroptosis in cancer. *Cell Death Discovery*, 6, 112.
18. Li, L., Wang, S., Zhou, W. (2023). Balance cell apoptosis and pyroptosis of caspase-3-activating chemotherapy for better antitumor therapy. *Cancers*, 15, 26.
19. Lazarevic-Pasti, T., Leskovac, A., Momic, T., et al. (2017). Modulators of acetylcholinesterase activity: from Alzheimer's disease to anti-cancer drugs. *Current Medicinal Chemistry*, 24, 3283–3309.
20. Smith, G.C., d'Adda di Fagagna, F., Lakin, N.D., et al. (1999). Cleavage and inactivation of ATM during apoptosis. *Molecular and Cellular Biology*, 19, 6076–6084.
21. Zhou, M., Liu, X., Li, Z., et al. (2018). Caspase-3 regulates the migration, invasion and metastasis of colon cancer cells. *International Journal of Cancer*, 143, 921–930.
22. Mohammed, R.N., Khosravi, M., Rahman, H.S., et al. (2022). Anastasis: cell recovery mechanisms and potential role in cancer. *Cell Communication and Signaling*, 20, 81.
23. Nano, M., Mondo, J.A., Harwood, J., et al. (2023). Cell survival following direct executioner-caspase activation. *Proceedings of the National Academy of Sciences of the United States of America*, 120, e2216531120.
24. Wang, Y., Wang, R., Liu, X., et al. (2023). Chemotherapy-induced executioner caspase activation increases breast cancer malignancy through epigenetic de-repression of CDH12. *Oncogenesis*, 12, 34.
25. Tang, H.M., Tang, H.L. (2018). Anastasis: recovery from the brink of cell death. *Royal Society Open Science*, 5, 180442.
26. Cheratta, A.R., Thayyullathil, F., Hawley, S.A., et al. (2022). Caspase cleavage and nuclear retention of the energy sensor AMPK-alpha1 during apoptosis. *Cell Reports*, 39, 110761.
27. Khalil, H., Peltzer, N., Walicki, J., et al. (2012). Caspase-3 protects stressed organs against cell death. *Molecular and Cellular Biology*, 32, 4523–4533.
28. Porta, C., Paglino, C., Mosca, A. (2014). Targeting PI3K/Akt/mTOR signaling in cancer. *Frontiers in Oncology*, 4, 64.
29. Medina, C.B., Mehrotra, P., Arandjelovic, S., et al. (2020). Metabolites released from apoptotic cells act as tissue messengers. *Nature*, 580, 130–135.
30. Tang, H.L., Tang, H.M., Hardwick, J.M., et al. (2015). Strategies for tracking anastasis, a cell survival phenomenon that reverses apoptosis. *Journal of Visualized Experiments*, e51964.
31. Tang, H.M., Talbot, C.C., Jr., Fung, M.C., et al. (2017). Molecular signature of anastasis for reversal of apoptosis. *F1000Res*, 6, 43.
32. Flanagan, L., Sebastia, J., Tuffy, L.P., et al. (2010). XIAP impairs Smac release from the mitochondria during apoptosis. *Cell Death Dis*, 1, e49.
33. Mirzayans, R., Murray, D. (2020). Intratumor heterogeneity and therapy resistance: contributions of dormancy, apoptosis reversal (Anastasis) and cell fusion to disease recurrence. *International Journal of Molecular Sciences*, 21, 1308.
34. Seervi, M., Sumi, S., Chandrasekharan, A., et al. (2019). Molecular profiling of anastatic cancer cells: potential role of the nuclear export pathway. *Cellular Oncology*, 42, 645–661.
35. Berthenet, K., Castillo Ferrer, C., Fanfone, D., et al. (2020). Failed apoptosis enhances melanoma cancer cell aggressiveness. *Cell Reports*, 31, 107731.

36. Sung, H., Ferlay, J., Siegel, R.L., et al. (2021). Global cancer statistics 2020: GLOBOCAN estimates of incidence and mortality worldwide for 36 cancers in 185 countries. *CA: A Cancer Journal for Clinicians*, 71, 209–249.
37. Kinnel, B., Singh, S.K., Oprea-Ilies, G., et al. (2023). Targeted therapy and mechanisms of drug resistance in breast cancer. *Cancers (Basel)*, 15, 1320.
38. Wu, M.-h., Huang, J.-c., Feng, Z.-b. (2012). Expression of ALDH1 protein and its relationship with cell apoptosis and caspase-3 in breast carcinoma. *Chinese Journal of Clinical and Experimental Pathology*, 28, 737–740.
39. Engels, C.C., Ruberta, F., de Kruijf, E.M., et al. (2013). The prognostic value of apoptotic and proliferative markers in breast cancer. *Breast Cancer Res Treat*, 142, 323–339.
40. Vegran, F., Boidot, R., Solary, E., et al. (2011). A short caspase-3 isoform inhibits chemotherapy-induced apoptosis by blocking apoptosome assembly. *PLOS ONE*, 6, e29058.
41. Yang, X.H., Sladek, T.L., Liu, X., et al. (2001). Reconstitution of caspase 3 sensitizes MCF-7 breast cancer cells to doxorubicin- and etoposide-induced apoptosis. *Cancer Research*, 61, 348–354.
42. Xu, Y., So, C., Lam, H.M., et al. (2018). Apoptosis reversal promotes cancer stem cell-like cell formation. *Neoplasia*, 20, 295–303.
43. Marima, R., Francies, F.Z., Hull, R., et al. (2021). MicroRNA and alternative mRNA splicing events in cancer drug response/resistance: potent therapeutic targets. *Biomedicines*, 9, 1818.
44. Yang, X., Zhong, D.N., Qin, H., et al. (2018). Caspase-3 over-expression is associated with poor overall survival and clinicopathological parameters in breast cancer: a meta-analysis of 3091 cases. *Oncotarget*, 9, 8629–8641.
45. Brown, M.F., Leibowitz, B.J., Chen, D., et al. (2015). Loss of caspase-3 sensitizes colon cancer cells to genotoxic stress via RIP1-dependent necrosis. *Cell Death Discovery*, 6, e1729.
46. Degterev, A., Yuan, J. (2008). Expansion and evolution of cell death programmes. *Nature Reviews Molecular Cell Biology*, 9, 378–390.
47. Wang, R., Wang, Y., Liu, X., et al. (2023). Anastasis enhances metastasis and chemoresistance of colorectal cancer cells through upregulating cIAP2/NFκB signaling. *Cell Death & Disease*, 14, 388.
48. Oberoi, T.K., Dogan, T., Hocking, J.C., et al. (2012). IAPs regulate the plasticity of cell migration by directly targeting Rac1 for degradation. *The EMBO Journal*, 31, 14–28.
49. Liu, J., Zhang, D., Luo, W., et al. (2012). E3 ligase activity of XIAP RING domain is required for XIAP-mediated cancer cell migration, but not for Its RhoGDI binding activity. *PLOS ONE*, 7, e35682.

7 Role of Serine Proteases in Gastric Cancer

Inhibition and Cure

Reena Kumari
Institute of Bioresources and Sustainable Development, Tadong-Gangtok, India
Manipur University, Canchipur-Imphal, India

Kshetrimayum Birla Singh and Sivakumar Vijayaraghavalu
Manipur University, Canchipur-Imphal, India

Amit Kumar Rai
Institute of Bioresources and Sustainable Development, Tadong-Gangtok, India

7.1 INTRODUCTION

Gastric cancer (GC) poses a significant global health burden, with alarming statistics as revealed by the GLOBOCAN 2020 report, ranking fifth in terms of morbidity and fourth in mortality rate worldwide. This malignancy exhibits a pronounced gender bias, afflicting men at a disproportionately higher rate than women, with a striking male-to-female ratio of 2:1 (Bray et al., 2018). Notably, its prevalence is most pronounced in countries such as China, Japan, and Korea, where GC has entrenched itself as a major public health concern (Lei et al., 2022; Takei et al., 2022). In contrast, Western nations have reported comparatively lower incidence rates of GC, underscoring the intricate interplay between socio-economic factors and disease prevalence (Alberts et al., 2003).

Historically, GC has left a trail of devastation, claiming countless lives due to the absence of comprehensive knowledge and effective treatment options. The turning point in the battle against this formidable adversary came in the late 19th century, thanks to the pioneering work of Prof. Theodor Billroth, who conducted the first successful surgery for GC. However, the earliest documented references to GC date back to the annals of medical history, with the Ebers Papyrus in 1600 BC and the contributions of Galen in 100 A.D, who notably introduced the term "carcinoma" (Guggenheim & Shah, 2013). The first anatomical illustration of GC was painstakingly rendered by Jean Cruveilhier (termed it as "hamartoma"), marking a significant milestone in our understanding of the disease (Cruveilhier, 1844).

Contemporary approaches to treating GC encompass a multifaceted arsenal of therapeutic modalities, including radiotherapy, chemotherapy, immunotherapy, surgical intervention, and targeted therapy (Lei et al., 2022; Gallanis and Davis, 2023; Pihlak et al., 2023). Among these, surgery and chemotherapy has emerged as the cornerstone of treatment, widely recognized for its efficacy and broad applicability in managing GC. The relentless pursuit of innovative strategies to combat GC continues to be fueled by the urgency of reducing its global disease burden.

DOI: 10.1201/9781003394693-9

In the intricate landscape of cancer biology, serine proteases assume a pivotal role as the most abundant and functionally diverse group of proteolytic enzymes, constituting 2–3% of the human genome (Vizovisek et al., 2021). These proteases exert profound influence over numerous biological processes and play a critical role in the host's defense mechanisms against cancer, including GC. Classification based on their structure and function has led to the recognition of four major clans of serine proteases, with clan PA representing chymotrypsin, clan SB representing subtilisin, and clans SB and SF encompassing a wide range of peptidases. Remarkably, clan PA alone accounts for nearly three-fourths of all serine proteases (Antalis et al., 2010).

The biological significance of serine proteases extends across an array of cellular functions, including apoptosis, cell differentiation, inflammation, and extracellular matrix (ECM) degradation. These enzymatic players are pivotal in maintaining the delicate balance of proteolytic activities required for normal physiological processes. However, despite the wealth of scientific literature, the enigma surrounding the role of serine proteases in the context of GC remains a compelling challenge for researchers. Factors contributing to this complexity include their ancient evolutionary origins, diverse signaling pathways, and intricate interplay within the evolving cancer microenvironment (Topi et al., 2020).

This chapter embarks on an exploration of the intricate world of serine proteases and their involvement in the context of GC, shedding light on their multifaceted roles, potential therapeutic implications, and the intriguing avenues that lie ahead for unravelling their contribution to GC progression and treatment response. In this quest, a comprehensive understanding of the enigmatic interplay between serine proteases and GC will undoubtedly pave the way for novel therapeutic strategies and insights into this relentless battle against a formidable adversary.

7.2 ETIOLOGY AND PATHOPHYSIOLOGY OF GASTRIC CANCER

The etiology GC is indeed a complex and multifaceted puzzle that continues to challenge our understanding. GC's progression involves a series of intricate events, including initiation, promotion, invasion, migration, and implantation. It is a disease influenced by a multitude of factors, encompassing genetic predisposition, environmental exposures, and lifestyle choices (Guggenheim & Shah, 2013).

GC's causative factors span a wide spectrum, ranging from dietary habits and hereditary factors to genetic variations. Notably, the consumption of heavily processed foods and beverages, tobacco use, smoking, and excessive alcohol consumption have emerged as significant contributors to GC (Guggenheim & Shah, 2013; Lei et al., 2022). Additionally, diets rich in salt and nitrates, along with excessive meat consumption and inadequate intake of vitamin A and C-rich foods, have the potential to disrupt gut health, elevating the risk of GC (Krejs, 2010; Piazuelo & Correa, 2013).

Historically, GC was primarily attributed to lifestyle factors and stress until a groundbreaking discovery reshaped our comprehension of this disease: the identification of *Helicobacter pylori*. GC is broadly categorized into cardia GC (arising from the upper stomach) and non-cardia GC (originating in the mid-distal stomach). Non-cardia GC is more prevalent in Asian populations, whereas, cardia GC is predominantly found in Western regions. Non-cardia GC is often associated with chronic *H. pylori* infection (Zabaleta, 2012; Lei et al., 2022). The ingestion of contaminated foods carrying *H. pylori* is a common mode of transmission. Prolonged consumption of *H. pylori*-infected food can lead to gastritis and gastric and duodenal ulcers, a discovery recognized by the Nobel Prize awarded to Australian researchers Barry J. Marshall and Robin Warren in 2005 (Topi et al., 2020). *H. pylori* thrives in the acidic gastric environment, resulting in increased nitrite and N-nitroso compound levels, contributing to carcinogenesis. *H. pylori* has been unequivocally classified as a class I human carcinogen by the World Health Organization (WHO) and the International Agency for Research on Cancer since 1994 (Zabaleta et al., 2004).

H. pylori infection manifests in various phenotypes. Corpus-predominant gastritis leads to atrophic gastritis and reduced stomach acid production (hypochlorhydria), a precursor to GC. The benign phenotype has a lesser impact on gastric acid secretion, while the duodenal ulcer phenotype is characterized by excessive gastric acid production and duodenal ulceration. The development of GC as a result of *H. pylori* infection is influenced by individual genetics and the host's unique response to the bacterium (Malfertheiner et al., 2023).

Matrix metalloproteinases (MMPs) are implicated in the pathophysiological development of GC. They exert their influence by modulating the inflammatory microenvironment and inducing architectural changes in the affected tissues. Genetic polymorphisms in proinflammatory cytokines, inflammation-related genes (TNF, IL-1RN, and IL-1B), interleukins (IL-1β), and tumor necrosis factor have also been identified as promoters of GC (Piazuelo and Correa, 2013). Upon entering the gastric mucosa *via* adhesion proteins like cagA and BabA, *H. pylori* accelerate the production of cytokines, leading to the infiltration of inflammatory cells and extensive tissue damage. This sets off a cascade of inflammatory reactions within the gastric epithelial tissues, eventually culminating in cancer (Zabaleta, 2012; Piazuelo and Correa, 2013). *H. pylori* also produces arginase, an enzyme that aids its survival within the host by evading T-cells and macrophages, effectively escaping both innate and acquired immune responses (Zabaleta et al., 2004). Furthermore, *H. pylori* interferes with nitric oxide (NO) production, a critical component of innate immunity. NO serves as an essential antimicrobial agent against various pathogens, including *Escherichia coli*, *Mycobacterium tuberculosis*, and *Leishmania major* through NO-dependent mechanisms (Gobert et al., 2001).

In addition to *H. pylori*, studies suggest that genes from the Epstein-Barr virus (EBV) play a role in GC development through mechanisms such as immune evasion, hypermethylation of tumor suppressor genes, and inflammation of the gastric mucosa (Lei et al., 2022). Individuals with conditions such as pernicious anemia, hereditary non-polyposis colon cancer, and Li-Fraumeni syndrome are at an increased risk of developing GC. Furthermore, those with blood type A and a family history of GC may face a higher likelihood of succumbing to this disease.

Genomic studies conducted by The Cancer Genome Atlas (TCGA) have classified GC mutations into four distinct types: genetically stable, chromosomally unstable, microsatellite unstable, and EBV-associated GC. Numerous somatic mutations can occur in the cancer genome, leading to DNA damage and impaired DNA repair pathways due to factors such as chemical exposure and DNA destruction. Molecular genetics research has identified specific genes, including MLH1, RUNX3, APC, DAPK, GSTP1, THBX-1, MGMT, TP53, TGFBR2, FGFR2/KSAM, DARPP32, MET, and CTNNB1, which, when dysregulated, can induce malignancy in epithelial cells by either activating oncogenes or silencing tumor suppressor genes. Other mutated or overexpressed genes like CDH1, PTGS2/COX2, EGF, HER2/NEU (gene amplification), and KRAS2 (point mutation) also contribute to GC and are potential targets for GC therapy (Lynch et al., 2005).

While the etiology of this multifactorial disease remains a formidable challenge, ongoing investigations in many Asian and South Asian countries are diligently seeking innovative ways to combat GC in our modern era of scientific and medical advancement. As we unravel the intricate web of GC causation, new avenues for prevention, early detection, and therapeutic interventions emerge, offering hope in the battle against this relentless adversary.

7.3 SERINE PROTEASES AND GASTRIC CANCER

Serine proteases, a formidable category of proteolytic enzymes, stand as the largest among the five main protease groups found in mammals, which also include metallo, aspartic, cysteine, and threonine proteases (Lei et al., 2012). These enzymes are initially synthesized and stored as zymogens in granules, bound to proteoglycans in an inactive state, until they undergo cleavage to become active. Serine proteases derive their name from the presence of serine (Ser) at their catalytic site, where they form an acyl-enzyme intermediate by attacking the peptide bond within the carbonyl moiety of their substrates. The nucleophilicity of the catalytic serine is often referred to as a charge relay system, owing to the involvement of Asp, Ser, and His residues, which are essential for its enzymatic activity. Serine proteases can be further categorized into 13 clans and 14 families (Di Cera, 2009). Table 7.1 provides an overview of the general properties of serine proteases.

Serine proteases play pivotal roles in various physiological and pathological activities, including digestion, fertilization, blood coagulation, wound healing, fibrinolysis, angiogenesis, and, cell differentiation and apoptosis (Qin et al., 2022). In the context of cancer, including GC, serine proteases assume a prominent role in facilitating neoplastic progression during the metastatic stages.

Tumor cells employ diverse cellular and biochemical mechanisms to spread, and serine proteases emerge as key contributors in this process. They actively participate in the degradation of the ECM, thereby promoting the metastasis of cancer cells within the host body (Wang et al., 2019). Some serine proteases, such as S1 serine proteases, directly puncture the host cell's plasma membrane through specific cytoplasmic domains, facilitating their entry into host cells. Conversely, others, like seprase, traverse the cell membrane through a series of reactions involving cell signaling pathways (Antalis et al., 2010).

Notably, certain serine proteases have been identified as promoters of GC malignancy. Among them, serpins (serine protease inhibitors) with pro-inflammatory activity have garnered attention for their role in cancer progression. For instance, SERPINB3 and SERPINB5 are associated with squamous cell carcinoma, while SERPINB2 and SERPINE1 have implications in breast and other cancers. Additionally, SERPIN11 and SERPIN12 have been linked to oral squamous carcinoma. Research by Lei et al. (2012) highlights genes such as SERPINA5, SERPINB5, HPN, SPINK1, KLK10, PRSS8, and KLK11 as contributors to cancer progression. KLK10 and KLK11 in particular are upregulated in colon, pancreatic, ovarian, and GCs. Studies exploring loss-of-function mutations in the PRSS23 gene have suggested its role as an oncogene in GC. Another serine protease gene has also been implicated as a potential contributor to malignancy (Wang et al., 2019; Qin et al., 2022).

TABLE 7.1
General Properties of Serine Proteases

Clan	No. of Families	Representative Members	Catalytic Residues
SB	2	Subtilisin	Asp, His, Ser
PA	12	Trypsin, Chymotrypsin	His, Asp, Ser
SP	3	Nucleoporin	His, Ser
ST	5	Rhomboid	His, Ser
SH	2	Cytomegalovirus Assemblin	His, Ser, His
SR	1	Lactoferrin	Lys, Ser
PB	1	Penicillin Amidohydrolase Precursor	Ser (N-terminal)
SQ	1	Aminopeptidase DmpA	Ser
SC	2	Prolyl oligopeptidase, Carboxypeptidase	Ser, Asp, His
SS	1	L, D- Carboxypeptidase	Ser, Glu, His
SK	2	CLP Peptidase and CLP Protease	Ser, His, Asp
SE	6	D-ALA-D-ALA- Carboxypeptidase A	Ser, Lys
SJ	1	Lon peptidase	Ser, Lys
SF	3	Lex-A Peptidase, Signal Peptidase I	Ser, Lys/His

Serine proteases, a diverse family of proteolytic enzymes, are integral players in various physiological and pathological processes, including cancer progression. In the context of GC, their involvement in ECM degradation and facilitation of metastasis underscores their significance in disease pathogenesis. Understanding the intricate roles of serine proteases in GC can pave the way for targeted therapeutic interventions and enhanced treatment strategies in the ongoing battle against this formidable disease.

7.4 ROLE OF MITOGEN-ACTIVATED PROTEIN KINASES (MAPKs) SIGNALING IN GASTRIC CANCER

The mitogen-activated protein kinases (MAPKs) are one of the families of serine/threonine kinases involved in the cellular response to multiple extracellular stimuli and undergo the most complicated signaling pathways for tumor invasion in GC (Cargnello and Roux, 2011). Different studies suggest that ERK/MAPK signaling is involved in GC regulation and cell motility. In fact, the regulation of matrix metalloproteins in GC is regulated by ERK, thereby affecting cell migration as well as invasion (Magnelli et al., 2020). The signaling pathway of MAPK is shared by five different cascades named after the components of each MAPK tier. The major signaling pathways involved in GC progression are illustrated in Figure 7.1. These include extracellular signal-related kinases ERK (ERK1/2), Jun amino terminal kinases (SAPK/JNK1,2,3), p38-MAPK (p38-α, β, γ, and δ), ERK5, and ERK3/4 (Lei et al., 2014). There are at least three core kinases (MAPK, MAPKK, and MAPKKK) present in each cascade of the canonical signal (et al., 2011). The onset of the MAPK/ERK pathway is done by binding of extracellular factors to the epidermal growth factor receptor (EGFR), G-protein couples receptors. and receptors and tyrosine kinases (RTK). It is also triggered by vascular endothelial growth factors and its receptors (VGEF and VGEFR). Under standard conditions, the activation of the RAS protein (KRAS, HRAS and NRAS) triggers MAPK signaling (Simanshu et al., 2017). Approximately 37% of GCs are found to be a result of the RTK signaling pathway (Deng et al., 2012). In the switch 1 and 2 regions, the RAS undergoes a conformational shift in its Guanosine-5'-triphosphate (GTP) bound activated form. This promotes interaction with the RAF family of kinases ARAF, BRAF, and CRAF (Pai et al., 1989). The RAF kinases, after being activated, phosphorylate and activate MEK1/2 kinases, which, in turn, are helpful in cell proliferation, cell differentiation, and survival signals (Roskoski, 2012; Lei et al., 2022). The activated ERK1/2 kinases then lead to a series of signaling cascades by phosphorylating a number of required substrates (Roskoski, 2012). The regulation of cell motility in GC is also done by MAPK/ERK pathway in coordination with the matrix metalloproteins, cell adhesion, and EGFR disassembly of focal adhesions that promotes cell migration and invasion (Yang & Huang, 2015).

The JNKs are another set of MAPKs that are encoded by three different genes viz. MAPK8, MAPK9, and MAPK10 encoding JNK1, JNK2, and JNK3, respectively. The activation of JNKs results in cell proliferation apoptosis and transformation (Weston and Davis, 2007). The JNK subtypes activate the p38-MAPK. The upstream MAPK kinases, MKK3 and MKK6 kinases specifically, activate the p38-MAPK, which has an important role in cancer regulation via progression, invasion, inflammation, and angiogenesis (Cargnello and Roux, 2011; Souma et al., 2012; Yan et al., 2015). A study performed by Guo et al. (2008) showed that p38-MAPK/AP-1 pathway activation can be effectively applied in GC chemotherapy.

Recent studies have shown that GC malignancy and metastasis can be epigenetically regulated via the MAPK/ERK pathway. In addition, overexpression of miR-592 also promotes cell proliferation, migration, and invasion in GC by targeting Sprouty-2 (SPRY2) via MAPK/ERK and PI3K/AKT pathways (He et al., 2018). Signaling pathways HER-2, PI3K/AKT/mTOR, p53, HGF/c-MET, Wnt/β-catenin, TGF-β signaling, and JAK2/STAT3 are also reported to be involved in the GC progression (Lei et al., 2022), mentioned in Figure 7.1

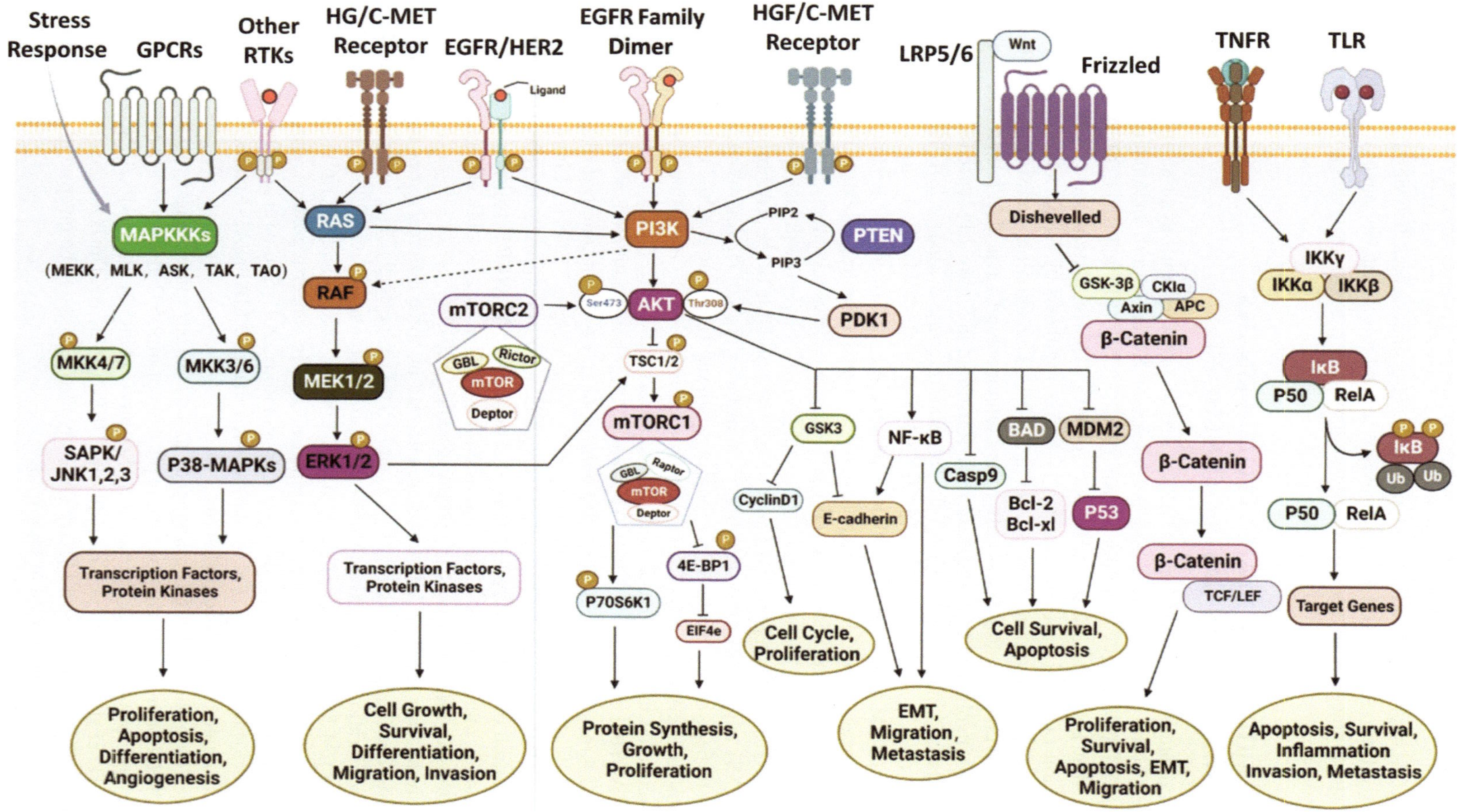

FIGURE 7.1 Signaling pathways associated with GC (Lei et al., 2022); key signaling pathways and regulators in gastric cancer. This diagram highlights the primary interactions and functions of pathways such as MAPK, HER2, PI3K/AKT/mTOR, HGF/c-Met, p53, Wnt/β-catenin, and NF-κB. Notable entities include G-protein-coupled receptors (GPCRs), epidermal growth factor receptor (EGFR), receptor tyrosine kinases (RTKs), RAS, RAF, MEK, ERK1/2, PI3K, AKT, mTORC1/2, PTEN, NF-κB, T-cell factor/lymphoid enhancer factor (TCF/LEF), and IκB kinase (IKK). The figure also highlights processes such as epithelial-mesenchymal transition (EMT). (Created using Biorender.com. Adapted after obtaining permission from Lei et al., 2022. Licensed under CC BY 4.0.)

7.5 SERINE PROTEASES IN GASTRIC CANCER INHIBITION AND CURE

Serine proteases play multifaceted roles in the intricate landscape of GC progression. Given their central involvement in various cell-signaling cascades associated with malignancy, researchers have increasingly explored these pathways to target specific intermediates or activators with the aim of curbing or controlling GC. Here, we delve into several promising avenues and strategies in the quest to inhibit and ultimately cure GC.

7.5.1 Targeting Serine Proteases in GC Progression

One avenue of research focuses on the understanding of how serine proteases contribute to the metastatic cascade of GC. By elucidating these mechanisms, researchers have identified potential targets for intervention. For instance, in colorectal carcinogenesis and metastasis, the serine protease PRSS8 has emerged as a crucial player. Inhibition of the EMT signaling pathway, mediated by PRSS8, has shown promise in suppressing cancer progression. Moreover, loss-of-function mutations, achieved through gene knockdown, have been explored as a means to reduce GC metastasis significantly (Qin et al., 2022).

PRSS23, another serine protease, has also garnered attention for its role in GC. Knockdown of PRSS23 leads to a decrease in FGF2 expression and secretion in GC cells. This inhibition of FGF2 can impact cancer progression. Additionally, PRSS23 has been implicated in blocking EIF2 signaling and hedgehog pathway, both of which are involved in inhibiting cell proliferation and inducing apoptotic pathways (Han et al., 2019).

7.5.2 Inhibiting Serine Protease Inhibitors

While serine proteases are themselves targets, their inhibitors also play significant roles in GC progression. SERPINE1, a member of serine protease inhibitor clade E, has been explored as a potential target for inhibition (Chen et al., 2022). Deletion of SERPINE1 affects the expression of VEGF and IL6 via the VEGF and JAK-STAT3 signaling pathways. This disruption hinders the invasion and migration of malignant cells in GC patients. Furthermore, SERPINE1 has been implicated in modulating immune cell infiltration, which can have a profound impact on the tumor microenvironment (Chen et al., 2022).

HAI-2/SPINT2, an inhibitor of the HGF/MET pathway, has also been investigated for its role in GC. Downregulation of HAI-2/SPINT2 has been associated with tumor suppression in GC patients. Understanding and manipulating these inhibitors hold promise for disrupting the molecular pathways that drive GC progression.

7.5.3 Exploring Synthetic Inhibitors

In addition to targeting specific serine proteases and their inhibitors, synthetic serine protease inhibitors have been evaluated for their potential in GC treatment. Nafamostat mesylate (NM), for example, has shown promise in inhibiting the progression of various cancers, including colorectal, pancreatic, gall bladder, and hepatocellular carcinoma. NM exerts its effects through multiple pathways, including the tumor necrosis factor receptor-1 (TNRF1) and nuclear factor-κB (NF-κB) signaling pathway (Iwase et al., 2013; Takada et al., 2021).

7.5.4 Natural Compounds as Potential Therapeutics

Nature also offers a wealth of potential therapeutic compounds for GC treatment. Sinulariolide, isolated from the marine organism *Sinularia flexibilis*, has demonstrated efficacy in curtailing GC (Wu et al., 2019). This natural compound acts by inhibiting cancer cell invasion and migration, partly by downregulating the EMT process. It also suppresses signaling pathways such as FAK/PI3K/AKT/mTOR and MAPK through distinct mechanisms, making it a promising candidate for GC therapy (Wu et al., 2019).

7.5.5 Reversing Silenced Tumor Suppressor Genes

Silenced tumor suppressor genes can be a key driver of GC progression. One such gene is tissue factor pathway inhibitor 2 (TFPI2), which belongs to the Kunitz-type serine protease inhibitor family (Takada et al., 2010). In GC, TFPI2 is often silenced, contributing to tumor growth and metastasis. Research has shown that the restoration of TFPI2 can inhibit the progression of GC, offering a potential avenue for therapeutic intervention.

7.5.6 Harnessing Overexpressed Serine Proteases

Interestingly, not all serine proteases promote GC progression. Some, like HtrA1, have been found to inhibit GC when overexpressed in gastric cells. HtrA1 limits invasion and cell migration, underscoring the complex and context-dependent roles of serine proteases in cancer (Kang et al., 2017).

7.5.7 Immunotherapeutic Approaches

Immunotherapeutic approaches using monoclonal antibodies have also entered the landscape of GC treatment. Monoclonal antibodies offer targeted therapies that can specifically interfere with key molecules and pathways involved in cancer progression. These emerging treatments hold promise for improving outcomes in GC patients (Takei et al., 2022).

Overall, serine proteases, while intricately involved in GC progression, also offer a rich landscape of potential targets for inhibiting and ultimately curing this complex disease. From targeting specific serine proteases and their inhibitors to exploring synthetic compounds and harnessing the power of natural compounds, researchers are making strides in understanding and combating GC. Additionally, the manipulation of silenced tumor suppressor genes and the utilization of immunotherapeutic approaches further expand the arsenal of strategies to tackle this formidable disease. As our knowledge deepens and therapeutic options continue to evolve, the future holds promise for improved treatments and outcomes for GC patients.

7.6 CHALLENGES AND FUTURE DIRECTIONS IN SERINE PROTEASES AND GASTRIC CANCER RESEARCH

While significant progress has been made in understanding the role of serine proteases in GC, numerous challenges remain, and promising avenues for future research are emerging. Here, we discuss some of the key challenges and potential future directions in this field.

7.6.1 Challenges

7.6.1.1 Heterogeneity of Gastric Cancer

One of the major challenges in GC research is its heterogeneity. GC comprises various subtypes with distinct molecular profiles, which can influence the role and expression of serine proteases differently (Qin et al., 2022). Understanding this heterogeneity and its implications for serine protease involvement is critical.

7.6.1.2 Identification of Biomarkers

Identifying reliable serine protease biomarkers for GC diagnosis, prognosis, and treatment response is challenging. Biomarkers are essential for early detection and personalized therapies, but the complexity of protease networks and their diverse functions in cancer make it difficult to pinpoint specific markers (Qin et al., 2022). In a recent study, LncRNAs have been recognized as a potential biomarker for GC tumor suppression due to its stability and specificity (Liu et al., 2021). Some other newly identified diagnostic biomarkers are members of SERPIN family (e.g., SERPINA1: α1antitrypsin, SERPINA3: α1antichymotrypsin, and SERPINC1: antithrombin-3) have been reported (Pączek and Mroczko, 2022). Some other circulating biomarkers for detecting GC in earlier stages are adipocytokine leptin, p3-autoantibody, M2-pyruvate kinase, olfactomedin-4, vascular adhesion protein-1, overexpressed genes (KRT17, COL10A1, and SALL4), gastrokine1, and many miRNAs (Jelski and Mroczko, 2022).

7.6.1.3 Therapeutic Resistance

Resistance to current treatments, such as chemotherapy, remains a significant obstacle in GC management. Serine proteases play a role in cancer progression and resistance mechanisms. Effective chemotherapy for advanced cancer stages remains elusive due to factors like genetic variation and diverse resistance mechanisms, including multidrug resistance, inhibited cell death, altered drug metabolism, epigenetic changes, enhanced DNA repair, and gene amplification. Developing targeted therapies that circumvent or exploit these mechanisms is a complex task.

7.6.1.4 Limited Clinical Translation

Despite promising preclinical findings, translating serine protease-based therapies from the bench to the bedside is challenging. Clinical trials are necessary to evaluate their safety and efficacy, but many hurdles must be overcome before these therapies become standard practice (Rudzińska et al., 2021).

7.6.1.5 Interplay with the Tumor Microenvironment

The interaction between serine proteases and the tumor microenvironment, including immune cells, stromal cells, and ECM components, is complex and poorly understood (Quail & Joyce, 2013). Elucidating this interplay is vital for designing effective therapies.

7.7 CONCLUSION AND FUTURE PERSPECTIVES

Advancements in genomics and proteomics allow for a more comprehensive understanding of individual tumors. Future research should focus on identifying patient-specific serine protease signatures to tailor treatments for improved outcomes.

Immunotherapies have shown promise in various cancers. Investigating how serine proteases influence immune responses within the tumor microenvironment and exploring immunotherapeutic interventions targeting these proteases could be a groundbreaking avenue.

Combining serine protease inhibitors with existing treatments, such as chemotherapy or targeted therapies, may overcome resistance mechanisms and enhance therapeutic efficacy. Synergistic approaches should be explored in preclinical and clinical settings.

Continued efforts to identify and validate serine protease biomarkers are essential. These biomarkers could aid in early detection, patient stratification, and monitoring treatment responses (Shen et al., 2019).

In-depth functional studies should be conducted to unravel the specific roles of different serine proteases in GC progression (Zhang et al., 2016). This knowledge will inform the development of targeted therapies and help predict treatment responses.

Initiating well-designed clinical trials to evaluate the safety and efficacy of serine protease-based therapies in GC patients is imperative. Collaborative efforts between researchers, clinicians, and pharmaceutical companies are needed to bring these therapies to the clinic.

Leveraging big data analytics and artificial intelligence in GC research can aid in identifying novel serine protease targets, predicting patient outcomes, and optimizing treatment regimens. Serine proteases play a multifaceted role in GC, offering both challenges and exciting opportunities for further research and therapeutic development. As our understanding of the intricate protease networks in cancer grows, we can expect more targeted and effective treatments, ultimately improving outcomes for GC patients. Collaborative efforts and innovative approaches will be key to addressing the challenges and advancing the field of serine proteases in GC research.

In conclusion, the intricate interplay between serine proteases and GC presents a multifaceted landscape for researchers and clinicians. Serine proteases, as a diverse family of proteolytic enzymes, exert significant influence on various stages of GC initiation, progression, and metastasis.

Their roles in modulating the tumor microenvironment, facilitating invasion and angiogenesis, and regulating immune responses make them central players in GC pathophysiology.

Despite substantial progress in understanding the involvement of serine proteases in GC, several challenges remain. Heterogeneity within GC subtypes, the identification of reliable biomarkers, therapeutic resistance, and the translation of promising preclinical findings into clinical practice are among the hurdles that must be addressed.

Looking ahead, the future of serine protease research in GC holds great promise. Precision medicine approaches, immunotherapies, and combination therapies are emerging as potential strategies to improve patient outcomes. The discovery and validation of serine protease biomarkers may revolutionize early diagnosis and treatment stratification.

Furthermore, in-depth functional studies of specific serine proteases and their roles in GC are essential to inform targeted therapies and predict treatment responses accurately. Clinical trials, driven by collaborative efforts, will be crucial in evaluating the safety and efficacy of serine protease-based interventions in GC patients.

As researchers harness the power of big data and artificial intelligence, new insights into serine protease targets and treatment optimization are expected. With continued dedication and innovative approaches, the complex web of serine proteases in GC can be unraveled, ultimately leading to improved therapeutic strategies and outcomes for patients with this challenging disease.

DECLARATIONS

Conflict of interest: The authors declare no conflicts of interest.

Consent for publishing: All authors have approved the final version of the manuscript and given consent for publication.

Funding: Not Applicable.

REFERENCES

Alberts S R. Cervantes A. van de Velde C J H. Gastric cancer: epidemiology, pathology and treatment. *Annals of oncology: official journal of the European Society for Medical Oncology.* 14 Suppl 2. 2003. ii31–6. 12810455.

Antalis, T. M., Bugge, T. H. & Wu, Q. (2010). Membrane-anchored serine proteases in health and disease. *Progress in Molecular Biology and Translational Science*, 99, 1–50. https://doi.org/10.1016/b978-0-12-385504-6.00001-4.

Bray, F., Ferlay, J., Soerjomataram, I., Siegel, R. L., Torre, L. A. & Jemal, A. (2018). Global cancer statistics 2018: GLOBOCAN estimates of incidence and mortality worldwide for 36 cancers in 185 countries. *CA: A Cancer Journal for Clinicians*, 68(6), 394–424. https://doi.org/10.3322/caac.21492.

Cargnello Marie. Roux Philippe P. Activation and function of the MAPKs and their substrates, the MAPK-activated protein kinases. *Microbiology and molecular biology reviews: MMBR.* 75 (1) 2011. 50–83. Mar 2011. 10.1128/MMBR.00031-10. 21372320.

Cargnello, M. & Roux, P. P. (2011). Activation and function of the MAPKs and their substrates, the MAPK-activated protein kinases. *Microbiology and Molecular Biology Reviews*, 75, 50–83. https://doi.org/10.1128%2FMMBR.00031-10.

Chen, S., Li, Y., Zhu, Y., Fei, J., Song, L., Sun, G., Guo, L. & Li, X. (2022). SERPINE1 overexpression promotes malignant progression and poor prognosis of gastric cancer. *Journal of Oncology*, 2647825. https://doi.org/10.1155/2022/2647825.

Cruveilhier, J. (1844). The Anatomy of the human body. New York, NY, Harper and Brothers. https://openlibrary.org/books/OL6949531M/The_anatomy_of_the_human_body.

Deng, N., Goh. L. K., Wang, H., Das, K., Tao, J., Tan, I. B., Zhang, S., Lee, M., Wu, J., Lim, K. H., Lei, Z., Goh, G., Lim, Q. Y., Tan, A. L., Poh, D. Y. S., Riahi, S, Bell, S., Shi, M. M., Linnartz, R., Zhu, F., Yeoh, K. G., Toh, H. C., Yong, W. P., Cheong, H. C., Rha, S. Y., Boussioutas, A., Grabsch, H., Rozen, S. & Tan, P. (2012). A comprehensive survey of genomic alterations in gastric cancer reveals systematic patterns of molecular exclusivity and co-occurrence among distinct therapeutic targets. *Gut*, 61(5), 673–684. https://doi.org/10.1136/gutjnl-2011-301839.

Di Cera Enrico. Serine proteases. IUBMB Life. 2009. 61 (5) 510–515. May 2009. 10.1002/iub.186.

Gallanis Amber F. Davis Jeremy L. Unique challenges of risk-reducing surgery for hereditary diffuse gastric cancer syndrome: a narrative review. *European journal of cancer prevention : the official journal of the European Cancer Prevention Organisation (ECP).* 32(4) 2023. 391–395. Jul 2023. 10.1097/CEJ.0000000000000798. 36977191.

Gobert, A. P, McGee, D. J., Akhtar, M., Mendz, G. L., Newton, J. C., Cheng, Y., Mobley, H. L. & Wilson, K. T. (2001). *Helicobacter pylori* arginase inhibits nitric oxide production by eukaryotic cells: a strategy for bacterial survival. *Proceedings of the National Academy of Sciences of the United States of America*, 98(24), 13844–13849. https://doi.org/10.1073/pnas.241443798.

Guo Xianling. Ma Nannan. Wang Jin. Song Jianrui. Bu Xinxin. Cheng Yue. Sun Kai. Xiong Haiyan. Jiang Guocheng. Zhang Baihe. Wu Mengchao. Wei Lixin. Increased p38-MAPK is responsible for chemotherapy resistance in human gastric cancer cells. BMC cancer. 8. 2008. 375. Dec 2008. 10.1186/1471-2407-8-375. 19091131.

Guggenheim, D. E. & Shah, M. A. (2013). Gastric cancer epidemiology and risk factors. *Journal of Surgical Oncology*, 107(3), 230–236. https://doi.org/10.1002/jso.23262.

Han, B., Yang, Y., Chen, J., He, X., Lv, N. & Yan, R. (2019). PRSS23 knockdown inhibits gastric tumorigenesis through EIF2 signaling. *Pharmacological Research*, 142, 50–57. https://doi.org/10.1016/j.phrs.2019.02.008.

He, Y., Ge, Y, Jiang, M., Zhou, J., Luo, D., Fan, H., Shi, L., Lin, L. & Yang, L. (2018). MiR-592 promotes gastric cancer proliferation, migration, and invasion through the PI3K/AKT and MAPK/ERK signaling pathways by targeting Spry2. *Cellular Physiology and Biochemistry*, 47, 1465–1481. https://doi.org/10.1159/000490839.

Iwase, R., Haruki, K., Fujiwara, Y., Furukawa, K., Shiba, H., Uwagawa, T., Misawa, T., Ohashi, T. & Yanaga K. (2013). Combination chemotherapy of nafamostat mesylate with gemcitabine for gallbladder cancer targeting nuclear factor-κB activation. *Journal of Surgical Research*, 184, 605–612. https://doi.org/10.1016/j.jss.2013.06.003.

Jelski Wojciech. Mroczko Barbara. Molecular and Circulating Biomarkers of Gastric Cancer. International Journal of Molecular Sciences. 2022. 23 (14) 7588. 10.3390/ijms23147588.

Kang, M. H., Jeong, G. S., Smoot, D. T., Ashktorab, H., Hwang, C. M., Kim, B. S., Kim, H. S. & Park, Y. Y. (2017). Verteporfin inhibits gastric cancer cell growth by suppressing adhesion molecule FAT1. *Oncotarget*, 8(58), 98887–98897.https://doi.org/10.18632/oncotarget.21946. Erratum in: Oncotarget. 2018 Dec 14;9(98):37268.

Krejs Guenter J. Gastric cancer: epidemiology and risk factors. Digestive diseases (Basel, Switzerland). 28 (4–5) 2010. 600–3. 10.1159/000320277. 21088409.

Lei, K. F., Liu, B. Y., Zhang, X. Q., Jin, X. L., Guo, Y., Ye, M. & Zhu Z. G. (2012). Development of a survival prediction model for gastric cancer using serine proteases and their inhibitors, *Experimental and Therapeutic Medicine*, 3(1), 109–116. https://doi.org/10.3892%2Fetm.2011.353.

Lei, Z. N., Teng, Q. X., Tian, Q., Chen, W., Xie, Y., Wu, K., Zeng, Q., Zeng, L., Pan, Y., Chen, Z. S. & He, Y. (2022). Signaling pathways and therapeutic interventions in gastric cancer. *Signal Transduction and Targeted Therapy*, 7, 358. https://doi.org/10.1038/s41392-022-01190-w.

Liu Ying. Ding Wei. Yu Wanpeng. Zhang Yuan. Ao Xiang. Wang Jianxun. Long noncoding RNAs: Biogenesis, functions, and clinical significance in gastric cancer. Molecular therapy oncolytics. 23. 2021. 458–476. Dec 2021. 10.1016/j.omto.2021.11.005. 34901389.

Lei, Y. Y., Wang, W. J., Mei, J. H. & Wang, C. L. (2014). Mitogen-activated protein kinase signal transduction in solid tumors. *Asian Pacific Journal of Cancer Prevention*, 15, 8539–8548. https://doi.org/10.7314/apjcp.2014.15.20.8539.

Lynch, H. T., Grady, W., Suriano, G. & Huntsman, D. (2005). Gastric cancer: new genetic developments, *Journal of Surgical Oncology*, 90(3), 114–133. https://doi.org/10.1002/jso.20214.

Magnelli, L., Schiavone, N., Staderini, F., Biagioni, A. & Papucci, L. (2020). MAP kinases pathways in gastric cancer. *International Journal of Molecular Sciences*, 21, 2893. https://doi.org/10.3390/ijms21082893.

Malfertheiner Peter. Camargo M. Constanza. El-Omar Emad. Liou Jyh-Ming. Peek Richard. Schulz Christian. Smith Stella I. Suerbaum Sebastian. Helicobacter pylori infection. Nature Reviews Disease Primers. 2023. 9 (1) 10.1038/s41572-023-00431-8.

Pączek Sara. Mroczko Barbara. The Role of Selected Serpins in Gastrointestinal (GI) Malignancies. *Journal of clinical medicine.* 11 (20) 2022. Oct 2022. 10.3390/jcm11206225. 36294546.

Pai, E. F., Kabsch, W., Krengel, U., Holmes, K. C., John, J. & Wittinghofer, A. (1989). Structure of the guanine-nucleotide-binding domain of the Ha-ras oncogene product p21 in the triphosphate conformation. *Nature*, 341(6239), 209–214. https://doi.org/10.1038/341209a0.

Qin, S., Wang, Z., Huang, C., Huang, P.& Li, D. (2022). Serine protease PRSS23 drives gastric cancer by enhancing tumor associated macrophage infiltration via FGF2. *Frontiers in Immunology*, 13, 955841. https://doi.org/10.3389/fimmu.2022.955841.

Quail, D. F. & Joyce, J. A. (2013). Microenvironmental regulation of tumor progression and metastasis. *Nature Medicine*, 19(11), 1423–1437. https://doi.org/10.1038/nm.3394.

Roskoski, R. Jr. (2012). ERK1/2 MAP kinases: structure, function, and regulation. *Pharmacological Research*, 66, 105–143. https://doi.org/10.1016/j.phrs.2012.04.005.

Rudzińska, M., Daglioglu, C., Savvateeva, L. V., Kaci, F. N., Antoine, R. & Zamyatnin, A. A., Jr. (2021). Current Status and Perspectives of Protease Inhibitors and Their Combination with Nanosized Drug Delivery Systems for Targeted Cancer Therapy. *Drug Design, Development and Therapy*, 15, 9–20. https://doi.org/10.2147/DDDT.S285852.

Shen, Q., Polom, K., Williams, C., de Oliveira, F. M. S., Guergova-Kuras, M., Lisacek, F., … Kamali-Moghaddam, M. (2019). A targeted proteomics approach reveals a serum protein signature as diagnostic biomarker for resectable gastric cancer. *EBioMedicine*, 44, 322–333. https://doi.org/10.1016/j.ebiom.2019.05.044.

Simanshu, D.K., Nissley, D.V. & Mc Cormick, F. (2017). RAS proteins and their regulators in human disease. *Cell*, 170(1), 17–33. https://doi.org/10.1016%2Fj.cell.2017.06.009.

Souma, Y., Nishida, T., Serada S., Iwahori, K., Takahashi, T., Fujimoto, M., Ripley, B., Nakajima, K., Miyazaki, Y., Mori, M., Doki, Y., Sawa, Y. & Naka, T. (2012). Antiproliferative effect of SOCS-1 through the suppression of STAT3 and p38 MAPK activation in gastric cancer cells. *International Journal of Cancer*, 131, 1287–1296. https://onlinelibrary.wiley.com/doi/10.1002/ijc.27350.

Takada, H., Wakabayashi, N., Dohi, O., Yasui, K., Sakakura, C., Mitsufuji, S., Taniwaki M. & Yoshikawa, T. (2010). Tissue factor pathway inhibitor 2 (TFPI2) is frequently silenced by aberrant promoter hypermethylation in gastric cancer. *Cancer Genetics and Cytogenetics*, 197(1), 16–24. https://doi.org/10.1016/j.cancergencyto.2009.11.004.

Takada Naoki. Sugano Hiroshi. Shirai Yoshihiro. Saito Nobuhiro. Hamura Ryoga. Taniai Tomohiko. Uwagawa Tadashi. Yanaga Katsuhiko. Ikegami Toru. Ohashi Toya. Eto Ken. Nafamostat mesilate, a nuclear factor kappa B inhibitor, enhances the antitumor action of radiotherapy on gallbladder cancer cells. PloS one. 16 (9) 2021. e0257019. 10.1371/journal.pone.0257019. 34473771.

Takei, S., Kawazoe, A. & Shitara, K. (2022). The New Era of Immunotherapy in Gastric Cancer. *Cancers*, 14(4), 1054. https://doi.org/10.3390/cancers14041054.

Topi, S., Santacroce, L., Bottalico, L., Ballini, A., Inchingolo, A. D., Dipalma, G., Charitos, I. A. & Inchingolo F. (2020). Gastric Cancer in History: A Perspective Interdisciplinary Study. *Cancers (Basel)*, 12(2), 264. https://doi.org/10.3390%2Fcancers12020264.

Vizovisek, Matej., Ristanovic, Dragana., Menghini Stefano., Christiansen, M.ichael G., Schuerle, Simone. The Tumor Proteolytic Landscape: A Challenging Frontier in Cancer Diagnosis and Therapy. International Journal of Molecular Sciences. 2021. 22 (5) 2514. https://doi.org/ 10.3390/ijms22052514

Wang, F., Hu, Y. L., Feng, Y., Guo, Y.B., Liu, Y. F., Mao, Q. S. & Xue, W. J. (2019). High-level expression of PRSS3 correlates with metastasis and poor prognosis in patients with gastric cancer. *Journal of Surgical Oncology*, 119(8):1108–1121. https://doi.org/10.1002/jso.25448.

Weston, C. R., and Davis, R. J. (2007). The JNK signal transduction pathway. *Current Opinion in Cell Biology*, 19(2), 142–149. https://doi.org/10.1016/j.ceb.2007.02.001.

Wu, Y.-J., Lin, S.-H., Din, Z.-H., Su, J.-H., Liu, C.-I. (2019). Sinulariolide Inhibits gastric cancer cell migration and invasion through downregulation of the emt process and suppression of FAK/PI3K/AKT/mTOR and MAPKs signaling pathways. *Drugs*, 17, 668. https://doi.org/10.3390/md17120668.

Yan, X., Rui, X. & Zhang, K. (2015). Baicalein inhibits the invasion of gastric cancer cells by suppressing the activity of the p38 signaling pathway. *Oncology Reports*, 33, 737–743. https://doi.org/10.3892/or.2014.3669.

Yang, M. & Huang, C. (2015). The Mitogen-activated protein kinase signaling pathway and invasion and metastasis of gastric cancer. *World Journal of Gastroenterology*, 21(41), 11673–11679. https://doi.org/10.3748%2Fwjg.v21.i41.11673.

Zabaleta, J. (2012). Multifactorial Etiology of Gastric Cancer. In: Dumitrescu, R., Verma, M. (eds) *Cancer Epigenetics. Methods in Molecular Biology*, 863, 411–435. Totowa, NJ, Humana Press. https://doi.org/10.1007/978-1-61779-612-8_26.

Zabaleta, J., McGee, D. J., Zea, A. H., Hernaandez, C. P., Rodriguez, P. C., Sierra, R. A., Correa, P. & Ochoa, A. C. (2004). *Helicobacter pylori* Arginase Inhibits T Cell Proliferation and Reduces the Expression of the TCRζ -Chain (CD3ζ). *Journal of Immunology*, 173(1), 586–593. https://doi.org/10.4049/jimmunol.173.1.586.

Zhang Mingming. Zhao Junjie. Tang Wenyi. Wang Yanru. Peng Peike. Li Lili. Song Shushu. Wu Hao. Li Can. Yang Caiting. Wang Xuefei. Zhang Chunyi. Gu Jianxin. High Hepsin expression predicts poor prognosis in Gastric Cancer. Scientific Reports. 2016. 6 (1) 10.1038/srep36902.

8 Serine Protease Inhibitors in Breast Cancer Therapy

Abinaya Elango and Arunkumar Radhakrishnan
Chettinad Hospital and Research Institute, Chettinad Academy of Research and Education, Chengalpattu, India

8.1 INTRODUCTION

Cancer is one of the leading causes of mortality worldwide. By 2030, the number of deaths due to cancer is expected to reach 11 million (Benson & Jatoi, 2012). In women, carcinoma breast is the most common cancer leading to death. Breast cancer affects a significant number of women globally, with an estimated annual diagnosis rate of 1.5 million, accounting for approximately 25% of all women affected with cancer. In the United States, breast tumors accounted for 30% of newly diagnosed cancer cases among women (Siegel et al., 2018).

Breast cancer is categorized as a metastatic malignancy, often exhibiting a propensity to spread to remote anatomical sites, including but not limited to the liver, bones, brain and lungs. This typical invasive pattern makes the treatment of breast cancer challenging. The timely identification of the disease can result in a favorable prognosis and a significant increase in the likelihood of survival. The 5-year survival rate in breast cancer exceeds 80% as a result of the prompt identification of this medical condition (DeSantis et al., 2016). Mammography is a commonly employed screening modality for the detection of breast cancer, which has demonstrated significant efficacy in reducing mortality rates. Additional screening techniques, including magnetic resonance imaging (MRI), which exhibits a higher level of sensitivity compared to mammography, have been utilized and investigated in the past ten years.

Though incidence is increasing, mortality due to breast cancer has started declining due to advancements in the diagnostics and treatment. This chapter discusses the risk factors, pathogenesis and types of breast cancer, currently available treatment options and the potential therapeutic role of serine protease inhibitors in breast cancer.

8.2 BREAST CANCER: RISK FACTORS

The risk factors of breast cancer include biological factors such as sex, age, and estrogen levels, as well as non-biological factors such as family history, gene mutations, and poor lifestyle choices (Majeed et al., 2014). The majority of breast cancer cases manifest in females, with the incidence rate being around 100 times greater in women compared to men (Siegel et al., 2018).

Besides female sex, one of the most important risk factors for breast cancer is advancing age, as the disease is strongly correlated with advancing age. Women over age 40 and 60 accounted for more than 99% and 71% respectively, of all breast cancer-related deaths (Siegel et al., 2018).

Family history is a significant contributing factor in nearly 25% of cases. Breast cancer is common in women if her mother or sibling has the disease. According to a cohort study involving more than 113,000 women, those who have a single first-degree cousin with breast cancer are 1.75 times more likely to develop cancer than those who do not have any affected relatives. Mutations in genes linked to breast cancer, including breast cancer antigen 1 and 2 (BRCA1 and 2), are also responsible for the inherited predisposition to breast cancer (Brewer et al., 2017).

DOI: 10.1201/9781003394693-10

There is a correlation between estrogen and the risk of breast cancer. Female reproductive factors such as attainment of menarche at an early age, late menopause, first pregnancy at a delayed age and low parity, which increases the exposure of endogenous estrogen, can increase breast cancer risk. A 1-year postponement of menopause results in a 3% rise in the risk (Dall & Britt, 2017; Washbrook, 2006).

Hormone replacement treatment (HRT) and oral hormonal contraceptives are the primary sources of exogenously administered estrogen. The risk of getting breast cancer is 1.5 times more with use of oral contraceptives (Soroush et al., 2016). Estrogen and other hormones are administered to menopausal or post-menopausal women as part of HRT. The use of HRT is found to be associated with increased incidence of breast cancer, according to several studies (Liu et al., 2016; Narod, 2011).

Modern lifestyle factors that raise the risk of breast cancer include drinking large amounts of alcohol and eating too much fat-containing food. Drinking alcohol can activate the estrogen receptor pathways, leading to cellular proliferation in the breast. A meta-analysis revealed that consuming more than 35 to 44 grams of alcohol daily can result in a 32% increased risk of breast cancer (Hamajima et al., 2002). Research studies indicate the positive association between smoking and breast cancer, particularly in young individuals (Catsburg et al., 2015; Kispert & McHowat, 2017).

8.3 BREAST CANCER: TYPES

Based on histology, breast cancer is broadly divided into *in situ* and invasive disease types. *In situ* comprises both ductal and lobular carcinomas. As per the World Health Organization (WHO) classification, invasive breast carcinoma has more than 21 subtypes, the most common of which is invasive carcinoma of no special type, which accounts for 40–75% of the cases. Some of the tumors with distinct morphology are categorized as special types that include invasive lobular, mucinous, tubular, metaplastic and cancers with apocrine, medullary and neuroendocrine characteristics (Vuong et al., 2014).

Based on immunohistochemistry (IHC) analysis, breast tumors are categorized as estrogen receptor (ER) positive (+)/negative (–), progesterone receptor (PR) positive (+)/negative (–) and human epidermal growth factor receptor 2 (HER-2) positive (+)/negative (–) (Vuong et al., 2014). The ER, PR and HER-2 assays are performed for all types of invasive carcinomas to decide on chemotherapy.

Other molecular classifications based on gene expression studies are not widely used in the management of breast tumors.

8.4 PATHOGENESIS OF BREAST CANCER

8.4.1 Breast cancer: Molecular Genetics

Germ-line mutations in genes such as BRCA1 and 2, tumor protein 53 (p53), and phosphatase and tensin (PTEN) cause 5–10% of all breast cancers and increase an individual's risk of acquiring hereditary breast cancer (Dumitrescu & Cotarla, 2005). Patients who are positive for BRCA genes 1 and 2 have an 80% chance of getting breast cancer, in the premenopausal age (Russell et al., 2000).

The polymorphisms in breast-cancer susceptibility genes when combined with exogenous lifestyle risk factors, as well as endogenous factors like estrogen and progesterone exposures, contribute to the pathogenesis of breast cancer (Weinberg et al., 2005). Most sporadic breast cancers account for more than 90% of all breast cancers in women. Familial malignancies are more common in younger individuals, more frequently bilateral, and have a worse outcome than sporadic occurrences (Russell et al., 2000).

8.4.2 HER-2/neu

HER-2/neu is a growth factor that is overexpressed in a variety of human malignancies, including breast, ovarian, lung and gastric cancers. The HER-2/neu gene was overexpressed in more than

20% of invasive breast tumors in 1987, and it was also linked to a poorer result and shorter survival (Ganecberg et al., 2000).

8.4.3 Estrogen and Progesterone

Through membrane receptors, steroid hormones such as estrogen, androgen and progesterone can rapidly regulate cell activity. Estrogen and progesterone play roles in the proliferation, differentiation, and invasion of cancer stem cells (CSCs), which have the capacity of self-renewal, initiate tumors, and contribute to tumor formation and recurrence. Cell-signaling transduction and the tumor microenvironment (TME) dynamically modulate their phenotype. Fibroblasts, macrophages and mesenchymal stem cells release interleukins 6 and 8, and chemokine ligand 7 in the TME, which promotes CSC self-renewal. Furthermore, the TME contains stromal cells that release cytokines, growth hormones, and proteases such as vascular endothelial growth factor, platelet-derived growth factor and matrix metalloproteinases (MMPs), which aid in tumor invasion and metastasis. Malignancies of the breast are especially sensitive to estrogen and progesterone via ERs and PRs, respectively, and alter CSC biology. Estrogen and progesterone also play a paracrine role in the regulation of CSC populations (Chen et al., 2020).

8.4.4 Triple Negative Breast Cancer (TNBC)

Triple negative breast cancer (TNBC) that is negative for estrogen, progesterone and HER-2/neu receptors has been observed to arise in specific groups of patients, most of whom are from lower socioeconomic classes or poor neighborhoods. It accounts for up to 20% of breast cancers and cannot be treated. In addition, these tumors have higher relapse rates, rapid growth, and worse outcomes when compared to other types of breast cancers (Wang et al., 2013).

8.5 CURRENTLY AVAILABLE TREATMENT OPTIONS FOR BREAST CANCER

The management of breast carcinoma involves a multidisciplinary approach combining surgery, neoadjuvant and adjuvant chemotherapy and other treatment methods including radiation and hormonal therapy. The treatment approach depends on various factors such as stage of cancer, metastasis, status of the hormone receptors (ER/PR) and the patient's overall health status.

8.5.1 Surgery

The type of surgery for breast cancer can be either lumpectomy i.e., excision of the tumor along with surrounding breast tissue or mastectomy i.e., removal of the entire breast, with or without axillary lymph node dissection, depending on the stage and spread of the tumor (Cardoso et al., 2019).

8.5.2 Chemotherapy and Hormonal Therapy

Chemotherapy may be needed in some patients even before the surgery to reduce the size of the tumor, which facilitates excision of the tumor completely. This type of chemotherapy is called neoadjuvant chemotherapy. Adjuvant chemotherapy i.e., chemotherapy post-surgery may be required in some patients depending on the type and stage of cancer combined with or without radiation therapy (Cardoso et al., 2019).

Patients with ER/PR-positive breast cancers should receive neoadjuvant and/or adjuvant therapy with drugs like tamoxifen (in pre- and post-menopausal females) and/or steroidal aromatase inhibitors like exemestane or non-steroidal aromatase inhibitors like anastrozole or letrozole (in postmenopausal females). Endocrine therapy is usually given for a period of 5 to 10 years to prevent

TABLE 8.1
Pharmacological Summary of Drugs Used in Breast Cancer

Type of Therapy	Drug (s)	Mechanism of Action	Common Adverse Effects
Chemotherapy	Alkylating agents • Platinum compounds (Cisplatin) • Cyclophosphamide	Binds with DNA irreversibly and interferes with its normal function, leading to cell death	• Bone-marrow suppression • Nausea and vomiting
	Anthracyclines • Adriamycin (doxorubicin) • Epirubicin	Blocks DNA replication and generates free radicals, which damage cancer cells	• Myelosuppression • Cardiotoxicity
	Taxanes • Paclitaxel • Docetaxel	Microtubule inhibitors, prevent cell division	• Myelosuppression • Hypersensitivity reactions
Endocrine Therapy	Aromatase inhibitors • Letrozole • Anastrozole • Exemestane	Inhibits aromatase enzyme, blocks conversion of androstenedione to estrone and testosterone to estradiol	• Hormonal side effects like hot flashes • Osteoporosis and bone fractures
	Selective estrogen receptor modulators • Tamoxifen • Raloxifene	Antagonists to estrogen receptors in breasts and prevent estrogen-induced cell proliferation in breast tissue	• Hot flashes • Thromboembolism • Risk for endometrial cancer
Immunotherapy	HER-2 targeted monoclonal antibodies • Pertuzumab • Trastuzumab	Binds to HER-2 receptor and interferes with the downstream signaling pathways	• Alopecia • Myelosuppression
	HER-2 tyrosine kinase inhibitor • Neratinib	Tyrosine kinase inhibitor of the HER-2 receptor and interferes with the downstream signaling pathways	• Fatigue • Skin rash • Abdominal pain

cancer recurrence. For HER-2 positive cancers, HER-2 targeted drugs like trastuzumab or pertuzumab are given (Cardoso et al., 2019).

For patients with TNBCs, the drugs used for neoadjuvant and/or adjuvant chemotherapy include doxorubicin, taxanes like paclitaxel, cyclophosphamide, methotrexate, 5-fluorouracil, pembrolizumab, olaparib, capecitabine, gemcitabine, vinorelbine and so on. Metastatic breast cancer is managed with palliative therapy (Won & Spruck, 2020).

A summary of the pharmacology of drugs commonly used in breast cancer is given in Table 8.1.

8.6 SERINE PROTEASES AND THEIR ROLE IN CANCER

The TME is composed of a diverse population of tumor cells and stromal cells, as well as the extracellular matrix (ECM). The structural support to tumor tissue and many cellular signaling pathways is regulated by ECM (Kim et al., 2011). As a major component of TME, tumor stroma actively promotes growth, angiogenesis, invasion and metastasis. The expression and secretion of numerous signaling molecules including cytokines and chemokines, growth factors and proteolytic enzymes regulate the activity of stromal cells and modulate their interaction with cancer cells inside the TME (De Groot et al., 2017).

Although once assumed to be restricted to ECM degradation, the involvement of proteases in malignancies is now known to be far more nuanced and crucial. In addition to cancer cells, the cells present in the TME secrete protease enzymes such as MMPs, cathepsins, and serine proteases contribute to tumor development. Proteases are engaged in ECM remodeling, growth factor and cytokine signaling and inflammatory and immunosuppressive responses. Depending on the biological setting and the type of tissue, proteases can either promote or inhibit cancer growth. In humans, serine proteases, which account for around one-third of proteases, play a vital role in carcinogenesis, with dysregulated expression of some of the serine proteases related with tumor formation and metastasis (Tagirasa & Yoo, 2022).

The process known as epithelial-mesenchymal transition (EMT) involves the reversible transformation of organized epithelial cells. During this process, these cells lose their polarity, reduce their adhesion to neighboring cells, and undergo cytoskeletal reorganization. Simultaneously, they gain mesenchymal characteristics, including a fibroblast-like morphology and the ability to invade surrounding tissues. Cells undergoing EMT exhibit a decrease in the activity of epithelial markers, including E-cadherin and specific cytokeratins. Conversely, there is an upregulation of a few markers of mesenchymal cells including vimentin, N-cadherin, fibronectin and integrins (Lu & Kang, 2019; Yang et al., 2020).

EMT is known to play pivotal roles in pathological processes, including cancer progression and organ fibrosis (Lu & Kang, 2019). EMT is initiated by paracrine signaling that originates from the reactive stroma within the tumor, which includes transforming growth factor-β, WNT proteins, hypoxia, cytokines and ECM-integrin adhesions (Dongre & Weinberg, 2019). As a result, these substances stimulate the functional activation of transcription factors that cause EMT. These transcription factors, known as EMT-inducing transcription factors (EMT-TFs), have the ability to coordinate various transcriptional alterations involved with EMT. The primary category of EMT-TFs comprises zinc-finger-binding transcription factors, specifically the SNAIL family, the TWIST family and the zinc-finger E-box-binding homeobox factors. Additional transcription factors, namely Forkhead box C and paired-related homeobox 1, have the ability to initiate EMT (Lambert & Weinberg, 2021). The aforementioned factors exhibit unique expression profiles and target gene patterns that vary based on the specific cell and tissue type.

Proteases play a role in the initiation of EMT. For instance, MMP-3 induces EMT by elevating the intracellular concentrations of reactive oxygen species (ROS), thus promoting the overexpression of SNAIL. The induction of EMT is facilitated by the stimulation of transforming growth factor-beta signaling by MMP-14 and MMP-8. The increase of interleukin-6 results in the upregulation of metalloprotease (ADAM) 9 expression, which subsequently leads to the generation of ROS and the initiation of EMT. The enzyme ADAM10 is responsible for cleaving the ectodomain of E-cadherin, hence influencing several cellular processes such as cell migration, β-catenin translocation and signaling and EMT (Mitschke et al., 2019). The induction of EMT in colon and hepatocellular carcinoma cells is facilitated by transmembrane protease, serine (TMPRSS) 4, concomitant with the overexpression of ZEB2, SNAIL and SLUG (Jung et al., 2008; Wang et al., 2015). The induction of EMT in gastric cancer cells is facilitated by the action of TMPRSS3, which regulates the signaling pathways of ERK1/2 and AKT. The induction of EMT and metastasic invasion in a prostate cancer model is attributed to TMPRSS2, which likely activates prohepatocyte growth factor (HGF) and subsequently triggers c-Met signaling (Li et al., 2018; Lucas et al., 2014).

In humans, the class of serine proteases has 175 members and, among those proteases, most are secreted proteases (Puente et al., 2005). A subset of serine proteases is tightly bound to cell membranes, called cell-surface anchored serine proteases. Based on the type of linkage to the cell membrane, the serine proteases are categorized as type I serine proteases (attached to the cell membrane via a carboxy-terminal transmembrane domain), type II serine proteases (attached to cell membrane via an amino-terminal) and GPI-linked proteases (attached to the cell membrane via a glycosyl-phosphatidylinositol linkage) (Caughey, 2007).

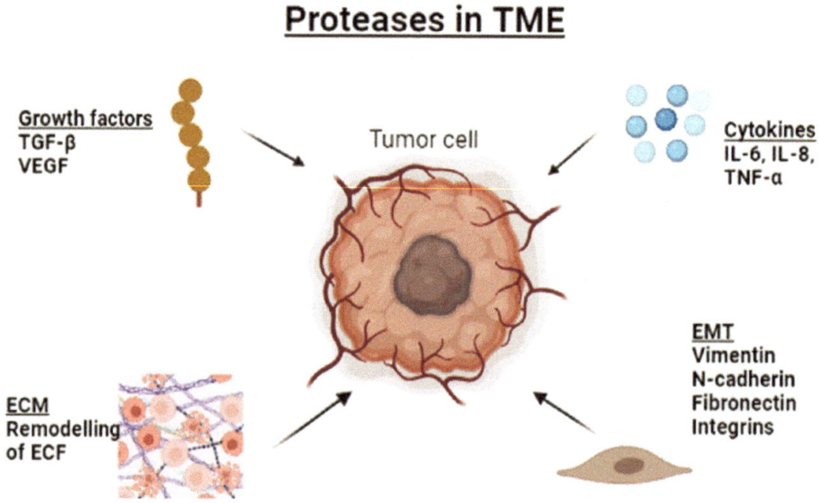

FIGURE 8.1 Role of proteases in carcinogenesis.

Matriptase, prostatin, TMPRSS, and heptin are cell surface-associated serine proteases that are increasingly expressed in the cancers of skin, head and neck, breast cancer, colorectal, pancreatic, prostate, endometrial, cervix and ovarian cancers and hematological malignancies (Martin & List, 2019).

The role of proteases in carcinogenesis is depicted in Figure 8.1.

8.7 SERINE PROTEASES AND BREAST CANCER

Several serine proteases are involved in the pathogenesis of breast cancer as follows:

8.7.1 Matriptase

Matriptase was the first identified novel serine protease associated with cancer in 1993. This enzyme exhibited gelatinolytic activity and was found to be expressed in cultured human breast cancer cells. Elevated matriptase expression in breast carcinomas exhibits a positive correlation with both tumor grade and stage, hence indicating a potential association between matriptase expression levels and disease severity (Siddiqui et al., 2005). Moreover, the presence of elevated matriptase expression serves as a predictor factor for unfavorable survival outcomes in affected individuals. The expression of the matriptase inhibitors, HAI-1 and 2, are notably reduced in breast cancers that are poorly differentiated. Additionally, there is an inverse correlation between the expression of HAI-2 and both nodal involvement and tumor propagation (Parr et al., 2004). In a study examining the long-term survival of breast cancer patients without lymph node involvement, it was shown that the presence of elevated levels of both matriptase and c-Met, the receptor for hepatocyte growth factor (HGF), was strongly correlated with a diminished likelihood of remaining disease-free. The involvement of matriptase in mammary carcinogenesis was demonstrated through the use of genetic mouse models. It was found that one of the molecular mechanisms by which matriptase promotes carcinogenesis is through the activation of pro-HGF on the surface of cancer cells. This activation leads to the initiation of the c-Met signaling pathway, resulting in the stimulation of mitogenic and invasive responses in breast cancer. The study utilized mice with reduced amounts of matriptase protein in their mammary glands, resulting in an approximate 75% decrease in matriptase expression. In the mouse model of breast carcinoma, it was observed that matriptase hypomorphic animals exhibited a notable delay in the initiation of tumors. Additionally, these mice demonstrated a reduction in tumor burden, which was attributed to the inhibition of tumor cell proliferation (Zoratti et al., 2015). To conduct mechanistic investigations, primary mammary cancer cells were created using tamoxifen-inducible Cre-loxP recombination, resulting in genetic disruption of the matriptase expressing gene. Cells lacking matriptase exhibited a diminished capacity to trigger the activation of the c-Met signaling pathway upon exposure to pro-HGF produced by fibroblasts. The proliferation and metastasis of inflammatory breast cancer (IBC) cell lines and non-IBC human TNBC cell lines are also mediated by the matriptase/c-Met signaling axis (Zoratti et al., 2016).

The substrate of matriptase that has been found to play a role in breast cancer cell migration and survival *in vitro* is the platelet-derived growth factor-C (PDGF-C). The MCF-7 luminal breast cancer cells were genetically modified to enhance the expression of PDGF-C, resulting in the production of proteases that had the ability to cleave PDGF-C into its biologically active form. The upregulation of PDGF-C expression resulted in the augmentation of cellular proliferation, anchorage-independent growth, and motility of tumor cells through autocrine signaling. The primary protease responsible for the digestion of PDGF-C in MCF-7 cells was identified as matriptase (Hurst et al., 2012).

8.7.2 TMPRSS6

The gene encoding matriptase-2, also known as TMPRSS6, exhibits significant expression levels in healthy mammary tissue, primarily localized in the epithelial cells. The levels of matriptase-2 protein in breast carcinomas exhibited a negative correlation with tumor grade, whereby higher tumor grades were associated with lower levels of matriptase-2. Particularly, undifferentiated ductal or lobular tumors displayed notably low levels of matriptase-2. There was a significant association observed between decreased levels of matriptase-2 in breast cancer tissues and an unfavorable prognosis. *In vitro* experiments were conducted to investigate the impact of stable expression of matriptase-2 in the MDA-MB-231 breast cancer cell line, which lacks endogenous matriptase-2 expression. The results revealed a significant decrease in both cell invasion and migration. In addition, it was shown that the introduction of matriptase-2 expressing cells into the subcutaneous tissue of nude mice resulted in a notable inhibition of tumor growth (Parr et al., 2007). The study examined 13 specific single nucleotide polymorphisms within the TMPRSS6 gene in relation to TNBC. The findings revealed that four of these variants were linked to decreased production of matriptase-2 and were related with worse survival outcomes (Tuhkanen et al., 2013).

8.7.3 Hepsin

The expression of hepsin is significantly higher in breast cancer cells compared to surrounding non-malignant breast cells. The expression of hepsin has also been observed to exhibit a favorable correlation with both the stage of the tumor and the occurrence of lymph node metastases. The upregulation of hepsin in mammary epithelial cells was found to be correlated with a decrease in the expression of HAI-1 and an increase in HGF/c-Met signaling, leading to the degradation of desmosomes and hemidesmosomes (Xing et al., 2011). Hepsin plays a crucial role in enhancing the ability of breast cancer cells to invade surrounding tissues by modifying the structure of the basement membrane. This modification is achieved through laminin-332 cleavage, a key constituent of hemidesmosomes located at the junctions between adjacent cells. The reduction of hepsin activity through the use of a specific inhibitor or the suppression of its expression via siRNA-mediated silencing resulted in a decrease in desmosome cleavage and hindered the multiplication and invasive nature of breast cancer cells cultivated *in vitro* (Pant et al., 2018).

8.7.4 TMPRSS3

In a study utilizing IHC, a comparison was made between tissue samples from breast cancer patients and healthy breast tissue. The findings revealed a statistically significant increase in the expression of TMPRSS3 in the malignant tissue. There was a significant correlation seen between the expression level of TMPRSS3 and various clinical parameters, including disease stage, lymph node positive and proliferation of the cancer cells. As a result, the upregulation of TMPRSS3 was associated with decreased rates of disease-free and overall survival (Rui et al., 2015). It was discovered that TMPRSS3 exhibits a favorable correlation with the occurrence of distant organ metastasis in cases of breast cancer. The study reported a diminished expression of TMPRSS3 in breast cancer samples, particularly in poorly differentiated tumors. A significant correlation was observed between low levels of TMPRSS3 expression and decreased overall survival rates (Pelkonen et al., 2015).

8.7.5 Prostatin

In a study, it was observed that the expression of prostasin was found to be absent in breast cancer cell lines that exhibited high invasiveness and metastatic potential. However, normal breast epithelial cells and breast cancer cells with minimal invasiveness were found to express this protease. The *in vitro* invasiveness of MDA-MB-231 and MDA-MB-435 breast cancer cell lines, which lack endogenous prostasin expression, was decreased upon transgenic expression of prostasin. The loss of prostasin expression in breast cancer cell lines can potentially be attributed to the methylation occurring in the promoter region of the prostasin gene (Bergum et al., 2012).

8.7.6 Other Serine Proteases

In silico analysis revealed the transmembrane serine proteases like TMPRSS13 and TMPRSS4 are associated with the invasive type of ductal carcinoma of the breast (Kim, 2023; Murray et al., 2020). An increase in the expression of TMPRSS13 could positively correlate with the tumor progression, overall tumor burden and growth of mammary tumors. Also, targeting TMPRSS13 makes the TNBC respond well to chemotherapy (Murray et al., 2020).

8.8 TARGETING SERINE PROTEASES FOR TREATMENT OF BREAST CANCER

Kempopeptin C, a novel serine protease inhibitor, was isolated from marine cyanobacteria. The antiproteolytic efficacy of kempopeptin C was assessed against trypsin and matriptase, resulting in the inhibition of these enzymes. Given the importance of both proteases in the advancement and spread of cancer, along with their functional overlap in terms of targeting similar substrates, the impact of kempopeptin C on the subsequent cellular substrates of matriptase, namely CDCP1 and desmoglein-2 (Dsg-2), was evaluated. *In vitro* experiments demonstrated the inhibitory effect of kempopeptin C on the cleavage of both substrates. The application of kempopeptin C resulted in a decrease in the cleavage of CDCP1 in MDA-MB-231 cells, with a maximum reduction observed at a concentration of 10 μM. The migratory behavior of MDA-MB-231 breast cancer cells was seen to be reduced by 37% and 60% at concentrations of 10 μM and 20 μM of kempopeptin C, respectively (Al-Awadhi et al., 2017).

Serine protease enzyme PRSS23 is a recently identified protein that has been linked to tumor growth in a variety of malignancies. PRSS23 is co-expressed with ER, and it is an important biomarker and drug target in human breast cancer. Estrogen signaling is also known to influence cell multiplication, cell death and survival, promoting cancer by controlling the synthesis of multiple downstream effector proteins (Chan et al., 2012).

A study aimed to determine the relationship of ER and PRSS23 in breast cancer found that the gene expression association between ER and PRSS23 is extremely significant among all other proteases in breast cancer. The levels of PRSS23 were then examined in primary breast cancer biopsies and breast cancer cell lines and the findings validated the relationship between PRSS23 and ER in breast cancer. *In vitro* experiments in MCF-7 breast cancer cells revealed that 17-estradiol-activated ER induces PRSS23 expression by interacting with the promoter region of the PRSS23 gene (Chan et al., 2012).

Proteases mediate carcinogenesis by degrading the ECM. Some of the serine protease inhibitors (serpins) are paradoxically overexpressed in human malignancies. It was discovered that the level of protease nexin-1 (PN-1), a serine protease inhibitor that inhibits the activity of several proteases, is much higher in ER-negative and high-grade breast cancer. This finding was supported by mechanistic studies on two breast cancer cell lines, PN-1 (–) 168FARN cells and PN-1(+) 4T1 cells, from primary breast tumors and 4T1 cancers that can metastasize to the lungs. It was observed that PN-1 administration increased extracellular signal-regulated kinase activation via low-density lipoprotein receptor-related protein-1 (LRP-1) binding, resulting in enhanced MMP-9 RNA proteins. MMP-9 levels are low in PN-1-silenced 4T1 cells. Upon injecting PN-1-silenced cells into mice, there was no effect on the formation of 4T1 primary mammary tumors; nevertheless, the tumors showed poor metastatic capability. An examination of 126 breast cancer patients revealed that individuals with raised PN-1 levels in their tumors had a considerably higher risk of developing lung metastasis on recurrence, but not metastasis to other locations (Fayard et al., 2009).

Maspin is a protease inhibitor related to plasminogen activator 1 and 2 and α-1-antitrypsin. Maspin was present in normal epithelial cells of the breast but was reduced and downregulated in breast cancer cells. The recombinant form of maspin inhibited the invasive phenotype of mammary cancer cells (MDA-MB 435). The treated breast cancer cells change their phenotype and morphology into a more benign and less invasive form. Maspin inhibits angiogenesis and invasion via interaction with the p53 tumor suppressor pathway (Maass et al., 2000).

Xioa et al. used differential cDNA sequencing to identify a serpin, a serine protease inhibitor, in normal mammary tissue. This serpin was discovered in myoepithelial cells of the breast on the epithelial side of the basement membrane by *in situ* hybridization and was thus called myoepithelium derived serine proteinase inhibitor (MEPI). MEPI cDNA was transfected into human breast cancer cells to test if MEPI could regulate the *in vivo* growth and progression of human breast malignancies, and the growth of tumor cells in MEPI-transfected as compared to control clones in the fat pad of mice breast was studied. MEPI overexpression decreased cell invasion in an *in vitro* experiment. When MEPI-transfected clones were injected into the athymic mice, the breast tumor sizes, axillary lymph node involvement and lung metastasis were considerably reduced compared to controls (Xiao et al., 1999).

Table 8.2 summarizes a few serine proteases, which are potential targets in breast cancer therapy and their *in vitro* and/or *in vivo* effects on breast cancer cells.

8.9 CONCLUSION AND FUTURE PERSPECTIVES

The currently employed treatment options for breast carcinoma including surgery, endocrine and chemotherapy are not devoid of serious adverse effects. The importance of targeted therapy for cancer has been considered recently for various cancers. Advancement in molecular profiling methods has led to the identification of several predictive biomarkers/proteins involved in cancer pathways. The discovery of novel biomarkers that are involved in breast cancer formation and progression would enable the development of drugs that target these proteins/genes and interfere with the downstream signaling pathways, thereby limiting the tumor growth and invasion. Serine proteases are among the potential biomarkers involved in breast cancer development, progression and metastasis. Several serine proteases are involved as discussed in this chapter. Targeting these proteases as a treatment modality for breast cancer has yielded promising results in the basic *in silico*, *in vitro* and preclinical research. Hence, targeted therapy against serine proteases could evolve into a potential therapeutic option for breast cancer in the future.

TABLE 8.2
Serine Protease Inhibitors for Breast Cancer

Target Serine Protease/Compound Inhibiting the Protease	*In vitro/In vivo* Effect on Breast Cancer Cells
Matriptase—inhibited by kempopeptin C	Decreases migration of the cancer cells and reduces the invasive nature (Al-Awadhi et al., 2017).
PRSS23	Inhibits breast cancer cell proliferation induced by estrogen (Chan et al., 2012).
Several proteases- inhibited by protease nexin-1	Decreases the levels of MMP-9 and prevents metastasis (Fayard et al., 2009).
Maspin—recombinant form (with reduced expression)	Inhibits angiogenesis and invasion by interacting with p53 tumor suppressor pathway (Maass et al., 2000).
Myoepithelium-derived serine protease inhibitor	Reduction in tumor size and tumor metastasis to axillary lymph nodes and lungs (Xiao et al., 1999).

DECLARATIONS

Conflict of interest: The authors declare no conflicts of interest.

Consent for publishing: All authors have approved the final version of the manuscript and given consent for publication.

Funding: Not Applicable.

REFERENCES

Al-Awadhi, F. H., Salvador, L. A., Law, B. K., Paul, V. J., & Luesch, H. (2017). Kempopeptin C, a novel marine-derived serine protease inhibitor targeting invasive breast cancer. *Marine Drugs*, *15*(9), 290.

Benson, J. R., & Jatoi, I. (2012). The global breast cancer burden. *Future Oncology*, *8*(6), 697–702.

Bergum, C., Zoratti, G., Boerner, J., & List, K. (2012). Strong expression association between matriptase and its substrate prostasin in breast cancer. *Journal of Cellular Physiology*, *227*(4), 1604–1609.

Brewer, H. R., Jones, M. E., Schoemaker, M. J., Ashworth, A., & Swerdlow, A. J. (2017). Family history and risk of breast cancer: an analysis accounting for family structure. *Breast Cancer Research and Treatment*, *165*, 193–200.

Cardoso, F., Kyriakides, S., Ohno, S., Penault-Llorca, F., Poortmans, P., Rubio, I., Zackrisson, S., & Senkus, E. (2019). Early breast cancer: ESMO Clinical Practice Guidelines for diagnosis, treatment and follow-up. *Annals of Oncology*, *30*(8), 1194–1220.

Catsburg, C., Miller, A. B., & Rohan, T. E. (2015). Active cigarette smoking and risk of breast cancer. *International Journal of Cancer*, *136*(9), 2204–2209.

Caughey, G. H. (2007). Mast cell tryptases and chymases in inflammation and host defense. *Immunological Reviews*, *217*(1), 141–154.

Chan, H.-S., Chang, S.-J., Wang, T.-Y., Ko, H.-J., Lin, Y.-C., Lin, K.-T., Chang, K.-M., & Chuang, Y.-J. (2012). Serine protease PRSS23 is upregulated by estrogen receptor α and associated with proliferation of breast cancer cells. *PLoS One*, *7*(1), e30397.

Chen, B., Ye, P., Chen, Y., Liu, T., Cha, J.-H., Yan, X., & Yang, W.-H. (2020). Involvement of the estrogen and progesterone axis in cancer stemness: elucidating molecular mechanisms and clinical significance. *Frontiers in Oncology*, *10*, 1657.

Dall, G. V., & Britt, K. L. (2017). Estrogen effects on the mammary gland in early and late life and breast cancer risk. *Frontiers in Oncology*, *7*, 110.

De Groot, A. E., Roy, S., Brown, J. S., Pienta, K. J., & Amend, S. R. (2017). Revisiting seed and soil: examining the primary tumor and cancer cell foraging in metastasis. *Molecular Cancer Research*, *15*(4), 361–370.

DeSantis, C. E., Fedewa, S. A., Goding Sauer, A., Kramer, J. L., Smith, R. A., & Jemal, A. (2016). Breast cancer statistics, 2015: convergence of incidence rates between black and white women. *CA: A Cancer Journal for Clinicians*, *66*(1), 31–42.

Dongre, A., & Weinberg, R. A. (2019). New insights into the mechanisms of epithelial–mesenchymal transition and implications for cancer. *Nature Reviews Molecular Cell Biology*, *20*(2), 69–84.

Dumitrescu, R., & Cotarla, I. (2005). Understanding breast cancer risk-where do we stand in 2005? *Journal of Cellular and Molecular Medicine*, *9*(1), 208–221.

Fayard, B., Bianchi, F., Dey, J., Moreno, E., Djaffer, S., Hynes, N. E., & Monard, D. (2009). The serine protease inhibitor protease nexin-1 controls mammary cancer metastasis through LRP-1–mediated MMP-9 expression. *Cancer Research*, *69*(14), 5690–5698.

Ganeċberg, D., Lespagnard, L., Rouas, G., Paesmans, M., Piccart, M., & Dileo, A. (2000). Sensitivity of HER-2/neu antibodies in archival tissue samples of invasive breast carcinomas. *American Journal of Clinical Pathology*, *113*, 675–682.

Hamajima, N., Hirose, K., Tajima, K., Rohan, T., Calle, E., Heath Jr, C., Coates, R., Liff, J., Talamini, R., & Chantarakul, N. (2002). Collaborative Group on Hormonal Factors in Breast Cancer Alcohol, tobacco and breast cancer–collaborative reanalysis of individual data from 53 epidemiological studies, including 58,515 women with breast cancer and 95,067 women without the disease. *British Journal of Cancer*, *87*(11), 1234–1245.

Hurst Jr., N. J., Najy, A. J., Ustach, C. V., Movilla, L., & Kim, H.-R. C. (2012). Platelet-derived growth factor-C (PDGF-C) activation by serine proteases: implications for breast cancer progression. *Biochemical Journal*, *441*(3), 909–919.

Jung, H., Lee, K., Park, S., Park, J., Jang, Y., Choi, S., Jung, J., Jo, K., Park, D., & Yoon, J. (2008). TMPRSS4 promotes invasion, migration and metastasis of human tumor cells by facilitating an epithelial–mesenchymal transition. *Oncogene*, *27*(18), 2635–2647.

Kim, S. (2023). TMPRSS4, a type II transmembrane serine protease, as a potential therapeutic target in cancer. *Experimental & Molecular Medicine*, *55*(4), 716–724.

Kim, S.-H., Turnbull, J., & Guimond, S. (2011). Extracellular matrix and cell signalling: the dynamic cooperation of integrin, proteoglycan and growth factor receptor. *The Journal of endocrinology*, *209*(2), 139–151.

Kispert, S., & McHowat, J. (2017). Recent insights into cigarette smoking as a lifestyle risk factor for breast cancer. *Breast Cancer: Targets and Therapy*, 127–132.

Lambert, A. W., & Weinberg, R. A. (2021). Linking EMT programmes to normal and neoplastic epithelial stem cells. *Nature Reviews Cancer, 21*(5), 325–338.

Li, S.-L., Chen, X., Wu, T., Zhang, X.-W., Li, H., Zhang, Y., & Ji, Z.-Z. (2018). Knockdown of TMPRSS3 inhibits gastric cancer cell proliferation, invasion and EMT via regulation of the ERK1/2 and PI3K/Akt pathways. *Biomedicine & Pharmacotherapy*, *107*, 841–848.

Liu, J.-Y., Chen, T.-J., & Hwang, S.-J. (2016). The risk of breast cancer in women using menopausal hormone replacement therapy in Taiwan. *International Journal of Environmental Research and Public Health*, *13*(5), 482.

Lu, W., & Kang, Y. (2019). Epithelial-mesenchymal plasticity in cancer progression and metastasis. *Developmental Cell*, *49*(3), 361–374.

Lucas, J. M., Heinlein, C., Kim, T., Hernandez, S. A., Malik, M. S., True, L. D., Morrissey, C., Corey, E., Montgomery, B., & Mostaghel, E. (2014). The androgen-regulated protease TMPRSS2 activates a proteolytic cascade involving components of the tumor microenvironment and promotes prostate cancer metastasis. *Cancer Discovery*, *4*(11), 1310–1325.

Maass, N., Hojo, T., Zhang, M., Sager, R., Jonat, W., & Nagasaki, K. (2000). Maspin-a novel protease inhibitor with tumor-suppressing activity in breast cancer. *Acta Oncologica*, *39*(8), 931–934.

Majeed, W., Aslam, B., Javed, I., Khaliq, T., Muhammad, F., Ali, A., & Raza, A. (2014). Breast cancer: major risk factors and recent developments in treatment. *Asian Pacific Journal of Cancer Prevention*, *15*(8), 3353–3358.

Martin, C. E., & List, K. (2019). Cell surface–anchored serine proteases in cancer progression and metastasis. *Cancer and Metastasis Reviews*, *38*, 357–387.

Mitschke, J., Burk, U. C., & Reinheckel, T. (2019). The role of proteases in epithelial-to-mesenchymal cell transitions in cancer. *Cancer and Metastasis Reviews*, *38*(3), 431–444.

Murray, A. S., Hyland, T. E., Sala-Hamrick, K. E., Mackinder, J. R., Martin, C. E., Tanabe, L. M., Varela, F. A., & List, K. (2020). The cell-surface anchored serine protease TMPRSS13 promotes breast cancer progression and resistance to chemotherapy. *Oncogene*, *39*(41), 6421–6436.

Narod, S. A. (2011). Hormone replacement therapy and the risk of breast cancer. *Nature Reviews Clinical Oncology*, *8*(11), 669–676.

Pant, S. M., Belitskin, D., Ala-Hongisto, H., Klefström, J., & Tervonen, T. A. (2018). Analyzing the type II transmembrane serine protease hepsin-dependent basement membrane remodeling in 3D cell culture. Proteases and Cancer: Methods and Protocols, pp. 169–178. Humana Press.

Parr, C., Sanders, A. J., Davies, G., Martin, T., Lane, J., Mason, M. D., Mansel, R. E., & Jiang, W. G. (2007). Matriptase-2 inhibits breast tumor growth and invasion and correlates with favorable prognosis for breast cancer patients. *Clinical Cancer Research*, *13*(12), 3568–3576.

Parr, C., Watkins, G., Mansel, R. E., & Jiang, W. G. (2004). The hepatocyte growth factor regulatory factors in human breast cancer. *Clinical Cancer Research*, *10*(1), 202–211.

Pelkonen, M., Luostari, K., Tengström, M., Ahonen, H., Berdel, B., Kataja, V., Soini, Y., Kosma, V.-M., & Mannermaa, A. (2015). Low expression levels of hepsin and TMPRSS3 are associated with poor breast cancer survival. *BMC Cancer*, *15*(1), 1–15.

Puente, X., Sanchez, L., Gutierrez-Fernandez, A., Velasco, G., & Lopez-Otin, C. (2005). A genomic view of the complexity of mammalian proteolytic systems. *Biochemical Society Transactions*, *33*(2), 331–334.

Rui, X., Li, Y., Jin, F., & Li, F. (2015). TMPRSS3 is a novel poor prognostic factor for breast cancer. *International Journal of Clinical and Experimental Pathology*, *8*(5), 5435.

Russell, R., Bulstrode, C., & Bailey, N. W. (2000). *Chapter on Breast Cancer.* Arnold.

Siddiqui, S. F., Pawelek, J., Handerson, T., Lin, C.-Y., Dickson, R. B., Rimm, D. L., & Camp, R. L. (2005). Coexpression of β1, 6-N-acetylglucosaminyltransferase V glycoprotein substrates defines aggressive breast cancers with poor outcome. *Cancer Epidemiology Biomarkers & Prevention*, *14*(11), 2517–2523.

Siegel, R. L., Miller, K. D., & Jemal, A. (2018). Cancer statistics, 2018. *CA: A Cancer Journal for Clinicians*, *68*(1), 7–30.

Soroush, A., Farshchian, N., Komasi, S., Izadi, N., Amirifard, N., & Shahmohammadi, A. (2016). The role of oral contraceptive pills on increased risk of breast cancer in Iranian populations: a meta-analysis. *Journal of Cancer Prevention*, *21*(4), 294.

Tagirasa, R., & Yoo, E. (2022). Role of serine proteases at the tumor-stroma interface. *Frontiers in Immunology*, *13*, 832418.

Tuhkanen, H., Hartikainen, J. M., Soini, Y., Velasco, G., Sironen, R., Nykopp, T. K., Kataja, V., Eskelinen, M., Kosma, V. M., & Mannermaa, A. (2013). Matriptase-2 gene (TMPRSS6) variants associate with breast cancer survival, and reduced expression is related to triple-negative breast cancer. *International Journal of Cancer*, *133*(10), 2334–2340.

Vuong, D., Simpson, P. T., Green, B., Cummings, M. C., & Lakhani, S. R. (2014). Molecular classification of breast cancer. *Virchows Archiv*, *465*, 1–14.

Wang, C., Pécot, T., Zynger, D. L., Machiraju, R., Shapiro, C. L., & Huang, K. (2013). Identifying survival associated morphological features of triple negative breast cancer using multiple datasets. *Journal of the American Medical Informatics Association*, *20*(4), 680–687.

Wang, C.-H., Guo, Z.-Y., Chen, Z.-T., Zhi, X.-T., Li, D.-K., Dong, Z.-R., Chen, Z.-Q., Hu, S.-Y., & Li, T. (2015). TMPRSS4 facilitates epithelial-mesenchymal transition of hepatocellular carcinoma and is a predictive marker for poor prognosis of patients after curative resection. *Scientific Reports*, *5*(1), 12366.

Washbrook, E. (2006). Risk factors and epidemiology of breast cancer. *Women's Health Medicine*, *3*(1), 8–14.

Weinberg, O. K., Marquez-Garban, D. C., & Pietras, R. J. (2005). New approaches to reverse resistance to hormonal therapy in human breast cancer. *Drug Resistance Updates*, *8*(4), 219–233.

Won, K. A., & Spruck, C. (2020). Triple-negative breast cancer therapy: current and future perspectives. *International Journal of Oncology*, *57*(6), 1245–1261.

Xiao, G., Liu, Y. E., Gentz, R., Sang, Q. A., Ni, J., Goldberg, I. D., & Shi, Y. E. (1999). Suppression of breast cancer growth and metastasis by a serpin myoepithelium-derived serine proteinase inhibitor expressed in the mammary myoepithelial cells. *Proceedings of the National Academy of Sciences*, *96*(7), 3700–3705.

Xing, P., Li, J.-G., Jin, F., Zhao, T.-T., Liu, Q., Dong, H.-T., & Wei, X.-L. (2011). Clinical and biological significance of hepsin overexpression in breast cancer. *Journal of Investigative Medicine*, *59*(5), 803–810.

Yang, J., Antin, P., Berx, G., Blanpain, C., Brabletz, T., Bronner, M., Campbell, K., Cano, A., Casanova, J., & Christofori, G. (2020). Guidelines and definitions for research on epithelial–mesenchymal transition. *Nature reviews Molecular Cell Biology*, *21*(6), 341–352.

Zoratti, G. L., Tanabe, L. M., Hyland, T. E., Duhaime, M. J., Colombo, É., Leduc, R., Marsault, E., Johnson, M. D., Lin, C.-Y., & Boerner, J. (2016). Matriptase regulates c-Met mediated proliferation and invasion in inflammatory breast cancer. *Oncotarget*, *7*(36), 58162.

Zoratti, G. L., Tanabe, L. M., Varela, F. A., Murray, A. S., Bergum, C., Colombo, É., Lang, J. E., Molinolo, A. A., Leduc, R., & Marsault, E. (2015). Targeting matriptase in breast cancer abrogates tumour progression via impairment of stromal-epithelial growth factor signalling. *Nature Communications*, *6*(1), 6776.

9 Serine Proteases in Cancer

Mechanisms and Treatment Options

Madhu Tanya Singh, Ammu V. V. V. Ravi Kiran, Garikapati Kusuma Kumari, and Praveen T. Krishnamurthy
JSS College of Pharmacy, JSS Academy of Higher Education & Research, Ooty, India

9.1 INTRODUCTION

Cancer is a heterogeneous disease characterized by uncontrolled growth and proliferation of cells that can infiltrate surrounding tissues and spread to distant parts of the body (1). The process of cancer cell invasion involves the adherence of tumour cells to the underlying basement membrane, localized breakdown of proteins through proteolysis and subsequent migration through the region that has been enzymatically modified (2).

Proteases are enzymes that catalyse the hydrolysis of peptide bonds in proteins and play an important role in various biological processes, such as digestion, blood clotting, immune response and cell signalling. Ever since Theodor Schwann's pioneering discovery of the digestive enzyme pepsin nearly two centuries ago, it has become increasingly evident that proteases are integral components in all biological signalling pathways and networks, fulfilling numerous indispensable functions across all forms of life (3). Proteases constitute approximately 3% of the human genes responsible for encoding proteins. Among these, half are secreted proteases, while the remaining are intracellular proteases, with a smaller subset being membrane-bound (4). A minute fraction of these membrane-bound proteases can undergo shedding, i.e., they can be released from the cell surface and facilitate activation of other proteases. They are also involved in the development and progression of cancer. Proteases can degrade the extracellular matrix (ECM), which provides structural and biochemical support to the cells and tissues. By breaking down the ECM, they enable cancer cells to detach from the primary tumour site, migrate through the surrounding stroma and invade the blood vessels or lymphatics (1). Each class of proteases has specific substrates and inhibitors that modulate their activity and function. The imbalance between proteases and their inhibitors can result in excessive or insufficient proteolysis, which can disrupt the normal homeostasis of the tissue microenvironment and promote cancer development (5).

Serine proteases constitute up to one-third of all the proteases involved in normal physiological processes and play a pivotal role in both physiological and various pathological conditions, including cancer (6). The set of peptides involved in cancer progression is denoted as the cancer "degradome," with serine proteases being significant constituents associated with cancer pathology (7). Serine proteases can influence the interaction between tumour and stromal cells, remodelling of ECM, activation of growth factors and cytokines and modulation of immune responses (1). These proteases can originate from both cancer and stromal cells, such as endothelial cells, fibroblasts and immune cells. Notable examples of serine proteases implicated in cancer include fibroblast activation protein (FAP), urokinase-type plasminogen activator (uPA), kallikrein-related peptidases

DOI: 10.1201/9781003394693-11

(KLKs) and granzymes (8). The effects of these serine proteases can vary as either tumour-promoting or tumour-suppressing effects depending on the context. For instance, FAP, expressed by cancer-associated fibroblasts can degrade collagen and activate growth factors, thereby promoting tumour invasion and angiogenesis (9). Conversely, granzymes, secreted by natural killer (NK) cells and cytotoxic T cells can induce apoptosis in tumour cells (10). Therefore, serine proteases play a vital role in cancer biology and can be potential targets or biomarkers for cancer diagnosis and therapy (7). In this chapter, we focus mainly the role of proteases in cancer, with an emphasis on serine proteases and their inhibitors.

9.2 BRIEFING ON SERINE PROTEASES: PHYSIOLOGICAL AND PATHOLOGICAL ASPECTS

Almost one-third of the proteases are serine proteases with nucleophilic serine residue at their active catalytic site. These serine proteases are grouped into 13 clans and 40 families with the stem name of nucleophile "ser" obtained from its active catalytic site. Nucleophile serine within these serine proteases attacks the substrate peptide bond at the carbonyl moiety to form an acyl intermediate (11). The nucleophilicity of serine typically relies on a catalytic triad involving Asp102, His57 and Ser195 residues, often known as the charge relay system.

9.2.1 Serine Proteases in Normal Physiology

Active site serine residues of serine proteases, enable them to participate in essential physiological processes such as food digestion, blood clotting, inflammation, development, fertilization, apoptosis and the complement system (12). Enzyme cascades governing the activation of serine proteases have been identified in the complement system and blood coagulation of vertebrates (13). Additionally, these proteases possess immunomodulatory functions by interacting with the components of the complement system such as C1r, C1s and C3 convertases (C4b, 2a, C3b and Bb). Serine proteases such as kallikreins and KLKs play an important role in regulating physiological functions such as skin desquamation, enamel formation, immune response, semen liquefaction and blood pressure homeostasis (14).

9.2.2 Serine Proteases in Cancer Pathology

The actions of serine proteases are intricately regulated by an endogenous anti-serine protease system referred to as serpins during normal physiological processes. Dysregulation of secreted and membrane-anchored serine proteases leads to an imbalance in signalling cascades and cellular homeostasis contributing to many pathophysiological implications including cancer. Secreted proteases such as the kallikrein system and uPA are involved in numerous physiological functions such as cell growth, signalling and tissue remodelling. Membrane-anchored type II serine proteases (TTSP) constitute a vital component of the human degradome. They actively participate in converting precursors into their active forms within the pericellular microenvironment. The dysregulation of these serine proteases promotes tumour growth, angiogenesis, invasion and metastasis, thereby fostering an environment favourable for cancer progression. Moreover, the dysregulation of serine proteases is linked to the remodelling of tumour microenvironment and immune response modulation. The overexpression of proteases is often associated with the aggressiveness of the cancer.

Human physiology is characterized by the presence of a dynamic three-dimensional ECM comprising collagen, enzymes and glycoproteins. The constituents of ECM are synthesized by resident cells and subsequently released into ECM through exocytosis. The ECM not only furnishes neighbouring cells with essential structural integrity but also provides crucial biochemical support. Furthermore, the ECM is involved in cellular processes, encompassing cell adhesion, cell–cell communication and differentiation (15).

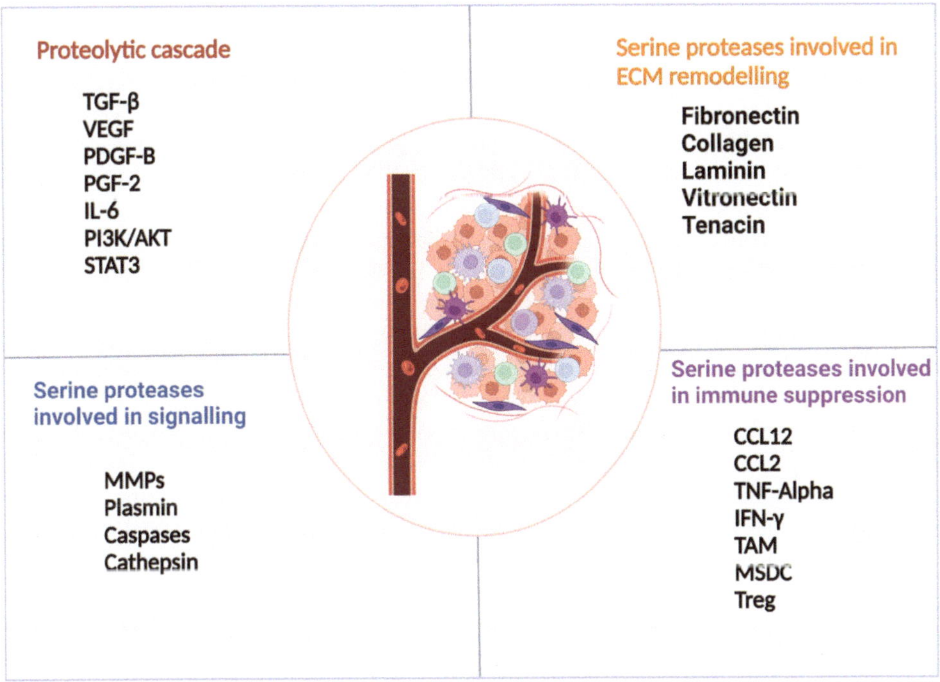

FIGURE 9.1 Various serine proteases involved in cancer pathology.

Serine proteases play a significant role in maintaining tissue homeostasis and activating signalling cascades that impact tissue integrity by degrading or disrupting ECM and basement membrane which are key players in tissue development, function, repair, and regeneration. Notably, a linear increase in the levels of serine proteases can be observed with the progression of cancer. This rise in serine proteases can cause cancer cells *in situ* to initiate migration into the surrounding stroma and further disseminate to distant organs, thereby causing the progression of carcinoma *in situ* to invasive and metastatic carcinoma (Figure 9.1).

Beyond their role in cellular migration, serine proteases also mediate the formation of the tumour microenvironment, acting as a catalyst for malignant transformation. Additionally, they increase the concentration of growth factors and adhesion molecules, thereby potentiating the overall tumour growth trajectory. Furthermore, these proteases exert a modulatory effect on immune reactions, potentially contributing to the immune evasion strategies employed by malignant cells (16). The complex interplay of serine proteases within the dynamic milieu of cancer pathophysiology underscores its roles as both tumour promoters as well as tumour suppressors, offering new avenues for understanding disease progression and fostering innovative therapeutic approaches.

9.3 VARIOUS SERINE PROTEASES AND THEIR ROLE IN CANCER

9.3.1 Granzymes

Granzymes constitute a unique family of serine proteases. Among these, granzyme A and B rank as the most prevalent granzymes and as a result, they have been subjects of extensive research concerning their role in cancer. Effector cells that do not possess granzyme B exhibit a significantly reduced rate of target-cell killing compared to wild-type cells. This highlights the crucial role played by granzyme B to facilitate the prompt elimination of tumourigenic cells (17). The efficiency of granzyme B-dependent killing primarily results from its capacity to activate the intrinsic cell death proteases within the target cell, namely, the caspases, either through direct or indirect means.

Granzyme B directly cleaves and activates caspases-3,7 and 8, initiating a cascade that results in caspase-mediated degradation of numerous cellular protein substrates. This process facilitates rapid and efficient apoptosis. The activation of caspase-3 is directly regulated by apoptotic protein inhibitor (18). Effective processing of pro-caspase-3 results in the cleavage of inhibitor of DNA-dependent protein kinase (DNA-PK), poly (ADP ribose) polymerase (PARP), nuclear-envelope intermediate-filament protein lamin B, caspase-activated DNase (ICAD), and nuclear mitotic apparatus protein (NuMA). These events lead to morphological changes associated with apoptosis, including fragmentation of DNA, condensation of chromatin, blebbing of membrane, shrinkage of cell and the formation of apoptotic bodies. The activation of the BH3-interacting domain death agonist (BID) pathway, either through direct action by GrB or indirectly through caspase-8, results in the fragmentation of mitochondrial BID, including BAX and BAK (19). The movement of truncated BID (tBID) to the mitochondria causes mitochondrial depolarization and the release of cytochrome c (cyt c) along with other pro-apoptotic molecules such as HtrA2/OMI and endonuclease G. The presence of cytochrome c in the cytosol is crucial for the formation of apoptosome, which involves the assembly of apoptotic protease activation factor (APAF-1). This assembly triggers the self-catalytic activation of pro-caspase-9 and the processing of caspase-3, thereby facilitating downstream caspase activation and the initiation of cell death. Apart from causing the loss of mitochondrial transmembrane potential, GrB can induce mitochondrial depolarization directly without the participation of cytosolic agents, although the precise mechanisms are not fully understood. GrB cleaves downstream aspartic or glutamic acid residues, which gives it a substrate specificity similar to that of caspases. Effector cells that lack granzyme B maintain their cytotoxic capability through the functions of the remaining granzymes, with granzyme A being the most prevalent among them. While granzyme A can induce target cell death through various pathways, the primary mechanism proposed for cell death by this granzyme involves the proteolysis of components associated with the endoplasmic reticulum-linked SET complex. There is limited evidence supporting granzyme C's involvement in promoting cell death, although it may exhibit cytotoxicity under specific circumstances (20). Researchers have demonstrated that metase (granzyme M) promotes a mechanism of cell death that operates independently of caspases and mitochondria (Figure 9.2).

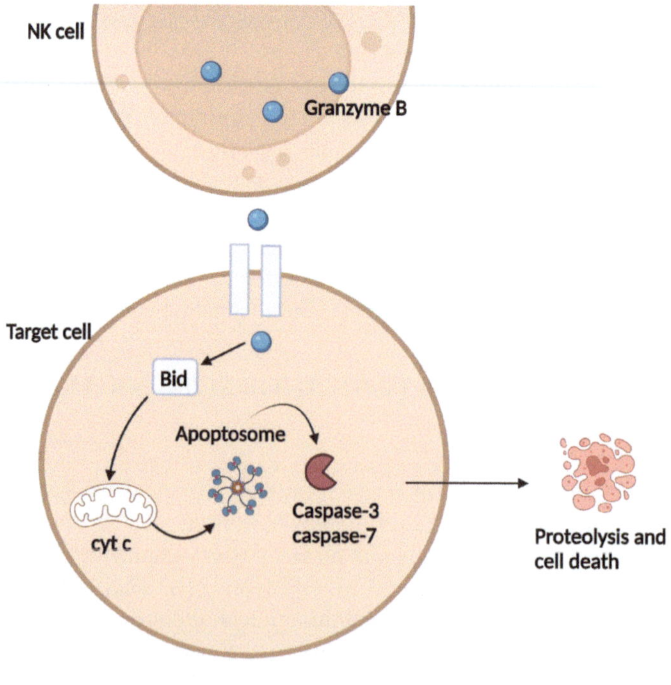

FIGURE 9.2 Involvement of granzyme B in cancer cell proteolysis and apoptosis.

9.3.2 Kallikreins and Related Proteases

Kallikreins and related proteases (KLKs) form the largest family of serine proteases expressed in various tissues, exerting significant influence on physiological functions and various pathophysiological conditions, particularly cancer (21). Within the context of cancer, KLKs exhibit a dual role, acting both as tumour suppressors, as seen in KLK10 downregulation in breast cancer and acute lymphoblastic leukaemia, and tumour promoters by impacting processes such as angiogenesis, invasiveness and metastasis (22). KLK3 dysregulation is linked to aggressive forms of prostate cancer. KLKs contribute to cancer progression by modulating apoptosis, affecting androgen receptor pathways and interacting with molecular networks, including protease-activated receptors (PARs), uPA and matrix metalloproteinases (MMPs). Specific KLKs can cleave IGF-binding proteins (IGFBPs), enhancing the availability of IGFs and influencing crucial cellular processes like cell survival and differentiation (23). KLKs also play a role in acquired resistance through the modulation of androgen receptor-dependent pathways (24). Activation of type IV collagenases and regulation of the uPA-uPAR system by KLKs play a vital role in the degradation of ECM components, promoting angiogenesis and releasing proangiogenic growth factors. Additionally, KLKs such as KLK6 and KLK7 are involved in cleaving E-cadherin, impacting tissue remodelling, tumour invasion and metastasis (25). Furthermore, certain KLKs, like KLK3 and plasmin, can activate TGF-beta, suggesting their indirect involvement in regulating tumourigenicity and immunosuppression within the tumour microenvironment. The intricate involvement of KLKs underscores their significance in cancer biology and therapeutic targeting (26) (Figure 9.3).

9.3.3 Urokinase-type Plasminogen Activator (uPA)

The uPA system is linked to several pro-carcinogenic effects, while its primary and widely recognized functions involve the proteolysis of basement membrane (BM) and ECM. The proteolysis of ECM and BM is a crucial step that promotes the migration, invasion and metastasis of cancer cells. The uPA system is also involved in initiating EMT and various signal transduction pathways, thereby promoting tumour angiogenesis, migration and metastasis. Plasminogen, specific plasminogen receptors and uPAR accumulate on the surface of tumour cells and catalyse the activation of plasminogen into plasmin. Thus, activated plasmin exhibits the ability to degrade substrates such as type IV collagen, laminins, myelin, vitronectin (Vn), perlecan and fibronectin. Additionally, it also plays a vital role in ECM degradation by activating MMPs and growth factors. ECM proteins are degraded binding sites for adhesion molecules such as collagen-IV, laminin-5, integrins and Vn

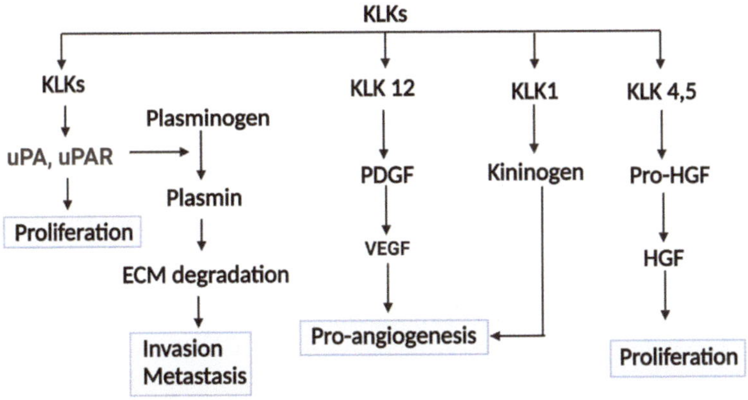

FIGURE 9.3 Involvement of other KLKs in proliferation, angiogenesis, invasion and metastasis.

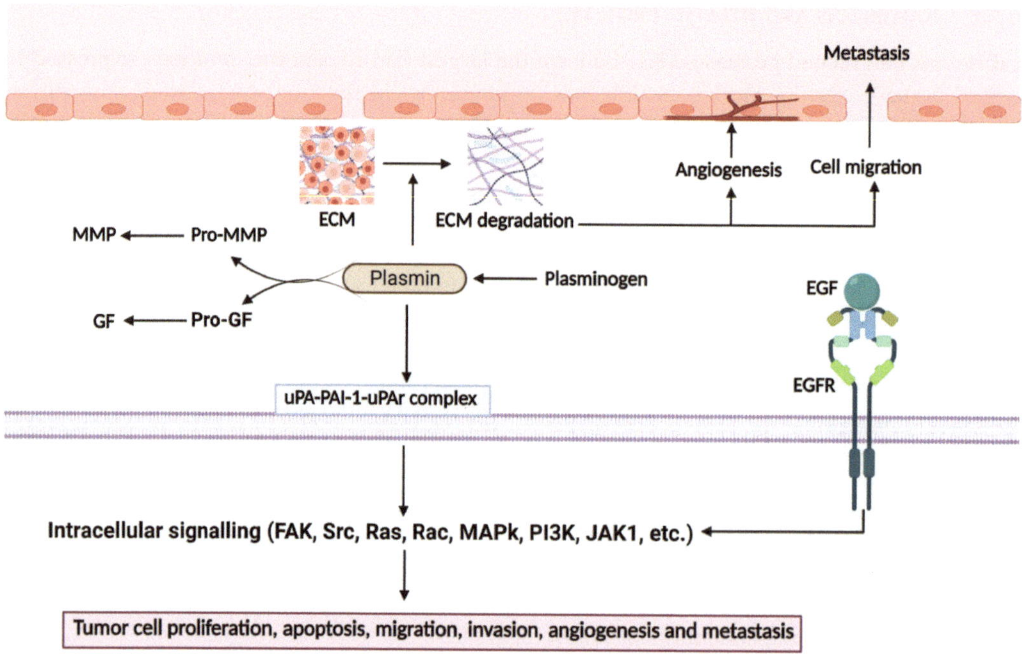

FIGURE 9.4 Involvement of urokinase-type plasminogen activator (uPA) in cancer proliferation, apoptosis, angiogenesis and metastasis.

which triggers metastasis. The interaction between uP, androgen receptor (AR) and Vn can modulate cytoskeleton rearrangements and enhance cell motility (27).

During EMT, cells undergo significant changes in phenotype and polarity, transitioning into a mobile spindle-shaped form. Hypoxia is a common occurrence in solid tumours that triggers EMT. Research findings have demonstrated that hypoxia induces EMT by upregulating the expression of uPAR (28). Furthermore, under hypoxic conditions, uPAR has been identified as an inducer of cellular signalling pathways that can promote EMT. It has been demonstrated that reversing EMT in cancer is possible through interventions such as reoxygenating tumour or targeting uPAR-mediated signalling transduction pathways. The association between uPA and uPAR can also trigger the activation of several cell-signalling proteins, including c-Src, focal adhesion kinase (FAK) and ERK/MAPK involved in cancer progression. Additionally, uPAR interacts with EGFR to initiate the ERK/MAPK signalling pathway, thereby playing a crucial role in regulating cell proliferation, growth and survival (29) (Figure 9.4).

9.3.4 Fibroblast Activation Protein (FAP)

FAP expression is generally undetectable in the majority of normal tissues, but it is elevated in nearly all carcinomas. Within tumour, multiple mesenchymal cell types exhibit FAP expression, including cancer-associated fibroblasts (CAFs), mesenchymal stem cells (MSCs), sarcoma cells and melanoma cells. Fibroblasts are a major component of the tumour microenvironment. They can enhance tumour cell growth and invasion by synthesizing, depositing and remodelling the ECM, thereby influencing angiogenesis and disrupting anti-tumour immune responses.

Numerous research findings have demonstrated the role of FAP in enhancing cell proliferation, invasion and metastasis in the cells where it is expressed. The high FAP expression is associated with the enrichment of gene pathways related to cell growth, differentiation, angiogenesis, migration and immune regulation. Stable overexpression of FAP in CAFs leads to significant alterations

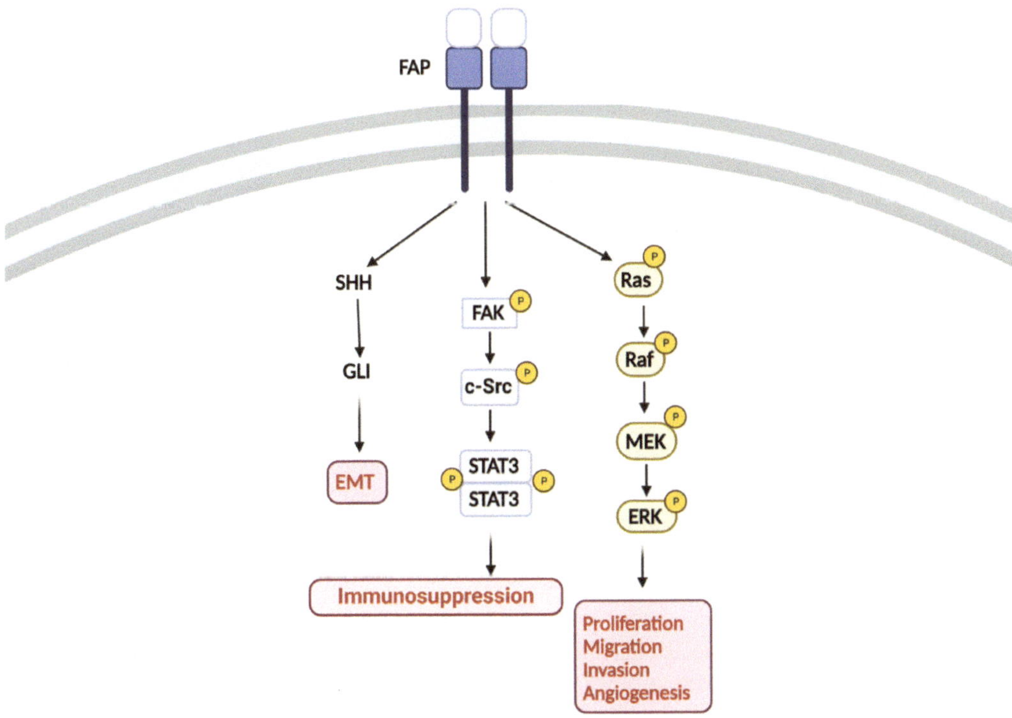

FIGURE 9.5 Role of FAP in cancer cell proliferation, migration, invasion, angiogenesis, EMT and immunosuppression.

in CAF secretome, including the upregulation of proliferative, inflammatory and ECM remodelling factors. To undergo metastasis, cancer cells must adhere to ECM adhesion proteins and enzymatically degrade ECM. This process involves extensive remodelling of ECM and alterations in the secretion of ECM proteins like laminin, tenascin, fibronectin and collagens. FAP serves as an active serine protease with the capacity to degrade type I collagen and possesses dipeptidyl-peptidase activity. FAP acts directly by participating in ECM degradation or indirectly by regulating proteases involved in the activation or modification of other ECM proteases and protease inhibitors. FAP contributes to the growth of tumours by promoting angiogenesis. Tumours originating from FAP exhibit a three-fold increase in microvessel density compared to tumours originating from FAP-non-expressing cells (Figure 9.5).

9.4 POTENTIAL OF SERINE PROTEASES AS DRUGGABLE TARGETS

9.4.1 Granzymes

Granzymes, particularly granzyme B (GrB), have emerged as potential druggable targets for cancer therapy due to their pivotal role in apoptosis and their involvement in immune surveillance against cancer cells. Granzymes play a n important role in the cytotoxic activity of immune cells like NK cells and cytotoxic T lymphocytes (CTLs) against cancer cells, initiating a cascade of events that leads to apoptosis and eventual destruction of cancer cells. One of the key advantages of targeting granzymes as potential cancer therapeutics lies in their ability to induce apoptosis selectively in cancer cells while sparing healthy cells. This targeted approach minimizes damage to surrounding tissues, reducing the side effects associated with conventional cancer treatments. Granzyme-based therapies hold great promise in precision medicine. Furthermore, cancer cells often employ various mechanisms to evade immune system destruction, including inhibition of granzyme activity.

Researchers are exploring strategies to enhance cancer cells by modulating granzyme activity or augmenting the delivery of granzymes to tumour sites. Targeting granzymes or the pathways they activate could potentially enhance the immune response against cancer cells, making them attractive candidates for the development of novel cancer therapies.

9.4.2 Kallikreins and Related Proteases

Given the vital roles of kallikreins in driving tumour progression, these serine proteases hold significant promise for developing novel therapeutics for cancer. Several therapeutic approaches are aimed at either inhibiting or harnessing the activity and tissue specificity of kallikrein. Based on numerous reports implicating the growth promoting activity of KLKs through proteolytic activity, the potential search for KLK inhibitors as therapeutic agents for treating cancer is actively ongoing. A novel method was identified for developing highly specific serpins targeting KLK2 by modulating reactive-site loops (30). Since serpins are naturally occurring proteins, their recombinant modified forms are likely to exhibit lower toxicity, reduced immunogenicity and slower clearance rates compared to small chemical inhibitors. Additionally, alternative therapeutic strategies leverage KLK activity and tissue specificity. KLK3 is highly expressed in diseased prostate tissue with largely unhindered enzymatic activity within tumour microenvironments. Efforts have been directed towards creating hidden "prodrugs" that encapsulate cytotoxic substances like doxorubicin, vinblastine and thapsigargin. These prodrugs are linked to a peptide carrier via a bond that can be cleaved by KLK3. Research conducted before clinical trials indicates that these prodrugs specifically attack cancer cells that produce KLK3 in laboratory settings and slow down tumour growth in animal models implanted with KLK3-producing LNCaP tumour cells, all while causing minimal to no harm or fatalities to the host (31). Moreover, the unique affinity of KLK3 for cancerous tissue has been utilized in cytoreductive gene therapy strategies devised to specifically target and eliminate prostate tumours. This approach involves constructing either non-viral (liposomal) or viral (adenoviral) vectors containing a suicide gene that encodes an enzyme capable of activating a prodrug. These vectors are designed with KLK3 promoter and/or enhancer elements regulating the expression of the suicide gene. Experimental studies have shown that the cancer-specific KLK3 promoter or enhancer effectively delivers and triggers the expression of suicide genes within KLK3-producing cancer cells, leading to the activation of prodrugs and subsequent apoptosis of cancer cells in laboratory settings, as well as inhibition of tumour growth in mice implanted with xenografted tumours (32). Furthermore, there is another prospective anti-cancer tactic which focuses on creating KLK3 vaccines for active immunotherapy. Phase I clinical trials have been carried out to assess vaccines that include dendritic cells transfected with KLK3 mRNA, recombinant vaccinia viruses expressing KLK3 and dendritic cells exposed to KLK3 in men with metastatic prostate cancer, as well as those experiencing biochemical relapse after prostatectomy (33). These vaccines have demonstrated the ability to activate KLK3-specific T-cell responses without causing any serious adverse effects. Thus, actively targeting kallikreins and related proteins can be a new strategy to target cancer.

9.4.3 Fibroblast Activation Protein (FAP)

The enzymatic activity of FAP in various carcinomas highlights its potential as a therapeutic target. Clinical inhibitor Val-boro-Pro (PT-100; talabostat) demonstrated efficacy in reducing tumour growth across mouse models, suggesting the potential of small molecule inhibitors to target FAP's protease activity (34). However, Phase II trials of talabostat in metastatic colorectal cancer and in combination with docetaxel in advanced non-small cell lung cancer did not yield significant clinical benefits (35). The more specific inhibitor PT630 displayed potent anti-cancer effects but may carry a higher risk of side effects affecting normal functions. Anti-FAP antibodies, like sibrotuzumab, showed limited efficacy in clinical trials, while the monoclonal antibody FAP5-DM1 demonstrated promising results in animal models of lung, head and neck and lung cancer (36). FAP-triggered photodynamic molecular beacons (FAP-PPB) and FAP-targeting prodrug of doxorubicin (FTPD) present novel strategies for cancer detection and treatment, showing promising results in preclinical studies (37). Nevertheless, further preclinical studies are essential for the development of novel FAP inhibitors for clinical applications.

9.5 CONCLUSION AND FUTURE PROSPECTS

The importance of serine proteases in cancer has emerged as a fascinating area of research, offering significant insights into cancer progression mechanisms and potential treatment strategies. Serine proteases are pivotal in various aspects of cancer biology, influencing tumour growth, invasion, angiogenesis and metastasis. Their impact on the tumour microenvironment and cancer-cell survival makes them appeal therapeutic targets. Although the field of serine protease inhibitors for cancer is evolving, promising treatment options have been explored. Personalized medicine approaches, considering individual genetic and molecular profiles, hold potential for optimizing serine protease inhibitor use and minimizing side effects. Ongoing research into the intricate network of serine proteases, their substrates and regulatory mechanisms is expected to reveal new insights, expanding our understanding of cancer pathogenesis and informing the development of more effective, targeted therapies. In summary, serine proteases offer a compelling avenue for cancer research and treatment, bringing us closer to personalized strategies for combating this disease and providing hope for those affected by cancer.

DECLARATIONS

Conflict of interest: The authors declare no conflicts of interest.

Consent for publishing: All authors have approved the final version of the manuscript and given consent for publication.

Funding: Not Applicable.

REFERENCES

1. Yuan Z, Li Y, Zhang S, et al (2023) Extracellular matrix remodeling in tumor progression and immune escape: from mechanisms to treatments. *Molecular Cancer.* 22(1), 48.
2. Nabeshima K, Inoue T, Shimao Y, et al (2002) Matrix metalloproteinases in tumor invasion: Role for cell migration. *Pathology International*, 52(4), 255–64.
3. Kousoulis AA, Tsoucalas G, Armenis I, et al (2012) From the "hungry acid" to pepsinogen: a journey through time in quest for the stomach's secretion. *Annals of Gastroenterology.* 25(2), 119–22.
4. Rawlings ND, Barrett AJ, Bateman A (2012) MEROPS: the database of proteolytic enzymes, their substrates and inhibitors. *Nucleic Acids Research.* 40(D1), D343–50.
5. Sevenich L, Joyce JA (2014) Pericellular proteolysis in cancer. *Genes & Development.* 28(21), 2331–47.
6. Owen CA. Serine proteases. In: Encyclopedia of Respiratory Medicine. Elsevier; 2006. pp. 1–10.
7. Tagirasa R, Yoo E. (2022) Role of serine proteases at the tumor-stroma interface. *Frontiers in Immunology*, 13, 832418.
8. Zhai BT, Tian H, Sun J, et al. (2022) Urokinase-type plasminogen activator receptor (uPAR) as a therapeutic target in cancer. *Journal of Translational Medicine.* 20(1), 135.
9. Puré E, Blomberg R (2018) Pro-tumorigenic roles of fibroblast activation protein in cancer: back to the basics. *Oncogene.* 37(32), 4343–57.
10. Cullen SP, Brunet M, Martin SJ (2010) Granzymes in cancer and immunity. *Cell Death & Differentiation.* 17(4), 616–23.
11. Hedstrom L (2002) Serine protease mechanism and specificity. *Chemical Reviews.* 102(12), 4501–24.
12. Sheida EV (2022) Adaptive responses of cattle digestive system as influenced by dietary ultrafine iron particles combined with fat diets. *Sel'skokhozyaistvennaya Biologiya.* 57(2), 328–42.
13. Oncul S, Afshar-Kharghan V (2020) The interaction between the complement system and hemostatic factors. *Current Opinion in Hematology.* 27(5), 341–52.

14. Prassas I, Eissa A, Poda G, et al. (2015) Unleashing the therapeutic potential of human kallikrein-related serine proteases. *Nature Reviews Drug Discovery.* 14(3), 183–202.
15. Abedin M, King N (2010) Diverse evolutionary paths to cell adhesion. *Trends in Cell Biology.* 20(12), 734–42.
16. Herszényi L (2014) Impact of proteolytic enzymes in colorectal cancer development and progression. *World Journal of Gastroenterology.* 20(37), 13246.
17. Pardo J, Bosque A, Brehm R, et al. (2004) Apoptotic pathways are selectively activated by granzyme A and/or granzyme B in CTL-mediated target cell lysis. *Journal of Cell Biology.* 167(3), 457–68.
18. Galvin JP, Spaeny-Dekking LHA, Wang B, et al. (1999) Apoptosis induced by granzyme B-glycosaminoglycan complexes: implications for granule-mediated apoptosis *in vivo. The Journal of Immunology.* 162(9), 5345–50.
19. Andrade F, Casciola-Rosen LA, Rosen A (2004) Granzyme B-induced cell death. *Acta Haematologica.* 11, 28–41.
20. Johnson H, Scorrano L, Korsmeyer SJ, et al (2003) Cell death induced by granzyme C. *Blood.* 101(8), 3093–3101.
21. Avgeris M, Mavridis K, Scorilas A (2010) Kallikrein-related peptidase genes as promising biomarkers for prognosis and monitoring of human malignancies. *Biological Chemistry.* 391, 501–11.
22. Matsumura M, Bhatt AS, Andress D, et al. (2005) Substrates of the prostate-specific serine protease prostase/KLK4 defined by positional-scanning peptide libraries. *The Prostate.* 62(1), 1–3.
23. Niu Y, Yeh S, Miyamoto H, et al. (2008) Tissue prostate-specific antigen facilitates refractory prostate tumor progression via enhancing ARA70-regulated androgen receptor transactivation. *Cancer Research.* 68(17), 7110–9.
24. Menashi S, Fridman R, Desrivieres S, et al. (1994) Regulation of 92-kDa Gelatinase B activity in the extracellular matrix by tissue kallikrein. *Annals of the New York Academy of Sciences.* 732(1), 466–8.
25. Kapadia C, Ghosh MC, Grass L, et al. (2004) Human kallikrein 13 involvement in extracellular matrix degradation. *Biochemical and Biophysical Research Communications.* 323(3), 1084–90.
26. Derynck R, Akhurst RJ, Balmain A (2001) TGF-β signaling in tumor suppression and cancer progression. *Nature Genetics.* 29, 117–29.
27. Chapin JC, Hajjar KA (2015) Fibrinolysis and the control of blood coagulation. *Blood Reviews.* 29(1), 17–24.
28. Lester RD, Jo M, Montel V, et al. (2007) uPAR induces epithelial-mesenchymal transition in hypoxic breast cancer cells. *The Journal of Cell Biology.* 178(3), 425–36.
29. Liu D, Aguirre Ghiso JA, Estrada Y, et al (2002) EGFR is a transducer of the urokinase receptor initiated signal that is required for *in vivo* growth of a human carcinoma. *Cancer Cell.* 1(5), 445–57.
30. Cloutier SM, Kündig C, Felber LM, et al (2003) Development of recombinant inhibitors specific to human kallikrein 2 using phage-display selected substrates. *European Journal of Biochemistry.* 271(3), 607–13.
31. Denmeade SR, Jakobsen CM, Janssen S, (2003) Prostate-specific antigen-activated thapsigargin prodrug as targeted therapy for prostate cancer. *Journal of the National Cancer Institute.* 95(13), 990–1000.
32. Suzuki S, Tadakuma T, Kunitomi M (2001) Liposome-mediated gene therapy using HSV-TK/ganciclovir under the control of human PSA promoter in prostate cancer cells. *Urological Internationalis.* 67(3), 216–23.
33. Heiser A, Maurice MA, Yancey DR (2001) Human dendritic cells transfected with renal tumor RNA stimulate polyclonal T-cell responses against antigens expressed by primary and metastatic tumors. *Cancer Research.* 61(8), 3388–93.
34. Jones B, Adams S, Miller GT (2003) Hematopoietic stimulation by a dipeptidyl peptidase inhibitor reveals a novel regulatory mechanism and therapeutic treatment for blood cell deficiencies. *Blood.* 102(5), 1641–8.
35. Hofheinz RD, Al-Batran SE, Hartmann F (2003) Stromal antigen targeting by a humanised monoclonal antibody: an early phase II trial of sibrotuzumab in patients with metastatic colorectal cancer. *Oncological Research and Treatment.* 26(1), 44–48.
36. Lo PC, Chen J, Stefflova K (2009) Photodynamic molecular beacon triggered by fibroblast activation protein on cancer-associated fibroblasts for diagnosis and treatment of epithelial cancers. *Journal of Medicinal Chemistry.* 52(2), 358–68.
37. Huang S, Fang R, Xu J (2011) Evaluation of the tumor targeting of a FAPα-based doxorubicin prodrug. *Journal of Drug Targeting.* 19(7), 487–96.

10 Survivin, a Moonlighting Caspase Inhibitor and Potential Oncotherapeutic Target

Abdul Basit Baba
Advanced Centre for Human Genetics, Sher-i-Kashmir Institute of Medical Sciences (SKIMS), Srinagar, India

Rajakishore Mishra
School of Natural Sciences, Central University of Jharkhand, Chery-Manaty, Ranchi, India

Siddavaram Nagini
Faculty of Science, Annamalai University, Annamalainagar, India

10.1 INTRODUCTION

Survivin, also known as baculoviral inhibitor of apoptosis protein (IAP) repeat-containing protein 5 (BIRC5), is a multitasking eukaryotic protein essential for stimulation of cell division and prevention of cell death (Wheatley & Altieri, 2019). Since its discovery by Ambrosini et al. (1997) as a novel anti-apoptosis gene, survivin has attracted significant attention as a promising cancer biomarker and potential oncotherapeutic target.

Survivin is expressed during embryonic development, barely detectable in normal adult somatic tissues, and is aberrantly overexpressed in malignant tumors. In actively proliferating adult normal cells such as melanocytes, gastrointestinal cells, neurons, hematopoietic cells, T cells, vascular endothelial cells, testes, ovary, and placenta, survivin has multifarious functions including regulation of cell cycle and angiogenesis, hematopoiesis, apoptosis inhibition, and self-renewal and differentiation of cryptic stem cells (Rafatmanesh et al., 2020; Warrier et al., 2020). Survivin, the fourth most common protein in malignant tumors, is implicated in various hallmarks of cancer as well as in resistance to cancer chemotherapy and radiation. In preclinical animal models, downregulation of survivin expression resulted in tumor regression, apoptosis, and chemosensitization. The involvement of survivin in cell-fate decision and differential expression in normal versus proliferating tissues renders it an important target for cancer therapy (Albadari & Li, 2023; Fang et al., 2024; Li et al., 2019a).

This chapter provides an overview of the structure, localization, physiological functions, regulation, and differential expression of survivin in cancer cells. The different strategies developed to target survivin for cancer therapy are also summarized.

10.2 STRUCTURE

The human survivin gene (14.5 kb) encoded by the BIRC5 gene is located in the telomeric end of chromosome 17q25. It has four exons and five introns (Figure 10.1). Ten different splice variants

DOI: 10.1201/9781003394693-12

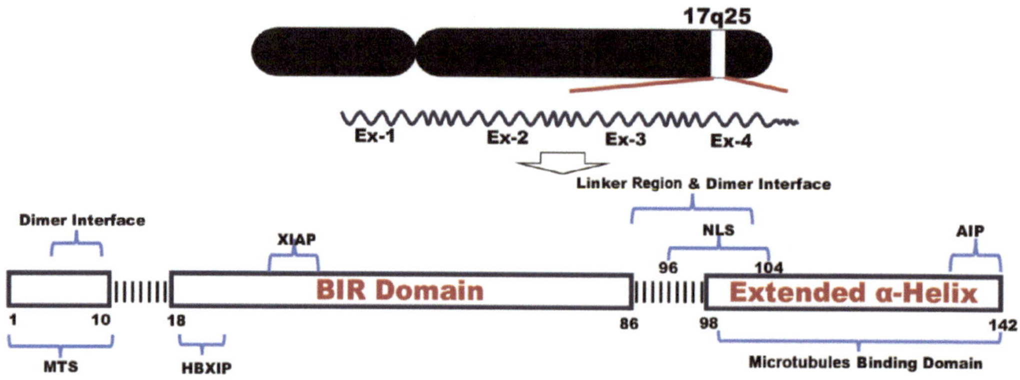

FIGURE 10.1 The structure of survivin. The survivin gene (14.5 kb) is located in the telomeric end of chromosome 17q25. It has four exons and five introns. The survivin protein, comprising 142 amino acids (Mr 16.5 kDa), contains a single baculovirus inhibitor of apoptosis repeat (BIR) domain in the N-terminal region (aa 18–88), a linker region (aa89–102), and an extended α-helix in the C-terminal end (aa 98–142). The first ten amino acids are proline-rich and required for mitochondrial importation. The BIR domain is important for protein–protein interactions. The linker region, together with the N-terminal residues 6–10, mediates dimerization. The nuclear localization signal extends between amino acid residues 96–104, between the BIR domain and the C-terminal region.

have been reported, some of which have not been well characterized. The wild type referred to as survivin, and the splice variants 2β and ΔEx3 are the common forms. Both the variants differ from the wild type at the exon2–exon3 junction. Survivin 2β contains a 26-amino-acid insert that renders it pro-apoptotic, whereas the ΔEx3 variant, which lacks exon 3, is anti-apoptotic (Sah & Seniya, 2015). Figure 10.2 provides a schematic representation of the common survivin splice variants.

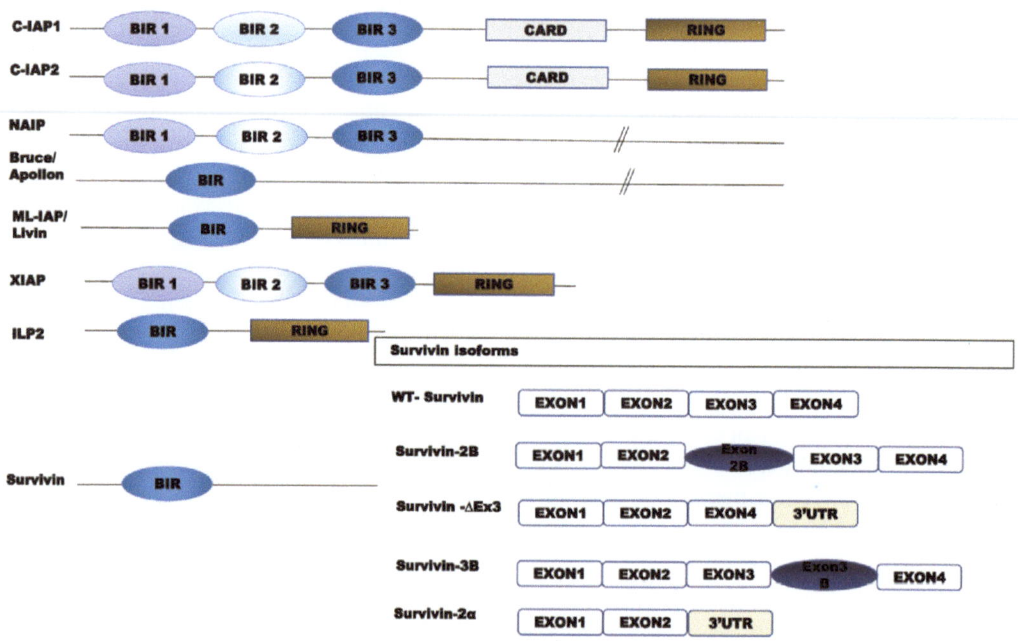

FIGURE 10.2 The wild type and common splice variants of survivin. Wild type survivin contains four exons. Survivin 2β contains a 26-amino acid insert between exon 2 and exon 3. The ΔEx3 variant lacks exon 3.

The human survivin protein is the smallest member of the IAP family of proteins containing 142 amino-acid residues with a molecular weight of 16.5 kDa (Figure 10.1). The protein structure comprises a single baculovirus inhibitor of apoptosis repeat (BIR) domain in the N-terminal region spanning amino acids 18–88 stabilized by a zinc finger involving Cys57, Cys60, His77, and Cys84 involved in protein–protein interactions, a linker region extending from amino acids 89–102, and an extended α-helix in the C-terminal end (amino acids 98–142). Survivin functions both as a monomer and a homodimer that forms via interactions between the linker region and residues 6–10 of the BIR domain (Verdecia et al., 2000). Post-translational modifications of survivin include phosphorylation, acetylation, and ubiquitylation. The half-life of survivin is 30–120 minutes and nuclear localization leads to decrease in the half-life of survivin. Survivin undergoes ubiquitylation and degradation by the 26S proteasome following mitosis (Nogueira-Ferreira et al., 2013; Wheatley & Altieri, 2019).

10.3 LOCALIZATION

Survivin is localized in multiple subcellular compartments, including the nucleus, cytoplasm, mitochondria, cell membrane, extracellular matrix, and exosomes. Nuclear export of survivin to the cytoplasm requires exportin-1, also known as chromosome region maintenance 1 (Rodriguez et al., 2002). Progressive cytoplasmic localization of survivin was reported in newly isolated fibroblasts with successive passages (Temme et al., 2005). Expression of survivin occurs during the G2/M phase of the cell cycle. During mitosis, survivin is localized in the mitotic spindle. Survivin was found to be expressed in the kinetochores, cleavage furrow, and midbody. Interestingly, in cancer cells, survivin was detectable in the mitochondria (Wheatley & Altieri, 2019).

The subcellular localization of survivin appears to dictate its cellular functions. While nuclear survivin regulates cell division, cytoplasmic and mitochondrial survivin play an important role in apoptosis resistance. Following apoptosis, mitochondrial survivin is released into the cytoplasm. The extracellular survivin secreted by cancer cells is found in exosomes and is implicated in intercellular communication. It is reported to re-enter tumor cells, where it is involved in proliferation, invasion, angiogenesis, and apoptosis inhibition (Wheatley & Altieri, 2019). There is evidence to indicate that localization of survivin in the nucleus has a pro-apoptotic function. Colnaghi et al. (2006) demonstrated that enforced nuclear localization of survivin destroyed its IAP function. On the contrary, some studies have also showed that nuclear survivin retains its IAP function (Krieg et al., 2022). Degradation of survivin by the 26S proteasome occurs in the nucleus following ubiquitination by the anaphase-promoting complex (Zhao et al., 2000).

10.4 BINDING PARTNERS

The interactions of survivin with other proteins influences stability. Survivin is stabilized by integrating with borealin, inner centromere protein (INCENP), and aurora B kinase to form the chromosomal passenger complex (CPC). On the other hand, the binding of survivin to the X-linked inhibitor of apoptosis (XIAP) leads to destabilization and increased turnover (Abad et al., 2019). The interaction of heat shock protein 90 (Hsp90) with survivin prevents proteasomal degradation of survivin. Survivin shares several biological functions with heat shock protein 60 (HSP60). Both the proteins are localized in the cytoplasm and mitochondria and are overexpressed in cancer cells. The binding of HSP60 to survivin through protein–protein complexes stabilizes survivin, leading to survival of cancer cells. The expression of both the proteins increases during G2/M phase of the cell cycle. In the mitochondria, both the proteins exhibit pro-survival and anti-apoptotic functions. However, in the cytosol, HSP60 and survivin may display either pro-survival or pro-apoptotic functions. In response to cell death signals, mitochondrial HSP60–survivin complex is released into the cytosol with degradation of survivin by ubiquitination (Huang & Yeh, 2019).

10.5 FUNCTIONS

Survivin is a multitasking protein that plays a critical role in cell division, apoptosis suppression, autophagy inhibition, angiogenesis, and stemness (Figure 10.3).

10.5.1 Cell Division

Survivin plays a central role in cell division as a component of the CPC, regulator of microtubule dynamics, segregation of sister chromatids, and cytokinesis. While the C-terminal region of survivin is required for cell division, the N-terminal region functions in apoptosis inhibition. Survivin is phosphorylated during prometaphase by aurora B kinase, which is essential for targeting the CPC to the centromeres before anaphase (Wheatley &Altieri, 2019). Survivin regulates microtubule dynamics in a process mediated by Ran-GTP and TPX2. Survivin plays a key role in coordinating mitosis with cytokinesis in all eukaryotes. Consequently, functional loss of survivin results in failure of cytokinesis, mitotic catastrophe, and enhanced apoptosis (Xia et al., 2008).

10.5.2 Cell Death

Survivin is documented to regulate apoptosis as well as autophagy. Survivin inhibits both the extrinsic and intrinsic pathways of apoptosis. Survivin, together with hepatitis B X-interacting protein (HBXIP) cofactor, suppresses pro-caspase-9 and inhibits the death receptor (extrinsic) pathway (Chen et al., 2016). Suvivin primarily inhibits the mitochondrial or intrinsic pathway of apoptosis by inhibiting caspase-3, -7, and -9 and via interactions with cyclin-dependent kinase (CDK)-2 and -4 (Wheatley & Altieri, 2019). Survivin interacts with XIAP, another IAP family protein to form the survivin–XIAP complex, which stabilizes XIAP and prevents ubiquitination (Hanifeh & Ataei, 2022). The survivin–XIAP complex prevents the activation of caspase-9 and consequently apoptosis. Survivin is recognized to enhance the anti-apoptotic function of XIAP by sequestration of second mitochondria-derived activator of caspases (Smac), also known as DIABLO, a negative regulator of XIAP, preventing it from being released from the mitochondria into the cytosol and subsequent caspase activation and apoptosis induction (Cheung et al., 2020). Additionally, survivin

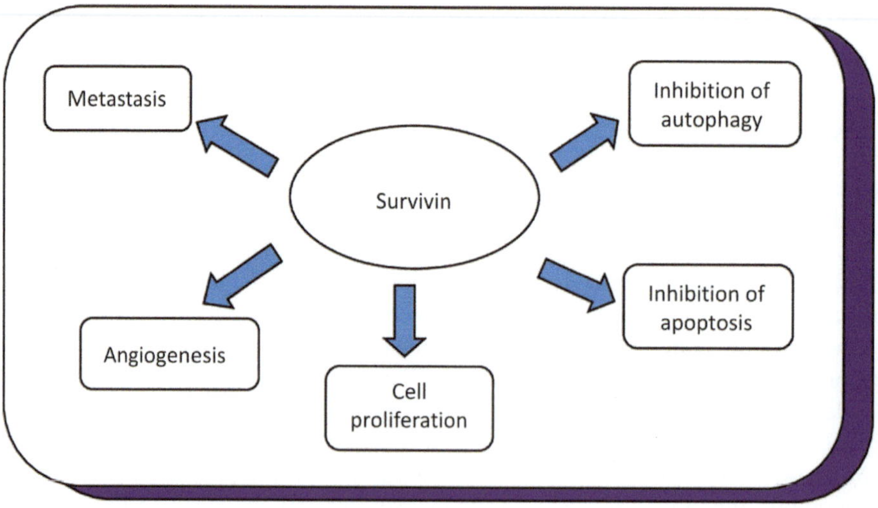

FIGURE 10.3 Important cellular functions of survivin. As an integral component of the chromosomal passenger complex, survivin coordinates mitosis with cytokinesis. Survivin inhibits both apoptosis and autophagy. Survivin also plays a key role in angiogenesis and stemness.

activates caspase-independent apoptosis by blocking the release of apoptosis-inducing factor (AIF) from the mitochondrial intermembrane space (Pavlyukov et al., 2011). However, the action of survivin on XIAP is reversed by XIAP-associated factor 1 (XAF1) that promotes survivin degradation via XIAPE3 ligase (Hanifeh & Ataei, 2022).

In addition to its IAP function, survivin has been reported to negatively regulate autophagy. While survivin inhibition by YM155 induces autophagic cell death, its overexpression inhibits autophagy. Survivin is believed to suppress autophagy by inhibiting the formation of autophagosomes. Furthermore, it inhibits complex formation between autophagy-related protein-5, -12, and -16L1 as well as the stability of ATG7 consequently blocking cellular autophagic flux (Cheung et al., 2020).

10.5.3 Angiogenesis

Survivin stimulates angiogenesis by enhancing the synthesis and release of vascular endothelial growth factor (VEGF) under hypoxic conditions, besides disruption of cell–cell and cell–matrix interactions (Sanhueza et al., 2015). In esophageal cancer, survivin was identified as the signal downstream of VEGF receptor-2 (VEGFR2) responsible for cell survival (Meng et al., 2018). Co-expression of survivin and VEGF-C was found to be predictive of poor prognosis in esophageal cancer (Zeng et al., 2021). In non-small cell lung cancer (NSCLC) and prostate cancer, survivin expression correlated positively with VEGF expression(Fang et al., 2021; Kong et al., 2021).

10.5.4 Stemness

Survivin is known to play a role in self-renewal and differentiation of cryptic stem cells. High survivin expression has been observed in embryonic stem (ES) cells (Warrier et al., 2020). Knockdown of survivin in ES cells was shown to downregulate transcription factors involved in pluripotency including octamer-binding transcription factor 4 (Oct4) and Nanog, suggesting its involvement in stemness (Mull et al., 2014). Constitutive survivin expression in cancer stem cells (CSCs) is promoted by the oncogenic transcription factor specificity protein 1 (Spl) and c-Myc as well as mitogen-activated protein kinase (MAPK) signaling (Wheatley &Altieri, 2019; Zhang et al., 2015).

10.5.5 Epigenetic Regulation

In a recent study, Jensen et al. (2023) showed that the interaction of survivin with the polycomb repressive complex 2 (PRC2) leads to epigenetic control of gene expression. PRC2 causes silencing of gene expression by trimethylating lysine 27 on histone 3. Using a panel of techniques including peptide microarray, machine learning, NMR, BLI, and docking, survivin was shown to interact with the EZH2 subunit of PRC2 to inhibit its H3K27 methylation activity.

10.6 REGULATION OF SURVIVIN EXPRESSION

Basal transcription of survivin requires Spl or Kruppel-like factor 5 (KLF5) (Zhu et al., 2006). Proteins that play a key role in the cell cycle influence survivin expression. The eukaryotic transcriptional activators E2F1, E2F2, and E2F3 of the E2F family of transcription factors bind to the survivin promoter to induce transcriptional activation via cell-cycle-dependent element/cellcycle genes homology region (CDE/CHE), whereas E2F4 and E2F5 downregulate survivin expression (Raj et al., 2008). The p53 tumor suppressor gene together with the retinoblastoma protein (pRB) negatively regulates survivin expression. The multifunctional transcription factor YY1 was shown to physically interact with the survivin promoter to downregulate survivin expression. However, the negative regulators of survivin are often mutated, downregulated, silenced, or dysfunctional in cancer cells (Galloway et al., 2017).

Many growth factors and their receptors as well as oncogenic transcription factors stimulate survivin expression. These include insulin-like growth factor-1, VEGF, epidermal growth factor receptor (EGFR), human epidermal growth factor receptor 2 (HER 2) and 3 (HER 3), nuclear factor kappa-light-chain-enhancer of activated B cells (NF-κB), hypoxia-inducible factor 1α (HIF-1α), and signal transducer and activator of transcription 3 (STAT3), (reviewed by Chen et al., 2016; Warrier et al., 2020). Several signaling pathways regulate survivin expression. The phosphoinositide-3 kinase (PI3K)/Akt pathway activates survivin expression by the downregulation of Forkhead box O (FOXO) that causes the silencing of survivin via phosphatase and tensin homolog (PTEN) (Guha et al., 2009). Activation of the Wnt/β-catenin, Janus-activated kinase (JAK)/STAT3, Notch, and Ras signaling pathways enhances survivin expression, whereas transforming growth factor β downregulates survivin expression (Chen et al., 2016; Fernández et al., 2014).

10.7 SURVIVIN AS A CANCER BIOMARKER

Overexpression of survivin has been documented in a wide range of malignant tumors including hematological malignancies, gynecological tumors, lung cancer, head and neck tumors, and gastrointestinal tumors to name a few (Chen et al., 2016; Jaiswal et al., 2015; Li et al., 2021b; Warrier et al., 2020). Increased expression of survivin positively correlated with invasion, lymph node metastasis, poor differentiation, decreased survival, and recurrence (Galloway et al., 2017). A pan-cancer analysis of approximately 9000 samples from 33 different cancer types and 11 pan-cancer organ systems obtained from The Cancer Genome Atlas (TCGA) revealed significantly high expression of survivin in 14/33 types of cancer that was linked to worse overall survival, especially for lung, gastric, and breast cancer. The findings also suggested that survivin is a promising prognostic biomarker (Beding et al., 2022).

Enhanced expression of survivin is associated with worse prognosis, apoptosis evasion, poor treatment response, and recurrence in a large number of malignant tumors including lung, oral, stomach, colorectal, breast, and ovarian cancer as well as melanoma. A meta-analysis of 18 publications based on data from 2233 ovarian cancer patients revealed an association between increased survivin expression and shorter overall survival, higher histological grade, and advanced disease stage, indicating that survivin may serve as a marker of poor prognosis (Gasowska-Bajger et al., 2021). In a cross-sectional study, survivin expression was higher in triple-negative breast cancer and lowest in receptor positive breast cancer (Quidute et al., 2023).

A strong association was seen between survivin expression and lung cancer staging, metastasis, and recurrence (Fung et al., 2021). Co-expression of survivin and cullin7 (Cul7), a core component of the cullin E3 ligase complex, was demonstrated to inhibit apoptosis, induce proliferation, and contribute to docetaxel chemoresistance in lung adenocarcinoma cells (Li et al., 2023b). A significant increase in survivin expression was seen in lung adenocarcinoma tissues that correlated with high mortality rates. Furthermore, knockdown of BIRC5 resulted in pyroptosis of A549 lung carcinoma cells via the caspase-3/GSDME pathway (Zhang et al., 2023).

Survivin together with the anti-apoptotic receptor B7-H1 is a reliable predictor of clear cell renal cell carcinoma (RCC) progression and mortality. A negative correlation was found between survivin expression and RCC survival using the TCGA database. Several studies have demonstrated that survivin is a potential target for RCC therapeutics (reviewed by Li et al., 2021a). Survivin positivity combined with overexpression of mutant p53 indicated poor prognosis in RCC (Radovanović et al., 2023). Survivin is recognized as a reliable diagnostic and prognostic marker in bladder cancer (Wolf, 2023).

Overexpression of HSP60 and survivin was found to predict poor prognosis in oral squamous cell carcinoma (OSCC) (Zhou et al., 2023). In a cohort of 32 patients with head and neck squamous cell carcinoma (HNSCC), a positive correlation was observed between survivin expression in the tumor tissues and serum glucose levels, indicating a potential role for survivin in rewiring cancer metabolism. Further, high tumoral survivin was associated with worse tumor stage and a low rate of loco-regional recurrence-free survival after radiotherapy. Moreover, radiation-induced survivin upregulation that stimulates a radioresistant phenotype in tumor cells was dependent on glucose levels (Benaiges et al., 2024).

Polymorphisms in the survivin gene have been reported in a wide range of malignancies (Jaiswal et al., 2015). Polymorphism in the survivin gene promoter region (–31G/C) was seen in advanced stage of OSCC (Cacina et al., 2023). In OSCC and oropharyngeal squamous cell carcinoma (OPSCC), 18 polymorphisms were identified, of which five (rs3764383, rs9904341, rs2071214, rs2239680, and rs2661694) showed a significant association with advanced clinical and pathological lymph node status and tumor node metastasis (TNM) stage (Mumlek et al., 2023). In NSCLC, the overall survival was found to be significantly associated with the *BIRC5* homozygous GG genotype (Szczyrek et al., 2022). The survivin rs1042489 polymorphism was associated with an increased risk of hepatitis C-induced hepatocellular carcinoma in an Egyptian population (Mohamed et al., 2021). Survivin –31G>C polymorphism has been suggested to contribute to the development of colorectal cancer in an Iranian population (Asadian et al., 2022). In a meta-analysis involving 9348 subjects, polymorphisms in BIRC5 rs9904341 and rs2071214 have been suggested to increase the risk of urinary cancer, whereas rs17878467 was associated with decreased risk (Xu et al., 2020). In breast cancer patients, polymorphisms were mostly located in the promoter and 3′UTR of *BIRC5*. Further, alleles of five *BIRC5* polymorphisms (c.–644T>C, c.–241C>T, c.9809T>C, c.–1547C>T, and c.9386T>C) were associated with younger age of onset (Sušac et al., 2019). Wang et al. (2019) assessed the association between polymorphisms of the BIRC5 gene on the development of esophageal squamous cell carcinoma (ESCC) in northern China. Analysis revealed that while the –31 C>G polymorphism may contribute to the development of ESCC, the –241 C>T and –625 G>C polymorphisms may serve as valuable prognostic markers.

10.8 THERAPEUTIC TARGETING

Survivin's lack of intrinsic catalytic activity and druggable sites makes it a difficult protein to target. However, several different approaches have been developed to inhibit survivin in cancer cells, some of which have entered clinical trials. These include antisense oligonucleotides, small molecule inhibitors (SMIs), dominant-negative mutants, small interfering RNAs, immunotherapy, nanoplatforms, and the clustered regularly interspaced short palindromic repeats/CRISPR-associated protein-9 (CRISPR/Cas9) technique.

10.8.1 Antisense Oligonucleotides

Ely Lilly & Co. developed the first antisense oligonucleotide (ASO), LY2181308, a second-generation, potent, ASO that binds to the 3′-untranslated region of the survivin transcript to inhibit translation and stimulate apoptosis (Coumar et al., 2013). In preclinical models, LY2181308 was shown to exert anti-cancer effects in combination with the chemotherapeutic agent docetaxel. ASOs such as LY2181308 are stable and exhibit low toxicity with an improved pharmacokinetic profile (Talbot et al., 2010). However, due to disappointing results from Phase II clinical trials, LY2181308 has been discontinued (Tanioka et al., 2011). SPC3042 (EZN-3042), another ASO that has reached clinical trials, targets exon 4 of the survivin transcript. SPC3042 initiates cell cycle arrest between metaphase and anaphase and activates caspase-3 and -7 (Coumar et al., 2013). Although promising, EZN-3042 was dropped after Phase I clinical trials owing to dose-limiting toxicity (Raetz et al., 2014).

10.8.2 Survivin SMIs

Many SMIs of survivin with potential anti-cancer activity have been identified. Albadari and Li (2023) provided a comprehensive and exhaustive review of SMIs of survivin. Here, we summarize the status of SMIs of survivin that are in preclinical and clinical stages of development. These can

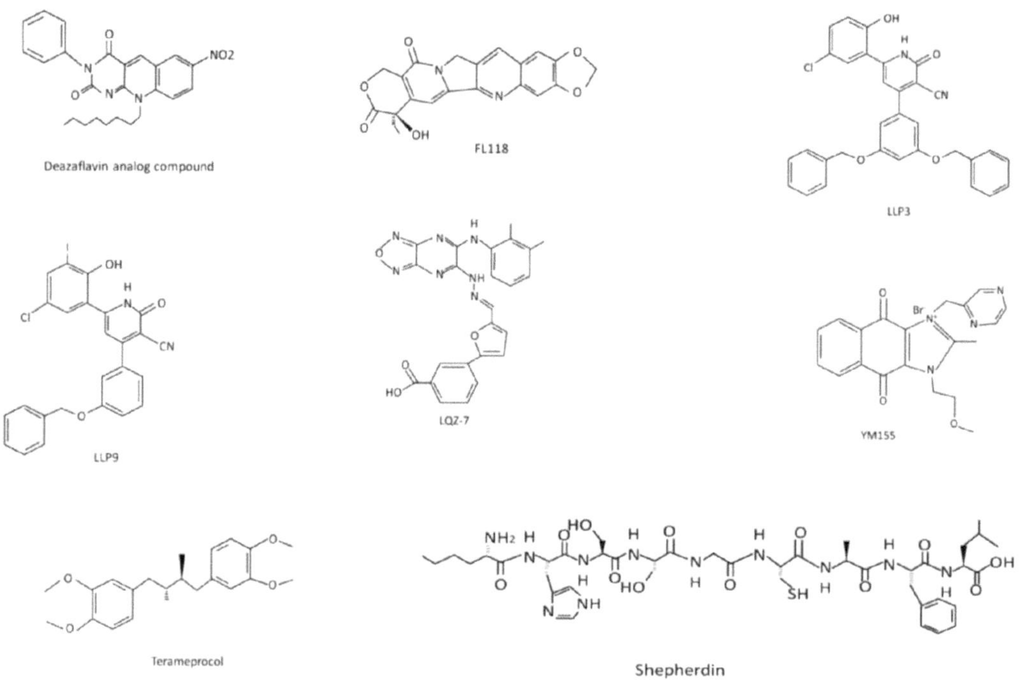

FIGURE 10.4 Some prototype SMIs of survivin. (*Source:* PubChem.)

be categorized into (i) SMIs that downregulate survivin expression, (ii) SMIs that prevent homodimerization, and (iii) SMIs that prevent interaction of survivin with partner proteins. The structures of some prototype SMIs are depicted in Figure 10.4.

10.8.2.1 SMIs that Downregulate Survivin Expression

Sepantronium bromide (YM155), a small-molecule inhibitor of survivin, which induces apoptosis of a wide panel of cancer cells primarily by repressing survivin transcription, generated immense attention as a promising anti-cancer agent. YM155 was first identified as an SMI of survivin by Astellas Pharma in 2007 during high-throughput screening (HTS) of chemical compound libraries (Nakahara et al., 2007). Infusion of YM155 induced tumor regression and chemosensitivity in a wide range of preclinical xenograft tumor models, both as a single agent and in combination with chemotherapy or immunotherapy (Albadari & Li, 2023; Li et al., 2023c). YM155 has been demonstrated to influence the expression of several genes to effectively inhibit proliferation, invasion, and metastasis, and induce apoptosis. YM155 also promotes the expansion of CD44+ CSCs and overcomes chemoresistance in tumors (reviewed by Albadari & Li, 2023; Cheng et al., 2016). YM155 inhibited survivin mRNA and protein synthesis at concentrations of 10–100 nM without affecting other IAPs or anti-apoptotic molecules other than Mcl-1. The finding that topoisomerase is a molecular target of YM155 precludes survivin as the sole target (Hong et al., 2017).

YM155 displayed a favorable outcome and a maximum tolerated dose of 4.8 mg/m^2. However, in multiple Phase II clinical trials, YM155 exhibited limited efficacy as an anti-cancer agent, both as a single agent and in combination with conventional chemotherapeutic agents that may be attributed to its short half-life, which necessitates continuous infusion, uncertain target specificity, and most importantly, being a substrate of P-glycoprotein, it could induce multidrug resistance (Albadari & Li, 2023). Recently, West et al. (2023) designed, synthesized, and characterized a prodrug form of YM155, termed aYM155 that exhibited exceptional cell-type specificity and induced apoptosis of patient-derived glioblastoma cancer stem-like cells and in EGFR-amplified and EGFR variant III-expressing (EGFRvIII) cells.

FL118, a camptothecin analog and topoisomerase inhibitor, is a non-specific SMI of survivin that also inhibits other IAPs. FL118 inhibits survivin from high pM to low nM concentrations and has a favorable toxicity profile (Li, 2014). Unlike YM155, FL118 is not a substrate of P-glycoprotein and can therefore overcome chemoresistance (Westover et al., 2015). FL118 has been demonstrated to directly target DEAD box RNA helicase DDX5, a positive regulator of survivin for proteasomal degradation. DDX5 (also known as p68), is regarded as a potential cancer biomarker and therapeutic target. Li et al. (2023a) provided evidence to show that FL118, a molecular glue DDX5 degrader, selectively targets therapy-resistant prostate cancer organoids/cells. FL118 has been documented to exert potent anti-cancer effects in pancreatic tumor orthotopic and patient-derived xenograft (PDX) animal models (Ling et al., 2018). Treatment with FL118 inhibited the growth of drug-resistant and metastatic lung cancer (Wang et al., 2017).

FL118 is endowed with drug-like attributes and exerts anti-cancer effects via multiple mechanisms, including inhibition of cell proliferation, invasion, and metastasis, induction of apoptosis, the ability to target CSCs, and chemoresistance. Moreover, the high stability and excellent pharmacokinetic profile render FL118 a choice therapeutic option (Li et al., 2019a). Zhang et al. (2022a) designed, synthesized, and evaluated a series of novel FL118 derivatives against drug-resistant small cell lung cancer. Compound 9c displayed the highest anti-cancer activity and inhibition of drug-resistant tumor growth by suppressing survivin. There is sufficient evidence from preclinical xenograft experiments to indicate the potent anti-tumor efficacy of FL118 that could eventually pave the way for clinical trials.

Terameprocol (tetra-O-methyl nordihydroguaiaretic acid), a semisynthetic plant lignan derivative, is a selective inhibitor of Sp1-regulated proteins such as CDK2, survivin, and VEGF expression to prevent proliferation and angiogenesis and promote apoptosis (Smolewski, 2008). In preclinical models, terameprocol, effectively suppressed the growth of several human tumor xenografts (Park et al., 2005). Terameprocol was also found to be very effective as an anti-cancer agent in combination with other drugs (Kimura et al., 2023). Phase I/II clinical trials with terameprocol showed promise in patients with recurrent high-grade glioma and advanced leukemia with strong safety profile (Grossman et al., 2012; Tibes et al., 2015).

10.8.2.2 SMIs that Prevent Homodimerization

Computational modeling and *in silico* analysis led to the discovery of several small molecules that prevent homodimerization of survivin. LLP3, LLP9, and LQZ-7 are some of the SMIs that inhibit survivin by binding to the dimerization interface of survivin (Albadari & Li, 2023). LLP3 and LLP9 blocked cell cycle and induced defects in mitosis in cancer cells even at low nanomolar concentrations besides enabling chemosensitivity (Chettiar et al., 2013; Steigerwald et al., 2018). LLP3 was demonstrated to disrupt the formation of the survivin–Ran complex, thereby affecting spindle formation in tumors (Guvenc et al., 2013). LQZ-7 prevented survivin dimerization by specifically targeting Leu98 and Phe101 in the dimerization interface with consequent proteasomal degradation (Qi et al., 2016). Analogs of LQZ-7 were shown to suppress prostate cancer cell growth in xenograft models (Peery et al., 2020).

10.8.2.3 SMIs that Prevent Interaction of Survivin with Partner Proteins

Deazaflavin analog compound 1 disrupts the interaction of survivin with Smac and stimulates apoptosis. PZ-6-QN was shown to enter the mitochondria and block the interaction of survivin with Smac with release of Smac as well as cytochrome c into the cytosol (Park et al., 2019). UC-112 and its analogs inhibit survivin via ubiquitin-mediated degradation (Wang et al., 2018a). MX-106 displayed greater selectivity to survivin relative to UC-112 (Albadari et al., 2021). Shepherdin, a small molecule peptidomimetic inhibitor and survivin-HSP complex antagonist is currently under clinical trial. Shepherdin functions as an HSP90 inhibitor via competitive inhibition with ATP, thereby preventing binding of HSP90 with survivin. Shepherdin also binds to and destabilizes HSP90 client proteins such as CDK4, CDK6, and Akt (Zhang et al., 2022b). The adenovirus-mediated gene transfer of shepherdin into gall bladder carcinoma cells and xenograft tumors exerted potent growth inhibitory and apoptosis inducing effects (Zhu et al., 2015). Co-delivery of 5-aminoimidazole-4-carboxamide-1-β-D-ribofuranoside (AICAR) a nonpeptidic SMI and doxorubicin conjugated multifunctional nanoparticles, increased cellular uptake and apoptosis, and inhibited proliferation in tumor-derived cell lines (Daglioglu & Okutucu, 2017).

10.8.3 Dominant Negative Mutants

Dominant negative mutants of survivin offer great promise as anti-cancer agents. These are competitive inhibitors of endogenous survivin that can induce apoptosis of cancer cells and CSCs as well as chemosensitization. Guo et al. (2021) developed a series of amino-acid mutants from survivin (TmSm34, TmSm48, TmSm84, TmSm34/48, TmSm34/84, and TmSm34/48/84)-fused cell-permeable peptide TATm at the N-terminus. The dominant negative mutant TmSm34/84 displayed more robust pro-apoptotic activity against A549 cells compared to the other mutants and reversed adriamycin resistance in A549 CSCs by decreasing the expression of survivin, Bcl-2, and P-glycoprotein and increasing caspase-3 expression.

10.8.4 Immunotherapy

Survivin-based immunotherapeutic agents that have been investigated include peptides, vaccines, and monoclonal antibodies. The survivin-based immunotherapy approaches are generally well tolerated and have generated good immunological responses both as single agents and in combination. For instance, EMD640744 is a multiepitope vaccine used for solid tumors (Lennerz et al., 2014; Zhenjiang et al., 2018), and SurVaxM for malignant glioblastoma multiformis (Fenstermaker & Ciesielski, 2014).

10.8.4.1 Survivin-Derived Peptides

When survivin-derived peptides are loaded on to the major histocompatibility complex (MHC), cytolytic T cell response is stimulated with resultant lysis of human leukocyte class I or II antigen (HLA)-matched target cancer cells (Garg et al., 2016). Survivin derived epitopes were found to provoke CD4+ and CD8+ T-cell responses in acute lymphocytic leukemia patients and decrease recurrence (Widenmeyer et al., 2012). Adoptive T cell therapy is an effective therapeutic modality that identifies survivin-specific clones and discriminates malignant from nonmalignant cells (Arber et al., 2015).

10.8.4.2 Survivin-Based Vaccines

Several survivin-based vaccines that are specific to malignant cells have entered clinical trials and reported to have a good safety profile (Warrier et al., 2020). A subunit vaccine containing survivin proved to be effective against lung cancer (Srivastava et al., 2012). Domínguez-Romero et al. (2020) generated multiepitope cancer vaccines from a novel class of vaccine immunogens called variable epitope libraries using three regions of survivin, comprising 40, 49, and 51 amino acids, together with the complete survivin protein. Vaccination of BALB/c mice harboring the highly aggressive 4T1 mammary cell line significantly suppressed tumor growth and metastasis following single immunization associated with enhanced cellular immune response. DPX-Survivac, as combination therapy with metronomic cyclophosphamide, generated an immune response ovarian cancer patients (Berinstein et al., 2015). Combination therapy with DPX-Survivac, pembrolizumab, and intermittent low-dose cyclophosphamide was effective and safe in patients with relapsed/refractory diffuse large B-cell lymphoma (Amitai et al., 2023). A Phase I first-in-human trial (NCT02851056) investigated the feasibility and safety of a dendritic cell (DC) vaccine transduced with an adenoviral vector encoding full-length survivin, harboring mutations that neutralized its IAP function in multiple myeloma. Two doses of the vaccine administered before and after autologous stem cell transplant generated a good immune response and durable clinical outcomes without causing any adverse side effects (Freeman et al., 2023).

10.8.4.3 Monoclonal Antibodies

Monoclonal antibodies that reduce survivin gene expression indirectly such as trastuzumab, which targets ErbB2, and Wnt-2 Ab that targets the Wnt pathway have been developed (You et al., 2004; Zhu et al., 2010). Combination therapy with involumab and autologous dendritic cell-based vaccines pulsed with survivin is under clinical investigation and appears to be well tolerated in Phase I and IIa trials (Clinical Trials Identifier: NCT04199559).

10.8.5 RNA Interference (RNAi)

Survivin expression was effectively silenced using the siRNA approach in neuroblastoma (Zhang et al., 2014). Survivin gene silencing was also achieved using microRNAs (miRNAs), such as miR-34a, miR-138p, miR-203, miR-214-3p, and miR-485-5p, in several cancers (Cao et al., 2013; Li et al., 2019b; Wang et al., 2018b; Xu et al., 2013; Yang et al., 2016).

10.8.6 Nanoparticle Platforms

Several nanotherapeutic approaches were developed for precise and controlled targeted delivery of survivin-based therapy. These include biocompatible nanoformulations such as liposomes, dendrimers, polymers, inorganic nanocarriers, nucleic acid nanotechnology, and exosomes to enable gene silencing or drug delivery (reviewed by George et al., 2022).

10.8.7 CRISPR/Cas-9 Gene Editing

The CRISPR/Cas9 gene-editing tool, was used to knock out the *BIRC5* gene in HL60 and KG1 leukemic cell lines. A 30-fold decrease of survivin mRNA was seen in the *BIRC5*-knocked out cells accompanied by apoptosis (Narimani et al., 2019). In SKOV3 and OVCAR3 ovarian cancer cells, lentiviral CRISPR/Cas9 nickase vector-based BIRC5 editing inhibited cell proliferation, migration, invasion, and chemosensitization (Zhao et al., 2017). Moreover, synergistic anti-cancer efficacy was achieved by co-delivery of doxorubicin and survivin CRISPR/Cas9-expressing plasmid into tumor cells (Li et al., 2022). Deng et al. (2023) combined CRISPR/Cas9-mediated gene editing and PROTACs (proteolysis-targeting chimeras) to target the survivin gene and control the fate of bladder cancer cells. They edited the survivin gene in the nucleus using photoactivatable split-Cas9-targeted gene editing with simultaneous degradation of the survivin protein in the cytoplasm using a nanobody-mediated target (paProtacL-Survivin). This approach prevented cell proliferation and migration, and stimulated apoptosis of bladder cancer cells *in vitro* and inhibited growth of UMUC-3 bladder cancer cell-derived xenograft *in vivo*. Recently, Li et al. (2024) combined photoactivation with the CRISPR/Cas12a-based sensing strategy for rapid detection of survivin that could have potential applications in biomarker imaging, point-of-care diagnosis, and treatment.

10.8.8 Targeting Survivin Expression in CSCs

The limited success of survivin in clinical trials has been attributed to nonspecific targeting in heterogeneous tumor-cell populations. Emerging evidence indicates that targeting survivin in CSCs is more promising. The strategies for targeting survivin in CSCs include combination therapy with survivin inhibitor and conventional treatment methods, targeting factors and pathways that promote self-renewal and stemness, RNAi-based methods using aptamer-survivin siRNA chimera, oncolytic virotherapy using competent replicating viruses that induce cell death and target chemoresistant CSCs, destruction of tumor vasculature, and targeted drug delivery by nanocomplexes (Alshamaileh et al., 2017; Kahm & Kim, 2023; Warrier et al., 2020).

10.9 CONCLUSION AND FUTURE PERSPECTIVES

Since its discovery more than 25 years ago, survivin has continued to attract the attention of oncologists as a potential cancer drug target given its high expression in a wide range of malignant tumors and absence from terminally differentiated, normal somatic cells. Survivin is a prototype member of the IAP family of proteins that suppresses apoptosis via caspase inhibition. Interestingly, survivin also inhibits autophagy, indicating its ability to concurrently regulate mitosis, apoptosis, and autophagy. With its rich interactome repertoire, survivin, despite its lack of catalytic activity, impacts myriad cellular signaling molecules and pathways that eventually culminate in neoplastic transformation. Further investigations on the mechanisms underlying protein–protein interactions that enable differential expression in normal and malignant cells, as well as under different conditions such as hypoxia and deprivation of nutrients and growth factors, are warranted.

Therapeutic targeting of survivin has generated mixed results, with some approaches that are not encouraging, while certain state-of-the art technologies and strategies have opened up opportunities as well as challenges. Therapeutic approaches should therefore envisage the complexity of the interplay between various factors that regulate survivin to achieve precise targeting. ASO therapy has met with limited success because of lack of target specificity and tumor heterogeneity. Although several SMIs of survivin have been investigated, only few have shown promise in clinical trials. Many natural products need to be screened for their ability to directly and precisely target survivin besides exploring the effects on regulators of survivin transcription/translation. Development of suitable assay systems to identify appropriate anti-survivin agents is imperative. Nanocomplexes and immunotherapeutic approaches have been promising. Nevertheless, given that anti-survivin therapeutics enhance apoptosis, a combinatorial approach with standard treatment modalities such as chemo- and radiotherapy appears to be a more pragmatic strategy to achieve synergy as well as to minimize toxicity and adverse side effects. Recently, Zhu et al. (2024) developed a novel therapeutic strategy for cutaneous squamous cell carcinomas using activable photosensitizers delivered by an aptamer-functionalized DNA probe for tumor targeting and a linear ASO for recognizing survivin mRNA. Precision tumor targeting and release of the ASO to downregulate survivin was achieved, resulting in enhanced efficacy of photodynamic therapy in both xenograft tumor-bearing mice and PDX models. This approach raises exciting possibilities for safe and efficient targeted therapy for other cancers.

DECLARATIONS

Conflict of interest: The authors declare no conflicts of interest.

Consent for publishing: All authors have approved the final version of the manuscript and given consent for publication.

Funding: Not Applicable.

REFERENCES

Abad MA, Ruppert JG, Buzuk L, et al. (2019) Borealin-nucleosome interaction secures chromosome association of the chromosomal passenger complex. J Cell Biol.218:3912–3925.

Albadari N, Deng S, Chen H, et al. (2021) Synthesis and biological evaluation of selective survivin inhibitors derived from the MX-106 hydroxyquinoline scaffold. Eur J Med Chem.224:113719.

Albadari N, Li W (2023) Survivin small molecules inhibitors: Recent advances and challenges. Molecules.28:1376.

Alshamaileh H, Wang T, Xiang D, et al. (2017). Aptamer-mediated survivin RNAi enables 5-fluorouracil to eliminate colorectal cancer stem cells. Sci Rep.7:5898.
Ambrosini G, Adida C, Altieri DC (1997). A novel anti-apoptosis gene, survivin, expressed in cancer and lymphoma. Nat Med.3:917–921.
Amitai I, Roos K, Rashedi I, et al. (2023) PD-L1 expression predicts efficacy in the phase II SPiReL trial with MVP-S, pembrolizumab, and low-dose CPA in R/R DLBCL. Eur J Haematol.111:191–200.
Arber C, Feng X, Abhyankar H, et al. (2015). Survivin-specific T cell receptor targets tumor but not T cells. J Clin Invest.125:157–168.
Asadian F, Ghadyani M, Antikchi MH, et al. (2022) Association of epidermal growth factor 61A>G, survivin -31G>C, and EFNA1 -1732G>A polymorphisms with susceptibility to colorectal cancer. J Gastrointest Cancer.53:78–83.
Beding AF, Larsson P, Helou K, et al. (2022) Pan-cancer analysis identifies BIRC5 as a prognostic biomarker. BMC Cancer.22:322.
Benaiges E, Ceperuelo-Mallafré V, Guaita S, et al. (2024) Survivin/BIRC5 as a novel molecular effector at the crossroads of glucose metabolism and radio resistance in head and neck squamous cell carcinoma. Head Neck. 46:1752–1765.
Berinstein NL, Karkada M, Oza AM, et al. (2015) Survivin-targeted immunotherapy drives robust polyfunctional T cell generation and differentiation in advanced ovarian cancer patients. Oncoimmunology. 4:e1026529.
Cacına C, Akgün A, Kayhan KB, et al. (2003) The analysis of Survivin promoter (-31G/C) gene variation in oral squamous cell carcinoma risk and prognosis. J Stomatol Oral Maxillofac Surg.8:101494.
Cao W, Fan R, Wang L, et al. (2013). Expression and regulatory function of miRNA-34a in targeting survivin in gastric cancer cells. Tumour Biology.34:963–971.
Chen X, Duan N, Zhang C, et al. (2016) Survivin and tumorigenesis: molecular mechanisms and therapeutic strategies. J Cancer.7:314–323.
Cheng XJ, Lin JC, Ding YF, et al. (2016) Survivin inhibitor YM155 suppresses gastric cancer xenograft growth in mice without affecting normal tissues. Oncotarget.7:7096–7109.
Chettiar SN, Cooley JV, Park IH, et al. (2013). Design, synthesis and biological studies of survivin dimerization modulators that prolong mitotic cycle. Bioorg Med Chem Lett.23:5429–5433.
Cheung CHA, Chang YC, Lin TY, et al. (2020) Anti-apoptotic proteins in the autophagic world: An update on functions of XIAP, Survivin, and BRUCE. J Biomed Sci. 27:31.
Colnaghi R, Connell CM, Barrett RM, et al. (2006) Separating the anti-apoptotic and mitotic roles of survivin. J Biol Chem. 281:33450–33456.
Coumar MS, Tsai F-Y, Kanwar JR, et al. (2013) Treat cancers by targeting survivin: just a dream or future reality? Cancer Treat Rev. 39:802–811.
Daglioglu C, Okutucu B (2017) Therapeutic effects of AICAR and DOX conjugated multifunctional nanoparticles in sensitization and elimination of cancer cells via survivin targeting. Pharm Res. 34:175–184.
Deng C, Li S, Liu Y, et al. (2023) Split-Cas9-based targeted gene editing and nanobody-mediated proteolysis-targeting chimeras optogenetically coordinated regulation of Survivin to control the fate of cancer cells. Clin Transl Med. 13:e1382.
Domínguez-Romero AN, Martínez-Cortés F, Munguía ME, et al. (2020) Generation of multiepitope cancer vaccines based on large combinatorial libraries of survivin-derived mutant epitopes. Immunology. 161:123–138.
Fang XL, Cao XP, Xiao J, et al. (2024) Overview of role of survivin in cancer: expression, regulation, functions, and its potential as a therapeutic target. J Drug Target.22:1–18.
Fang C, Kang B, Zhao P, et al. (2021) Targeting neuregulin 1 (NRG1): A novel biomarker for non-small-cell lung cancer. Environ Pathol Toxicol Oncol.40:61–72.
Fenstermaker RA, Ciesielski MJ (2014). Challenges in the development of a survivin vaccine (SurVaxM) for malignant glioma. Expert Rev Vaccines.13:377–385.
Fernández JG, Rodríguez DA, Valenzuela M, et al. (2014) Survivin expression promotes VEGF-induced tumor angiogenesis via PI3K/Akt enhanced β-catenin/Tcf-Lef dependent transcription. Mol Cancer.13:209.
Freeman CL, Atkins R, Varadarajan I, et al. (2023) Survivin dendritic cell vaccine safely induces immune responses and is associated with durable disease control after autologous transplant in patients with myeloma. Clin Cancer Res. 22:OF1–OF11.
Fung S, Knoefel WT, Krieg A (2021) Clinicopathological and prognostic significance of Inhibitor of Apoptosis Protein (IAP) family members in lung cancer: A meta-analysis. Cancers (Basel). 13:4098.
Galloway NR, Ball KF, Stiff T, et al. (2017) Yin Yang 1 (YY1): Regulation of survivin and its role in invasion and metastasis. Crit Rev Oncog.22:23–36.

Garg H, Suri P, Gupta JC, et al. (2016). Survivin: a unique target for tumor therapy. Cancer Cell Int.16:49.
Gasowska-Bajger B, Gasowska-Bodnar A, Knapp P, et al. (2021) Prognostic significance of survivin expression in patients with ovarian carcinoma: a meta-analysis. J Clin Med.10:879.
George R, Hehlgans S, Fleischmann M, et al. (2022) Advances in nanotechnology-based platforms for survivin-targeted drug discovery. Expert Opin Drug Discov. 17:733–754.
Grossman SA, Ye X, Peereboom D, et al. (2012) Adult Brain Tumor, C. Phase I study of terameprocol in patients with recurrent high-grade glioma. Neuro Oncol.14:511–517.
Guha M, Plescia J, Leav I, et al. (2009) Endogenous tumor suppression mediated by PTEN involves survivin gene silencing. Cancer Res.69:4954–4958.
Guo W, Ma X, Fu Y, et al. (2021) Discovering and characterizing of survivin dominant negative mutants with stronger pro-apoptotic activity on cancer cells and CSCs. Front Oncol. 11:635233.
Guvenc H, Pavlyukov MS, Joshi K, et al. (2013) Impairment of glioma stem cell survival and growth by a novel inhibitor for survivin-Ran protein complex. Clin Cancer Res. 19:631–642.
Hanifeh M, Ataei F (2022) XIAP as a multifaceted molecule in cellular signaling. Apoptosis27:441–453.
Hong M, Ren MQ, Silva J, et al. (2017) YM155 inhibits topoisomerase function. Anti-Cancer Drugs.28:142–152.
Huang YH, Yeh CT (2019) Functional compartmentalization of HSP60-survivin interaction between mitochondria and cytosol in cancer cells. Cells9:23.
Jaiswal PK, Goel A, Mittal RD (2015) Survivin: a molecular biomarker in cancer. Indian J Med Res.141:389–397.
Jensen M, Chandrasekaran V, García-Bonete MJ, et al. (2023) Survivin prevents the polycomb repressor complex 2 from methylating histone 3 lysine 27. iScience. 26:106976.
Kahm YJ, Kim RK (2023) BIRC5: a novel therapeutic target for lung cancer stem cells and glioma stem cells. Biochem Biophys Res Commun. 682:141–147.
Kimura K, Chun JH, Lin YL, et al. (2023) Tetra-O-methyl-nordihydroguaiaretic acid inhibits energy metabolism and synergistically induces anticancer effects with temozolomide on LN229 glioblastoma tumors implanted in mice while preventing obesity in normal mice that consume high-fat diets. PLoS One.18:e0285536.
Kong L, Qi R, Zhou G, et al. (2021) Correlation analysis of survivin, ING4, CXCL8 and VEGF expression in prostate cancer tissue. Am J Transl Res.13:13784–13790.
Krieg S, Roderburg C, Fung S, et al. (2022) Nuclear survivin is A prognosticator in gastroenteropancreatic neuroendocrine neoplasms: A meta-analysis. J Cancer Res Clin Oncol.148:2235–2246.
Lennerz V, Gross S, Gallerani E, et al. (2014) Immunologic response to the survivin-derived multi-epitope vaccine EMD640744 in patients with advanced solid tumors. Cancer Immunol Immunother. 63:381–394.
Li F (2014) Anticancer drug FL118 is more than a survivin inhibitor: where is the Achilles' heel of cancer? Am J Cancer Res.4:304–311.
Li F, Aljahdali I, Ling X (2019a) Cancer therapeutics using survivin BIRC5 as a target: what can we do after over two decades of study? J Exp Clin Cancer Res.38:368.
Li F, Aljahdali IAM, Zhang R, et al. (2021a) Kidney cancer biomarkers and targets for therapeutics: survivin (BIRC5), XIAP, MCL-1, HIF1α, HIF2α, NRF2, MDM2, MDM4, p53, KRAS and AKT in renal cell carcinoma. J Exp Clin Cancer Res.40:254.
Li Y, He W, Gao X, et al. (2023b) Cullin7 induces docetaxel resistance by regulating the protein level of the antiapoptotic protein survivin in lung adenocarcinoma cells. J Thorac Dis.15:5006–5019.
Li F, Ling X, Chakraborty S, et al. (2023a) Role of the DEAD-box RNA helicase DDX5 (p68) in cancer DNA repair, immune suppression, cancer metabolic control, virus infection promotion, and human microbiome (microbiota) negative influence. J Exp Clin Cancer Res. 42:213.
Li Y, Lu W, Yang J, et al. (2021b) Survivin as a biological biomarker for diagnosis and therapy. Expert Opin Biol Ther.21:1429–1441.
Li Q, Lv X, Tang C, et al. (2022) Co-delivery of doxorubicin and CRISPR/Cas9 or RNAi-expressing plasmid by chitosan-based nanoparticle for cancer therapy. Carbohydr Polym.287:119315.
Li S, Sun Y, Tian T, et al. (2019b). MicroRNA-214-3p modified tetrahedral framework nucleic acids target survivin to induce tumor cell apoptosis. Cell Prolif.53:e12708.
Li QN, Wang DX, Chen DY, et al. (2024) Photoactivatable CRISPR/Cas12a sensors for biomarkers imaging and point-of-care diagnostics. Anal Chem.96:2692–2701.
Li X, Yang F, He N, et al. (2023c) YM155 inhibits neuroblastoma growth through degradation of MYCN: A new role as A USP7 inhibitor. Eur J Pharm Sci. 181:106343.
Ling X, Wu W, Fan C, et al. (2018) An ABCG2 nonsubstrate anticancer agent FL118 targets drug-resistant cancer stem-like cells and overcomes treatment resistance of human pancreatic cancer. J Exp Clin Cancer Res.37:240.
Meng L, Fuhao Z, Xiaoming Z, et al. (2018) Survivin is critically involved in VEGFR2 signaling-mediated esophageal cancer cell survival. Biomed Pharmacother.107:139–145.

Mohamed AA, Yassin AS, Gomaa BS, et al. (2021) Association of polymorphism in survivin gene and the risk of liver cancer resulting from hepatitis c virus among Egyptian patients. Curr Cancer Drug Targets. 21:536–543.

Mull AN, Klar A, Navara CS (2014) Differential localization and high expression of SURVIVIN splice variants in human embryonic stem cells but not in differentiated cells implicate a role for SURVIVIN in pluripotency. Stem Cell Res.12:539–549.

Mumlek I, Ozretić P, Sabol M, Leović M, Glavaš-Obrovac L, Leović D, Musani V (2023) *BIRC5* gene polymorphisms are associated with a higher stage of local and regional disease in oral and oropharyngeal squamous cell carcinomas. Int J Mol Sci2417490.

Nakahara T, Kita A, Yamanaka K, et al. (2007) YM155, a novel small-molecule survivin suppressant, induces regression of established human hormone-refractory prostate tumor xenografts. Cancer Res. 67:8014–8021.

Narimani M, Sharifi M, Jalili A (2019) Knockout of *BIRC5* Gene by CRISPR/Cas9 induces apoptosis and inhibits cell proliferation in leukemic cell lines, HL60 and KG1. Blood Lymphat Cancer.9:53–61.

Nogueira-Ferreira R, Vitorino R, Ferreira-Pinto MJ, et al. (2013) Exploring the role of post-translational modifications on protein-protein interactions with survivin. Arch Biochem Biophys.538:64–70.

Park R, Chang CC, Liang YC, et al. (2005) Systemic treatment with tetra-O-methyl nordihydroguaiaretic acid suppresses the growth of human xenograft tumors. Clin Cancer Res.11:4601–4609.

Park SH, Shin I, Park SH, et al. (2019) An inhibitor of the interaction of survivin with Smac in mitochondria promotes apoptosis. Chem Asian J.14: 4035–4041.

Pavlyukov MS, Antipova NV, Balashova MV, et al. (2011) Survivin monomer plays an essential role in apoptosis regulation. J Biol Chem.286:23296–23307.

Peery R, Kyei-Baffour K, Dong Z, et al. (2020) Synthesis and identification of a novel lead targeting survivin dimerization for proteasome-dependent degradation. J Med Chem.63:7243–7251.

Qi J, Dong Z, Liu J, et al. (2016) Effective targeting of the survivin dimerization interface with small-molecule inhibitors. Cancer Res.76:453–462.

Quidute P, Quidute R, Perez MM, et al. (2023) Survivin expression as a prognostic marker for breast cancer. Rev Assoc Med Bras.69:e20230167.

Radovanović M, Petrović M, Šantrić V, et al. (2023) P53 and survivin expression in renal cell carcinoma. Urol Ann.15:186–190.

Raetz EA, Morrison D, Romanos-Sirakis E, et al. (2014) A phase I study of EZN-3042, A novel survivin messenger ribonucleic acid (mRNA) antagonist, administered in combination with chemotherapy in children with relapsed acute lymphoblastic leukemia (ALL): a report from the therapeutic advances in childhood leukemia and lymphoma (TACL) consortium. J Pediatr Hematol Oncol. 36:458–463

Rafatmanesh A, Behjati M, Mobasseri N, et al. (2020) The survivin molecule as a double-edged sword in cellular physiologic and pathologic conditions and its role as a potential biomarker and therapeutic target in cancer. J Cell Physiol.235:725–744.

Raj D, Liu T, Samadashwily G, et al. (2008) Survivin repression by p53, Rb and E2F2 in normal human melanocytes. Carcinogenesis.29: 194–201.

Rodriguez JA, Span SW, Ferreira CG, et al. (2002) CRM1-mediated nuclear export determines the cytoplasmic localization of the antiapoptotic protein survivin. Exp Cell Res.275:44–53.

Sah NK, Seniya C (2015) Survivin splice variants and their diagnostic significance. Tumour Biol. 36:6623–6631.

Sanhueza C, Wehinger S, Castillo Bennett J, et al. (2015) The twisted survivin connection to angiogenesis. Mol Cancer.14:198.

Smolewski P (2008) Terameprocol, a novel site-specific transcription inhibitor with anticancer activity. IDrugs.11:204–214.

Srivastava AK, Sharma RK, Yolcu ES, et al. (2012) Prime-boost vaccination with SA-4-1BBL costimulatory molecule and survivin eradicates lung carcinoma in CD8+ T and NK cell dependent manner. PLoS One. 7:e48463.

Steigerwald C, Rasenberger B, Christmann M, et al. (2018) Sensitization of colorectal cancer cells to irinotecan by the survivin inhibitor LLP3 depends on XAF1 proficiency in the context of mutated p53. Arch Toxicol.92:2645–2648.

SušacI, Ozretić P, Gregorić M, et al. (2019) Polymorphisms in survivin (*BIRC5* Gene) are associated with age of onset in breast cancer patients. J Oncol.2019:3483192.

Szczyrek M, Mlak R, Szudy-Szczyrek A, et al. (2022) Polymorphism of baculoviral inhibitor of apoptosis repeat-containing 5 (BIRC5) can be associated with clinical outcome of non-small cell lung cancer. Cells.11:956.

Talbot DC, Ranson M, Davies J, et al. (2010) Tumor survivin is downregulated by the antisense oligonucleotide LY2181308: A proof-of-concept, first-in-human dose study. Clin Cancer Res.16:6150–6158.

Tanioka M, Nokihara H, Yamamoto N, et al. (2011) Phase I study of LY2181308, an antisense oligonucleotide against survivin, in patients with advanced solid tumors. Cancer Chemother Pharmacol.68:505–511.

Temme A, Diestelkoetter-Bachert P, Schmitz M, et al. (2005) Increased p21(ras) activity in human fibroblasts transduced with survivin enhances cell proliferation. Biochem Biophys Res Commun.327:765–773.

Tibes R, McDonagh KT, Lekakis L, et al. (2015) Phase I study of the novel Cdc2/CDK1 and AKT inhibitor terameprocol in patients with advanced leukemias. Investig New Drugs.33:389–396.

Verdecia MA, Huang H, Dutil E, et al. (2000) Structure of the human anti-apoptotic protein survivin reveals a dimeric arrangement. Nat Struct Biol.7:602–608.

Wang Q, Arnst KE, Xue Y, Lei ZN, Ma D, Chen ZS, Miller DD, Li W (2018a) Synthesis and biological evaluation of indole-based UC-112 analogs as potent and selective survivin inhibitors. Eur J Med Chem.149:211–224.

Wang M, Cai WR, Meng R, Chi J-R, et al. (2018b). miR-485-5p suppresses breast cancer progression and chemosensitivity by targeting survivin. Biochem Biophys Res Commun.501:48–54.

Wang N, Li Y, Zhou RM, et al. (2019) The effect of polymorphisms in the promoter of the BIRC5 gene on the risk of oesophageal squamous cell carcinoma and patient's outcomes. Mutagenesis.34:307–313.

Wang J, Liu Z, Zhang D, et al. (2017) FL118, a novel survivin inhibitor, wins the battle against drug-resistant and metastatic lung cancers through inhibition of cancer stem cell-like properties. Am J Transl Res.9:3676–3686.

Warrier NM, Agarwal P, Kumar P (2020) Emerging importance of survivin in stem cells and cancer: the development of new cancer therapeutics. Stem Cell Rev Rep.16:828–852.

West TJ, Bi J, Martínez-Peña F, et al. (2023) A cell type selective YM155 prodrug targets receptor-interacting protein kinase 2 to induce brain cancer cell death. J Am Chem Soc. doi:10.1021/jacs.2c11715.

Westover D, Ling X, Lam H, et al. (2015) FL118, a novel camptothecin derivative, is insensitive to ABCG2 expression and shows improved efficacy in comparison with irinotecan in colon and lung cancer models with ABCG2-induced resistance. Mol Cancer.14:92.

Wheatley SP, Altieri DC (2019) Survivin at a glance. J Cell Sci.132:jcs223826.

Widenmeyer M, Griesemann H, Stevanovic S, et al. (2012) Promiscuous survivin peptide induces robust CD41 T-cell responses in the majority of vaccinated cancer patients. Int J Cancer.131: 140–149.

Wolf P (2023) Inhibitor of apoptosis proteins as therapeutic targets in bladder cancer. Front Oncol.13:1124600.

Xia F, Canovas PM, Guadagno TM, et al. (2008) A survivin-Ran complex regulates spindle formation in tumor cells. Mol Cell Biol.28:5299–5311.

Xu M, Hu X, Zhang M, et al. (2020) What is the impact of BIRC5 gene polymorphisms on urinary cancer susceptibility? Evidence from 9348 subjects. Gene.733:144268.

Xu D, Wang Q, An Y, et al. (2013). MiR-203 regulates the proliferation, apoptosis and cell cycle progression of pancreatic cancer cells by targeting survivin. Mol Med Rep.8:379–384.

Yang R, Liu M, Liang H, et al. (2016) miR-138-5p contributes to cell proliferation and invasion by targeting survivin in bladder cancer cells. Mol Cancer.15:82.

You L, He B, Xu Z, et al. (2004). Inhibition of Wnt-2-mediated signaling induces programmed cell death in non-small-cell lung cancer cells. Oncogene.23:6170–6174.

Zeng YZ, Zhang YQ, Lin XQ, et al. (2021) Co-expression of VEGF-C and survivin predicts poor prognosis in esophageal squamous cell carcinoma. Transl Cancer Res.10:210–222.

Zhang Y, Chen HX, Zhou SY, et al. (2015) Sp1 and c-Myc modulate drug resistance of leukemia stem cells by regulating survivin expression through the ERK-MSK MAPK signaling pathway. Mol Cancer.14:56.

Zhang J, Li H, Liu Y, et al. (2022b) Targeting HSP90 as a novel therapy for cancer: mechanistic insights and translational relevance. Cells.11:2778.

Zhang L, Liang H, Cao W, et al. (2014) Downregulation of survivin by siRNA inhibits invasion and promotes apoptosis in neuroblastoma SH-SY5Y cells. Braz J Med Biol Res.47:548–553.

Zhang Q, Chen X, Hu Y, et al. (2023) BIRC5 inhibition is associated with pyroptotic cell death via caspase3-GSDME pathway in lung adenocarcinoma cells. Int J Mol Sci.24:14663.

Zhang G, Yin R, Dai X, et al. (2022a) Design, synthesis, and biological evaluation of novel 7-substituted 10,11-methylenedioxy-camptothecin derivatives against drug-resistant small-cell lung cancer *in vitro* and *in vivo*. Eur J Med Chem.241:114610.

Zhao J, Tenev T, Martins LM, et al. (2000) The ubiquitin-proteasome pathway regulates survivin degradation in a cell cycle-dependent manner. J Cell Sci.113(Pt 23):4363–4371.

Zhao G, Wang Q, Gu Q, et al. (2017) Lentiviral CRISPR/Cas9 nickase vector mediated BIRC5 editing inhibits epithelial to mesenchymal transition in ovarian cancer cells. Oncotarget.8:94666–94680.

Zhenjiang L, Rao M, Luo X, et al. (2018) Cytokine networks and survivin peptide-specific cellular immune responses predict improved survival in patients with glioblastoma multiforme. EBioMed.33:49–56.

Zhou Y, Tang Y, Luo J, Yang Y, Zang H, Ma J, Fan S, Wen Q (2023) High expression of HSP60 and survivin predicts poor prognosis for oral squamous cell carcinoma patients. BMC Oral Health.23:629.

Zhu N, Gu L, Findley HW, et al. (2006) KLF5 interacts with p53 in regulating survivin expression in acute lymphoblastic leukemia. J Biol Chem.281:14711–14718.

Zhu A, Ren Y, Wang N, et al. (2015) Adeno-associated virus mediated gene transfer of Shepherdin inhibits gallbladder carcinoma growth *in vitro* and *in vivo*. Gene.572:87–94.

Zhu H, Zhang G, Wang Y, et al. (2010) Inhibition of ErbB2 by herceptin reduces survivin expression via the ErbB2–β-catenin TCF4-survivin pathway in ErbB2-overexpressed breast cancer cells. Cancer Sci.101: 1156–1162.

Zhu J, Peng L, Jehan S, Wang H, Chen X, Zhao S, Zhou W. (2024) Activable photodynamic DNA probe with an "AND" logic gate for precision skin cancer therapy. Research (Wash D C). 7:0295.

11 Ubiquitin-specific Proteases for the Treatment of Breast Cancer

Suchetana Pal, Subhasish Sarkar, Rajlakshmi Laha, and Somasri Dam
The University of Burdwan, Burdwan, India

11.1 INTRODUCTION

Breast cancer is the most common cancer in women, representing an 18% rise over lung cancer (Siegel et al., 2020). The controlled regulation of protein turnover is a critical component. With a half-life of 10 minutes, nearly 30% of newly synthesised proteins in the body are rapidly destroyed. Alterations in the ubiquitin proteasome system (UPS) cause disruptions in the regulation of protein levels, protein interactions, and signalling pathways, ultimately playing a significant role in the development and progression of breast cancer (Xiao et al., 2016). The UPS degrades more than 80% of cellular proteins, highlighting the relevance of this route in the control of several cellular activities. Mutations or altered expression of certain components within the ubiquitination cascade have been linked to cancer. Ubiquitination is an essential post-translational modification, mediated by three classes of enzymes: ubiquitin activating enzymes (E1), ubiquitin conjugating enzymes (E2), and ubiquitin ligases (E3). E1 activates ubiquitin and transfers it to E2, whereas E3 brings the substrate proteins to facilitate the transfer of ubiquitin moieties. In breast cancer, ubiquitin E3 ligases can behave as oncoproteins or tumour suppressors. Breast cancer 1 (BRCA1) cancer-predisposing missense mutations, for example, inactivate its ubiquitin ligase activity, implying that the tumour suppressor role of BRCA1 is linked to its E3 ligase function (Ruffner et al., 2001). Deubiquitinases (DUBs), a class of enzymes encompassing both cysteine proteases and metalloproteases, are involved in the dismantling of ubiquitin-protein associations, leading to the reversal of ubiquitination. About 100 DUBs are encoded in the human genome and they have been divided into six families: ubiquitin-specific proteases (USPs), ubiquitin carboxy-terminal hydrolases (UCHs), ovarian tumour proteases (OTUs), Machado-Joseph disease protein domain proteases (MJDs), JAMM/MPN domain-associated metallopeptidases (JAMMs), and the monocyte chemotactic protein-induced protein (MCPIP) family (Nijman et al., 2005). With more than60 proteases in humans, the USPs are the biggest DUB family. DUBs influence protein stability, signal transduction, and other non-proteasomal activities, thereby playing an important role in the control of critical cancer proteins and pathways. Successful use of proteasome inhibitors in models of human malignancies demonstrated significant therapeutic efficacy in cancer. In this review, we discuss the mechanisms of USPs in the promotion of breast cancer and address the role of the small-molecule USP inhibitors in breast cancer therapy.

11.2 STRUCTURE OF USP

The deubiquitinase activity of USPs is directly correlated with their structural features. Every USP features three conserved subdomains resembling the palm, thumb, and fingers of the right hand. The finger domain is in charge of interactions with distal ubiquitin, whereas the catalytic site is positioned between the palm and thumb domains (Komander et al., 2009). Given their structural variety with

 DOI: 10.1201/9781003394693-13

extra domains and terminal extensions, the majority of USPs has a common feature: a characteristic conformational shift. When certain portions of USP domains are incapable of effectively interacting with the target protein, they assume a non-operational catalytic configuration. When ubiquitin binds, it triggers a transition in USPs from an inactive state to an enzymatically active state. The catalytic centre of USP7/HAUSP was the first USP protein to have an X-ray structure revealed.

11.3 USPs SHOWING ALTERED EXPRESSION IN BREAST CANCER

USPs represent a diverse family of proteins having a deubiquitinating activity. USPs have received particular attention as a cancer target since many of them exhibit differential expression or activation in tumours or their microenvironment, making them excellent candidates for therapeutic development. USPs implicated in every stage of breast cancer development are listed and displayed in Table 11.1.

TABLE 11.1

Roles of USPs in Breast Cancer Progression

USP Family Member	Functions in Breast Cancer	Substrates	Deubiquitination-Induced Change	Pathways Involved	References
USP1	• Promotes metastasis in breast cancer cells • Regulates pro-metastatic genes	KPNA2, TAZ, ERα, TGF-β	Protein stabilisation	KPNA2, ERα signalling, Hippo signalling pathway, TGF-β signalling	Han et al., 2021; Ma et al., 2019; Mussell et al., 2020; Niu et al., 2020
USP2	• Localises and stabilises EGFR • Promotes Mdm2-mediated p53 degradation	EGFR, BMI1	Protein stability or activity	ERα signalling	Liu et al., 2013; Stevenson et al., 2007
USP4	Promotes tumour growth and metastasis	HDAC2, TβRI	Protein stabilisation	TGF-β and AKT signalling	Zhang et al., 2012
USP7	• Upregulation of cyclin A2 • Forms a positive feedback loop with ECT2, promoting carcinogenesis	ERα, PHF8, ECT2, RNF169, PHF8, CHK1	Protein stabilisation and localisation		Wang et al., 2016; Xia et al., 2019b; Zhang et al., 2020c
USP9X	• Overexpresses in breast cancer and is correlated with CEP131 • TRB3 upregulation and induction of tumour promoting activities of Notch • Stabilises β-catenin and mediates TRAIL-resistance • Upregulation of cyclin D1 and downregulation of cyclin-dependent inhibitor kinase 1A	β-catenin, BCL9	Protein stabilisation and interaction	CEP131, Hippo pathway, Notch signalling, cyclin D1, Wnt signalling	Izrailit et al., 2017; Li et al., 2017; Li & Zheng, 2019; Ouyang et al., 2016

(Continued)

TABLE 11.1 *(Continued)*
Roles of USPs in Breast Cancer Progression

USP Family Member	Functions in Breast Cancer	Substrates	Deubiquitination-Induced Change	Pathways Involved	References
USP10	• Deubiquitinates and modifies TCF4 • Promotes breast cancer metastasis	TCF4, IGF2BP1	Protein stabilisation	–	Shi et al., 2023; Yang et al., 2023
USP11	• Promotes epithelial-mesenchymalplasticity induced by TGFβ • Interacts with BRCA2, potentially promoting cell survival in response to DNA damage	BRCA2, X-linked inhibitor of apoptosis protein (XIAP)	Protein stabilisation	TGF-β signalling, DNA damage, XIAP	Garcia et al., 2018; Zhou et al., 2017
USP14	Promotes tumour growth, promotes chemoresistance	Cyclin B1, androgen receptor	Protein stabilisation	–	Liu et al., 2019; Zhu et al., 2016
USP15	Influences cancer cell response to PARPi by regulating DNA repair processes	BARD1	Protein stabilisation	DNA repair, ERα signalling	Peng et al., 2019
USP18	Promotes breast cancer cell proliferation and inhibits apoptosis via the EGFR/AKT/Skp2 pathway	EGFR	–	AKT/Skp2 pathway	Tan et al., 2018
USP20	Correlated with SNAI2 protein, a metastasis-promoting transcription factor	SNAI2	Protein stabilisation	–	Li et al., 2020b
USP22	Overexpresses in breast cancer cells and positively regulates c-Myc and ERα	c-Myc, ERα	Protein stabilisation	ERα signalling	Wang et al., 2020
USP28	Stabilises proteins like MYC and LSD1	MYC, LSD1	Protein stabilisation	Wnt/β-catenin pathway	Popov et al., 2007; Wu et al., 2013
USP32	Prognostic gene with copy number alterations in estrogen receptor-positive human breast tumour	CCDC49	–	–	Schulte et al., 2012
USP37	• Regulates stemness, epithelial-mesenchymal transition and sensitivity to cisplatin • Stabilises cyclin A and is overexpressed in adriamycin-resistant breast cancer cells	Cyclin A	Protein stabilisation	Hedgehog pathway	Qin et al., 2018; Huang et al., 2011
USP51	Regulates ZEB1 and is associated with epithelial-to-mesenchymal transition	–	Protein stabilisation	–	Zhang et al., 2020d

11.3.1 USP1

USP1 regulates the expression of several pro-metastatic genes (Ma et al., 2019). For example, S100A4 induces metastasis in breast cancer. In basal-like breast cancers, LOXL2 regulates metastasis (Moreno-Bueno et al., 2011). Karyopherin Subunit Alpha 2 (KPNA2) is a nuclear protein and participates in cancer metastasis. USP1 stabilises KPNA2 by deubiquitination, thereby promoting metastasis (Ma et al., 2019). YAP/TAZ function as effectors in the Hippo signalling pathway, and the presence of USP1 enhances the stability of TAZ, leading to its increased expression in breast cancer (Mussell et al., 2020). USP1 is connected to estrogen receptor α (ERα) and improves ERα stability in breast cancer cells by preventing ER K48-linked poly-ubiquitination (Niu et al., 2020). Triple-negative breast cancer (TNBC) stands out as a highly fatal form of breast cancer, marked by a poor prognosis and a lack of tailored therapeutic options. In the case of TNBC, USP1 has been found to promote epithelial-mesenchymal plasticity (EMP) induced by TGF-β (Han et al., 2021).

11.3.2 USP2

USP2a plays a pivotal role in breast cancer, especially breast cancer by preventing epidermal growth factor receptor's (EGFR) ubiquitination and subsequent degradation, contributing to its sustained activity (Liu et al., 2013). USP2 overexpression is linked to poor clinical outcomes in breast cancer patients, particularly those with TNBC (Qu et al., 2015). USP2a also interacts with fatty acid synthase and the Mdm2-p53 axis, suggesting its role in tumourigenesis and cellular processes (Stevenson et al., 2007).

11.3.3 USP4

The genome-wide gain of function screening identifies that USP4 strongly induces TGF-β signalling. It deubiquitinates and stabilises TβRI in the plasma membrane. Metastasis and epithelial-mesenchymal transition (EMT) caused by TGF-β are reduced by USP4 depletion. It's significantly considered that AKT which is also called protein kinase B, has been linked to a poor diagnosis of breast cancer, reacts with and phosphorylates USP4 directly. Nuclear USP4 is transported to the cytoplasm and membrane via AKT-mediated phosphorylation, which is important for maintaining the protein's stability. Blocking TβRI kinase activity and lowering USP4 levels likewise hindered the metastatic spread of breast cancer cells induced by AKT. USP4 is a crucial factor in the interaction of the TGF-β and AKT signalling pathways (Zhang et al., 2012). USP4 triggers the expression of relaxin (RLX) that regulates TGF-β. The intracellular Smad2 proteins play important roles in the anti-mitogenic effect of TGF-β-1 and can induce metastasis and invasion of breast cancer cells. It upregulates MMP-9, suggesting that USP4 induces breast cancer cells through RLX-mediated TGF-β/Smad2/MMP-9 pathway (Cao et al., 2016).

11.3.4 USP7

USP7 regulates several cellular pathways like Mdm2, p53, and NF-κB. USP7 overexpression in many types of cancer indicates its therapeutic potential as a target (Wang et al., 2016). A positive feedback loop formed by USP7 and an oncogene epithelial cell transforming sequence 2 (ECT2), promotes breast cancer. USP7 deubiquitinates and stabilises ECT2, while ECT2 promotes USP7 intermolecular self-association, and stabilisation. These circuits increase breast cancer cell survival by maintaining the expression of the cancer-causing protein Mdm2 (Zhang et al., 2020c). The expression of USP7 and geminin has been found to have a strong correlation in breast cancer. The presence of USP7 contributes to the advancement of ERα positive breast cancer. Deletion of USP7 leads to ERα inhibition and cell death (Xia et al., 2019b).

11.3.5 USP9X

USP9, X-linked (USP9X) is physically associated with the centriolar satellite protein CEP131 and plays important roles in different types of malignancies. Overexpression of USP9X in breast cancer is correlated with CEP131. Thus, USP9X/CEP131 can be a potential target for breast cancer treatment (Li et al., 2017). In human breast cancer cell line, USP9X has a vital role in stabilising BRCA1, a tumour suppressor protein. This stabilisation results in resistance to chemotherapy or radiation therapy in cancer cells. This suggests USP9X to be a key cause of treatment failure in cancer cells. USP9X mediates resistance to apoptosis induced by TRAIL, a cell death-inducing ligand through the regulation of β-catenin. This interaction can resist apoptosis in breast cancer cells, offering insights into novel therapeutic approaches (Ouyang et al., 2016).

11.3.6 USP10

The N6-methyladenosine (m6A) modification and its related regulators are critical in the progression of breast cancer. Patients with an overexpressed m6A reader IGF2BP1 showed a clinically significant correlation with metastasis. IGF2BP1's ability to attach to USP10 enhances its stability by deubiquitination, resulting in increased expression levels. Additionally, the interaction of IGF2BP1 with m6A sites on CPT1A mRNA boosts mRNA stability, ultimately contributing to the progression of breast cancer metastasis (Shi et al., 2023). A pronounced overexpression of USP10 was also found in TNBC. USP10 aids in the deubiquitination of TCF4 and subsequently increases the protein's stability and contributes to the development of TNBC (Yang et al., 2023).

11.3.7 USP11

USP11 mediates the process of EMP, potentially facilitating their spread to other parts of the body (Garcia et al., 2018). USP11 interacts with BRCA2, which is a tumour suppressor gene having a pro-survival role in the cellular response to DNA damage. This interaction with USP11 suggests a mechanism by which BRCA2 may promote cell survival in case of DNA damage, thereby contributing to genomic stability (Schoenfeld et al., 2004). The X-linked inhibitor of the apoptosis protein (XIAP) has been identified as a target of USP11. XIAP belongs to the inhibitor of apoptosis (IAP) family and is distinguished for its role in blocking apoptosis by inhibiting caspase-9, 3, and 7 through protein degradation. USP11 promotes tumourigenesis by modulating XIAP levels by influencing XIAP turnover (Zhou et al., 2017). USP11 acts as a transcriptional regulator of ERα, thereby impacting the course of breast cancer and to be a therapeutic target for ERα positive breast cancer (Dwane et al., 2020).

11.3.8 USP14

USP14 plays a crucial role in promoting the rapid growth of breast cancer cells, a hallmark of cancer progression. Reduced expression of USP14 in breast cancer cells inhibits the multiplication and metastasis of cancer cells. By targeting USP14, it may be possible to slow down or even halt the uncontrolled division of cancer cells (Zhu et al., 2016).

In androgen receptor positive (AR+) human breast cancer cells, inhibition of androgen receptor deubiquitination by USP14 arrests the proliferation of cancer cells and induces apoptosis to stop cancer. USP14-mediated deubiquitination of cyclin B1 plays an important role in breast cancer (Liu et al., 2019).

11.3.9 USP18

Overexpression of USP18 suppresses apoptotic signalling in breast cancer cells. USP18 inhibition upregulates miR-7, which in turn suppresses the translation of EGFR messenger RNA and induces apoptosis (Duex et al., 2011).

11.3.10 USP22

USP22 has been reported to overexpress in breast cancer. In breast cancer cells, USP22 has been shown to positively regulate c-Myc level, and USP22 depletion affects the level and stability of c-Myc by deubiquitination. USP22 enhances ERα's stability and activity, which contributes to breast cancer progression by promoting the expression of genes associated with cell growth and tumour development (Wang et al., 2020).

11.3.11 USP37

An increase in the USP37 level has been associated with the regulation of stemness, EMT, and sensitivity to the chemotherapy drug cisplatin. This suggests that USP37 could play a significant role in breast cancer progression and response to treatment, making it a potential target for further research and therapeutic development (Qin et al., 2018).

Overexpression of USP37 stabilises cyclin A by deubiquitination, which promotes the transformation of cells from the G1 phase to the S phase (Huang et al., 2011). USP37 expression is upregulated in CD44+/CD24$^-$ breast cancer stem cells and controls stemness (Qin et al., 2018).

11.3.12 USP28

USP28 stabilises critical proteins, including Myc, which promotes cell proliferation (Popov et al., 2007). Additionally, USP28 contributes to angiogenesis by preventing the degradation of proteins like HIF-1α. Moreover, it enhances the stem cell-like properties of breast cancer cells through the stabilisation of LSD1, a chromatin modulator overexpressed in tumours. The high expression of USP28 in breast cancer tumours, along with its correlation with LSD1 levels, underscores its potential role as a target for therapeutic intervention in breast cancer (Wu et al., 2013).

11.3.13 USP32

USP32 has been reported to be highly expressed in breast cancer (Akhavantabasi et al., 2010). It has been identified as a prognostic gene that exhibits copy number alterations in ER-positive human breast tumours. The fusion gene USP32-CCDC49 has been identified as one of the nine expressed fusion genes during a structural analysis of the genome of ZR-75-30, an ER-positive breast cancer cell line. For the study of ER-positive breast tumours that fail to respond to tamoxifen therapy, this cell line has been suggested as a model system (Schulte et al., 2012).

11.3.14 USP51

USP51 is a well-known target of CDK4/6. CDK4/6 phosphorylates and activates USP51 which later deubiquitinates and stabilises zinc-finger E-box binding homeobox 1 (ZEB1). ZEB1 is associated with EMT and is also involved in breast cancer metastasis. This indicates a correlation between USP51 and breast cancer development suggesting this is a potential therapeutic target (Zhang et al., 2020d).

11.3.15 Others

An exome sequencing study suggests that the variants of USP39 are significantly associated with breast cancer development (Kuligina et al., 2020). USP20 is positively correlated with the level of SNAI2 protein, a metastasis-promoting transcription factor in breast tumour samples (Li et al., 2020b). USP15 influences cancer-cell response to poly-ADP-ribose polymerase inhibitors(PARPi) by regulating DNA repair processes in several cancer cells including breast cancer (Peng et al., 2019).

USP18 promotes breast cancer cell proliferation and inhibits apoptosis via the EGFR/AKT/Skp2 pathway. High USP18 levels are associated with advanced cancer stages and poorer prognosis, suggesting its potential as a therapeutic target in breast cancer (Tan et al., 2018). USP17L2, also known as DUB3, stabilises CDC25A and its overexpression induces oncogenic transformation in cases of breast and other cancers (Pereg et al., 2010).

11.4 USPs INHIBITORS IN BREAST CANCER TREATMENT

An alternative way to address cancer efficiently could involve directing attention towards USP inhibitors that might hold promise for treating various forms of malignancies. Several small molecule DUB inhibitors are known, ranging from general inhibitors to selective inhibitors that target certain DUB enzymes.

The development of USP inhibitors has been ongoing during the last several years (Huang et al., 2022). The inhibition of USPs emerges as a potential therapeutic possibility with a special targetable specificity, which proteasome inhibitors do not provide. The pan inhibitory activity of several USP inhibitors is a concerning factor. USP9X, USP5, and USP14 may all be directly blocked by WP1130, which has a broad pan-enzymatic DUB profile. This can be attributed to the similarity observed in the preserved catalytic region within the USP family. Several USP inhibitors have been explored for breast cancer, however none of them has received clinical use approval (Table 11.2).

TABLE 11.2
USP Inhibitors and Their Effects on Specific Cancer Subtype

USP Inhibitors	Structures	Breast cancer subtype	Targets	Effects
Pimozide	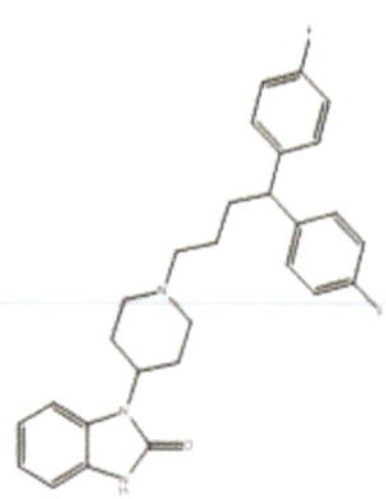	ER-negative BC, TNBC	USP1, USP 46	Suppression of angiogenesis and paracrine stimulation through cell cycle, AKT signalling pathway, EMT, MMP-9, vimentin, STAT3
ML323	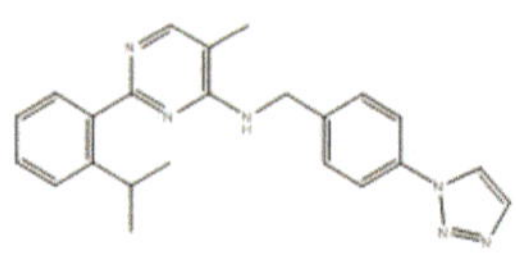	–	USP1	Supresses breast cancer metastasis
Rottlerin	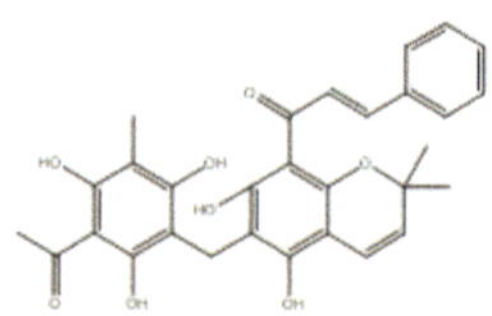	ER-positive BC, TNBC, CSCs	USP1	Induces autophagy, which results in apoptosis for breast cancer stem cells
Trifluoperazin	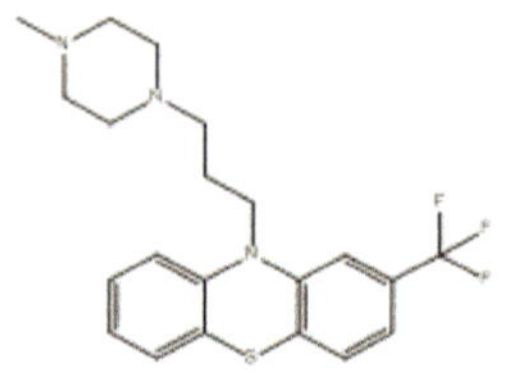	TNBC	USP1	Inhibits the growth of TNBC tumours and brain metastasis, G0/G1 arrest

TABLE 11.2 *(Continued)*

USP Inhibitors and Their Effects on Specific Cancer Subtype

USP Inhibitors	Structures	Breast cancer subtype	Targets	Effects
6-TG	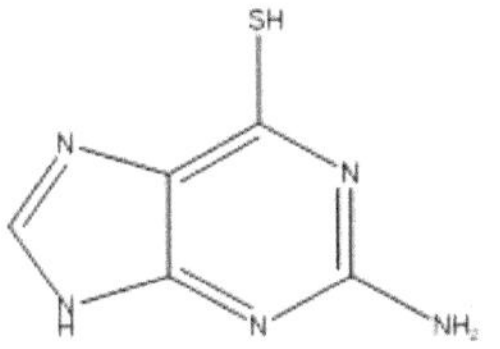	BRCA2-defective PARP inhibitor-resistant BC, BRCA1-mutant BC, TNBC	USP2	Selectively kill BRCA2-defective tumours and overcomes PARP inhibitor resistance, DNA repair
ML364	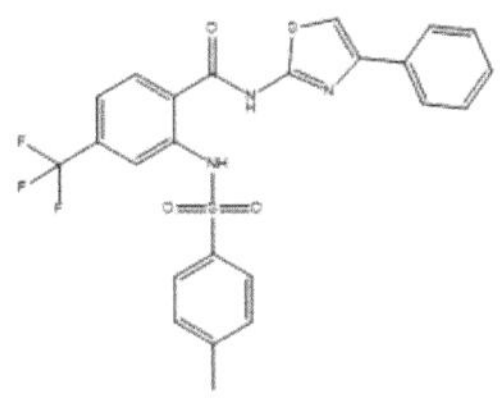	ER-positive BC	USP2	Inhibition of USP2 through endocytic degradation
WP1130	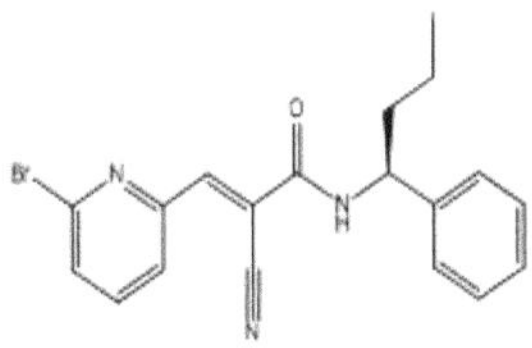	ER-negative BC	USP5, USP9X	Increases the cytotoxicity ofcisplatin in ER-negative breast cancer cells
P5091	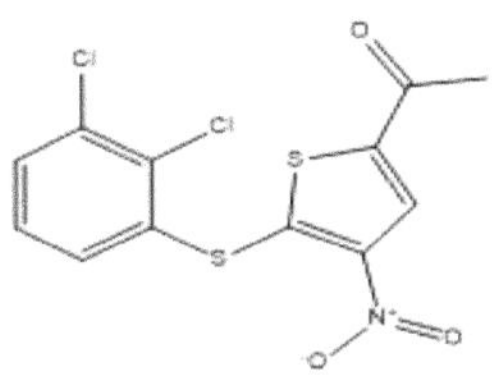	–	USP7/47	Blockage of deubiquitination, reduces cell proliferation, colony formation, migration, and sphere dissemination for breast cancer cell lines
Costunolide	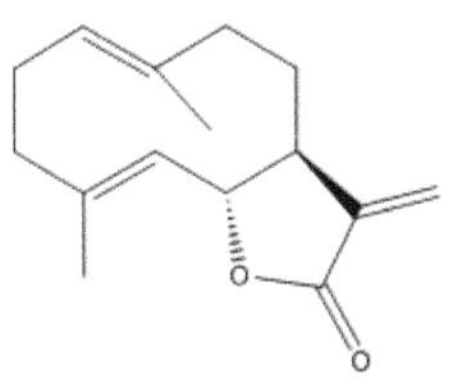	Metastatic TNBC	USP7	Induces breast cancer cell apoptosis
RA-9	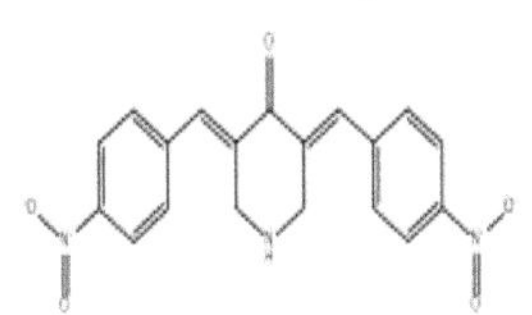	TNBC	USP2, USP14	Induces breast cancer cell apoptosis
Spautin-1	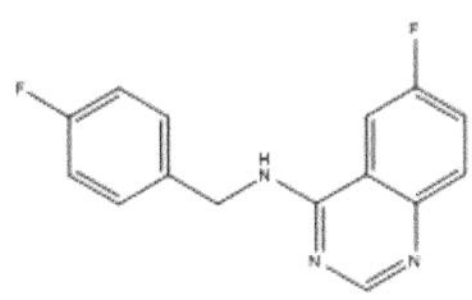	TNBC	USP10, USP13	Autophagy inhibition
Mitoxantrone	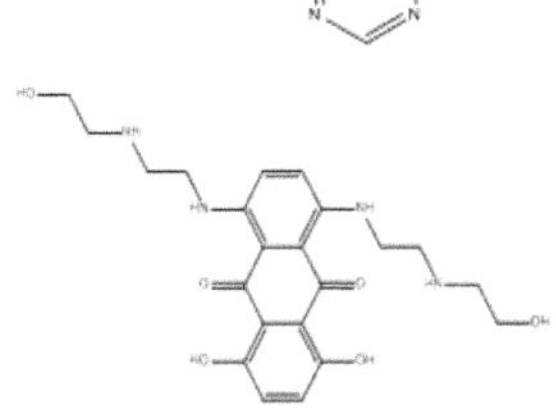	–	USP11	Increases apoptosis

(Continued)

TABLE 11.2 *(Continued)*
USP Inhibitors and Their Effects on Specific Cancer Subtype

USP Inhibitors	Structures	Breast cancer subtype	Targets	Effects
b-AP15	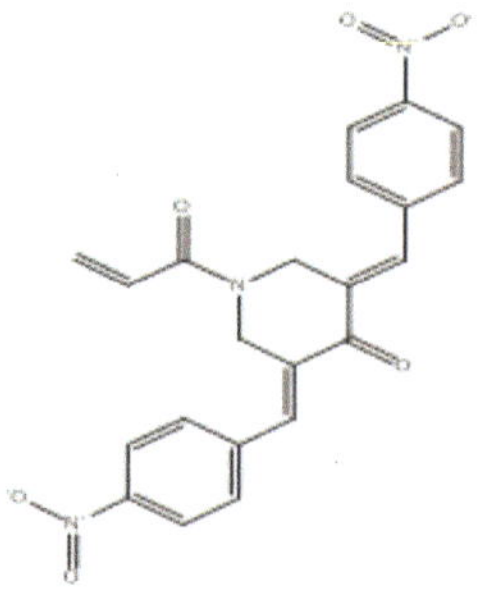	ER-positive BC, TNBC	USP14	Inhibits tumour progression
IU1	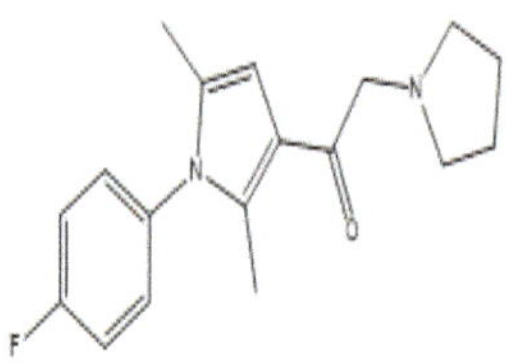	AR-positive BC	USP14	Suppresses cell proliferation and induces apoptosis in breast cancer cell lines
Auranofin	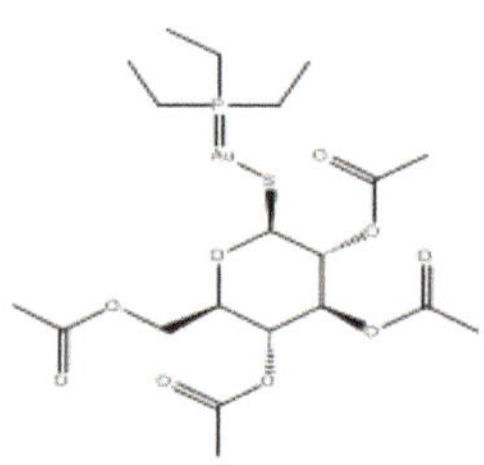	ER-positive BC, TNBC	USP14	Synergistic breast cancer inhibition

Abbreviations: AR: Androgen receptor; CSC: Cancer stem cell; ER: Estrogen receptor; TNBC: Triple negative breast cancer.

11.4.1 USP1 Inhibitors

The ability of pimozide to suppress angiogenesis and stimulate paracrine signalling has been demonstrated to inhibit breast cancer development (Dakir et al., 2018). In TNBC, pimozide can significantly reduce invasion and migration by activating signal transducer and activator of transcription 3 (STAT3) through phosphorylation. Pimozide's main mechanism of action is to induce apoptosis, thereby preventing the growth and formation of spheres in breast and lung cancer cells. Pimozide showed genotoxic activity by enhancing the occurrence of double strand breaks, as indicated by an elevation in the phosphorylation of histone H2AX.

11.4.1.1 Mechanism of Action of Pimozide in Breast Cancer

The actual anti-neoplastic activity of pimozide against breast cancer is yet unclear. However, inhibition of USP1 and USP46 can be achieved by STAT3 pathway and destabilising USP1/WDR48 complex.

11.4.1.2 STAT3 Pathway

One transcription factor that is frequently active in cancer is STAT. Tyrosine kinases, both receptor and non-receptor types, phosphorylate a tyrosine residue to activate the STATs found in the plasma membrane. Through a reciprocal phosphotyrosine-SH2 interaction between two STAT molecules,

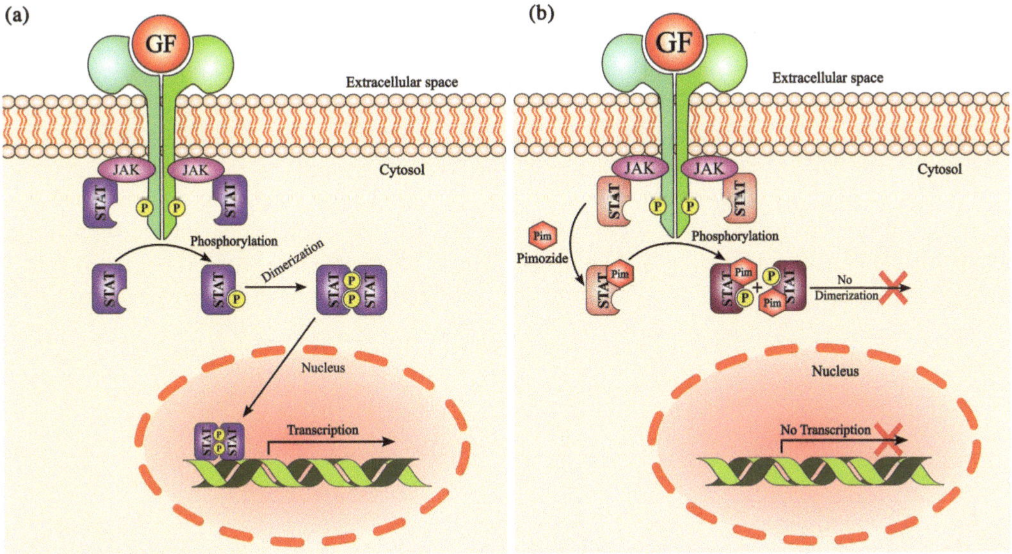

FIGURE 11.1 (a) Non-receptors and receptors tyrosine kinases activate the plasma membrane located signal transducer and activator of transcription (STAT) by phosphorylating a tyrosine residue. STAT activation leads to dimerisation via the reciprocal phosphotyrosine-SH2 interaction of two STAT molecules. The STAT dimers are then translocated to the nucleus and activate the transcription of their target genes. (b) Pimozide, on the other hand, suppresses STAT phosphorylation and inhibits transcription of STAT signalling downstream genes BCL-xL, MCL1, and MYC. *Abbreviations*: Pim: Pimozide, GF: Growth factors or ligand.

dimerisation is induced by STAT activation. The target genes' transcription is then activated by the translocation of STAT dimers to the nucleus. These genes produce the proteins needed for cell survival, proliferation, and self-renewal (Figure 11.1). Because of this, cancer stem cells exhibit upregulation of STAT3 and activation of the STAT pathway. Targeting STATs directly may be a helpful approach in the therapy of cancer, since they may serve as important mediators of the behaviour of malignant cells.

After phosphorylation, STAT3 dimerises and moves into the nucleus where it binds to DNA response elements to trigger the transcription of target genes downstream, including MCL1 and BCL-xL. STAT3 phosphorylation and luciferase activity, both are suppressed by pimozide, while it also diminishes the transcription of genes downstream of STAT3 signalling, including BCL-xL, MCL1, and MYC (Cai et al., 2017). Pimozide brings down the phosphorylation of STAT3 at Tyr 705,consequently downregulating STAT3 phosphorylation (Dakir et al., 2018).

11.4.1.3 Destabilising USP1/WDR48 Complex

Pimozide inhibits USP1/WDR48 complex leading to the ubiquitination of inhibitors of DNA binding (ID) proteins and regulates the development and differentiation by basic helix–loop–helix (bHLH) transcription factors. Deregulated growth during cancer is linked to a number of pathways connected to the HLH (helix-loop-helix) gene. Two categories of bHLH proteins play roles in governing the processes of development and differentiation. The activation of anti-proliferative genes, such as cyclin-dependent kinase inhibitor (CDKN1A, CDKN2A, and CDKN2B), is encouraged by the class I of bHLH homodimers, whereas class II bHLH proteins create heterodimers in conjunction with class I bHLH proteins, regulating genes specific to particular tissues.

ID protein prevents the HLH transcription factor from binding to DNA. It is made feasible by the formation of high affinity heterodimers that block this binding and, as a result, the transcription of genes linked to differentiation. ID1, ID2, and ID3, but not ID4, are polyubiquitinated before being degraded by the 26S proteasome. As a result, most IDs in cells have a brief lifetime. On the other hand, the abundance and ID stability may indicate ubiquitin modification.

When USP1 is alone, its deubiquitinating activity is minimal. Thus, this protein functions in concert with UAF1, also known as WDR48, an 80 kDa protein. The ubiquitin protease is stimulated by the formation of this complex, which increases the catalytic activity of WDR48. IDs continue to antagonise the bHLH transcription factors in the following ways: the USP1/WDR48 complex prevents IDs from being ubiquitinated, enabling these proteins to continue preventing the binding of bHLH transcription factors to DNA. As a result, these proteins prevent development and keep the cell in an undifferentiated condition, identifying these cells as cancer stem cells (Williams et al., 2011). Figure 11.2a depicts the process of maintaining the undifferentiated and proliferative cell state.

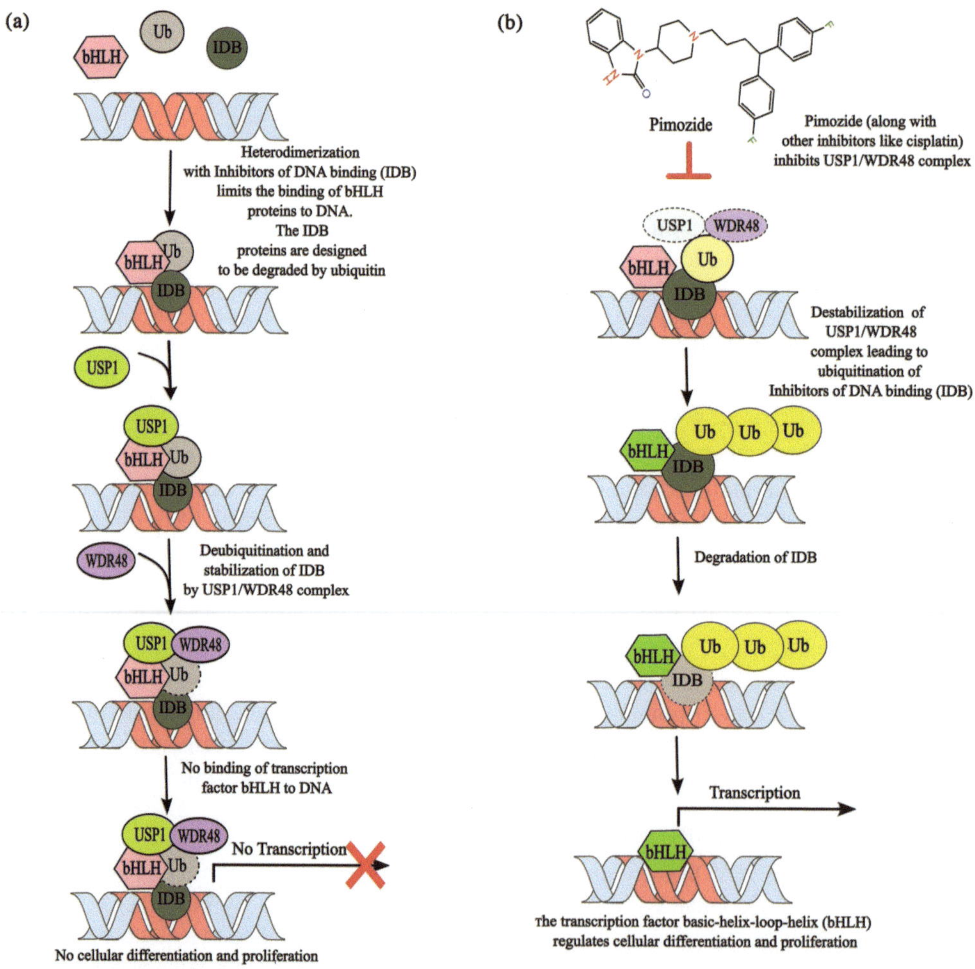

FIGURE 11.2 (a) In the absence of pimozide, USP1/WDR48 complex stabilises the IDs by degrading ubiquitin which leads to heterodimerisation of bHLH with IDs. Thus, ID proteins prevent transcription factor bHLH to bind the DNA response element region and blocks the transcription and activation of the anti-proliferative genes. (b) In the presence of pimozide, destabilisation of the USP1/WDR48 complex leads to the ubiquitination of IDs. Subsequently, the transcription factor bHLH becomes free to bind the DNA in the response element region and, thus, promotes transcription and activation of anti-proliferative genes. *Abbreviations*: bHLH: Basic Helix-Loop-Helix, ID: Inhibitors of DNA binding, Ub: Ubiquitin.

Pimozide destabilises the USP1/WDR48 complex, leading to the ubiquitination of IDs. Degradation of IDs promotes the binding of transcription factor bHLH to the DNA response-element region of DNA, followed by transcription and activation of anti-proliferative genes (Figure 11.2b).

Trifluoperazine (TFP) hydrochloride hinders the proliferation of TNBC tumours and brain metastasis by triggering G0/G1 cell-cycle arrest and apoptosis. *In vitro* research demonstrates that it triggered mitochondria-mediated apoptosis in TNBC cell lines MDA-MB-468, MDA-MB-231, and 4T1 and caused G0/G1 cell-cycle arrest by reducing the expression of both cyclinD1/CDK4 and cyclin E/CDK2 (Feng et al., 2018).

Rottlerin, a naturally occurring substance from *Mallotus philippensis,* has been shown to have anti-angiogenic effects in breast cancer cells. The induction of autophagy by rottlerin leads to apoptosis in breast cancer stem cells. Rottlerin was observed to cause growth cessation in MCF-7 breast cancer cells and human immortalised keratinocytes by suppressing NF-κB and lowering cyclin D-1 levels (Valacchi et al., 2011). Limited research has been conducted on ML323 in breast cancer, but KPNA2 may be a target of ML323 in suppressing breast cancer metastasis (Ma et al., 2019).

11.4.2 USP2 Inhibitors

A purine analogue6-Thioguanine (6-TG) was found to selectively eliminate tumours with BRCA2 defects and overcome resistance to PARP inhibitors (Issaeva et al., 2010). Through downregulating phosphatase and tensin homolog (PTEN's) DNA methylation level, 6-TG suppressed the PI3K-AKT pathway, which in turn prevented cell proliferation and tumour cell growth. Additionally, apoptosis was brought on by activating target tuberous sclerosis 1 (TSC1), a downstream target of PI3K-AKT, and downregulating the methylation of DAXX [X-linked nuclear protein (ATRX)-Fas death domain-associated protein], tumour necrosis factor (TNF), Fas-associated death domain (FADD), and caspase-8 (CASP8), among other molecules (Zhang et al., 2020a). 6-TG may cause p21-dependent G2/M arrest and FAS-mediated exogenous apoptosis in MCF-7 breast cancer cells via reducing the activity of maintenance DNA methyltransferase 1 (DNMT1) (Li et al., 2020a). ML364 ubiquitinates erythroblastic oncogene B2(ErbB2), leading to its increased turnover. ErbB2-positive breast cancer cells become more sensitive to heat shock protein 90 (HSP90) inhibition as a result of ML364's potentiation of the pro-degradation effects of HSP90 inhibitors on ErbB2 (Zhang et al., 2020b).

11.4.3 USP7, USP7/47 Inhibitors

The USP7 inhibitor costunolide inhibits breast cancer growth and spread and shows promise as a potential anti-cancer treatment, especially for advanced breast cancer. Costunolide, a sesquiterpene lactone extracted and purified from *Saussurea Lappa* Clarke (SLC) suppressed MDA-MB-231 (an epithelial, human breast cancer cell line) tumour development and metastasis. Costunolide is reported to downregulate TNF-α-induced invasion and migration of MDA-MB-231 cells. It also hindered the activation of TNF-α-induced NF-κB signalling, resulting in diminished expression of MMP-9, a well-established NF-κB dependent gene associated with breast cancer cell proliferation and metastasis (Choi et al., 2013). Both the MCF-7 and MDA-MB-231 cell lines' viability was significantly decreased by costunolide, which also caused cell-cycle arrest in the G2/M phase in both cell types. After treatment with costunolide, apoptosis inducers (caspase-3 and caspase-9) and cell cycle regulatory proteins (cyclin D1, D3, CDK-4, CDK-6, p18 INK4c, p21 CIP1/Waf-1, and p27 KIP1) exhibit remarkable differences in their expression compared to that of normal breast cell line (MCF-10A). Combining costunolide with dehydrocostus lactone can inhibit breast cancer by halting cell division and inducing cell death (Peng et al., 2017).

P5091, a USP47 inhibitor, has the capability to revert alterations in the cellular morphology of MCF-10A cells and diminish the expression of EMT markers (Silvestrini et al., 2020). Blocking deubiquitination with P5091 can decrease cell growth, formation of colonies, migration, and spread of breast cancer cells (Hayal et al., 2020).

11.4.4 USP9X Inhibitors

The application of USP9X blocker is very limited. WP1130 enhances the cell-killing effects of cisplatin in breast cancer cells that lacked estrogen receptors. Additionally, combining WP1130 with cisplatin treatment may enhance the sensitivity of ER negative breast cancer cells to the drug, and this effect is dependent on the presence of USP9X (Fu et al., 2017).

11.4.5 USP10 Inhibitors

The documented blockers of USP10 mainly consist of P22077, HBX19818, and Spautin-1. P22077 and HBX19818 were first recognised as irreversible inhibitors of USP7. Nevertheless, a recent study unveiled that these compounds could also impede the deubiquitinase function of USP10 and lead to the suppression of cell proliferation in cancer cells harbouring mutant-FLT3 (FLT3-ITD) (Weisberg et al., 2017). Treatment with Spautin-1 effectively hinders the deubiquitinase activity of both USP10 and USP13 as a potent small molecule inhibitor, leading to an increase in ubiquitination and hastening the degradation of Beclinl in Vps34 complexes, ultimately inhibiting autophagy (Liu et al., 2011).

11.4.6 USP11 Inhibitors

Advanced breast cancer can be effectively treated with the USP11 inhibitor mitoxantrone. However, mild side effects like cardiotoxicity, nausea, vomiting, and hair loss are associated with it (Landys et al., 1985).

11.4.7 USP14 Inhibitors

The inhibitor b-AP15 targets USP14 and UCHL5 and can prevent the formation of tumours in the MCF-7 breast cancer cell line. b-AP15 and RA-9 have a significant impact on TNBC survival and trigger autophagy as a cellular response to manage stress on the UPS. Additionally, b-AP15 and PtPT may be able to treat breast cancer that is ER positive. These inhibitors cause endoplasmic reticulum stress, caspase activation, and the downregulation of ER, which results in cell-cycle arrest and death. Additionally, PtPT and b-AP15 disrupt ER signalling by preventing ER transcription and promoting ubiquitin-mediated ER degradation (Xia et al., 2018).

Auranofin, a USP14 inhibitor, synergistically suppresses breast cancer, particularly TNBC (Hatem et al., 2019). Auranofin combined with vitamin C demonstrates effectiveness against TNBC. An innovative approach involves combining trametinib and auranofin for breast cancer therapy, preventing cancer cell growth. The combined therapy induces cell-cycle arrest and apoptosis by cleaving poly (ADP-ribose) polymerase and activating caspase-3/7. The notable effects include increased phosphorylation of the p38 MAPK subunit, leading to mitochondrial stress and promotion of death via the release of apoptosis inducing factor (AIF) in cancer cells apoptosis induction in cancer cells, as observed in MCF-7 cells (Joo et al., 2021).

Enzalutamide hinders cell growth and triggers apoptosis by suppressing USP14. It significantly downregulates AR proteins and suppresses signalling pathways related to AR which increases the anti-proliferative and pro apoptotic ability of breast cancer cell making the treatment successful (Xia et al., 2019a). Additionally, IU1, another USP14 inhibitor, can enhance enzalutamide's ability to restrain cell proliferation and induce apoptosis in breast cancer cell lines both *in vitro* and *in vivo* (Xia et al., 2019a).

There is the need for additional research on the pharmacological properties of these substances in order to be taken into consideration for clinical application, particularly in regard to pharmacodynamics and unfavourable side effects (which may be connected to toxicity or off-target activity).

11.5 CONCLUSION AND FUTURE PERSPECTIVES

Post-translational modification is essential for sustaining physiological functions of a protein. USPs are crucial and highly specialised class of DUB with rising therapeutic potentialities for breast cancer. The constant and rapid advancement of small molecule USP inhibitors may increase the prospects of successfully treating a variety of malignant tumours. Future research should concentrate on uncovering synergistic DUB inhibition mechanisms and devising combinatorial therapeutic methods for cancer treatment. The emergence of innovative bioassays and disease models will also help this process.

Therapeutic strategies that target specific DUBs may be better tolerated than addressing the entire UPS. However, with no inhibitors in clinical trials, the route to establishing them as therapy choices is not straightforward. While pan-inhibition may result in undesirable side effects, finding non-conserved areas is helpful for developing particular USP inhibitors. More study into the interactions between drugs and the USP catalytic site is required.

ABBREVIATIONS

6-TG 6-thioguanine
AIF Apoptosis-inducing factor
BRCA Breast cancer gene
Daxx Death domain associated proteins
DUB Deubiquitinase
ECT2 Epithelial cell transforming sequence 2
EGFR Epidermal growth factor receptor
EMP Epithelial-mesenchymal plasticity
EMT Epithelial-mesenchymal transition
ER Estrogen receptor
KPNA2 Karyopherin Subunit Alpha 2
MAPK Mitogen activated protein kinase
TFP Trifluoperazine
TGF Transforming growth factor
USP Ubiquitin specific proteases
UPS Ubiquitin proteasome system
XIAP X-linked inhibitor of apoptosis

DECLARATIONS

Conflict of interest: The authors declare no conflicts of interest.

Consent for publishing: All authors have approved the final version of the manuscript and given consent for publication.

Funding: Not Applicable.

REFERENCES

Akhavantabasi, S., Akman, H. B., Sapmaz, A., Keller, J., Petty, E. M., & Erson, A. E. (2010). USP32 is an active, membrane-bound ubiquitin protease overexpressed in breast cancers. *Mammalian Genome*, *21*(7–8), 388–397. https://doi.org/10.1007/s00335-010-9268-4

Cai, N., Zhou, W., Ye, L.-L., Chen, J., Liang, Q.-N., Chang, G., & Chen, J.-J. (2017). The STAT3 inhibitor pimozide impedes cell proliferation and induces ROS generation in human osteosarcoma by suppressing catalase expression. *American Journal of Translational Research*, *9*(8), 3853–3866.

Cao, W.-H., Liu, X.-P., Meng, S.-L., Gao, Y.-W., Wang, Y., Ma, Z.-L., Wang, X.-G., & Wang, H.-B. (2016). USP4 promotes invasion of breast cancer cells via Relaxin/TGF-β1/Smad2/MMP-9 signal. *European Review for Medical and Pharmacological Sciences*, *20*(6), 1115–1122.

Choi, Y. K., Cho, S.-G., Woo, S.-M., Yun, Y. J., Jo, J., Kim, W., Shin, Y. C., & Ko, S.-G. (2013). Saussurea lappa Clarke-derived costunolide prevents TNF α-induced breast cancer cell migration and invasion by inhibiting NF-κB activity. *Evidence-Based Complementary and Alternative Medicine: eCAM, 2013*, 936257. https://doi.org/10.1155/2013/936257

Dakir, E.-H., Pickard, A., Srivastava, K., McCrudden, C. M., Gross, S. R., Lloyd, S., Zhang, S.-D., Margariti, A., Morgan, R., Rudland, P. S., & El-Tanani, M. (2018). The anti-psychotic drug pimozide is a novel chemotherapeutic for breast cancer. *Oncotarget*, *9*(79), 34889–34910. https://doi.org/10.18632/oncotarget.26175

Duex, J. E., Comeau, L., Sorkin, A., Purow, B., & Kefas, B. (2011). Usp18 regulates epidermal growth factor (EGF) receptor expression and cancer cell survival via microRNA-7. *The Journal of Biological Chemistry*, *286*(28), 25377–25386. https://doi.org/10.1074/jbc.M111.222760

Dwane, L., O'Connor, A. E., Das, S., Moran, B., Mulrane, L., Pinto-Fernandez, A., Ward, E., Blümel, A. M., Cavanagh, B. L., Mooney, B., Dirac, A. M., Jirström, K., Kessler, B. M., NíChonghaile, T., Bernards, R., Gallagher, W. M., & O'Connor, D. P. (2020). A Functional genomic screen identifies the deubiquitinase USP11 as a novel transcriptional regulator of ERα in breast cancer. *Cancer Research*, *80*(22), 5076–5088. https://doi.org/10.1158/0008-5472.CAN-20-0214

Feng, Z., Xia, Y., Gao, T., Xu, F., Lei, Q., Peng, C., Yang, Y., Xue, Q., Hu, X., Wang, Q., Wang, R., Ran, Z., Zeng, Z., Yang, N., Xie, Z., & Yu, L. (2018). The antipsychotic agent trifluoperazine hydrochloride suppresses triple-negative breast cancer tumor growth and brain metastasis by inducing G0/G1 arrest and apoptosis. *Cell Death & Disease*, *9*(10), 1006. https://doi.org/10.1038/s41419-018-1046-3

Fu, P., Du, F., Liu, Y., Yao, M., Zhang, S., Zheng, X., & Zheng, S. (2017). WP1130 increases cisplatin sensitivity through inhibition of usp9x in estrogen receptor-negative breast cancer cells. *American Journal of Translational Research*, *9*(4), 1783–1791.

Garcia, D. A., Baek, C., Estrada, M. V., Tysl, T., Bennett, E. J., Yang, J., & Chang, J. T. (2018). USP11 enhances TGFβ-induced epithelial-mesenchymal plasticity and human breast cancer metastasis. *Molecular Cancer Research: MCR*, *16*(7), 1172–1184. https://doi.org/10.1158/1541-7786.MCR-17-0723

Han, D., Wang, L., Chen, B., Zhao, W., Liang, Y., Li, Y., Zhang, H., Liu, Y., Wang, X., Chen, T., Li, C., Song, X., Luo, D., Li, Z., & Yang, Q. (2021). USP1-WDR48 deubiquitinase complex enhances TGF-β induced epithelial-mesenchymal transition of TNBC cells via stabilizing TAK1. *Cell Cycle (Georgetown, Tex.)*, *20*(3), 320–331. https://doi.org/10.1080/15384101.2021.1874695

Hatem, E., Azzi, S., El Banna, N., He, T., Heneman-Masurel, A., Vernis, L., Baïlle, D., Masson, V., Dingli, F., Loew, D., Azzarone, B., Eid, P., Baldacci, G., & Huang, M.-E. (2019). Auranofin/Vitamin C: A novel drug combination targeting triple-negative breast cancer. *Journal of the National Cancer Institute*, *111*(6), 597–608. https://doi.org/10.1093/jnci/djy149

Hayal, T. B., DoĞan, A., ŞİŞlİ, H. B., Kiratli, B., & Şahİn, F. (2020). Ubiquitin-specific protease 7 downregulation suppresses breast cancer *in vitro*. *Turkish Journal of Biology = Turk Biyoloji Dergisi*, *44*(4), 145–157. https://doi.org/10.3906/biy-1912-83

Huang, M.-L., Shen, G.-T., & Li, N.-L. (2022). Emerging potential of ubiquitin-specific proteases and ubiquitin-specific proteases inhibitors in breast cancer treatment. *World Journal of Clinical Cases*, *10*(32), 11690–11701. https://doi.org/10.12998/wjcc.v10.i32.11690

Huang, X., Summers, M. K., Pham, V., Lill, J. R., Liu, J., Lee, G., Kirkpatrick, D. S., Jackson, P. K., Fang, G., & Dixit, V. M. (2011). Deubiquitinase USP37 is activated by CDK2 to antagonize APC (CDH1) and promote S phase entry. *Molecular Cell*, *42*(4), 511–523. https://doi.org/10.1016/j.molcel.2011.03.027

Issaeva, N., Thomas, H. D., Djureinovic, T., Jaspers, J. E., Stoimenov, I., Kyle, S., Pedley, N., Gottipati, P., Zur, R., Sleeth, K., Chatzakos, V., Mulligan, E. A., Lundin, C., Gubanova, E., Kersbergen, A., Harris, A. L., Sharma, R. A., Rottenberg, S., Curtin, N. J., & Helleday, T. (2010). 6-thioguanine selectively kills BRCA2-defective tumors and overcomes PARP inhibitor resistance. *Cancer Research*, *70*(15), 6268–6276. https://doi.org/10.1158/0008-5472.CAN-09-3416

Izrailit, J., Jaiswal, A., Zheng, W., Moran, M. F., & Reedijk, M. (2017). Cellular stress induces TRB3/USP9x-dependent Notch activation in cancer. *Oncogene*, *36*(8), 1048–1057. https://doi.org/10.1038/onc.2016.276

Joo, M.-K., Shin, S., Ye, D.-J., An, H.-G., Kwon, T.-U., Baek, H.-S., Kwon, Y.-J., & Chun, Y.-J. (2021). Combined treatment with auranofin and trametinib induces synergistic apoptosis in breast cancer cells. *Journal of Toxicology and Environmental Health. Part A*, *84*(2), 84–94. https://doi.org/10.1080/15287394.2020.1835762

Komander, D., Clague, M. J., & Urbé, S. (2009). Breaking the chains: Structure and function of the deubiquitinases. *Nature Reviews. Molecular Cell Biology*, *10*(8), 550–563. https://doi.org/10.1038/nrm2731

Kuligina, E. S., Sokolenko, A. P., Bizin, I. V., Romanko, A. A., Zagorodnev, K. A., Anisimova, M. O., Krylova, D. D., Anisimova, E. I., Mantseva, M. A., Varma, A. K., Hasan, S. K., Ni, V. I., Koloskov, A. V., Suspitsin, E. N., Venina, A. R., Aleksakhina, S. N., Sokolova, T. N., Milanović, A. M., Schürmann, P., … Imyanitov, E. N. (2020). Exome sequencing study of Russian breast cancer patients suggests a predisposing role for USP39. *Breast Cancer Research and Treatment*, *179*(3), 731–742. https://doi.org/10.1007/s10549-019-05492-6

Landys, K., Borgstrom, S., Andersson, T., & Noppa, H. (1985). Mitoxantrone as a first-line treatment of advanced breast cancer. *Investigational New Drugs*, *3*(2), 133–137. https://doi.org/10.1007/BF00174160

Li, H., An, X., Zhang, D., Li, Q., Zhang, N., Yu, H., & Li, Z. (2020a). Transcriptomics analysis of the tumor-inhibitory pathways of 6-thioguanine in MCF-7 cells via silencing DNMT1 activity. *OncoTargets and Therapy*, *13*, 1211–1223. https://doi.org/10.2147/OTT.S236543

Li, H., & Zheng, B. (2019). Overexpression of the ubiquitin-specific peptidase 9 X-linked (USP9X) gene is associated with upregulation of cyclin D1 (CCND1) and downregulation of cyclin-dependent inhibitor kinase 1A (CDKN1A) in breast cancer tissue and cell lines. *Medical Science Monitor*, *25*, 4207–4216. https://doi.org/10.12659/MSM.914742

Li, W., Shen, M., Jiang, Y.-Z., Zhang, R., Zheng, H., Wei, Y., Shao, Z.-M., & Kang, Y. (2020b). Deubiquitinase USP20 promotes breast cancer metastasis by stabilizing SNAI2. *Genes & Development*, *34*(19–20), 1310–1315. https://doi.org/10.1101/gad.339804.120

Li, X., Song, N., Liu, L., Liu, X., Ding, X., Song, X., Yang, S., Shan, L., Zhou, X., Su, D., Wang, Y., Zhang, Q., Cao, C., Ma, S., Yu, N., Yang, F., Wang, Y., Yao, Z., Shang, Y., & Shi, L. (2017). USP9X regulates centrosome duplication and promotes breast carcinogenesis. *Nature Communications*, *8*, 14866. https://doi.org/10.1038/ncomms14866

Liu, B., Liu, Y., Wang, Y., Xie, C., Gan, M., Han, T., Cao, J., & Wang, J. (2019). CyclinB1 deubiquitination by USP14 regulates cell cycle progression in breast cancer. *Pathology, Research and Practice*, *215*(10), 152592. https://doi.org/10.1016/j.prp.2019.152592

Liu, J., Xia, H., Kim, M., Xu, L., Li, Y., Zhang, L., Cai, Y., Norberg, H. V., Zhang, T., Furuya, T., Jin, M., Zhu, Z., Wang, H., Yu, J., Li, Y., Hao, Y., Choi, A., Ke, H., Ma, D., & Yuan, J. (2011). Beclin1 controls the levels of p53 by regulating the deubiquitination activity of USP10 and USP13. *Cell*, *147*(1), 223–234. https://doi.org/10.1016/j.cell.2011.08.037

Liu, Z., Zanata, S. M., Kim, J., Peterson, M. A., Di Vizio, D., Chirieac, L. R., Pyne, S., Agostini, M., Freeman, M. R., & Loda, M. (2013). The ubiquitin-specific protease USP2a prevents endocytosis-mediated EGFR degradation. *Oncogene*, *32*(13), 1660–1669. https://doi.org/10.1038/onc.2012.188

Ma, A., Tang, M., Zhang, L., Wang, B., Yang, Z., Liu, Y., Xu, G., Wu, L., Jing, T., Xu, X., Yang, S., & Liu, Y. (2019). USP1 inhibition destabilizes KPNA2 and suppresses breast cancer metastasis. *Oncogene*, *38*(13), 2405–2419. https://doi.org/10.1038/s41388-018-0590-8

Moreno-Bueno, G., Salvador, F., Martín, A., Floristán, A., Cuevas, E. P., Santos, V., Montes, A., Morales, S., Castilla, M. A., Rojo-Sebastián, A., Martínez, A., Hardisson, D., Csiszar, K., Portillo, F., Peinado, H., Palacios, J., & Cano, A. (2011). Lysyl oxidase-like 2 (LOXL2), a new regulator of cell polarity required for metastatic dissemination of basal-like breast carcinomas. *EMBO Molecular Medicine*, *3*(9), 528–544. https://doi.org/10.1002/emmm.201100156

Mussell, A., Shen, H., Chen, Y., Mastri, M., Eng, K. H., Bshara, W., Frangou, C., & Zhang, J. (2020). USP1 regulates TAZ protein stability through ubiquitin modifications in breast cancer. *Cancers*, *12*(11), 3090. https://doi.org/10.3390/cancers12113090

Nijman, S. M. B., Luna-Vargas, M. P. A., Velds, A., Brummelkamp, T. R., Dirac, A. M. G., Sixma, T. K., & Bernards, R. (2005). A genomic and functional inventory of deubiquitinating enzymes. *Cell*, *123*(5), 773–786. https://doi.org/10.1016/j.cell.2005.11.007

Niu, Z., Li, X., Feng, S., Huang, Q., Zhuang, T., Yan, C., Qian, H., Ding, Y., Zhu, J., & Xu, W. (2020). The deubiquitinating enzyme USP1 modulates ERα and modulates breast cancer progression. *Journal of Cancer*, *11*(23), 6992–7000. https://doi.org/10.7150/jca.50477

Ouyang, W., Zhang, S., Yang, B., Yang, C., Zhang, J., Zhou, F., & Xie, C. (2016). β-catenin is regulated by USP9x and mediates resistance to TRAIL-induced apoptosis in breast cancer. *Oncology Reports*, *35*(2), 717–724. https://doi.org/10.3892/or.2015.4463

Peng, Y., Liao, Q., Tan, W., Peng, C., Hu, Z., Chen, Y., Li, Z., Li, J., Zhen, B., Zhu, W., Li, X., Yao, Y., Song, Q., Liu, C., Qi, X., He, F., & Pei, H. (2019). The deubiquitylating enzyme USP15 regulates homologous recombination repair and cancer cell response to PARP inhibitors. *Nature Communications*, *10*(1), 1224. https://doi.org/10.1038/s41467-019-09232-8

Peng, Z., Wang, Y., Fan, J., Lin, X., Liu, C., Xu, Y., Ji, W., Yan, C., & Su, C. (2017). Costunolide and dehydrocostuslactone combination treatment inhibit breast cancer by inducing cell cycle arrest and apoptosis through c-Myc/p53 and AKT/14-3-3 pathway. *Scientific Reports*, *7*, 41254. https://doi.org/10.1038/srep41254

Pereg, Y., Liu, B. Y., O'Rourke, K. M., Sagolla, M., Dey, A., Komuves, L., French, D. M., & Dixit, V. M. (2010). Ubiquitin hydrolase Dub3 promotes oncogenic transformation by stabilizing Cdc25A. *Nature Cell Biology*, *12*(4), 400–406. https://doi.org/10.1038/ncb2041

Popov, N., Wanzel, M., Madiredjo, M., Zhang, D., Beijersbergen, R., Bernards, R., Moll, R., Elledge, S. J., & Eilers, M. (2007). The ubiquitin-specific protease USP28 is required for MYC stability. *Nature Cell Biology*, *9*(7), 765–774. https://doi.org/10.1038/ncb1601

Qin, T., Li, B., Feng, X., Fan, S., Liu, L., Liu, D., Mao, J., Lu, Y., Yang, J., Yu, X., Zhang, Q., Zhang, J., Song, B., Li, M., & Li, L. (2018). Abnormally elevated USP37 expression in breast cancer stem cells regulates stemness, epithelial-mesenchymal transition and cisplatin sensitivity. *Journal of Experimental & Clinical Cancer Research: CR*, *37*(1), 287. https://doi.org/10.1186/s13046-018-0934-9

Qu, Q., Mao, Y., Xiao, G., Fei, X., Wang, J., Zhang, Y., Liu, J., Cheng, G., Chen, X., Wang, J., & Shen, K. (2015). USP2 promotes cell migration and invasion in triple negative breast cancer cell lines. *Tumour Biology*, *36*(7), 5415–5423. https://doi.org/10.1007/s13277-015-3207-7

Ruffner, H., Joazeiro, C. A., Hemmati, D., Hunter, T., & Verma, I. M. (2001). Cancer-predisposing mutations within the RING domain of BRCA1: Loss of ubiquitin protein ligase activity and protection from radiation hypersensitivity. *Proceedings of the National Academy of Sciences of the United States of America*, *98*(9), 5134–5139. https://doi.org/10.1073/pnas.081068398

Schoenfeld, A. R., Apgar, S., Dolios, G., Wang, R., & Aaronson, S. A. (2004). BRCA2 is ubiquitinated *in vivo* and interacts with USP11, a deubiquitinating enzyme that exhibits prosurvival function in the cellular response to DNA damage. *Molecular and Cellular Biology*, *24*(17), 7444–7455. https://doi.org/10.1128/MCB.24.17.7444-7455.2004

Schulte, I., Batty, E. M., Pole, J. C. M., Blood, K. A., Mo, S., Cooke, S. L., Ng, C., Howe, K. L., Chin, S.-F., Brenton, J. D., Caldas, C., Howarth, K. D., & Edwards, P. A. W. (2012). Structural analysis of the genome of breast cancer cell line ZR-75-30 identifies twelve expressed fusion genes. *BMC Genomics*, *13*, 719. https://doi.org/10.1186/1471-2164-13-719

Shi, J., Zhang, Q., Yin, X., Ye, J., Gao, S., Chen, C., Yang, Y., Wu, B., Fu, Y., Zhang, H., Wang, Z., Wang, B., Zhu, Y., Wu, H., Yao, Y., Xu, G., Wang, Q., Wang, S., & Zhang, W. (2023). Stabilization of IGF2BP1 by USP10 promotes breast cancer metastasis via CPT1A in an m6A-dependent manner. *International Journal of Biological Sciences*, *19*(2), 449–464. https://doi.org/10.7150/ijbs.76798

Siegel, R. L., Miller, K. D., & Jemal, A. (2020). Cancer statistics, 2020. *CA: A Cancer Journal for Clinicians*, *70*(1), 7–30. https://doi.org/10.3322/caac.21590

Silvestrini, V. C., Thomé, C. H., Albuquerque, D., de Souza Palma, C., Ferreira, G. A., Lanfredi, G. P., Masson, A. P., Delsin, L. E. A., Ferreira, F. U., de Souza, F. C., de Godoy, L. M. F., Aquino, A., Carrilho, E., Panepucci, R. A., Covas, D. T., & Faça, V. M. (2020). Proteomics analysis reveals the role of ubiquitin specific protease (USP47) in Epithelial to Mesenchymal Transition (EMT) induced by TGFβ2 in breast cells. *Journal of Proteomics*, *219*, 103734. https://doi.org/10.1016/j.jprot.2020.103734

Stevenson, L. F., Sparks, A., Allende-Vega, N., Xirodimas, D. P., Lane, D. P., & Saville, M. K. (2007). The deubiquitinating enzyme USP2a regulates the p53 pathway by targeting Mdm2. *The EMBO Journal*, *26*(4), 976–986. https://doi.org/10.1038/sj.emboj.7601567

Tan, Y., Zhou, G., Wang, X., Chen, W., & Gao, H. (2018). USP18 promotes breast cancer growth by upregulating EGFR and activating the AKT/Skp2 pathway. *International Journal of Oncology*, *53*(1), 371–383. https://doi.org/10.3892/ijo.2018.4387

Valacchi, G., Pecorelli, A., Sticozzi, C., Torricelli, C., Muscettola, M., Aldinucci, C., & Maioli, E. (2011). Rottlerin exhibits antiangiogenic effects *in vitro*. *Chemical Biology & Drug Design*, *77*(6), 460–470. https://doi.org/10.1111/j.1747-0285.2011.01121.x

Wang, Q., Ma, S., Song, N., Li, X., Liu, L., Yang, S., Ding, X., Shan, L., Zhou, X., Su, D., Wang, Y., Zhang, Q., Liu, X., Yu, N., Zhang, K., Shang, Y., Yao, Z., & Shi, L. (2016). Stabilization of histone demethylase PHF8 by USP7 promotes breast carcinogenesis. *The Journal of Clinical Investigation*, *126*(6), 2205–2220. https://doi.org/10.1172/JCI85747

Wang, S., Zhong, X., Wang, C., Luo, H., Lin, L., Sun, H., Sun, G., Zeng, K., Zou, R., Liu, W., Sun, N., Song, H., Liu, W., Zhang, Q., Liao, Z., Teng, X., Zhou, T., Sun, X., & Zhao, Y. (2020). USP22 positively modulates ERα action via its deubiquitinase activity in breast cancer. *Cell Death and Differentiation*, *27*(11), 3131–3145. https://doi.org/10.1038/s41418-020-0568-2

Weisberg, E. L., Schauer, N. J., Yang, J., Lamberto, I., Doherty, L., Bhatt, S., Nonami, A., Meng, C., Letai, A., Wright, R., Tiv, H., Gokhale, P. C., Ritorto, M. S., De Cesare, V., Trost, M., Christodoulou, A., Christie, A., Weinstock, D. M., Adamia, S., … Buhrlage, S. J. (2017). Inhibition of USP10 induces degradation of oncogenic FLT3. *Nature Chemical Biology*, *13*(12), 1207–1215. https://doi.org/10.1038/nchembio.2486

Williams, S. A., Maecker, H. L., French, D. M., Liu, J., Gregg, A., Silverstein, L. B., Cao, T. C., Carano, R. A. D., & Dixit, V. M. (2011). USP1 deubiquitinates ID proteins to preserve a mesenchymal stem cell program in osteosarcoma. *Cell*, *146*(6), 918–930. https://doi.org/10.1016/j.cell.2011.07.040

Wu, Y., Wang, Y., Yang, X. H., Kang, T., Zhao, Y., Wang, C., Evers, B. M., & Zhou, B. P. (2013). The deubiquitinase USP28 stabilizes LSD1 and confers stem-cell-like traits to breast cancer cells. *Cell Reports*, *5*(1), 224–236. https://doi.org/10.1016/j.celrep.2013.08.030

Xia, X., Huang, C., Liao, Y., Liu, Y., He, J., Guo, Z., Jiang, L., Wang, X., Liu, J., & Huang, H. (2019a). Inhibition of USP14 enhances the sensitivity of breast cancer to enzalutamide. *Journal of Experimental & Clinical Cancer Research: CR*, *38*(1), 220. https://doi.org/10.1186/s13046-019-1227-7

Xia, X., Liao, Y., Guo, Z., Li, Y., Jiang, L., Zhang, F., Huang, C., Liu, Y., Wang, X., Liu, N., Liu, J., & Huang, H. (2018). Targeting proteasome-associated deubiquitinases as a novel strategy for the treatment of estrogen receptor-positive breast cancer. *Oncogenesis*, *7*(9), 75. https://doi.org/10.1038/s41389-018-0086-y

Xia, X., Liao, Y., Huang, C., Liu, Y., He, J., Shao, Z., Jiang, L., Dou, Q. P., Liu, J., & Huang, H. (2019b). Deubiquitination and stabilization of estrogen receptor α by ubiquitin-specific protease 7 promotes breast tumorigenesis. *Cancer Letters*, *465*, 118–128. https://doi.org/10.1016/j.canlet.2019.09.003

Xiao, Z., Zhang, P., & Ma, L. (2016). The role of deubiquitinases in breast cancer. *Cancer Metastasis Reviews*, *35*(4), 589–600. https://doi.org/10.1007/s10555-016-9640-2

Yang, J.-R., Lu, Y.-B., Su, H.-X., Xiao, Y., Pan, Q., Su, F., Zhang, X.-B., Zhu, K.-L., Guan, Q.-L., & Ling, X.-L. (2023). USP10 promotes the progression of triple-negative breast cancer by enhancing the stability of TCF4 protein. *Biochemical Pharmacology*, 115864. https://doi.org/10.1016/j.bcp.2023.115864

Zhang, D., An, X., Li, Q., Man, X., Chu, M., Li, H., Zhang, N., Dai, X., Yu, H., & Li, Z. (2020a).Thioguanine induces apoptosis in triple-negative breast cancer by regulating PI3K-AKT pathway. *Frontiers in Oncology*, *10*, 524922. https://doi.org/10.3389/fonc.2020.524922

Zhang, J., Liu, S., Li, Q., Shi, Y., Wu, Y., Liu, F., Wang, S., Zaky, M. Y., Yousuf, W., Sun, Q., Guo, D., Wang, T., Zhang, Y., Wang, Y., Li, M., & Liu, H. (2020b).The deubiquitylase USP2 maintains ErbB2 abundance via counteracting endocytic degradation and represents a therapeutic target in ErbB2-positive breast cancer. *Cell Death and Differentiation*, *27*(9), 2710–2725. https://doi.org/10.1038/s41418-020-0538-8

Zhang, L., Zhou, F., Drabsch, Y., Gao, R., Snaar-Jagalska, B. E., Mickanin, C., Huang, H., Sheppard, K.-A., Porter, J. A., Lu, C. X., & Ten Dijke, P. (2012). USP4 is regulated by AKT phosphorylation and directly deubiquitylates TGF-β type I receptor. *Nature Cell Biology*, *14*(7), 717–726. https://doi.org/10.1038/ncb2522

Zhang, Q., Cao, C., Gong, W., Bao, K., Wang, Q., Wang, Y., Bi, L., Ma, S., Zhao, J., Liu, L., Tian, S., Zhang, K., Yang, J., Yao, Z., Song, N., & Shi, L. (2020c). A feedforward circuit shaped by ECT2 and USP7 contributes to breast carcinogenesis. *Theranostics*, *10*(23), 10769–10790. https://doi.org/10.7150/thno.46878

Zhang, Z., Li, J., Ou, Y., Yang, G., Deng, K., Wang, Q., Wang, Z., Wang, W., Zhang, Q., Wang, H., Sun, W., Sun, P., & Yang, S. (2020d). Correction to: CDK4/6 inhibition blocks cancer metastasis through a USP51-ZEB1-dependent deubiquitination mechanism. *Signal Transduction and Targeted Therapy*, *5*(1), 103. https://doi.org/10.1038/s41392-020-00212-9

Zhou, Z., Luo, A., Shrivastava, I., He, M., Huang, Y., Bahar, I., Liu, Z., & Wan, Y. (2017). Regulation of XIAP turnover reveals a role for USP11 in promotion of tumorigenesis. *EBioMedicine*, *15*, 48–61. https://doi.org/10.1016/j.ebiom.2016.12.014

Zhu, L., Yang, S., He, S., Qiang, F., Cai, J., Liu, R., Gu, C., Guo, Z., Wang, C., Zhang, W., Zhang, C., & Wang, Y. (2016). Downregulation of ubiquitin-specific protease 14 (USP14) inhibits breast cancer cell proliferation and metastasis, but promotes apoptosis. *Journal of Molecular Histology*, *47*(1), 69–80. https://doi.org/10.1007/s10735-015-9650-3

12 Microbial Secretory Proteases

A New Insight in the Field of Cancer Therapeutics

Nanda Singh, Rima Tapader, and Amit Pal
ICMR-National Institute of Cholera and Enteric Diseases (NICED), Kolkata, India

12.1 INTRODUCTION

Growth and reproduction are two fundamental characteristics of all living organisms. Eukaryotic organisms grow by a process called cell division, where a parent cell divides and produces two new daughter cells (Cooper, 2000). Cell division occurs through a series of events called the cell cycle. The cell cycle is a tightly regulated process, and an abnormality in this process caused by genetic or environmental factors may lead to uncontrolled cell proliferation. The uncontrolled proliferation of cells further changes the molecular and genetic makeup of normal cells, and as a result, a normal cell becomes a cancerous cell (Amy, 2010). There are different types of cancer that can be classified into five groups: carcinomas, sarcomas, hematopoietic malignancies, neuroectodermal malignancies and atypical tumors (Link, 2019). Among them, carcinoma is the leading cause of death worldwide (Figure 12.1). The incident of cancer increases with age, as the cellular repair machinery becomes less effective. Worldwide, cancer continues to be one of the leading causes of death each year (Are et al., 2019). In 2022, a total of 1,918,030 new cases and 609,360 deaths related to cancer were reported in the United States (Siegel et al., 2022). In India, the adjusted mortality-to-incident ratio of cancer was 26.7 million in 2021, which may increase to 29.8 million by 2025, as predicted by the Indian Council of Medical Research (ICMR) (Kulothungan et al., 2022). It is evident from these statistics that more research is required to develop better cancer treatments.

Prevention of cancer depends on various factors like lifestyle changes, early detection of cancers and developing better cancer treatments that can reduce the risk of cancer progression in individuals (LoConte et al., 2018; Loomans-Kropp & Umar, 2019). A conventional approach to cancer treatment involves chemotherapy, radiation therapy and surgery (Tannock, 1998). Treatments with these conventional approaches can prevent cancer and save lives, but there is a high level of toxicity and disfigurement associated with these treatments (Figure 12.2). Surgery relies on the removal of cancer-containing mass of tissue, which works well with solid tumor removal. Surgery may not be able to remove all cancerous tumors, as it may cause damage to normal tissue if the entire region is removed (Dare et al., 2015). There is also some risk associated with surgery, such as pain, infection and recurrence of cancer after surgery (Demicheli et al., 2008). Chemotherapy relies on the principle that a drug administered in the body can kill tumor cells more effectively than normal cells. Treatment for cancer remains largely based on chemotherapy. In some cases, chemotherapy can be effective, but sometimes its side effects can kill cancer patients. A major risk associated with chemotherapy is that along with the cancer cells, it kills the normal/healthy cells in the body, which can have serious consequences for the patient and may even change their genetic makeup in such a way that it becomes resistant to the treatment(Bukowski et al., 2020; Chabner & Roberts, 2005). Similarly, radiation therapy

 DOI: 10.1201/9781003394693-14

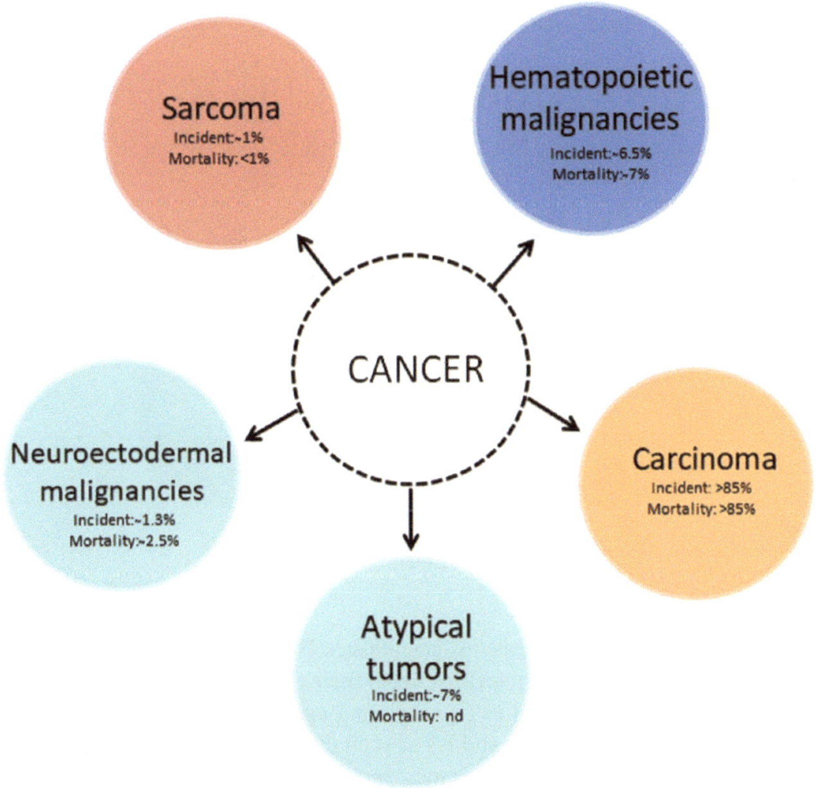

FIGURE 12.1 Cancer types and their incidence and mortality.

reduces the size of tumors or kills cancer cells more effectively than normal cells when administered to a cancer-affected area of the body. In radiation therapy, different components of the cell can be targeted with ionizing radiation, including proteins, nucleic acids, and lipids, but the major target is the cellular DNA. Similar to chemotherapy, radiation therapy also has major side effects. It is highly destructive to normal tissue and it may induce new types of cancer in the body (Vilalta et al., 2016). The fundamental problem of these conventional therapies is specificity against cancer cells. Due to the limitation of the conventional therapy, scientists have discovered alternate approaches to treat cancer that target cancer cells more effectively without affecting normal cells. One alternate approach is to find novel anti-cancer molecules/drugs that can specifically target cancer cells while minimizing the effect on healthy cells (Dehelean et al., 2021; Khalifa et al., 2019). In targeted therapy, advances in cancer research help to identify the cancer-specific targeted molecule for therapeutics specificity. Identification and discovery of drugs against a targeted molecule reduce the burden of toxicity and side effects found in conventional therapy.

In anti-cancer research, one of the objectives is to discover novel molecules/drugs that can specifically inhibit cancer. Molecules discovered from natural resources are the major sources for drug development in today's world. Over the years, research has been conducted to isolate molecules from plants, animals and microbes and study their anti-cancer effect. About 25% of currently approved anti-cancer drugs are discovered from these sources (Gordaliza, 2007; Huang et al., 2018; Newman & Cragg, 2020). In addition to the pharmaceutical industry, research conducted by academics, governments and private laboratories has also contributed significantly to the discovery and development of anti-cancer agents isolated directly or indirectly from natural sources (Cragg et al., 2011). Currently, anti-cancer research has led to the discovery of a large and diverse group

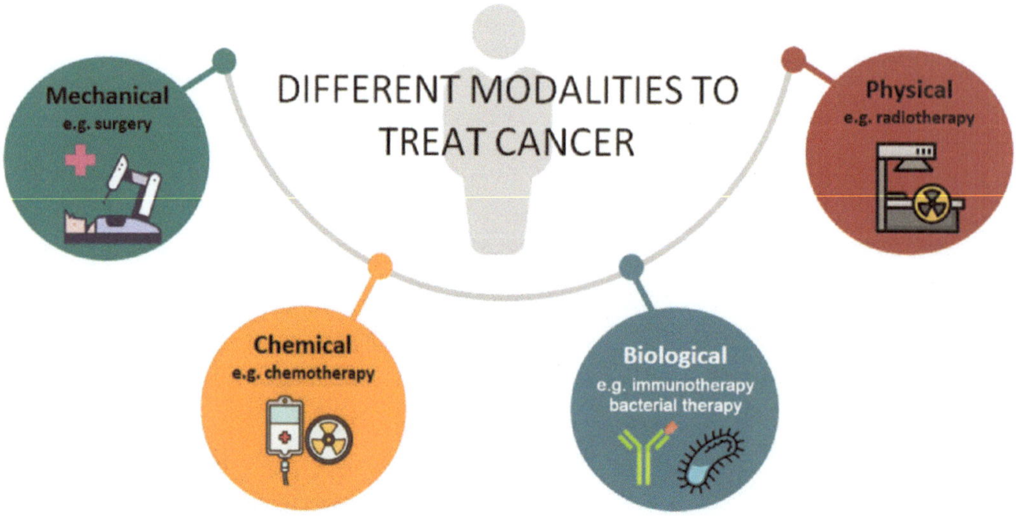

FIGURE 12.2 Different modalities of cancer treatment.

of anti-cancer molecules that are not easy to categorize. There are many factors that can be considered when classifying anti-cancer molecules, from their chemical nature to their mechanisms of action. Traditionally they are classified as (i) alkylating agents, (ii) natural products, (iii) antimetabolites, (iv) hormones and antagonists and (v) miscellaneous. The miscellaneous group in recent years has grown with novel anti-cancer molecules, including bacterial proteins, antibodies and toxins. Recently, proteins and peptides from microorganisms have shown to be promising anti-cancer agents. Among them, it has been shown that bacterial antibiotics, toxins, and proteases are effective against cancer cells. Some of these are already used in cancer treatment, and some are in experimental stages(Baindara & Mandal, 2020).

12.2 THERAPEUTIC POTENTIAL OF MICROBIAL ORIGIN MOLECULES

Microorganisms remain great sources of molecules that are useful to humans. Advancement in biotechnology has found microbes a valuable source of bioactive molecules that can be used as therapeutics against various diseases such as cancer, cardiovascular disease and antimicrobial therapy, and these molecules can be purified in a relatively cost-effective manner (Du et al., 2011; Karpiński & Adamczak, 2018; Pham et al., 2019; Weiman, 2014). The use of microorganisms as a source of anti-cancer drugs is, therefore, very promising. As early as the 1800s, molecules isolated from microbes were evaluated as anti-cancer agents. Bacteria and their byproducts were first used in cancer treatment by physician William B. Coley and are referred to as 'Coley's toxins' (Natarajan & Bhatt, 2020). The principle of Coley's toxin has been proven to be effective by recent discoveries. A variety of anti-cancer agents derived from microbes are being evaluated for the treatment of cancer. In 1940, the antibiotic actinomycin was isolated from *Streptomyces parvulus* and it was the first used antibiotic shown to have anti-cancer properties. Actinomycin-D was later approved by US Food and Drug Administration (FDA) and has been widely used as an anti-cancer molecule in the treatment of various types of tumors (McCarthy, 2006; Waksman & Woodruff, 1940). Microbial natural products such as bleomycin, epirubicin, doxorubicin and rapamycin were isolated from different microbial sources and were approved by the FDA for the treatment of cancer (Giddings & Newman, 2013; Hollstein, 1974). More than 60% of currently used anti-cancer compounds are of microbial origin and this number will increase further with the discovery of different natural products from microbes.

In addition to these microbial natural products, current research identifies the importance of microbial enzymes in disease therapeutics. Microbes are the source of many useful enzymes. Microbiological enzymes are widely used in textile, food, detergent, leather, biofuels, and other industry (da Rocha, 2001). In a survey in 2019, the microbial enzymes have shown to gather a market of about $9.9 billion (USD), and this growing demand encourages researchers to screen for more of these enzymes (Yari et al., 2017). The use of enzymes in the biopharmaceutical industry has grown steadily in recent years (Demain & Adrio, 2008). The advantages of the production of therapeutic enzymes from microbes are that it is cost effective, easy to isolate and environmentally friendly. The concept of therapeutic enzymes is not new. It has been used as a replacement therapy since the 1960 (Dhankhar et al., 2021). With the development of modern biotechnology, scientists realize the importance of microbiological enzymes in different disease therapeutics, including cancer.

12.3 MICROBIAL ENZYMES IN CANCER THERAPEUTICS

A large number of microbial enzymes have been screened for cancer therapeutics in recent years. Though very few of them are in clinical trials, the numbers are increasing daily. The fundamental characteristic of these enzymes is cytotoxicity against cancer and their use of different modes of action for the induction of cytotoxicity in cancer cells. For example, cytosine deaminase (CD) from bacteria and fungus has been used for pro-drug activation in cancer gene therapy. CD enzymes convert 5-flurocytosine to toxic 5-flurouracil, which kills cancer cells(Negroni et al., 2007; Wei et al., 2007). It has been found that cancer cells use extracellular amino acids and metabolites to sustain their growth. Arginine is one such amino acid that cancer cells uptake from its extracellular environment to use in different metabolic pathways. This metabolic vulnerability targeted by the use of the bacterial origin arginine deaminase enzyme that degrades arginine and converts it to ammonia and citrulline. Anti-cancer arginine deaminase could be isolated from different microbial origin, for example *Pseudomonas* sp., *Lactococcus* sp., *Enterococcus* sp., *Mycoplasma* sp., etc. Similarly, L-asparaginase causes the depletion of L-asparagine amino acid, which is a non-essential amino acid and can take part in cellular metabolism, glycoprotein bio-synthesis and epithelial-mesenchymal transition (EMT) of cancer. Therefore, depletion of asparagine by microbial L-asparaginase potentially inhibits cancer by targeting its metabolism. L-asparaginase from microbial origin is already an FDA-approved enzyme used in acute lymphoblastic leukemia (ALL) (Dhankhar et al., 2020; Radadiya et al., 2020). Another microbial enzyme heparinase showed anti-tumor activity by targeting angiogenesis and it was first isolated from bacteria *Pedobacter heparinus* (Korn &Payza, 1956; Yang et al., 2007). There are several other examples where microbial enzymes were studied for anticancer properties.

12.4 MICROBIAL SECRETORY PROTEASES IN CANCER THERAPEUTICS

Among various types of enzymes, proteases are the most abundant enzymes that hydrolyze peptide bonds in proteins. Proteases act as an important enzyme in every organism. It plays many roles, including maturation of intracellular proteins, digestion of nutrients, and taking part in immune cascades, etc. (Shankar et al., 2021). Due to their unique function, proteases have been used in commercial industries, such as the detergent industry, leather industry and food-processing industry. Outside of other industrial enzymes markets, proteases serve 60% of the total worldwide sale of enzymes (Figure 12.3) and a significant part of this protease market comes from microbial proteases, which represent 40% of the total (Rao et al., 1998). A significant step toward the use of proteases from industry to therapeutics was made when the FDA approved the enzyme 'Activase' for the treatment of myocardial damage in 1978 (Baker et al., 2005; Vellard, 2003). Following that, identification of proteases from diverse origins and the study of their therapeutic effects on different diseases took the protease industry into a new direction. Today, proteolytic enzymes are established drugs, prescribed

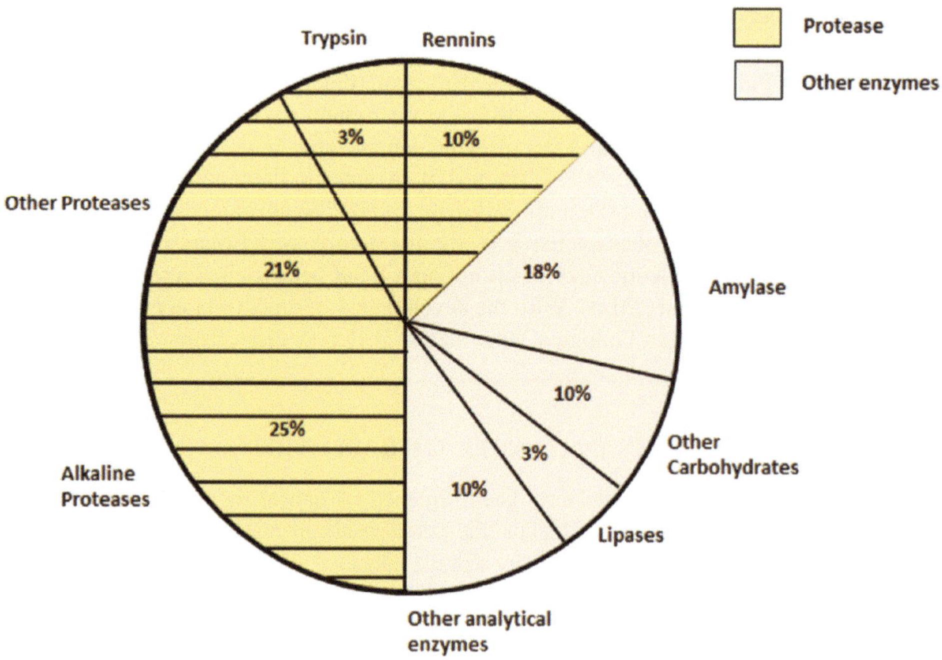

FIGURE 12.3 Distribution of the commercial enzyme market.

to reduce symptoms of irritable bowel syndrome (IBS), decrease inflammation and reduce muscle soreness for faster recovery after surgery and for others diseases(Baker et al., 2005).

12.4.1 Overview and Therapeutic Potential of Microbial Secretory Proteases

Protease is a group of enzymes that hydrolyze protein polypeptide chains into smaller peptides. Proteases are classified into two groups, endopeptidase and exopeptidase. Exopeptidase cleaves proteins at their C or N terminal end, while endopeptidase cleaves the proteins anywhere in between the two ends of the polypeptide chain. When the structure and mechanism of protease catalysis were discovered, they were classified into six groups: metalloproteases, serine proteases, cysteine proteases, aspartic proteases, glutamic proteases and threonine proteases (López-Otín & Bond, 2008). The name of each protease type is related to the interacting amino acid present at the active site with the substrate. Protease can be found in variety of organisms from single-cell to multicellular organisms. Among these, proteases from microorganisms are popular because of its easy availability, genetic manipulation and cost-effective purification. Microbial proteases are classified into alkaline, acidic and neutral proteases depending on their activity at different pH levels. The activity of analkaline protease is optimum in an alkaline environment, similarly, the activity of a neutral protease is optimum at a neutral pH and the activity of an acidic protease is optimum at an acidic pH (Razzaq et al., 2019).

Apart from the previously mentioned classification, depending upon cellular physiology, microbes process two types of proteases, intracellular and extracellular. Intracellular proteases mainly carry out a specific function related to cellular biochemistry. They are involved in different biological events required for maintaining normal physiology of the microbes such as, maturation of newly synthesized protein, regulation of signal transduction, protection against foreign proteins, etc. Extracellular or secretory proteases are secreted by the microbes in their extracellular environment. Bacteria and other microorganisms use secretory proteases for their survival in response to different environmental conditions. These proteases are mainly toxins or other factors that help in the survival of the microbes within their surroundings. Many pathogenic bacteria need secretory proteases to infect their hosts (Culp & Wright, 2017; Figaj et al., 2019). These types of proteases are of particular interest to researchers and are studied extensively for humankind (Wandersman, 1989).

Microbial secretory proteases have been used in industry for years, but their use in the biopharmaceutical industry has been growing in recent years. Microbial secretory proteases have been used as therapeutic candidates for different diseases, such as the treatment of sepsis, cardiovascular disease, inflammation, cystic fibrosis, retinal disorder, cancer, and other diseases. The FDA has approved a variety of secretory proteases for therapeutic applications such as botulinum toxin B, botulinum toxin A, reteplase and urokinase and few more are still in clinical trials (Shankar et al., 2021). Surprisingly, there are few reports available where anti-cancer activity of microbial secretory proteases has been studied. In reports that are available, it has been found they are very promising in inhibiting cell proliferation, inducing apoptosis and cytotoxicity in cancer cells. The field is quite unexplored and further thorough research is needed to identify novel bacterial proteases having anti-cancer properties.

12.4.2 Bacterial Serratia Protease as an Anti-Cancer Agent

In 1987, Maeda et al., from Japan, isolated protease from *Serratia marcescens* and showed that it has a potent anti-tumor activity when injected into radiation-induced syngeneic mouse tumors (Maeda et al., 1987). It has been shown that the *Serratia* protease in low doses can completely eradicate tumors locally, thus, for regional tumor treatment, this protease can be a good option. In high doses, this protease is toxic to animals and the anti-tumor activity is completely inhibited when incubated with ovomacroglobulin (an inhibitor of the *Serratia* protease). The *Serratia* protease is inhibited *in vivo* by alpha-2 macroglobulin, but after 5–6 hours, protease activity is regained. It has been found that the *Serratia* protease forms a complex with alpha-2 macroglobulin (E/I complex) after becoming internalized via the alpha-2 macroglobulin receptor. Following internalization, cellular integrity was lost and eventually the cell died (Maeda et al., 1989).

12.4.3 Serralysin-Extracellular Protease from *S. marcescens* Showed Cytotoxic Effects on Colorectal Cell

Another extracellular metalloprotease, serralysin, from *S. marcescens* was isolated and purified, which showed a cytotoxic effect on the Caco-2 cell line (Araghi et al., 2019). The MTT assay showed 24 hours' exposure of 24.8 μg/ml serralysin to the Caco-2 cell line can kill 50% of the cell population. Inhibition of the protease by EDTA showed no cytotoxic effects on cancer cells. However, the subcellular target and mode of action of this protease has not been explored yet.

12.4.4 Hemagglutinin Protease from *Vibrio cholerae* is a Potent Anti-Cancer Molecule

In our laboratory, we have shown that *Vibrio cholerae* hemagglutinin protease (HAP) is effective against various cancer cells. It showed significant cytotoxic effects on colon cancer (HT-29) and breast cancer (MCF-7) cell lines. Solid tumors developed in a syngeneic mouse model were completely abolished when treated with HAP (Ray et al., 2016). Importantly, HAP can induce cytotoxic effects on cancer cells but not on normal cells (MCF-10A, normal breast epithelial cell). The molecular mechanism revealed that HAP acts on the PAR-1 receptor of the cell surface and activates it. The activated PAR-1 receptor further activates its downstream signaling of NF-κB and MAPK (Ray & Pal, 2016). This signaling pathway enhanced reactive oxygen species (ROS) generation in cancer cells and induced the intrinsic pathway of apoptosis. Detailed analysis showed that *V. cholerae* HAP cleaved PAR-1 at a site different from its conventional thrombin-mediated cleavage site. The

N-terminal site of cleaved PAR-1 interacts with one of its transmembrane domains and activates its downstream signaling pathways. HAP exposed a new N-terminal sequence 'PFISED,' which by itself could be used as an activating peptide for PAR-1 (Ray et al., 2018).

12.4.5 Subtilisin from *Bacillus amyloliquefaciens* Acts as an Anti-Cancer Agent

In a separate discovery, our lab has shown that extracellular protease subtilisin has anti-cancer properties. Subtilisin is a well-studied protein and widely used in the detergent industry, but surprisingly, we found it could induce apoptosis specifically in cancer cells. Our result showed, subtilisin activates ubiquitination and proteasomal-mediated tubulin degradation in MCF-7 cells. Furthermore, we showed PARKIN acts as an E3 ligase of tubulin and caspase-7/PARP activation executes the finale stage of apoptosis in subtilisin-treated MCF-7 cells. Why subtilisin could activate tubulin degradation only in cancer cells and not in normal cells is yet to be determined, but preliminary findings suggest that subtilisin has some receptor specificity toward cancer cells. This study indicates that the secreted protease subtilisin could be used as a tubulin-targeted molecule to treat cancer (Singh et al., 2022).

12.5 STRATEGY TO DISCOVER NOVEL MICROBIAL SECRETORY PROTEASE FOR ANTI-CANCER RESEARCH

Microbes are found everywhere in nature, from the desert to a volcanic eruption, and the number of microbial taxa that are currently known is only 0.001% (Mathur & Hoskins, 2017). By exploring the environmental microbiome, we will be able to discover novel secretory proteases with anti-cancer properties. To fulfil this, the general and straightforward approach would be: microbial sample collection and screening for protease secretion, purification and identification of the extracellular protease and screening for anti-cancer potential of the protease using cell lines or animal models.

Microbial samples could be collected from different sources, such as environmental sources, biological sources, air sources, chemical sources, contaminant site, etc. The procedure involves using a cotton or non-cotton sterile swab on the surface of interest (such as a solid or liquid surface) and then keeping those swabs in a sterile zip-seal packet or vial containing growth media, generally in nutrient broth or tryptic soy broth media for the transportation purposes(Jansson et al., 2020; Wise et al., 2021). In the laboratory, the collected samples are either streaked directly to a media plate, generally, a nutrient agar plate, or streaked after serially diluting the collected sample in sterile PBS or media. The plates are then kept in an incubator at 37°C for the growth. Grown bacterial culture can be stored in glycerol or as a stab (a sterile vial with solidified agar) culture. For longer storage, bacterial culture can be dehydrated through lyophilization (Heckly, 1978).

Protease secretions from microbial sample could be detected by various methods such as fluorescence, mass spectrometry, calorimetry, immunoassay, etc. (Hu et al., 2016; Luo et al., 2018; Pinto & Schanze, 2004; Quinn et al., 2002; Zhang et al., 2021). However, the simple and straightforward approach is to grow the microbial colony in a 1% casein or skim milk and 1% agar-containing plate. If the sample secretes any active protease, the protease will digest the casein protein, resulting in the formation of a clear zone surrounding the colony, which will be visible in a white opaque casein agar plate. This observation confirms the presence of protease from the microbial colony. Another experiment that could be used for routine screening of a large number of microbial samples for protease secretion is the azocasein assay(Syngkon et al., 2010). In this assay, microbial culture supernatant is incubated with the azocasein solution and (similar with a casein agar assay) if any protease presents in the supernatant, it will cleave the casein protein. As a result, azo dye will be released from the casein and, in presence of 0.5 M NaOH, it will turn an orange color that will give absorbance at 440 nm(Coêlho et al., 2016).

Following screening, different methods can be applied for the purification of the secreted protease from the microbial culture. One cost-effective approach would be salting out methods—where

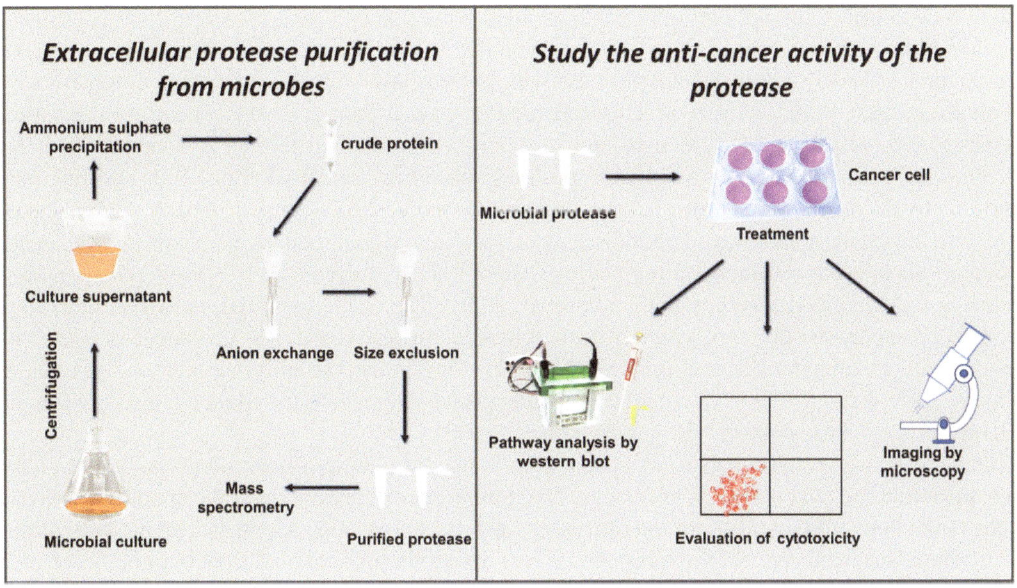

FIGURE 12.4 Schematic representation–strategy to discover microbial proteases for anti-cancer research.

proteins present in the microbial culture supernatant are precipitated out by the addition of salt–– such as adding ammonium sulphate to the culture supernatant(Wingfield, 1998). The protease from the precipitated proteins can then be isolated and purified by the column chromatography technique (Burden & Whitney, 1995; Coskun, 2016). Once purified, mass spectrometry would be a standard choice for the identification of the purified protease. After successful identification of the protease, the gene of the protease can be cloned for better characterization and future application (Figure 12.4).

The basic principle in studying anti-cancer activity is by evaluating the cytotoxicity of the purified protease on the cancer cell line. The cytotoxicity of the cancer cell can be evaluated by different types of assays. Among them MTT assay, comet assay, lactate dehydrogenase (LDH) assay and Annexin-V/PI staining assays for apoptosis are popular, and the commercial availability of these assay kits also makes them popular among researchers(Terry et al., 2019). To gain a deeper understanding of the signaling pathway related to the anti-cancer activity of microbial proteases, researcher used different molecular biology tools and techniques. Depending on the specific research interest, mice models or other animal models may be used for studying *in vivo* anti-cancer activity of the proteases.

12.6 APOPTOSIS IS THE KEY TARGET OF MICROBIAL PROTEASES HAVING ANTI-CANCER PROPERTIES

Apoptosis or programmed cell death is the cellular physiological process by which cells activate their own death program (Elmore, 2007). A series of pro-apoptotic or anti-apoptotic proteins can be regulated in this process that leads to ultimate death of the cells (O'Brien & Kirby, 2008). Apoptosis occurs to eliminate the unwanted or abnormal cells of an organism. Thus, this process is important for maintaining cellular homeostasis (Delong, 1998). Apoptosis in the cells can be activated through two major pathways, the intrinsic or mitochondrial pathway and the extrinsic or death receptor pathway. Intracellular signals such as DNA damage, cytokine deprivation and unfolded protein response can activate the intrinsic pathway of apoptosis and extracellular signals such as cellular damage and pathogen infections can induce the extrinsic pathway of apoptosis (Igney & Krammer, 2002; Xu et al., 2015; Zaman et al., 2014). In the intrinsic pathway, pro-apoptotic proteins induce damage of the mitochondria or disrupt the mitochondrial membrane potential and, as a result, cytochrome-c is released from the mitochondria to the cytoplasm that in-turn activates apoptotic protease activating factor 1 (APAF1) and forms apoptosome with pro-caspase-9. This apoptosome complex helps to cleave caspase-9 and activate it. This activated caspase-9 then activates executioner caspases, caspase-3/7/6, which leads to the execution of apoptosis in the cell (Green & Llambi, 2015; Xu & Shi, 2007). Some of the pro-apoptotic proteins, including Bax, Bad, Bim, Bid, and p53 play important role in the regulation of these proteins (Elmore, 2007; Lomonosova & Chinnadurai, 2008). In the extrinsic pathway of apoptosis, an extracellular signal induces apoptosis through death receptors such as Fas receptors, tumor necrosis factor (TNF) receptors and TNF-related apoptosis-inducing ligand (TRAIL) receptors (Goldar et al., 2015). The death domain of these receptors then recruits pro-apoptotic proteins which in-turn activates initiator caspase-8. Caspase-8 activates its downstream executioner caspase, pro-caspase-3, resulting in the execution of apoptosis (Cullen & Martin, 2009; Nagata, 1999). During the execution phase of apoptosis crosstalk occurs between the intrinsic and extrinsic pathway (Roy & Nicholson, 2000).

One of the major characteristics of cancer cells is uncontrolled proliferation of the cells and, for this, the inhibition of apoptosis is necessary. The inhibition or deregulation of the apoptosis pathway leads to the uncontrolled proliferation of cancer cells. In cancer cells, apoptosis can be deregulated by different mechanisms. Over-expression of anti-apoptotic proteins and downregulation of pro-apoptotic proteins is common in most cancer types. For example, anti-apoptotic protein bcl-2 is upregulated and loss of bax or a mutation in the *Bax* gene is reported in various cancers (Adams & Cory, 1998; Elkholi et al., 2014; Yip & Reed, 2008).

In anti-cancer research, different strategies are employed to target cancer cells and targeting apoptosis is one of the common approaches (Pfeffer & Singh, 2018). Many naturally derived anti-cancer molecules have been shown to induce apoptosis in cancer cells. For example, curcumin found in turmeric is reported to have anti-cancer activity through inhibiting bcl-2 and the x-linked inhibitor of apoptosis protein (XIAP) (Shanmugam et al., 2015; Yallapu et al., 2015). Quercetin, found in apples and onions, activates executioner caspases and induces apoptosis (Ranganathan et al., 2015). Cytotoxic necrotizing factor Y (CNFy) from *Yersinia pseudotuberculosis* induces PARP cleavage and activates caspase-3-mediated apoptosis in cancer cells (Augspach et al., 2013). Several other examples exist in which anti-cancer molecules isolated from microbes, plants and other sources have the potential to induce apoptosis in cancer cells. Similarly, most of the reported microbial proteases were shown to induce apoptosis in cancer cells. Serralysin secreted by *S. marcescens* induces cytotoxicity and apoptosis in the colon carcinoma Caco-2 cell line (Araghi et al., 2019). HAP from *V. cholerae* activates the PAR-1-mediated intrinsic pathway of apoptosis in colon and breast cancer cells (Ray & Pal, 2016; Ray et al., 2016). Similarly, secreted protease subtilisin induces apoptosis in breast cancer cells by targeting the tubulin of the cells (Singh et al., 2022). Overall, these reports indicate that apoptosis, or programmed cell death, could be an important area of interest while searching for anti-cancer proteases from microbial sources.

12.7 CONCLUSION AND FUTURE PERSPECTIVES

Protease is an enzyme found in every living organism, including plants, animals and microbes. Microbial proteases have been studied for decades for their importance in commercial industrial applications. Besides their industrial use, proteases isolated from microbial sources have shown promising results against various diseases. The discovery of the therapeutic benefit of microbial protease takes the protease industry in a new direction. Though the activity of microbial proteases on different diseases have been observed, there are very few reports available where the anti-cancer activity of microbial proteases has been studied. However, the reported studies indicate that microbial protease has great potential against cancer. So, there is a clear need for further studies to better understand the therapeutic potential of microbial proteases against cancer. A straightforward and simple strategy is vital in identifying more novel proteases with anti-cancer properties.

Microbial extracellular proteases hold significant potential in future anti-cancer research due to their ease of extraction, higher yields, high stability, cost-effective production and ease of genetic manipulation. Recent advances in protein engineering brightens the future use of these proteases as more promising in cancer therapeutics. Novel microbial proteases can be manipulated through protein engineering to target altered or mutated cell-surface receptors of cancer or different tumor-associated proteins. Research in this area could further explore how these proteases can be used to selectively target cancer cells. The goal is to develop more effective and prompt treatment of cancer with minimal side-effects.

DECLARATIONS

Conflict of interest: The authors declare no conflicts of interest.

Consent for publishing: All authors have approved the final version of the manuscript and given consent for publication.

Funding: Not Applicable.

REFERENCES

Adams, J. M., & Cory, S. (1998). The Bcl-2 protein family: Arbiters of cell survival. *Science*, *281*(5381), 1322–1326. https://doi.org/10.1126/science.281.5381.1322

Amy, Y. C. (2010). Cell cycle control by oncogenes and tumor suppressors: Driving the transformation of normal cells into cancerous cells. *Nature Education*, *3*(9), 7.

Araghi, A., Hashemi, S., Sepahi, A. A., Faramarzi, M. A., & Amin, M. (2019). Purification and study of anti-cancer effects of Serratia marcescens serralysin. *Iranian Journal of Microbiology*, *11*(4), 320–327.

Are, C., McMasters, K. M., Giuliano, A., Yanala, U., Balch, C., Anderson, B. O., Berman, R., Audisio, R., Kovacs, T., Savant, D., Toprani, R., Amira, G., Sallam, I., Baek, J. H., Baek, M.-J., Park, D. J., Beulo, G. Q., Rocha, E. B., Said, H. M., … Mori, M. (2019). Global forum of cancer surgeons: Perspectives on barriers to surgical care for cancer patients and potential solutions. *Annals of Surgical Oncology*, *26*(6), 1577–1582. https://doi.org/10.1245/s10434-019-07301-2

Augspach, A., List, J., Wolf, P., Bielek, H., Schwan, C., Elsässer-Beile, U., Aktories, K., & Schmidt, G. (2013). Activation of Rho A, B, C by Yersinia cytotoxic necrotizing factor (CNFy) induces apoptosis in LNCaP prostate cancer cells. *Toxins*, *5*(11), 2241–2257. https://doi.org/10.3390/toxins5112241

Baindara, P., & Mandal, S. M. (2020). Bacteria and bacterial anticancer agents as a promising alternative for cancer therapeutics. *Biochimie*, *177*, 164–189. https://doi.org/10.1016/j.biochi.2020.07.020

Baker, S. S., Borowitz, D., Duffy, L., Fitzpatrick, L., Gyamfi, J., & Baker, R. D. (2005). Pancreatic enzyme therapy and clinical outcomes in patients with cystic fibrosis. *The Journal of Pediatrics*, *146*(2), 189–193. https://doi.org/10.1016/j.jpeds.2004.09.003

Bukowski, K., Kciuk, M., & Kontek, R. (2020). Mechanisms of multidrug resistance in cancer chemotherapy. *International Journal of Molecular Sciences*, *21*(9). https://doi.org/10.3390/ijms21093233

Burden, D. W., & Whitney, D. B. (1995). Protein Purification by Column Chromatography. In *Biotechnology Proteins to PCR* (pp. 93–124). Birkhäuser Boston. https://doi.org/10.1007/978-1-4612-4278-9_5

Chabner, B. A., & Roberts, T. G. (2005). Chemotherapy and the war on cancer. *Nature Reviews Cancer*, *5*(1), 65–72. https://doi.org/10.1038/nrc1529

Coêlho, D. F., Saturnino, T. P., Fernandes, F. F., Mazzola, P. G., Silveira, E., & Tambourgi, E. B. (2016). Azocasein substrate for determination of proteolytic activity: Reexamining a traditional method using bromelain samples. *BioMed Research International*, *2016*, 1–6. https://doi.org/10.1155/2016/8409183

Cooper, G. M. (2000). The Eukaryotic Cell Cycle. In *The Cell: A Molecular Approach.* (2nd ed.). Sinauer Associates.

Coskun, O. (2016). Separation techniques: Chromatography. *Northern Clinics of Istanbul*. https://doi.org/10.14744/nci.2016.32757

Cragg, G. M., Kingston, D. G. I., & Newman, D. J. (Eds.). (2011). *Anticancer Agents from Natural Products*. CRC Press. https://doi.org/10.1201/b11185

Cullen, S. P., & Martin, S. J. (2009). Caspase activation pathways: Some recent progress. *Cell Death & Differentiation*, *16*(7), 935–938. https://doi.org/10.1038/cdd.2009.59

Culp, E., & Wright, G. D. (2017). Bacterial proteases, untapped antimicrobial drug targets. *The Journal of Antibiotics*, *70*(4), 366–377. https://doi.org/10.1038/ja.2016.138

da Rocha, A. (2001). Natural products in anticancer therapy. *Current Opinion in Pharmacology*, *1*(4), 364–369. https://doi.org/10.1016/S1471-4892(01)00063-7

Dare, A. J., Anderson, B. O., Sullivan, R., Pramesh, C. S., Yip, C.-H., Ilbawi, A., Adewole, I. F., Badwe, R. A., & Gauvreau, C. L. (2015). Surgical Services for Cancer Care. In *Disease Control Priorities, Third Edition (Volume 3): Cancer* (pp. 223–238). The World Bank. https://doi.org/10.1596/978-1-4648-0349-9_ch13

Dehelean, C. A., Marcovici, I., Soica, C., Mioc, M., Coricovac, D., Iurciuc, S., Cretu, O. M., & Pinzaru, I. (2021). Plant-derived anticancer compounds as new perspectives in drug discovery and alternative therapy. *Molecules*, *26*(4), 1109. https://doi.org/10.3390/molecules26041109

Delong, M. J. (1998). Apoptosis: A modulator of cellular homeostasis and disease states. *Annals of the New York Academy of Sciences*, *842*(1 SALIVARY GLAN), 82–90. https://doi.org/10.1111/j.1749-6632.1998.tb09635.x

Demain, A. L., & Adrio, J. L. (2008). Contributions of microorganisms to industrial biology. *Molecular Biotechnology*, *38*(1), 41–55. https://doi.org/10.1007/s12033-007-0035-z

Demicheli, R., Retsky, M. W., Hrushesky, W. J. M., Baum, M., & Gukas, I. D. (2008). The effects of surgery on tumor growth: A century of investigations. *Annals of Oncology*, *19*(11), 1821–1828. https://doi.org/10.1093/annonc/mdn386

Dhankhar, R., Gupta, V., Kumar, S., Kapoor, R. K., & Gulati, P. (2020). Microbial enzymes for deprivation of amino acid metabolism in malignant cells: Biological strategy for cancer treatment. *Applied Microbiology and Biotechnology*, *104*(7), 2857–2869. https://doi.org/10.1007/s00253-020-10432-2

Dhankhar, R., Kawatra, A., Mohanty, A., & Gulati, P. (2021). Microbial enzymes used in prodrug activation for cancer therapy: Insights and future perspectives. *Current Protein & Peptide Science*, *22*(7), 514–525. https://doi.org/10.2174/1389203721666201207231932

Du, J., Shao, Z., & Zhao, H. (2011). Engineering microbial factories for synthesis of value-added products. *Journal of Industrial Microbiology & Biotechnology*, *38*(8), 873–890. https://doi.org/10.1007/s10295-011-0970-3

Elkholi, R., Renault, T. T., Serasinghe, M. N., & Chipuk, J. E. (2014). Putting the pieces together: How is the mitochondrial pathway of apoptosis regulated in cancer and chemotherapy? *Cancer & Metabolism*, *2*(1), 16. https://doi.org/10.1186/2049-3002-2-16

Elmore, S. (2007). Apoptosis: A review of programmed cell death. *Toxicologic Pathology*, *35*(4), 495–516. https://doi.org/10.1080/01926230701320337

Figaj, D., Ambroziak, P., Przepiora, T., & Skorko-Glonek, J. (2019). The role of proteases in the virulence of plant pathogenic bacteria. *International Journal of Molecular Sciences*, *20*(3). https://doi.org/10.3390/ijms20030672

Giddings, L.-A., & Newman, D. J. (2013). Microbial natural products: Molecular blueprints for antitumor drugs. *Journal of Industrial Microbiology and Biotechnology*, *40*(11), 1181–1210. https://doi.org/10.1007/s10295-013-1331-1

Goldar, S., Khaniani, M. S., Derakhshan, S. M., & Baradaran, B. (2015). Molecular mechanisms of apoptosis and roles in cancer development and treatment. *Asian Pacific Journal of Cancer Prevention*, *16*(6), 2129–2144. https://doi.org/10.7314/APJCP.2015.16.6.2129

Gordaliza, M. (2007). Natural products as leads to anticancer drugs. *Clinical and Translational Oncology*, *9*(12), 767–776. https://doi.org/10.1007/s12094-007-0138-9

Green, D. R., & Llambi, F. (2015). Cell death signaling. *Cold Spring Harbor Perspectives in Biology*, *7*(12), a006080. https://doi.org/10.1101/cshperspect.a006080

Heckly, R. J. (1978). Preservation of Microorganisms (pp. 1–53). Academic Press. https://doi.org/10.1016/S0065-2164(08)70635-X

Hollstein, U. (1974). Actinomycin. Chemistry and mechanism of action. *Chemical Reviews*, *74*(6), 625–652. https://doi.org/10.1021/cr60292a002

Hu, J., Liu, F., Feng, N., & Ju, H. (2016). Peptide codes for multiple protease activity assay via high-resolution mass spectrometric quantitation. *Rapid Communications in Mass Spectrometry*, *30*(S1), 196–201. https://doi.org/10.1002/rcm.7631

Huang, M.-Y., Zhang, L.-L., Ding, J., & Lu, J.-J. (2018). Anticancer drug discovery from Chinese medicinal herbs. *Chinese Medicine*, *13*(1), 35. https://doi.org/10.1186/s13020-018-0192-y

Igney, F. H., & Krammer, P. H. (2002). Death and anti-death: Tumour resistance to apoptosis. *Nature Reviews Cancer*, *2*(4), 277–288. https://doi.org/10.1038/nrc776

Jansson, L., Akel, Y., Eriksson, R., Lavander, M., & Hedman, J. (2020). Impact of swab material on microbial surface sampling. *Journal of Microbiological Methods, 176*, 106006. https://doi.org/10.1016/j.mimet.2020.106006

Karpiński, T., & Adamczak, A. (2018). Anticancer activity of bacterial proteins and peptides. *Pharmaceutics, 10*(2), 54. https://doi.org/10.3390/pharmaceutics10020054

Khalifa, S. A. M., Elias, N., Farag, M. A., Chen, L., Saeed, A., Hegazy, M.-E. F., Moustafa, M. S., Abd El-Wahed, A., Al-Mousawi, S. M., Musharraf, S. G., Chang, F.-R., Iwasaki, A., Suenaga, K., Alajlani, M., Göransson, U., & El-Seedi, H. R. (2019). Marine natural products: A source of novel anticancer drugs. *Marine Drugs, 17*(9), 491. https://doi.org/10.3390/md17090491

Korn, E. D., & Payza, A. N. (1956). The degradation of heparin by bacterial enzymes. I. Adaptation and lyophilized cells. *The Journal of Biological Chemistry, 223*(2), 853–858.

Kulothungan, V., Sathishkumar, K., Leburu, S., Ramamoorthy, T., Stephen, S., Basavarajappa, D., Tomy, N., Mohan, R., Menon, G. R., & Mathur, P. (2022). Burden of cancers in India – estimates of cancer crude incidence, YLLs, YLDs and DALYs for 2021 and 2025 based on National Cancer Registry Program. *BMC Cancer, 22*(1), 527. https://doi.org/10.1186/s12885-022-09578-1

Link, W. (2019). *Principles of Cancer Treatment and Anticancer Drug Development*. Springer International Publishing. https://doi.org/10.1007/978-3-030-18722-4

LoConte, N. K., Gershenwald, J. E., Thomson, C. A., Crane, T. E., Harmon, G. E., & Rechis, R. (2018). Lifestyle modifications and policy implications for primary and secondary cancer prevention: Diet, exercise, sun safety, and alcohol reduction. *American Society of Clinical Oncology Educational Book, 38*, 88–100. https://doi.org/10.1200/EDBK_200093

Lomonosova, E., & Chinnadurai, G. (2008). BH3-only proteins in apoptosis and beyond: An overview. *Oncogene, 27*(S1), S2–S19. https://doi.org/10.1038/onc.2009.39

Loomans-Kropp, H. A., & Umar, A. (2019). Cancer prevention and screening: The next step in the era of precision medicine. *NPJ Precision Oncology, 3*, 3. https://doi.org/10.1038/s41698-018-0075-9

López-Otín, C., & Bond, J. S. (2008). Proteases: Multifunctional enzymes in life and disease. *The Journal of Biological Chemistry, 283*(45), 30433–30437. https://doi.org/10.1074/jbc.R800035200

Luo, Q., Chen, D., Boom, R. M., & Janssen, A. E. M. (2018). Revisiting the enzymatic kinetics of pepsin using isothermal titration calorimetry. *Food Chemistry, 268*, 94–100. https://doi.org/10.1016/j.foodchem.2018.06.042

Maeda, H., Matsumura, Y., & Molla, A. (1987). Antitumor activity of some bacterial proteases: Eradication of solid tumors in mice by intratumor injection. *Cancer Research, 47*(2), 563–566.

Maeda, H., Molla, A., Sakamoto, K., Murakami, A., & Matsumura, Y. (1989). Cytotoxicity of bacterial proteases in various tumor cells mediated through alpha 2-macroglobulin receptor. *Cancer Research, 49*(3), 660–664.

Mathur, S., & Hoskins, C. (2017). Drug development: Lessons from nature. *Biomedical Reports, 6*(6), 612–614. https://doi.org/10.3892/br.2017.909

McCarthy, E. F. (2006). The toxins of William B. Coley and the treatment of bone and soft-tissue sarcomas. *The Iowa Orthopaedic Journal, 26*, 154–158.

Nagata, S. (1999). Biddable death. *Nature Cell Biology, 1*(6), E143–E145. https://doi.org/10.1038/14094

Natarajan, A., & Bhatt, A. S. (2020). Microbes and microbiomes in 2020 and beyond. *Nature Communications, 11*(1), 4988. https://doi.org/10.1038/s41467-020-18850-6

Negroni, L., Samson, M., Guigonis, J.-M., Rossi, B., Pierrefite-Carle, V., & Baudoin, C. (2007). Treatment of colon cancer cells using the cytosine deaminase/5-fluorocytosine suicide system induces apoptosis, modulation of the proteome, and Hsp90beta phosphorylation. *Molecular Cancer Therapeutics, 6*(10), 2747–2756. https://doi.org/10.1158/1535-7163.MCT-07-0040

Newman, D. J., & Cragg, G. M. (2020). Natural products as sources of new drugs over the nearly four decades from 01/1981 to 09/2019. *Journal of Natural Products, 83*(3), 770–803. https://doi.org/10.1021/acs.jnatprod.9b01285

O'Brien, M. A., & Kirby, R. (2008). Apoptosis: A review of pro-apoptotic and anti-apoptotic pathways and dysregulation in disease. *Journal of Veterinary Emergency and Critical Care, 18*(6), 572–585. https://doi.org/10.1111/j.1476-4431.2008.00363.x

Pfeffer, C., & Singh, A. (2018). Apoptosis: A target for anticancer therapy. *International Journal of Molecular Sciences, 19*(2), 448. https://doi.org/10.3390/ijms19020448

Pham, J. V., Yilma, M. A., Feliz, A., Majid, M. T., Maffetone, N., Walker, J. R., Kim, E., Cho, H. J., Reynolds, J. M., Song, M. C., Park, S. R., & Yoon, Y. J. (2019). A review of the microbial production of bioactive natural products and biologics. *Frontiers in Microbiology, 10*. https://doi.org/10.3389/fmicb.2019.01404

Pinto, M. R., & Schanze, K. S. (2004). Amplified fluorescence sensing of protease activity with conjugated polyelectrolytes. *Proceedings of the National Academy of Sciences*, *101*(20), 7505–7510. https://doi.org/10.1073/pnas.0402280101

Quinn, C. P., Semenova, V. A., Elie, C. M., Romero-Steiner, S., Greene, C., Li, H., Stamey, K., Steward-Clark, E., Schmidt, D. S., Mothershed, E., Pruckler, J., Schwartz, S., Benson, R. F., Helsel, L. O., Holder, P. F., Johnson, S. E., Kellum, M., Messmer, T., Thacker, W. L., ... Perkins, B. A. (2002). Specific, sensitive, and quantitative enzyme-linked immunosorbent assay for human immunoglobulin G antibodies to anthrax toxin protective antigen. *Emerging Infectious Diseases*, *8*(10), 1103–1110. https://doi.org/10.3201/eid0810.020380

Radadiya, A., Zhu, W., Coricello, A., Alcaro, S., & Richards, N. G. J. (2020). Improving the treatment of acute lymphoblastic leukemia. *Biochemistry*, *59*(35), 3193–3200. https://doi.org/10.1021/acs.biochem.0c00354

Ranganathan, S., Halagowder, D., & Sivasithambaram, N. D. (2015). Quercetin suppresses twist to induce apoptosis in MCF-7 breast cancer cells. *PLoS One*, *10*(10), e0141370. https://doi.org/10.1371/journal.pone.0141370

Rao, M. B., Tanksale, A. M., Ghatge, M. S., & Deshpande, V. V. (1998). Molecular and biotechnological aspects of microbial proteases. *Microbiology and Molecular Biology Reviews*, *62*(3), 597–635. https://doi.org/10.1128/MMBR.62.3.597-635.1998

Ray, T., Chakrabarti, M. K., & Pal, A. (2016). Hemagglutinin protease secreted by V. cholerae induced apoptosis in breast cancer cells by ROS mediated intrinsic pathway and regresses tumor growth in mice model. *Apoptosis*, *21*(2), 143–154. https://doi.org/10.1007/s10495-015-1194-1

Ray, T., Kar, D., Pal, A., Mukherjee, S., Das, C., & Pal, A. (2018). Molecular targeting of breast and colon cancer cells by PAR1 mediated apoptosis through a novel pro-apoptotic peptide. *Apoptosis*, *23*(11–12), 679–694. https://doi.org/10.1007/s10495-018-1485-4

Ray, T., & Pal, A. (2016). PAR-1 mediated apoptosis of breast cancer cells by V. cholerae hemagglutinin protease. *Apoptosis*, *21*(5), 609–620. https://doi.org/10.1007/s10495-016-1229-2

Razzaq, A., Shamsi, S., Ali, A., Ali, Q., Sajjad, M., Malik, A., & Ashraf, M. (2019). Microbial proteases applications. *Frontiers in Bioengineering and Biotechnology*, *7*, 110. https://doi.org/10.3389/fbioe.2019.00110

Roy, S., & Nicholson, D. W. (2000). Cross-talk in cell death signaling. *The Journal of Experimental Medicine*, *192*(8), F21–5.

Shankar, R., Upadhyay, P. K., & Kumar, M. (2021). Protease enzymes: Highlights on potential of proteases as therapeutics agents. *International Journal of Peptide Research and Therapeutics*, *27*(2), 1281–1296. https://doi.org/10.1007/s10989-021-10167-2

Shanmugam, M., Rane, G., Kanchi, M., Arfuso, F., Chinnathambi, A., Zayed, M., Alharbi, S., Tan, B., Kumar, A., & Sethi, G. (2015). The multifaceted role of curcumin in cancer prevention and treatment. *Molecules*, *20*(2), 2728–2769. https://doi.org/10.3390/molecules20022728

Siegel, R. L., Miller, K. D., Fuchs, H. E., & Jemal, A. (2022). Cancer statistics, 2022. *CA: A Cancer Journal for Clinicians*, *72*(1), 7–33. https://doi.org/10.3322/caac.21708

Singh, N., Tapader, R., Chatterjee, S., Pal, A., & Pal, A. (2022). Subtilisin from *Bacillus amyloliquefaciens* induces apoptosis in breast cancer cells through ubiquitin-proteasome-mediated tubulin degradation. *International Journal of Biological Macromolecules*, *220*, 852–865. https://doi.org/10.1016/j.ijbiomac.2022.08.086

Syngkon, A., Elluri, S., Koley, H., Rompikuntal, P. K., Saha, D. R., Chakrabarti, M. K., Bhadra, R. K., Wai, S. N., & Pal, A. (2010). Studies on a novel serine protease of a ΔhapAΔprtV vibrio cholerae O1 strain and its role in hemorrhagic response in the rabbit ileal loop model. *PLoS ONE*, *5*(9), e13122. https://doi.org/10.1371/journal.pone.0013122

Tannock, I. F. (1998). Conventional cancer therapy: Promise broken or promise delayed? *The Lancet*, *351*(Suppl 2), SII9–SII16. https://doi.org/10.1016/s0140-6736(98)90327-0

Terry, R., Andrew, N., Rich, M., Natashia, K., & Jolanta, V. (2019). *Cytotoxicity Assays: In Vitro Methods to Measure Dead Cells* (Markossian S, Grossman A, Brimacombe K, et al. Eds.). Eli Lilly & Company and the National Center for Advancing Translational Sciences.

Vellard, M. (2003). The enzyme as drug: Application of enzymes as pharmaceuticals. *Current Opinion in Biotechnology*, *14*(4), 444–450. https://doi.org/10.1016/S0958-1669(03)00092-2

Vilalta, M., Rafat, M., & Graves, E. E. (2016). Effects of radiation on metastasis and tumor cell migration. *Cellular and Molecular Life Sciences*, *73*(16), 2999–3007. https://doi.org/10.1007/s00018-016-2210-5

Waksman, S. A., & Woodruff, H. B. (1940). Bacteriostatic and bactericidal substances produced by a soil actinomyces. *Experimental Biology and Medicine*, *45*(2), 609–614. https://doi.org/10.3181/00379727-45-11768

Wandersman, C. (1989). Secretion, processing and activation of bacterial extracellular proteases. *Molecular Microbiology*, *3*(12), 1825–1831. https://doi.org/10.1111/j.1365-2958.1989.tb00169.x

Wei, J., Wahl, J., Knauss, H., Zeller, S., Jarmy, G., Fitze, G., Debatin, K.-M., & Beltinger, C. (2007). Cytosine deaminase/5-fluorocytosine gene therapy and Apo2L/TRAIL cooperate to kill TRAIL-resistant tumor cells. *Cancer Gene Therapy*, *14*(7), 640–651. https://doi.org/10.1038/sj.cgt.7701051

Weiman, S. (2014). Cover of Harnessing the Power of Microbes as Therapeutics: Bugs as Drugs Harnessing the Power of Microbes as Therapeutics: Bugs as Drugs. American Society for Microbiology. https://www.ncbi.nlm.nih.gov/books/NBK519801/

Wingfield, P. (1998). Protein precipitation using ammonium sulfate. *Current Protocols in Protein Science*, *13*(1). https://doi.org/10.1002/0471140864.psa03fs13

Wise, N. M., Wagner, S. J., Worst, T. J., Sprague, J. E., & Oechsle, C. M. (2021). Comparison of swab types for collection and analysis of microorganisms. *Microbiology Open*, *10*(6). https://doi.org/10.1002/mbo3.1244

Xu, G., & Shi, Y. (2007). Apoptosis signaling pathways and lymphocyte homeostasis. *Cell Research*, *17*(9), 759–771. https://doi.org/10.1038/cr.2007.52

Xu, W., Jing, L., Wang, Q., Lin, C.-C., Chen, X., Diao, J., Liu, Y., & Sun, X. (2015). Bax-PGAM5L-Drp1 complex is required for intrinsic apoptosis execution. *Oncotarget*, *6*(30), 30017–30034. https://doi.org/10.18632/oncotarget.5013

Yallapu, M. M., Nagesh, P. K. B., Jaggi, M., & Chauhan, S. C. (2015). Therapeutic applications of curcumin nanoformulations. *The AAPS Journal*, *17*(6), 1341–1356. https://doi.org/10.1208/s12248-015-9811-z

Yang, Y., MacLeod, V., Dai, Y., Khotskaya-Sample, Y., Shriver, Z., Venkataraman, G., Sasisekharan, R., Naggi, A., Torri, G., Casu, B., Vlodavsky, I., Suva, L. J., Epstein, J., Yaccoby, S., Shaughnessy, J. D., Barlogie, B., & Sanderson, R. D. (2007). The syndecan-1 heparan sulfate proteoglycan is a viable target for myeloma therapy. *Blood*, *110*(6), 2041–2048. https://doi.org/10.1182/blood-2007-04-082495

Yari, M., Ghoshoon, M. B., Vakili, B., & Ghasemi, Y. (2017). Therapeutic enzymes: Applications and approaches to pharmacological improvement. *Current Pharmaceutical Biotechnology*, *18*(7). https://doi.org/10.2174/1389201018666170808150742

Yip, K. W., & Reed, J. C. (2008). Bcl-2 family proteins and cancer. *Oncogene*, *27*(50), 6398–6406. https://doi.org/10.1038/onc.2008.307

Zaman, S., Wang, R., & Gandhi, V. (2014). Targeting the apoptosis pathway in hematologic malignancies. *Leukemia & Lymphoma*, *55*(9), 1980–1992. https://doi.org/10.3109/10428194.2013.855307

Zhang, X., Shuai, Y., Tao, H., Li, C., & He, L. (2021). Novel method for the quantitative analysis of protease activity: The casein plate method and its applications. *ACS Omega*, *6*(5), 3675–3680. https://doi.org/10.1021/acsomega.0c05192

13 The Emerging Role of the γ-Secretase Inhibitors in Cancer Management

Anuj Kumar Borah, Chongtham Sovachandra Singh, Babiangshisha Kharbuli and Amlan Das
Royal School of Bio-Sciences, The Assam Royal Global University, Betkuchi, Guwahati, India

Anupam Chatterjee
Dean, Royal School of Bio-Sciences, The Assam Royal Global University, Betkuchi, Guwahati, India

Subhendu Chakrabarty
M.U.C. Women's College, Burdwan, India

13.1 INTRODUCTION

Gamma (γ) secretase (γ-secretase or GS) is a multi-subunit aspartyl protease complex responsible for cleaving single-pass transmembrane proteins within the cell membrane. The stimulated activity of GS, a hetero-tetrameric aspartyl enzyme complex, has been known to be responsible for several physiological disturbances, mainly neurological disorders and several types of malignancies. GS comprises four subunits, such as presenilin (PS), anterior pharynx-defective 1 (APH-1), nicastrin (NCT), and presenilin enhancer 2 (PEN2). The subunit PS contains the catalytic center—the aspartyl residues in the transmembrane helical domains six and seven. Subunits APH-1 and PEN2 participate in conjunction with the maturation of the enzyme complex. The other subunit, NCT, is reportedly involved in substrate recognition (Carroll and Li 2016) (Figure 13.1).

Once correctly folded, all four subunits assembled and then glycosylated, GS becomes active and carries out its aspartyl protease action on more than 90 integral membrane proteins after ectodomain shedding (Carroll and Li 2016). Amyloid precursor protein (APP) is reportedly the most common substrate of γ-secretase action, which, when cleaved by β and γ-secretase, forms the β-amyloid plaques in Alzheimer's disease. Notch receptors (Notch 1–4), canonical Notch transmembrane ligand, E-Cadherin, deleted in colorectal cancer (DCC), insulin-like growth factor receptor (IGF1R), vascular endothelial growth factor receptor (VEGF-1R), interleukin 6 and interleukin1 receptor (IL6R and IL1R), which are involved in different cellular signaling pathways, are reportedly prime substrates for the γ-secretase complex (Pine 2018). The γ-secretase complex activity is also assisted by associated proteins, which provide specificity to the aspartyl protease. One such associated protein, GSAP, which interacts with both GS and APP, exerts influence on GS so that APP is preferred to Notch as substrate (Carroll and Li 2016). Nevertheless, as stated previously, Notch receptors (Notch 1–4) remain a prime substrate, and with today's technology, the similar Notch–gamma secretase interaction and modulation would receive major attention.

 DOI: 10.1201/9781003394693-15

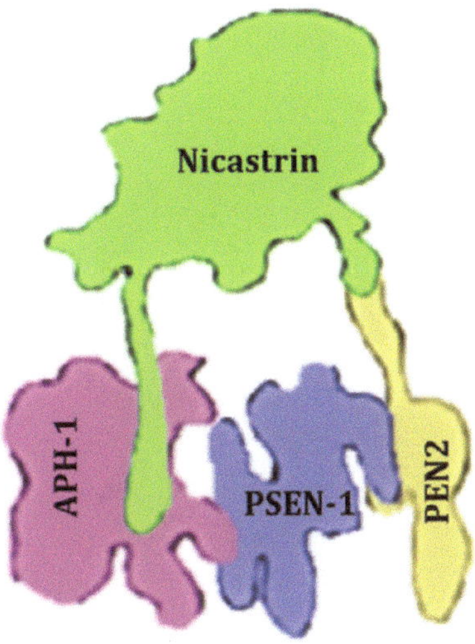

FIGURE 13.1 Subunit composition of γ-secretase: GS comprises the subunits Presenilin (PS), Nicastrin (NCT), anterior pharynx-defective 1 (APH-1), and Presenilin enhancer 2 (PEN2).

Deregulation of Notch signaling, which is associated with arrays of cellular events such as cellular growth, differentiation, signal transduction, and homeostasis, often reportedly culminates in dreadful consequences including T-cell leukemia, breast cancer, colorectal cancer, lung cancer, esophageal cancers, and neuro-malignancies, to name a few (Yuan et al. 2015). GS, the major activator of Notch-signaling, unarguably stands as the major target to tackle such consequences. GS action and activation can be regulated by antibodies for the extracellular domain of NCT, which would prevent substrate recognition by the enzyme, while inhibitors of the catalytic activity by GS inhibitors (GSI) represent the alternative but sound measure. This discrepancy is mainly because of reports suggesting that NCT is not critical for GS action (Carroll and Li 2016). Subsequent sections of this chapter aim to demonstrate extant investigations with GSI for the management of Notch activation and cancers of various kinds, in particular.

13.2 ROLE OF GAMA SECRETASE IN NOTCH ACTIVATION AND IMPLICATION IN CANCER

Binding of specific ligands to the Notch-receptors leads to the activation of the Notch-signaling cascade (Figure 13.2). There are four different mammalian Notch receptors viz Notch 1–4; such as Jagged-1 (JAG1), Jagged-2 (JAG2), delta-like ligand 1 (DLL1), delta-like ligand 3 (DLL3), and delta-like ligand 4 (DLL4). The transmembrane heterodimeric Notch receptors demonstrate an extracellular, a transmembrane, and a cytoplasmic domain. Upon ligand-receptor interaction, Notch, which has already received an initial cleavage (S1 cleavage), undergoes a further cleavage by "A Disintegrin and Metalloproteinase" (ADAM) family of protease (S2 cleavage), releasing the Notch extracellular domain. The remaining Notch receptor then undergoes cleavage by the γ-secretase complex (S3 cleavage), leading to the formation of the Notch Intracellular Domain (NICD). The activated NICD further translocates into the nucleus and interacts with the transcription factor CBF-1/suppressor of hairless/Lag1 (CSL) (Kopan 2012). CSL is an adapter molecule that may interact with transcriptional repressors, including histone deacetylase (HDAC) or NICD, once Notch receives S3 cleavage by γ-secretase (GS). The NICD-CSL interaction recruits histone acetyltransferase (open-up chromatin) and Mastermind-like adapter protein (MAML). MAML further recruits HAT, p300, and other transcription machinery, ultimately resulting in the expression of Hairy Enhance of Split (HES), Hairy/Enhancer of Spit related with YRPW motif (HEY), Akt, and PTEN (Al-Hussaini

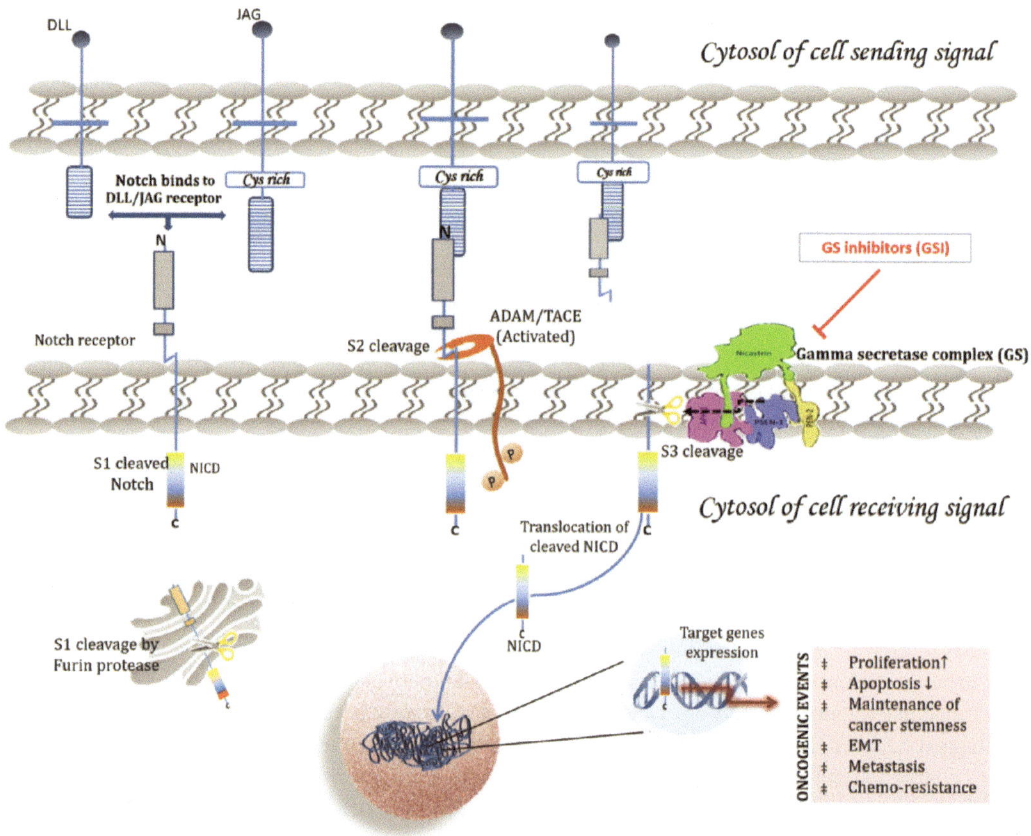

FIGURE 13.2 Schematic representation of the activation and subsequent mechanism of action of the Notch pathway.

et al. 2011). This signaling cascade regulates numerous other factors that include p27cip1/WAF1, c-Myc, cyclinD1, Nuclear factor-kappa B (NF-κB), p21, Slug, Nanog, etc (Kuramoto et al. 2012).

In normal physiological conditions, a Notch receptor not bound to a ligand usually becomes recycled or undergoes lysosomal degradation. However, Notch receptor activation may occur in the absence of ligands in scenarios such as exposure to calcium chelators, overexpression of Notch with ADAM proteases, and mutation in the NRR domain and ESCRT domain of Notch. The Notch receptors, after synthesis and receiving S1 cleavage, are glycosylated in the Golgi complex by Fringe (a type of glycosyl transferase), which governs the ligand-specific responses, timing, and amplitude of activation of the receptor (Kopan 2012). The E3 ubiquitin ligases (e.g., Deltex), Numb protein, and α-adaptin are involved in the maintenance of Notch receptor content at the cell surface. On the other hand, the ubiquitinylated Notch ligand undergoes Epsin-mediated endocytosis to activate and interact with the receptor. Ligands, after activation of the receptor, may undergo recycling or degradation (Kopan 2012).

Notch signaling has immense physiological significance, contributing to neuronal, cardiovascular, endocrine as well and embryogenic development; consequently, its deregulation is linked to a number of health issues, including Alagille syndrome, spondylocostal dysostosis, endothelial dysfunctions, and cancers of various tissues. Among the diverse events and effects resulting from deregulated Notch activation proliferation, apoptosis, epithelial-to-mesenchymal transition (EMT), invasion and migration, angiogenesis, stemness, and chemoresistance are the most discussed oncogenic events (Table 13.1). Available evidence suggests the involvement of Notch ligands, receptors,

TABLE 13.1
Impact of Notch Receptors/Ligands in Cancer

Ligand/Receptor	Oncogenic Events	Cancer Types	References
JAG1	Proliferation, angiogenesis, and chemosensitivity	Breast cancer, prostate cancer, adrenocortical carcinoma, etc.	Zheng et al. (2017)
JAG2	Proliferation, cell survival, and metastasis.	Lung adenocarcinoma, bladder cancer, etc.	Chen et al. (2020)
DLL1	Proliferation, migration, metastasis, chemoresistance, radiation resistance, and mammosphere formation (in breast cancer)	Breast cancer, medulloblastoma tumors, colorectal cancer, etc.	Xiu, Liu, and Kuang (2020)
DLL3	Metastasis, cell survival, reduction in apoptosis	Small cell lung cancer, melanoma, glioblastoma multiforme, small cell bladder cancer, prostate cancer, etc.	Xiu et al. (2020)
DLL4	Survival, tumor angiogenesis and maintenance of cancer stem cells (CSC), metastasis, and epithelial-to-mesenchymal transition (EMT)	Ovarian cancers, liver cancer, breast cancer, colorectal cancer, etc.	Kuramoto et al. (2012); Xiu et al. (2020)
Notch1	Tumor invasion and metastasis, suppression of apoptosis, EMT, maintenance of CSC, and chemoresistance	Lung cancer, ovarian and colorectal cancer cells, breast cancer, prostate cancer, esophageal cancer, etc.	Aleksic and Feller (2008); Giuli et al. (2019)
Notch2	Tumor invasion metastasis, maintenance of CSC, drug resistance, and EMT	Glioma, breast cancer, pancreatic cancer, etc.	Giuli et al. (2019); Wang et al. (2009)
Notch3	Maintenance of CSC, chemoresistance, inhibition of apoptosis, and tumor invasion and metastasis	Breast cancer, lung cancer, B-cell acute lymphoblastic leukemia (B-ALL), esophageal cancer, ovarian cancer, etc.	Giuli et al. (2019); Yao and Qian (2010)
Notch4	Tumor invasion and metastasis	Breast cancer, B-cell acute lymphoblastic leukemia (B-ALL), etc.	Giuli et al. (2019)

and target genes in the pathogenesis of cancers of various types, including breast cancer, lung cancer, pancreatic cancer, leukemia, melanoma, and esophageal cancer (Table 13.1). On the other hand, Notch signaling is reported to also exhibit anti-tumorigenic activity, such as in the case of neuroendocrine tumors like small-cell lung cancer (SCLC) where Notch and p53 mediate tumor-suppressing effects (Zhou et al. 2022).

13.3 GSI AS NOTCH-TARGETED ANTI-CANCER THERAPY

It is well established now that Notch signaling contributes to cancer initiation, progression, and metastasis through a range of routes. Consequently, regulation of Notch signaling attracts scientific attention as a critical anti-cancer measure. Out of the several measures taken up to modulate Notch signaling, the S1 cleavage inhibitors (SERCA inhibitors), S2 cleavage inhibitors (ADAM10/17 inhibitors), and S3 cleavage modulation (γ-secretase inhibitors and modulators) are notable (Zhou et al. 2022).

Sarcoendoplasmic reticulum Ca^{2+}-ATPase (SERCA) activity is required for the initial S1 cleavage of the inactive Notch receptor in the Golgi appratus. Loss of SERCA activity thus blocks oncogenic Notch1 signaling. Consequently, SERCA inhibitors such as thapsigargin have seen preclinical investigation for the management of breast cancer and leukemia (Jackisch et al. 2000; Treiman, Caspersen, and Christensen 1998). Similarly, inhibitors S2 cleavage enzymes, namely,

ADAM10 or tumor necrosis factor-alpha convertase (TACE or ADAM17), have been investigated to halt Notch signaling in the management of non-small-cell lung cancer (NSCLC), renal carcinoma, breast cancer, and hepatocellular carcinoma (HCC) in *in vitro* and *in vivo* experiments, but are still waiting clinical investigations (Zhou et al. 2022). The transduction of Notch signaling relies significantly on S3 cleavage, induced by the GS complex. Consequently, modulators as well as inhibitors of γ-secretase are recognized as promising therapeutic approaches to tackle issues related to deregulated Notch activation (Zhou et al. 2022). While γ-secretase modulators (GSM) are explored for efficacies against Alzheimer's disease, GSIs are investigated as potent anti-tumorigenic/anti-cancer agents, many being tested clinically. The current research delves into demonstrating extant investigations with GSIs for the management of cancers of various kinds.

GSIs originally tested as anti-Alzheimer's disease measures reached a short duration of clinical trials because of serious Notch-related adverse issues, including gastrointestinal problems, non-melanoma skin cancers, and increased infections. Since the oncogenic effects of Notch became understood, modulation of Notch activation through GSIs has become a mainstay in anti-cancer research. GSIs can modulate various cellular processes, including cell growth and differentiation, cell division, proliferation, motility and migration, communication with the neighboring cells, and different modes of cell death (Yamashita et al. 2013). As evident in preclinical studies, GSIs display anti-tumor effects against various cancers, including the malignancies of the breast, lung, cervix, colon, prostate, colorectal, liver, and also gliomas. Several GSIs are presently undergoing clinical trials to assess their efficacy and safety profiles. These trials span various phases, reflecting the progression of these agents through the rigorous process of medical validation. While specific details may evolve as new data emerges, some notable GSIs under investigation include RO4929097 (RG-4733), nirogacestat (PF-03084014), AL102, and crenigacestat (LY3039478) (Table 13.2A).

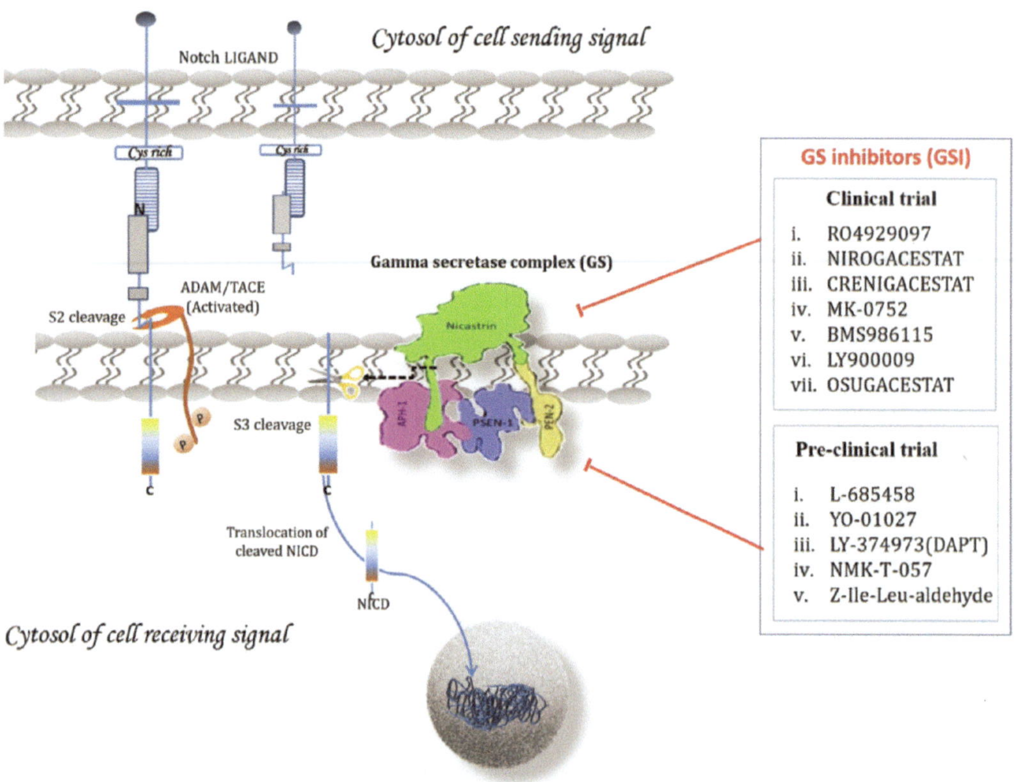

FIGURE 13.3 GSIs that have been used in clinical trials as well as preclinical studies.

TABLE 13.2
Implication of GSIs in Cancer Management

GSI	Structure
A. GSIs in Clinical Trials	
Name of the GSI: RO4929097 *Cancer and Study Cohort*: Colorectal cancer, melanoma, epithelial ovarian cancer, fallopian tube cancer, primary peritoneal cancer, non-small cell lung cancer (NSCLC), pancreatic adenocarcinoma *Reference*: Lee et al. 2015	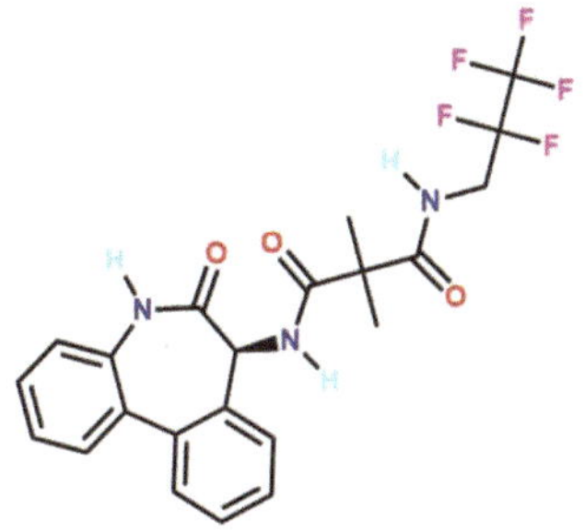
Name of the GSI: Nirogacestat *Cancer and Study Cohort*: Colorectal cancer, desmoid tumor, breast cancer, thyroid cancer, pancreatic cancer, NSCLC, ovarian granulosa cell tumor *Reference*: "Nirogacestat in Ovarian Granulosa Cell Tumors" 2023	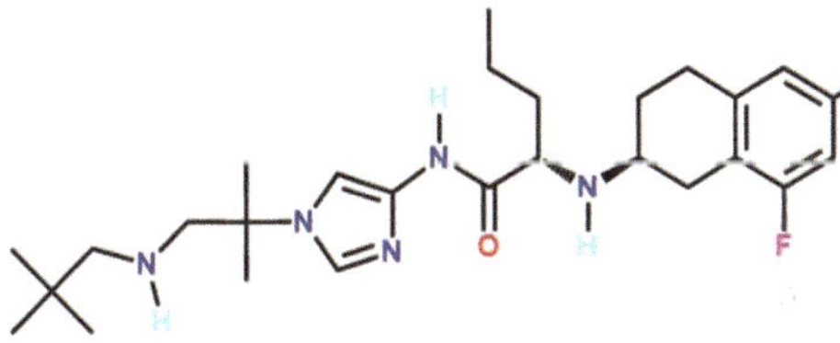
Name of the GSI: Crenigacestat *Cancer and Study Cohort*: Breast cancer, leiomyosarcoma, adenoid cystic carcinoma, T-ALL/T-LBL *Reference*: Doi et al. 2021	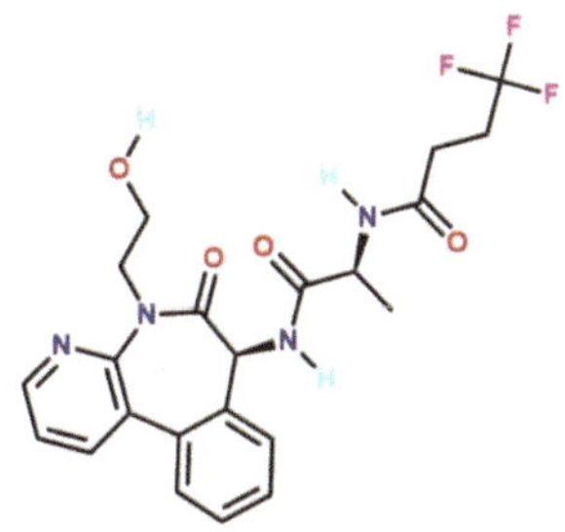
Name of the GSI: Osugacestat *Cancer and Study Cohort*: T-ALL/T-LBL, Triple-negative breast cancer, adenoid cystic cancer *References*: El-Khoueiry et al. 2018; "AL101 Before Surgery for the Treatment of Notch Activated Adenoid Cystic Cancer" 2023	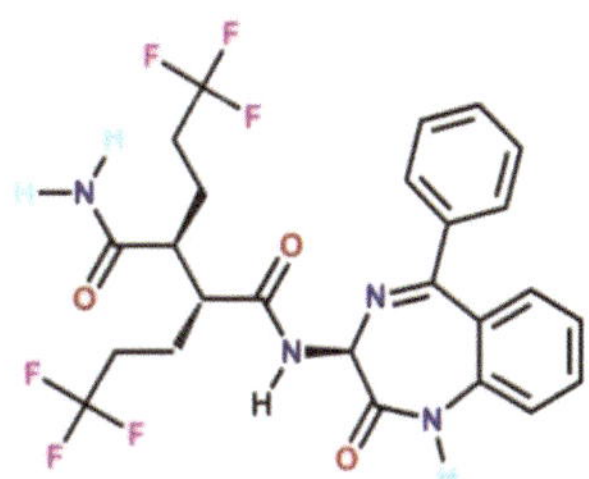
Name of the GSI: BMS986115 *Cancer and Study Cohort*: Colorectal cancer and desmoid tumors *Reference*: Federman 2022	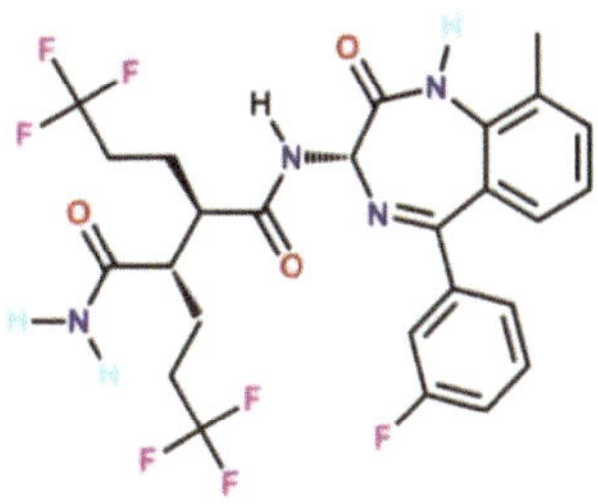

(Continued)

TABLE 13.2 *(Continued)*
Implication of GSIs in Cancer Management

GSI	Structure
Name of the GSI: MK-0752 *Cancer and Study Cohort*: T-ALL, acute myeloid leukemia (AML), ovarian cancer, colorectal cancer, glioma, breast cancer, NSCLC, pancreatic cancer, melanoma, CNS malignancies *Reference*: Krop et al. 2012	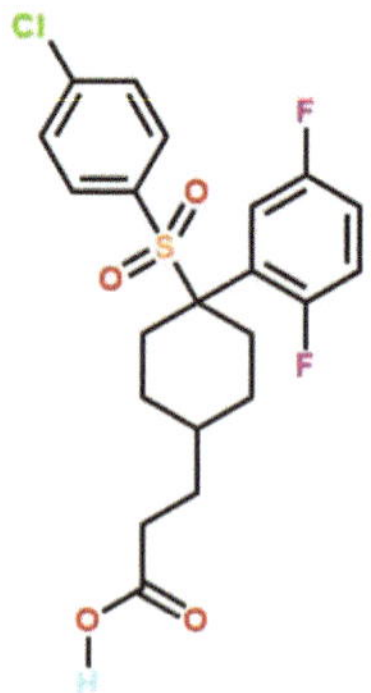
Name of the GSI: LY900009 *Cancer and Study Cohort*: Ovarian cancer and leiomyosarcoma *Reference*: Pant et al. 2016	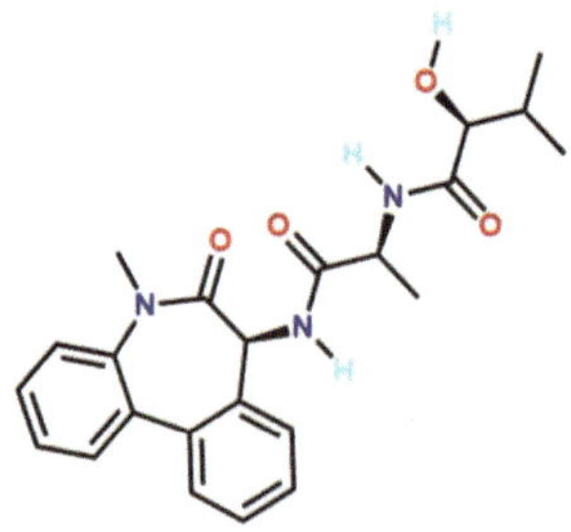
B. GSIs in Preclinical Trial/Experimentation	
Name of the GSI: L-685458 *Cancer Type/Level of investigation*: In vitro *cell lines*: Neuroblastoma, renal cancer, hepatoma In vivo *mouse model*: Hepatoma *References*: (Kawaguchi et al. 2016; Sjölund et al. 2008)	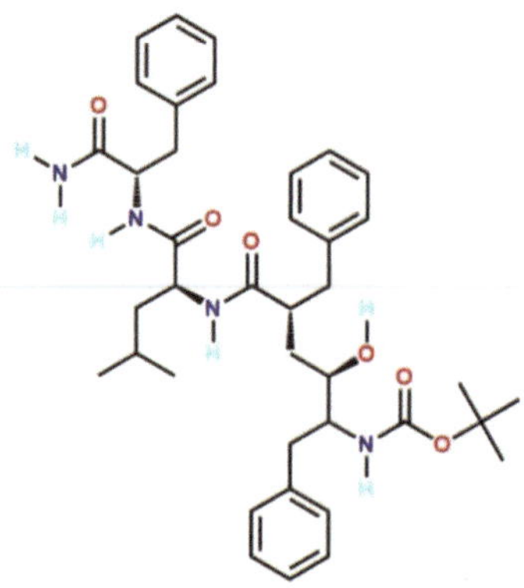
Name of the GSI: YO-01027 *Cancer Type/Level of investigation*: Cell-free assay In vitro *cell lines*: Breast cancer In vivo *mouse model*: Breast cancer *References*: Groth et al. 2010; Harrison et al. 2010	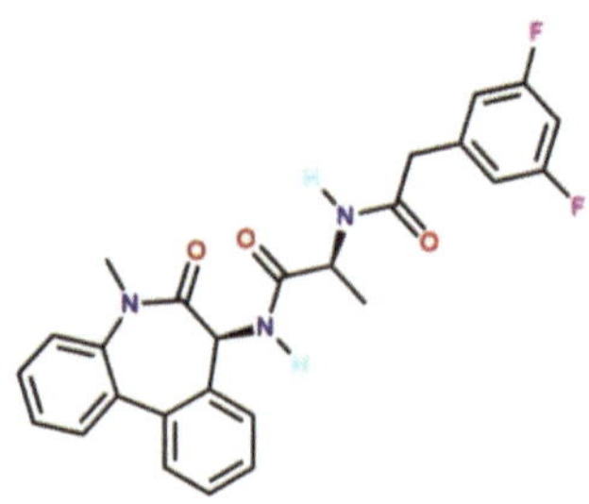

TABLE 13.2 *(Continued)*

Implication of GSIs in Cancer Management

GSI	Structure
Name of the GSI: LY-374973 *Cancer Type/Level of investigation*: In vitro *cell lines*: NSCLC Neuroblastoma, colorectal cancer *Reference*: Cao et al. 2012	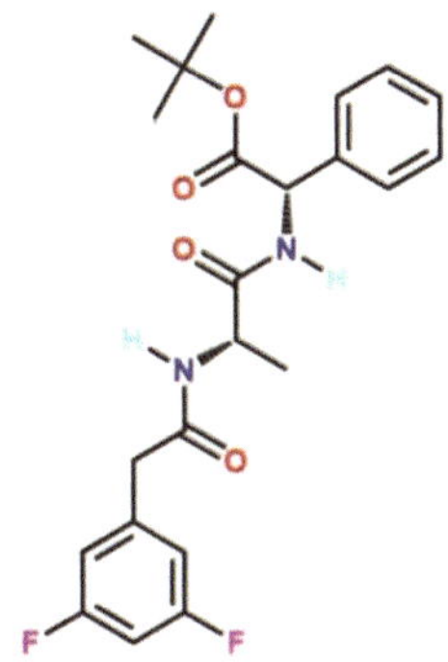
Name of the GSI: Z-Ile-Leu-aldehyde *Cancer Type/Level of investigation*: In vitro *cell lines*: Myeloma cell *Reference*: Schwarzer et al. 2014	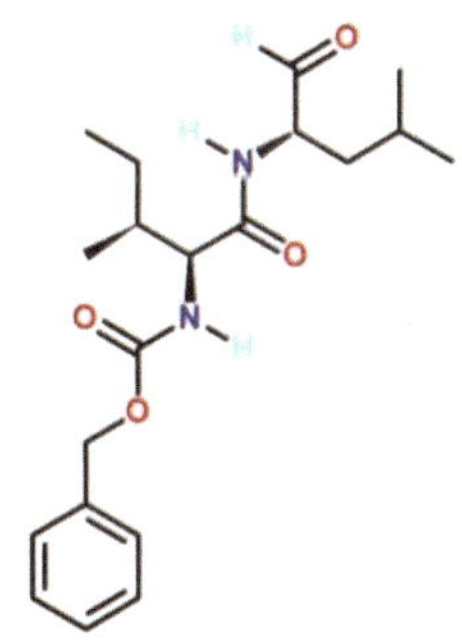
Name of the GSI: NMK-T-057 *Cancer Type/Level of investigation*: In vitro *cell lines*: Breast cancer (triple negative) In vivo *mouse model*: Breast cancer *Reference*: Das et al. 2019	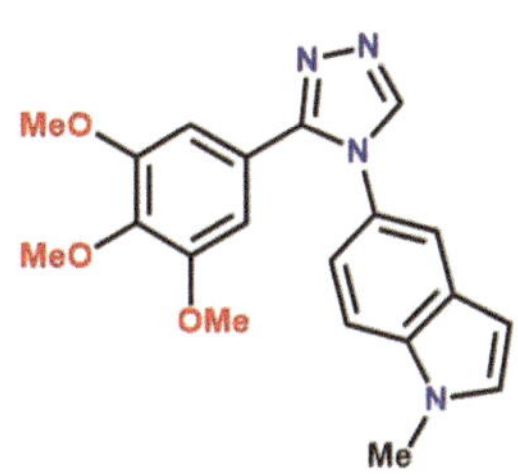
Name of the GSI: Rutin *Cancer Type/Level of investigation*: In vitro *cell lines*: Cervical cancer and colon cancer *Reference*: Singh et al. 2022	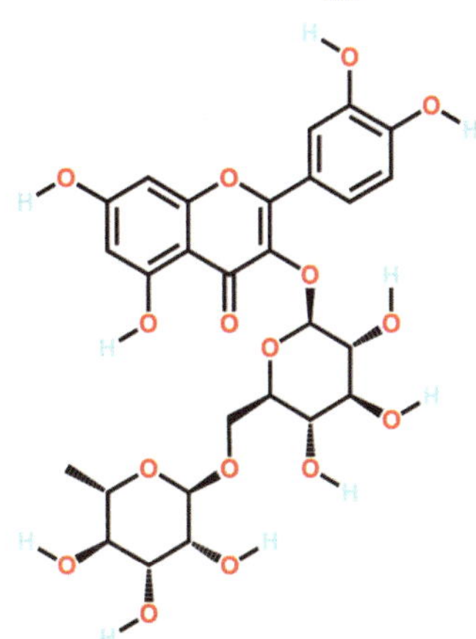

(Continued)

TABLE 13.2 *(Continued)*
Implication of GSIs in Cancer Management

GSI	Structure
Name of the GSI: Epigallocatechin Gallate *Cancer Type/Level of investigation*: In vitro *cell lines*: HNSC In silico *model*: Breast cancer *References*: Lee et al. 2013; Shekar, Vuong, and Kaur 2024	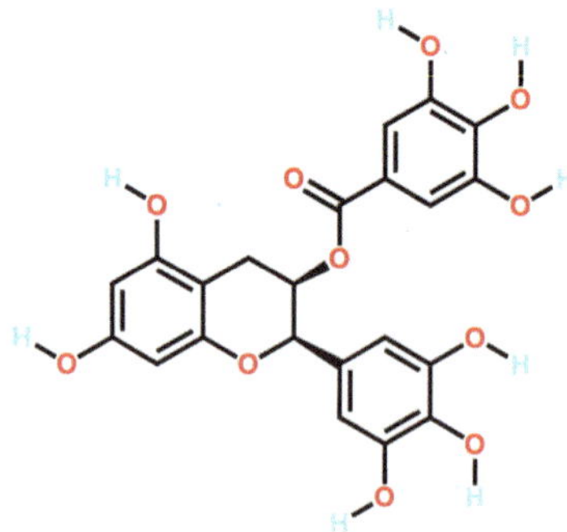
Name of the GSI: Curcumin *Cancer Type/Level of investigation*: In vitro *cell lines*: Esophageal cancer *Reference*: Subramaniam et al. 2012	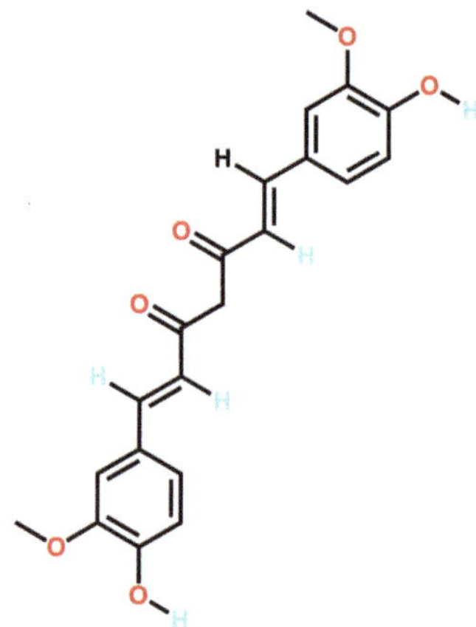
Name of the GSI: Resveratrol *Cancer Type/Level of investigation*: In vitro *cell lines*: T-ALL, cervical cancer, ovarian cancer, breast cancer *Reference*: Kiesel and Stan 2022	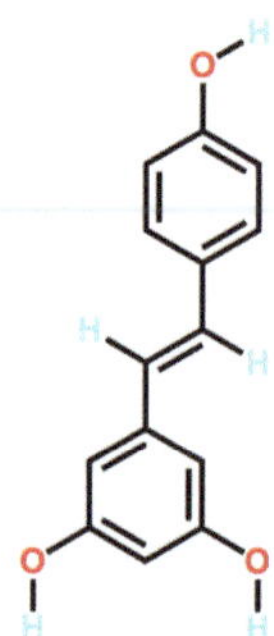
Name of the GSI: Quercetin *Cancer Type/Level of investigation*: In vitro *cell lines*: Colon cancer *Reference*: Li et al. 2020	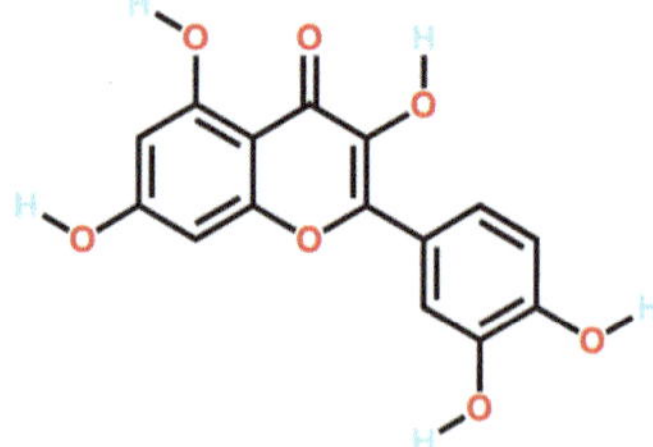

Similarly, preclinical data on various other natural as well as synthetic GSIs demonstrating anti-cancer effects have been generated, providing scope to scale up and conduct clinical experimentation (Table 13.2B).

13.3.1 GSIs in Clinical Trials

13.3.1.1 RO4929097

RO4929097, a dibenzoazepine, is a small-molecule selective inhibitor targeting γ-secretase, which can inhibit Notch-activation and downregulate HES-1 in NSCLC cells (Luistro et al. 2009). RO4929097 demonstrated an anti-proliferative effect and suppressed the oncogenic potential of cancer cells (Huynh et al. 2011). While Phase III trials are yet to be made, RO4929097 has been investigated in Phase I/II clinical trials for various cancers, including recurrent and/or metastatic colorectal cancer, melanoma, NSCLCs, ovarian cancer, etc., with many of the studies exhibiting an extended period of stable disease lasting at least 3 months (Tolcher et al. 2012). This small molecule has shown favorable tolerability among patients with advanced cancer (Lee et al. 2015). Furthermore, in combination with drugs such as capecitabine, gemcitabine, and Cediranib, RO4929097 showed favorable outcomes in clinical trials in patients with refractory cancer of the cervix and colon, primary cancers of the pancreas, trachea, and breast, and advanced solid tumors with well-tolerated, clinical anti-tumor effectiveness, resulting in prolonged stabilization of disease among most patients (LoConte et al. 2015; Sahebjam et al. 2013). Interestingly, when RO4929097 was co-administered with temsirolimus, an mTOR inhibitor, the drug–drug interaction led to increased clearance and decreased exposure to temsirolimus in refractory advanced solid tumors (Diaz-Padilla et al. 2013).

13.3.1.2 Nirogacestat

Nirogacestat (PF-03084014) is a reversible, non-competitive GSI, which resulted in a significant decrease in the Notch1 intracellular domain (NICD) level, reducing phosphorylation of Stat3, suppressing Akt signaling, and reversion of EMT, thereby targeting the cancer stem cells (CSCs), respectively, in a preclinical model of T-ALL (Wei et al. 2010). A Phase III, double-blind, randomized, and placebo-controlled clinical trial of nirogacestat was conducted in patients with progressing desmoid cancer, and satisfactory responses were recorded. In this study, the patients were administered orally with nirogacestat (150 mg) or placebo two times a day. The study revealed a significant enhancement in progression-free survival compared to the placebo group. Specifically, the probability of remaining free from events after 2 years was 76% for those treated with nirogacestat, in contrast to 44% for those who received the placebo (Gounder, Ratan, et al. 2023).

13.3.1.3 Crenigacestat

Crenigacestat (LY3039478) selectively inhibits the γ-secretase enzyme, prevents the release of NICD, and subsequently demonstrates anti-cancer activity (Mäemets-Allas, Belitškin, and Jaks 2016). Crenigacestat has active as well as completed Phase I clinical trials against advanced or metastatic solid tumors. In a first-in-human clinical study, 110 cancer patients (advanced stage with metastasis) were treated with crenigacestat (45–100 mg dose). An effective suppression of Notch signaling (~80% inhibition of plasma Aβ) by crenigacestat was observed as well as promising clinical outcomes in the treated patients, such as tumor reduction and imporved metabolic responses (Massard et al. 2018). In combination trials, crenigacestat with different anti-cancer agents, such as dexamethasone, gemcitabine, cisplatin, and carboplatin, was poorly tolerated and demonstrated limited and unsatisfactory clinical outcomes in patients with T-ALL/T-LBL and advanced or metastatic solid tumors (Borthakur et al. 2021).

13.3.1.4 Osugacestat

Osugacestat (AL101) is a best-in-class GSI that can inhibit all four Notch receptors (1–4) to a similar degree. In the preclinical model, osugacestat inhibits the growth of tumors in adenoid cystic carcinoma, leukemia (T-ALL) xenografts, and triple-negative breast cancer (TNBC) (MDA-MB-468) cells (Ferrarotto et al. 2022). A Phase 1 clinical trial was conducted to evaluate the safety of osugacestat among patients with relapsed or refractory T-ALL and T-LBL; 32% of the patients exhibited a decrease of at least 50% in bone marrow (BM) blasts, showing efficacy with one complete response (CR) and one partial response (PR) (Zweidler-McKay et al. 2014). Currently, osugacestat is in Phase II monotherapy trials in patients with Notch-activated ACC and TNBC, a subtype associated with poor prognosis.

13.3.1.5 BMS986115

BMS986115 (AL102) is another GSI that can inhibit all Notch receptors (1–4) receptors. In a multi-arm, Phase I dose-escalating trial in patients with advanced solid tumors, BMS986115 demonstrated safe and tolerable daily dosing with biological activity (continuous target engagement and Notch signaling inhibition) (Federman 2022). BMS986115 is undergoing Phase II/III trials in patients with progressing desmoid tumors. The group receiving 1.2 mg once a day (QD) showed early clinical activity, such as tumor response, reduced volume, and decreased T2 signal with deeper and maintained treatment responses. This dosage (1.2 mg QD) has been chosen for investigation in the ongoing enrollment of Phase III across several countries (Gounder, Jones, et al. 2023).

13.3.1.6 MK-0752

MK-0752 is a moderately potent, orally active, and specific GSI, which reduces Aβ40 in a dose-dependent manner by blocking NICD with an IC50 of 5 nM in human SH-SY5Y cells (Harrison et al. 2010). In a Phase I clinical trial, it exhibited favorable tolerance levels in leukemia patients, administered at doses below 300 mg/m^2 (Deangelo et al. 2006). In another study conducted with patients with advanced solid tumors, MK-0752 was well tolerated, and significant inhibition of Notch signaling was observed (1.8–4.2 g weekly doses). These findings strongly validate the compound's engagement with its intended target at these specific dosage levels. Furthermore, the study yielded clinical benefits, with a subset of patients affected by high-grade gliomas showed promising outcomes and 10 patients exhibiting sustained stable disease exceeding 4 months (Krop et al. 2012).

A preclinical and clinical study of GSIs with docetaxel, a combination study, was conducted on human breast tumors. Treatment of GSI reduced CSCs within MC1 and BCM-2147 tumor grafts through the inhibition of the Notch pathway, and MK-0752 enhanced the efficacy of docetaxel ("Phase I/II Study of MK-0752 Followed by Docetaxel in Advanced or Metastatic Breast Cancer" 2014). In the clinical trial, 30 patients diagnosed with advanced breast cancer underwent treatment involving progressively higher doses of MK-0752 alongside docetaxel. The tumors of patients who underwent successive biopsies displayed notable reductions in CD44(+)/CD24(-), ALDH (+), and mammosphere-forming efficiency (Schott et al. 2013). Furthermore, MK-0752, in combination with tamoxifen or letrozole, has completed a Phase IV clinical trial to ascertain the feasibility, safety/tolerance, and impact on biomarkers in early breast cancers. Early study results showed anti-proliferative and proapoptotic effects and inhibition of Notch4 processing, a significant player in breast tumor-initiating cells (TIC) with favorable mild-to-moderate toxicity among the patients. These results indicate a potential anti-TIC effect of MK-0752 and tamoxifen or letrozole and their role in overcoming endocrine resistance (Albain et al. 2010).

13.3.1.7 LY900009

LY900009 is a potent and orally bioavailable compound and first-in-class GSI that inhibits the Notch signaling pathway. LY900009 has demonstrated notable inhibition of Notch signaling, with an IC50 range of 0.005–20 nM in cancer cells and endothelial cells (Pant et al. 2016). In a Phase I study of LY900009, 35 patients with advanced leiomyosarcoma and ovarian cancer showed significant inhibition of plasma Aβ by 80–90% in a dose-responsive manner (dose: 30–60 mg). LY900009 was well tolerated, and an maximum tolerated dose (MTD) schedule of 30 mg twice weekly was recommended for future studies (Pant et al. 2016).

13.3.2 GSIs in Preclinical Research

Researchers are actively exploring novel GSIs, and it's worth noting that alongside the GSIs currently undergoing clinical trials, there are also several other compounds in earlier stages of development. These compounds are still being investigated in basic and preclinical research stages, laying the foundation for potential future therapeutic advancements targeted against the Notch signaling pathways such as L-685458, YO-01027, LY- 374973 (DAPT), Z-Ile-Leu-aldehyde, and a novel triazole, NMK-T-057 (Table 13.2B).

13.3.2.1 L-685458

L-685458 is a potent transition state analog of GSI that suppresses γ-secretase by inhibiting and modulating its activities with IC50 of 17 nM (Shearman et al. 2000) and induces significant downregulation of the Notch-targeting gene HES-1 in clear cell renal cell carcinoma (CCRC) cell line, 786-O cells (Sjölund et al. 2008). It has demonstrated anti-tumor effects in HCC cell lines such as Huh7, HepG2, HLE, and SKHep1 cells and mouse hepatoma models by inhibiting EpCAM production and reducing HES1 expression in the nucleus (Kawaguchi et al. 2016).

13.3.2.2 YO-01027

YO-01027 (a dibenzazepine) is a dipeptide that showed potent GSI activity with IC_{50} of 2.6 nM and 2.9 nM in cell-free assays for amyloid precursor protein like protein (APPL) and Notch cleavage (Groth et al. 2010). This compound directly binds to the GS complex and targets the N-terminal of the PS1 subunit (Fuwa et al. 2007). YO-01027 was also found to inhibit the CSCs in breast cancer cells (Harrison et al. 2010). The same team injected MCF7 cells in nude mice by intraperitoneal (IP) injection along with an IP injection of 1 mg/mL on the same day and every subsequent 3 days and observed that YO-01027 significantly decreased the developed MCF7 tumors and increased latency compared with controlled mice (Harrison et al. 2010).

13.3.2.3 LY-374973

LY-374973 (DAPT) is a novel and orally active GSI that inhibits the production of Aβ(40) and Aβ(42) with an IC50 of 115 nM and 200 nM, respectively, in HEK 293 cells (Dovey et al. 2001). A study on SK-MES-1 cells showed the anti-proliferative properties of DAPT and induction of the caspase-dependent and caspase-independent apoptosis in NSCLC cells by inhibiting the Notch signaling pathway (Cao et al. 2012). Similarly, in rat cerebral cortexes, LY-374973 induced cell apoptosis with prolonged neuroinflammation by suppressing γ-secretase activity (Nasoohi et al. 2012).

13.3.2.4 Z-Ile-Leu-Aldehyde and NMK-T-057

Z-Ile-Leu-aldehyde and NMK-T-057 are new GSIs, with potent inhibitory activity against γ-secretase receptors and the Notch signaling pathway, which are still in the early development phase. Z-Ile-Leu-aldehyde is a peptide aldehyde GSI that competitively inhibits GS activity and subsequent Notch activation and, finally, induces apoptosis in murine MOPC315.BM myeloma cells (Schwarzer et al. 2014).

NMK-T-057 is a triazole-based compound that binds to γ-secretase, with IC50 8–12 μM, displayed an ability to inhibit proliferation, colony-forming capacity, and mobility across different types of breast cancer cells, including TNBCs such MDA-MB-231, MDA-MB-468, 4T1, and ER-positive MCF7 cells (Das et al. 2019). Further, NMK-T-057 induced notable apoptosis and inhibited EMT and stemness in breast cancer cells. Notably, these observations were substantiated by the discovery that NMK-T-057 also effectively curtailed tumor advancement in a 4T1-BALB/c mouse model. These findings inferred that NMK-T-057 could emerge as a promising contender for targeted drug development against breast cancer. Now, it has been known that NMK-T-057 targeted the γ-secretase, thereby modulating the Notch signaling pathway and initiating autophagy-mediated cell death (Das et al. 2019).

While synthetic GSIs have potent efficacies, their off-target effects and potential toxicological aspects led scientific focus to natural compounds as alternative means. Consequently, numerous natural compounds (mostly plant-derived) have been explored as modulators and inhibitors of γ-secretase and subsequent anti-cancer effects.

13.3.2.5 Rutin

Rutin, a natural flavonol, is generally found in citrus fruits and a variety of foods such as fruits, vegetables, and cereals, etc. Rutin exerts anti-cancer effects via modulating various signaling pathways, including Notch. Rutin binds to the γ-secretase catalytic site and forms a stable complex, thereby inhibiting NICD formation and downregulating the expression of Notch-responsive genes at the mRNA level in colon cancer cells (Singh et al. 2022).

13.3.2.6 Epigallocatechin gallate

Epigallocatechin gallate (EGCG), the active polyphenolic compound present in green tea, is known to inhibit the oncogenic properties of multiple carcinomas. EGCG exerts an anti-cancer effect by inhibiting the γ-secretase-mediated Notch activation and also targets cancer stem cells (Lee et al. 2013).

13.3.2.7 Curcumin

Curcumin, a phyto-polyphenolic compound extracted from turmeric (*Curcuma longa*), has been proven to possess a variety of anti-cancer properties including suppression of cell proliferation and formation of new blood vessels (angiogenesis), inducing apoptosis and inhibiting the activity of DNA topoisomerase II. Recent research demonstrated the inhibitory action of Curcumin against γ-secretase by inhibiting Notch activation and downregulating γ-secretase complex in esophageal cancer cells (Subramaniam et al. 2012).

13.3.2.8 Resveratrol

Resveratrol, a natural plant-derived polyphenol, demonstrated anti-cancer effects, inhibiting various stages of carcinogenesis (initiation, promotion, and progression). It serves as a chemo-preventive and chemotherapeutic agent, exhibiting anti-inflammatory, antioxidant, proapoptotic, and anti-proliferative actions by targeting intracellular signaling pathways and key regulators of cell survival, apoptosis, inflammation, angiogenesis, and metastasis. In the past few years, there has been a growing emphasis on the use of resveratrol synergistically due to the swift advancement in drug efficacy and the rise of anti-drug resistance. Recently, the Notch suppressive effects of resveratrol have been demonstrated in cell models of T-ALL, cervical, ovarian, and breast cancer and also showed a decrease in the expression of HES-1 protein when the ovarian and cervical cancer cells were treated with resveratrol and a GSI (Kiesel and Stan 2022).

13.3.2.9 Quercetin

Quercetin and its glycosides, present as primary polyphenolic flavonoid in various plant foods and vegetables including onion, broccoli, apple, tea, and red wine, demonstrated antioxidant, proapoptotic, anti-inflammatory, and anti-angiogenic properties, as well as anti-cancer activity. Recently, a study demonstrated the γ-secretase inhibition activity of quercetin in colon cancer cells. A dramatic decline in the level of NICD was observed in cells treated with quercetin along with ionizing radiation (Li et al. 2020).

13.4 FUTURE PROSPECTIVES AND CONCLUSION

Given the impact of Notch signaling on cellular physiology and functioning, a total shutdown of Notch signaling demonstrates several adverse effects. Thus, antibody-drug conjugates specific for Notch ligands and receptors are gaining popularity to exert a precision effect in specific cancers. However, the majority of these antibody-drug conjugates (ADC) show only mediocre effects, possibly because of cell heterogeneity. ADCs against DLL1/JAG2 (that influence a majority of pro-oncogenic events) are promising but still in development. On the other hand, the blocking of Notch-mediated transcription and Notch agonists for cancer management has also been explored. However, as far as effectiveness is concerned, GSIs have the most favorable outlook among the existing Notch targeted therapeutics (Zhou et al. 2022). GSIs offer a promising avenue for targeted cancer therapy by disrupting the Notch-mediated major oncogenic processes such as proliferation, anti-apoptosis, EMT, metastasis, maintenance of cancer stem cells, and chemoresistance. Clinical trials and preclinical research continue to evaluate the efficacy and safety of these agents, with the potential to revolutionize cancer treatment strategies.

DECLARATIONS

Conflict of interest: The authors declare no conflicts of interest.

Consent for publishing: All authors have approved the final version of the manuscript and given consent for publication.

Funding: Not Applicable.

REFERENCES

'AL101 Before Surgery for the Treatment of Notch Activated Adenoid Cystic Cancer', 2023. https://classic.clinicaltrials.gov/show/NCT04973683.

Al-Hussaini, H., D. Subramanyam, M. Reedijk, and S. S. Sridhar. 2011. 'Notch signaling pathway as a therapeutic target in breast cancernotch in breast cancer', *Molecular Cancer Therapeutics*, 10: 9–15.

Albain, K. S., C. Czerlanis, P. Rajan, A. Zlobin, C. Godellas, D. Bova, S. S. Lo, P. Robinson, S. Sarker, and E. R. Gaynor. 2010. 'Abstract PD05-12: combination of notch inhibitor MK-0752 and endocrine therapy for early stage ERα+ breast cancer in a presurgical window pilot Study', *Cancer Research*, 70: PD05–12.

Aleksic, T., and S. M. Feller. 2008. 'Gamma-secretase inhibition combined with platinum compounds enhances cell death in a large subset of colorectal cancer cells', *Cell Communication and Signaling*, 6: 1–13.

Borthakur, G., G. Martinelli, E. Raffoux, P. Chevallier, J. Chromik, A. Lithio, C. L. Smith, E. Yuen, G. J. Oakley 3rd, K. A. Benhadji, and D. J. DeAngelo. 2021. 'Phase 1 study to evaluate Crenigacestat (LY3039478) in combination with dexamethasone in patients with T-cell acute lymphoblastic leukemia and lymphoma', *Cancer*, 127: 372–80.

Cao, H., Y. Hu, P. Wang, J. Zhou, Z. Deng, and J. Wen. 2012. 'Down-regulation of Notch receptor signaling pathway induces caspase-dependent and caspase-independent apoptosis in lung squamous cell carcinoma cells', *APMIS*, 120: 441–50.

Carroll, C. M, and Y.-M. Li. 2016. 'Physiological and pathological roles of the γ-secretase complex', *Brain Research Bulletin*, 126: 199–206.

Chen, Y.-T., C.-R. Huang, C.-L. Chang, J. Y. Chiang, C.-W. Luo, H.-H. Chen, and H.-K. Yip. 2020. 'Jagged2 progressively increased expression from Stage I to III of Bladder Cancer and Melatonin-mediated down-regulation of Notch/Jagged2 suppresses the Bladder Tumorigenesis via inhibiting PI3K/AKT/mTOR/MMPs signaling', *International Journal of Biological Sciences*, 16: 2648.

Das, A., M. K. Narayanam, S. Paul, P. Mukhnerjee, S. Ghosh, D. G. Dastidar, S. Chakrabarty, A. Ganguli, B. Basu, M. Pal, U. Chatterji, S. K. Banerjee, P. Karmakar, D. Kumar, and G. Chakrabarti. 2019. 'A novel triazole, NMK-T-057, induces autophagic cell death in breast cancer cells by inhibiting γ-secretase-mediated activation of Notch signaling', *Journal of Biological Chemistry*, 294: 6733–50.

Deangelo, D. J., R. M. Stone, L. B. Silverman, W. Stock, E. C. Attar, I. Fearen, A. Dallob, C. Matthews, J. Stone, S. J. Freedman, and J. Aster. 2006. 'A phase I clinical trial of the notch inhibitor MK-0752 in patients with T-cell acute lymphoblastic leukemia/lymphoma (T-ALL) and other leukemias', *Journal of Clinical Oncology*, 24: 6585–85.

Diaz-Padilla, I., H. Hirte, A. M. Oza, B. A. Clarke, B. Cohen, M. Reedjik, T. Zhang, S. Kamel-Reid, S. P. Ivy, S. J. Hotte, A. A. Razak, E. X. Chen, I. Brana, M. Wizemann, L. Wang, L. L. Siu, and P. L. Bedard. 2013. 'A phase Ib combination study of RO4929097, a gamma-secretase inhibitor, and temsirolimus in patients with advanced solid tumors', *Investigational New Drugs*, 31: 1182–91.

Doi, T., M. Tajimi, J. Mori, H. Asou, K. Inoue, K. A. Benhadji, and Y. Naito. 2021. 'A phase 1 study of crenigacestat (LY3039478), the Notch inhibitor, in Japanese patients with advanced solid tumors', *Investigational New Drugs*, 39: 469–76.

Dovey, H. F., V. John, J. P. Anderson, L. Z. Chen, P. de Saint Andrieu, L. Y. Fang, S. B. Freedman, B. Folmer, E. Goldbach, E. J. Holsztynska, K. L. Hu, K. L. Johnson-Wood, S. L. Kennedy, D. Kholodenko, J. E. Knops, L. H. Latimer, M. Lee, Z. Liao, I. M. Lieberburg, R. N. Motter, L. C. Mutter, J. Nietz, K. P. Quinn, K. L. Sacchi, P. A. Seubert, G. M. Shopp, E. D. Thorsett, J. S. Tung, J. Wu, S. Yang, C. T. Yin, D. B. Schenk, P. C. May, L. D. Altstiel, M. H. Bender, L. N. Boggs, T. C. Britton, J. C. Clemens, D. L. Czilli, D. K. Dieckman-McGinty, J. J. Droste, K. S. Fuson, B. D. Gitter, P. A. Hyslop, E. M. Johnstone, W. Y. Li, S. P. Little, T. E. Mabry, F. D. Miller, and J. E. Audia. 2001. 'Functional gamma-secretase inhibitors reduce beta-amyloid peptide levels in brain', *Journal of Neurochemistry*, 76: 173–81.

El-Khoueiry, A. B., J. Desai, S. P. Iyer, S. M. Gadgeel, S. S. Ramalingam, L. Horn, P. LoRusso, G. Bajaj, G. Kollia, Z. Qi, S. Basak, B. S. Fischer, M. Davis, and P. L. Bedard. 2018. 'A phase I study of AL101, a pan-NOTCH inhibitor, in patients (pts) with locally advanced or metastatic solid tumors', *Journal of Clinical Oncology*, 36: 2515–15.

Federman, N. 2022. 'Molecular pathogenesis of desmoid tumor and the role of γ-secretase inhibition', *NPJ Precision Oncology*, 6: 62.

Ferrarotto, R., V. Mishra, E. Herz, A. Yaacov, O. Solomon, R. Rauch, A. Mondshine, M. Motin, T. Leibovich-Rivkin, M. Davis, J. Kaye, C. R. Weber, L. Shen, A. T. Pearson, A. J. Rosenberg, X. Chen, A. Singh, J. C. Aster, N. Agrawal, and E. Izumchenko. 2022. 'AL101, a gamma-secretase inhibitor, has potent antitumor activity against adenoid cystic carcinoma with activated NOTCH signaling', *Cell Death & Disease*, 13: 678.

Fuwa, H., Y. Takahashi, Y. Konno, N. Watanabe, H. Miyashita, M. Sasaki, H. Natsugari, T. Kan, T. Fukuyama, T. Tomita, and T. Iwatsubo. 2007. 'Divergent synthesis of multifunctional molecular probes to elucidate the enzyme specificity of dipeptidic gamma-secretase inhibitors', *American Chemical Society Chemical Biology*, 2: 408–18.

Giuli, M. V., E. Giuliani, I. Screpanti, D. Bellavia, and S. Checquolo. 2019. 'Notch signaling activation as a hallmark for triple-negative breast cancer subtype', *Journal of Oncology*, 2019: 8707053.

Gounder, M., R. Ratan, T. Alcindor, P. Schöffski, W. T. van der Graaf, B. A. Wilky, R. F. Riedel, A. Lim, L. M. Smith, S. Moody, S. Attia, S. Chawla, G. D'Amato, N. Federman, P. Merriam, B. A. Van Tine, B. Vincenzi, C. Benson, N. Q. Bui, R. Chugh, G.Tinoco, J. Charlson, P. Dileo, L. Hartner, L. Lapeire, F. Mazzeo, E. Palmerini, P. Reichardt, S. Stacchiotti, H. H. Bailey, M. A. Burgess, G. M. Cote, L. E. Davis, H. Deshpande, H. Gelderblom, G. Grignani, E. Loggers, T. Philip, J. G. Pressey, S. Kummar, and B. Kasper. 2023. 'Nirogacestat, a γ-secretase inhibitor for desmoid Tumors', *The New England Journal of Medicine*, 388: 898–912.

Gounder, M. M., R. L. Jones, R. Chugh, M. Agulnik, A. S. Singh, B. A. Van Tine, V. Andelkovic, E. Choy, J. Howard Lewin, R. Ratan, G. B. Gordon, J. Yovell, A. A. Gutierrez, and B. Kasper. 2023. 'RINGSIDE Phase 2/3 trial of AL102 for treatment of desmoid tumors (DT): Phase 2 results', *Journal of Clinical Oncology*, 41: 11515–15.

Groth, C., W. G. Alvord, O. A. Quiñones, and M. E. Fortini. 2010. 'Pharmacological analysis of Drosophila melanogaster gamma-secretase with respect to differential proteolysis of Notch and APP', *Molecular Pharmacology*, 77: 567–74.

Harrison, H., G. Farnie, S. J. Howell, R. E. Rock, S. Stylianou, K. R. Brennan, N. J. Bundred, and R. B. Clarke. 2010. 'Regulation of breast cancer stem cell activity by signaling through the Notch4 receptor', *Cancer Research*, 70: 709–18.

Huynh, C., L. Poliseno, M. F. Segura, R. Medicherla, A. Haimovic, S. Menendez, S. Shang, A. Pavlick, Y. Shao, F. Darvishian, J. F. Boylan, I. Osman, and E. Hernando. 2011. 'The novel gamma secretase inhibitor RO4929097 reduces the tumor initiating potential of melanoma', *PLoS One*, 6: e25264.

Jackisch, C., H. A. Hahm, B. Tombal, D. McCloskey, K. Butash, N. E. Davidson, and S. R. Denmeade. 2000. 'Delayed micromolar elevation in intracellular calcium precedes induction of apoptosis in thapsigargin-treated breast cancer cells', *Clinical Cancer Research*, 6: 2844–50.

Kawaguchi, K., M. Honda, T. Yamashita, H. Okada, T. Shirasaki, M. Nishikawa, K. Nio, K. Arai, Y. Sakai, T. Yamashita, E. Mizukoshi, and S. Kaneko. 2016. 'Jagged1 DNA copy number variation is associated with poor outcome in liver Cancer', *The American Journal of Pathology*, 186: 2055–67.

Kiesel, V. A., and S. D. Stan. 2022. 'Modulation of Notch signaling pathway by bioactive dietary agents', *International Journal of Molecular Sciences*, 23: 3532.

Kopan, R. 2012. 'Notch signaling', *Cold Spring Harbor Perspectives in Biology*, 4: a011213.

Krop, I., T. Demuth, T. Guthrie, P. Y. Wen, W. P. Mason, P. Chinnaiyan, N. Butowski, M. D. Groves, S. Kesari, S. J. Freedman, S. Blackman, J. Watters, A. Loboda, A. Podtelezhnikov, J. Lunceford, C. Chen, M. Giannotti, J. Hing, R. Beckman, and P. Lorusso. 2012. 'Phase I pharmacologic and pharmacodynamic study of the gamma secretase (Notch) inhibitor MK-0752 in adult patients with advanced solid tumors', *Journal of Clinical Oncology*, 30: 2307–13.

Kuramoto, T., H. Goto, A. Mitsuhashi, S. Tabata, H. Ogawa, H. Uehara, A. Saijo, S. Kakiuchi, Y. Maekawa, and K. Yasutomo. 2012. 'Dll4-Fc, an inhibitor of Dll4-notch signaling, suppresses liver metastasis of small cell lung cancer cells through the downregulation of the NF-κB activity', *Molecular Cancer Therapeutics*, 11: 2578–87.

Lee, S. H., H. J. Nam, H. J. Kang, H. W. Kwon, and Y. C. Lim. 2013. 'Epigallocatechin-3-gallate attenuates head and neck cancer stem cell traits through suppression of Notch pathway', *European Journal of Cancer*, 49: 3210–18.

Lee, S. M., J. Moon, B. G. Redman, T. Chidiac, L. E. Flaherty, Y. Zha, M. Othus, A. Ribas, V. K. Sondak, T. F. Gajewski, and K. A. Margolin. 2015. 'Phase 2 study of RO4929097, a gamma-secretase inhibitor, in metastatic melanoma: SWOG 0933', *Cancer*, 121: 432–40.

Li, Y., Z. Wang, J. Jin, S. X. Zhu, G. Q. He, S. H. Li, J. Wang, and Y. Cai. 2020. 'Quercetin pretreatment enhances the radiosensitivity of colon cancer cells by targeting Notch-1 pathway', *Biochemical and Biophysical Research Communications*, 523: 947–53.

LoConte, N. K., A. R. A. Razak, P. Ivy, A. Tevaarwerk, R. Leverence, J. Kolesar, L. Siu, S. J. Lubner, D. L. Mulkerin, W. R. Schelman, D. A. Deming, K. D. Holen, L. Carmichael, J. Eickhoff, and G. Liu. 2015. 'A multicenter phase 1 study of γ -secretase inhibitor RO4929097 in combination with capecitabine in refractory solid tumors', *Investigational New Drugs*, 33: 169–76.

Luistro, L., W. He, M. Smith, K. Packman, M. Vilenchik, D. Carvajal, J. Roberts, J. Cai, W. Berkofsky-Fessler, and H. Hilton. 2009. 'Preclinical profile of a potent γ-secretase inhibitor targeting notch signaling with *in vivo* efficacy and pharmacodynamic properties', *Cancer Research*, 69: 7672–80.

Mäemets-Allas, K., D. Belitškin, and V. Jaks. 2016. 'The inhibition of Akt-Pdpk1 interaction efficiently suppresses the growth of murine primary liver tumor cells', *Biochemical and Biophysical Research Communications*, 474: 118–25.

Massard, C., A. Azaro, J. C. Soria, U. Lassen, C. Le Tourneau, D. Sarker, C. Smith, U. Ohnmacht, G. Oakley, B. K. R. Patel, E. S. M. Yuen, K. A. Benhadji, and J. Rodon. 2018. 'First-in-human study of LY3039478, an oral Notch signaling inhibitor in advanced or metastatic cancer', *Annals of Oncology*, 29: 1911–17.

Nasoohi, S., A. A. Hemmati, F. Moradi, and A. Ahmadiani. 2012. 'The γ-secretase blocker DAPT impairs recovery from lipopolysaccharide-induced inflammation in rat brain', *Neuroscience*, 210: 99–109.

'Nirogacestat in Ovarian Granulosa Cell Tumors', 2023. https://classic.clinicaltrials.gov/show/NCT05348356.

Pant, S., S. F. Jones, C. D. Kurkjian, J. R. Infante, K. N. Moore, H. A. Burris, D. S. McMeekin, K. A. Benhadji, B. K. R. Patel, M. J. Frenzel, J. D. Kursar, M. J. Zamek-Gliszczynski, E. S. M. Yuen, E. M. Chan, and J. C. Bendell. 2016. 'A first-in-human phase I study of the oral Notch inhibitor, LY900009, in patients with advanced cancer', *European Journal of Cancer*, 56: 1–9.

'Phase I/II Study of MK-0752 Followed by Docetaxel in Advanced or Metastatic Breast Cancer', 2014. https://classic.clinicaltrials.gov/show/NCT00645333.

Pine, S. R. 2018. 'Rethinking gamma-secretase inhibitors for treatment of Non–small-cell lung cancer: Is notch the target? Rethinking gamma secretase inhibitors', *Clinical Cancer Research*, 24: 6136–41.

Sahebjam, S., P. L. Bedard, V. Castonguay, Z. Chen, M. Reedijk, G. Liu, B. Cohen, W. J. Zhang, B. Clarke, T. Zhang, S. Kamel-Reid, H. Chen, S. P. Ivy, A. R. A. Razak, A. M. Oza, E. X. Chen, H. W. Hirte, A. McGarrity, L. Wang, L. L. Siu, and S. J. Hotte. 2013. 'A phase I study of the combination of ro4929097 and cediranib in patients with advanced solid tumours (PJC-004/NCI 8503)', *British Journal of Cancer*, 109: 943–49.

Schott, A. F., M. D. Landis, G. Dontu, K. A. Griffith, R. M. Layman, I. Krop, L. A. Paskett, H. Wong, L. E. Dobrolecki, M. T. Lewis, A. M. Froehlich, J. Paranilam, D. F. Hayes, M. S. Wicha, and J. C. Chang. 2013. 'Preclinical and clinical studies of gamma secretase inhibitors with docetaxel on human breast tumors', *Clinical Cancer Research*, 19: 1512–24.

Schwarzer, R., N. Nickel, J. Godau, B. M. Willie, G. N. Duda, R. Schwarzer, B. Cirovic, A. Leutz, R. Manz, B. Bogen, B. Dörken, and F. Jundt. 2014. 'Notch pathway inhibition controls myeloma bone disease in the murine MOPC315.BM model', *Blood Cancer Journal*, 4: e217.

Shearman, M. S., D. Beher, E. E. Clarke, H. D. Lewis, T. Harrison, P. Hunt, A. Nadin, A. L. Smith, G. Stevenson, and J. L. Castro. 2000. 'L-685,458, an aspartyl protease transition state mimic, is a potent inhibitor of amyloid beta-protein precursor gamma-secretase activity', *Biochemistry*, 39: 8698–704.

Shekar, N., P. Vuong, and P. Kaur. 2024. 'Analysing potent biomarkers along phytochemicals for breast cancer therapy: an in silico approach', *Breast Cancer Research and Treatment*, 203: 29–47.

Singh, A. K., M. Shuaib, K. S. Prajapati, and S. Kumar. 2022. 'Rutin potentially binds the gamma secretase catalytic site, down regulates the Notch signaling pathway and reduces sphere formation in colonospheres', *Metabolites*, 12: 926.

Sjölund, J., M. Johansson, S. Manna, C. Norin, A. Pietras, S. Beckman, E. Nilsson, B. Ljungberg, and H. Axelson. 2008. 'Suppression of renal cell carcinoma growth by inhibition of Notch signaling *in vitro* and *in vivo*', *The Journal of Clinical Investigation*, 118: 217–28.

'Study Of MK-0752 In Combination With Tamoxifen Or Letrozole to Treat Early Stage Breast Cancer', 2019. https://classic.clinicaltrials.gov/show/NCT00756717.

Subramaniam, D., S. Ponnurangam, P. Ramamoorthy, D. Standing, R. J. Battafarano, S. Anant, and P. Sharma. 2012. 'Curcumin induces cell death in esophageal cancer cells through modulating Notch signaling', *PLoS One*, 7: e30590.

Tolcher, A. W., W. A. Messersmith, S. M. Mikulski, K. P. Papadopoulos, E. L. Kwak, D. G. Gibbon, A. Patnaik, G. S. Falchook, A. Dasari, G. I. Shapiro, J. F. Boylan, Z. X. Xu, K. Wang, A. Koehler, J. Song, S. A. Middleton, J. Deutsch, M. Demario, R. Kurzrock, and J. J. Wheler. 2012. 'Phase I study of RO4929097, a gamma secretase inhibitor of Notch signaling, in patients with refractory metastatic or locally advanced solid tumors', *Journal of Clinical Oncology*, 30: 2348–53.

Treiman, M., C. Caspersen, and S. B. Christensen. 1998. 'A tool coming of age: thapsigargin as an inhibitor of sarco-endoplasmic reticulum Ca^{2}+-ATPases', *Trends in Pharmacological Sciences*, 19: 131–35.

Wang, Z., Y. Li, D. Kong, S. Banerjee, A. Ahmad, A. S. Azmi, S. Ali, J. L. Abbruzzese, G. E. Gallick, and F. H. Sarkar. 2009. 'Acquisition of epithelial-mesenchymal transition phenotype of gemcitabine-resistant pancreatic cancer cells is linked with activation of the notch signaling pathway', *Cancer Research*, 69: 2400–07.

Wei, P., M. Walls, M. Qiu, R. Ding, R. H. Denlinger, A. Wong, K. Tsaparikos, J. P. Jani, N. Hosea, and M. Sands. 2010. 'Evaluation of selective γ-secretase inhibitor PF-03084014 for its antitumor efficacy and gastrointestinal safety to guide optimal clinical trial design', *Molecular Cancer Therapeutics*, 9: 1618–28.

Xiu, M.-X., Y.-M. Liu, and B.-H. Kuang. 2020. 'The role of DLLs in cancer: a novel therapeutic target', *OncoTargets and Therapy*, 13: 3881–901.

Yamashita, A. S., M. V. Geraldo, C. S. Fuziwara, M. A. Kulcsar, C. U. Friguglietti, R. B. da Costa, G. S. Baia, and E. T. Kimura. 2013. 'Notch pathway is activated by MAPK signaling and influences papillary thyroid cancer proliferation', *Translational Oncology*, 6: 197–205.

Yao, J., and C. Qian. 2010. 'Inhibition of Notch3 enhances sensitivity to gemcitabine in pancreatic cancer through an inactivation of PI3K/Akt-dependent pathway', *Medical Oncology*, 27: 1017–22.

Yuan, X., H. Wu, H. Xu, H. Xiong, Q. Chu, S. Yu, G. S. Wu, and K. Wu. 2015. 'Notch signaling: an emerging therapeutic target for cancer treatment', *Cancer Letters*, 369: 20–27.

Zheng, H., Y. Bae, S. Kasimir-Bauer, R. Tang, J. Chen, G. Ren, M. Yuan, M. Esposito, W. Li, and Y. Wei. 2017. 'Therapeutic antibody targeting tumor-and osteoblastic niche-derived Jagged1 sensitizes bone metastasis to chemotherapy', *Cancer Cell*, 32: 731–47. e6.

Zhou, B., W. Lin, Y. Long, Y. Yang, H. Zhang, K. Wu, and Q. Chu. 2022. 'Notch signaling pathway: Architecture, disease, and therapeutics', *Signal Transduction and Targeted Therapy*, 7: 95.

Zweidler-McKay, P. A., D. J. DeAngelo, D. Douer, H. Dombret, O. G. Ottmann, N. Vey, D. A. Thomas, L. Zhu, F. Huang, and G. Bajaj. 2014. 'The safety and activity of BMS-906024, a gamma secretase inhibitor (GSI) with anti-notch activity, in patients with relapsed T-cell acute lymphoblastic leukemia (T-ALL): initial results of a phase 1 trial', *Blood*, 124: 968.

14 Cathepsins in Cancer Therapy and Precision Medicine

Namrata Singh and Asma Mohin Rojewale
Glenmark Pharmaceuticals Limited, Glenmark Research Centre, Mumbai, India

Aqsa Khan and Shinjinee Sengupta
Amity Institute of Molecular Medicine and Stem Cell Research (AIMMSCR), Amity University, Noida, India

14.1 INTRODUCTION

The word "cathepsin" was first coined from the Greek word *kathepsins*, which means "to digest", mentioned in 1929 by Richard Willstätter, a Nobel Prize winner, to define proteases that are found in the endosomal/lysosomal system and are active in a mildly acidic environment (Yadati et al., 2020). Cathepsins are intracellular globular proteases that may function at neutral to acidic pH levels and are typically, though not always, present in lysosomes. They are initially synthesized as inactive zymogens, which are then carried to endosomes to be transformed into active zymogens (through autocatalysis or the action of other proteases), and then sent to lysosomes (Jiang et al., 2023; Turk et al., 2012; Yadati et al., 2020).

Cathepsins play a crucial role by participating in immune responses via processing antigens and breaking down various proteases and chemokines to ensure cellular balance. Additionally, they play a significant role in essential physiological processes, including digestion, blood coagulation, bone resorption, ion channel function, apoptosis, vesicular trafficking, autophagy, angiogenesis, proliferation, and metastasis (Turk et al., 2012). They are typically found in lysosomes, and exhibit their greatest activity within the acidic pH environment. Nevertheless, certain cathepsins have been observed to retain activity even outside their optimal pH range of 5 (Turk et al., 2012; Yadati et al., 2020).

Mammalian proteases are categorized into five distinct families, such as serine, metallo, aspartic, threonine, and cysteine proteases, based on the specific amino acid found at their active sites (Soond et al., 2019). Among these families, cathepsins belong to three different protease families—aspartic proteases (cathepsin D and E), serine proteases (cathepsins A and G), and 11 cysteine cathepsins (cathepsins B, C, F, H, K, L, O, S, V, X, and W) (Turk et al., 2012; Yadati et al., 2020). Some cathepsins, such as cathepsin B, L, H, and C, are expressed everywhere, but the recently discovered cathepsins K, W, and X are expressed by certain cells and tissues (Soond et al., 2019). In the past 20 years, several distinct and specific roles for the cathepsin family have been discovered utilizing gene knockout models (Gocheva et al., 2006). For example, cathepsin L is involved in heart functions, reproduction, and keratinocyte differentiation; cathepsin C stimulates granzymes and mast cell proteases, and cathepsin K plays a significant role in bone remodeling (Yadati et al., 2020). Reports have stated that pycnodysostosis and Papillon-Lefevre syndrome are genetic illnesses caused by mutations in the cathepsin K or C genes, respectively (Vasiljeva et al., 2007). Cathepsins also play significant roles in neurodegenerative disorders by activating or breaking down various crucial neuronal proteins. For instance, cathepsin D is involved in maintaining neuronal cell balance, and its dysfunction impairs the proteolysis of important target proteins like huntingtin,

DOI: 10.1201/9781003394693-16

tau, α-synuclein, lipofuscin, and apolipoprotein E (apoE) (Yadati et al., 2020). This dysfunction contributes to the development of neurological disorders such as Huntington's disease, Parkinson's disease, and others. Cathepsins also play a role in cardiovascular disorders such as cardiomyopathy, myocardial infarction, hypertension, atherosclerosis, and aortic aneurysms (Yadati et al., 2020). Additionally, apart from cathepsin D, multiple cathepsins are associated with neurological diseases like Alzheimer's disease, Niemann-Pick type C (NPC) disease, and neuronal ceroid lipofuscinosis (NCL). These conditions are characterized by extensive degradation and remodeling of the extracellular matrix (ECM), which is primarily mediated by cathepsins (Buck et al., 1992; Fonović & Turk, 2014).

14.2 CATHEPSINS: SYNTHESIS, FAMILY, STRUCTURE, AND FUNCTIONS

Cathepsins are a family of protease enzymes synthesized within cells through a tightly regulated process. They are initially synthesized as inactive zymogens to prevent premature digestion of cellular components. Activation often involves specific cleavage events or changes in pH within cellular compartments like lysosomes, where many cathepsins are active (Rudzińska et al., 2019; Turk et al., 2012; Yadati et al., 2020) (Figure 14.1).

Cathepsins have various enzymatic characteristics and functions, with their distribution depending on their specific functions within the cell. They are involved in proteolytic activities in lysosomes, where they play roles in protein turnover, metabolism, antigen processing, and tissue remodeling.

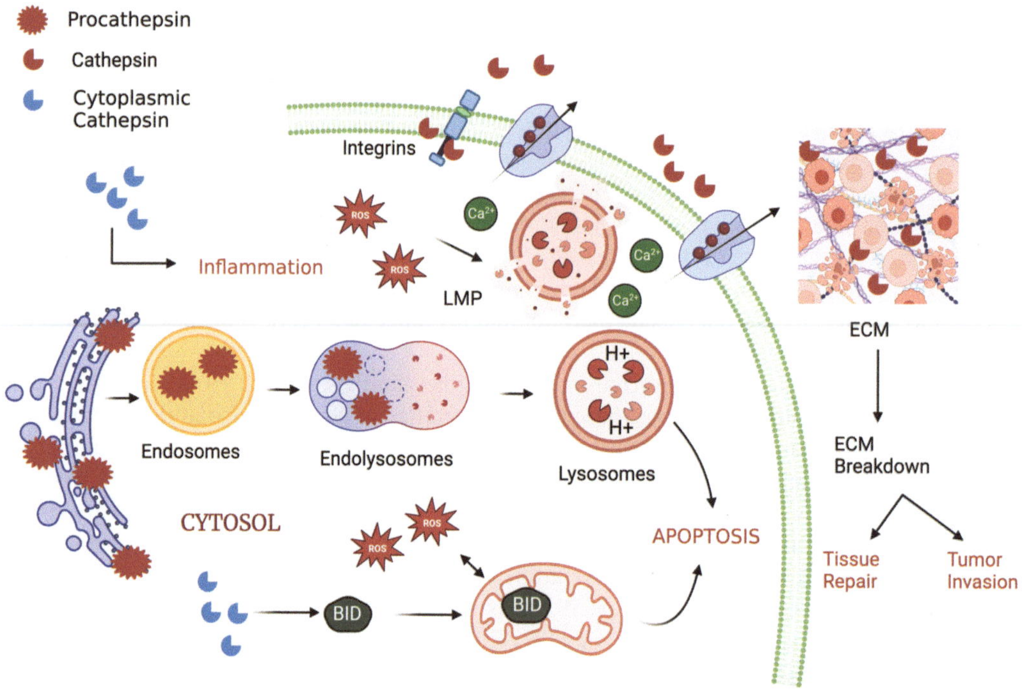

FIGURE 14.1 General functions of cathepsins. This schematic portrays the central role of cathepsins as proteolytic enzymes within lysosomes in regulating inflammation. Integrins, essential cell adhesion molecules, link extracellular signals to intracellular events, modulating inflammatory responses. Cathepsin can trigger inflammation, influence apoptotic pathways involving BID (a BCL-2 family member). Concurrently, reactive oxygen species (ROS) and Calcium (Ca^{2+}) act as secondary messengers, further shaping the inflammatory response. Cathepsins, by orchestrating these processes, play a pivotal role in inflammation, connecting lysosomal functions with integrin signaling, apoptotic pathways, and redox/calcium regulation.

They also play roles in innate and adaptive immune responses, autophagy, and regulation of growth factors and hormones. Dysregulation of lysosomal cathepsins can lead to diseases such as lysosomal storage disorders (LSD) (Turk et al., 2012; Yadati et al., 2020). In the cytosol, cathepsins mediate various forms of programmed cell death and inflammatory responses. They are found in the nucleus, where they participate in processes like autophagy, chromatin remodeling, activation of transcription factors, and histone processing (Chevriaux et al., 2020; Yadati et al., 2020). In the extracellular space, cathepsins participate in tissue remodeling, inflammation, and immune responses (Stoka et al., 2016; Turk et al., 2012; Yadati et al., 2020). Dysregulation of extracellular cathepsins is associated with diseases like cancer, arthritis, and cardiovascular conditions. Cathepsins' presence in different cellular compartments and diverse functions highlight their importance in maintaining cellular homeostasis and regulating various physiological processes. Dysregulation of cathepsin activity can lead to pathological conditions, making them potential targets for therapeutic interventions in diseases related to protease dysfunction. To name a few, for example, cathepsin B is released due to impaired autophagy in Sjögren syndrome, a chronic Kawasaki disease (Xie et al., 2023).

Cathepsins are proteases that play a role in the nucleus, primarily in lysosomes and endosomes. They are involved in processes like autophagy, chromatin remodeling, and nuclear envelope breakdown (Tamhane et al., 2016; Yadati et al., 2020). They are also involved in apoptosis and DNA damage in prostate cancer cells (Kenig et al., 2011).

In the extracellular space, cathepsins can have various functions often associated with tissue remodeling, inflammation, and immune responses. They can be released into the extracellular space by damaged or dying cells during inflammatory responses, contributing to the processing of pro-inflammatory cytokines and chemokines. They can also be secreted by immune cells, such as macrophages and dendritic cells, and involved in antigen processing, activating the immune system (Turk et al., 2012; Vidak et al., 2019; Yadati et al., 2020).

During an infection, cathepsin secretion to the extracellular space is often high, leading to the release of antimicrobial peptides and cathepsin, which have antimicrobial properties. These proteins are abundantly expressed in intestinal goblet cells and paneth cells, helping to manage pathogens and microbiota and repair epithelial barriers (Anes et al., 2022; Yan et al., 2020). Dysregulation of cathepsins in the extracellular space can have both beneficial and detrimental effects depending on the context. Their dysregulation has been associated with various diseases, including cancer, arthritis, and cardiovascular conditions. Researchers continue to study the roles of cathepsins in various diseases and their potential role in cancer treatment (Fonović & Turk, 2014; Vidak et al., 2019).

14.3 CATHEPSIN'S ROLE IN CANCER

Elevated levels and activity of cathepsins have been observed in the blood and tumors of many cancer patients. In breast cancer, the expression and activity of cathepsins B and L have been found to correlate with disease progression and relapse after treatment. Similar correlations have been found in colon, lung, brain, and head and neck cancers, where higher concentrations of cathepsins B and L within tumors are associated with poorer survival rates (Joyce & Hanahan, 2004; Linders et al., 2023; Ratna & Das, 2017; Rudzińska et al., 2019). An increase in levels of cathepsin L have also been detected in the blood of patients with pancreatic and liver cancers. Additionally, elevated levels of cathepsin H have been identified in lung, melanoma, and colorectal cancers. Although there are some discrepancies in the findings of cathepsin expression and activity in cancer, recent studies suggest that cathepsins hold promise as potential biomarkers for various diseases, including coronary artery disease, aneurysms, adiposity, peripheral arterial disease, non-alcoholic fatty liver disease (NAFLD), and type 2 diabetes (Fang et al., 2020). Cathepsin D levels have been linked to metabolic changes in the liver during NAFLD and type 2 diabetes, while cathepsins Z and K are being explored as diagnostic markers for osteoporosis. Furthermore, cathepsins have shown potential as fluorescent probes for non-invasive diagnostic imaging in preclinical mouse models (Joyce & Hanahan, 2004; Linders et al., 2023; Ratna & Das, 2017; Rudzińska et al., 2019; Yadati et al., 2020).

14.3.1 Cysteine Cathepsin and Its Role in Cancer

Cysteine cathepsins, which include the lysosomal protease cathepsin B, have been the subject of the most research among all cathepsins (Pizzorno, 2001). Cathepsin B can work as endopeptidases at neutral pH and as exopeptidases (carboxypeptidases) at low pH. Normally, the levels of cathepsin B are carefully controlled in the body in a coordinated manner. However, during the development of cancer, the regulation of cathepsin B can be disrupted at multiple levels, leading to an excessive production of this enzyme (Gerber et al., 2002; Joyce & Hanahan, 2004; Linders et al., 2023; Rudzińska et al., 2019; Tan et al., 2013; Yoo et al., 2022). High levels of cathepsin B have been observed in various types of cancer, including brain, lung, prostate, breast, and colorectal cancer, which can be seen based on the analysis using The Cancer Genome Atlas Program (TCGA) database when compared to healthy individuals (Figure 14.2). Interestingly, cathepsin B was observed to be downregulated in head and neck cancer, however, the exact mechanism is still unknown (Figure 14.2c). Cathepsin B is located on chromosome 8p22, and interestingly, this specific region has also been identified as having an important role in preventing tumor formation according to several studies (Aggarwal & Sloane, 2014; Gondi & Rao, 2013; Tan et al., 2013). In cervical cancer, the presence of cathepsin B was found to be linked to the depth of tumor invasion and the spread of cancer cells to lymph nodes (Figure 14.2b). When the cathepsin B gene was turned off in cervical cancer cells, their ability to invade decreases significantly. In brain tumors, cathepsin B levels increase with tumor progression, and higher levels were found to be associated with aggressive forms of cancer (Gondi & Rao, 2013; Tan et al., 2013). Cathepsin B has also been shown to bind directly to a protein related to hepatitis B, thereby promoting the movement and invasion of liver cancer cells (Figure 14.2d). In studies where cathepsin B was removed, cell growth and tumor invasion were slowed. Cathepsin B plays a prominent role in promoting tumor progression by breaking down components of the basement membrane and ECM. Upon inhibition of cathepsin B activity, ECM degradation and invasion were found to be reduced in inflammatory breast cancer (Gondi & Rao, 2013; Jiang et al., 2023; Tan et al., 2013). Knocking down cathepsin B in tumor cells through RNA interference decreases collagen degradation *in vitro* and bone metastasis *in vivo*. Administering a specific cathepsin B inhibitor also reduces metastasis in animals with tumors (Gondi & Rao, 2013; Mijanović et al., 2019; Withana et al., 2012; Wu et al., 2012). Thyroid carcinoma with extracapsular invasion and metastasis has high levels of cathepsin B activity and showed an increase in the degradation of type I and IV collagen. In non-invasive tumors, cathepsin B was found in distinct cytoplasmic granules, while in invasive tumors it appeared to be diffusely distributed in the cytosol (Tan et al., 2013). Analyzing cellular fractions confirmed that invasive cells have active cathepsin B localized in the plasma membrane, while non-invasive cells confine cathepsin B to lysosomes. The mature form of cathepsin B, with a molecular weight of 25 kDa, is predominantly present in invasive cells, whereas non-invasive cells primarily have precursor forms with molecular weights ranging from 30 to 35 kDa (Turk et al., 2012; Victor et al., 2011). *In vitro* experiments using plasma membrane fractions from invasive and non-invasive cells demonstrated that the plasma membrane of invasive cells can degrade laminin, like the degradation caused by purified cathepsin B. Experimental and clinical studies have also associated cathepsin B with tumor invasion and metastasis (PMID: 14587299). In several types of human cancers, cathepsin B expression was found to be markedly increased at the mRNA, protein, and activity levels (Aggarwal & Sloane, 2014; Gondi & Rao, 2013) (Figure 14.2; Table 14.1). This overexpression has also been observed in premalignant lesions, suggesting its involvement in the early stages of cancer invasion. Cathepsin B's expression regulation occurs at multiple levels, including gene amplification, alternative promoters, increased transcription, alternative splicing, and enhanced stability of transcripts. During the transition to malignancy, cathepsin B changes its localization, appearing in vesicles at the cell periphery and basal pole of polarized cells (PMID: 14587299). This altered localization results in increased secretion of both pro-cathepsin B and active cathepsin B from tumors, facilitated by lysosomal exocytosis or extracellular processing. Cathepsin B on the tumor surface binds to annexin II heterotetramer and activates downstream proteolytic cascades

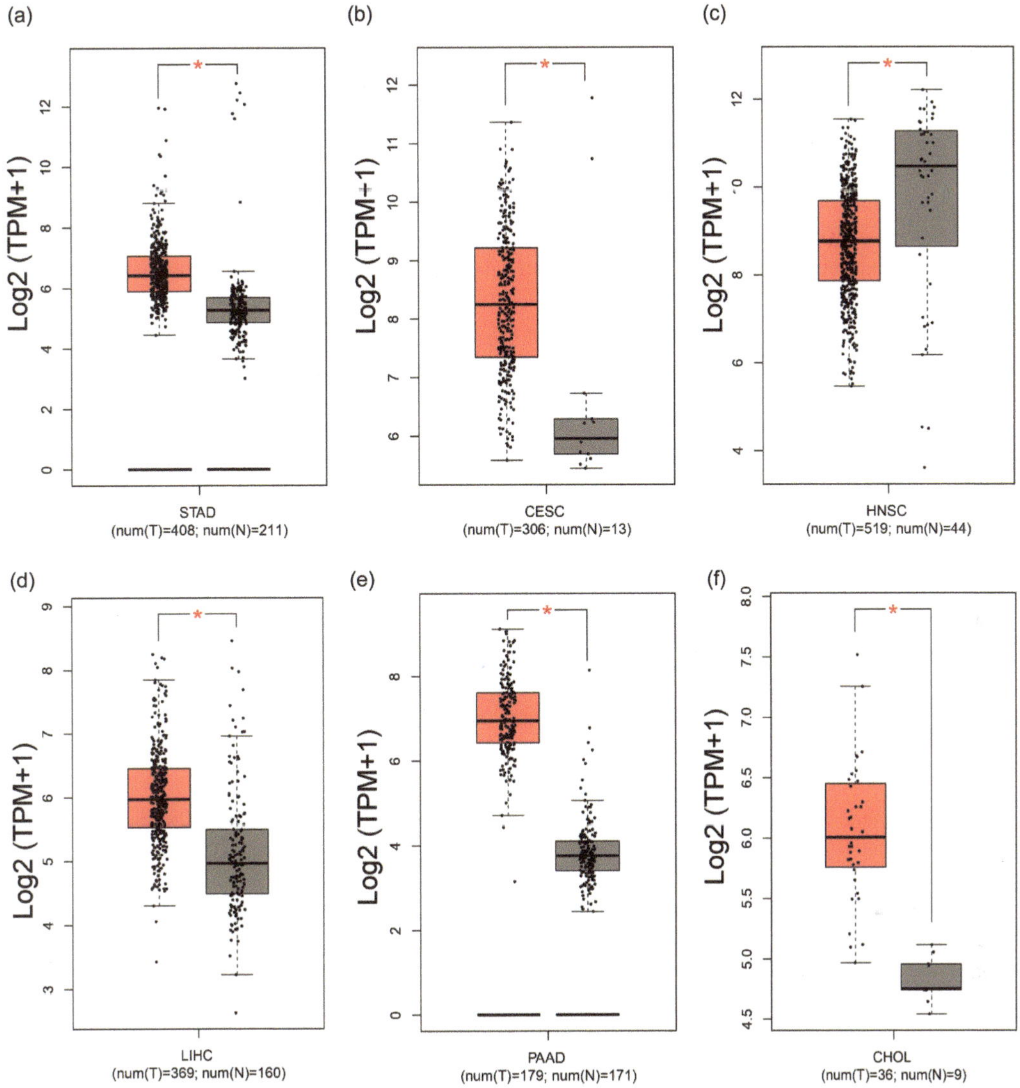

FIGURE 14.2 Expression profile of cathepsin B in various human cancers from TCGA database. Cathepsin B is encoded by the CTSB gene, which is observed to be upregulated in tumor samples from patients of (a) stomach cancer (STAD), (b) cervical squamous cell carcinoma (CESC), (c) head and neck cancer (HNSC), (d) liver hepatocellular carcinoma (LIHC), (e) pancreatic adenocarcinoma (PAAD), and (f) cholangiocarcinoma (CCA). Except for HNSC, in all the other cancer types, Cathepsin B is upregulated in the tumor samples as compared to the healthy samples.

(Aggarwal & Sloane, 2014; Gondi & Rao, 2013; Joyce & Hanahan, 2004; Mijanović et al., 2019; Tan et al., 2013; Victor et al., 2011; Wu et al., 2012; Yan & Sloane, 2003).

Cathepsin L, a type of cysteine protease, is known to possess a stronger endoproteolytic activity compared to cathepsin B. In the acidic tumor microenvironment created by anaerobic glycolysis in cancer cells, cathepsin L is particularly active. It effectively breaks down collagen, fibronectin, and laminin (Turk et al., 2012; Vidak et al., 2019; Yadati et al., 2020). Studies using mass spectrophotometry in a mouse model of pancreatic cancer have shown that cathepsin L and cathepsin S enzymes are released from cancer cells and target the extracellular region of cell adhesion molecules on the cancer cell surface. Cathepsin L also contributes to invasiveness by cleaving E-cadherin, a cell–cell adhesion molecule (Sobotič et al., 2015; Turk et al., 2012). Additionally, cathepsin L has been investigated for its role in apoptosis, as it cleaves BID, a pro-apoptotic molecule involved in apoptotic pathways. In a transgenic mouse model of pancreatic islet cell cancer, known as the RIP-Tag2 model, cathepsin L has been found to promote tumor cell proliferation, tumor growth, and angiogenesis. It is also involved in the processing of the transcription factor CDP/Cux, which aids in the rapid transition of cancer cells into the S phase of the cell cycle. Recently, cathepsin L has been implicated in Myc-driven tumorigenesis in neuroendocrine pancreatic tumors (Gerber et al., 2002; Han et al., 2016).

In squamous cell carcinomas, cathepsin K levels in the surrounding tissue were significantly higher compared to other skin tumors, suggesting its role in promoting the breakdown of the ECM and aiding tumor growth and invasion (Jiang et al., 2023; Tan et al., 2013). Cathepsin K was also present in tumor cells of basal cell carcinoma and perivascular epithelioid cell neoplasms. The roles of cathepsin H in tumor angiogenesis have been discovered recently. Under *in vivo* conditions, cathepsin H deletion inhibited angiogenic switching in premalignant hyperplastic islets and RIP-Tag2 mouse model tumor growth (Gocheva et al., 2010). Additionally, it was discovered that a homozygous cathepsin H deletion increased apoptosis (Gocheva et al., 2006).

Human cathepsin V, initially named cathepsin U, was first isolated from cDNA libraries sourced from the corneal epithelium, thymus, and testis by independent research groups (Lecaille et al., 2022). The cathepsin V gene (CTSL2/CTSV) was mapped to human chromosome 9q22.2 using Fluorescence in situ hybridization (FISH) analysis, adjacent to the cathepsin L locus (CTSL1) (Lecaille et al., 2022). Cathepsin V and human cathepsin L share significant genetic similarities and are positioned next to each other on chromosome 9, suggesting they originated from a gene duplication event about 100 million years ago (Lecaille et al., 2022). While attempts were made to find a mouse counterpart for human cathepsin V, it was found that cathepsin V, not human cathepsin L, functionally corresponded to mouse cathepsin L. The cathepsin V gene comprises 8813 base pairs, with transcription beginning at exon 2 (Lecaille et al., 2022).Unlike other cathepsins, cathepsin V lacks specific cis-regulatory elements, which may explain its limited expression pattern compared to ubiquitously expressed cathepsin B (Lecaille et al., 2022; Turk et al., 2012; Yadati et al., 2020). Interestingly, unlike some other cathepsins like C, F, K, and H, no human diseases caused by mutations in the cathepsin V gene have been reported yet. While cathepsin V is frequently elevated in human cancers and correlates with tumor aggressiveness, little is known about its epigenetic and post-transcriptional regulation compared to other cathepsins like B, C, K, and L. Cysteine cathepsins, including cathepsin V, have specific regions in their DNA that might be controlled by methylation, suggesting potential epigenetic regulation. Certain factors like the Spl transcription factor may boost cathepsin V in tumors, while the ETS family factors, commonly found in invasive tumors, might also influence its expression. Cathepsin V increase, often through the ERK pathway, might lead to lysosomal membrane damage and its release into the cell, impacting tumor growth and invasion. Although cathepsin V can be involved in cancer cell death, its overproduction alone doesn't induce apoptosis effectively. Furthermore, a specific transcription factor called E2F1 can regulate cathepsin V, influencing cancer cell apoptosis. Silencing cathepsin V reduces apoptosis, suggesting its role in cancer cell death during chemotherapy, but its overexpression alone doesn't trigger cell death (Lecaille et al., 2022; Wang et al., 2019; Yadati et al., 2020).

14.3.2 Aspartate Cathepsin and Its Role in Cancer

In breast and laryngeal cancers, higher levels of cathepsin D were associated with the recurrence of cancer and the spread of cancer cells to nearby lymph nodes (Table 14.1). Increased levels of cathepsin D and the presence of lymphatic vessels were linked to the spread of cancer in laryngeal tumors (Tan et al., 2013). Pro-cathepsin D, a precursor form of the enzyme, secreted by cancer cells, was found to promote the growth, spread, and progression of breast and lung cancers. In high-grade

TABLE 14.1
Major Functions of All the Different Types of Cathepsins

Cathepsin	Substrate	Function	References
		Cysteine protease	
Cathepsin B	Collagen, Bcl-2, Mcl-1, XIAP	Protein catabolism, bone turnover, antigen hormone activation	Cavallo-Medved et al., 2011 Droga-Mazovec et al., 2008 Yadati et al., 2020
Cathepsin F	Elastolytic activity	Antigen presentation	Andrault et al., 2019 Yadati et al., 2020
Cathepsin H	Elastolytic activity	Endopeptidase activity	Turk et al., 2012 Yadati et al., 2020
Cathepsin K	Collagen, elastin	Bone modelling, collagenolytic activity	Dai et al., 2020a Garnero et al., 1998
Cathepsin L	MHC-II, trysinogen, collagen	Antigen presentation, keratinocyte differentiation, collagenolytic activity	Han et al., 2016 Sudhan & Siemann, 2015 Yadati et al., 2020 Turk et al., 2012
		Aspartate Protease	
Cathepsin D	BID, BAX, caspase-8	Protein degradation in lysosomes	Turk et al., 2012 Yadati et al., 2020
		Serine Protease	
Cathepsin G	Fibronectin, collagen, PARs	Degradation of ECM, immune complex-mediated inflammation	Lu et al., 2011 Morimoto-Kamata et al., 2022 Palesch et al., 2016
Cathepsin A	Substance P, angiotensin 1, bioactive peptides	Platelet activation, autophagy	Patel et al., 2018

carcinomas, strong cathepsin D expression was linked to the invasive potential and aggressive behavior of the tumors. Patients with positive lymph nodes and cathepsin-D-positive tumor cells were at higher risk of cancer recurrence. Cathepsin D, a protein localized on the surface of breast fibroblasts, appears to function as a mitogen that induces stromal proliferation (Aggarwal & Sloane, 2014; Gocheva et al., 2006; Tan et al., 2013).

Recent studies have demonstrated that cathepsin D is essential for fibroblast invasive growth in a 3D co-culture assay with cancer cells, regardless of whether they secrete pro-cathepsin D or not. Ectopic expression of cathepsin D in cathepsin D-deficient fibroblasts led to significant increases in fibroblast proliferation, survival, motility, and invasive capacity, along with activation of the ras-MAP kinase pathway. Surprisingly, these stimulatory effects were independent of cathepsin D's proteolytic activity. The mechanism by which cathepsin D induces these changes in fibroblast behavior is still unclear. While the M6P/IGF2 receptor is known to interact with the secreted pro-cathepsin D, its ability to stimulate cellular responses remains controversial (Tan et al., 2013; Yadati et al., 2020). An alternative hypothesis is that pro-cathepsin D interacts with an unidentified cell surface receptor associated with the MAP kinase transduction pathway. This concept has been proposed in both cancer and endothelial cells. It is theorized that cathepsin D might enhance tumor progression by influencing the epithelial compartment and encouraging fibroblast outgrowth through a paracrine loop. In this scenario, cancer cells overexpressing and hyper-secreting cathepsin D could influence stromal cells, promoting their proliferation, survival, motility, and invasion, thereby enhancing tumor–host homeostasis. Initially, it was believed that proteases played a major role in metastasis by aiding the invasion of cancer cells through the digestion of ECM and basement membrane components (Benes et al., 2008; Tan et al., 2013; Yadati et al., 2020). However, emerging evidence indicates that proteases also act as key regulators of tumor growth at both the primary and metastatic sites. The mechanism of their mitogenic activity has been attributed to their proteolytic actions, enabling growth factor release, degradation of growth inhibitors, or activation of mitogenic receptors. Similar to neutral proteases, lysosomal protease precursors, such as pro-cathepsin B, D, and L, are also overexpressed and hypersecreted by carcinoma cells. Hypersecreted cathepsin D may act either as a protease upon extracellular activation in an acidic environment or as a binding protein, functioning extracellularly due to its high secretion, or intracellularly following endocytosis via membrane receptors (Benes et al., 2008; Laurent-Matha et al., 2005; Lecaille et al., 2022). It is unlikely that cathepsin D's action primarily involves degrading the basement membrane, as its activation requires an acidic pH found within endosomes, lysosomes, or phagosomes. Recent findings suggest that cathepsin D may function as a secreted binding protein, independent of its proteolytic activity, interacting with unknown membrane receptors on cancer, endothelial, and fibroblastic cells (Benes et al., 2008; Laurent-Matha et al., 2005; Mijanović et al., 2021; Yadati et al., 2020). Additionally, pro-cathepsin D may act as a ligand, triggering a transmembrane receptor through its pro-peptide. Another potential mechanism is that cathepsins-D could act as a mitogen on cancer cells by competing with IGFII on the M6P/IGFII receptor, leading to IGFI receptor activation. These mechanisms may vary based on local *in vivo* tumor conditions, particularly pH and oxygenation. In contrast to other proteases, cathepsin D appears to have a greater mitogenic than proteolytic role, facilitating cancer cell invasion through the basement membrane and connective tissue, or entering the bloodstream. It has been identified as a significant mediator of apoptosis induced by various apoptotic agents, including IFN-gamma, FAS/APO, TNF-alpha, oxidative stress, and several chemotherapeutic drugs (Benes et al., 2008; Laurent-Matha et al., 2005; Máša et al., 2006; Mijanović et al., 2021; Vetvicka, 2010). Its role in apoptosis is linked to the release of mature cathepsin D from lysosomes into the cytosol, leading to cytochrome c release from mitochondria, activation of pro-caspases 9 and 3, and cleavage of BID. While previous studies have suggested that cathepsin D's catalytic activity is crucial for its pro-apoptotic function, it is important to note that cathepsin D's proteolytic activity is significantly impaired at cytosolic pH levels lower than 6.2. Therefore, the ability of cathepsin D to cleave cytosolic substrates implicated in the apoptotic cascade under adverse pH conditions remains uncertain. Recent findings have shown that pepstatin A, an aspartic protease inhibitor, did not prevent cell death induced by certain apoptotic agents, and a deficiency in cathepsin D activity did not alter cell death induction in certain cell types. These results raise questions about the exclusive role of cathepsin D's proteolytic activity in apoptosis, suggesting other mechanisms may be at play (Lecaille et al., 2022; Máša et al., 2006; Mijanović et al., 2021; Turk et al., 2012; Yadati et al., 2020).

14.3.3 Serine Cathepsins and Its Role in Cancer

Cathepsin G is located in the endocytic compartment of antigen-presenting cells (APCs), and to a lesser extent, cathepsin D and cathepsin S, actively degrade MHC I molecules (Palesch et al., 2016). To initiate an adaptive immune response, MHC I molecules play a crucial role by presenting antigenic peptides to CTLs. Immune evasion strategies often involve reducing the transcription of MHC I molecules, impairing the detection of infected or cancerous cells by CTLs (Palesch et al., 2016). In the absence of MHC I, Natural Killer (NK) cells target and eliminate these cells. Consequently, infected or cancerous cells maintain a limited presence of MHC I at their surface to evade NK cell recognition (Patil et al., 2021). Inhibition of cathepsin G resulted in an increased expression of MHC I on the surface of primary human immune cells. Conversely, human glioblastoma cells lack active cathepsin G and may have lost the ability to degrade MHC I during tumorigenesis to evade NK cell-mediated destruction (Palesch et al., 2016). Studies have highlighted that introducing cathepsin G into glioblastoma cells led to a rapid and efficient degradation of MHC I (Palesch et al., 2016). Thus, cathepsin G plays a crucial role as a key protease in regulating MHC I molecules. Modulating cathepsin G activity could represent a promising avenue for therapeutic intervention, offering new possibilities for immune response regulation in the context of infections and cancer. Cathepsin G is also reported to induce cell-cell adhesion in MCF-7 human breast cancer cells, inhibiting their movement specifically on fibronectin but not on type IV collagen (Kudo et al., 2009). This effect leads to the formation of highly compact cell colonies due to enhanced E-cadherin-mediated cell-cell adhesion. Importantly, this action of cathepsin G relies on its protease activity and is hindered by serine protease inhibitors. Furthermore, cathepsin G promotes the formation of the E-cadherin/catenin complex and activates Rap1 in MCF-7 cells, both of which are known to regulate E-cadherin-based cell–cell junctions. Additionally, cathepsin G facilitates the formation of the E-cadherin/protein kinase D1 (PKD1) complex. Notably, the cathepsin G-induced cell condensation is suppressed by Go6976, a selective PKD1 inhibitor, indicating the involvement of PKD1 in this process (Kudo et al., 2009). These findings represent the initial evidence demonstrating that cathepsin G plays a role in modulating E-cadherin function, suggesting a novel regulatory mechanism in tumor cell-cell adhesion (Kudo et al., 2009). This discovery opens new avenues for understanding the intricate dynamics of cancer cell interactions and offers potential targets for therapeutic intervention (Morimoto-Kamata et al., 2022).

14.4 ROLE OF CATHEPSINS IN PROTEIN AGGREGATION

Proteolysis catalyzed by the major lysosomal aspartyl protease cathepsin D is crucial for maintaining proteostasis in the central nervous system and is implicated in neurodegenerative diseases (Yu et al., 2022). The absence of cathepsin D leads to defective autophagic flux and pathological protein accumulation, as observed in NCL type 10. Previous studies demonstrated therapeutic clearance of protein aggregates in NCL10 mice and α-synuclein-accumulating cells and mice after treatment with recombinant human pro-cathepsin D. Recent work investigated the ability of cathepsin D to cleave Alzheimer's disease-related amyloid beta (Aβ) peptides and to understand the potential efficacy of pro-cathepsin D in clearing toxic Aβ species. Although it was shown that cathepsin D's proteolytic cleavage of human Aβ1–42,it did not reveal a significant role for cathepsin D in β-amyloid precursor protein (APP) processing or Aβ degradation (Gallwitz et al., 2022). These results shed light on the complexity of lysosomal proteases' interactions in Aβ metabolism, indicating the need for a comprehensive understanding of their collective functions for effective therapeutic interventions in AD.

TP53, which has a known functional role as a tumor suppressor gene, might be a possible regulator of cathepsins (Soond et al., 2019). Though a transcriptional factor, several post-translational modifications regulate p53 functioning and contribute to its cytoplasmic translocation, where it arbitrates the mitochondria or lysosomal-mediated cell death (Aubrey et al., 2016). Phosphorylated p53 can directly induce apoptosis via MOMP, by activating apoptosis-inducing genes BAX or BAK or by replacing anti-apoptotic Bcl-2 proteins with their pro-apoptotic counterparts. Nevertheless, another potential contributing mechanism involves the action of p53 in mediating lysosome-mediated permeabilization (LMP) directly, serving as the initial step for lysosomal-mediated cell death through mitochondrial outer membrane permeabilization (MOMP) induction (Yu & Zhang, 2008). P53 has been shown to directly trigger LMP in myeloid leukemia cells. Upon nuclear stabilization, p53 translocates to the cytoplasm in a monoubiquitinated form, where it undergoes phosphorylation, inducing LMP (Soond et al., 2019). This process results in the release of cathepsin proteases from the lysosome, subsequently leading to MOMP and triggering caspase-dependent apoptosis. Monoubiquitinated p53 can also be deubiquitinated by herpesvirus-associated ubiquitin-specific protease (HAUSP), allowing p53, phosphorylated p53 (P-p53), or acetylated forms to reside in the outer mitochondrial membrane, inducing

MOMP, or releasing apoptosis regulating factor (AIF) and EndoG. These proteins mediate caspase-independent apoptosis through PARP-1 activity. Cathepsin genes are either translated and secreted for ECM regulation or transported to the lysosome as mature cathepsin proteases. Upon LMP, lysosomal cathepsins are released into the cytoplasm, leading to MOMP and caspase activation. Simultaneously, leakage of cathepsins into the cytoplasm via BID-mediated MOMP can also activate the classic apoptosis pathway (Aubrey et al., 2016; Soond et al., 2019; Wang et al., 2023; Yu & Zhang, 2008). Furthermore, they can cleave and release mitochondrial AIF, a crucial step in PARP-mediated caspase-independent cell death. Wu et al. (1998) reported the presence of two p53 DNA binding sites in the promoter sequence of the cathepsin D gene, indicating that either site could be specifically bound by the p53 protein. Studies have provided evidence supporting a direct correlation between p53 protein and cathepsin D expression in esophageal cancer. In breast cancer prognosis, traditional indicators like lymph node involvement and estrogen receptor status often fail to predict metastatic relapse effectively. This study aimed to identify additional prognostic markers beyond individual oncogenic factors. When a series of ~217 breast cancer cases were analyzed, it revealed that p53, Bcl-2, and cathepsin D, in combination, significantly influenced the risk of disease relapse. Mutations in TP53 and Ha-RAS were exclusively detected in relapsed cases, emphasizing their association with unfavorable outcomes. Such studies highlight the possible interaction with mutant p53, however detailed study is still lacking (Soond et al., 2019).

14.5 TARGETING CATHEPSIN AS A STRATEGY FOR CANCER THERAPY

Targeting cathepsins as a strategy for cancer therapy is an area of active research and drug development. Cathepsins are a family of proteases that are overexpressed in many types of cancer and play a role in tumor progression, invasion, and metastasis. Several facets of cathepsin functions are controlled by their localization (endo, lysosomal, cytosolic, or extracellular space). The functional characteristics and differential expression of cathepsins in various pathologies have been extensively studied, making them promising therapeutic targets and potential diagnostic markers. Therefore, targeting cathepsins offers several potential therapeutic strategies for cancer treatment (Joyce & Hanahan, 2004; Mijanović et al., 2019; Rudzińska et al., 2019; Sudhan & Siemann, 2015; Withana et al., 2012).

14.5.1 Targeting Cathepsin B

Increased cysteine cathepsin enzymatic activity promotes tumor growth, invasion, and metastasis in a variety of cancer types through a number of distinct mechanisms. It is conceivable that cancer cells use cathepsins to meet their increased metabolic needs given their critical role in protein catabolism. Furthermore, secreted cathepsins mediate the breakdown of ECM to promote cancer invasion and spread. Additionally, there is mounting proof that cathepsin's proteolytic products, which include receptors and cell adhesion molecules, might activate signaling cascades that promote cancer. Of the various subtypes, cathepsin K (CK) plays a critical role in metastatic breast and prostate cancer, specifically in the bone tumor microenvironment, where it catalyzes collagen I degradation and contributes to bone resorption and remodeling (Gerber et al., 2002; Dai et al., 2020a; Gondi & Rao, 2013; Linders et al., 2023; Rudzińska et al., 2019; Tan et al., 2013; Wang et al., 2023; Withana et al., 2012; Wu et al., 2012; Xie et al., 2023; Yadati et al., 2020).

Cysteine cathepsin B is crucial to the treatment of cancer, because it plays a role in a number of human diseases and oncogenic processes. Numerous studies have shown a strong correlation between the invasive and metastatic properties of cancer and the overexpression of cathepsin B. It has been observed that when cathepsin B is not adequately regulated, it can lead to the transformation of cells into a carcinogenic state. While cathepsin B is primarily associated with lysosomes, which play a crucial role in processes like autophagy and the immune response, its abnormal production has been identified as a significant factor in the development and progression of cancer (Chevriaux et al., 2020; Gondi & Rao, 2013; Wang et al., 2023; Wu et al., 2012; Xie et al., 2023).

Targeting cathepsin B in cancer is an approach aimed at inhibiting or modulating the activity of this protease to impede tumor growth, invasion, and metastasis. Cathepsin B has been implicated in various aspects of cancer progression, making it an attractive target for cancer therapy (Ruan et al., 2015). In an effort to prevent protease-mediated metastatic infiltration, a number of strategies for targeting cathepsin B have been developed to decrease the proteolytic activity of proteases. Chemical inhibitors, monoclonal antibodies, RNA Interference (RNAi), nanoparticles, combination therapy, protease-activated prodrugs, and immune therapy techniques are among the tactics that have made progress. Small molecule inhibitors have been developed to specifically target its enzymatic activity. These inhibitors work by blocking its ability to degrade ECM components and promote tumor invasion. Several cathepsin B inhibitors are in preclinical and clinical development (Veselov et al., 2022). The marine bacterium *Pseudomonas*, marine sponges, and other species have all been used as sources for cathepsin B inhibitors. Additionally, scientists have found that non-peptidic substances such as flavonoids have potent cathepsin B inhibitory characteristics. Among natural inhibitors, examples are peptidyl aldehydes, aziridinyl peptides, and epoxysuccinyl peptides (Frlan & Gobec, 2006; Kos et al., 2014; Vidal-Albalat & González, 2016). Epoxysuccinyl peptide E-64 (isolated from *Aspergillus japonicas*) can make parasites undergo oxidative stress and apoptosis (Xie et al., 2023). Endothelial cell damage brought on by scleroderma is lessened by the aldehyde inhibitor leupeptin. Despite many compounds having been isolated and identified, none have actually demonstrated any profound medicinal potential. Hence, researchers are also working on monoclonal antibodies that can be designed to target and neutralize cathepsin B, effectively reducing its proteolytic activity. These antibodies may be used to inhibit tumor invasion and disrupt metastasis. Dai et al. (2020b) developed a humanized antibody inhibitor with high efficacy and specificity for human cathepsin B by genetically fusing the pro-peptide of pro-cathepsin B into the Herceptin heavy chain complementarity determining region 3. The recombinant chimeric analog of murine mAb 2A2 effectively decreased the ECM breakdown and tumor cell invasion *in vitro*. Cathepsin B inhibitors have the potential to complement various cancer treatment modalities, including chemotherapy, radiation therapy, and immunotherapy, with the aim of improving overall treatment effectiveness (Gondi et al., 2013). In a study conducted by Shree et al., (2011), it was revealed that cathepsin B expression within breast tumors serves a protective role for the tumor. Moreover, their research demonstrated that employing a combination strategy involving cathepsin B inhibition, through the use of the compound JPM, in conjunction with the chemotherapy agent Taxol, resulted in a substantial deceleration of tumor growth (Shree et al., 2011). This suggests that cathepsin B inhibitors may offer promise as part of combination therapies to combat cancer.

Recently, disease-associated protease activatable pro-drugs have received a lot of attention and have led to ground-breaking treatments in the field of antibody-drug conjugation (ADC). For targeted delivery, the majority of ADCs are made by coupling a cytotoxic chemical to an antibody via a protease-sensitive module (Fu et al., 2022). For instance, the development of US Food and Drug Administration (FDA)-approved treatments like Brentuximab vedotin (Adcetris®) for CD30-positive recurrent or refractory Hodgkin's lymphoma was made possible by the effective implementation of a cathepsin B-specific module for this function (Jin et al., 2020; Scott, 2017). Additionally, the cathepsin B selective probe was created as a result of the protease cleavable pro-drug method, which also made it possible to monitor drug release in real time. It's important to note that while cathepsin B is a promising target for cancer therapy, research in this field is ongoing, and the clinical efficacy and safety of cathepsin B-targeted treatments are subjects of ongoing investigations in clinical trials (Park et al., 2023). The development of cathepsin B-targeted therapies offers the potential to improve the treatment options for various types of cancer. A list of cathepsin B inhibitors is presented in Table 14.2.

TABLE 14.2
Inhibitors of Cathepsin B toward Cancer Therapy

Inhibitors	Cellular Mechanism	References
Nitroxoline APL-1202 HLX-1502	Angiogenesis inhibitors Epigenetic modifier modulators	Milan Bonotto et al., 2023
Allicin DADSO	Antioxidants Signal transduction modulators	Ge et al., 2016
VBY-825	Angiogenesis inhibitors Viral replication inhibitors	Elie et al., 2010
[177Lu]A-AG [177Lu]E-AG [177Lu]A-ANT [177Lu]E-AN	Signal transduction modulators	Zhang et al., 2020
125I-CaBInIV-3	Radio iodinated inhibitors as tumor imaging probes	Edem et al., 2014
Viral entry inhibitors	Blocking the protease	Padmanabhan & Dixit, 2023
CA-074Me CA-074	Anti-amyloidogenic agents	Yoon et al., 2022
BPC-11	p53-independent apoptosis	Bechara et al., 2014
Z-FA-FMK Z-Phe-Ala-CH2F	Prevents LPS-induced production of cytokines	Lawrence et al., 2006 Schotte et al., 2001 Smith et al., 1988
pGAEMA-PTX-DOTA-Gd	Angiogenesis inhibitors Antimitotic drugs Apoptosis inducers	Zhou et al., 2022
Z-Arg-Lys-AOMK z-FA-CMK	Apoptosis inducers	Liow & Chow, 2013 Yoon et al., 2021
FA-GFLG-SN-38	Antiamyloidogenic agents Apoptosis inducers	Jin et al., 2020
17,18-dihydrobarnesin Barnesin A	Natural inhibitor	Rischer et al., 2018
Shishicrellastatin A & B	Natural inhibitor	Murayama et al., 2011
4‴-methylamentoflavone Podocarpusflavone A Hinokiflavone	Apoptosis Inducers Viral Replication Inhibitors	Pan et al., 2005
ANO-2	Angiogenesis Inhibitors	Kostoulas et al., 2000
Miraziridine A	Natural Inhibitor	Konno et al., 2007
Z-Phe-Gly-NHO-Bz	Apoptosis Inducers	Zhu & Uckun, 2000

14.5.2 Targeting Cathepsin L

Cathepsin L2s participation in a variety of highly invasive malignancies is particularly remarkable, as detailed in recent reviews (Liu et al., 2006; Zhao et al., 2021). It has been shown that the extracellular cathepsin L levels in transformed cells can grow up to 200-fold. They can rapidly destroy many structural components in the ECM environment, dramatically increasing the migratory/invasive potential of these cells and encouraging metastasis. Indeed, the dysregulation of cathepsin L function has been implicated in several human diseases, making it an appealing target for drug development. The unchecked activity of cathepsin L often arises from an imbalance between the enzyme and the levels of endogenous inhibitors (Fonović & Turk, 2014; Gomes et al., 2020; Liu et al., 2006; Sudhan & Siemann, 2015). To counteract its proteolytic function, various studies have explored approaches involving the introduction or overexpression of endogenous inhibitors, such as cystatins and stefins. These strategies aim to restore the balance and mitigate the adverse effects associated with cathepsin L dysregulation in diseases (Sudhan & Siemann, 2015).

Similar to this, cathepsin L downregulation through RNA interference consistently inhibited the tumorigenicity and invasiveness of neoplastic cells in glioma, osteosarcoma, myeloma, and melanoma tumor models (Qin et al., 2016; Zhang et al., 2015). Although anti-sense and genetic knockout strategies support the anti-metastatic significance of cathepsin L abrogation, the high structural homology between various cathepsin family members has hindered the development of specific cathepsin L inhibitors. In fact, the majority of cathepsin inhibitors now under investigation exhibit wide-ranging activity. Despite the challenges imposed by the high degree of homology between the S2 substrate recognition domains of cysteine cathepsins, a few cathepsin L-specific inhibitors have been developed and tested in preclinical models. KGP94 significantly retarded the growth of both recently implanted and established tumors. However, the prolonged use of selective inhibitors of cathepsin L (CTSL) carries the potential risk of therapeutic resistance, as compensation by other members of the cathepsin family can occur over time (Katunuma, 2011; Kishore Kumar et al., 2010; Qin et al., 2016; Song et al., 2013; Zhang et al., 2015). Therefore, a more effective approach might involve simultaneous inhibition of cathepsin L and other critical tumor-promoting cathepsins. For instance, cathepsins B and K have been shown to significantly influence different aspects of tumor progression, including metastasis and bone resorption (Tan et al., 2013). Testing dual targeting strategies that involve cathepsin L in combination with cathepsins B or K could potentially yield better treatment outcomes. While preclinical models have shown promising results with cathepsin L inhibition, the lessons learned from clinical trials of matrix metalloproteinase (MMP) inhibitors have raised questions about protease targeting in cancer therapy (Coussens et al., 2002). One key issue is the development of reliable tools for identifying patients who would benefit most from anti-cathepsin-L therapy. Proposed methodologies, such as cathepsin L immunohistochemistry of tumor tissue and the measurement of CTSL levels or activity in serum samples, need to be rigorously evaluated for their reliability. A recent non-cell permeable cathepsin L-specific photo-affinity-based probe has been taken into consideration, which can be employed to measure secreted CTSL activity *ex vivo* (Sudhan & Siemann, 2015). Stratifying patients based on the expression levels and activity of cathepsin L can help determine the most appropriate treatment approach, ensuring that therapy is tailored to the individual characteristics of the patient's cancer. Research into targeting cathepsin L for cancer therapy and personalized medicine is ongoing, and the effectiveness of such treatments may vary depending on the specific cancer type and the patient's profile. Personalized approaches are gaining importance in cancer treatment, as they offer the potential for more precise and effective therapies with reduced side effects (Gerber et al., 2002) Gomes et al., 2020; Kenig et al., 2011; Kishore Kumar et al., 2010; Qin et al., 2016; Simmons et al., 2005; Song et al., 2013; Sudhan & Siemann, 2015).

14.5.3 Targeting Cathepsin S

Cathepsin S has been demonstrated to be differently expressed in a range of malignancies, including hepatocellular, pancreatic, and prostate carcinomas, which contrasts with the usual distribution of cathepsin S under normal settings. It is highly expressed in the TME's myeloid-derived suppressor cells (MDSC), dendritic cells (DC), and tumor-associated M2-type macrophages (TAM). Patients with higher cathepsin S immunohistochemistry scores had significantly worse survival rates, according to a survival analysis, and a multivariable analysis demonstrated that cathepsin S was a significant prognostic factor. Thus, cathepsin S is another promising target for cancer therapy (Turk et al., 2012; Yadati et al., 2020).

14.5.4 Targeting Nuclear Cathepsins

The role of nuclear cathepsins, particularly cathepsin L, is an area of ongoing research, and their functions in the nucleus are not as well established as their roles in other cellular compartments.

Some of the proposed roles of nuclear cathepsins include DNA repair, transcriptional regulation, chromatin remodeling, cell proliferation and differentiation, and immune response. The role of nuclear cathepsin L and its interaction with various inhibitors like stefin B and cystatin D in colorectal cancer cells is of significant interest. In colon cancer cells, cystatin D can also translocate to the nucleus and influence gene transcription processes related to the production of FGF4 and CXC3CL1. This research into nuclear proteases and their interactions with nuclear substrates can also provide insights into targeting systems like PROTACs for the selective degradation of nuclear proteins (Tamhane et al., 2016; Turk et al., 2012; Yadati et al., 2020). Understanding how nuclear proteases work in these contexts may lead to innovative approaches in the development of targeted therapies and the treatment of various diseases, including cancer and neurodegenerative disorders. To create efficient therapies and treatments that can fully utilize the potential of nuclear proteases, it will be essential to comprehend the mechanism of action of these enzymes, their selectivity, and the effects of their activity. It's important to emphasize that research in this area is ongoing, and the functions of nuclear cathepsins are still being explored. While these proposed roles are intriguing, more studies are needed to confirm and elucidate the exact mechanisms and significance of nuclear cathepsins in various cellular processes. Additionally, the specific functions of nuclear cathepsins may vary between different cell types and contexts.

14.6 CONCLUSION AND FUTURE PERSPECTIVES

Recent advancements in proteomics, *in vivo* imaging, and the use of *in vivo* models have greatly advanced research on proteases. However, there are challenges and opportunities in developing effective treatments targeting cathepsins in medicine. Achieving efficacy and safety in human patients has been challenging, leading to setbacks in the translation of cathepsin-targeted therapies. Additionally, understanding the involvement of cathepsins in immune responses and diseases like cancer is a significant challenge. The development of targeted medication delivery systems, such as ADCs, holds promise in overcoming these challenges. By delivering drugs specifically to cancer cells, ADCs minimize off-target effects and enhance the precision and effectiveness of treatments. However, the transition from animal models to human subjects has been marked by variability due to differences in cysteine and cathepsin structures and specificity across species. Advances in genetic profiling and biomarker identification will enable personalized therapies targeting specific cathepsins based on individual patient characteristics. Despite challenges, the integration of cutting-edge technologies, personalized medicine strategies, and targeted medication delivery systems offer hope for the future of cathepsins as therapeutic targets. However, addressing challenges related to specificity, resistance, and delivery will be crucial for the successful development and application of cathepsin-targeted treatments in various diseases, including cancer.

ACKNOWLEDGMENT

The authors wish to acknowledge the DBT/Wellcome Trust India Alliance Fellowship [IA/E/17/1/503663] awarded to Dr. Shinjinee Sengupta for financial support.

DECLARATION

Conflict of interest: The authors declare no conflicts of interest. The manuscript, including figures and tables, is entirely original and has not been taken from any published paper.

Consent for publishing: All authors have approved the final version of the manuscript and given consent for publication.

REFERENCES

Aggarwal, N., & Sloane, B. F. (2014). Cathepsin B: Multiple roles in cancer. *PROTEOMICS – Clinical Applications*, *8*(5–6), 427–437. https://doi.org/10.1002/prca.201300105

Andrault, P.-M., Panwar, P., Mackenzie, N. C. W., et al (2019). Elastolytic activity of cysteine cathepsins K, S, and V promotes vascular calcification. *Scientific Reports*, *9*(1), 9682. https://doi.org/10.1038/s41598-019-45918-1

Anes, E., Pires, D., Mandal, M., et al (2022). Spatial localization of cathepsins: Implications in immune activation and resolution during infections. *Frontiers in Immunology*, *13*. https://doi.org/10.3389/fimmu.2022.955407

Aubrey, B. J., Strasser, A., & Kelly, G. L. (2016). Tumor-suppressor functions of the TP53 pathway. *Cold Spring Harbor Perspectives in Medicine*, *6*(5). https://doi.org/10.1101/cshperspect.a026062

Bechara, A., Barbosa, C. M. V, Paredes-Gamero, E. J., et al (2014). Palladacycle (BPC) antitumour activity against resistant and metastatic cell lines: the relationship with cytosolic calcium mobilisation and cathepsin B activity. *European Journal of Medicinal Chemistry*, *79*, 24–33. https://doi.org/10.1016/j.ejmech.2014.03.073

Benes, P., Vetvicka, V., & Fusek, M. (2008). Cathepsin D–many functions of one aspartic protease. *Critical Reviews in Oncology/Hematology*, *68*(1), 12–28. https://doi.org/10.1016/j.critrevonc.2008.02.008

Buck, M. R., Karustis, D. G., Day, N. A. et al (1992). Degradation of extracellular-matrix proteins by human cathepsin B from normal and tumour tissues. *Biochemical Journal*, *282*(1), 273–278. https://doi.org/10.1042/bj2820273

Cavallo-Medved, D., Moin, K., & Sloane, B. (2011). Cathepsin B: Basis Sequence: Mouse. *The AFCS-Nature Molecule Pages*, *2011*.

Chevriaux, A., Pilot, T., Derangère, V., et al (2020). Cathepsin B is required for NLRP3 inflammasome activation in macrophages, through NLRP3 interaction. *Frontiers in Cell and Developmental Biology*, *8*. https://doi.org/10.3389/fcell.2020.00167

Coussens, L. M., Fingleton, B., & Matrisian, L. M. (2002). Matrix metalloproteinase inhibitors and cancer—trials and tribulations. *Science*, *295*(5564), 2387–2392. https://doi.org/10.1126/science.1067100

Dai, R., Wu, Z., Chu, H. Y., et al (2020a). Cathepsin K: The action in and beyond bone. *Frontiers in Cell and Developmental Biology*, *8*. https://doi.org/10.3389/fcell.2020.00433

Dai, Z., Cheng, Q., & Zhang, Y. (2020b). Rational design of a humanized antibody inhibitor of cathepsin B. *Biochemistry*, *59*(14), 1420–1427. https://doi.org/10.1021/acs.biochem.0c00046

Droga-Mazovec, G., Bojič, L., Petelin, A., et al (2008). Cysteine cathepsins trigger caspase-dependent cell death through cleavage of bid and antiapoptotic Bcl-2 homologues. *Journal of Biological Chemistry*, *283*(27), 19140–19150. https://doi.org/10.1074/jbc.M802513200

Edem, P. E., Czorny, S., & Valliant, J. F. (2014). Synthesis and evaluation of radioiodinated acyloxymethyl ketones as activity-based probes for cathepsin B. *Journal of Medicinal Chemistry*, *57*(22), 9564–9577. https://doi.org/10.1021/jm501357r

Elie, B. T., Gocheva, V., Shree, T., et al (2010). Identification and pre-clinical testing of a reversible cathepsin protease inhibitor reveals anti-tumor efficacy in a pancreatic cancer model. *Biochimie*, *92*(11), 1618–1624. https://doi.org/10.1016/j.biochi.2010.04.023

Fang, W., Deng, Z., Benadjaoud, F., et al (2020). Cathepsin B deficiency ameliorates liver lipid deposition, inflammatory cell infiltration, and fibrosis after diet-induced nonalcoholic steatohepatitis. *Translational Research*, *222*, 28–40. https://doi.org/10.1016/j.trsl.2020.04.011

Fonović, M., & Turk, B. (2014). Cysteine cathepsins and extracellular matrix degradation. *Biochimica et Biophysica Acta (BBA) - General Subjects*, *1840*(8), 2560–2570. https://doi.org/10.1016/j.bbagen.2014.03.017

Frlan, R., & Gobec, S. (2006). Inhibitors of cathepsin B. *Current Medicinal Chemistry*, *13*(19), 2309–2327. https://doi.org/10.2174/092986706777935122

Fu, Z., Li, S., Han, S., et al (2022). Antibody drug conjugate: The "biological missile" for targeted cancer therapy. *Signal Transduction and Targeted Therapy*, *7*(1), 93. https://doi.org/10.1038/s41392-022-00947-7

Gallwitz, L., Schmidt, L., Marques, A. R. A., et al (2022). Cathepsin D: Analysis of its potential role as an amyloid beta degrading protease. *Neurobiology of Disease*, *175*, 105919. https://doi.org/10.1016/j.nbd.2022.105919.

Garnero, P., Borel, O., Byrjalsen, I., et al (1998). The Collagenolytic Activity of Cathepsin K Is Unique among Mammalian Proteinases. *Journal of Biological Chemistry*, *273*(48), 32347–32352. https://doi.org/10.1074/jbc.273.48.32347

Ge, L., Xu, Y., Jiang, X., et al (2016). Broad-spectrum inhibition of proteolytic enzymes by allicin and application in mitigating textural deterioration of ice-stored grass carp (*Ctenopharyngodon idella*) fillets. *International Journal of Food Science & Technology, 51*(4), 902–910. https://doi.org/10.1111/ijfs.13047

Gerber, A., Welte, T., Ansorge, S., Bühling, F. (2002). Expression of Cathepsins B and L in human lung epithelial cells is regulated by cytokines BT – Cellular peptidases in immune functions and diseases 2, in: Langner, J., Ansorge, S. (Eds.). Springer USA, Boston, MA, pp. 287–292. https://doi.org/10.1007/0-306-46826-3_31

Gocheva, V., Chen, X., Peters, C., et al (2010). Deletion of cathepsin H perturbs angiogenic switching, vascularization and growth of tumors in a mouse model of pancreatic islet cell cancer. *Biological Chemistry, 391*(8). https://doi.org/10.1515/bc.2010.080

Gocheva, V., Zeng, W., Ke, D., et al (2006). Distinct roles for cysteine cathepsin genes in multistage tumorigenesis. *Genes & Development, 20*(5), 543–556. https://doi.org/10.1101/gad.1407406

Gomes, C. P., Fernandes, D. E., Casimiro, F., et al (2020). Cathepsin L in COVID-19: From Pharmacological Evidences to Genetics. *Frontiers in Cellular and Infection Microbiology, 10*. https://doi.org/10.3389/fcimb.2020.589505

Gondi, C. S., & Rao, J. S. (2013). Cathepsin B as a cancer target. *Expert Opinion on Therapeutic Targets, 17*(3), 281–291. https://doi.org/10.1517/14728222.2013.740461

Han, M., Zhao, Y., Tan, C., et al (2016). Cathepsin L upregulation-induced EMT phenotype is associated with the acquisition of cisplatin or paclitaxel resistance in A549 cells. *Acta Pharmacologica Sinica, 37*(12), 1606–1622. https://doi.org/10.1038/aps.2016.93

Jiang, H., Dong, Z., Xia, X., et al (2023). Cathepsins in oral diseases: Mechanisms and therapeutic implications. *Frontiers in Immunology, 14*. https://doi.org/10.3389/fimmu.2023.1203071

Jin, X., Zhang, J., Jin, X., et al (2020). Folate receptor targeting and cathepsin B-sensitive drug delivery system for selective cancer cell death and imaging. *ACS Medicinal Chemistry Letters, 11*(8), 1514–1520. https://doi.org/10.1021/acsmedchemlett.0c00031

Joyce, J. A., & Hanahan, D. (2004). Multiple roles for cysteine cathepsins in cancer. *Cell Cycle, 3*(12), 1516–1519. https://doi.org/10.4161/cc.3.12.1289

Katunuma, N. (2011). Structure-based development of specific inhibitors for individual cathepsins and their medical applications. *Proceedings of the Japan Academy, Series B, 87*(2), 29–39. https://doi.org/10.2183/pjab.87.29

Kenig, S., Frangež, R., Pucer, A., et al (2011). Inhibition of cathepsin L lowers the apoptotic threshold of glioblastoma cells by up-regulating p53 and transcription of caspases 3 and 7. *Apoptosis, 16*(7), 671–682. https://doi.org/10.1007/s10495-011-0600-6

Kishore Kumar, G. D., Chavarria, G. E., Charlton-Sevcik, A. K., et al (2010). Design, synthesis, and biological evaluation of potent thiosemicarbazone based cathepsin L inhibitors. *Bioorganic & Medicinal Chemistry Letters, 20*(4), 1415–1419. https://doi.org/10.1016/j.bmcl.2009.12.090

Konno, H., Kubo, K., Makabe, H., et al (2007). Total synthesis of miraziridine A and identification of its major reaction site for cathepsin B. *Tetrahedron, 63*(38), 9502–9513. https://doi.org/10.1016/j.tet.2007.06.082

Kos, J., Mitrović, A., & Mirković, B. (2014). The current stage of cathepsin B inhibitors as potential anticancer agents. *Future Medicinal Chemistry, 6*(11), 1355–1371. https://doi.org/10.4155/fmc.14.73

Kostoulas, G., Lang, A., Nagase, H., et al (2000). Erratum to: Stimulation of angiogenesis through cathepsin B inactivation of the tissue inhibitors of matrix metalloproteinases (FEBS 22347). *FEBS Letters, 466*(2–3), 394–394. https://doi.org/10.1016/S0014-5793(99)01266-1

Kudo, T., Kigoshi, H., Hagiwara, T., et al (2009). Cathepsin G, a neutrophil protease, induces compact cell-cell adhesion in MCF-7 human breast cancer cells. *Mediators of Inflammation, 2009*, 1–11. https://doi.org/10.1155/2009/850940

Laurent-Matha, V., Maruani-Herrmann, S., Prébois, C., et al (2005). Catalytically inactive human cathepsin D triggers fibroblast invasive growth. *The Journal of Cell Biology, 168*(3), 489–499. https://doi.org/10.1083/jcb.200403078

Lawrence, C. P., Kadioglu, A., Yang, A.-L., et al (2006). The cathepsin B inhibitor, z-FA-FMK, inhibits human T cell proliferation *in vitro* and modulates host response to pneumococcal infection *in vivo*. *The Journal of Immunology, 177*(6), 3827–3836. https://doi.org/10.4049/jimmunol.177.6.3827

Lecaille, F., Chazeirat, T., Saidi, A., et al (2022). Cathepsin V: Molecular characteristics and significance in health and disease. *Molecular Aspects of Medicine, 88*, 101086. https://doi.org/10.1016/j.mam.2022.101086

Linders, D. G. J., Bijlstra, O. D., Fallert, L. C., et al (2023). Cysteine cathepsins in breast cancer: Promising targets for fluorescence-guided surgery. *Molecular Imaging and Biology, 25*(1), 58–73. https://doi.org/10.1007/s11307-022-01768-4

Liow, K. Y., & Chow, S. C. (2013). The cathepsin B inhibitor, z-FA-CMK is toxic and readily induced cell death in human T lymphocytes. *Toxicology and Applied Pharmacology*, *272*(3), 559–567. https://doi.org/10.1016/j.taap.2013.07.022

Liu, J., Sukhova, G. K., Yang, J.-T., et al (2006). Cathepsin L expression and regulation in human abdominal aortic aneurysm, atherosclerosis, and vascular cells. *Atherosclerosis*, *184*(2), 302–311. https://doi.org/10.1016/j.atherosclerosis.2005.05.012

Lu, P., Takai, K., Weaver, V. M., et al (2011). Extracellular matrix degradation and remodeling in development and disease. *Cold Spring Harbor Perspectives in Biology*, *3*(12), a005058. https://doi.org/10.1101/cshperspect.a005058

Máša, M., Marešová, L., Vondrášek, J., et al (2006). Cathepsin D propeptide: Mechanism and regulation of its interaction with the catalytic core. *Biochemistry*, *45*(51), 15474–15482. https://doi.org/10.1021/bi0614986

Mijanović, O., Branković, A., Panin, A. N., et al (2019). Cathepsin B: A sellsword of cancer progression. *Cancer Letters*, *449*, 207–214. https://doi.org/10.1016/j.canlet.2019.02.035

Mijanović, O., Petushkova, A. I., Brankovic, A., Turk, B., Solovieva, A. B., Nikitkina, A. I., Bolevich, S., Timashev, P. S., Parodi, A., Zamyatnin, A. A. J. (2021). Cathepsin D-managing the delicate balance. *Pharmaceutics*, 13. https://doi.org/10.3390/pharmaceutics13060837

Milan Bonotto, R., Mitrović, A., Sosič, I., et al (2023). Cathepsin inhibitors nitroxoline and its derivatives inhibit SARS-CoV-2 infection. *Antiviral Research*, *216*, 105655. https://doi.org/10.1016/j.antiviral.2023.105655

Morimoto-Kamata, R., Matsuki, S., Ohkura, N., et al (2022). Cathepsin G-induced cell aggregation of breast cancer MCF-7 decreases doxorubicin sensitivity in a hypoxia-inducible factor-independent mechanism. *Biological and Pharmaceutical Bulletin*, *45*(12), 1772–1783. https://doi.org/10.1248/bpb.b22-00447

Murayama, S., Imae, Y., Takada, K., et al (2011). Shishicrellastatins, inhibitors of cathepsin B, from the marine sponge Crella (Yvesia) spinulata. *Bioorganic & Medicinal Chemistry*, *19*(22), 6594–6598. https://doi.org/10.1016/j.bmc.2011.06.052

Padmanabhan, P., & Dixit, N. M. (2023). Modelling how increased Cathepsin B/L and decreased TMPRSS2 usage for cell entry by the SARS-CoV-2 Omicron variant may affect the efficacy and synergy of TMPRSS2 and Cathepsin B/L inhibitors. *Journal of Theoretical Biology*, *572*, 111568. https://doi.org/10.1016/j.jtbi.2023.111568

Palesch, D., Wagner, J., Meid, A., et al (2016). Cathepsin G-mediated proteolytic degradation of MHC class I molecules to facilitate immune detection of human glioblastoma cells. *Cancer Immunology, Immunotherapy: CII*, *65*(3), 283–291. https://doi.org/10.1007/s00262-016-1798-5

Pan, X., Tan, N., Zeng, G., et al (2005). Amentoflavone and its derivatives as novel natural inhibitors of human Cathepsin B. *Bioorganic & Medicinal Chemistry*, *13*(20), 5819–5825. https://doi.org/10.1016/j.bmc.2005.05.071

Park, S.-H., Lee, J.-H., Yang, S.-B., et al (2023). Development of a peptide-based nano-sized cathepsin B inhibitor for anticancer therapy. *Pharmaceutics*, *15*(4), 1131. https://doi.org/10.3390/pharmaceutics15041131

Patel, S., Homaei, A., El-Seedi, H. R., et al (2018). Cathepsins: Proteases that are vital for survival but can also be fatal. *Biomedicine & Pharmacotherapy*, *105*, 526–532. https://doi.org/10.1016/j.biopha.2018.05.148

Patil, N. D., Theresine, M., & Zimmer, J. (2021). Human natural killer cells in major histocompatibility complex class I deficiency. *Scandinavian Journal of Immunology*, *94*(2). https://doi.org/10.1111/sji.13029

Pizzorno, M. C. (2001). Nuclear cathepsin B-like protease cleaves transcription factor YY1 in differentiated cells. *Biochimica et Biophysica Acta (BBA) - Molecular Basis of Disease*, *1536*(1), 31–42. https://doi.org/10.1016/S0925-4439(01)00032-1

Qin, G., Cai, Y., Long, J., et al (2016). Cathepsin L is involved in proliferation and invasion of breast cancer cells. *Neoplasma*, *63*(01), 30–36. https://doi.org/10.4149/neo_2016_004

Ratna, A., & Das, S. K. (2017). Role of Cathepsins, in Particular Cathepsins B and D in Breast Cancer: Mechanisms and Clinical Implications. In *Pathophysiological Aspects of Proteases* (pp. 171–189). Springer Singapore. https://doi.org/10.1007/978-981-10-6141-7_8

Rischer, M., Raguž, L., Guo, H., et al (2018). Biosynthesis, synthesis, and activities of barnesin A, a NRPS-PKS hybrid produced by an anaerobic epsilonproteobacterium. *ACS Chemical Biology*, *13*(8), 1990–1995. https://doi.org/10.1021/acschembio.8b00445

Ruan, H., Hao, S., Young, P., & Zhang, H. (2015). Targeting cathepsin B for cancer therapies. *Horizons in Cancer Research*, *56*, 23–40.

Rudzińska, M., Parodi, A., Soond, S. M., et al (2019). The role of cysteine cathepsins in cancer progression and drug resistance. *International Journal of Molecular Sciences*, *20*(14), 3602. https://doi.org/10.3390/ijms20143602

Schotte, P., Schauvliege, R., Janssens, S., & Beyaert, R. (2001). The cathepsin B inhibitor z-FA.fmk inhibits cytokine production in macrophages stimulated by lipopolysaccharide. *Journal of Biological Chemistry*, *276*(24), 21153–21157. https://doi.org/10.1074/jbc.M102239200

Scott, L. J. (2017). Brentuximab vedotin: A review in CD30-positive Hodgkin lymphoma. *Drugs*, *77*(4), 435–445. https://doi.org/10.1007/s40265-017-0705-5

Shree, T., Olson, O. C., Elie, B. T., et al (2011). Macrophages and cathepsin proteases blunt chemotherapeutic response in breast cancer. *Genes & Development*, *25*(23), 2465–2479. https://doi.org/10.1101/gad.180331.111

Simmons, G., Gosalia, D. N., Rennekamp, A. J., et al (2005). Inhibitors of cathepsin L prevent severe acute respiratory syndrome coronavirus entry. *Proceedings of the National Academy of Sciences*, *102*(33), 11876–11881. https://doi.org/10.1073/pnas.0505577102

Smith, R. E., Rasnick, D., Burdick, C. O., et al (1988). Visualization of time-dependent inactivation of human tumor cathepsin B isozymes by a peptidyl fluoromethyl ketone using a fluorescent print technique. *Anticancer Research*, *8*(4), 525–529.

Sobotič, B., Vizovišek, M., Vidmar, R., et al (2015). Proteomic identification of cysteine cathepsin substrates shed from the surface of cancer cells. *Molecular & Cellular Proteomics*, *14*(8), 2213–2228. https://doi.org/10.1074/mcp.M114.044628

Song, J., Jones, L. M., Chavarria, G. E., et al (2013). Small-molecule inhibitors of cathepsin L incorporating functionalized ring-fused molecular frameworks. *Bioorganic & Medicinal Chemistry Letters*, *23*(9), 2801–2807. https://doi.org/10.1016/j.bmcl.2012.12.025

Soond, S. M., Kozhevnikova, M. V., Townsend, P. A., et al (2019). Cysteine cathepsin protease inhibition: An update on its diagnostic, prognostic and therapeutic potential in cancer. *Pharmaceuticals*, *12*(2), 87. https://doi.org/10.3390/ph12020087

Soond, S. M., Kozhevnikova, M. V, & Zamyatnin, A. A. J. (2019). "Patchiness" and basic cancer research: unravelling the proteases. *Cell Cycle*, *18*, 1687–1701. https://doi.org/10.1080/15384101.2019.1632639

Stoka, V., Turk, V., & Turk, B. (2016). Lysosomal cathepsins and their regulation in aging and neurodegeneration. *Ageing Research Review*, *32*, 22–37. https://doi.org/https://doi.org/10.1016/j.arr.2016.04.010

Sudhan, D. R., & Siemann, D. W. (2015). Cathepsin L targeting in cancer treatment. *Pharmacology & Therapeutics*, *155*, 105–116. https://doi.org/10.1016/j.pharmthera.2015.08.007

Tamhane, T., Illukkumbura, R., Lu, S., et al (2016). Nuclear cathepsin L activity is required for cell cycle progression of colorectal carcinoma cells. *Biochimie*, *122*, 208–218. https://doi.org/10.1016/j.biochi.2015.09.003

Tan, G.-J., Peng, Z.-K., Lu, J.-P., et al (2013). Cathepsins mediate tumor metastasis. *World Journal of Biological Chemistry*, *4*(4), 91. https://doi.org/10.4331/wjbc.v4.i4.91

Turk, V., Stoka, V., Vasiljeva, O., et al (2012). Cysteine cathepsins: From structure, function and regulation to new frontiers. *Biochimica et Biophysica Acta*, *1824*(1), 68–88. https://doi.org/10.1016/j.bbapap.2011.10.002

Vasiljeva, O., Reinheckel, T., Peters, C., et al (2007). Emerging roles of cysteine cathepsins in disease and their potential as drug targets. *Current Pharmaceutical Design*, *13*(4), 387–403. https://doi.org/10.2174/138161207780162962

Veselov, V. V, Nosyrev, A. E., Jicsinszky, L., et al (2022). Targeted delivery methods for anticancer drugs. *Cancers*, *14*(3). https://doi.org/10.3390/cancers14030622

Vetvicka, V. (2010). Procathepsin D and cancer: From molecular biology to clinical applications. *World Journal of Clinical Oncology*, *1*(1), 35. https://doi.org/10.5306/wjco.v1.i1.35

Victor, B. C., Anbalagan, A., Mohamed, M. M., et al (2011). Inhibition of cathepsin B activity attenuates extracellular matrix degradation and inflammatory breast cancer invasion. *Breast Cancer Research*, *13*(6), R115. https://doi.org/10.1186/bcr3058

Vidak, E., Javoršek, U., Vizovišek, M., et al (2019). Cysteine cathepsins and their extracellular roles: Shaping the microenvironment. *Cells*, *8*(3), 264. https://doi.org/10.3390/cells8030264

Vidal-Albalat, A., & González, F. V. (2016). Natural Products as Cathepsin Inhibitors. In *Studies in Natural Products Chemistry* (pp. 179–213). Elsevier. https://doi.org/10.1016/B978-0-444-63749-9.00006-2

Wang, J., Zheng, M., Yang, X., et al (2023). The role of cathepsin B in pathophysiologies of non-tumor and tumor tissues: A systematic review. *Journal of Cancer*, *14*(12), 2344–2358. https://doi.org/10.7150/jca.86531

Wang, Y., Niu, H., Hu, Z., et al (2019). Targeting the lysosome by an aminomethylated Riccardin D triggers DNA damage through cathepsin B-mediated degradation of BRCA1. *Journal of Cellular and Molecular Medicine*, *23*(3), 1798–1812. https://doi.org/10.1111/jcmm.14077

Withana, N. P., Blum, G., Sameni, M., Slaney, C., Anbalagan, A., Olive, M.B., Bidwell, B. N., Edgington, L., Wang, L., Moin, K., Sloane, B. F., Anderson, R. L., Bogyo, M. S., & Parker, B. S. (2012). Cathepsin B inhibition limits bone metastasis in breast cancer. *Cancer Research*, 72, 1199–1209. https://doi.org/10.1158/0008-5472.CAN-11-2759

Wu, D., Wang, H., Li, Z., et al (2012). Cathepsin B may be a potential biomarker in cervical cancer. *Histology and Histopathology*, *27*(1), 79–87. https://doi.org/10.14670/HH-27.79
Wu, G. S., Saftig, P., Peters, C., et al (1998). Potential role for cathepsin D in p53-dependent tumor suppression and chemosensitivity. *Oncogene*, *16*(17), 2177–2183. https://doi.org/10.1038/sj.onc.1201755
Xie, Z., Zhao, M., Yan, C., et al (2023). Cathepsin B in programmed cell death machinery: mechanisms of execution and regulatory pathways. *Cell Death & Disease*, *14*(4), 255. https://doi.org/10.1038/s41419-023-05786-0
Yadati, T., Houben, T., Bitorina, A., et al (2020). The Ins and Outs of cathepsins: Physiological function and role in disease management. *Cells*, *9*(7). https://doi.org/10.3390/cells9071679
Yan, S., & Sloane, B. F. (2003). Molecular regulation of human cathepsin B: Implication in pathologies. *Biological Chemistry*, *384*(6). https://doi.org/10.1515/BC.2003.095
Yan, X., Wu, Z., Wang, B., et al (2020). Involvement of cathepsins in innate and adaptive immune responses in periodontitis. *Evidence-Based Complementary and Alternative Medicine*, *2020*, 1–9. https://doi.org/10.1155/2020/4517587
Yoo, Y., Choi, E., Kim, Y., et al (2022). Therapeutic potential of targeting cathepsin S in pulmonary fibrosis. *Biomedicine & Pharmacotherapy*, *145*, 112245. https://doi.org/10.1016/j.biopha.2021.112245
Yoon, M. C., Christy, M. P., Phan, V. V., et al (2022). Molecular features of CA-074 pH-dependent inhibition of cathepsin B. *Biochemistry*, *61*(4), 228–238. https://doi.org/10.1021/acs.biochem.1c00684
Yoon, M. C., Solania, A., Jiang, Z., et al (2021). Selective neutral pH inhibitor of cathepsin B designed based on cleavage preferences at cytosolic and lysosomal pH conditions. *ACS Chemical Biology*, *16*(9), 1628–1643. https://doi.org/10.1021/acschembio.1c00138
Yu, D. M. T., Dauth, S., Margineanu, M. B., et al (2022). Characterization of cysteine cathepsin expression in the central nervous system of aged wild-type and cathepsin-deficient mice. *Applied Sciences*, *12*(5), 2608. https://doi.org/10.3390/app12052608
Yu, J., & Zhang, L. (2008). PUMA, a potent killer with or without p53. *Oncogene*, *27*(Suppl 1), S71–S83. https://doi.org/10.1038/onc.2009.45
Zhang, Q., Han, M., Wang, W., et al (2015). Downregulation of cathepsin L suppresses cancer invasion and migration by inhibiting transforming growth factor-β-mediated epithelial-mesenchymal transition. *Oncology Reports*, *33*(4), 1851–1859. https://doi.org/10.3892/or.2015.3754
Zhang, W., Fan, W., Ottemann, B. M., Alshehri, S., & Garrison, J. C. (2020). Development of improved tumor-residualizing, GRPR-targeted agents: Preclinical comparison of an endolysosomal trapping approach in agonistic and antagonistic constructs. *Journal of Nuclear Medicine*, *61*, 443–450. https://doi.org/10.2967/jnumed.119.231282
Zhao, M.-M., Yang, W.-L., Yang, F.-Y., et al (2021). Cathepsin L plays a key role in SARS-CoV-2 infection in humans and humanized mice and is a promising target for new drug development. *Signal Transduction and Targeted Therapy*, *6*(1), 134. https://doi.org/10.1038/s41392-021-00558-8
Zhou, M., Wen, L., Wang, C., et al (2022). Recent advances in stimuli-sensitive amphiphilic polymer-paclitaxel prodrugs. *Frontiers in Bioengineering and Biotechnology*, *10*. https://doi.org/10.3389/fbioe.2022.875034
Zhu, D. M., & Uckun, F. M. (2000). Z-Phe-Gly-NHO-Bz, an inhibitor of cysteine cathepsins, induces apoptosis in human cancer cells. *Clinical Cancer Research: An Official Journal of the American Association for Cancer Research*, *6*(5), 2064–2069.

15 The Emerging Role of Calpain 2 Protease as a Therapeutic Target in Breast Cancer

Praveen Kumar Chandra Sekar and Ramakrishnan Veerabathiran
Human Cytogenetics and Genomics Laboratory, Faculty of Allied Health Sciences, Chettinad Hospital and Research Institute, Chettinad Academy of Research and Education, Kelambakkam, India

Jayakumar Subramaniyan
SRM Arts and Science College, Kattankulathur, India

15.1 INTRODUCTION

Calpain 2 protease, an essential component of the calpain family, plays a pivotal role in the intricate realm of cancer biology, with a notable focus on studies on breast cancer [Storr et al., 2016]. Worldwide, 2.1 million women experience breast cancer each year, making it the biggest cause of death for women in 2018. Similarly, in 2022 breast cancer accounts for more than 627,000 women and 15% of all cancer-related deaths. The prevalence of breast cancer is notably higher in developed countries, with global incidence rates increasing [Pandi et al., 2022]. This versatile enzyme, also known as the calpain 2 large subunit or calpain 2 non-catalytic subunit, exerts its influence on a wide spectrum of cellular processes that are deeply interconnected with cancer development and progression [Ho et al., 2012]. Given its calcium-dependent cysteine protease nature, its significance extends beyond that of breast cancer as it operates in various cancer types [MacLeod et al., 2018].

Calpain 2 actively participates in the proteolytic cleavage of a diverse range of substrates, including essential proteins that govern processes such as apoptosis, cell-cycle regulation, and DNA repair. However, in this chapter, we will specifically focus on calpain 2's role in breast cancer, a common and fatal type of cancer that affects women worldwide. The contribution of calpain is prominent within the realm of breast cancer research. It is a key player in crucial processes essential for the advancement of breast cancer, notably, metastasis, cell migration, and invasion, and it has garnered significant attention [Leloup & Wells, 2011]. Through its impact on cytoskeletal dynamics and disruption of cell adhesion, calpain 2 empowers cancer cells to breach tissue barriers, thereby facilitating their spread throughout the body [Niessen et al., 2011]. These characteristics make calpain 2 a compelling candidate for understanding and potentially targeting the aggressive nature of breast cancer.

Additionally, breast cancer tissues' calpain 2 expression levels have a significant impact on prognosis. Elevated calpain 2 expression has been linked to more advanced disease stages and poorer outcomes, indicating its potential as a diagnostic marker [Potz et al., 2016]. Monitoring calpain 2 levels can provide valuable insights into the aggressiveness of breast cancer and guide clinical decision-making. As researchers delve deeper into the intricate interaction between calpain

DOI: 10.1201/9781003394693-17

2 and breast cancer, it has become increasingly evident that this enzyme's role transcends mere proteolysis [Zhang et al., 2019]. Its effect on hormone receptor (HR) signaling, specifically the estrogen receptor (ER) and progesterone receptor (PR), highlights its ability to shape the molecular landscape of breast cancer [García-Trevijano et al., 2023]. By regulating the degradation and proteolysis of these receptors, Calpain 2 can profoundly influence the signaling pathways that drive breast cancer [Wang et al., 2020]. This chapter is dedicated to uncovering the multifaceted involvement of calpain 2 in breast cancer research. The objective of this study was to investigate the complex mechanisms underlying its role in tumor invasion, metastasis, HR signaling, and prognosis. Through this exploration, we aimed to contribute to the ongoing search for novel therapeutic strategies and diagnostic markers specific to breast cancer, offering fresh insights into the enigma of this formidable disease.

15.2 CALPAIN 2 PROTEASE AND BREAST CANCER

Several studies have unequivocally confirmed a substantial association between calpain 2 and breast cancer, with both m-calpain and μ-calpain being linked to diverse aggressive characteristics in this ailment, as demonstrated in both *in vitro* and *in vivo* experiments. Calpain-2-deficient mouse mammary carcinoma cells were injected into the mouse mammary fat pad, where they inhibited tumor growth and promoted the expression of the tumor suppressor p27kip1. This result emphasizes the critical role of calpain 2 in promoting the development and aggressiveness of breast tumors. [Storr et al., 2015]. In addition, in mouse models of HER2+ breast cancer, genetic disruption of both calpain 1 and calpain 2 resulted in a reduced incidence of tumorigenesis. Furthermore, this disruption made cancer cells more responsive to chemotherapeutic agents such as doxorubicin and lapatinib, highlighting the multifaceted role of calpain 2 in breast cancer development and advancement and emphasizing its potential as a therapeutic target [MacLeod et al., 2018].

In the context of breast cancer, there is clear evidence of upregulated calpain 2 expression, which is closely linked to the increased invasive traits of tumors. It has been shown that calpain 2 is essential for the development of invadopodia, specialized membrane protrusions that increase the invasiveness of breast cancer cells [Cortesio et al., 2008]. Elevated calpain expression in breast cancer has been linked to a less favorable prognosis, and this association is believed to be attributed to increased calpain activity, which fosters tumor progression and metastasis [Storr et al., 2016]. In summary, these studies emphasize the intricate involvement of calpain 2 in breast cancer, spanning its contributions to tumor growth and invasiveness, its potential impact on the response to chemotherapy, and the overall prognosis of this devastating disease.

15.3 STRUCTURE AND FUNCTION OF CALPAIN 2

Calpain 2 is a heterodimer composed of a regulatory subunit known as CAPNS1 and a catalytic subunit encoded by the CAPN2 gene. The protease core 1 domain (PC1), protease core 2 domain (PC2), a calpain-type beta-sandwich-like domain (CBSW), and a penta EF-hand domain (PEF(L)) comprise the catalytic subunit. The catalytic cleft is created by the interaction between the PC1 and PC2 domains when calcium binds. Within this cleft, the catalytic triad comprises the amino-acid residues C105, H262, and N286 [Condrat et al., 2021]. The C2 domain is essential for the calcium-dependent activation of calpain 2. When intracellular calcium levels fluctuate, calcium ions bind to this domain, leading to conformational changes that facilitate enzyme activation and enhance its ability to recognize specific targets [Hosfield et al., 1999].

Calpain 2 comprises calpain homology domains that are pivotal for its interaction with calpain 1, leading to the formation of a functional calpain protease complex. This interaction is vital for regulating the proteolytic activity of calpain 1 [Luo et al., 2015]. Calpain 2, while primarily serving as a regulatory subunit, retains the ability to directly cleave specific substrates. Important proteins involved in a variety of cellular processes, including cell-cycle regulation, apoptosis, and intracellular signaling pathways, may degrade as a result of this ability. In the context of breast cancer and other hormone-related conditions, calpain 2 plays a substantial role in regulating HRs and affects the stability and activity of HRs such as the ER and PR, which are pivotal for the advancement of the disease and therapeutic approaches [Kovacs & Su, 2014].

15.4 MECHANISMS OF CALPAIN 2 IN BREAST CANCER

By influencing cell-cycle regulation, destroying anti-apoptotic proteins, and fostering oncogenic pathways, calpain 2 is deeply involved in promoting the unchecked growth and proliferation of breast cancer cells.

15.4.1 Cell-Cycle Regulation

The cell cycle is a tightly regulated process that governs the orderly progression of cells through growth, DNA replication, and division [Sever & Brugge, 2015]. Calpain 2 exerts its influence on this essential regulatory network, calpain 2 can cleave proteins that are essential cell cycle regulators and checkpoints. Cyclins, cyclin-dependent kinases (CDKs), and CDK inhibitors are all integral parts of the CDK regulatory system. By cleaving cyclins and activating CDKs, calpain 2 facilitates the transition from one phase of the cell cycle to the next. This leads to an expedited cell cycle, with cells progressing through these phases at a rapid pace [Łukasik et al., 2021]. Calpain 2 can degrade CDK inhibitors such as p21 and p27, which typically function as brakes in the cell cycle. Their degradation enables cancer cells to circumvent crucial checkpoints, promote uncontrolled cell division, and ultimately contribute to tumor growth [Abbas & Dutta, 2009].

15.4.2 Apoptosis

Apoptosis, also referred to as programmed cell death, is a natural mechanism for eliminating damaged or unwanted cells. Anti-apoptotic proteins protect cells from undergoing this process, and calpain 2 can selectively target and degrade anti-apoptotic proteins, including Bcl-2, which inhibits the activation of apoptotic pathways [Wang et al., 2018]. Through the degradation of anti-apoptotic proteins, calpain 2 reduces the cell's capacity to resist apoptosis. Consequently, cancer cells can evade programmed cell death and persist during proliferation [Eldeeb et al. 2018].

15.4.3 Oncogenic Pathways

Oncogenic pathways are signaling cascades that are often hyperactive in cancer cells, driving their uncontrolled growth. The proteolytic activity of calpain 2 can activate oncogenic pathways by cleaving inhibitors of these pathways or by directly activating signaling molecules. Calpain 2 has the potential to cleave proteins that hinder the Ras or PI3K/Akt pathways, resulting in increased activation and cell proliferation. This activation contributes to unrestrained proliferation of breast cancer cells [Ho et al., 2012]. Calpain 2 is capable of activating transcription factors, such as NF-κB

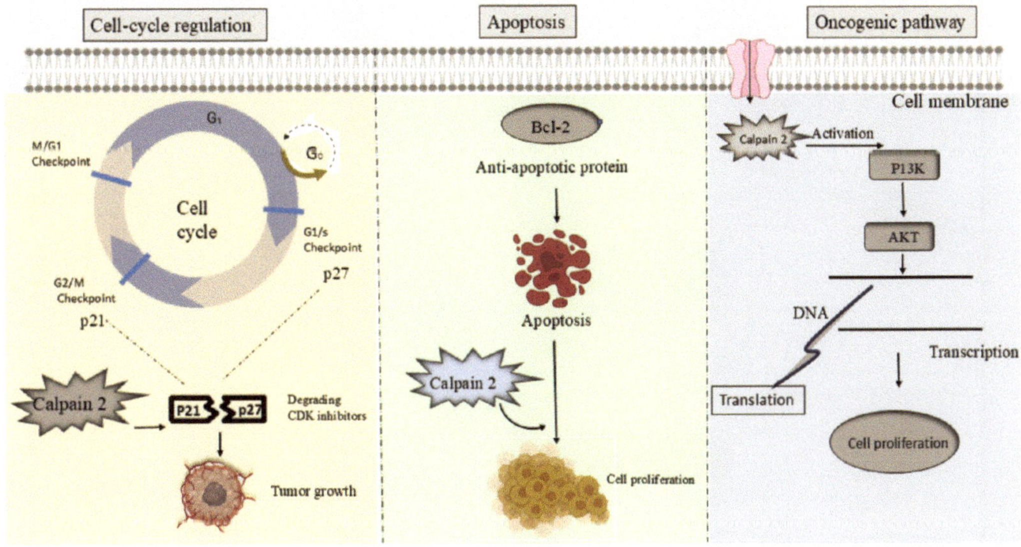

FIGURE 15.1 Mechanisms of calpain 2 in breast cancer. Calpain 2 plays a crucial role in promoting the uncontrolled growth and multiplication of breast cancer cells by regulating the cell cycle, breaking down proteins that inhibit apoptosis, and supporting pathways that drive cancer development.

or STAT3, which in turn stimulates the expression of genes that support cell growth and division [Zhang et al., 2017]. Figure 15.1 illustrates this process.

15.5 REGULATION OF CALPAIN 2 IN BREAST CANCER

The regulation of calpain 2 in breast cancer is intricately intertwined with various signaling pathways, shaping the complex landscape of this disease. Several key players in these pathways impact calpain 2's activity, contributing to the multifaceted role of this enzyme in breast cancer. Akt, a serine/threonine kinase that is crucial for cell survival and proliferation, is a central component of these pathways. Cleaved by calpain 2, FoxO is a transcription factor that regulates the expression of genes linked to DNA repair, apoptosis, and cell-cycle arrest. This cleavage leads to the degradation of FoxO and the suppression of p27Kip1 expression, ultimately encouraging cell proliferation [Ho et al., 2012]. In breast cancer, the interplay between the non-receptor tyrosine kinase Src, the regulatory protein calpain 2, and the protein tyrosine phosphatase PTP1B is pivotal for cell adhesion, migration, and invasion. Calpain 2 disassembles adhesions by targeting substrates, such as talin and FAK. It also cleaves PTP1B, leading to PTP1B degradation and Src activation, thereby promoting cell invasion. PTP1B is critical for invadopodia formation and underscores its role in breast cancer progression. This intricate interplay regulates key aspects of breast cancer invasion and metastasis [Cortesio et al., 2008]. An illustration of this is shown in Figure 15.2.

Hypoxia, which is characterized by low oxygen levels in cancer cells, initiates amoeboid cancer cell dissemination through the action of HIF. Calpain 2 plays a vital role in this process by deactivating β1 integrins through the cleavage of talin-1, facilitating its transformation into an amoeboid state. Blocking calpain 2 can reverse this transformation by restoring cells to their elongated form and preventing hypoxia-induced metastasis (Figure 15.3). This pathway is a potential target for the control of metastasis and energy efficiency. Furthermore, calpain 2 is involved in degrading

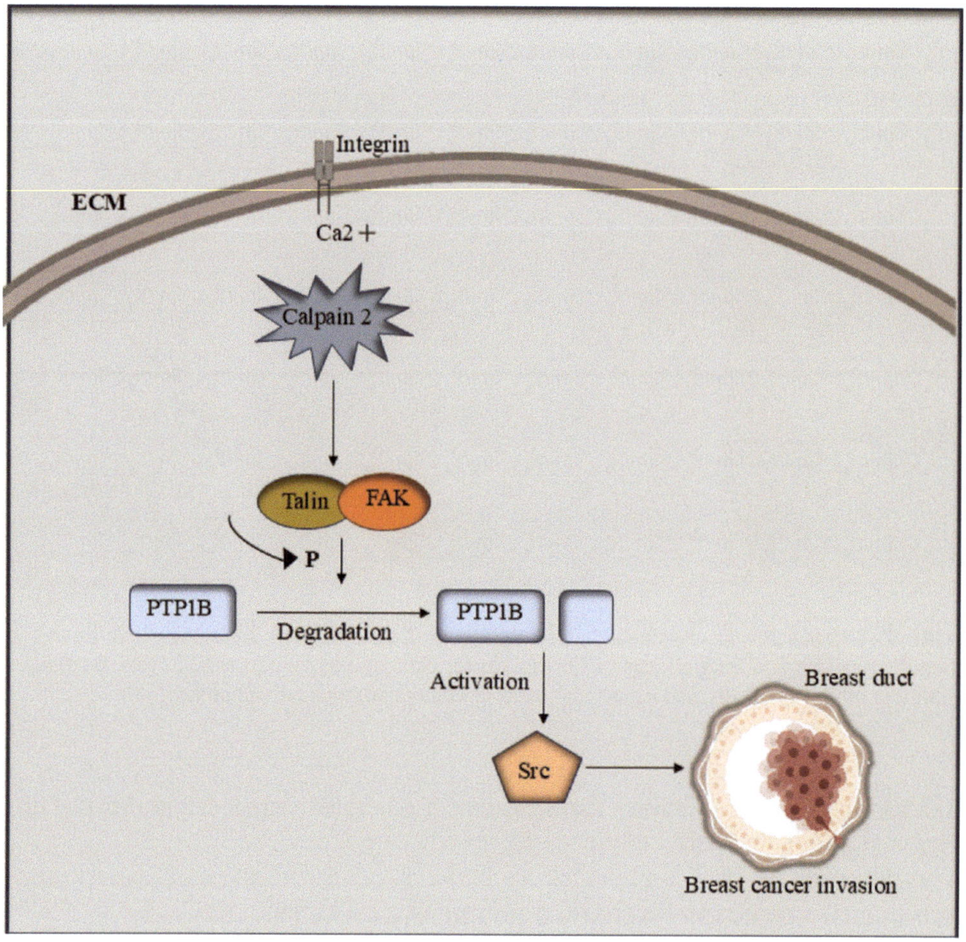

FIGURE 15.2 Regulation of calpain 2 in breast cancer. In breast cancer, the interplay between Src, calpain 2, and PTP1B is pivotal in regulating cell adhesion, migration, and invasion. Calpain 2 dismantles adhesions and activates Src through the cleavage of PTP1B, thereby promoting cell invasion and metastasis. This underscores the importance of calpain 2 in the progression of breast cancer progression.

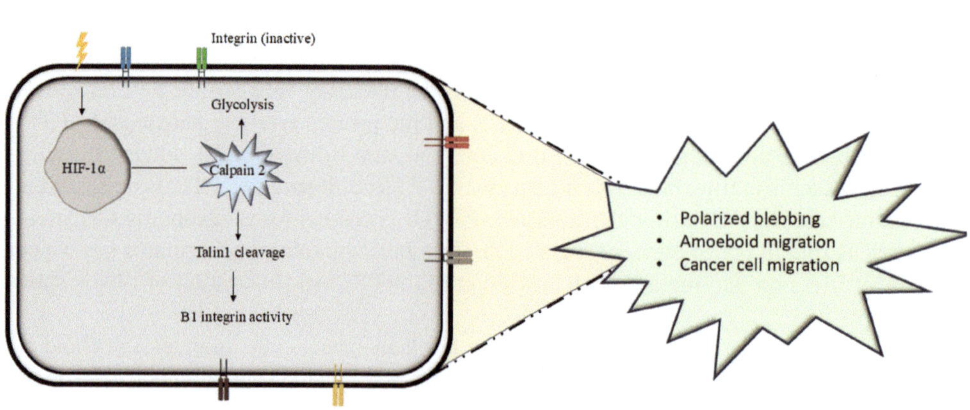

FIGURE 15.3 Regulation of calpain 2 in breast cancer. Calpain 2 plays a critical role in this process, as it inactivates β1 integrins through the cleavage of talin-1, thus promoting the transition to an amoeboid state. Inhibiting calpain 2 can reverse this transformation, returning cells to their elongated form and preventing hypoxia-induced metastasis.

HIF-1α, further hindering hypoxia-induced amoeboid cancer cell migration and metastasis [Te Boekhorst et al., 2022].

RNA-binding proteins have emerged as a new category of potential therapeutic drug targets in breast cancer. This is especially pertinent in the context of triple-negative breast cancer (TNBC), in which many other molecular drug targets are absent. Among these RNA-binding proteins, YTHDF2 has garnered considerable attention from RNAs that fine-tune gene expression and has also been associated with the development and advancement of breast cancer [Zhang et al., 2022]. Multiple miRNA-based therapeutic strategies have been investigated, including the use of anti-sense oligonucleotides to inhibit oncomiRs, the reintroduction of tumor suppressors through miRNA mimics, and chemical modifications of miRNAs. These approaches offer promising avenues for the treatment of breast cancer [Nagini, 2017].

15.6 CALPAIN 2 INHIBITORS AS THERAPEUTIC AGENTS

The exploration of calpain 2 inhibitors as potential therapeutic agents offers promising avenues for cancer research and various other diseases. It has been demonstrated that selective calpain 2 inhibitors improve learning, tasks in particular, and offer neuroprotection in animal models of acute neurodegeneration. Among these inhibitors, NA-184 is particularly noteworthy and is currently being developed for its potential use in the management of traumatic brain injury with plans for future clinical trials. These inhibitors also demonstrate promise for potential applications in the treatment of diseases such as breast cancer [Baudry et al., 2023].

In the context of breast cancer, researchers have employed RNA interference (RNAi), which uses small RNA molecules to suppress gene expression. This technique has been used to reduce calpain-2 expression in breast cancer cells, leading to a marked decrease in cell proliferation and invasion. RNAi-based strategies are promising for potential therapeutic applications in breast cancer [Wang et al., 2018]. Proteolytic inhibitors, including calpain inhibitors, are being considered as alternative medicinal options for treating diseases caused by trypanosomatids such as Chagas disease and leishmaniasis [Ennes-Vidal et al., 2022]. Notably, calpain inhibitors have demonstrated effectiveness in laboratory settings against *Trypanosoma cruzi,* the pathogen responsible for Chagas disease. This suggests potential for the development of novel therapeutic interventions for these diseases [Ennes-Vidal et al., 2022]. Calpain 2 inhibitors are specifically designed to target the proteolytic activity of calpain 2, a protease implicated in breast cancer progression. These inhibitors aim to mitigate the harmful effects of calpain 2 by interfering with its function at various levels. They primarily function by impeding the proteolytic activity of the enzyme, typically by attaching to its active site, thereby preventing the cleavage of its substrates.

By inhibiting the proteolytic activity of calpain 2, these inhibitors effectively reduce the breakdown of critical cellular proteins essential for functions, such as cell-cycle regulation and apoptosis, ultimately slowing unregulated cell proliferation, a prominent characteristic of cancer [Kovacs & Su, 2014]. Moreover, certain calpain 2 inhibitors work by disrupting the interaction between calpain 2 and its substrates. Calpain 2 identifies specific sequences or patterns within target proteins, and inhibitors can mimic these sequences or bind to crucial regions of calpain 2 responsible for substrate recognition. This interference prevents the enzyme from cleaving its substrates, thus ensuring the stability and preservation of vital proteins [Leloup & Wells, 2011]. Certain inhibitors disrupt the interaction between calpain 2 and its substrates. Calpain 2 identifies particular sequences or patterns within target proteins. Inhibitors can mimic these sequences or bind to crucial regions of calpain 2, which is responsible for substrate recognition. This interference prevents the enzyme from cleaving its substrates, thereby preserving protein stability (Table 15.1).

TABLE 15.1
Calpain 2 Inhibitors as Therapeutic Agents

Disease	Calpain Inhibitors	Study/Experiment Details	Novel Approaches	Reference
Traumatic brain injury	NA-184	Calpain 2 inhibitors improve learning and offer neuroprotection in animal models of acute neurodegeneration.	Plans are underway for upcoming clinical trials.	Sever and Brugge (2015)
Trypanosomatid diseases	PD150606	Calpain inhibitors have demonstrated effectiveness in laboratory settings against *Trypanosoma cruzi*, the pathogen responsible for Chagas disease.	This suggests the potential for the development of novel therapeutic interventions for these diseases.	Storr et al. (2016)
	PC-3	RNAi has been used to suppress calpain 2 expression in breast cancer cells, leading to decreased cell proliferation and invasion.	RNAi-based strategies are promising for therapeutic applications in breast cancer treatment.	Storr et al. (2015)
Breast cancer	c-Myc	Calpain-2 inhibitors target the proteolytic activity of calpain 2.	They primarily function by impeding the enzyme's proteolytic activity, typically by attaching to its active site.	Tatarova et al. (2022)
	17-AAG	Certain calpain 2 inhibitors work by disrupting the interaction between calpain 2 and its substrates.	This can be achieved by mimicking the sequences of target proteins or binding to crucial regions of calpain 2 responsible for substrate recognition.	Te Boekhorst et al. (2022)

Abbreviations: 17-AAG, 17-allylamino-17-demethoxygeldanamycin; c-Myc, cellular Myc; PC3, prostate cancer.

15.7 CLINICAL TRIALS AND TRANSLATIONAL RESEARCH

Clinical trials represent a pivotal stage in the drug development process and serve as a critical bridge between laboratory research and potential therapeutic applications. To date, no clinical trials have specifically focused on the utilization of calpain 2 inhibitors as therapeutic remedies. Nevertheless, numerous ongoing clinical trials are actively investigating the use of calpain 2 inhibitors in the treatment of a range of ailments, including cancer and traumatic brain injury. These trials rigorously evaluated the safety and effectiveness of calpain 2 inhibitors in human subjects, offering invaluable insights that can shape the future development of these inhibitors as potential therapeutic agents. These trials represent a vital step in bringing the promise of calpain 2 inhibitors closer to practical clinical use.

Currently, calpain 2 inhibitors are in the developmental stages for potential use in the management of traumatic brain injury, which holds promise for addressing this critical medical need. Additionally, the use of RNAi to suppress calpain 2 expression in breast cancer cells has yielded encouraging results, leading to reduced cell proliferation and invasion. This method holds promise for potential development as a therapeutic approach for breast cancer, offering new hope for patients in need of effective treatments [Austin, 2021]. Translational research is a crucial process that involves translating fundamental scientific findings into innovative therapies and treatments. In the context of calpain 2 inhibitors as potential therapeutic agents, there are both challenges and opportunities. A significant challenge is the intricate nature of the calpain system, which poses difficulties in developing selective inhibitors that exclusively target calpain 2 while sparing other members of the calpain family. Additionally, there is a need for biomarkers that can predict the effectiveness of calpain 2 inhibitors in individual patients, allowing for more personalized treatment approaches [Yu, 2011]. Addressing these challenges is crucial to fully harness the therapeutic potential of calpain 2 inhibitors and advance them to the forefront of clinical practice.

15.8 COMBINATION THERAPIES

Combination therapies hold great promise as approaches for treating complex diseases, including cancer. Calpain 2 inhibitors have emerged as potential therapeutic agents for various diseases, including breast cancer and acute neuronal injury, through the binding of NR2B to RasGRF1, establishing a connection between NMDAR activation and ERK activation. ERK activation leads to the phosphorylation and subsequent activation of calpain 2. Therefore, this pathway is likely responsible for the sustained activation of calpain 2 following stimulation of the extrasynaptic NMDA receptor. Numerous studies have demonstrated that calpain activity causes striatal-enriched tyrosine phosphatase (STEP) to cleave, producing inactive fragments. This then causes p38 to become active and initiates downstream cell death signaling pathways [Wang et al., 2018], and their potential for use in combination with conventional therapies such as chemotherapy and radiation therapy to enhance treatment outcomes. For instance, in the context of breast cancer, combining calpain 2 inhibitors with CDK4/6 inhibitors, abemaciclib, palbociclib, and ribociclib, which are often used alongside endocrine therapy (ET), have shown synergistic effects. This suggests that a combined treatment approach may offer improved therapeutic benefits, potentially leading to more effective treatments for patients [Ayoub, 2021] (Figure 15.4).

Despite the potential advantages of combination therapies, several challenges remain associated with their development and implementation. One such challenge relates to the potential for drug interactions and pharmacokinetics of concurrently administered agents. These factors may affect the effectiveness of the treatment regimen, and careful consideration and monitoring are necessary to ensure the safety and efficacy of combination therapies. Another significant challenge is the

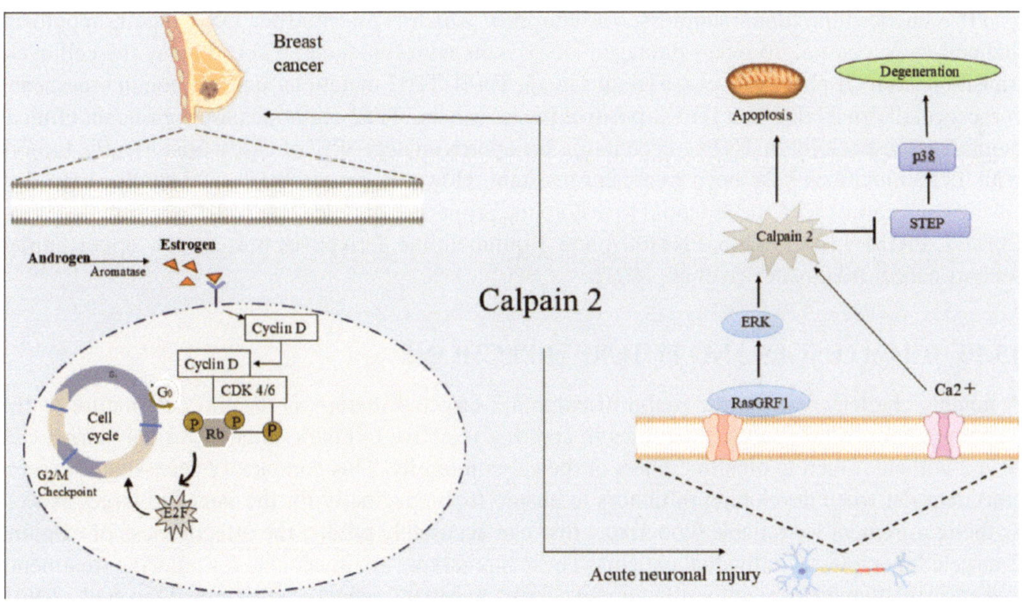

FIGURE 15.4 Combination Therapies. Calpain 2 inhibitors have emerged as potential therapeutic agents for various diseases, including breast cancer and acute neuronal injury. The interaction between NR2B and RasGRF1 connects NMDAR activation to ERK activation, subsequently triggering calpain 2 activation and ultimately initiating cell death pathways. When calpain 2 inhibitors are combined with CDK4/6 inhibitors, they are frequently used for breast cancer treatment.

inherent heterogeneity of cancer, which makes it difficult to predict how individual patients will respond to combination treatment. Personalized medical approaches that consider a patient's unique profile may be crucial for addressing this challenge and optimizing treatment outcomes [Schmucker et al., 2021]. Moreover, combining calpain 2 inhibitors with immunotherapies offers a promising avenue for augmenting treatment efficacy. This combination approach has the potential to enhance immunogenicity and counteract resistance mediated by the tumor microenvironment, thereby providing enhanced therapeutic benefits. Harnessing the power of the immune system alongside calpain 2 inhibitors may lead to more effective treatments that leverage the body's defense against cancer [Tatarova et al., 2022].

15.9 BIOMARKERS FOR PATIENT STRATIFICATION

Biomarkers are invaluable tools for tailoring patient treatment, especially in the realm of calpain-2-targeted therapies. This enables healthcare professionals to predict patient responses and optimize treatment strategies. For instance, Ki-67, which is encoded by the MKI-67 gene on chromosome 10q26.2, is associated with cell proliferation. Elevated Ki-67 levels in breast cancer indicate greater aggressiveness [Menon et al., 2019]. Ki-67 staining is prognostically significant, aiding in survival and recurrence rate predictions and tumor classification when combined with other markers [Remnant et al., 2021]. In clinical practice, Ki-67 is particularly useful when HRs are positive [Hilton et al., 2018]. The ER status is a pivotal breast cancer marker that guides classification and treatment decisions [Rakha et al., 2022]. ER-positive cases benefit from ET with a treatment choice based on ER expression levels. Postmenopausal patients often receive aromatase inhibitors (AIs) such as anastrozole, whereas premenopausal patients may use selective estrogen receptor modulators (SERMs) such as tamoxifen [Hernando et al., 2021]. Selective estrogen receptor degraders (SERDs) such as fulvestrants are also employed. Resistance, often caused by ESR1 mutations, can emerge after prolonged AI treatment. Sequential use of different ET classes and targeted therapies (e.g., CDK4/6, PI3K, and mTORC1) can be considered for ER-positive breast cancer [Brett et al., 2021].

TP53 encodes the tumor suppressor protein p53, which is essential for DNA repair, apoptosis, and cell-cycle control. To repair damaged DNA, cells with functional p53 can delay the cell cycle transition from G1 phase to S phase [Radha et al., 2014]. TP53 mutations are common in breast cancer, especially in TNBC and HER2-positive breast cancer. TP53 mutations are significant clinical biomarkers, especially in TNBC, accounting for approximately 80% of cases. Historically, tumors with TP53 mutations have been treatment-resistant. However, recent preclinical studies have discovered substances with the potential to restore the properties of wild-type p53. Compounds such as COTI-2, PRIMA-1, APR-246, PK11007, and 3-quinuclidine derivatives present new opportunities for anti-cancer treatment [Xu et al., 2021].

15.10 CHALLENGES AND FUTURE DIRECTIONS

A notable challenge within the realm of calpain-2-targeted therapy is the intricate nature of the calpain system, which poses difficulties in creating selective inhibitors that exclusively target calpain 2 without affecting other members of the calpain family. This complexity necessitates careful consideration when developing inhibitors to ensure their specificity for the intended target as well as the requirement for reliable biomarkers that can accurately predict the effectiveness of calpain-2-targeted therapies in individual patients. These biomarkers are invaluable for tailoring treatments and ensuring that they are optimally effective for each patient's specific condition [Ono et al., 2016]. Furthermore, the inherent heterogeneity of cancer presents a considerable challenge in predicting how patients respond to calpain 2-targeted therapies. This diversity in cancer types and individual patient profiles makes the accurate prediction of treatment outcomes challenging [Te Boekhorst et al., 2022]. Despite these challenges, there are numerous promising avenues for future research and innovation in the field of calpain 2-targeted therapy.

Ongoing research and development of selective calpain 2 inhibitors, which have shown promise in animal models of acute neurodegeneration, is one of these strategies. These inhibitors are currently being developed, offer a promising strategy to deal with this condition. Another promising direction is the exploration of combination therapies involving the use of two or more therapeutic agents in tandem to enhance treatment outcomes. This approach has the potential to enhance the effectiveness of calpain-2-targeted therapy significantly by addressing the complexity and diversity of diseases such as cancer and neurodegenerative disorders [Wang et al., 2018].

15.11 CONCLUSION AND CLINICAL IMPLICATIONS

In conclusion, the multifaceted involvement of calpain 2 in breast cancer research highlights its pivotal role in tumor growth, invasion, HR signaling, and prognosis. Hope exists for improving treatment outcomes for diseases such as breast cancer and other conditions, owing to the development of selective calpain 2 inhibitors and the investigation of combination therapies. However, challenges related to specificity, patient heterogeneity, and the need for biomarkers must be addressed to fully harness the therapeutic potential of calpain 2 inhibitors. The clinical implications of this study are significant. By better understanding calpain 2's role in breast cancer, we can potentially develop more effective therapeutic strategies. Selective calpain 2 inhibitors may offer new treatment options, particularly for patients with aggressive breast cancer subtypes. Combination therapies that incorporate calpain 2 inhibitors along with existing treatments can enhance the overall effectiveness of cancer therapies. Additionally, identifying and validating biomarkers associated with calpain 2 activity may enable personalized treatment approaches, ensuring that patients receive the most appropriate and effective care. Continued research and translational efforts in the field of calpain 2 and its inhibitors have the potential to significantly impact cancer treatment and other diseases, ultimately improving patient outcomes and quality of life.

ACKNOWLEDGMENT

The authors thank the Chettinad Academy of Research and Education for their support and encouragement.

DECLARATIONS

Conflict of interest: The authors declare no conflicts of interest.

Consent for publishing: All authors have approved the final version of the manuscript and given consent for publication.

Funding: Not Applicable.

REFERENCES

Abbas, T., & Dutta, A. (2009). p21 in cancer: intricate networks and multiple activities. *Nat Rev Cancer*, 9(6), 400–414. https://doi.org/10.1038/nrc2657

Austin C. P. (2021). Opportunities and challenges in translational science. *Clin Transl Sci*, 14(5), 1629–1647. https://doi.org/10.1111/cts.13055

Ayoub N. M. (2021). Editorial: Novel Combination Therapies for the Treatment of Solid Cancers. *Front Oncol*, 11, 708943. https://doi.org/10.3389/fonc.2021.708943

Baudry, M., Luo, Y. L., & Bi, X. (2023). Calpain-2 Inhibitors as Therapy for Traumatic Brain Injury. Neurotherapeutics: *NeuroTherapeutics*, 20, 1592–1602. https://doi.org/10.1007/s13311-023-01407-y

Brett, J. O., Spring, L. M., Bardia, A., & Wander, S. A. (2021). ESR1 mutation as an emerging clinical biomarker in metastatic hormone receptor-positive breast cancer. *Breast Cancer Res*: BCR, 23(1), 85. https://doi.org/10.1186/s13058-021-01462-3

Condrat, C. E., Barbu, M. G., Thompson, D. C., Dănilă, C. A., Boboc, A. E., Suciu, N., ... & Voinea, S. C. (2021). Roles and distribution of telocytes in tissue organization in health and disease. In *Tissue Barriers in Disease, Injury and Regeneration* (pp. 1–41). Elsevier. https://doi.org/10.1016/B978-0-12-818561-2.00001-1.

Cortesio, C. L., Chan, K. T., Perrin, B. J., Burton, N. O., Zhang, S., Zhang, Z. Y., & Huttenlocher, A. (2008). Calpain 2 and PTP1B function in a novel pathway with Src to regulate invadopodia dynamics and breast cancer cell invasion. *J Cell Biol*, 180(5), 957–971. https://doi.org/10.1083/jcb.200708048

Eldeeb, M. A., Fahlman, R. P., Esmaili, M., & Ragheb, M. A. (2018). Regulating Apoptosis by Degradation: The N-End Rule-Mediated Regulation of Apoptotic Proteolytic Fragments in Mammalian Cells. *Int J Mol Sci*, 19(11), 3414. https://doi.org/10.3390/ijms19113414

Ennes-Vidal, V., Dos Santos, A. L. S., Branquinha, M. H., & d'Avila-Levy, C. M. (2022). Proteolytic inhibitors as alternative medicines to treat trypanosomatid-caused diseases: experience with calpain inhibitors. *Mem Inst Oswaldo Cruz*, 117, e220017. https://doi.org/10.1590/0074-02760220017

García-Trevijano, E. R., Ortiz-Zapater, E., Gimeno, A., Viña, J. R., & Zaragozá, R. (2023). Calpains, the proteases of two faces controlling the epithelial homeostasis in mammary gland. *Front Cell Dev Biol*, 11, 1249317. https://doi.org/10.3389/fcell.2023.1249317

Hernando, C., Ortega-Morillo, B., Tapia, M., Moragón, S., Martínez, M. T., Eroles, P.,Garrido-Cano, I., Adam-Artigues, A., Lluch, A., Bermejo, B., & Cejalvo, J. M. (2021). Oral Selective Estrogen Receptor Degraders (SERDs) as a Novel Breast Cancer Therapy: Present and Future from a Clinical Perspective. *Int J Mol Sci*, 22(15), 7812. https://doi.org/10.3390/ijms22157812

Hilton, H. N., Clarke, C. L., & Graham, J. D. (2018). Estrogen and progesterone signalling in the normal breast and its implications for cancer development. *Mol Cell Endocrinol*, 466, 2–14. https://doi.org/10.1016/j.mce.2017.08.011

Ho, W. C., Pikor, L., Gao, Y., Elliott, B. E., & Greer, P. A. (2012). Calpain 2 regulates Akt-FoxO-p27(Kip1) protein signaling pathway in mammary carcinoma. *J Biol Chem*, 287(19), 15458–15465. https://doi.org/10.1074/jbc.M112.349308

Hosfield, C. M., Elce, J. S., Davies, P. L., & Jia, Z. (1999). Crystal structure of calpain reveals the structural basis for Ca(2+)-dependent protease activity and a novel mode of enzyme activation. *EMBO J*, 18(24), 6880–6889. https://doi.org/10.1093/emboj/18.24.6880

Kovacs, L., & Su, Y. (2014). The Critical Role of Calpain in Cell Proliferation, *J Biomol Res Ther*, 3(3), 1000112. https://doi.org/10.4172/2167-7956.1000112

Leloup, L., & Wells, A. (2011). Calpains as potential anti-cancer targets. *Expert Opin Ther Targets*, 15(3), 309–323. https://doi.org/10.1517/14728222.2011.553611

Łukasik, P., Załuski, M., & Gutowska, I. (2021). Cyclin-Dependent Kinases (CDK) and Their Role in Diseases Development-Review. *Int J Mol Sci*, 22(6), 2935. https://doi.org/10.3390/ijms22062935

Luo, Y., D.F. Sellitti, and K. Suzuki. (2015). The calpain proteolytic system. In *Encyclopedia of Cell Biology* (pp. 670–680). Academic Press. Accessed October 16, 2023. https://doi.org/10.1016/B978-0-12-394447-4.10075-6

MacLeod, J. A., Gao, Y., Hall, C., Muller, W. J., Gujral, T. S., & Greer, P. A. (2018). Genetic disruption of calpain-1 and calpain-2 attenuates tumorigenesis in mouse models of HER2+ breast cancer and sensitizes cancer cells to doxorubicin and lapatinib. *Oncotarget*, 9(70), 33382–33395. https://doi.org/10.18632/oncotarget.26078

Menon, S. S., Guruvayoorappan, C., Sakthivel, K. M., & Rasmi, R. R. (2019). Ki-67 protein as a tumour proliferation marker. Clinica chimica acta; *Clin Chim Acta*, 491, 39–45. https://doi.org/10.1016/j.cca.2019.01.011

Nagini S. (2017). Breast Cancer: Current Molecular Therapeutic Targets and New Players. *Anticancer Agents Med Chem*, 17(2), 152–163. https://doi.org/10.2174/1871520616666160502122724

Niessen, C. M., Leckband, D., & Yap, A. S. (2011). Tissue organization by cadherin adhesion molecules: dynamic molecular and cellular mechanisms of morphogenetic regulation. *Physiol Rev*, 91(2), 691–731. https://doi.org/10.1152/physrev.00004.2010

Ono, Y., Saido, T. C., & Sorimachi, H. (2016). Calpain research for drug discovery: challenges and potential. *Nat Rev Drug Discov*, 15(12), 854–876. https://doi.org/10.1038/nrd.2016.212

Pandi, S., Kulanthaivel, L., Subbaraj, G. K., Rajaram, S., & Subramanian, S. (2022). Screening of Potential Breast Cancer Inhibitors through Molecular Docking and Molecular Dynamics Simulation. *Biomed Res Int*, 2022, 3338549. https://doi.org/10.1155/2022/3338549

Potz, B. A., Abid, M. R., & Sellke, F. W. (2016). Role of Calpain in Pathogenesis of Human Disease Processes. *J Nat Sci.*, 2(9), e218.

Radha, R. K., P, V., & B, K. (2014). Histopathology and Prognostic Indices of Carcinoma Breast with Special Reference to p53 Marker. *J Clin Diagn Res: JCDR*, 8(7), FC04–FC8. https://doi.org/10.7860/JCDR/2014/9114.4609

Rakha, E. A., Chmielik, E., Schmitt, F. C., Tan, P. H., Quinn, C. M., & Gallagy, G. (2022). Assessment of Predictive Biomarkers in Breast Cancer: Challenges and Updates. *Pathobiology*: 9(5), 263–277. https://doi.org/10.1159/000525092

Remnant, L., Kochanova, N. Y., Reid, C., Cisneros-Soberanis, F., & Earnshaw, W. C. (2021). The intrinsically disorderly story of Ki-67. *Open Biol*, 11(8), 210120. https://doi.org/10.1098/rsob.210120

Schmucker, R., Farina, G., Faeder, J., Fröhlich, F., Saglam, A. S., & Sandholm, T. (2021). Combination treatment optimization using a pan-cancer pathway model. *PLoS Comput Biol*, 17(12), e1009689. https://doi.org/10.1371/journal.pcbi.1009689

Sever, R., & Brugge, J. S. (2015). Signal transduction in cancer. *Spring Harb Perspect Med*, 5(4), a006098. https://doi.org/10.1101/cshperspect.a006098

Storr, S. J., Thompson, N., Pu, X., Zhang, Y., & Martin, S. G. (2015). Calpain in Breast Cancer: Role in Disease Progression and Treatment Response. *Pathobiology*, 82(3-4), 133–141. https://doi.org/10.1159/000430464

Storr, S. J., Zhang, S., Perren, T., Lansdown, M., Fatayer, H., Sharma, N., Gahlaut, R., Shaaban, A., & Martin, S. G. (2016). The calpain system is associated with survival of breast cancer patients with large but operable inflammatory and non-inflammatory tumours treated with neoadjuvant chemotherapy. *Oncotarget*, 7(30), 47927–47937. https://doi.org/10.18632/oncotarget.10066

Tatarova, Z., Blumberg, D. C., Korkola, J. E., Heiser, L. M., Muschler, J. L., Schedin, P. J., Ahn, S. W., Mills, G. B., Coussens, L. M., Jonas, O., & Gray, J. W. (2022). A multiplex implantable microdevice assay identifies synergistic combinations of cancer immunotherapies and conventional drugs. *Nat Biotechnol*, 40(12), 1823–1833. https://doi.org/10.1038/s41587-022-01379-y

Te Boekhorst, V., Jiang, L., Mählen, M., Meerlo, M., Dunkel, G., Durst, F. C., Yang, Y., Levine, H., Burgering, B. M. T., & Friedl, P. (2022). Calpain-2 regulates hypoxia/HIF-induced plasticity toward amoeboid cancer cell migration and metastasis. *Curr Biol: CB*, 32(2), 412–427.e8. https://doi.org/10.1016/j.cub.2021.11.040

Wang, Y., Bi, X., & Baudry, M. (2018). Calpain-2 as a therapeutic target for acute neuronal injury. *Expert Opin Ther Targets*, 22(1), 19–29. https://doi.org/10.1080/14728222.2018.1409723

Wang, Y., Liu, Y., Bi, X., & Baudry, M. (2020). Calpain-1 and Calpain-2 in the Brain: New Evidence for a Critical Role of Calpain-2 in Neuronal Death. *Cells*, 9(12), 2698. https://doi.org/10.3390/cells9122698

Xu, Z., Wu, W., Yan, H., Hu, Y., He, Q., & Luo, P. (2021). Regulation of p53 stability as a therapeutic strategy for cancer. *Biochem Pharmacol*, 185, 114407. https://doi.org/10.1016/j.bcp.2021.114407

Yu D. (2011). Translational research: current status, challenges and future strategies. *Am J Transl Res*, 3(5), 422–433.

Zhang, S., Deen, S., Storr, S. J., Chondrou, P. S., Nicholls, H., Yao, A., Rungsakaolert, P., & Martin, S. G. (2019). Calpain system protein expression and activity in ovarian cancer. *J Cancer Res Clin Oncol*, 145(2), 345–361. https://doi.org/10.1007/s00432-018-2794-2

Zhang, W., Liu, L., Zhao, S., Chen, L., Wei, Y., Chen, W., & Ge, F. (2022). Research progress on RNA-binding proteins in breast cancer. *Oncol Lett*, 23(4), 121. https://doi.org/10.3892/ol.2022.13241

Zhang, Y., Liu, N. M., Wang, Y., Youn, J. Y., & Cai, H. (2017). Endothelial cell calpain as a critical modulator of angiogenesis. *Biochim Biophys Acta Mol Basis Dis*, 1863(6), 1326–1335. https://doi.org/10.1016/j.bbadis.2017.03.021

Section III

Epigenetic Tools and Stem Cell Therapy

16 Calpain, miRNA, and Cancer

An Overview

Sajal Chakraborti, Mahasweta Ghosh,
Dibyapriya Roy Chowdhury, and Tapati Chakraborti
University of Kalyani, Kalyani, India

Rajabrata Bhuyan
Banasthali Vidyapith, Banasthali, Tonk, India

16.1 INTRODUCTION

Calpains, belonging to the calcium-dependent neutral cysteine protease family, are an integral part of cytoskeletal remodelling, cell movement, proliferation, and gene expression. They assist in the catalysis of the selective proteolysis of a wide variety of specific substrates in response to calcium signals (Kar et al., 2007; Samanta et al., 2007; Kar et al., 2008; Kar et al., 2009; Kar et al., 2010; Samanta et al., 2010; Shaikh et al., 2010). Sixteen isoforms of the calpain genes have been characterized in humans. Some isoforms of the calpain family, including calpain-1 (μ-calpain: CAPN1) and calpain-2 (m-calpain: CAPN2) are expressed throughout the body, whilst others have a more restricted expression. For instance, calpain-8 and calpain-9 are expressed in the gastrointestinal tract and stomach mucosa (Cai et al., 2014). Several studies suggest that there is a possible involvement of abnormal calpain expression and the initiation and progression of cancer (Goll et al., 2003)

Calpains are enzymes that catalyse the proteolysis of numerous substrates involved in a broad range of cellular functions, like cell migration, invasion, and proliferation (Croall and DeMartino, 1991; Hanna et al., 2008; Zhuang et al., 2014). Calpain-4 (calpain small subunit 1, known as CAPNS1 and CAPN4), the tiny regulatory component of calpains, is essential for controlling the stability and activity of calpains (Bai et al., 2009; Storr et al., 2011).

CAPN4 plays a critical role in a number of physiological and pathological processes. It has been demonstrated that cellular dysregulation of CAPN4 is linked to carcinogenesis in various human malignancies (Bai et al., 2009; Zhang et al., 2013). Numerous cancer forms such as hepatocellular, colorectal, lung, and adrenal malignancies, have been found to express CAPN4 at high levels (Goll et al., 2003; Bai et al., 2009; Zhang et al., 2013; Zhuang et al., 2014).

Human glioma tissues displayed greater CAPN4 expression when compared to normal brain tissue. *In vitro* invasion and migration have decreased as a result of CAPN4 in glioma cells being downregulated. CAPN4 could serve as a potential target in therapeutic strategies for treating human glioblastoma because its overexpression represents a unique negative prognostic marker (Cai et al., 2014).

It was established that overexpression of CAPN4 acts as the key player in the metastasis of intrahepatic cholangiocarcinoma (ICC, Zhang et al., 2013). Bai et al. (2009) have established a link between the expression of CAPN4 and an intrusive phenotype of epithelial hepatocellular cancer. Additionally, the expression of CAPN4 was found to be negatively correlated with the overall survival (OS) of people with clear cell renal cell carcinoma (ccRCC) (Zhuang et al.,

DOI: 10.1201/9781003394693-19

2014). Also, OS and progression-free survival (PFS) were poorer in glioblastoma multiforme (GBM) patients which can be tied to the CAPN4 aberrant expression in glioma tissues (Cai et al., 2016). In schwannomas and meningiomas, abnormal expression of calpain-1 was spotted (Kimura et al., 1998). It is seen that there is overexpression of m-calpain in colorectal adenocarcinoma and endometrial cancer (Goll et al., 2003; Saez et al., 2006). Additionally, CAPN4 has been observed to be upregulated in hepatocellular carcinoma (HCC) (Bai et al., 2009) and ICC (Zhang et al., 2013).

Overexpression of CAPN4 can promote tumourigenesis, metastasis, and invasiveness of HCC that are associated with the focal adhesion kinase (FAK)-Src signalling pathways (Dai et al., 2014). The hepatitis B virus X protein's upregulation of CAPN4 aids in enhancing the HCC cells migration (Zhang et al., 2010; Zhang et al., 2012b; Li et al., 2018). Overexpression of CAPN4 is a characteristic feature of nasopharyngeal carcinoma (NPC) and lung cancer tissues, whereas there is a reduction in cancer cell invasion and metastasis when CAPN4 is silenced via downregulation of MMP-2 (Gu et al., 2015; Shi et al., 2017). According to a study, CAPN4 stimulated the growth of tumours, lowered their apoptotic death, and stimulated the Wnt/β-catenin-mediated signalling pathway, which enhanced epithelial-mesenchymal transition (EMT) in human melanoma cells (Esquela-Kerscher and Slack, 2006; Zhang et al., 2010; Rabbani et al., 2013; Wang et al., 2017a). Moreover, it has been discovered that, in cases of prostate cancer, CAPN4 is abundantly expressed and may facilitate the proliferation and invasion of cells, pointing to an oncogenic role of CAPN4 in this disease (Esquela-Kerscher and Slack, 2006; Zhang et al., 2010; Rabbani et al., 2013; Wang et al., 2017a).

In melanoma cells, the expression of β-catenin inside the nucleus is downregulated when CAPN4 is silenced (Wang et al., 2017a). Wnt/β-catenin signalling has been shown to be inhibited upon CAPN4 silencing, but CAPN4 overexpression was found to facilitate Wnt/β-catenin signalling. Therefore, CAPN4 seems to behave as a positive regulator of Wnt/β-catenin signalling (Dai et al., 2008; Rabbani et al., 2013).

In prostate cancer, abnormal Wnt/β-catenin pathway activation is linked to growth of tumours and their metastasis (Dai et al., 2008; Rabbani et al., 2013). Preclinical research has shown that inhibiting the Wnt receptor complex or preventing β-catenin-mediated activation of transcription effectively stops the spread of prostate cancer (Murillo-Garzón and Kypta, 2017). Therefore, CAPN4 might be a promising target for the creation of novel Wnt signalling inhibitors for the treatment of cancer as CAPN4 overexpression prevents β-catenin from being degraded by the ubiquitin proteasome resulting in its accumulation in cells (Cheng et al., 2018).

M-calpain has also been found to be localized to focal adhesion components that are integrin linked and can cleave the focal adhesion protein talin. So, it has been suggested that m-calpain could play a crucial role in the regulation of integrin-mediated cell adhesion (Carragher et al., 2004). Through pharmacological inhibition and RNA interference (RNAi) to restrict calpain activity, m-calpain has been observed to modulate focal adhesion turnover and cell migration in a range of tumour-derived and oncogenic transformed cells, for example, U87 GBM cells. The functional mutation analysis of the substrates of focal adhesions showed that the regulation of focal adhesion components are involved in the proteolytic cleavage of focal adhesion kinase and talin by calpain (Jang et al., 2010; Ma et al., 2012). Notably, CAPN4 has been linked to integrin-mediated cell migration, and its overexpression may contribute to the malignancy of human gliomas (Kimura et al., 1998; Dourdin et al., 2001; Cuevas et al., 2003; Franco et al., 2004; Chan et al., 2010).

16.2 MicroRNA (miRNA)

MicroRNA (miRNA), a family of endogenous short non-coding RNAs with about 23 nucleotides, controls gene expression post-transcriptionally (Cai et al., 2016). MiRNAs can suppress gene expression and negatively regulate it by binding to the 3′-untranslated region (UTR) of the target gene, for example, CAPN4 in a base-pairing manner (Bartel, 2004; Winter et al., 2009). About 30–40% of human proteins are regulated by miRNAs (Van Kouwenhove et al., 2011), indicating that miRNAs are associated with various physiological and pathological processes (Kloosterman and Plasterk, 2006). A study suggested that miRNAs could act as aberrant regulators of gene expression, and thereby elicit a pivotal role in cell proliferation and invasion (Esquela-Kerscher and Slack, 2006). MiRNAs are currently used in cancer detection, prognosis, and treatment of many types of cancer (Li et al., 2018).

16.2.1 MicroRNA-520b

Prostate cancer is the second most common cause of cancer-related mortality in men and is one of the most common tumours of the urinary system (Siegel et al., 2018). Compared to normal cells, prostate cancer cells have higher levels of CAPN4 expression. While CAPN4 overexpression had an oncogenic effect, CAPN4 gene silencing significantly reduced the proliferation, invasion, and Wnt/β-catenin signalling of prostate cancer cells. Additionally, the overexpression of CAPN4 has an oncogenic effect which was greatly reduced when β-catenin was silenced (Ren et al., 2018).

Functional studies demonstrated that CAPN4 encourages prostate cancer cells' *in vitro* proliferation, invasion, and activation of Wnt/β-catenin signalling. Notably, CAPN4 was found to be a gene that miR-520b targets. Prostate cancer tissues and cell lines exhibit decreased miR-520b expression, and miR-520b suppresses prostate cancer cell growth and invasion via targeting CAPN4. Accordingly, miR-520-b is a crucial CAPN4 regulator that supports the carcinogenic properties of prostate cancer as shown in Figure 16.1 (Ren et al., 2018).

A bioinformatics study revealed that CAPN4 is a possible target gene of miRNA-520b, which functions as a tumour suppressor miRNA in several forms of cancer. The 3′-UTR of CAPN4 is directly bound by miRNA-520b. Prostate cancer cells' expression of CAPN4 is negatively regulated

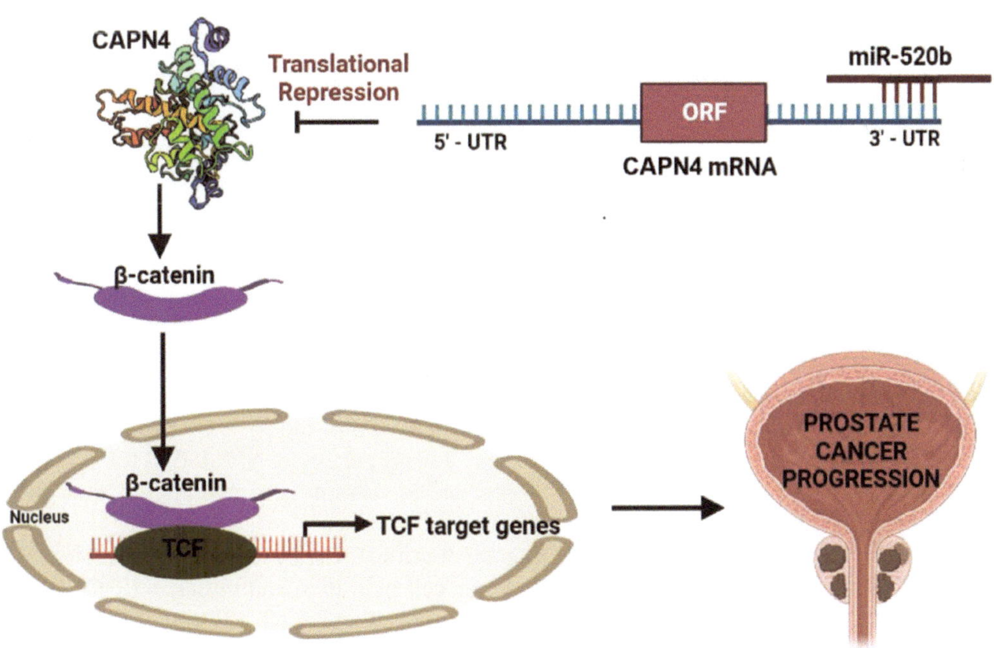

FIGURE 16.1 The miR-520b/CAPN4/Wnt/β-catenin regulatory axis in prostate cancer progression is schematically represented in this figure. The transcription of β-catenin/TCF-dependent genes is accelerated by CAPN4 overexpression, which also speeds up the accumulation of β-catenin target genes, which causes prostate cancer to spread. Also, miR-520b has the ability to bind to the CAPN4 3′-UTR and limit CAPN4 expression, preventing CAPN4 from activating Wnt/β-catenin signalling.

by miR-520b. While miR-520b inhibition had an oncogene effect, miR-520b overexpression mimicked CAPN4 siRNA's tumour-suppressive effects. Restoring CAPN4 effectively counteracted miR-520b's anti-tumour effects in prostate cancer cells. Thus, miR-520b can epigenetically control CAPN4's oncogenic involvement in prostate cancer. A significant function for the miR-520b-CAPN4-Wnt-β-catenin signalling axis in the molecular pathogenesis of prostate cancer has been demonstrated by studying the inhibition of prostate cancer cells' ability to proliferate and invade when CAPN4 is suppressed by miR-520b (Ren et al., 2018). Thus, for treatment of prostate cancer, miR-520b and CAPN4 may serve as potential and novel therapeutic targets.

As previously mentioned, miR-520b has tumour-suppressive properties in different cancers. Due to the downregulation of the oncoprotein hepatitis B X-interacting protein, ten-eleven translocation 1, mitogen-activated protein kinase 2, and cyclin D1 S1, MiR-520b operates as a new tumour suppressive gene in HCC (Zhang et al., 2012a; Zhang et al., 2014; Zhang et al., 2015). MiR-520b has also been shown to target interleukin (IL)-8 and hepatitis B X-interacting protein to prevent breast cancer cells from migrating (Hu et al., 2011). MiR-520b elicits tumour suppressive properties in a number of cancer types such as colorectal, head and neck, gastric, osteosarcoma, and glioma. It does this by targeting a number of oncogenes, such as the epidermal growth factor receptor (Li et al., 2016; Lu et al., 2017; Wang et al., 2017b; Xiao et al., 2018). Notably, in HCC and breast cancer, miR-520b downregulates CD46 and autophagy-related protein 7, making cancer cells more responsive to anti-cancer treatments (Cui et al., 2010; Sonntag et al., 2012).

16.2.2 MicroRNA-124

In healthy brain tissue, miR-124 is highly expressed and is essential for the development and operation of neurons. It also has a role in the pathophysiology of diseases of the human brain (Sonntag et al., 2012). Numerous studies demonstrated a link between the tumour grading of GBM and miR-124's downregulation (Li et al., 2011; Xia et al., 2012a; Zhao et al., 2013). These investigations demonstrated that GBMs with significant invasiveness also had decreased miR-124 expression (An et al., 2013). The expression of CAPN4 is also increased in NPC cells when miR-124 was downregulated (Shi et al., 2017).

It has been demonstrated that miR-124 can suppress CAPN4 in colorectal cancer (CRC) cells and tissues to decrease glioma cell growth and invasion. MiR-124 inhibited the growth of NPC by targeting CAPN4 (Cai et al., 2016; Hu et al., 2017; Shi et al., 2017).

MiR-124 and CAPN4 are abnormally expressed in GBM tissues. GBM is one of the most severe forms of malignant tumour, which is difficult to treat (Van Meir et al., 2010). In GBM tissues, miR-124 has been found to have a markedly reduced expression. In GBM tissues, CAPN4 expression was significantly upregulated, according to immunoblotting.

The migration and invasion of glioma cells were inhibited by miR-124 mimics or by downregulation of CAPN4. The nucleotide 440–446 region of CAPN4's 3′-UTR was also found to include a miR-124 binding site. Vimentin, MMP-2, phospho-FAK, and N-cadherin levels in U87 glioblastoma cells were decreased by miR-124 mimics or by downregulating CAPN4. CAPN4 is therefore a potential miR-124 target gene, and miR-124 inhibited *in vitro* invasion and migration of glioma cells via modifying CAPN4.

It has been noted that ROC1, a protein that plays a critical part in the control of cell motility, mediates the inhibitory action of miR-124 on glioma cell migration and invasion. ROC1 is a key catalytic subunit of the cullin-RING E3 ligase. ROC1 may affect the NF-κB signalling pathway to induce glioma cancer invasion and metastasis through ubiquitination (An et al., 2013). By contrast, Zhao et al. (2013) observed that miR-124 inhibits the invasiveness of glioma cells by targeting protein phosphatase 1 regulatory subunit 13 like (PPP1R3L), an oncoprotein that has only recently been discovered and which is also an inhibitory member of the p53 family of apoptosis-stimulating proteins (IASPP) that regulates growth, cell-cycle progression and metastasis of various types of cancer (Cai et al., 2016).

CAPN4 promotion of the metastasis of HCC cells has been shown to occur via activation of the FAK-Src signalling pathway and MMP-2 (Dai et al., 2014). CAPN4 downregulation by short hairpin RNA (shRNA) was found to be linked with lower levels of phosphor-FAK and MMP-2 (Dai et al., 2014). Similarly, the levels of phosphor-FAK and MMP-2 can also be reduced by miR-124 mimics in U87 cells, and these effects are inhibited by cotransfection with plasmid vectors expressing CAPN4. Invasion of glioma involves multiple steps that includes detachment from the tumour mass, remodelling of the extracellular matrix, and the relocation and migration of glioma cells (Cuddapah et al., 2014). Several factors, including CD44 (Pietras et al., 2014), cadherins (Barami et al., 2006), and MMPs (Könnecke and Bechmann, 2013) were shown to mediate these processes.

The treatment of cancer presently faces a significant problem due to drug resistance. However, the fundamental mechanisms of miRNA control that underlie renal cell carcinoma (RCC) resistance are not entirely known. Numerous diseases, including glioma, ovarian cancer, gastric cancer, non-small cell lung cancer, and liver cancer have been linked to the anti-cancer mechanism of miR-124 (Cui et al., 2017; Shi et al., 2017; Li et al., 2017b). It is also thought to play an oncogenic role in RCC (Gebauer et al., 2013; Adelaiye-Ogala et al., 2017).

Similarly, R-Ras is an oncogene homolog of the Ras virus and neuroblastoma. It is established that miR-124 targets N-Ras, the oncogene homolog of the Ras virus. miR-124 decreased glioma cell proliferation, invasion, and tumour growth and increased chemosensitivity during temozolomide (TMZ) treatment by negatively regulating the Ras family and its downstream signalling pathways, i.e., phosphatidylinositol-3 kinase/Akt and Raf/extracellular signal regulated kinase 1/2. Moreover, R- and N-Ras overexpression restored the inhibitory effects of miR-124. Clinical glioma specimens had higher amounts of R-Ras and N-Ras proteins, which were inversely correlated with miR-124 expression levels (Shi et al., 2014).

One of the most successful chemotherapy treatments is cisplatin. Despite decades of research into the underlying causes of cisplatin resistance, there are currently no effective pharmaceutical treatments for this type of complicated resistance (Li et al., 2017a; Liu et al., 2020a). MiRNAs have been largely reported to control cisplatin resistance in a variety of malignancies (Li et al., 2017a; Tian et al., 2017; Zeng et al., 2019), but have rarely been reported in RCC. In RCC, miR-124 was found to be markedly elevated by cisplatin. Functionally, in cases of overexpression of miR-124, the sensitivity of renal cancer cells to cisplatin was markedly reduced. CAPN4 negatively regulates CAPN4 expression and thereby acts as a direct target of miR-124. CAPN4's abnormal expression mitigated miR-124's negative effects on cisplatin sensitivity and necroptosis (Mao et al., 2021).

All eukaryotes include the multi-subunit CCR4-NOT complex, which helps to control gene expression. CCR4-NOT Transcription Complex Subunit 3 (CNOT3) controls mRNA turnover in the cytoplasm and transcription and RNA processing in the nucleus (Collart and Panasenko, 2012; Villanyi and Collart, 2015; Collart and Panasenko, 2017). Research on endometriotic stromal cells (ESCs) in the past revealed that the transcription factors CNOT3, cMYC, and ZFX cooperate to formulate a transcriptional regulatory module that can bind to a range of gene promoters in ESCs (Hu et al., 2009). CNOT3 has also been shown to have tumour-suppressing as well as tumour-promoting characteristics (De Keersmaecker et al., 2013; Cejas et al., 2017). The MAPK/ERK signalling pathway and the cell-cycle kinase Aurora B are connected by CNOT3, which functions as an adapter protein (Sarkar et al., 2021). The nuclear localization of CNOT3 and increased interaction between ERK and CNOT3 are promoted by Aurora B phosphorylation of the protein at sites near a nuclear localization sequence (NLS), which appears to be the plausible mechanism for the tumour-promoting effects of CNOT3 (Mao et al., 2021; Sarkar et al., 2021). CAPN4 activates and promotes RCC cell proliferation via the FAK signalling pathway (Shen et al., 2017). MiR-124 targets the CAPN4 3′-UTR to prevent the enzyme from being expressed. CAPN4 could directly interact with CNOT3 and encourage its breakdown, and upregulating CNOT3 could reverse the cisplatin resistance phenotype. It is interesting to note that cisplatin-increased miR-124 prevents CNOT3 from being degraded by CAPN4 in RCC cells, which reduces necroptosis

(Mao et al., 2021). As a result, miR-124 has become a significant inhibitor of cisplatin-induced necroptosis, and the miR-124-CAPN4-CNOT3 signalling axis is essential for the development of cisplatin resistance.

miR-124 also acts as a tumour suppressor in gastric cancer (Xia et al., 2012b) and HCC (Furuta et al., 2010). Moreover, miR-124 expression was downregulated in NPC and overexpression of miR-124 inhibited proliferation, migration, and invasion of NPC cells (Peng et al., 2014; Xu et al., 2016). However, how miR-124 expression was regulated in NPC remains unclear.

Increased levels of lncRNAs appear to function as an oncogene or tumour suppressor in many different forms of cancer. Among these lncRNAs, metastasis-associated lung adenocarcinoma transcript 1 (MALAT1), which acts as an oncogene, is significantly expressed in a number of malignancies (Gutschner et al., 2013). For instance, increased MALAT1 expression significantly accelerates the development and invasion of cervical cancer cells that are HR-HPV positive (Liu et al., 2016b). Adenosine monophosphate-activated (PRKA) kinase anchor protein 9 is involved in the upregulation of MALAT1 which results in the proliferation, migration, and invasion of CRC cells (Yang et al., 2015b). MALAT1 promoted brain metastasis of cancer cells in non-small cell lung cancer by triggering EMT (Shen et al., 2015).

In the pathophysiology of NPC, lncRNAs have been postulated to function as competing endogenous RNAs (ceRNA) for post-transcriptional control. By competitively binding to miRNAs, CeRNA shields miRNAs from their target mRNAs or depresses the expression of those mRNAs (Yoon et al., 2014).

Given that MALAT1 contains two miR-124 target sites and that miR-124 and MALAT1 expression levels in NPC cells were opposite, MALAT1 may behave as miRNA sponge that interacts with miR-124. While MALAT1 upregulation decreased miR-124 expression, MALAT1 knockdown increased it. Additionally, the inhibition of miR-124 on NPC cell proliferation, invasion, and EMT has been observed to be reversed by MALAT1 overexpression. MALAT1 may encourage NPC cell proliferation, invasion, and EMT by sponging miR-124 to inhibit CAPN4. Therefore, it appears that the MALTA/miR-124/CAPN4 axis is involved in the pathogenesis of NPC (Shi et al., 2017).

The CAPN4 gene has been shown to causes cancer in a variety of malignancies. CAPN4 downregulation by miR-124 inhibited the invasion and migration of glioma cells (Kimura et al., 1998). CAPN4 enhances the invasion ability of non-small cell lung cancer cells by upregulating the expression of MMP-2 (Gu et al., 2015).

16.2.3 miR-1271

miR-1271 may operate as a tumour suppressive miRNA in a variety of malignancies, including gastric cancer (Yang et al., 2014; Xiang et al., 2015b), ovarian cancer (Liu et al., 2015a), lung cancer (Zhou et al., 2015), pancreatic cancer (Liu et al., 2016a), and prostate cancer (Zhong et al., 2017) according to the data collected from various studies. These investigations have shown that miR-1271 suppresses cancer growth by targeting a number of oncogenes. Uncertainty exists on whether miR-1271 targets and controls CAPN4 expression.

One of the most common cancers and the main reason for cancer-related deaths worldwide is CRC (Brenner et al., 2014; Siegel et al., 2016). Numerous elements, including genetic and epigenetic changes, have a role in the pathogenesis of CRC (De Sousa E Melo et al., 2013). The function of CAPN4 in CRC is poorly understood. In CRC cells, CAPN4 expression was markedly elevated. SiRNA-mediated CAPN4 knockdown drastically reduced the ability of CRC cell lines to proliferate and invade. Additionally, CAPN4 knockdown reduced Wnt signalling in CRC cells. Interestingly, the tumour-suppressive miRNA miR-1271 was shown to have CAPN4 as a target gene as depicted in Figure 16.2. CAPN4 expression in CRC cells is adversely regulated by an epigenetic change caused by miR-1271. Thus, miR-1271 reduces CAPN4 to prevent CRC cells from proliferating and invasively spreading (Li et al., 2018).

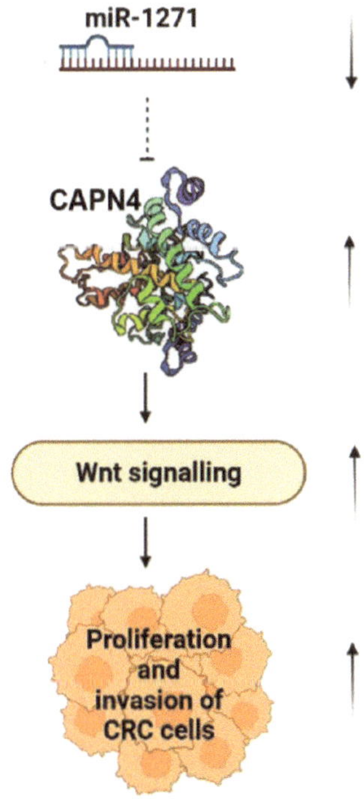

FIGURE 16.2 The schematic representation of how miR-1271/CAPN4/Wnt signalling controls the growth and invasion of colorectal cancer (CRC) cells. The increased CAPN4 expression and activation of Wnt signalling in CRC, which results in higher proliferation and invasion of CRC cells, may be caused by the decreased miR-1271 expression.

16.2.4 miR-585-3p

OC contributes the most in human fatality rates of all gynaecological tumours (Jayson et al., 2014). OC manifests a diverse range of histopathological, clinical, and molecular heterogeneity. Nearly 70% of OC leads to high-grade serous ovarian cancer (HGSOC) subtypes (Berns and Bowtell, 2012). Amidst these, 80% of the cases are detected at advanced stages, mainly because the clinical symptoms evinced are generally non-specific and the non-availability of reliable biomarkers that can be detected at early stages (Jayson et al., 2014).

Downregulation of miR-585-3p has been shown to promote cell development and growth in colon cancer upon dysregulation of proteasome activator subunit 3 (PSME3) (Liu et al., 2019). By specifically targeting MDM2, microRNA-585 reduces the growth of human glioma cells (Chen et al., 2020). In gastric cancer, miRNA that directly targets MAPK1 inhibits the growth and migration of the tumour (Hu et al., 2018).

The expression of calpain-9 (CAPN9), a calpain family member, was once believed to be reliant on the tissues of the digestive tract. In gastric cancer, there is lower amount of CAPN9 gene expression (Peng et al., 2016). This belief was proved wrong when CAPN9 expression was reflected in invasive breast cancer, hence showing that it is not expressed solely in a digestive-tract-specific manner (Davis et al., 2014). CAPN9 has been observed to be expressed at considerably elevated levels in tissues and cells of tumorous growths and in normal ovarian tissues and cells.

Aberrant expression of miRNA-585-3p leads to the development and progression of HGSOC. The expression of miR-5850-3p was found to be considerably lower in HGSOC tissues and cells than that in the normal ovarian tissues and cells. In HGSOC tissues, expression of CAPN9 has been observed to be contrariwise associated with miR-585-3p expression. miR-585-3p suppresses CAPN9 expression by attaching to the 3′-UTR of the gene. When CAPN9 was overexpressed, miR-585-3p's inhibitory effect on HGSOC cells was observed to be lessened. Lnc01436 accelerates the development of gastric cancer as it upregulates MAPK1 expression and also inhibits miR-585-3p expression (Lu et al., 2021; Xu et al., 2021).

The FBXO11 gene in humans encodes a protein known as F-box only protein 11. The SKP1-cullin-F-box (SCF) complex, which is in charge of the ubiquitination of the substrates and its subsequent degradation of that further contributes to the preservation of genome stability, is encoded by FBXO11 (also known as VIT1 and PRMT9) (Li et al., 2018). Lnc01436, an oncogenic long non-coding RNA, controls the expression of miR-585 and FBOX11 and encourages the growth and spread of gastric cancer cells (Zhang et al., 2020).

Currently, it is unknown how miR-585/CAPN9 and lncRNAs are related. miR-585-3p promoted HGSOC migration and proliferation by directly acting on the target gene CAPN9. Compared to healthy ovarian tissues and cell lines, miR-585-3p expression was found to be decreased OC tissue and cell lines. HGSOC's growth and migration were diminished when miR-585-3p was downregulated. Treatment with a specific focus on miR-585/CAPN9 to encourage HGSOC migration and proliferation is in review. The therapeutic success of this targeted treatment might prove to be beneficial for HGSOC patients. (Lu et al., 2021).

16.2.5 miR-203

HCC is a high mortality disease and, according to literature, it has been identified as the third largest cause of cancer-related death globally, after stomach and oesophageal malignancies (Fan et al., 1995; Jemal et al., 2011; Zheng et al., 2017). The most frequent current therapeutic treatment for HCC is liver resection, because liver regeneration encourages HCC growth in the regenerated liver. The consequent aftereffect of liver resection for HCC patients is the high rate of HCC recurrence/metastasis, which ultimately leads to death in majority of cases. Diethyl nitrosamine was used to induce HCC in a rat model of liver cirrhosis to acquire insight into the aforementioned clinicopathology (Zheng et al., 2017). After partial (30%) hepatectomy (PH), recombinant miR-203 adenovirus was used to promote hepatic miR-203 overexpression. Through inhibiting the expression of CAPN4, miR-203 was found to significantly reduce the proliferation, invasion, and metastasis of the residual HCC, ultimately leading to lung metastases. Aside from that, inhibiting IL-1, Snail1, and Twist1, miR-203 overexpression prevented the EMT brought on by hepatectomy. Due to the increased liver regeneration after PH, miR-203 overexpression prevented the heightened proliferation and lung metastasis of the residual HCC in part through controlling the EMT. (Zheng et al., 2017).

miRNA serves as a tumour promoter or suppressor, regulating a wide range of biologic processes such as proliferation, invasion, and metastasis. Of significance was the observation that miR-203 could possibly act as a target gene product that suppresses tumours in HCC (Chen et al., 2012; Wei et al., 2013; Peng et al., 2014; Liu et al., 2015b; Yang et al., 2015a; Wang et al., 2016). Higher miR-203 expression may indicate a better prognosis, according to research on the clinical consequences and biological significance of aberrant miR-203 expression in HCC (Chen et al., 2012; Liu et al., 2015b). In HCC cells, miR-203 overexpression inhibited growth and metastasis of HCC by targeting surviving EZH2, Bmi-1, and ATP binding cassette E1 (Chen et al., 2012; Wei et al., 2013; Yang et al., 2015a).

The prototypical anti-inflammatory molecule A20 has been shown to be connected to a number of illnesses in humans, including cancer. A20 functions as an oncogene in breast malignancies, gastric cancers, and melanomas, but performs anti-tumour roles in CRC and HCC (Shi et al., 2021). da Silva et al. (2013) demonstrated that miR-203 could function at some point in the A20 enhanced IL-6/STAT3 pro-proliferative signals by downregulating suppressors of cytokine signalling (SOCS) proteins, which serve as feedback inhibitors of the JAK/STAT signalling pathway and could promote liver regeneration and repair after PH. By targeting SOCS3 and then boosting IL-6/STAT3 signalling following partial hepatectomy (~70% PH) in an animal model of CCl4-mediated liver cirrhosis, miR-203 overexpression may directly accelerate the beginning of cirrhotic liver regeneration.

By upregulating the expression of Ki67, a protein that, in humans, is encoded by the MKI67 gene, a well-known proliferation marker for the evaluation of cell proliferation, and by enhancing IL-5/SOCS3/STAT3 proliferative signals, miR-203 overexpression may promote the regeneration of the non-tumorous liver. Importantly, miR-203 overexpression suppresses the expression of Ki67 and CAPN4 and lung metastasis, which significantly reduces the proliferation, invasion, and metastasis of hepatic residual HCC (Zheng et al., 2017).

Furthermore, it has been reported that by targeting IL 1, Snail1, and Twist1, miR-203 overexpression prevented the EMT caused by hepatectomy. Possibly by controlling EMT, miR-203 overexpression prevents the increased proliferation and lung metastasis of the remaining HCC brought on by the increase in liver regeneration after PH. After liver resection, miR-203 may both enhance liver regeneration and limit the increased proliferation and spread of remaining HCC (Zheng et al., 2017).

16.2.6 miR-142-5p

The KCNQ1OT1 (604115)—potassium voltage gated channel subfamily Q member 1 opposite strand/antisense transcript 1—a long non-coding RNA found at chromosome 11p15.5, has been proven to promote the expression of colon and rectal cancer (Zhang et al., 2019; Liu et al., 2020b).

In individuals with colon adenocarcinoma, high KCNQ1OT1 was found to be a predictor of worse OS and recurrence-free survival (Zhang et al., 2019; Liu et al., 2020b). Also, the response of cancer cells to chemo-reagents could be regulated by KCNQ1OT1. For example, KCNQ1OT1 was found to be overexpressed in methotrexate-resistant CRC cells (Xian and Zhao, 2019; Liu et al., 2020b). Also, It has also been was observed that by targeting protein phosphatase 1 regulatory subunit 1B through sponging miR-760, KCNQ1OT1 has the ability to affect the chemo sensitivity of CRC cells (Xian and Zhao, 2019; Liu et al., 2020b). In colon cancer, KCNQ1OT1 expression was found to be elevated and that could be correlated with poor overall cancer survivors (Li et al., 2019).

In vitro studies have shown that KCNQ1OT1 may encourage OC cell migration and proliferation. Furthermore, miR-142 expression was observed to be downregulated, whilst CAPN10 expression was upregulated in OC cells compared to the normal cells. Low miR-142-5p or high CAPN10 expression were markers known to modulate OC patients' survival. KCNQ1OT1 could control OC growth through the miR-142-5p-CAPN10 signalling axis. KCNQ1OT1, therefore, stimulates miR-142-5p by upregulating CAPN10 expression, causing OC to proliferate and migrate (Liu et al., 2020b).

Examples of RNA molecules that are essential for several physiological activities, such as metastasis, drug resistance and cell proliferation are LncRNA and miRNAs (Li and Zhan, 2019). According to research, KCNQ1OT1 expression levels in OC cells was shown to be greater than in normal cells, suggesting oncogenic function of KCNQ1OT1 in OC (Liu et al., 2020b).

In vitro studies have revealed that KCNQ1OT1 promotes OC cell migration and proliferation. In addition, miR-142 expression was reduced and CAPN10 expression was increased in OC cells compared to normal cells. As a result, lower OS rates for OC patients were indicated by low miR-142-5p or high CAPN10 expression. OC development could be regulated by KCNQ1OT1 via the miR-142-5p/CAPN10 axis. Thus, KCNQ1OT1 upregulates CAPN10 expression via sponging miR-142-5p, thereby inducing the proliferation and migration of OC (Liu et al., 2020b).

A bioinformatics research revealed that miR-142-5p seems to be a potential KCNQ1OT1 target. Reduced miR-142-5p expression was discovered in the tissues and cells of non-small cell lung cancer (Wang et al., 2017c). *In vitro* and *in vivo*, overexpression of miR-142-5p inhibited the advancement of non-small cell lung cancer via controlling the catalytic subunit of phosphatidylinositol 4,5-bisphosphate 3 kinase (Wang et al., 2017c). Furthermore, miR-142-5p may act as a link between KCNQ1OT1 and cyclin-dependent kinase 5, which control lung adenocarcinoma cell invasion, migration, and EMT (Jia et al., 2019). miR-142-5p expression was shown to be downregulated in OC cells and was associated with a bad prognosis for people with OC. The association between miR-142-5p and KCNQ10T1 was detected using a reporter assay for luciferase activity. Furthermore, it was discovered that miR-142-5p and CAPN10 do their many functions upon association. CAPN10 was discovered to be significantly expressed in OC cells and demonstrated a significant decline in cancer patient survival. The oncogene amplified in oesophageal cancer 1 (GAEC1), which promotes the growth of tumours, was found to influence the expression of CAPN10, a member of the mitochondrial calpain system (Chan et al., 2013).

To influence the development of OC, miR-142-5p is sponged by KCNQ10T1 and targets CAPN10. The clinical importance of KCNQ10T1/miR-142-5p/CAPN10 in OC requires additional research (Liu et al., 2020b).

16.2.7 miR-181, miR-125, and miR-143

Osteosarcoma accounts for 5% of childhood cancers and the mortality rate is about 9% of all deaths in children that are linked to cancer (Yang and Zhang, 2013). Therefore, it is of utmost importance that novel approaches are taken to treat patients suffering from osteosarcoma (Jones et al., 2012). High expression of miR-181a, miR-181b, and miR-181c are part of the miRNA profile linked with osteosarcoma. Particularly, miR-181b expression has been found to be confined to specific cell types in cancer tissues, and patients who acquire metastatic illness had high levels of miR-181c expression in biopsy samples (Yang and Zhang, 2013). Osteosarcoma samples and cell lines usually have lower levels of miR-125b (Liu et al., 2011). Through STAT3 downregulation, miR-125b reduced osteosarcoma cell proliferation and migration, indicating that ectopic restoration of miR-125b expression may be a promising therapeutic strategy. Additionally, the downregulation of miR-143 encourages cellular invasion, conceivably through m-calpain and consequent MMP-13 overexpression, and are associated with osteosarcoma lung metastasis (Osaki et al., 2011).

16.2.8 miRNA 337-3p

The most prevalent subtype of renal cancer is ccRCC. It has been proposed that miRNAs could assist as potential biomarkers for cancer diagnosis, prognosis, and tumour management (Zhuang et al., 2018b).

It was discovered that miR-337-3p regulates CAPN4 negatively and that overexpressing miR-337-3p reduced its effects on ccRCC cellular activity. In ccRCC, miR-337-3p prevents the transition of the epithelium to the mesenchyme. When there was overexpression of miR-337-3p, the growth of the tumour *in vivo* was significantly inhibited. Data from clinical samples revealed that miR-337-3p was downregulated and CAPN4 mRNA protein was upregulated in ccRCC tissues. The expression levels of miR-337-3p and CAPN4 mRNA were also inversely associated in ccRCC tissues (Zhuang et al., 2018b).

It has been discovered that miR-337-3p directly targets CAPN4. CAPN4 plays a crucial role in tumour metastasis. CAPN4 was discovered to be upregulated in ICC and ccRCC and to have a positive correlation with lymphatic metastasis and the stage of tumour node metastasis. The talin-focal adhesion kinase signalling pathway was modulated by CAPN4 which seems to promote tumour invasion and metastasis in ccRCC cells (Zhuang et al., 2018a). EMT in ccRCC was inhibited by miR-337-3p. Importantly, the overexpression of miR-337-3p resulted in the downregulation of CAPN4 mRNA expression in mouse cancer tissues. The level of miR-337 expression in ccRCC tissues was inversely linked with the degree of CAPN4 mRNA expression in clinical samples. Together, these findings suggested that miR-337-3p may have reduced the development of ccRCC by targeting CAPN4 to have a better understanding the pathophysiology of ccRCC. Therefore, it is important to further investigate other miR-337-3p targets in the ccRCC development as miR-337-3p could target other genes that govern ccRCC advancement (Xiang et al., 2015a).

16.2.9 MicroRNA 122

In advanced human malignancies, a discernible decrease in skeletal muscle mass and reduced muscular strength have been observed (Tkach and Théry, 2016; Aweida et al., 2018). Breast cancer has been shown to inhibit the modification of O-linked N-acetylglucosamine (O-GlcNAc) protein in muscle by targeting O-GlcNAc transferase (OGT) in extracellular vesicle-encapsulated miR-122. It has been demonstrated that the ryanodine receptor-1 (RyR1) protein expression increases noticeably in response to the miR-122-mediated reduction in OGT (Ruan et al., 2017; Zou et al., 2021). Exercise has been shown to increase O-GlcNAcylation of muscle proteins, which is known to be regulated by hypoxia and lactate through hypoxia inducing factor-1A (HIF1A)-dependent OGT promoter activation (Favier et al., 2015; Yan et al., 2022). Through an increase in (RyR1), suppression of O-GlcNcylation in the context of cancer caused greater calpain protease activation along with increase in cytosolic calcium levels, which resulted in the cleavage of desmin filaments and disintegration of myofibrils. In tumour-bearing mice, this was linked to a marked decrease in the skeletal muscle mass and contractility (Cai et al., 2004; Favier et al., 2015; Cohen, 2020; Yan et al., 2022). Thus, O-GlcNAccylation seems to be linked to homeostasis and contractility of muscle proteins, which suggests a prominent mechanism exists with the dysregulation of muscles linked to cancer.

Our bioinformatics survey on the above-mentioned miRNA interactions with calpain were performed using the miRANDA tool and the results are given in Tables 16.1 and 16.2.

TABLE 16.1
Interaction of miRNAs with CAPN4 Gene Leading to Cancer

miRNAs	Cancer Types	miRANDA Score	Interaction Energy
miR-520b	Prostate cancer	161	−18.51
miR-1271-3p	Colon cancer	160	−24.72
miR-491	Colon cancer	152	−22.15
miR-203	Hepatocellular carcinoma (HCC)	148	−20.13
miR-181a	Human glioma	162	−16.35
miR-181b	Human glioma	154	−16.26
miR-16	Human glioma	154	−18.28
miR-29b	Human glioma	154	−22.41

Note: *Score Score threshold = 130, Minimum free energy (MFE) threshold = (−15.00 kcal/mol)

TABLE 16.2
Interaction of miRNAs with CAPN9, CAPN10, and m-calpain Leading to Cancer

miRNAs	Gene Name	Cancer Types	miRANDA Score	Interaction Energy
miR-585-3p	CAPN9	High-grade serous ovarian cancer (HGSOC)	141	−22.95
miR-142-5p	CAPN10	Ovarian cancer (OC)	143	−15.8
miR-125b	m-calpain	Osteosarcoma	152	−21.34
miR-143	m-calpain	Osteosarcoma	179	−30.99

Note: *Score threshold = 130; Minimum free energy (MFE) threshold (=−15.00 kcal/mol)

The list of genes regulated by those miRNAs were identified by the TargetScan (McGeary et al., 2019) database and pathway enrichments were done using the KEGG tool under the DAVID web server (Sherman et al., 2022). The significant pathways activated in the cell due to the regulation of miRNAs are given in Figures 16.3 and 16.4, suggesting the leading cancer related pathways.

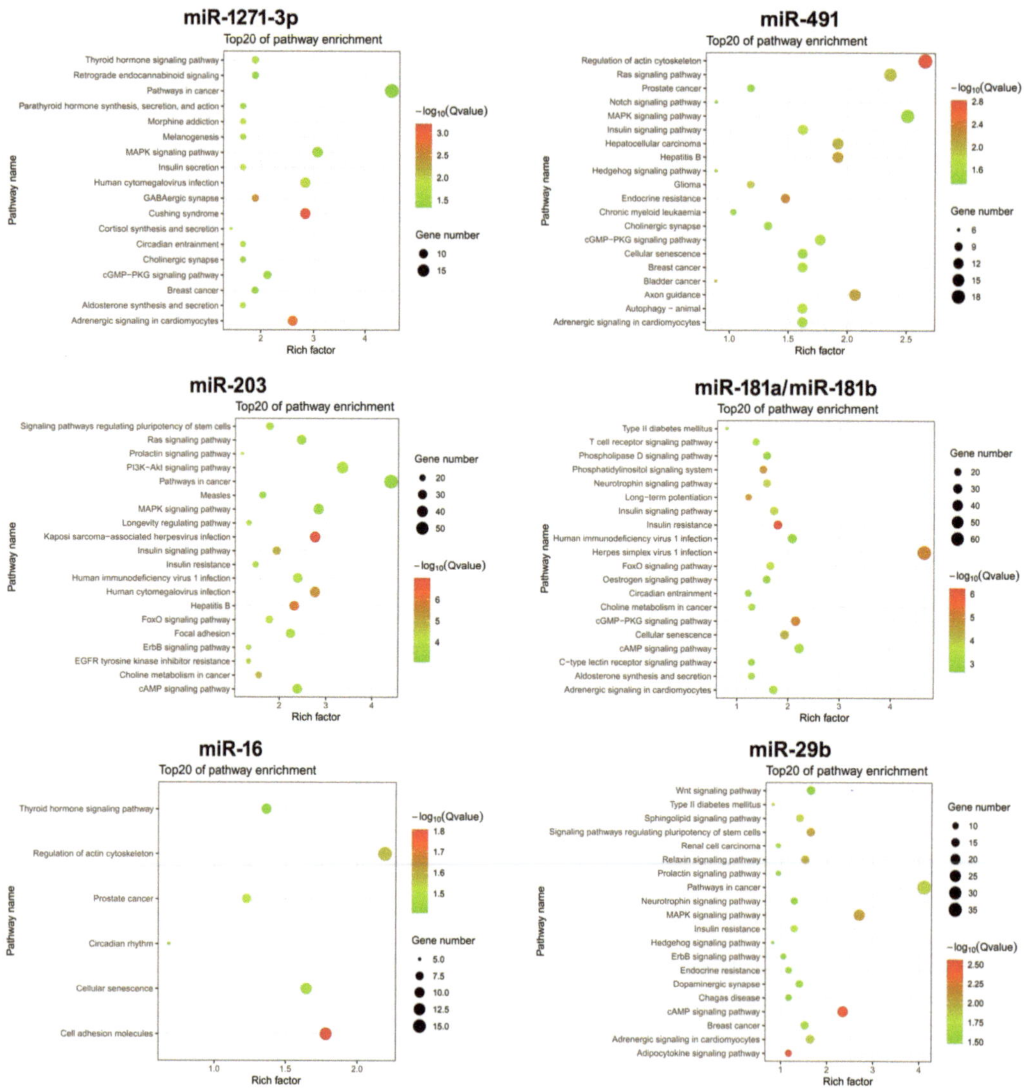

FIGURE 16.3 Pathway enrichment due to interaction of miRNAs with CAPN4.

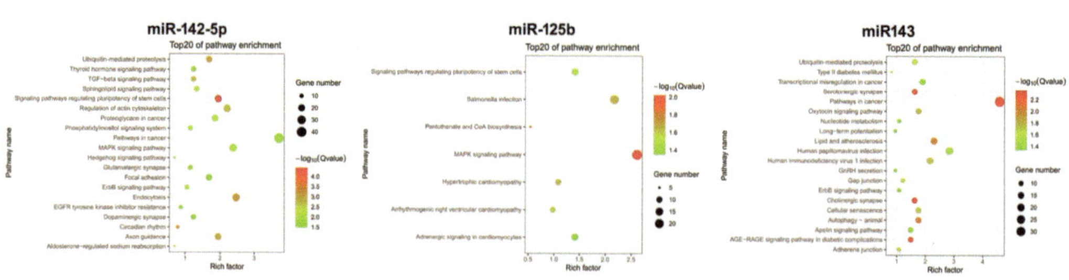

FIGURE 16.4 Pathway enrichment due to interaction of miRNAs with CAPN9, CAPN10, and m-calpain.

16.3 CONCLUSION AND FUTURE PERSPECTIVES

While studying *Caenorhabditis elegans*, Victor Ambros and Gary Ruvkun discovered a correlation between the amount of LIN14 protein and 22-nucleotide RNA, expressed by the LIN-4 gene, which is important for *C. elegans* development. The discovery that tiny RNA molecules (with a length of 19–25 nucleotides) post-transcriptionally silence target mRNA altered our understanding of the mechanism regulating the expression of genetic material (Lee et al., 1993; Smolarz et al., 2022). The relationship between miRNA molecules and cancer was revealed in 2005 (Mitchell et al., 2008; Smolarz et al., 2022). The miRNAs are divided into two groups based on their role in the growth of tumours: (i) suppressor miRNAs, which prevent the expression of oncogenes or genes that cause apoptosis; and (ii) oncogenic miRNAs, which either activate oncogenes or prevent the expression of suppressor genes (Smolarz et al., 2022).

Interest in miRNA molecules has grown since they were discovered. Almost all diseases exhibit dysregulation of the expression of miRNAs, which are prevalent in tissues and blood and play a critical role in pathological processes. It is very promising to use circulating miRNAs as non-invasive biomarkers, because they are present and stable in readily accessible blood and sputum. Based on initial research and current outcomes, miRNA molecules have the potential to assist in the creation of early cancer detection tests and greatly enhance the standard of care for cancer patients by evaluating risk, which is essential in the era of personalized medicine (Mitchell et al., 2008).

The miRNAs have the potential to function as tumour suppressors or oncogenes depending on the intracellular conditions. The dysregulated miRNAs have been connected to the features of cancer such as maintenance of proliferative signalling, evading growth inhibitors, resisting cell death, initiating invasion and metastasis, and inducing angiogenesis.

Expression of genes necessary for tumour progression, such as anti-apoptotic proteins or transcription factors, is boosted when suppressor miRNAs are expressed less or not at all. The process of DNA methylation is crucial in controlling the expression of suppressor miRNAs in cancer cells (Piao et al., 2017). Hyper methylation of miRNAs (miR-34, miR-342, miR-345, miR-9, miR-129, miR-137) play major roles in the development of CRC as hypermethylation reduces their expression in cells. Reduced expression of miR-143 in CRC cells results in a marked increase in the activity of methyltransferase 3A DNA (DNMT3A) and increased cancer cell proliferation (Wang et al., 2021). It is important to emphasize that DNA methylation is necessary for miRNA production, and that miRNA expression can influence the activity of epigenetic regulators such DNA methyltransferases and histone deacetylases.

CAPN4 may accelerate prostate cancer cell growth and invasion via the Wnt-β-catenin pathway. Notably, miR-520b functions as a tumour suppressor by targeting the 3′UTR of prostate cancer cells, hence inhibiting the synthesis of CAPN4. The results of this study provide credence to the theory that the miR-520b-CAPN4-Wnt-catenin signalling axis could play a significant role in the advancement of prostate cancer. Additionally, they enhance our comprehension of the molecular mechanism responsible for the constitutive activation of Wnt-β-catenin signalling in the course of prostate cancer development and progression. The miR-520b and CAPN4 have been proposed as potential therapeutic targets for the development of prostate cancer prevention techniques.

CRC progression is critically dependent on CAPN4. The proliferation, invasion and Wnt signalling of CRC cells were all suppressed when CAPN4 was knocked down, indicating an oncogenic role for CAPN4 in CRC. Furthermore, it was discovered that miR-1271 regulates CAPN4 expression in CRC. The epigenetic modulation of CAPN4 caused by miR-1271 may be crucial in the pathogenesis of CRC. As a result, miR-1271's inhibition of CAPN4 may represent a promising therapeutic approach for the management of CRC. The miR-124-CAPN4 signalling axis may be important in both glioma cell invasion and NPC. In ccRCC, miR-337-3p reduced *in vitro* cell proliferation, metastasis, and tumour growth. Additionally, mechanistic research showed that CAPN4 served as a mediator, at least partly, of miR-337-3p's impacts on ccRCC cellular activities.

Compared to healthy ovarian tissues and cell lines, miR-585-3p expression was decreased in OC tissues and cell lines. HGSOC could be prevented by miR-585. The miR-585-3p has been shown to influence the target gene CAPN9 to encourage HGSOC migration and proliferation. So, for ovarian, nasopharyngeal, colorectal, and prostate cancers, targeted treatment based on miR-5853p-CAPN9, miR-124-CAPN4, miR-1271-CAPN4, miR-203, miR-337-3p-CAPN4, and miR-520-b-CAPN4 offers a promising therapeutic target.

The miRNAs have an impact on the cancer progression that seems crucial for the proper operation of the organism. Numerous malignancies have been linked to altered miRNA expression, which may be a sign of the molecules suppressive or oncogenic potential. In this chapter, we have highlighted some mechanisms as well as applications of miRNA as diagnostic markers and potential cancer treatment targets.

DECLARATIONS

Conflict of interest: The authors declare no conflicts of interest.

Consent for publishing: All authors have approved the final version of the manuscript and given consent for publication.

Funding: Not Applicable.

REFERENCES

Adelaiye-Ogala, R., Budka, J., Damayanti, N. P., Arrington, J., Ferris, M., Hsu, C. C., Chintala, S., Orillion, A., Miles, K. M., Shen, L., Elbanna, M., Ciamporcero, E., Arisa, S., Pettazzoni, P., Draetta, G. F., Seshadri, M., Hancock, B., Radovich, M., Kota, J., Buck, M., … Pili, R. (2017). EZH2 modifies sunitinib resistance in renal cell carcinoma by kinome reprogramming. *Cancer Research*, *77*(23), 6651–6666.

An, L., Liu, Y., Wu, A., & Guan, Y. (2013). microRNA-124 inhibits migration and invasion by down-regulating ROCK1 in glioma. *PLoS One*, *8*(7), e69478.

Aweida, D., Rudesky, I., Volodin, A., Shimko, E., & Cohen, S. (2018). GSK3-β promotes calpain-1-mediated desmin filament depolymerization and myofibril loss in atrophy. *The Journal of Cell Biology*, *217*(10), 3698–3714.

Bai, D. S., Dai, Z., Zhou, J., Liu, Y. K., Qiu, S. J., Tan, C. J., Shi, Y. H., Huang, C., Wang, Z., He, Y. F., & Fan, J. (2009). CAPN4 overexpression underlies tumor invasion and metastasis after liver transplantation for hepatocellular carcinoma. *Hepatology (Baltimore, Md.)*, *49*(2), 460–470.

Barami, K., Lewis-Tuffin, L., & Anastasiadis, P. Z. (2006). The role of cadherins and catenins in gliomagenesis. *Neurosurgical Focus*, *21*(4), E13.

Bartel, D. P. (2004). MicroRNAs: Genomics, biogenesis, mechanism, and function. *Cell*, *116*(2), 281–297.

Berns, E. M., & Bowtell, D. D. (2012). The changing view of high-grade serous ovarian cancer. *Cancer Research*, *72*(11), 2701–2704.

Brenner, H., Kloor, M., & Pox, C. P. (2014). Colorectal cancer. *Lancet (London, England)*, *383*(9927), 1490–1502.

Cai, D., Frantz, J. D., Tawa, N. E., Jr, Melendez, P. A., Oh, B. C., Lidov, H. G., Hasselgren, P. O., Frontera, W. R., Lee, J., Glass, D. J., & Shoelson, S. E. (2004). IKKbeta/NF-kappaB activation causes severe muscle wasting in mice. *Cell*, *119*(2), 285–298.

Cai, J. J., Qi, Z. X., Chen, L. C., Yao, Y., Gong, Y., & Mao, Y. (2016). miR-124 suppresses the migration and invasion of glioma cells *in vitro* via CAPN4. *Oncology Reports*, *35*(1), 284–290.

Cai, J. J., Qi, Z. X., Hua, W., Zhu, J. J., Zhang, X., Yao, Y., & Mao, Y. (2014). Increased expression of CAPN4 is associated with the malignancy of human glioma. *CNS Neuroscience & Therapeutics*, *20*(6), 521–527.

Carragher, N. O., Fonseca, B. D., & Frame, M. C. (2004). Calpain activity is generally elevated during transformation but has oncogene-specific biological functions. *Neoplasia (New York, N.Y.)*, *6*(1), 53–73.

Cejas, P., Cavazza, A., Yandava, C. N., Moreno, V., Horst, D., Moreno-Rubio, J., Burgos, E., Mendiola, M., Taing, L., Goel, A., Feliu, J., & Shivdasani, R. A. (2017). Transcriptional regulator CNOT3 defines an aggressive colorectal cancer subtype. *Cancer Research*, *77*(3), 766–779.

Chan, D., Tsoi, M. Y., Liu, C. D., Chan, S. H., Law, S. Y., Chan, K. W., Chan, Y. P., Gopalan, V., Lam, A. K., & Tang, J. C. (2013). Oncogene GAEC1 regulates CAPN10 expression which predicts survival in esophageal squamous cell carcinoma. *World Journal of Gastroenterology*, *19*(18), 2772–2780.

Chan, K. T., Bennin, D. A., & Huttenlocher, A. (2010). Regulation of adhesion dynamics by calpain-mediated proteolysis of focal adhesion kinase (FAK). *The Journal of Biological Chemistry, 285*(15), 11418–11426.

Cheng, F., Mohanmed, M. M., Li, Z., Zhu, L., Zhang, Q., Huang, Q., & Shen, W. (2018). CAPN4 promotes colorectal cancer cell proliferation by increasing MAPK7 through activation of the Wnt/β-Catenin pathway. *Experimental Cell Research, 363*(2), 235–242.

Chen, H. Y., Han, Z. B., Fan, J. W., Xia, J., Wu, J. Y., Qiu, G. Q., Tang, H. M., & Peng, Z. H. (2012). miR-203 expression predicts outcome after liver transplantation for hepatocellular carcinoma in cirrhotic liver. *Medical Oncology (Northwood, London, England), 29*(3), 1859–1865.

Chen, W., Hong, L., Hou, C., Wang, Y., Wang, F., & Zhang, J. (2020). MicroRNA-585 inhibits human glioma cell proliferation by directly targeting MDM2. *Cancer Cell International, 20*, 469.

Cohen, S. (2020). Role of calpains in promoting desmin filaments depolymerization and muscle atrophy. *Biochimica et Biophysica Acta Molecular Cell Research, 1867*(10), 118788.

Collart, M. A., & Panasenko, O. O. (2012). The Ccr4–not complex. *Gene, 492*(1), 42–53.

Collart, M. A., & Panasenko, O. O. (2017). The Ccr4-not complex: Architecture and structural insights. *Sub-Cellular Biochemistry, 83*, 349–379.

Croall, D. E., & DeMartino, G. N. (1991). Calcium-activated neutral protease (calpain) system: Structure, function, and regulation. *Physiological Reviews, 71*(3), 813–847.

Cuddapah, V. A., Robel, S., Watkins, S., & Sontheimer, H. (2014). A neurocentric perspective on glioma invasion. *Nature Reviews. Neuroscience, 15*(7), 455–465.

Cuevas, B. D., Abell, A. N., Witowsky, J. A., Yujiri, T., Johnson, N. L., Kesavan, K., Ware, M., Jones, P. L., Weed, S. A., DeBiasi, R. L., Oka, Y., Tyler, K. L., & Johnson, G. L. (2003). MEKK1 regulates calpain-dependent proteolysis of focal adhesion proteins for rear-end detachment of migrating fibroblasts. *The EMBO Journal, 22*(13), 3346–3355.

Cui, M., Wang, J., Li, Q., Zhang, J., Jia, J., & Zhan, X. (2017). Long non-coding RNA HOXA11-AS functions as a competing endogenous RNA to regulate ROCK1 expression by sponging miR-124-3p in osteosarcoma. *Biomedicine & Pharmacotherapy = Biomedicine & Pharmacotherapie, 92*, 437–444.

Cui, W., Zhang, Y., Hu, N., Shan, C., Zhang, S., Zhang, W., Zhang, X., & Ye, L. (2010). miRNA-520b and miR-520e sensitize breast cancer cells to complement attack via directly targeting 3′UTR of CD46. *Cancer Biology & Therapy, 10*(3), 232–241.

da Silva, C. G., Studer, P., Skroch, M., Mahiou, J., Minussi, D. C., Peterson, C. R., Wilson, S. W., Patel, V. I., Ma, A., Csizmadia, E., & Ferran, C. (2013). A20 promotes liver regeneration by decreasing SOCS3 expression to enhance IL-6/STAT3 proliferative signals. *Hepatology (Baltimore, Md.), 57*(5), 2014–2025.

Dai, J., Hall, C. L., Escara-Wilke, J., Mizokami, A., Keller, J. M., & Keller, E. T. (2008). Prostate cancer induces bone metastasis through Wnt-induced bone morphogenetic protein-dependent and independent mechanisms. *Cancer Research, 68*(14), 5785–5794.

Dai, Z., Zhou, S. L., Zhou, Z. J., Bai, D. S., Xu, X. Y., Fu, X. T., Chen, Q., Zhao, Y. M., Zhu, K., Yu, L., Yang, G. H., Wang, Z., Wu, W. Z., Zhou, J., & Fan, J. (2014). CAPN4 contributes to tumour growth and metastasis of hepatocellular carcinoma by activation of the FAK-Src signalling pathways. *The Journal of Pathology, 234*(3), 316–328.

Davis, J., Martin, S. G., Patel, P. M., Green, A. R., Rakha, E. A., Ellis, I. O., & Storr, S. J. (2014). Low calpain-9 is associated with adverse disease-specific survival following endocrine therapy in breast cancer. *BMC Cancer, 14*, 995.

De Keersmaecker, K., Atak, Z. K., Li, N., Vicente, C., Patchett, S., Girardi, T., Gianfelici, V., Geerdens, E., Clappier, E., Porcu, M., Lahortiga, I., Lucà, R., Yan, J., Hulselmans, G., Vranckx, H., Vandepoel, R., Sweron, B., Jacobs, K., Mentens, N., Wlodarska, I., … Cools, J. (2013). Exome sequencing identifies mutation in CNOT3 and ribosomal genes RPL5 and RPL10 in T-cell acute lymphoblastic leukemia. *Nature Genetics, 45*(2), 186–190.

De Sousa E Melo, F., Wang, X., Jansen, M., Fessler, E., Trinh, A., de Rooij, L. P., de Jong, J. H., de Boer, O. J., van Leersum, R., Bijlsma, M. F., Rodermond, H., van der Heijden, M., van Noesel, C. J., Tuynman, J. B., Dekker, E., Markowetz, F., Medema, J. P., & Vermeulen, L. (2013). Poor-prognosis colon cancer is defined by a molecularly distinct subtype and develops from serrated precursor lesions. *Nature Medicine, 19*(5), 614–618.

Dourdin, N., Bhatt, A. K., Dutt, P., Greer, P. A., Arthur, J. S., Elce, J. S., & Huttenlocher, A. (2001). Reduced cell migration and disruption of the actin cytoskeleton in calpain-deficient embryonic fibroblasts. *The Journal of Biological Chemistry, 276*(51), 48382–48388.

Esquela-Kerscher, A., & Slack, F. J. (2006). Oncomirs - microRNAs with a role in cancer. *Nature Reviews. Cancer, 6*(4), 259–269.

Fan, S. T., Lai, E. C., Lo, C. M., Ng, I. O., & Wong, J. (1995). Hospital mortality of major hepatectomy for hepatocellular carcinoma associated with cirrhosis. *Archives of Surgery (Chicago, Ill.: 1960)*, *130*(2), 198–203.

Favier, F. B., Britto, F. A., Freyssenet, D. G., Bigard, X. A., & Benoit, H. (2015). HIF-1-driven skeletal muscle adaptations to chronic hypoxia: Molecular insights into muscle physiology. *Cellular and Molecular Life Sciences: CMLS*, *72*(24), 4681–4696.

Franco, S. J., Rodgers, M. A., Perrin, B. J., Han, J., Bennin, D. A., Critchley, D. R., & Huttenlocher, A. (2004). Calpain-mediated proteolysis of talin regulates adhesion dynamics. *Nature Cell Biology*, *6*(10), 977–983.

Furuta, M., Kozaki, K. I., Tanaka, S., Arii, S., Imoto, I., & Inazawa, J. (2010). miR-124 and miR-203 are epigenetically silenced tumor-suppressive microRNAs in hepatocellular carcinoma. *Carcinogenesis*, *31*(5), 766–776.

Gebauer, K., Peters, I., Dubrowinskaja, N., Hennenlotter, J., Abbas, M., Scherer, R., Tezval, H., Merseburger, A. S., Stenzl, A., Kuczyk, M. A., & Serth, J. (2013). Hsa-mir-124-3 CpG island methylation is associated with advanced tumours and disease recurrence of patients with clear cell renal cell carcinoma. *British Journal of Cancer*, *108*(1), 131–138.

Goll, D. E., Thompson, V. F., Li, H., Wei, W., & Cong, J. (2003). The calpain system. *Physiological Reviews*, *83*(3), 731–801.

Gutschner, T., Hämmerle, M., & Diederichs, S. (2013). MALAT1 – A paradigm for long noncoding RNA function in cancer. *Journal of Molecular Medicine (Berlin, Germany)*, *91*(7), 791–801.

Gu, J., Xu, F. K., Zhao, G. Y., Lu, C. L., Lin, Z. W., Ding, J. Y., & Ge, D. (2015). CAPN4 promotes non-small cell lung cancer progression via upregulation of matrix metalloproteinase 2. *Medical Oncology (Northwood, London, England)*, *32*(3), 51.

Hanna, R. A., Campbell, R. L., & Davies, P. L. (2008). Calcium-bound structure of calpain and its mechanism of inhibition by calpastatin. *Nature*, *456*(7220), 409–412.

Hu, G., Kim, J., Xu, Q., Leng, Y., Orkin, S. H., & Elledge, S. J. (2009). A genome-wide RNAi screen identifies a new transcriptional module required for self-renewal. *Genes & Development*, *23*(7), 837–848.

Hu, H., Wang, G., & Li, C. (2017). miR-124 suppresses proliferation and invasion of nasopharyngeal carcinoma cells through the Wnt/β-catenin signaling pathway by targeting CAPN4. *OncoTargets and Therapy*, *10*, 2711–2720.

Hu, L., Wu, H., Wan, X., Liu, L., He, Y., Zhu, L., Liu, S., Yao, H., & Zhu, Z. (2018). MicroRNA-585 suppresses tumor proliferation and migration in gastric cancer by directly targeting MAPK1. *Biochemical and Biophysical Research Communications*, *499*(1), 52–58.

Hu, N., Zhang, J., Cui, W., Kong, G., Zhang, S., Yue, L., Bai, X., Zhang, Z., Zhang, W., Zhang, X., & Ye, L. (2011). miR-520b regulates migration of breast cancer cells by targeting hepatitis B X-interacting protein and interleukin-8. *The Journal of Biological Chemistry*, *286*(15), 13714–13722.

Jang, H. S., Lal, S., & Greenwood, J. A. (2010). Calpain 2 is required for glioblastoma cell invasion: Regulation of matrix metalloproteinase 2. *Neurochemical Research*, *35*(11), 1796–1804.

Jayson, G. C., Kohn, E. C., Kitchener, H. C., & Ledermann, J. A. (2014). Ovarian cancer. *Lancet (London, England)*, *384*(9951), 1376–1388.

Jemal, A., Bray, F., Center, M. M., Ferlay, J., Ward, E., & Forman, D. (2011). Global cancer statistics. *CA: A Cancer Journal for Clinicians*, *61*(2), 69–90.

Jia, Y., Duan, Y., Liu, T., Wang, X., Lv, W., Wang, M., Wang, J., & Liu, L. (2019). LncRNA TTN-AS1 promotes migration, invasion, and epithelial mesenchymal transition of lung adenocarcinoma via sponging miR-142-5p to regulate CDK5. *Cell Death & Disease*, *10*(8), 573.

Jones, K. B., Salah, Z., Del Mare, S., et al. (2012). miRNA signatures associate with pathogenesis and progression of osteosarcoma. *Cancer Research*, *72*(7), 1865–1877.

Kar, P., Chakraborti, T., Roy, S., Choudhury, R., & Chakraborti, S. (2007). Identification of calpastatin and mu-calpain and studies of their association in pulmonary smooth muscle mitochondria. *Archives of Biochemistry and Biophysics*, *466*(2), 290–299.

Kar, P., Chakraborti, T., Samanta, K., & Chakraborti, S. (2008). Submitochondrial localization of associated mu-calpain and calpastatin. *Archives of Biochemistry and Biophysics*, *470*(2), 176–186.

Kar, P., Chakraborti, T., Samanta, K., & Chakraborti, S. (2009). mu-Calpain mediated cleavage of the Na+/Ca2+ exchanger in isolated mitochondria under A23187 induced Ca2+ stimulation. *Archives of Biochemistry and Biophysics*, *482*(1–2), 66–76.

Kar, P., Samanta, K., Shaikh, S., Chowdhury, A., Chakraborti, T., & Chakraborti, S. (2010). Mitochondrial calpain system: An overview. *Archives of Biochemistry and Biophysics*, *495*(1), 1–7.

Kimura, Y., Koga, H., Araki, N., Mugita, N., Fujita, N., Takeshima, H., Nishi, T., Yamashima, T., Saido, T. C., Yamasaki, T., Moritake, K., Saya, H., & Nakao, M. (1998). The involvement of calpain-dependent proteolysis of the tumor suppressor NF2 (merlin) in schwannomas and meningiomas. *Nature Medicine*, *4*(8), 915–922.

Kloosterman, W. P., & Plasterk, R. H. (2006). The diverse functions of microRNAs in animal development and disease. *Developmental Cell*, *11*(4), 441–450.

Könnecke, H., & Bechmann, I. (2013). The role of microglia and matrix metalloproteinases involvement in neuroinflammation and gliomas. *Clinical & Developmental Immunology*, *2013*, 914104.

Lee, R. C., Feinbaum, R. L., & Ambros, V. (1993). The C. elegans heterochronic gene lin-4 encodes small RNAs with antisense complementarity to lin-14. *Cell*, *75*(5), 843–854.

Li, B., Wang, W., Li, Z., Chen, Z., Zhi, X., Xu, J., Li, Q., Wang, L., Huang, X., Wang, L., Wei, S., Sun, G., Zhang, X., He, Z., Zhang, L., Zhang, D., Xu, H., El-Rifai, W., & Xu, Z. (2017a). MicroRNA-148a-3p enhances cisplatin cytotoxicity in gastric cancer through mitochondrial fission induction and cyto-protective autophagy suppression. *Cancer Letters*, *410*, 212–227.

Li, D., Chen, P., Li, X. Y., Zhang, L. Y., Xiong, W., Zhou, M., Xiao, L., Zeng, F., Li, X. L., Wu, M. H., & Li, G. Y. (2011). Grade-specific expression profiles of miRNAs/mRNAs and docking study in human grade I-III astrocytomas. *Omics: A Journal of Integrative Biology*, *15*(10), 673–682.

Li, J., Xu, J., Yan, X., Jin, K., Li, W., & Zhang, R. (2018). Suppression of CAPN4 by microRNA-1271 impedes the proliferation and invasion of colorectal cancer cells. *Biomedicine & Pharmacotherapy = Biomedecine & Pharmacotherapie*, *99*, 162–168.

Li, N., & Zhan, X. (2019). Identification of clinical trait-related lncRNA and mRNA biomarkers with weighted gene co-expression network analysis as useful tool for personalized medicine in ovarian cancer. *The EPMA Journal*, *10*(3), 273–290.

Li, S., Zhang, H., Ning, T., Wang, X., Liu, R., Yang, H., Han, Y., Deng, T., Zhou, L., Zhang, L., Bai, M., Wang, X., Ge, S., Ying, G., & Ba, Y. (2016). MiR-520b/e regulates proliferation and migration by simultaneously targeting EGFR in gastric cancer. *Cellular Physiology and Biochemistry: International Journal of Experimental Cellular Physiology, Biochemistry, and Pharmacology*, *40*(6), 1303–1315.

Li, S. L., Gao, H. L., Lv, X. K., Hei, Y. R., Li, P. Z., Zhang, J. X., & Lu, N. (2017b). MicroRNA-124 inhibits cell invasion and epithelial-mesenchymal transition by directly repressing Snail2 in gastric cancer. *European Review for Medical and Pharmacological Sciences*, *21*(15), 3389–3396.

Li, Y., Li, C., Li, D., Yang, L., Jin, J., & Zhang, B. (2019). lncRNA KCNQ1OT1 enhances the chemoresistance of oxaliplatin in colon cancer by targeting the miR-34a/ATG4B pathway. *OncoTargets and Therapy*, *12*, 2649–2660.

Liu, C., Yang, J., Wu, H., & Li, J. (2019). Downregulated miR-585-3p promotes cell growth and proliferation in colon cancer by upregulating *PSME3*. *OncoTargets and Therapy*, *12*, 6525–6534.

Liu, D., Luo, M., Hu, J., Chen, C., & Mei, H. (2020a). Osteopontin enhances cisplatin resistance of human A549 lung cancer cells via stimulating the PI3K signaling pathway and upregulating ERCC1 expression. *Translational Cancer Research*, *9*(5), 3258–3265.

Liu, H., Chen, R., Kang, F., Lai, H., & Wang, Y. (2020b). KCNQ1OT1 promotes ovarian cancer progression via modulating MIR-142-5p/CAPN10 axis. *Molecular Genetics & Genomic Medicine*, *8*(2), e1077.

Liu, H., Wang, H., Liu, X., & Yu, T. (2016a). miR-1271 inhibits migration, invasion and epithelial-mesenchymal transition by targeting ZEB1 and TWIST1 in pancreatic cancer cells. *Biochemical and Biophysical Research Communications*, *472*(2), 346–352.

Liu, K., Yao, H., Lei, S., Xiong, L., Qi, H., Qian, K., Liu, J., Wang, P., & Zhao, H. (2017). The miR-124-p63 feedback loop modulates colorectal cancer growth. *Oncotarget*, *8*(17), 29101–29115.

Liu, L. H., Li, H., Li, J. P., Zhong, H., Zhang, H. C., Chen, J., & Xiao, T. (2011). miR-125b suppresses the proliferation and migration of osteosarcoma cells through down-regulation of STAT3. *Biochemical and Biophysical Research Communications*, *416*(1–2), 31–38.

Liu, S., Song, L., Zeng, S., & Zhang, L. (2016b). MALAT1-miR-124-RBG2 axis is involved in growth and invasion of HR-HPV-positive cervical cancer cells. *Tumour Biology: The Journal of the International Society for Oncodevelopmental Biology and Medicine*, *37*(1), 633–640.

Liu, X., Ma, L., Rao, Q., Mao, Y., Xin, Y., Xu, H., Li, C., & Wang, X. (2015a). MiR-1271 inhibits ovarian cancer growth by targeting cyclin G1. *Medical Science Monitor: International Medical Journal of Experimental and Clinical Research*, *21*, 3152–3158.

Liu, Y., Ren, F., Rong, M., Luo, Y., Dang, Y., & Chen, G. (2015b). Association between under expression of microrna-203 and clinicopathological significance in hepatocellular carcinoma tissues. *Cancer Cell International*, *15*, 62.

Lu, Y. C., Cheng, A. J., Lee, L. Y., You, G. R., Li, Y. L., Chen, H. Y., & Chang, J. T. (2017). MiR-520b as a novel molecular target for suppressing stemness phenotype of head-neck cancer by inhibiting CD44. *Scientific Reports*, *7*(1), 2042.

Lu, X., Li, G., Liu, S., Wang, H., & Chen, B. (2021). MiR-585-3p suppresses tumor proliferation and migration by directly targeting CAPN9 in high grade serous ovarian cancer. *Journal of Ovarian Research*, *14*(1), 90.

Ma, J., Cui, W., He, S. M., Duan, Y. H., Heng, L. J., Wang, L., & Gao, G. D. (2012). Human U87 astrocytoma cell invasion induced by interaction of βig-h3 with integrin α5β1 involves calpain-2. *PLoS One*, *7*(5), e37297.

Mao, Q., Zhuang, Q., Shen, J., Chen, Z., Xue, D., Ding, T., & He, X. (2021). MiRNA-124 regulates the sensitivity of renal cancer cells to cisplatin-induced necroptosis by targeting the CAPN4-CNOT3 axis. *Translational Andrology and Urology*, *10*(9), 3669–3683.

McGeary, S. E., Lin, K. S., Shi, C. Y., Pham, T. M., Bisaria, N., Kelley, G. M., &Bartel, D. P. (2019). The biochemical basis of microRNA targeting efficacy. *Science*, *366*(6472), eaav1741.

Mitchell, P. S., Parkin, R. K., Kroh, E. M., Fritz, B. R., Wyman, S. K., Pogosova-Agadjanyan, E. L., Peterson, A., Noteboom, J., O'Briant, K. C., Allen, A., Lin, D. W., Urban, N., Drescher, C. W., Knudsen, B. S., Stirewalt, D. L., Gentleman, R., Vessella, R. L., Nelson, P. S., Martin, D. B., & Tewari, M. (2008). Circulating microRNAs as stable blood-based markers for cancer detection. *Proceedings of the National Academy of Sciences of the United States of America*, *105*(30), 10513–10518.

Murillo-Garzón, V., & Kypta, R. (2017). WNT signalling in prostate cancer. *Nature Reviews. Urology*, *14*(11), 683–696.

Osaki, M., Takeshita, F., Sugimoto, Y., Kosaka, N., Yamamoto, Y., Yoshioka, Y., Kobayashi, E., Yamada, T., Kawai, A., Inoue, T., Ito, H., Oshimura, M., & Ochiya, T. (2011). MicroRNA-143 regulates human osteosarcoma metastasis by regulating matrix metalloprotease-13 expression. *Molecular Therapy: The journal of the American Society of Gene Therapy*, *19*(6), 1123–1130.

Peng, P., Wu, W., Zhao, J., Song, S., Wang, X., Jia, D., Shao, M., Zhang, M., Li, L., Wang, L., Duan, F., Zhao, R., Yang, C., Wu, H., Zhang, J., Shen, Z., Ruan, Y., & Gu, J. (2016). Decreased expression of Calpain-9 predicts unfavorable prognosis in patients with gastric cancer. *Scientific Reports*, *6*, 29604.

Peng, X. H., Huang, H. R., Lu, J., Liu, X., Zhao, F. P., Zhang, B., Lin, S. X., Wang, L., Chen, H. H., Xu, X., Wang, F., & Li, X. P. (2014). MiR-124 suppresses tumor growth and metastasis by targeting Foxq1 in nasopharyngeal carcinoma. *Molecular Cancer*, *13*, 186.

Piao, Y., Piao, M., & Ryu, K. H. (2017). Multiclass cancer classification using a feature subset-based ensemble from microRNA expression profiles. *Computers in Biology and Medicine*, *80*, 39–44.

Pietras, A., Katz, A. M., Ekström, E. J., Wee, B., Halliday, J. J., Pitter, K. L., Werbeck, J. L., Amankulor, N. M., Huse, J. T., & Holland, E. C. (2014). Osteopontin-CD44 signaling in the glioma perivascular niche enhances cancer stem cell phenotypes and promotes aggressive tumor growth. *Cell Stem Cell*, *14*(3), 357–369.

Rabbani, S. A., Arakelian, A., & Farookhi, R. (2013). LRP5 knockdown: Effect on prostate cancer invasion growth and skeletal metastasis *in vitro* and *in vivo*. *Cancer Medicine*, *2*(5), 625–635.

Ren, W., Wang, D., Li, C., Shu, T., Zhang, W., & Fu, X. (2018). CAPN4 expression is modulated by microRNA-520b and exerts an oncogenic role in prostate cancer cells by promoting Wnt/β-catenin signaling. *Biomedicine & Pharmacotherapy = Biomedecine & Pharmacotherapie*, *108*, 467–475.

Ruan, H. B., Ma, Y., Torres, S., Zhang, B., Feriod, C., Heck, R. M., Qian, K., Fu, M., Li, X., Nathanson, M. H., Bennett, A. M., Nie, Y., Ehrlich, B. E., & Yang, X. (2017). Calcium-dependent O-GlcNAc signaling drives liver autophagy in adaptation to starvation. *Genes & Development*, *31*(16), 1655–1665.

Saez, M. E., Ramirez-Lorca, R., Moron, F. J., & Ruiz, A. (2006). The therapeutic potential of the calpain family: New aspects. *Drug Discovery Today*, *11*(19–20), 917–923.

Samanta, K., Kar, P., Chakraborti, T., & Chakraborti, S. (2010). Calcium-dependent cleavage of the Na(+)/Ca(2+) exchanger by m-calpain in isolated endoplasmic reticulum. *Journal of Biochemistry*, *147*(2), 225–235.

Samanta, K., Kar, P., Ghosh, B., Chakraborti, T., & Chakraborti, S. (2007). Localization of m-calpain and calpastatin and studies of their association in pulmonary smooth muscle endoplasmic reticulum. *Biochimica et Biophysica Ata*, *1770*(9), 1297–1307.

Sarkar, M., Martufi, M., Roman-Trufero, M., Wang, Y. F., Whilding, C., Dormann, D., Sabbattini, P., & Dillon, N. (2021). CNOT3 interacts with the Aurora B and MAPK/ERK kinases to promote survival of differentiating mesendodermal progenitor cells. *Molecular Biology of the Cell*, *32*(22), ar40.

Shaikh, S., Samanta, K., Kar, P., Roy, S., Chakraborti, T., & Chakraborti, S. (2010). m-Calpain-mediated cleavage of Na+/Ca2+ exchanger-1 in caveolae vesicles isolated from pulmonary artery smooth muscle. *Molecular and Cellular Biochemistry*, *341*(1–2), 167–180.

Shen, J., Zhuang, Q., Chen, Z., Fan, M., Lu, H., Ding, T., & He, X. (2017). CAPN4 induces human renal cancer cell proliferation by activating NF-κB signaling pathway through FAK phosphorylation. *International Journal of Clinical and Experimental Pathology*, *10*(7), 7466–7474.

Shen, L., Chen, L., Wang, Y., Jiang, X., Xia, H., & Zhuang, Z. (2015). Long noncoding RNA MALAT1 promotes brain metastasis by inducing epithelial-mesenchymal transition in lung cancer. *Journal of Neuro-Oncology*, *121*(1), 101–108.

Sherman, B. T., Hao, M., Qiu, J., Jiao, X., Baseler, M. W., Lane, H. C., & Chang, W. (2022). DAVID: A web server for functional enrichment analysis and functional annotation of gene lists (2021 update). *Nucleic Acids Research*, *50*(W1), W216–W221.

Shi, B., Wang, Y., & Yin, F. (2017). MALAT1/miR-124/CAPN4 axis regulates proliferation, invasion and EMT in nasopharyngeal carcinoma cells. *Cancer Biology & Therapy*, *18*(10), 792–800.

Shi, Y., Wang, X., Wang, J., Wang, X., Zhou, H., & Zhang, L. (2021). The dual roles of A20 in cancer. *Cancer Letters*, *511*, 26–35.

Shi, Z., Chen, Q., Li, C., Wang, L., Qian, X., Jiang, C., Liu, X., Wang, X., Li, H., Kang, C., Jiang, T., Liu, L. Z., You, Y., Liu, N., & Jiang, B. H. (2014). MiR-124 governs glioma growth and angiogenesis and enhances chemosensitivity by targeting R-Ras and N-Ras. *Neuro-Oncology*, *16*(10), 1341–1353.

Siegel, R. L., Miller, K. D., & Jemal, A. (2016). Cancer statistics, 2016. *CA: A Cancer Journal for Clinicians*, *66*(1), 7–30.

Siegel, R. L., Miller, K. D., & Jemal, A. (2018). Cancer statistics, 2018. *CA: A Cancer Journal for Clinicians*, *68*(1), 7–30.

Smolarz, B., Durczyński, A., Romanowicz, H., Szyłło, K., & Hogendorf, P. (2022). miRNAs in cancer (Review of Literature). *International Journal of Molecular Sciences*, *23*(5), 2805.

Sonntag, K. C., Woo, T. U., & Krichevsky, A. M. (2012). Converging miRNA functions in diverse brain disorders: A case for miR-124 and miR-126. *Experimental Neurology*, *235*(2), 427–435.

Storr, S. J., Carragher, N. O., Frame, M. C., Parr, T., & Martin, S. G. (2011). The calpain system and cancer. *Nature Reviews. Cancer*, *11*(5), 364–374.

Tian, J., Xu, Y. Y., Li, L., & Hao, Q. (2017). MiR-490-3p sensitizes ovarian cancer cells to cisplatin by directly targeting ABCC2. *American Journal of Translational Research*, *9*(3), 1127–1138.

Tkach, M., & Théry, C. (2016). Communication by extracellular vesicles: Where we are and where we need to go. *Cell*, *164*(6), 1226–1232.

Van Kouwenhove, M., Kedde, M., & Agami, R. (2011). MicroRNA regulation by RNA-binding proteins and its implications for cancer. *Nature Reviews Cancer*, *11*(9), 644–656.

Van Meir, E. G., Hadjipanayis, C. G., Norden, A. D., Shu, H. K., Wen, P. Y., & Olson, J. J. (2010). Exciting new advances in neuro-oncology: The avenue to a cure for malignant glioma. *CA: A Cancer Journal for Clinicians*, *60*(3), 166–193.

Villanyi, Z., & Collart, M. A. (2015). Ccr4-Not is at the core of the eukaryotic gene expression circuitry. *Biochemical Society Transactions*, *43*(6), 1253–1258.

Wang, D., Feng, M., Ma, X., Tao, K., & Wang, G. (2021). Transcription factor SP1-induced microRNA-146b-3p facilitates the progression and metastasis of colorectal cancer via regulating FAM107A. *Life Sciences*, *277*, 119398.

Wang, E., Wang, D., Li, B., Ma, H., Wang, C., Guan, L., Zhang, H., Yi, L., & Li, S. (2017a). CAPN4 promotes epithelial-mesenchymal transition in human melanoma cells through activation of the Wnt/β-catenin pathway. *Oncology Reports*, *37*(1), 379–387.

Wang, F., Qiang, Y., Zhu, L., Jiang, Y., Wang, Y., Shao, X., Yin, L., Chen, J., & Chen, Z. (2016). MicroRNA-7 downregulates the oncogene VDAC1 to influence hepatocellular carcinoma proliferation and metastasis. *Tumour Biology: The Journal of the International Society for Oncodevelopmental Biology and Medicine*, *37*(8), 10235–10246.

Wang, J., Pang, W., Zuo, Z., Zhang, W., & He, W. (2017b). MicroRNA-520b suppresses proliferation, migration, and invasion of spinal osteosarcoma cells via downregulation of Frizzled-8. *Oncology Research*, *25*(8), 1297–1304.

Wang, Z., Liu, Z., Fang, X., & Yang, H. (2017c). MiR-142-5p suppresses tumorigenesis by targeting PIK3CA in non-small cell lung cancer. *Cellular Physiology and Biochemistry: International Journal of Experimental Cellular Physiology, Biochemistry, and Pharmacology*, *43*(6), 2505–2515.

Wei, W., Wanjun, L., Hui, S., Dongyue, C., Xinjun, Y., & Jisheng, Z. (2013). miR-203 inhibits proliferation of HCC cells by targeting survivin. *Cell Biochemistry and Function*, *31*(1), 82–85.

Winter, J., Jung, S., Keller, S., Gregory, R. I., & Diederichs, S. (2009). Many roads to maturity: microRNA biogenesis pathways and their regulation. *Nature Cell Biology*, *11*(3), 228–234.

Xia, H., Cheung, W. K. C., Ng, S. S., Jiang, X., Jiang, S., Sze, J., Leung, G. K. K., Lu, G., Chan, D. T. M., Bian, X. W., Kung, H. F., Poon, W. S., & Lin, M. C. (2012a). Loss of brain-enriched miR-124 microRNA enhances stem-like traits and invasiveness of glioma cells. *The Journal of Biological Chemistry*, *287*(13), 9962–9971.

Xia, J., Wu, Z., Yu, C., He, W., Zheng, H., He, Y., Jian, W., Chen, L., Zhang, L., & Li, W. (2012b). miR-124 inhibits cell proliferation in gastric cancer through down-regulation of SPHK1. *The Journal of Pathology*, *227*(4), 470–480.

Xiang, X., Mei, H., Zhao, X., Pu, J., Li, D., Qu, H., Jiao, W., Zhao, J., Huang, K., Zheng, L., & Tong, Q. (2015a). miRNA-337-3p suppresses neuroblastoma progression by repressing the transcription of matrix metalloproteinase 14. *Oncotarget*, *6*(26), 22452–22466.

Xiang, X. J., Deng, J., Liu, Y. W., Wan, L. Y., Feng, M., Chen, J., & Xiong, J. P. (2015b). MiR-1271 inhibits cell proliferation, invasion and EMT in gastric cancer by targeting FOXQ1. *Cellular Physiology and Biochemistry: International Journal of Experimental Cellular Physiology, Biochemistry, and Pharmacology*, *36*(4), 1382–1394.

Xian, D., & Zhao, Y. (2019). LncRNA KCNQ1OT1 enhanced the methotrexate resistance of colorectal cancer cells by regulating miR-760/PPP1R1B via the cAMP signalling pathway. *Journal of Cellular and Molecular Medicine*, *23*(6), 3808–3823.

Xiao, J., Li, G., Zhou, J., Wang, S., Liu, D., Shu, G., Zhou, J., & Ren, F. (2018). MicroRNA-520b functions as a tumor suppressor in colorectal cancer by inhibiting defective in cullin neddylation 1 domain containing 1 (DCUN1D1). *Oncology Research*, *26*(4), 593–604.

Xu, Y., Dong, M., Wang, J., Zhao, W., & Jiao, M. (2021). LINC01436 inhibited miR-585-3p expression and upregulated MAPK1 expression to promote gastric cancer progression. *Digestive Diseases and Sciences*, *66*(6), 1885–1894.

Xu, S., Zhao, N., Hui, L., Song, M., Miao, Z. W., & Jiang, X. J. (2016). MicroRNA-124-3p inhibits the growth and metastasis of nasopharyngeal carcinoma cells by targeting STAT3. *Oncology Reports*, *35*(3), 1385–1394.

Yan, W., Cao, M., Ruan, X., Jiang, L., Lee, S., Lemanek, A., Ghassemian, M., Pizzo, D. P., Wan, Y., Qiao, Y., Chin, A. R., Duggan, E., Wang, D., Nolan, J. P., Esko, J. D., Schenk, S., & Wang, S. E. (2022). Cancer-cell-secreted miR-122 suppresses O-GlcNAcylation to promote skeletal muscle proteolysis. *Nature Cell Biology*, *24*(5), 793–804.

Yang, F., Lv, L. Z., Cai, Q. C., & Jiang, Y. (2015a). Potential roles of EZH2, Bmi-1 and miR-203 in cell proliferation and invasion in hepatocellular carcinoma cell line Hep3B. *World Journal of Gastroenterology*, *21*(47), 13268–13276.

Yang, J., & Zhang, W. (2013). New molecular insights into osteosarcoma targeted therapy. *Current Opinion in Oncology*, *25*(4), 398–406.

Yang, M., Shan, X., Zhou, X., Qiu, T., Zhu, W., Ding, Y., Shu, Y., & Liu, P. (2014). miR-1271 regulates cisplatin resistance of human gastric cancer cell lines by targeting IGF1R, IRS1, mTOR, and BCL2. *Anti-Cancer Agents in Medicinal Chemistry*, *14*(6), 884–891.

Yang, M. H., Hu, Z. Y., Xu, C., Xie, L. Y., Wang, X. Y., Chen, S. Y., & Li, Z. G. (2015b). MALAT1 promotes colorectal cancer cell proliferation/migration/invasion via PRKA kinase anchor protein 9. *Biochimica et Biophysica Acta*, *1852*(1), 166–174.

Yoon, J. H., Abdelmohsen, K., & Gorospe, M. (2014). Functional interactions among microRNAs and long noncoding RNAs. *Seminars in Cell & Developmental Biology*, *34*, 9–14.

Zeng, S. G., Xie, J. H., Zeng, Q. Y., Dai, S. H., Wang, Y., Wan, X. M., & Zhou, X. L. (2019). MicroRNA-497-5p negatively regulates the proliferation and cisplatin resistance of non-small cell lung cancer cells by targeting YAP1 and TEAD1. *Translational Cancer Research*, *8*(6), 2470–2480.

Zhang, C., Bai, D. S., Huang, X. Y., Shi, G. M., Ke, A. W., Yang, L. X., Yang, X. R., Zhou, J., & Fan, J. (2013). Prognostic significance of CAPN4 overexpression in intrahepatic cholangiocarcinoma. *PLoS One*, *8*(1), e54619.

Zhang, F., Wang, Q., Ye, L., Feng, Y., & Zhang, X. (2010). Hepatitis B virus X protein upregulates expression of calpain small subunit 1 via nuclear factor-kappaB/p65 in hepatoma cells. *Journal of Medical Virology*, *82*(6), 920–928.

Zhang, K., Yan, J., Yi, B., Rui, Y., & Hu, H. (2019). High KCNQ1OT1 expression might independently predict shorter survival of colon adenocarcinoma. *Future Oncology (London, England)*, *15*(10), 1085–1095.

Zhang, W., Kong, G., Zhang, J., Wang, T., Ye, L., & Zhang, X. (2012a). MicroRNA-520b inhibits growth of hepatoma cells by targeting MEKK2 and cyclin D1. *PLoS One*, *7*(2), e31450.

Zhang, W., Lu, Z., Gao, Y., Ye, L., Song, T., & Zhang, X. (2015). MiR-520b suppresses proliferation of hepatoma cells through targeting ten-eleven translocation 1 (TET1) mRNA. *Biochemical and Biophysical Research Communications*, *460*(3), 793–798.

Zhang, W., Lu, Z., Kong, G., Gao, Y., Wang, T., Wang, Q., Cai, N., Wang, H., Liu, F., Ye, L., & Zhang, X. (2014). Hepatitis B Virus X protein accelerates hepatocarcinogenesis with partner survivin through modulating miR-520b and HBXIP. *Molecular Cancer*, *13*, 128.

Zhang, Y., Yang, G., He, X., Chen, S., Zhang, F., & Fang, X. (2020). LINC01436, regulating miR-585 and FBXO11, is an oncogenic lncRNA in the progression of gastric cancer. *Cell Biology International*, *44*(3), 882–893.

Zhang, X., You, X., Wang, Q., Zhang, T., Du, Y., Lv, N., Zhang, Z., Zhang, S., Shan, C., Ye, L., & Zhang, X. (2012b). Hepatitis B virus X protein drives multiple cross-talk cascade loops involving NF-κB, 5-LOX, OPN and CAPN4 to promote cell migration. *PLoS One*, *7*(2), e31458.

Zhao, W. H., Wu, S. Q., & Zhang, Y. D. (2013). Downregulation of miR-124 promotes the growth and invasiveness of glioblastoma cells involving upregulation of PPP1R13L. *International Journal of Molecular Medicine*, *32*(1), 101–107.

Zheng, X. B., Chen, X. B., Xu, L. L., Zhang, M., Feng, L., Yi, P. S., Tang, J. W., & Xu, M. Q. (2017). miR-203 inhibits augmented proliferation and metastasis of hepatocellular carcinoma residual in the promoted regenerating liver. *Cancer Science*, *108*(3), 338–346.

Zhong, J., Liu, Y., Xu, Q., Yu, J., & Zhang, M. (2017). Inhibition of DIXDC1 by microRNA-1271 suppresses the proliferation and invasion of prostate cancer cells. *Biochemical and Biophysical Research Communications*, *484*(4), 794–800.

Zhou, Z., Niu, X., Li, C., Sheng, S., & Lu, S. (2015). Inhibition of the growth of non-small cell lung cancer by miRNA-1271. *American Journal of Translational Research*, *7*(10), 1917–1924.

Zhuang, Q., Luo, W., Zhang, M., Fan, M., Lu, H., Xu, R., & He, X. (2018a). CAPN4 contributes to tumor invasion and metastasis in clear cell renal cell carcinoma cells via modulating talin-focal adhesion kinase signaling pathway. *Acta biochimica et biophysica Sinica*, *50*(5), 465–472.

Zhuang, Q., Qian, X., Cao, Y., Fan, M., Xu, X., & He, X. (2014). CAPN4 mRNA level is correlated with tumour progression and clinical outcome in clear cell renal cell carcinoma. *The Journal of International Medical Research*, *42*(2), 282–291.

Zhuang, Q., Shen, J., Chen, Z., Zhang, M., Fan, M., Xue, D., Lu, H., Xu, R., He, X., & Hou, J. (2018b). MiR-337-3p suppresses the proliferation and metastasis of clear cell renal cell carcinoma cells via modulating CAPN4. *Cancer Biomarkers: Section A of Disease Markers*, *23*(4), 515–525.

Zou, L., Collins, H. E., Young, M. E., Zhang, J., Wende, A. R., Darley-Usmar, V. M., & Chatham, J. C. (2021). The identification of a novel calcium-dependent link between NAD+ and glucose deprivation-induced increases in protein O-GlcNAcylation and ER stress. *Frontiers in Molecular Biosciences*, *8*, 780865.

17 The Roles of Ubiquitin-Specific Proteases (USPs) in Cancer Stem Cell Maintenance

Kuan Ping Ang
School of Biosciences, Faculty of Health and Medical Sciences
Taylor's University, Subang Jaya, Selangor, Malaysia

Won Fen Wong
Faculty of Medicine, Universiti Malaya, Kuala Lumpur, Malaysia

Chung Yeng Looi
School of Biosciences, Faculty of Health and Medical Sciences
Taylor's University, Subang Jaya, Selangor, Malaysia
Digital Health and Medical Advancement Impact Lab
Taylor's University, Subang Jaya, Selangor, Malaysia

17.1 INTRODUCTION

Cancer stem cells (CSC) present with stemness characteristics and a high level of genetic lesions compared to normal embryonic and somatic stem cells, which allow them to regenerate themselves for tumour differentiation and other carcinogenesis events to sustain oncogenesis (Abad et al., 2020). On the other hand, stemness is regulated by stemness factors (Aponte & Caicedo, 2017), which are the transcription factors that suppress the cell's differentiation program and allow it to regain pluripotency and self-renewal capabilities (Ervin et al., 2022). These stemness factors are activated by the upstream signalling pathway, usually involving the Wnt, Notch, Hedgehog (Hh), and Hippo/Yap/Taz pathways.

The Wnt pathway controls cellular developmental and differentiation events (Liu et al., 2022). It is an ancient conserved pathway, which is including the canonical and noncanonical (planar cell polarity and calcium) pathways (Qin et al., 2023). The dysregulated Wnt pathway involves CSC development (Katoh & Katoh, 2022). In addition to the Wnt pathway, the Notch pathway is well-recognised for its roles in multiple cellular development events (Zhou et al., 2022). Notch activity has been studied for its roles in CSC development in liposarcoma (Tien et al., 2023) and breast cancer (Yousefi et al., 2022). Meanwhile, the Hh pathway is critical in vertebrate embryogenesis, tumorigenesis, and some tumours' treatment resistance (Briscoe & Thérond, 2013). The related Shh-Gli (glioma-associated oncogene homologue) crosstalk was studied for its role in CSC development (Heiden et al., 2014). Furthermore, the evolutionary high-conserved Hippo signalling pathway involves cellular fate determination and organ size regulation through Yap/Taz activities (Moya & Halder, 2019). Evidence shows that Yap/Taz is the stemness regulator for multiple CSCs, including glioblastoma (Castellan et al., 2021) and breast cancer (Luo et al., 2023).

 DOI: 10.1201/9781003394693-20

There are several pathways regulating the CSC stemness signalling pathways post-translationally, including ubiquitination (Zubarev et al., 2022). Since ubiquitination is a mutable process, it can be regulated negatively by deubiquitylation through the action of deubiquitinases, e.g., ubiquitin-specific proteases (USPs) (Komander et al., 2009). USPs are proteases with a core catalytic domain of 350–400 amino acids, on average (Komander et al., 2009). They also consist of a modular structure and localisation domains (Komander et al., 2009). Certain USPs have been studied extensively for their deubiquitylation in the CSC stemness regulations (Haq et al., 2018), which will be summarised in this chapter with updated findings.

17.2 USPs ACT ON THE STEMNESS SIGNALLING PATHWAYS

Among the known USPs, USP7, USP9x, and USP21 have been identified for their regulatory roles in the most stemness-related signal transductions. As shown in Figure 17.1, USP7 stabilises the β-catenin from ubiquitination in colorectal cancer, sustains Wnt signalling transduction continuously and the stem cell markers expression (e.g. Lgr5 and Olfm4) during the colorectal tumorigenesis (Novellasdemunt et al., 2023). USP7 also stabilises T-ALL's Notch1 (Shan et al., 2018). Interestingly, the promoter region of the USP7 gene is occupied then regulated by Notch1, suggesting a positive feedback relation between USP7 and Notch1 (Jin et al., 2019). For the Hh pathway, USP7 regulates Hh signalling positively (Zhou et al., 2015) as it interrupts ITCH-mediated Gli1's protein level degradation in Hh signalling (Zhou et al., 2015). This leads to the overexpression of the Gli protein to enhance the stemness characteristics of various cancers. Furthermore, USP7 regulates the Hippo/Yap/Taz pathway via the deubiquitylation of the Hippo's downstream effector Yap. This event will promote hepatocellular carcinoma (HCC) tumorigenesis and the stemness acquisition of HCC (Sun et al., 2019).

Meanwhile, the role of USP9x in CSC maintenance is illustrated in Figure 17.2. USP9x promotes the canonical Wnt pathway through the deubiquitylation of DVL2 and BCL9, contributing to breast cancer tumorigenesis (Nielsen et al., 2019). In addition, USP9x regulates the Notch pathway by promoting ITCH ubiquitination (Premarathne et al., 2017). USP9x also deubiquitylates Mind Bomb (MIB) 1 to prevent Jagged-1 from ubiquitination and sustains Notch activation (Izrailit et al., 2017). The inhibition of such deubiquitylation benefits triple-negative breast cancer treatment (Jaiswal et al., 2021). For the Hippo signalling pathway, USP9x is known for handling angiomotin (AMOT), a junction-associated Hippo signalling-associated protein in a renal precise cell carcinoma study. The deubiquitylation of AMOT will inhibit YAP/TAZ from excessive activity (Thanh Nguyen

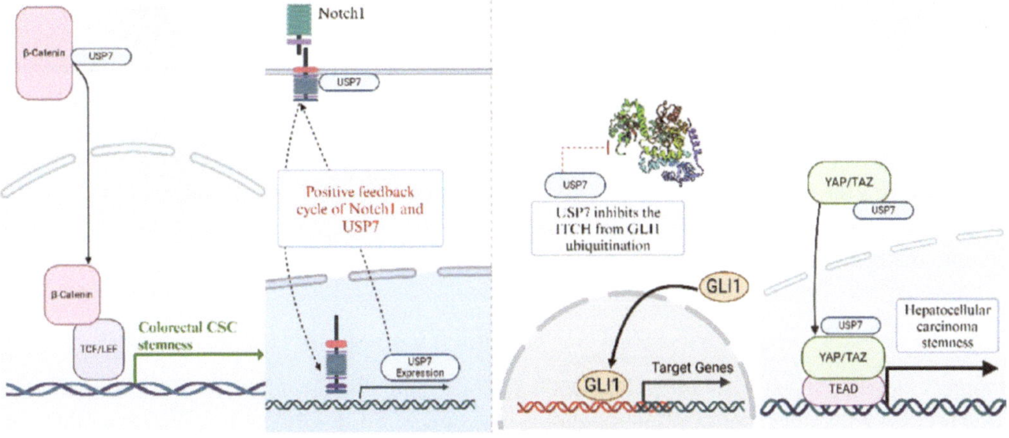

FIGURE 17.1 The regulations of USP7 in stemness signalling pathways.

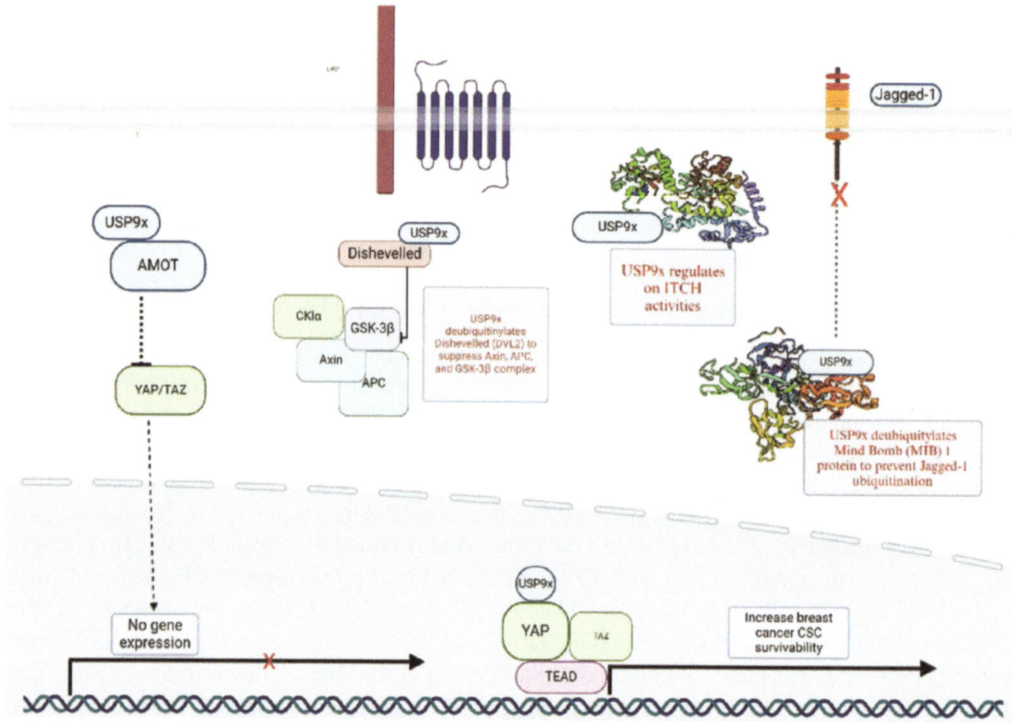

FIGURE 17.2 The regulation of USP9x in stemness signalling pathways.

et al., 2016). However, USP9x can stabilise the YAP1 protein to ensure the survivability of breast cancer cell (Li et al., 2018).

USP21 belongs to the C19 peptidase family. It is involved in signal transduction, DNA repair, apoptosis, as well as cancer cell stemness acquisition and maintenance. Its expression is regulated by LIF/STAT3, in which both transcription factors act on the USP21 promoter (Jin et al., 2016). USP21 is also involved in pancreatic adenocarcinoma stemness characterisation through interacting with TCF7 and promotes Wnt pathway activation for pancreatic CSC development (Hou et al., 2019). USP21 was also identified as an interacting partner to Gli1, the downstream of Hh pathway. It regulates the Gli1 activities through the transcriptional regulation and phosphorylated Gli1 stabilisation (Heride et al., 2016). Furthermore, the USP21 regulates YAP via interacting with microtubule affinity-regulating kinase (MARK), a regulator for LATS1/2 kinases (Nguyen et al., 2017). As shown in Figure 17.3, USP21 stabilises MARK through deubiquitylation and enhances the degradation of YAP/TAZ through the phosphorylation cascade of MARK to LATS1/2. The depletion of USP21 leads to the oncogenesis of renal cell carcinoma due to the ubiquitin-mediated degradation of MARK, prevents YAP/TAZ from phosphorylation and degradation, and YAP/TAZ will translocate into the nucleus.

17.3 CSCs ACT ON STEMNESS AND OTHER TRANSCRIPTION FACTORS

Stemness factors (Nanog, KLF members, Oct4, and Sox family members) drive the stemness acquisition process with other transcription regulators (e.g., YB-1, YY-1, DNMT-1, EZH2, and BMI-1) to establish the CSC identity. Their USPs regulatory interactions will be discussed next.

Nanog appears in CSCs for stemness reprogramming and acquisition (Vasefifar et al., 2022). The dysregulation of Nanog activity is observed in the CSC development (e.g., squamous cell carcinoma

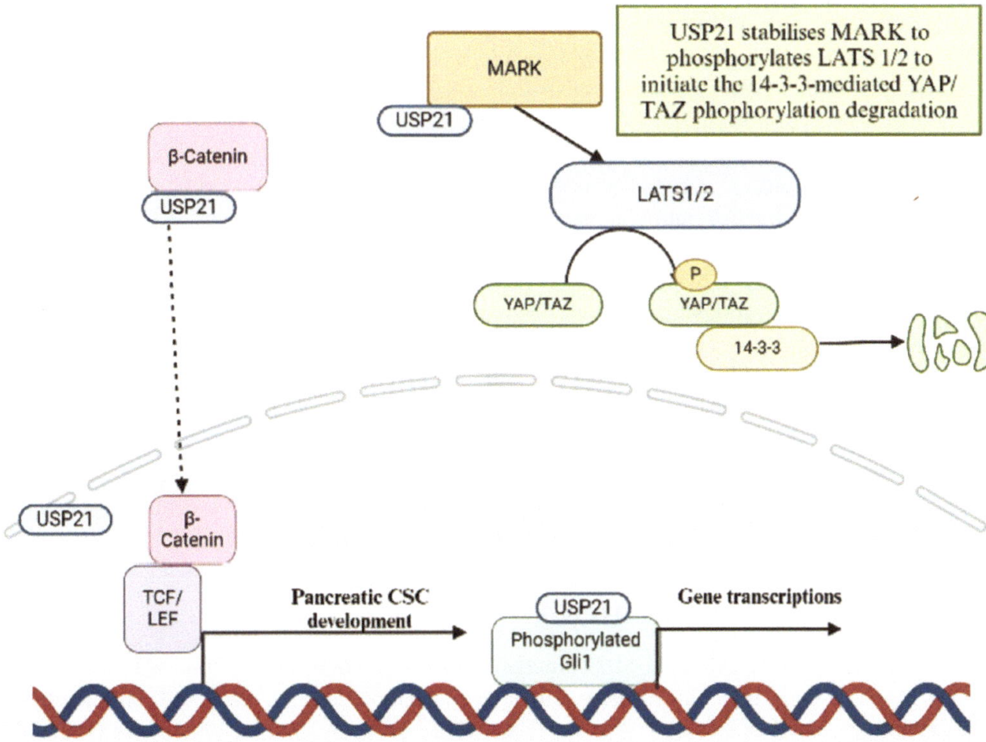

FIGURE 17.3 The regulations of USP21 in stemness-related signalling pathways.

of the head and neck region) (Huang et al., 2020). Nanog activity is regulated by USP7 (herpesvirus-associated USP) in glioma proliferation via the stabilisation of Nanog from the digestion of the 26-s proteasome pathway (Cheng et al., 2020). Additionally, USP21 is the critical enzyme stabilising the Nanog protein, sustaining stemness and pluripotency by interacting with Nanog and deubiquitylating the polyubiquitination of Nanog's lysine-48 (Liu et al., 2016; Kwon et al., 2017). In addition, the Kruppel-like family (KLF) members are the transcription factors that approach the promoters' and enhancers' consensus sequences (CACCC-, GC or GT box elements) (Zhang et al., 2015). KLF4 is known for the formation of colorectal, breast, lung, and pancreatic cancers CSCs (Zhang et al., 2015). Through a study on HCC, USP11 regulates KLF4 deubiquitylation via removing K63-polyubiquitination (Yang et al., 2021).

Furthermore, Oct4 appears abundantly in various CSCs, such as prostate cancer (Kosaka et al., 2016) and bladder cancer (Siddiqui et al., 2019). The USP28-mediated deubiquitylation of lysine-specific demethylase 1 (LSD1) will cause the upregulation of Oct4 (Ren et al., 2023), since LSD1 is upstream of Oct4 (Ren et al., 2023). Other stemness-related transcription factors are Sox family members (Pouremamali et al., 2022). Regarding the USP activity on Sox regulation, USP34 stabilises Sox2 from polyubiquitination in laryngeal squamous cell carcinoma's stem cell development (Dai et al., 2020). Meanwhile, Sox9 has been identified as a CSC marker in HCC (Kawai et al., 2016). It also engages in colorectal tumorigenesis via the USP47 promoter binding activity upon the colorectal cancer hypoxic condition. Sox9 is also regulated by USP18 to stabilise it in reactive astrogliosis (Liu et al., 2021).

Regarding the regulation of USPs on other CSC stemness transcription regulators (e.g., YB-1, YY-1, DNMT-1, EZH2, and BMI-1), Y-box binding protein 1 (YB-1) enforces the CSC's stemness, where it drives the CSC stemness expression via Notch and Wnt pathways, upregulates the GINS1 expression for CSC proliferation, and suppresses CSC apoptosis via p21 and GLP-1 regulations (Alkrekshi et al., 2021). It can re-establish differentiated cancer cell stemness characteristics, then dedifferentiate the cancer cell into CSCs (Yang et al., 2019). Meanwhile, a chronic myelogenous leukaemia study proved that USP47 was highly expressed during DNA damage to sustain the YB-1 activity post-translationally and the YB-1 is a substrate for USP47 (Lei et al., 2021). On the other hand, Yin-Yang 1 (YY-1) is a GLI-Kruppel class transcription regulator with four C2H2 zinc finger domains. It regulates the expression of genes through chromatin modifications. YY-1 is also a transcription factor (Verheul et al., 2020). Through the deubiquitylation of USP21 on YY-1, YY-1 drives epithelial-mesenchymal transition, and enables non-small cell lung cancer metastasis (Xu et al., 2020).

Furthermore, abnormal DNA methyltransferase 1 (DNMT-1) activities promote CSC development via histone modifications (activating H3K4me3 and repressive H3K27me3) and DNA hypermethylation in breast cancer (Pathania et al., 2015), in which the DNMT-1 is stabilised by USP7 to sustain its activity from ubiquitination (Cheng et al., 2015). Also, the polycomb repressive complex 2 (PRC2)'s component enhancer of zeste homolog 2 (EZH2) is a histone methyltransferase that plays a role in CSC development (Balinth et al., 2022). It can be stabilised by US7 on the EZH2's C-terminal domain (Gagarina et al., 2020). Moreover, the polycomb repressive complex 1 (PRC1) component B lymphoma Mo-MLV insertion region 1 homolog (BMI-1) can be deubiquitylated then stabilised by USP22 to promote the formation of glioma CSC (Qiu et al., 2018).

17.4 CONCLUSION AND OUTLOOK

In conclusion, USP7, USP9x, and USP21 emerge as pivotal regulators of stemness in cancer stem cells, each contributing uniquely to the maintenance and modulation of stem cell characteristics. USP7, with its broad influence on various stemness-related pathways and factors, plays a crucial role in sustaining the stem cell phenotype and promoting tumour progression. USP9x and USP21 also significantly impact stemness by regulating key signalling pathways and transcriptional networks that govern cancer stem cell behaviour. Despite these advancements, a deeper understanding of the precise mechanisms by which these USPs influence stemness and their interactions with other molecular players is essential. Continued research into these deubiquitinases will not only elucidate their roles in cancer stem cell biology but also aid in developing targeted therapies aimed at disrupting stemness and improving cancer treatment outcomes.

DECLARATIONS

Conflict of interest: The authors declare no conflicts of interest.

Consent for publishing: All authors have approved the final version of the manuscript and given consent for publication.

Funding: Not Applicable.

REFERENCES

Abad, E., Graifer, D., & Lyakhovich, A. (2020). DNA damage response and resistance of CSCs. Cancer Letters, 474, 106–117. https://doi.org/10.1016/j.canlet.2020.01.008

Alkrekshi, A., Wang, W., Rana, P. S., Markovic, V., & Sossey-Alaoui, K. (2021). A comprehensive review of the functions of YB-1 in cancer stemness, metastasis and drug resistance. Cellular Signalling, 85, 110073. https://doi.org/10.1016/j.cellsig.2021.110073

Aponte, P. M., & Caicedo, A. (2017). Stemness in cancer: Stem cells, CSCs, and their microenvironment. Stem Cells International, 2017, 5619472. https://doi.org/10.1155/2017/5619472

Balinth, S., Fisher, M. L., Hwangbo, Y., Wu, C., Ballon, C., Sun, X., & Mills, A. A. (2022). EZH2 regulates a SETDB1/ΔNp63α axis via RUNX3 to drive a cancer stem cell phenotype in squamous cell carcinoma. Oncogene, 41(35), 4130–4144. https://doi.org/10.1038/s41388-022-02417-4

Briscoe, J., & Thérond, P. P. (2013). The mechanisms of hedgehog signalling and its roles in development and disease. Nature Reviews. Molecular Cell Biology, 14(7), 416–429. https://doi.org/10.1038/nrm3598

Castellan, M., Guarnieri, A., Fujimura, A., Zanconato, F., Battilana, G., Panciera, T., Sladitschek, H. L., Contessotto, P., Citron, A., Grilli, A., Romano, O., Bicciato, S., Fassan, M., Porcù, E., Rosato, A., Cordenonsi, M., & Piccolo, S. (2021). Single-cell analyses reveal YAP/TAZ as regulators of stemness and cell plasticity in glioblastoma. Nature Cancer, 2(2), 174–188. https://doi.org/10.1038/s43018-020-00150-z

Cheng, C. D., Dong, Y. F., Niu, W. X., & Niu, C. S. (2020). HAUSP promoted the growth of glioma cells *in vitro* and *in vivo* via stabilising NANOG. Pathology, Research and Practice, 216(4), 152883. https://doi.org/10.1016/j.prp.2020.152883

Cheng, J., Yang, H., Fang, J., Ma, L., Gong, R., Wang, P., Li, Z., & Xu, Y. (2015). Molecular mechanism for USP7-mediated DNMT1 stabilization by acetylation. Nature Communications, 6, 7023. https://doi.org/10.1038/ncomms8023

Dai, W. L., Yuan, S. X., & Cao, J. P. (2020). The deubiquitinase USP34 stabilizes SOX2 and induces cell survival and drug resistance in laryngeal squamous cell carcinoma. The Kaohsiung Journal of Medical Sciences, 36(12), 983–989. https://doi.org/10.1002/kjm2.12285

Ervin, E. H., French, R., Chang, C. H., & Pauklin, S. (2022). Inside the stemness engine: Mechanistic links between deregulated transcription factors and stemness in cancer. Seminars in Cancer Biology, 87, 48–83. https://doi.org/10.1016/j.semcancer.2022.11.001

Gagarina, V., Bojagora, A., Lacdao, I. K., Luthra, N., Pfoh, R., Mohseni, S., Chaharlangi, D., Tan, N., & Saridakis, V. (2020). Structural basis of the interaction between ubiquitin specific protease 7 and enhancer of zeste homolog 2. Journal of Molecular Biology, 432(4), 897–912. https://doi.org/10.1016/j.jmb.2019.12.026

Haq, S., Suresh, B., & Ramakrishna, S. (2018). Deubiquitylating enzymes as CSC therapeutics. Biochimica et biophysica Acta. Reviews on Cancer, 1869(1), 1–10. https://doi.org/10.1016/j.bbcan.2017.10.004

Heiden, K. B., Williamson, A. J., Doscas, M. E., Ye, J., Wang, Y., Liu, D., Xing, M., Prinz, R. A., & Xu, X. (2014). The sonic hedgehog signaling pathway maintains the CSC self-renewal of anaplastic thyroid cancer by inducing snail expression. The Journal of Clinical Endocrinology and Metabolism, 99(11), E2178–E2187. https://doi.org/10.1210/jc.2014-1844

Heride, C., Rigden, D. J., Bertsoulaki, E., Cucchi, D., De Smaele, E., Clague, M. J., & Urbé, S. (2016). The centrosomal deubiquitylase USP21 regulates Gli1 transcriptional activity and stability. Journal of Cell Science, 129(21), 4001–4013. https://doi.org/10.1242/jcs.188516

Hou, P., Ma, X., Zhang, Q., Wu, C. J., Liao, W., Li, J., Wang, H., Zhao, J., Zhou, X., Guan, C., Ackroyd, J., Jiang, S., Zhang, J., Spring, D. J., Wang, Y. A., & DePinho, R. A. (2019). USP21 deubiquitinase promotes pancreas cancer cell stemness via Wnt pathway activation. Genes & Development, 33(19–20), 1361–1366. https://doi.org/10.1101/gad.326314.119

Huang, C., Yoon, C., Zhou, X. H., Zhou, Y. C., Zhou, W. W., Liu, H., Yang, X., Lu, J., Lee, S. Y., & Huang, K. (2020). ERK1/2-Nanog signaling pathway enhances CD44(+) cancer stem-like cell phenotypes and epithelial-to-mesenchymal transition in head and neck squamous cell carcinomas. Cell Death & Disease, 11(4), 266. https://doi.org/10.1038/s41419-020-2448-6

Izrailit, J., Jaiswal, A., Zheng, W., Moran, M. F., & Reedijk, M. (2017). Cellular stress induces TRB3/USP9x-dependent Notch activation in cancer. Oncogene, 36(8), 1048–1057. https://doi.org/10.1038/onc.2016.276

Jaiswal, A., Murakami, K., Elia, A., Shibahara, Y., Done, S. J., Wood, S. A., Donato, N. J., Ohashi, P. S., & Reedijk, M. (2021). Therapeutic inhibition of USP9x-mediated notch signaling in triple-negative breast cancer. Proceedings of the National Academy of Sciences of the United States of America, 118(38), e2101592118. https://doi.org/10.1073/pnas.2101592118

Jin, Q., Martinez, C. A., Arcipowski, K. M., Zhu, Y., Gutierrez-Diaz, B. T., Wang, K. K., Johnson, M. R., Volk, A. G., Wang, F., Wu, J., Grove, C., Wang, H., Sokirniy, I., Thomas, P. M., Goo, Y. A., Abshiru, N. A., Hijiya, N., Peirs, S., Vandamme, N., Berx, G., … Ntziachristos, P. (2019). USP7 cooperates with NOTCH1 to drive the oncogenic transcriptional program in T-cell leukemia. Clinical Cancer Research: An Official Journal of the American Association for Cancer Research, 25(1), 222–239. https://doi.org/10.1158/1078-0432.CCR-18-1740

Jin, J., Liu, J., Chen, C., Liu, Z., Jiang, C., Chu, H., Pan, W., Wang, X., Zhang, L., Li, B., Jiang, C., Ge, X., Xie, X., & Wang, P. (2016). The deubiquitinase USP21 maintains the stemness of mouse embryonic stem cells via stabilization of Nanog. Nature Communications, 7, 13594. https://doi.org/10.1038/ncomms13594

Katoh, M., & Katoh, M. (2022). WNT signaling and cancer stemness. Essays in Biochemistry, 66(4), 319–331. https://doi.org/10.1042/EBC20220016

Kawai, T., Yasuchika, K., Ishii, T., Miyauchi, Y., Kojima, H., Yamaoka, R., Katayama, H., Yoshitoshi, E. Y., Ogiso, S., Kita, S., Yasuda, K., Fukumitsu, K., Komori, J., Hatano, E., Kawaguchi, Y., & Uemoto, S. (2016). SOX9 is a novel CSC marker surrogated by osteopontin in human hepatocellular carcinoma. Scientific Reports, 6, 30489. https://doi.org/10.1038/srep30489

Komander, D., Clague, M. J., & Urbé, S. (2009). Breaking the chains: Structure and function of the deubiquitinases. Nature Reviews. Molecular Cell Biology, 10(8), 550–563. https://doi.org/10.1038/nrm2731

Kosaka, T., Mikami, S., Yoshimine, S., Miyazaki, Y., Daimon, T., Kikuchi, E., Miyajima, A., & Oya, M. (2016). The prognostic significance of OCT4 expression in patients with prostate cancer. Human Pathology, 51, 1–8. https://doi.org/10.1016/j.humpath.2015.12.008

Kwon, S. K., Lee, D. H., Kim, S. Y., Park, J. H., Choi, J., & Baek, K. H. (2017). Ubiquitin-specific protease 21 regulating the K48-linked polyubiquitination of NANOG. Biochemical and Biophysical Research Communications, 482(4), 1443–1448. https://doi.org/10.1016/j.bbrc.2016.12.055

Lei, H., Xu, H. Z., Shan, H. Z., Liu, M., Lu, Y., Fang, Z. X., Jin, J., Jing, B., Xiao, X. H., Gao, S. M., Gao, F. H., Xia, L., Yang, L., Liu, L. G., Wang, W. W., Liu, C. X., Tong, Y., Wu, Y. Z., Zheng, J. K., Chen, G. Q., … Wu, Y. L. (2021). Targeting USP47 overcomes tyrosine kinase inhibitor resistance and eradicates leukemia stem/progenitor cells in chronic myelogenous leukemia. Nature Communications, 12(1), 51. https://doi.org/10.1038/s41467-020-20259-0

Li, L., Liu, T., Li, Y., Wu, C., Luo, K., Yin, Y., Chen, Y., Nowsheen, S., Wu, J., Lou, Z., & Yuan, J. (2018). The deubiquitinase USP9X promotes tumor cell survival and confers chemoresistance through YAP1 stabilization. Oncogene, 37(18), 2422–2431. https://doi.org/10.1038/s41388-018-0134-2

Liu, J., Xiao, Q., Xiao, J., Niu, C., Li, Y., Zhang, X., … Yin, G. (2022). Wnt/β-catenin signalling: Function, biological mechanisms, and therapeutic opportunities. Signal Transduction and Targeted Therapy, 7(1), 3. https://doi.org/10.1038/s41392-021-00762-6

Liu, W., Ge, X., Zhou, Z., Jiang, D., Rong, Y., Wang, J., Ji, C., Fan, J., Yin, G., & Cai, W. (2021). Deubiquitinase USP18 regulates reactive astrogliosis by stabilizing SOX9. Glia, 69(7), 1782–1798. https://doi.org/10.1002/glia.23992

Liu, X., Yao, Y., Ding, H., Han, C., Chen, Y., Zhang, Y., Wang, C., Zhang, X., Zhang, Y., Zhai, Y., Wang, P., Wei, W., Zhang, J., & Zhang, L. (2016). USP21 deubiquitylates Nanog to regulate protein stability and stem cell pluripotency. Signal Transduction and Targeted Therapy, 1, 16024. https://doi.org/10.1038/sigtrans.2016.24

Luo, J., Zou, H., Guo, Y., Tong, T., Chen, Y., Xiao, Y., Pan, Y., & Li, P. (2023). The oncogenic roles and clinical implications of YAP/TAZ in breast cancer. British Journal of Cancer, 128(9), 1611–1624. https://doi.org/10.1038/s41416-023-02182-5

Moya, I. M., & Halder, G. (2019). Hippo-YAP/TAZ signalling in organ regeneration and regenerative medicine. Nature Reviews. Molecular Cell Biology, 20(4), 211–226. https://doi.org/10.1038/s41580-018-0086-y

Nguyen, H. T., Kugler, J. M., Loya, A. C., & Cohen, S. M. (2017). USP21 regulates Hippo pathway activity by mediating MARK protein turnover. Oncotarget, 8(38), 64095–64105. https://doi.org/10.18632/oncotarget.19322

Nielsen, C. P., Jernigan, K. K., Diggins, N. L., Webb, D. J., & MacGurn, J. A. (2019). USP9X deubiquitylates DVL2 to regulate WNT pathway specification. Cell Reports, 28(4), 1074–1089.e5. https://doi.org/10.1016/j.celrep.2019.06.083

Novellasdemunt, L., Kucharska, A., Baulies, A., Hutton, C., Vlachogiannis, G., Repana, D., Rowan, A., Suárez-Bonnet, A., Ciccarelli, F., Valeri, N., & Li, V. S. W. (2023). USP7 inactivation suppresses APC-mutant intestinal hyperproliferation and tumor development. Stem Cell Reports, 18(2), 570–584. https://doi.org/10.1016/j.stemcr.2022.12.013

Pathania, R., Ramachandran, S., Elangovan, S., Padia, R., Yang, P., Cinghu, S., Veeranan-Karmegam, R., Arjunan, P., Gnana-Prakasam, J. P., Sadanand, F., Pei, L., Chang, C. S., Choi, J. H., Shi, H., Manicassamy, S., Prasad, P. D., Sharma, S., Ganapathy, V., Jothi, R., & Thangaraju, M. (2015). DNMT1 is essential for mammary and cancer stem cell maintenance and tumorigenesis. Nature Communications, 6, 6910. https://doi.org/10.1038/ncomms7910

Pouremamali, F., Vahedian, V., Hassani, N., Mirzaei, S., Pouremamali, A., Kazemzadeh, H., Faridvand, Y., Jafari-Gharabaghlou, D., Nouri, M., & Maroufi, N. F. (2022). The role of SOX family in CSC maintenance: With a focus on SOX2. Pathology, Research and Practice, 231, 153783. https://doi.org/10.1016/j.prp.2022.153783

Premarathne, S., Murtaza, M., Matigian, N., Jolly, L. A., & Wood, S. A. (2017). Loss of Usp9x disrupts cell adhesion, and components of the Wnt and Notch signaling pathways in neural progenitors. Scientific Reports, 7(1), 8109. https://doi.org/10.1038/s41598-017-05451-5

Qin, K., Yu, M., Fan, J., Wang, H., Zhao, P., Zhao, G., … Shi, L. L. (2023). Canonical and noncanonical Wnt signaling: A comprehensive review of multilayered mediators, signaling mechanisms and cross-talk with major signaling pathways. Genes & Diseases, 11(1), 103–134. https://doi.org/10.1016/j.gendis.2023.01.030

Qiu, G. Z., Mao, X. Y., Ma, Y., Gao, X. C., Wang, Z., Jin, M. Z., Sun, W., Zou, Y. X., Lin, J., Fu, H. L., & Jin, W. L. (2018). Ubiquitin-specific protease 22 acts as an oncoprotein to maintain glioma malignancy through deubiquitinating B cell-specific Moloney murine leukemia virus integration site 1 for stabilization. Cancer Science, 109(7), 2199–2210. https://doi.org/10.1111/cas.13646

Ren, X., Jiang, M., Ding, P., Zhang, X., Zhou, X., Shen, J., Liu, D., Yan, X., & Ma, Z. (2023). Ubiquitin-specific protease 28: The decipherment of its dual roles in cancer development. Experimental Hematology & Oncology, 12(1), 27. https://doi.org/10.1186/s40164-023-00389-z

Shan, H., Li, X., Xiao, X., Dai, Y., Huang, J., Song, J., Liu, M., Yang, L., Lei, H., Tong, Y., Zhou, L., Xu, H., & Wu, Y. (2018). USP7 deubiquitinates and stabilizes NOTCH1 in T-cell acute lymphoblastic leukemia. Signal Transduction and Targeted Therapy, 3, 29. https://doi.org/10.1038/s41392-018-0028-3

Siddiqui, Z., Srivastava, A. N., Sankhwar, S. N., Zaidi, N., Fatima, N., Singh, S., & Yusuf, M. (2019). Oct-4: A prognostic biomarker of urinary bladder cancer in North India. Therapeutic Advances in Urology, 11, 1756287219875576. https://doi.org/10.1177/1756287219875576

Sun, X., Ding, Y., Zhan, M., Li, Y., Gao, D., Wang, G., Gao, Y., Li, Y., Wu, S., Lu, L., Liu, Q., & Zhou, Z. (2019). Usp7 regulates Hippo pathway through deubiquitinating the transcriptional coactivator Yorkie. Nature Communications, 10(1), 411. https://doi.org/10.1038/s41467-019-08334-7

Thanh Nguyen, H., Andrejeva, D., Gupta, R., Choudhary, C., Hong, X., Eichhorn, P. J., Loya, A. C., & Cohen, S. M. (2016). Deubiquitylating enzyme USP9x regulates hippo pathway activity by controlling angiomotin protein turnover. Cell Discovery, 2, 16001. https://doi.org/10.1038/celldisc.2016.1

Tien, P. C., Chen, X., Elzey, B. D., Pollock, R. E., & Kuang, S. (2023). Notch signaling regulates a metabolic switch through inhibiting PGC-1α and mitochondrial biogenesis in dedifferentiated liposarcoma. Oncogene. Advance online publication. https://doi.org/10.1038/s41388-023-02768-6

Vasefifar, P., Motafakkerazad, R., Maleki, L. A., Najafi, S., Ghrobaninezhad, F., Najafzadeh, B., Alemohammad, H., Amini, M., Baghbanzadeh, A., & Baradaran, B. (2022). Nanog, as a key CSC marker in tumor progression. Gene, 827, 146448. https://doi.org/10.1016/j.gene.2022.146448

Verheul, T. C. J., van Hijfte, L., Perenthaler, E., & Barakat, T. S. (2020). The Why of YY1: Mechanisms of transcriptional regulation by Yin Yang 1. Frontiers in Cell and Developmental Biology, 8, 592164. https://doi.org/10.3389/fcell.2020.592164

Xu, P., Xiao, H., Yang, Q., Hu, R., Jiang, L., Bi, R., Jiang, X., Wang, L., Mei, J., Ding, F., & Huang, J. (2020). The USP21/YY1/SNHG16 axis contributes to tumor proliferation, migration, and invasion of non-small-cell lung cancer. Experimental & Molecular Medicine, 52(1), 41–55. https://doi.org/10.1038/s12276-019-0356-6

Yang, F., Cui, P., Lu, Y., & Zhang, X. (2019). Requirement of the transcription factor YB-1 for maintaining the stemness of CSCs and reverting differentiated cancer cells into CSCs. Stem Cell Research & Therapy, 10(1), 233. https://doi.org/10.1186/s13287-019-1360-4

Yang, H., Park, D., Ryu, J., & Park, T. (2021). USP11 degrades KLF4 via its deubiquitinase activity in liver diseases. Journal of Cellular and Molecular Medicine, 25(14), 6976–6987. https://doi.org/10.1111/jcmm.16709

Yousefi, H., Bahramy, A., Zafari, N., Delavar, M. R., Nguyen, K., Haghi, A., Kandelouei, T., Vittori, C., Jazireian, P., Maleki, S., Imani, D., Moshksar, A., Bitaraf, A., & Babashah, S. (2022). Notch signaling pathway: A comprehensive prognostic and gene expression profile analysis in breast cancer. BMC Cancer, 22(1), 1282. https://doi.org/10.1186/s12885-022-10383-z

Zhang, Y., Hao, J., Zheng, Y., Jing, D., Shen, Y., Wang, J., & Zhao, Z. (2015). Role of Krüppel-like factors in CSCs. Journal of Physiology and Biochemistry, 71(1), 155–164. https://doi.org/10.1007/s13105-015-0381-4

Zhou, B., Lin, W., Long, Y., Yang, Y., Zhang, H., Wu, K., & Chu, Q. (2022). Notch signaling pathway: Architecture, disease, and therapeutics. Signal Transduction and Targeted Therapy, 7(1), 95. https://doi.org/10.1038/s41392-022-00934-y

Zhou, Z., Yao, X., Li, S., Xiong, Y., Dong, X., Zhao, Y., Jiang, J., & Zhang, Q. (2015). Deubiquitination of Ci/Gli by Usp7/HAUSP regulates hedgehog signaling. Developmental Cell, 34(1), 58–72. https://doi.org/10.1016/j.devcel.2015.05.016

Zubarev, I. V., Podenkova, U. I., Tomilin, A. N., & Tsimokha, A. S. (2022). 'Ubiquitin–Proteasome system in the induction and maintenance of cellular pluripotency'. Russian Journal of Developmental Biology, 53(6), 373–388. https://doi.org/10.1134/s1062360422060091

18 Modulatory Effects of Stem Cell Therapy on Cysteine Proteases and Matrix Metaloproteases in Breast Cancer

Nour Y. S. Yassin, Noha A. Ahmed,
Manal N. Abdel Azeem, and Osama M. Ahmed
Faculty of Science, Beni-Suef University, Beni-Suef, Egypt

18.1 INTRODUCTION

Proteases, which are also referred to as peptidases, proteinases, and proteolytic enzymes, catalyze the irreversible hydrolysis of peptide bonds by specifically targeting the carbonyl group within the peptide bond. Proteases are grouped into clans due to the active site amino acids, such as, glutamic, aspartic, serine, cysteine, or threonine. In contrast to the asparagine peptide lyses, there are also proteases for which these amino acids are combined or unknown (Rawlings *et al.*, 2016). A variety of important physiological processes, including cell-cycle progression, homeostasis, wound healing, cell death, tissue remodeling, and immunological responses, are controlled by both total protein breakdown and restricted proteolysis carried out by intracellular and secreted proteases (López-Otín and Bond, 2008). Protease activity is closely controlled at many levels, from epigenetic and genetic variables regulating gene expression and protein synthesis, and post-translational alterations that impact trafficking of protein and compartmentalization to zymogen activation, to preserve homeostasis under normal physiological conditions. Finally, the quantity of selective endogenous inhibitors of protease also regulates activity (Turk *et al.*, 2012).

18.2 CYSTEINE PROTEASES

Cysteine proteases, in contrast to metalloproteases, can have either an intracellular or an extracellular function. They either perform numerous regulatory functions within the cells or participate in general intracellular degradation in lysosomal/endosomal vesicles. It is now recognized that cysteine proteases, including caspases and papain-like cathepsins (CTS), are active in both extracellularly and intracellularly in diverse compartments. Caspases and cysteine proteases are associated with cancer initiation and progression. The 11 members of the human cysteine CTS family are CTSB, C, F, H, K, L/V, O, S, Z, X, and W (Rawlings *et al.*, 2016).

Studies have demonstrated that upregulation of CTSX promotes the rapid proliferation of various malignancies by conferring resistance to apoptosis, facilitating epithelial-mesenchymal transition (EMT), promoting invasion through interaction with integrin receptors $\alpha v\beta 3$ and $\alpha v\beta 5$, and cleaving the tumor suppressor protein profilin-1 (Kos *et al.*, 2022). The primary role

 DOI: 10.1201/9781003394693-21

of CTSX is significant in the immune response against tumors. Earlier studies have indicated that reduced activity of CTSX affects the migration and adhesion of dendritic cells and T cells. It is proposed that CTSX may sequentially cleave the C-terminus of leukocyte-specific β2 integrin receptors, such as LFA-1. This alteration modifies their interaction with adaptor proteins and modulates their responsiveness to external ligands. Overexpression of CTSX has been experimentally shown to significantly enhance T-cell migration through the LFA-1 receptor (Jevnikar *et al.*, 2011).

A variety of biochemically important signaling cascades, including apoptosis, can be cleaved and modulated by CTS. Concerning apoptosis, CTS can control both intrinsic and extrinsic signaling pathways as well as cancer pro- and anti-apoptotic activities. Extensive documentation exists regarding the role of B-cell lymphoma (Bcl-2) family proteins as substrates for CTS enzymes (Kalkavan and Green, 2018). Soond *et al.* (2021) provided a comprehensive overview of the therapeutic targeting of CTS, highlighting the potential strategies to either preserve their activity for degrading anti-apoptotic Bcl-2 proteins or promote their activity for degrading pro-apoptotic Bcl-2 proteins. The main objective is to induce apoptosis in cancer cells, thereby inhibiting tumor growth. Resistance to apoptosis is a hallmark feature of cancer. Apart from CTS and Bcl-2 proteins, caspases, which are a group of cysteine-aspartic peptidases, play a crucial role in initiating and executing apoptosis. Caspases have a complex role in cancer, and like many other proteins, they may either inhibit or promote carcinogenesis. The key research on the dual involvement of caspases in cancer in *Drosophila* was reviewed by Hounsell and Fan (2021). A thorough understanding of caspase role could enhance current anti-cancer therapeutic strategies because the induction of apoptosis is the objective of cancer therapies, such as radiotherapy and chemotherapy.

18.3 MATRIX METALLOPROTEASES (MMPs)

Matrix metalloproteases (MMPs), a class of zinc-dependent endopeptidases, have been extensively investigated as the most studied group of proteases in the context of cancer. MMPs can be categorized into four main groups based on their protein domain structure and sequence homology. These groups include gelatinases (MMP-9 and MMP-2), collagenases (MMP-13, MMP-8, and MMP-1), membrane-type MMPs (MT1-MMP to MT6-MMP, including MMP-25, MMP-24, and MMP14-17), and stromelysin subclass MMPs (MMP-26, MMP-11, MMP-10, MMP-7, and MMP-3) (Rawlings *et al.*, 2016). MMPs process a wide variety of cell surface molecules and function in several crucial biological processes. Cancer cells can degrade extracellular matrix (ECM) by activating MMPs, which promotes invasion and migration into organs. The primary enzymes responsible for breaking down the basement membrane are MMP-2 and MMP-9, largely due to their capacity to degrade type IV collagen. These two types of MMPs are crucial for cancer cell invasion and metastasis (Brown and Murray, 2015). In various pathological conditions like atherosclerosis, rheumatoid arthritis, inflammation, and tumor invasion, MMP-9 can be released and activated by different triggers such as cytokines and phorbol 12-myristate 13-acetate (PMA), while MMP-2 is typically expressed continuously. The endogenous tissue inhibitors of MMPs, known as TIMPs (1–4), block MMP activity, and nearly all of them have a specific function in the proliferation of cancer (Yamamoto *et al.*, 2015). The activity of MMPs outside the cells is controlled not just by TIMPs, but also by how they interact with proteins in the ECM and on the cell surface, like integrins (specifically the β1 subunit) and tetraspanins, and by endocytic trafficking (Yamamoto *et al.*, 2015). ADAMTSs, a family of proteases with disintegrin and metalloprotease with thrombospondin motifs, which operate mainly in the area surrounding cells. They belong to a comparable protease family and have the potential to alter the tumor microenvironment through various mechanisms associated with both promoting and protecting tumors. These proteases can be secreted by both cancerous and stromal cells (Cal and López-Otín, 2015).

18.4 PROTEASES IN BREAST CANCER

Breast cancer (BC) has emerged as the most widespread ailment among women worldwide and ranks as the second most fatal form of cancer following lung cancer, despite notable progress in diagnostic and therapeutic approaches (Bahrami *et al.*, 2018). Based on a worldwide cancer projection, BC exhibits a morbidity rate of 27.8% and a mortality rate of 15% among women on a global scale (Bray *et al.*, 2018). Despite recent advancements in BC diagnosis and treatment methods, there hasn't been a discernible decline in the disease's mortality (Dihge *et al.*, 2019). It has the trait of gradually accumulating mutations, which cause the uncontrolled growth of breast tissue cells. The primary risk factors for BC include age, using oral contraceptives, female gender, family history, and alcohol use (Bardaweel *et al.*, 2019). In recent decades, BC mortality has steadily decreased, and a better patient outcome depends on an early diagnosis (DeSantis *et al.*, 2019). Nevertheless, the identification of serum markers, understanding hormone receptor status, histologic grading, and lymph node involvement all offer advantages in the early detection of BC, thereby enhancing prognosis, assessment, and treatment guidance (Teufelsbauer *et al.*, 2019). Numerous functional changes take place during the malignant transformation of tumor cells, leading to the induction of various gene products with therapeutic significance. These gene products can serve as biological markers and targets for therapy, facilitating the development of innovative multi-target approaches to enhance current cancer therapies and prevent disease relapse (Mimeault and Batra, 2014). Consequently, numerous researchers are consistently seeking molecular markers to facilitate early BC diagnosis, identify molecular targets for its treatment, and establish molecular indicators for early BC detection and prognosis assessment. Localized early-stage breast tumors typically undergo treatment with surgery, radiotherapy, and estrogen-depleting therapy. However, treatment efficacy is often limited once BC has advanced (Curigliano *et al.*, 2017).

CTS are a class of proteins that have been linked to cancer metastasis and are highly expressed in a variety of human malignancies (Tan *et al.*, 2013). Several CTS family members have been investigated in BC previously (Bengsch *et al.*, 2014). Moreover, previous studies have demonstrated that CTSD exhibited higher expression levels in invasive ductal carcinoma compared to lobular carcinoma of the mammary glands (Dian *et al.*, 2014). Furthermore, there exists a correlation between the overexpression of CTSD and lymph node metastasis in breast invasive ductal carcinoma (Dian *et al.*, 2014). It also has a bad prognosis and significantly increases angiogenesis in both ovarian and breast malignancies (Pranjol and Whatmore, 2020). Under acidic conditions, extracellularly released CTSD is automatically activated, and active CTSD negatively impacts the progression of BC by promoting the growth of fibroblasts, boosting the growth of BC cells, tumor angiogenesis, tumor growth, and metastasis (Ashraf *et al.*, 2019).

MMPs potentially enhance the growth and proliferation of BC cells through various mechanisms, including (a) their capacity to degrade proteins that bind growth factors, thereby increasing the availability of these factors. One such mechanism involves the cleavage of perlecan by MMPs, leading to the liberation of fibroblast growth factors (FGFs) (Manes *et al.*, 1999); (b) MMPs break down the ECM, releasing cryptic proteins like integrins that are necessary for MMPs to control growth signals; (c) By producing molecules that promote cell survival, such as the ErbB4 receptor tyrosine kinase (RTK), proheparin-binding EGF-like growth factor (proHB-EGF), and insulin-like growth factors (IGFs), MMPs, particularly MMP-11 and MMP-9 with partial involvement from MMP-7, can inhibit cancer cell apoptosis and enhance cell viability; and (d) MMPs have the capability to activate growth factor receptors, such as by cleaving the ectodomain of FGF receptor 1. For example, MMP-9 and MMP-2 can induce the activation of FGF receptor 1, thereby facilitating the binding of FGF to this receptor (Manes *et al.*, 1999). Additionally, MMP-2 enhances angiogenesis by degrading collagen type IV, thus creating binding sites for $\alpha v\ \beta 3$. Earlier research has indicated that MMP-9 elevates the availability of two essential pro-angiogenic proteins, basic fibroblast growth factor (bFGF) and vascular endothelial growth factor (VEGF), by breaking down ECM components such as collagen types IV, VIII, X, and perlecan (Iozzo *et al.*, 2009). MMP-14 promotes the invasion of endothelial cells by cleaving the fibrin matrix surrounding blood vessels (Hiraoka *et al.*, 1998). Therefore, inhibiting MMPs would be a successful treatment strategy for breast carcinomas.

18.5 CYTOKINES ACTIVATING DIFFERENT SIGNALING PATHWAYS THAT AFFECT PROTEASES ACTIVITY AND PROMOTE BREAST CANCER

BC is the leading cause of cancer-related death among women worldwide (Barrios, 2022). The main cause contributing to BC mortality is metastasis. Tumor metastasis involves a multi-faceted process comprising cell proliferation, degradation of ECM matrix, cell migration, and subsequent tumor growth at distant metastatic sites (Tan *et al.*, 2021). Morphologically, tumor invasion is distinguished by an irregular margin of the primary tumor, where individual or groups of tumor cells invade the surrounding tissue, including the ECM, in an aggressive manner (Im *et al.*, 2014). MMPs have been recognized as a key molecule in processing mammary tumors invasion and metastasis (Duffy *et al.*, 2000). In previous reports, it was shown that phorbol esters, growth factors, or inflammatory cytokines increase the activity of MMPs in BC by activating several intracellular-signaling pathways (Siveen *et al.*, 2014). Because various tumor cells may be prompted by potential inducers of invasion and migration, nuclear factor-kappa B (NF-κB), signal transducers and activators of transcription (STAT3), phosphoinositide 3-kinase/protein kinase B (PI3K/Akt), transforming growth factor β (TGF-β), and Wnt/β-catenin signaling pathways can become activated. The upregulation and activation of these pathways are strongly correlated with heightened protease activity and increased invasiveness in BC (Villar *et al.*, 2011; Pal *et al.*, 2014; Cichon *et al.*, 2015; Dong *et al.*, 2019). It is crucial to inhibit the expression of these factors to develop an experimental therapeutic model for BC metastasis.

18.6 NF-κB SIGNALING PATHWAY

The stress-inducible transcription factor NF-κB is thought to regulate cancer cell survival, proliferation, cellular invasion, angiogenesis, chemoresistance, and metastasis (Aggarwal and Sung, 2011). Previous research has suggested that IκB is an inhibitor that stops NF-κB from translocating into the nucleus, whereas NF-κB may be the transcriptional regulatory factor of MMPs and other genes. Moreover, the transcription factor NF-κB regulates the expression binding sites of MMP-9 on its promoter (Kim *et al.*, 2007). NF-κB represents a group of inducible transcription factors that regulate the inflammatory and immune responses of the host. Various signal transduction pathways govern the activation of the NF-κB pathway (Yamamoto and Gaynor, 2001). It was stated that PI3K stimulates the IκB kinases (IKKs) to phosphorylate IκB, which causes it to be ubiquitinated and proteasomal degradation by proteases, releasing the NF-κB/Rel complex.

Studies have shown that in cultured mammary epithelial cells, MMP-3 (stromelysin-1) can initiate EMT, provoke a premalignant phenotype, and stimulate spontaneous tumor growth upon expression in mouse mammary glands (Sternlicht *et al.*, 1999). A prior investigation revealed that MMP-3 prompts EMT in mammary epithelial cells through the activation of Rac1b, which is a splice variant of Rac1 Rho GTPase (Fiegen *et al.*, 2004). Moreover, it was observed that MMP-3/Rac1b-induced EMT relies on the generation of cellular reactive oxygen species (ROS)(Radisky *et al.*, 2005). In a separate investigation, the role of external signals from the ECM in controlling the synthesis and positioning of Rac1b, as well as the generation of ROS in BC cells, has been clarified. (Chen *et al.*, 2013). Another study outlined the precise mechanism by which mammary epithelial cell EMT is stimulated by MMP-3/Rac1b-induced ROS (Cichon and Radisky, 2014). The researchers additionally illustrated that the heightened expression of the essential EMT transcription factor, Snail, is induced by the direct interaction of the NF-κB subunits p65 and cRel with the promoter region of Snail. This mechanism is vital for MMP-3-triggered EMT and is reliant on the activation of p65/cRel by ROS generated by MMP-3. They also established that that transcriptional changes linked to human BC subtypes with poor prognosis are activated by MMP-3/Rac1b-induced EMT (Cichon and Radisky, 2014).

CTSL, a cysteine proteinase primarily localized in lysosomes, has been observed to exhibit increased expression levels in various cancer types such as those affecting the neck, head, breast, colon, lung, gliomas, melanomas, and gastric carcinomas (Jedeszko and Sloane, 2004; Gocheva and Joyce, 2007). Research indicates that heightened CTSL expression levels play a role in tumor advancement by influencing cellular processes like the cell cycle and apoptosis. This effect stems from CTSL's capability to degrade NF-κB (Brix *et al.*, 2008). Suggesting the targeting of CTSL using small-molecule inhibitors as a potentially effective therapeutic approach for cancer treatment (Strojnik *et al.*, 2005). Research has shown that inhibiting CTSL can inhibit the nuclear translocation of NF-κB. These findings highlight the potential of targeting CTSL to modulate NF-κB signaling and potentially inhibit cancer progression (Wang *et al.*, 2013).

Hence, it is imperative to develop novel treatments capable of inhibiting the NF-κB signaling pathway-mediated expression of MMPs, thereby thwarting the invasion and metastasis of breast tumors.

18.7 STAT3 SIGNALING PATHWAY

The STAT family, comprising seven closely related and conserved members (STAT1, STAT2, STAT3, STAT4, STAT5a, STAT5b, and STAT6), holds considerable importance as a transcription factor family in BC (Furtek *et al.*, 2016). High activity of STAT3 has been associated with increased MMP-2 activity, which is a target gene of STAT3. Treatment with Janus kinases (JAKs) inhibitors or dominant-negative STAT3 has shown to significantly decrease MMP activity (Huang *et al.*, 2010). Furthermore, the expression of MMP-7 is carefully controlled through the transcriptional level. In human BC cell lines, it was observed that the activation of STAT3 and AP-1 by HER2 is accountable for controlling the production of MMP-7 induced by heregulin-β (Yuan *et al.*, 2008). Moreover, they discovered specific binding sites for AP-1 and STAT3 in the first 350 base pairs of the MMP-7 promoter region, which are crucial for its transcriptional activity (Yuan *et al.*, 2008). Another investigation revealed that FGF-2 can directly stimulate the expression of MMP-7 in human tumor cell lines and umbilical vein endothelial cells by activating STAT3 and AP-1. (Holnthoner *et al.*, 2006). Another research indicated that STAT3 serves as a central focal point for multiple cancer-causing signaling pathways (Yu *et al.*, 2009). In a distinct investigation, scientists illustrated that an abnormally active variant of STAT3 in immortalized human mammary epithelial cells prompts the production of MMP-9 mRNA and protein. Moreover, the expression of MMP-9 is associated with nuclear pSTAT3 expression in BC tissues (Dechow *et al.*, 2004). Additionally, studies have demonstrated that STAT3 increases the expression of the lysosomal enzyme CTSB, which is implicated in breast tumor progression (Vasiljeva *et al.*, 2008).

18.8 PI3K/Akt SIGNALING PATHWAY

Numerous cancers exhibit abnormal activation of the PI3K/Akt pathway, which is associated with neoplastic characteristics such as increased cell proliferation, drug resistance, and stem-cell-like phenotypes (Jiang *et al.*, 2020). The PI3K/Akt/mTOR pathway is commonly disturbed during the progression of BC, resulting in unfavorable patient outcomes and impacting crucial cellular functions such as survival, growth, invasion, and metabolism (Manning and Toker, 2017). In numerous cases, the activation of the PI3K/Akt/mTOR pathway is initiated by the interaction of PI3K with either G-protein coupled receptors (GPCRs) or mTOR (Murga *et al.*, 2000) as well as RTKs like the IGF-1 receptor (Zhang *et al.*, 2018a) and ErbB family receptors (Schoeberl *et al.*, 2009). The expression of MMPs can be controlled by the constant activation of the PI3K/Akt pathway, and several anti-cancer medications have shown the capability to hinder cancer spread by regulating Akt-mediated MMPs (Chen *et al.*, 2016). Previous decreased levels of MMP-9 and MMP-2 in BC cells are closely associated with the inhibition of PI3K/Akt and ERK, both of which are downstream targets of focal adhesion kinase (FAK) (Cao *et al.*, 2015). PI3K participates in numerous cellular signaling pathways through the activation of Akt, and the activation of Akt can enhance the invasion and spread of cancer cells by inducing the release of MMPs (Jung *et al.*, 2013). The involvement of MAP family kinases in signaling pathways that regulate MMPs, particularly MMP-9, has been widely acknowledged (O-charoenrat *et al.*, 2004). MMP-13 plays a critical role in regulating tumor invasion and spread, with multiple studies emphasizing its involvement in bone metastasis in BC (Pivetta *et al.*, 2011). The P2X7 receptor, a subtype of ligand-gated ion channel receptors, is linked to the development and spread of tumors. It has been identified to trigger numerous intracellular signaling pathways, such as the Akt and ERK1/2 pathways, which stimulate the proliferation of ovarian carcinoma cells (Vázquez-Cuevas *et al.*, 2014). Furthermore, research indicates that the receptor is capable of activating the PI3K/Akt and AMPK-PRAS40-mTOR signaling pathways, resulting in the death of tumor cells. These discoveries imply that targeting the P2X7 receptor could be a promising approach for cancer treatment (Bian *et al.*, 2013). In another investigation, it was shown that the activation of the P2X7 receptor by ATP can augment the invasion and migration of BC cells, possibly through the activation of the Akt pathway and modulation of E-cadherin and MMP-13 expression (Xia *et al.*, 2015). Therefore, blocking the Akt pathway could mitigate the impact of MMPs on the invasion and migration of BC cells.

18.9 Wnt/β-CATENIN SIGNALING PATHWAY

The Wnt/β-catenin pathway is essential in the progression and growth of BC (Lamb *et al.*, 2013). Anomalous activation of β-catenin and upregulation of Wnt receptors and co-receptors have been detected in BC tissues. The aberrant activation of the Wnt/β-catenin pathway has been associated with the emergence of resistance to radiation and chemotherapy in BC (Zhang *et al.*, 2012). In the absence of Wnt binding to its receptor, β-catenin forms a complex with glycogen synthase kinase-3 beta (GSK-3β), adenomatous polyposis coli (APC), and Axin, resulting in its degradation. GSK-3β and casein kinase 1 (CK1) phosphorylate β-catenin, tagging it for degradation by β-TrCP1. PPARα and PPARβ positively regulate this mechanism. Conversely, upon Wnt binding to its receptor and co-receptors, Dishevelled (Dsh) proteins undergo phosphorylation and activation. Activated Dsh then recruits GSK-3β, resulting in the release of β-catenin and facilitating its translocation into the nucleus. Within the nucleus, β-catenin interacts with Tcf and Lef transcription factors, initiating the transcription of genes implicated in tumorigenesis and metastasis, including cyclin D1, c-Myc, MMP-9, MMP-7, and MMP-2 (Tejeda-Muñoz and Robles-Flores, 2015; Wang *et al.*, 2021).

The process of EMT has been found to enhance the stemness of BC cells, as well as promote their invasion and metastasis (Dittmer, 2018). The Wnt/β-catenin signaling pathway has been associated with the regulation of EMT in multiple cancer types, including BC (Wu *et al.*, 2012). Moreover, in the process of tumor invasion and metastasis, the Wnt/β-catenin signaling pathway governs the expression of MMP-2, MMP-7, and MMP-9, enzymes that degrade the ECM and basement membrane, thereby promoting tumor cell detachment, invasion, and metastasis (Villar *et al.*, 2011).

Previous studies conducted on cultured cells and animal models have implicated the contributor role of MMP3 in breast premalignancy and cancer. However, limited information is available about the relationship between MMP3 expression in human tumors and disease progression and overall survival (Cichon *et al.*, 2015).

In another investigation, it was demonstrated that the overexpression of MMP3 in transgenic mice leads to the development of preneoplastic and malignant tumors in the mammary glands, whereas inhibiting MMP3 with antisense oligodeoxynucleotides prevents invasion of an artificial basement membrane in mouse mammary cancer cell lines (Sternlicht *et al.*, 1999). Moreover, MMP3 acts as an extracellular modulator of the Wnt signaling pathway (Anastas and Moon, 2013).

18.10 TGF-β PATHWAY

TGF-β, a crucial factor in various physiological processes like wound healing and tissue development, has also been associated with cancer (Tsuda, 2018). It plays a role in regulating events such as immune suppression, cell migration, and angiogenesis (Chen *et al.*, 2020). BC is characterized by the expression of TGF-β, with TGF-β1 being the most prevalent form found in the plasma of BC patients (Chiechi *et al.*, 2013). In a mouse model of BC, TGF-β signaling was shown to promote both lung and bone metastasis (Imamura *et al.*, 2012).

The TGF-β pathway, widely investigated for its function in restraining cell proliferation during the initial phases of tumor development yet stimulating EMT and invasion in advanced BC, is regarded as a double-edged sword in cancer (Zhang *et al.*, 2021). In normal cells, activation of the TGF-β signaling pathway results in cell-cycle arrest and apoptosis (while, in the early stages of BC, hyperplastic breast ducts have been observed to have an elevated likelihood of progressing into invasive BC (Zhang *et al.*, 2021). Research has shown that the TGF-β signaling pathway, particularly involving G proteins, PI3K, p38, c-Jun-NH2-kinase, and mitogen-activated protein kinase, could have a significant impact on the oncogenic effects observed during advanced-stage tumorigenesis (Zhang *et al.*, 2021).

The biological effects of TGF-β occur through its interaction with transmembrane receptors called TβRI and TβRII. The canonical intracellular TGF-β signal transduction process operates via the Smad pathway. This entails the phosphorylation of receptor-regulated Smads 2 and -3 (R-Smad2–3) by the type I receptor, which then form heteromeric complexes with the common mediator Smad4. These complexes then move into the nucleus to govern transcriptional responses. Various factors, including pH, ROS, thrombospondin-1, and proteases trigger the activation of TGF-β. Notably, MMP-9 and MMP-2 possess the ability to cleave latent TGF-β (Yu and Stamenkovic, 2000).

Multiple research endeavors have shown that pericellular proteases and their receptors closely interact with TGF-β1, whether within the ECM or the tumor microenvironment, exerting a notable influence on ECM remodeling. Additionally, it has been observed that urokinase-type plasminogen activator (uPA) is the first protease known to activate TGF-β1 when membrane-bound uPAR is present in peritoneal macrophages (Kitaw, 2019). On the other hand, the rapid increase in MMP-2 and MMP-9 levels triggered by TGF-β1 have proven to enhance the invasion and growth of preneoplastic breast epithelial cells (Kim *et al.*, 2007).

Research indicates that CTSL is heightened in various cancer forms, such as colon, gastric, lung, breast, gliomas, and melanomas. Moreover, the degree of CTSL expression correlates with the malignancy's severity (Skrzydlewska *et al.*, 2005). Increased extracellular activity of CTSL leads to reduced cell adhesion and heightened degradation of the ECM (Gocheva and Joyce, 2007). Moreover, the elevation of CTSL expression was observed to contribute to the heightened malignant characteristics subsequent to Ras-induced transformation of NIH/3T3 cells (Chambers *et al.*, 1992). These investigations revealed that CTSL facilitated the invasion and migration of ovarian cancer cells, suggesting its viability as a molecular target for combating cancer metastasis (Wang *et al.*, 2010).

In an earlier investigation, the levels of MMP-9 and TGF-β expression were analyzed in samples of canine BC, revealing significant overexpression of both proteins in malignant tissues. Transient expression of mouse MMP-9 in three distinct breast carcinoma cell lines resulted in heightened cell colony formation, migration, and induction of EMT. The study additionally noted a notable activation of Smad signaling, evidenced by increased expressions of Smad4, Smad2/3, and phosphorylated Smad2, in the cell lines expressing MMP-9. Furthermore, the authors showed that inhibiting the activity of either MMP-9 or TGF-β effectively countered the MMP-9-induced activation of Smad signaling, thereby reducing invasion in the breast carcinoma cell lines (Dong *et al.*, 2019).

18.11 STEM CELLS REGULATE CYSTEINE PROTEASES AND MMPs IN BC PROGRESSION

Cysteine proteases and MMPs play a crucial role in angiogenesis by regulating angiogenic factors and synthesizing endogenous angiogenic inhibitors. These proteases specifically break down basement membrane components like type-IV collagen, revealing hidden sites in ECM components that are biologically relevant. They also influence angiogenic factors and contribute to the production of natural inhibitors of angiogenesis. Researchers have explored the role of proteases in tumor angiogenesis using mouse models and synthetic inhibitors (Ohno-Matsui *et al.*, 2003). These proteases and inhibitors have become the focus of anti-cancer research and have been utilized in clinical studies to inhibit tumor angiogenesis and metastasis (Trisciuoglio *et al.*, 2005).

Interestingly, researchers have previously focused on targeting cysteine proteases and MMPs in the treatment of human cancer and other diseases (Vasiljeva *et al.*, 2007). Regenerative medicine shows great potential in utilizing stem cells and microenvironmental factors to induce differentiation into different lineages. Recent findings suggest that stem cells possess a specific inhibitory influence on the growth of tumor cells (Leng *et al.*, 2014). Mesenchymal stem cells (MSCs), originating from the mesodermal germ layer, represent a form of pluripotent stem cell capable of self-renewal and differentiation into various lineages. They are present in diverse tissues such as bone marrow, skin, umbilical cord, placenta, adipose tissue, and blood, as outlined by Brown *et al.* (2019).

A recent investigation revealed that MSCs can significantly reduce inflammatory markers such as IL-6 and NF-κB in cancer cells, indicating their anti-inflammatory effect (Ahmed *et al.*, 2021). Moreover, the study also demonstrated a significant decrease in cell proliferation of BC cells (MCF-7) when cocultured with MSCs compared to untreated cells (Ahmed *et al.*, 2021). These discoveries indicate that MSCs possess the capacity to reduce inflammation and impede the growth of tumor cells.

Several studies have reached the same conclusion that MSCs have the ability to hinder tumor growth and progression (Rhee *et al.*, 2015; Lan *et al.*, 2021). Although MSCs influence the biology of BC, the exact mechanism through which MSCs home in on the tumor site remains unclear and appears to be influenced by the specific tumor microenvironment (Pittenger *et al.*, 2019; Andrzejewska *et al.*, 2019). Divergent results have been documented concerning the effects of MSCs on BC (Vakhshiteh *et al.*, 2019). For instance, studies have shown that MSCs markedly decreased the proliferation and migration of MDA-MB231 cell lines in both 2D and 3D co-culture setups (Maumus *et al.*, 2020). These results indicate the potential of MSCs to impede tumor growth. Another study proposes that MSCs can diminish tumor growth by suppressing angiogenesis (Maffey *et al.*, 2017). Observations have indicated that the exchange of growth factors between co-cultured endothelial and BC cells can regulate the angiogenic response and tumor invasion capability. These investigations underscore the intricate interplay among MSCs, tumor cells, and endothelial cells within the framework of BC (Zhang *et al.*, 2018b). Moreover, prior studies have shown that MSCs possess the capacity to inhibit the PI3K/Akt pathway, resulting in the initiation of apoptosis in cancer cells and constraining their proliferation potential (Lu *et al.*, 2019).

Certain MMPs, including MMP-1, MMP-2, and MMP-9, have been directly linked to the dissemination of cancer to distant sites in the body, a process referred to as metastasis (Walsh *et al.*, 2012). TIMPs are endogenous molecules that bind to the active site of MMPs, leading to their inactivation. However, the relationship between TIMP expression and cancer prognosis is still not fully understood, as TIMPs are both overexpressed and silenced in tumors (Bandapalli *et al.*, 2012). In research conducted by Clarke *et al.* (2015), a co-culture model involving BC cells and MSCs or MSC-conditioned media was established. The results showed that MSCs release proteins that efficiently reduce the migration of cancer cells. Over time, the levels of TIMP subtypes, specifically TIMP-1 and TIMP-2, were observed to accumulate in the medium and were primarily responsible for the reduction in cancer cell migration and invasion. This suggests that MSC-produced TIMP-1 and TIMP-2 in the tumor stroma could potentially have anti-metastatic properties, counteracting the pro-tumorigenic effects of the stroma. Based on the increasing body of evidence, stem cells have the ability to stimulate natural repair mechanisms and regenerative processes through the release of paracrine substances like exosomes, miRNAs, and secretomes (Lu *et al.*, 2022). Exosomes, a subset of extracellular vesicles, are integral to intercellular communication as they ferry various biomolecules (such as oncogenes, oncomiRs, proteins, and therapeutic agents). These bioactive components can alter the transcriptome of recipient cells, thereby influencing signaling pathways linked to tumor progression (Wang *et al.*, 2021). An earlier investigation presented proof that miRNA-100, transferred from human MSCs to BC via exosomes, efficiently suppressed angiogenesis in the recipient cells. This process was accomplished by modulating the mTOR/HIF-1 α/VEGF signaling pathway in BC cells (Pakravan *et al.*, 2017).

Notably, the Wnt/β-catenin pathway has been recognized as one of the activated pathways facilitated by SCL9A1 (Sun *et al.*, 2018). This pathway is crucial for embryonic and tissue development, as well as the maintenance of stemness and the progression and metastasis of tumors. It serves as a hallmark in various types of tumors, including BC (Prosperi and Goss, 2010). Prior research documented the aberrant activation of the Wnt/β-catenin pathway in BC (VanKlompenberg *et al.*, 2015). This activation was linked to resistance to cisplatin (DDP), a common chemotherapy drug. In BC, it was discovered that the Wnt/β-catenin pathway was activated in DDP-resistant cells but can be suppressed by adMSC-Exos. Nevertheless, this suppression was reversed when miR-1236 was inhibited or SLC9A1 was overexpressed. Moreover, MSCs have been demonstrated to modulate Wnt signaling in cancer (Li *et al.*, 2021).

Another study explored the function of MSC-derived exosomal miR-551b-3p in BC. The investigators found that elevation of miR-551b-5p levels suppressed BC growth and metastasis while inducing apoptosis through decreased expression of tripartite motif-containing protein 31 (TRIM31). Additionally, they utilized a mouse xenograft model to assess the effects of MSC-derived exosomal miR-551b-3p *in vivo*. The outcomes demonstrated that MSC-derived exosomal miR-551b-3p significantly impeded tumor growth in the mouse xenograft model of BC. These results collectively indicate that MSC-derived exosomal miR-551b-3p holds promise in suppressing BC progression by reducing TRIM31 expression. Consequently, miR-551b-3p emerges as a potential therapeutic target for BC treatment (Yang *et al.*, 2022).

18.12 CONCLUSION

In conclusion, this chapter underscores the importance of investigating novel biological and physiological mechanisms in BC. The NF-κB, STAT3, PI3K/Akt, TGF-β, and Wnt/β-catenin signaling pathways function as transcriptional activators in BC. They regulate various genes, including those of cysteine proteases and MMPs, influencing BC progression, proliferation, apoptosis, metastasis, and chemo-resistance. Considering the involvement of proteases in tumor promotion, it is imperative to devise therapeutic approaches that selectively address these enzymes. Stem cell therapy effectively exhibits potential in suppressing proteases and thwarting BC probably through inhibitory effects on these signaling pathways (Figure 18.1).

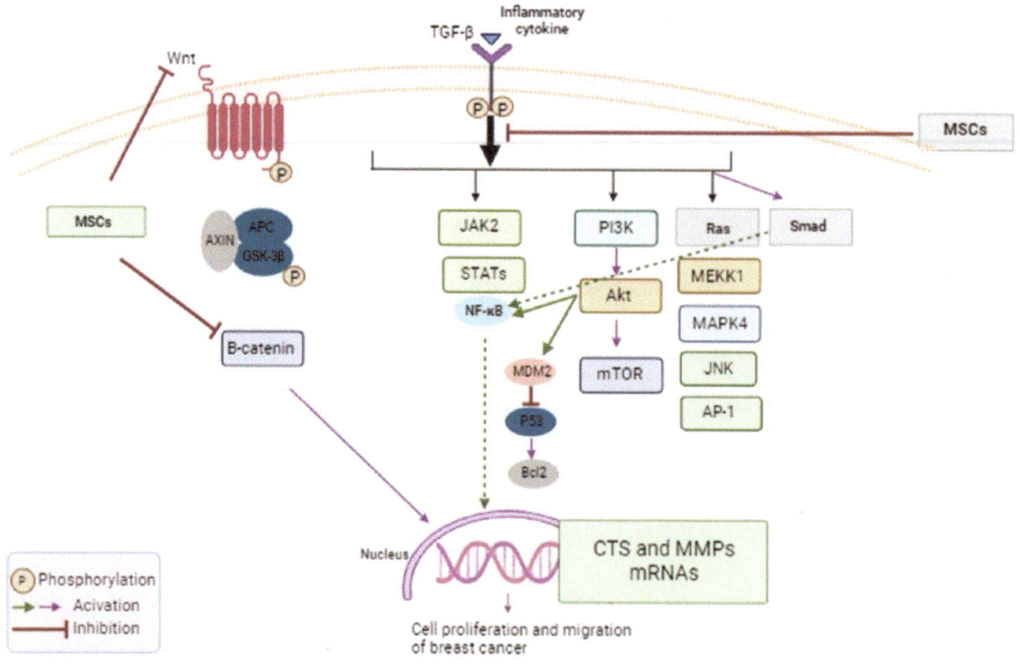

FIGURE 18.1 A summary of TGF-β, NF-κB, STAT3, PI3K/Akt, and Wnt/β-catenin signaling pathways-induced activation of CTS and MMPs, which affects BC progression, proliferation, apoptosis, and metastasis.

DECLARATIONS

Conflict of interest: The authors declare no conflicts of interest.

Consent for publishing: All authors have approved the final version of the manuscript and given consent for publication.

Funding: Not Applicable.

REFERENCES

Aggarwal, B. B., & Sung, B. (2011). NF-κB in cancer: A matter of life and death. *Cancer Discovery*, *1*(6), 469–471. https://doi.org/10.1158/2159-8290.CD-11-0260

Ahmed, S. H., Mostafa, A., & Abou-Elalla, A. A. (2021). Antiproliferative effect of mesenchymal stem cells on human breast carcinoma: New insight on FOXO/lncRNA-AF085935 axis. *Open Access Macedonian Journal of Medical Sciences* [Internet], *9*(A):748–752. https://doi.org/10.3889/oamjms.2021.6814

Anastas, J. N., & Moon, R. T. (2013). WNT signalling pathways as therapeutic targets in cancer. *Nature Reviews Cancer*, *13*(1), 11–26. https://doi.org/10.1038/nrc3419

Andrzejewska, A., Lukomska, B., & Janowski, M. (2019). Concise review: Mesenchymal stem cells: From roots to boost. *Stem Cells*, *37*(7), 855–864. https://doi.org/10.1002/stem.3016

Ashraf, Y., Mansouri, H., Laurent-Matha, V., Alcaraz, L. B., Roger, P., Guiu, S., & Liaudet Coopman, E. (2019). Immunotherapy of triple-negative breast cancer with cathepsin D-targeting antibodies. *Journal for Immunotherapy of Cancer*, *7*, 1–17. https://doi.org/10.1186/s40425-019-0498-z

Bahrami, A., Aledavood, A., Anvari, K., Hassanian, S. M., et al (2018). The prognostic and therapeutic application of microRNAs in breast cancer: Tissue and circulating microRNAs. *Journal of Cellular Physiology*, *233*(2), 774–786. https://doi.org/10.1002/jcp.25813

Bandapalli, O. R., Paul, E., Schirmacher, P., et al (2012). Opposite effects of tissue inhibitor of metalloproteinases-1 (TIMP-1) over-expression and knockdown on colorectal liver metastases. *BMC Research Notes*, *5*, 1–7. https://doi.org/10.1186/1756-0500-5-14

Bannoud, N., Carvelli, F. L., Troncoso, M., et al (2018). Cation-dependent mannose-6-phosphate receptor expression and distribution are influenced by estradiol in MCF-7 breast cancer cells. *PLoS One*, *13*(8), e0201844. https://doi.org/10.1371/journal.pone.0201844

Bardaweel, S. K., Akour, A. A., Al-Muhaissen, S., et al (2019). Oral contraceptive and breast cancer: Do benefits outweigh the risks? A case–control study from Jordan. *BMC Women's Health*, *19*(1), 1–7. https://doi.org/10.1186/s12905-019-0770-x

Barrios, C. H. (2022). Global challenges in breast cancer detection and treatment. *The Breast*, *62*, S3–S6. https://doi.org/10.1016/j.breast.2022.02.003

Bengsch, F., Buck, A., Günther, S. C., et al (2014). Cell type-dependent pathogenic functions of overexpressed human cathepsin B in murine breast cancer progression. *Oncogene*, *33*(36),4474–4484. https://doi.org/10.1038/onc.2013.395

Bian, S., Sun, X., Bai, A., et al (2013). P2X7 integrates PI3K/AKT and AMPK-PRAS40-mTOR signaling pathways to mediate tumor cell death. *PLoS One*, *8*(4), e60184. https://doi.org/10.1371/journal.pone.0060184

Bray, F., Ferlay, J., Soerjomataram, I., et al (2018). Global cancer statistics 2018: GLOBOCAN estimates of incidence and mortality worldwide for 36 cancers in 185 countries. *CA: A Cancer Journal for Clinicians*, *68*(6), 394–424. https://doi.org/10.3322/caac.21492

Brix, K., Dunkhorst, A., Mayer, K., et al (2008). Cysteine cathepsins: Cellular roadmap to different functions. *Biochimie*, *90*(2), 194–207. https://doi.org/10.1016/j.biochi.2007.07.024

Brown, C., McKee, C., Bakshi, S., et al (2019). Mesenchymal stem cells: Cell therapy and regeneration potential. *Journal of Tissue Engineering and Regenerative Medicine*, *13*(9), 1738–1755. https://doi.org/10.1002/term.2914

Brown, G. T., & Murray, G. I. (2015). Current mechanistic insights into the roles of matrix metalloproteinases in tumour invasion and metastasis. *The Journal of Pathology*, *237*(3), 273–281. https://doi.org/10.1002/path.4586

Cal, S., & López-Otín, C. (2015). ADAMTS proteases and cancer. *Matrix Biology*, *44*, 77–85. https://doi.org/10.1016/j.matbio.2015.01.013

Cao, W., Zheng, W., & Chen, T. (2015). Ruthenium polypyridyl complex inhibits growth and metastasis of breast cancer cells by suppressing FAK signaling with enhancement of TRAIL-induced apoptosis. *Scientific reports*, *5*(1), 9157. https://doi.org/10.1038/srep09157

Chambers, A. F., Colella, R., Denhardt, D. T., et al (1992). Increased expression of cathepsins L and B and decreased activity of their inhibitors in metastatic, ras-transformed NIH 3T3 cells. *Molecular Carcinogenesis*, *5*(3), 238–245. https://doi.org/10.1002/mc.2940050311

Chen, Q. K., Lee, K., Radisky, D. C., et al (2013). Extracellular matrix proteins regulate epithelial–mesenchymal transition in mammary epithelial cells. *Differentiation*, *86*(3), 126–132. https://doi.org/10.1016/j.diff.2013.03.003

Chen, Y., Di, C., Zhang, X., et al (2020). Transforming growth factor β signaling pathway: A promising therapeutic target for cancer. *Journal of Cellular Physiology*, *235*(3), 1903–1914. https://doi.org/10.1002/jcp.29108

Chen, Y. J., Lin, K. N., Jhang, L. M., et al (2016). Gallic acid abolishes the EGFR/Src/Akt/Erk-mediated expression of matrix metalloproteinase-9 in MCF-7 breast cancer cells. *Chemico-Biological Interactions*, *252*, 131–140. https://doi.org/10.1016/j.cbi.2016.04.025

Chiechi, A., Waning, D. L., Stayrook, K. R., et al (2013). Role of TGF-β in breast cancer bone metastases. *Advances in Bioscience and Biotechnology (Print)*, *4*(10C), 15. https://doi.org/10.4236/abb.2013.410A4003

Cichon, M. A., & Radisky, D. C. (2014). ROS-induced epithelial-mesenchymal transition in mammary epithelial cells is mediated by NF-κB-dependent activation of Snail. *Oncotarget*, *5*(9), 2827. https://doi.org/10.18632/oncotarget.1940

Cichon, M. A., Nelson, C. M., & Radisky, D. C. (2015). Regulation of epithelial-mesenchymal transition in breast cancer cells by cell contact and adhesion. *Cancer Informatics*, *14*, CIN-S18965.

Clarke, M. R., Imhoff, F. M., & Baird, S. K. (2015). Mesenchymal stem cells inhibit breast cancer cell migration and invasion through secretion of tissue inhibitor of metalloproteinase-1 and-2. *Molecular Carcinogenesis*, *54*(10), 1214–1219. https://doi.org/10.1002/mc.22178

Curigliano, G., Burstein, H. J., Winer, E. P., et al (2017). De-escalating and escalating treatments for early-stage breast cancer: The St. Gallen International Expert Consensus Conference on the Primary Therapy of Early Breast Cancer 2017. *Annals of Oncology*, *28*(8), 1700–1712. https://doi.org/10.1093/annonc/mdx308

Dechow, T. N., Pedranzini, L., Leitch, A., Leslie, K., Gerald, W. L., Linkov, I., & Bromberg, J. F. (2004). Requirement of matrix metalloproteinase-9 for the transformation of human mammary epithelial cells by Stat3-C. *Proceedings of the National Academy of Sciences*, *101*(29), 10602–10607. https://doi.org/10.1073/pnas.0404100101

DeSantis, C. E., Ma, J., Gaudet, M. M., et al (2019). Breast cancer statistics, 2019. *CA: A Cancer Journal for Clinicians*, *69*(6), 438–451. https://doi.org/10.3322/caac.21583

Dian, D., Heublein, S., Wiest, I., Barthell, L., Friese, K., & Jeschke, U. (2014). Significance of the tumor protease cathepsin D for the biology of breast cancer. https://doi.org/10.14670/HH-29.10.433

Dihge, L., Ohlsson, M., Edén, P., et al (2019). Artificial neural network models to predict nodal status in clinically node-negative breast cancer. *BMC Cancer*, *19*(1), 1–12. https://doi.org/10.1186/s12885-019-5827-6

Dittmer, J. (2018, December). Breast cancer stem cells: Features, key drivers and treatment options. *Seminars in Cancer Biology*, *53*, 59–74. Academic Press. https://doi.org/10.1016/j.semcancer.2018.07.007

Dong, H., Diao, H., Zhao, Y., et al (2019). Overexpression of matrix metalloproteinase-9 in breast cancer cell lines remarkably increases the cell malignancy largely *via* activation of transforming growth factor beta/SMAD signalling. *Cell Proliferation*, *52*(5), e12633. https://doi.org/10.1111/cpr.12633

Duffy, M. J., Maguire, T. M., Hill, A., McDermott, E., & O'Higgins, N. (2000). Metalloproteinases: role in breast carcinogenesis, invasion and metastasis. *Breast Cancer Research*, *2*(4), 1–6. https://doi.org/10.1186/bcr65

Fiegen, D., Haeusler, L. C., Blumenstein, L., Herbrand, U., Dvorsky, R., Vetter, I. R., & Ahmadian, M. R. (2004). Alternative splicing of Rac1 generates Rac1b, a self-activating GTPase. *Journal of Biological Chemistry*, *279*(6), 4743–4749. https://doi.org/10.1074/jbc.M310281200

Furtek, S. L., Backos, D. S., Matheson, C. J., et al (2016). Strategies and approaches of targeting STAT3 for cancer treatment. *ACS Chemical Biology*, *11*(2), 308–318. https://doi.org/10.1021/acschembio.5b00945

Gocheva, V., & Joyce, J. A. (2007). Cysteine cathepsins and the cutting edge of cancer invasion. *Cell Cycle*, *6*(1), 60–64. https://doi.org/10.4161/cc.6.1.3669

Hiraoka, N., Allen, E., Apel, I. J., et al (1998). Matrix metalloproteinases regulate neovascularization by acting as pericellular fibrinolysins. *Cell*, *95*(3), 365–377. https://doi.org/10.1016/S0092-8674(00)81768-7

Holnthoner, W., Kerenyi, M., Gröger, M., et al (2006). Regulation of matrilysin expression in endothelium by fibroblast growth factor-2. *Biochemical and Biophysical Research Communications*, *342*(3), 725–733. https://doi.org/10.1016/j.bbrc.2006.02.011

Hounsell, C., & Fan, Y. (2021). The duality of caspases in cancer, as told through the fly. *International Journal of Molecular Sciences*, *22*(16), 8927. https://doi.org/10.3390/ijms22168927

Huang, C., Yang, G., Jiang, T., et al (2010). Effects of IL-6 and AG490 on regulation of Stat3 signaling pathway and invasion of human pancreatic cancer cells *in vitro*. *Journal of Experimental & Clinical Cancer Research*, *29*, 1–8. https://doi.org/10.1186/1756-9966-29-51

Im, N. K., Jang, W. J., Jeong, C. H., et al (2014). Delphinidin suppresses PMA-induced MMP-9 expression by blocking the NF-κB activation through MAPK signaling pathways in MCF-7 human breast carcinoma cells. *Journal of Medicinal Food*, *17*(8), 855–861. https://doi.org/10.1089/jmf.2013.3077

Imamura, T., Hikita, A., & Inoue, Y. (2012). The roles of TGF-β signaling in carcinogenesis and breast cancer metastasis. *Breast Cancer*, *19*, 118–124. https://doi.org/10.1007/s12282-011-0321-2

Iozzo, R. V., Zoeller, J. J., & Nyström, A. (2009). Basement membrane proteoglycans: Modulators Par Excellence of cancer growth and angiogenesis. *Molecules and Cells*, *27*, 503–513. https://doi.org/10.1007/s10059-009-0069-0

Jedeszko, C., & Sloane, B. F. (2004). Cysteine cathepsins in human cancer. https://doi.org/10.1515/BC.2004.132

Jevnikar, Z., Obermajer, N., Doljak, B., et al (2011). Cathepsin X cleavage of the β2 integrin regulates talin-binding and LFA-1 affinity in T cells. *Journal of Leukocyte Biology*, *90*(1), 99–109. https://doi.org/10.1189/jlb.1110622

Jiang, N., Dai, Q., Su, X., et al (2020). Role of PI3K/AKT pathway in cancer: The framework of malignant behavior. *Molecular Biology Reports*, *47*(6), 4587–4629. https://doi.org/10.1007/s11033-020-05435-1

Jung, C. H., Kim, E. M., Park, J. K., et al (2013). Bmal1 suppresses cancer cell invasion by blocking the phosphoinositide 3 kinase-Akt-MMP-2 signaling pathway. *Oncology Reports*, *29*(6), 2109–2113. https://doi.org/10.3892/or.2013.2381

Kalkavan, H., & Green, D. R. (2018). MOMP, cell suicide as a BCL-2 family business. *Cell Death & Differentiation*, *25*(1), 46–55. https://doi.org/10.1038/cdd.2017.179

Kim, H. S., Kim, H. J., Park, K. G., et al (2007). α-Lipoic acid inhibits matrix metalloproteinase-9 expression by inhibiting NF-κB transcriptional activity. *Experimental & Molecular Medicine*, *39*(1), 106–113. https://doi.org/10.1038/emm.2007.12

Kitaw, M. E. (2019). Regulation of urokinase receptor (uPAR) cleavage by cancer associated fibroblasts (CAFs) in oral cancer (Master's thesis, UiT Norges arktiske universitet).

Kos, J. (2022). Proteases: Role and function in cancer. *International Journal of Molecular Sciences*, *23*(9), 4632. https://doi.org/10.3390/ijms23094632

Lamb, R., Ablett, M. P., Spence, K., et al (2013). Wnt pathway activity in breast cancer sub-types and stem-like cells. *PLoS One*, *8*(7), e67811. https://doi.org/10.1371/journal.pone.0067811

Lan, T., Luo, M., & Wei, X. (2021). Mesenchymal stem/stromal cells in cancer therapy. *Journal of Hematology & Oncology*, *14*(1), 1–16. https://doi.org/10.1186/s13045-021-01208-w

Leng, L., Wang, Y., He, N., et al (2014). Molecular imaging for assessment of mesenchymal stem cells mediated breast cancer therapy. *Biomaterials*, *35*(19), 5162–5170. https://doi.org/10.1016/j.biomaterials.2014.03.014

Li Y, Zhong X, Zhang Y, et al (2021). Mesenchymal stem cells in gastric cancer: Vicious but hopeful. *Frontiers in Oncology*, *11*, 617677. https://doi.org/10.3389/fonc.2021.617677

López-Otín, C., & Bond, J. S. (2008). Proteases: Multifunctional enzymes in life and disease. *Journal of Biological Chemistry*, *283*(45), 30433–30437. https://doi.org/10.1074/jbc.R800035200

Lu, L., Chen, G., Yang, J., et al (2019). Bone marrow mesenchymal stem cells suppress growth and promote the apoptosis of glioma U251 cells through downregulation of the PI3K/AKT signaling pathway. *Biomedicine & Pharmacotherapy*, *112*, 108625. https://doi.org/10.1016/j.biopha.2019.108625

Lu, L., Xu, A., Gao, F., et al (2022). Mesenchymal stem cell-derived exosomes as a novel strategy for the treatment of intervertebral disc degeneration. *Frontiers in Cell and Developmental Biology*, *9*, 770510. https://doi.org/10.3389/fcell.2021.770510

Maffey, A., Storini, C., Diceglie, C., et al (2017). Mesenchymal stem cells from tumor microenvironment favour breast cancer stem cell proliferation, cancerogenic and metastatic potential, via ionotropic purinergic signalling. *Scientific Reports*, *7*(1), 13162. https://doi.org/10.1038/s41598-017-13460-7

Manes, S., Llorente, M., Lacalle, R. A., et al (1999). The matrix metalloproteinase-9 regulates the insulin-like growth factor-triggered autocrine response in DU-145 carcinoma cells. *Journal of Biological Chemistry*, *274*(11), 6935–6945. https://doi.org/10.1074/jbc.274.11.6935

Manning, B. D., & Toker, A. (2017). AKT/PKB signaling: Navigating the network. *Cell*, *169*(3), 381–405. https://doi.org/10.1016/j.cell.2017.04.001

Maumus, M., Rozier, P., Boulestreau, J., et al (2020). Mesenchymal stem cell-derived extracellular vesicles: opportunities and challenges for clinical translation. *Frontiers in Bioengineering and Biotechnology*, *8*, 997. https://doi.org/10.3389/fbioe.2020.00997

Mimeault, M., & Batra, S. K. (2014). Altered gene products involved in the malignant reprogramming of cancer stem/progenitor cells and multitargeted therapies. *Molecular Aspects of Medicine, 39*, 3–32. https://doi.org/10.1016/j.mam.2013.08.001

Murga, C., Fukuhara, S., & Gutkind, J. S. (2000). A novel role for phosphatidylinositol 3-kinase β in signaling from G protein-coupled receptors to Akt. *Journal of Biological Chemistry, 275*(16), 12069–12073. https://doi.org/10.1074/jbc.275.16.12069

O-charoenrat, P., Wongkajornsilp, A., Rhys-Evans, P. H., et al (2004). Signaling pathways required for matrix metalloproteinase-9 induction by betacellulin in head-and-neck squamous carcinoma cells. *International Journal of Cancer, 111*(2), 174–183. https://doi.org/10.1002/ijc.20228

Ohno-Matsui, K., Uetama, T., Yoshida, T., et al (2003). Reduced retinal angiogenesis in MMP-2–deficient mice. *Investigative Ophthalmology & Visual Science, 44*(12), 5370–5375. https://doi.org/10.1167/iovs.03-0249

Pakravan, K., Babashah, S., Sadeghizadeh, M., et al (2017). MicroRNA-100 shuttled by mesenchymal stem cell-derived exosomes suppresses *in vitro* angiogenesis through modulating the mTOR/HIF-1α/VEGF signaling axis in breast cancer cells. *Cellular Oncology, 40*, 457–470. https://doi.org/10.1007/s13402-017-0335-7

Pal, S., Moulik, S., Dutta, A., & Chatterjee, A. (2014). Extracellular matrix protein laminin induces matrix metalloproteinase-9 in human breast cancer cell line mcf-7. *Cancer Microenvironment, 7*, 71–78. https://doi.org/10.1007/s12307-014-0146-6

Pittenger, M. F., Discher, D. E., Péault, B. M., et al (2019). Mesenchymal stem cell perspective: Cell biology to clinical progress. *NPJ Regenerative Medicine, 4*(1), 22. https://doi.org/10.1038/s41536-019-0083-6

Pivetta, E., Scapolan, M., Pecolo, M., et al (2011). MMP-13 stimulates osteoclast differentiation and activation in tumour breast bone metastases. *Breast Cancer Research, 13*, 1–15. https://doi.org/10.1186/bcr3047

Pranjol, Z. I., & Whatmore, J. L. (2020). Cathepsin D in the tumor microenvironment of breast and ovarian cancers. *Tumor Microenvironment: Molecular Players–Part A*, 1–16. https://doi.org/10.1007/978-3-030-43093-1_1

Prosperi, J. R., & Goss, K. H. (2010). A Wnt-ow of opportunity: Targeting the Wnt/beta-catenin pathway in breast cancer. *Curr Drug Targets, 11*(9), 1074–1088. https://doi.org/10.2174/138945010792006780

Radisky, D. C., Levy, D. D., Littlepage, L. E., et al (2005). Rac1b and reactive oxygen species mediate MMP-3-induced EMT and genomic instability. *Nature, 436*(7047), 123–127. https://doi.org/10.1038/nature03688

Rawlings, N. D., Barrett, A. J., & Finn, R. (2016). Twenty years of the MEROPS database of proteolytic enzymes, their substrates and inhibitors. *Nucleic Acids Research, 44*(D1), D343–D350. https://doi.org/10.1093/nar/gkv1118

Rhee, K. J., Lee, J. I., & Eom, Y. W. (2015). Mesenchymal stem cell-mediated effects of tumor support or suppression. *International Journal of Molecular Sciences, 16*(12), 30015–30033. https://doi.org/10.3390/ijms161226215

Schoeberl, B., Pace, E. A., Fitzgerald, J. B., et al (2009). Therapeutically targeting ErbB3: A key node in ligand-induced activation of the ErbB receptor–PI3K axis. *Science Signaling, 2*(77), ra31. https://doi.org/10.1126/scisignal.2000352

Siveen, K. S., Sikka, S., Surana, R., Dai, X., Zhang, J., Kumar, A. P., & Bishayee, A. (2014). Targeting the STAT3 signaling pathway in cancer: role of synthetic and natural inhibitors. *Biochimica et Biophysica Acta (BBA)-Reviews on Cancer, 1845*(2), 136–154. https://doi.org/10.1016/j.bbcan.2013.12.005

Sternlicht, M. D., Lochter, A., Sympson, C. J., Huey, B., Rougier, J. P., Gray, J. W., & Werb, Z. (1999). The stromal proteinase MMP3/stromelysin-1 promotes mammary carcinogenesis. *Cell, 98*(2), 137–146. https://doi.org/10.1016/S0092-8674(00)81009-0

Skrzydlewska, E., Sulkowska, M., Koda, M., et al (2005). Proteolytic-antiproteolytic balance and its regulation in carcinogenesis. *World Journal of Gastroenterology: WJG, 11*(9), 1251. https://doi.org/10.3748/wjg.v11.i9.1251

Soond, S. M., Kozhevnikova, M. V., Savvateeva, L. V., et al (2021). Intrinsically connected: Therapeutically targeting the cathepsin proteases and the Bcl-2 family of protein substrates as co-regulators of apoptosis. *International Journal of Molecular Sciences, 22*(9), 4669. https://doi.org/10.3390/ijms22094669

Strojnik, T., Kavalar, R., Trinkaus, M., et al. (2005). Cathepsin L in glioma progression: Comparison with cathepsin B. *Cancer Detection and Prevention, 29*(5), 448–455. https://doi.org/10.1016/j.cdp.2005.07.006

Sun, D. I., Tasca, A., Haas, M., et al. (2018). Na+/H+ exchangers are required for the development and function of vertebrate mucociliary epithelia. *Cells Tissues Organs, 205*(5–6), 279–292. https://doi.org/10.1159/000492973

Tan, Y., Luo, X., Lv, W., et al. (2021). Tumor-derived exosomal components: The multifaceted roles and mechanisms in breast cancer metastasis. *Cell Death & Disease*, *12*(6), 547. https://doi.org/10.1038/s41419-021-03825-2

Tan, G. J., Peng, Z. K., Lu, J. P., et al. (2013). Cathepsins mediate tumor metastasis. *World Journal of Biological Chemistry*, *4*(4), 91. https://doi.org/10.4331/wjbc.v4.i4.91

Tejeda-Muñoz, N., & Robles-Flores, M. (2015). Glycogen synthase kinase 3 in Wnt signaling pathway and cancer. *IUBMB Life*, *67*(12), 914–922. https://doi.org/10.1002/iub.1454

Teufelsbauer, M., Rath, B., Moser, D., et al (2019). Interaction of adipose-derived stromal cells with breast cancer cell lines. *Plastic and Reconstructive Surgery*, *144*(2), 207e–217e. https://doi.org/10.1097/PRS.0000000000005839

Trisciuoglio, D., Iervolino, A., Zupi, G., et al (2005). Involvement of PI3K and MAPK signaling in bcl-2-induced vascular endothelial growth factor expression in melanoma cells. *Molecular Biology of the Cell*, *16*(9), 4153–4162. https://doi.org/10.1091/mbc.e04-12-1087

Tsuda, T. (2018). Extracellular interactions between fibulins and transforming growth factor (TGF)-β in physiological and pathological conditions. *International Journal of Molecular Sciences*, *19*(9), 2787. https://doi.org/10.3390/ijms19092787

Turk, B., Turk, D., & Turk, V. (2012). Protease signalling: The cutting edge. *The EMBO Journal*, *31*(7), 1630–1643. https://doi.org/10.1038/emboj.2012.42

Vakhshiteh, F., Atyabi, F., & Ostad, S. N. (2019). Mesenchymal stem cell exosomes: A two-edged sword in cancer therapy. *International Journal of Nanomedicine*, 2847–2859. https://doi.org/10.2147/IJN.S200036

VanKlompenberg, M. K., Bedalov, C. O., Soto, K. F., et al (2015) APC selectively mediates response to chemotherapeutic agents in breast cancer. *BMC Cancer*, *15*, 457. https://doi.org/10.1186/s12885-015-1456-x

Vasiljeva, O., Korovin, M., Gajda, M., et al (2008). Reduced tumour cell proliferation and delayed development of high-grade mammary carcinomas in cathepsin B-deficient mice. *Oncogene*, *27*(30), 4191–4199. https://doi.org/10.1038/onc.2008.59

Vasiljeva, O., Reinheckel, T., Peters, C., et al (2007). Emerging roles of cysteine cathepsins in disease and their potential as drug targets. *Current Pharmaceutical Design*, *13*(4), 387–403. https://doi.org/10.2174/138161207780162962

Vázquez-Cuevas, F. G., Martínez-Ramírez, A. S., Robles-Martínez, L., et al (2014). Paracrine stimulation of P2X7 receptor by ATP activates a proliferative pathway in ovarian carcinoma cells. *Journal of Cellular Biochemistry*, *115*(11), 1955–1966. https://doi.org/10.1002/jcb.24867

Villar, J., Cabrera, N. E., Valladares, F., Casula, M., Flores, C., Blanch, L., & Slutsky, A. S. (2011). Activation of the Wnt/β-catenin signaling pathway by mechanical ventilation is associated with ventilator-induced pulmonary fibrosis in healthy lungs. *PloS One*, *6*(9), e23914. https://doi.org/10.1371/journal.pone.0023914

Walsh, L. A., Cepeda, M. A., & Damjanovski, S. (2012). Analysis of the MMP-dependent and independent functions of tissue inhibitor of metalloproteinase-2 on the invasiveness of breast cancer cells. *Journal of Cell Communication and Signaling*, *6*, 87–95. https://doi.org/10.1007/s12079-011-0157-8

Wang, S. M., Li, L., Zhang, W., et al (2010). Relationship between cathepsin L and invasion and metastasis of ovarian carcinoma cells. *Zhonghua Fu Chan Ke Za Zhi*, *45*(8), 598–602.

Wang, Y. R., Qin, S., Han, R., et al (2013). Cathepsin L plays a role in quinolinic acid-induced NF-κβ activation and excitotoxicity in rat striatal neurons. *PLoS One*, *8*(9), e75702. https://doi.org/10.1371/journal.pone.0075702

Wang, C., Zhang, R., Wang, X., et al (2021). Silencing of KIF3B suppresses breast cancer progression by regulating EMT and Wnt/β-catenin signaling. *Frontiers in Oncology*, *10*, 597464. https://doi.org/10.3389/fonc.2020.597464

Wu, Z. Q., Li, X. Y., Hu, C. Y., et al (2012). Canonical Wnt signaling regulates Slug activity and links epithelial–mesenchymal transition with epigenetic Breast Cancer 1, Early Onset (BRCA1) repression. *Proceedings of the National Academy of Sciences*, *109*(41), 16654–16659. https://doi.org/10.1073/pnas.1205822109

Xia, J., Yu, X., Tang, L., et al (2015). P2X7 receptor stimulates breast cancer cell invasion and migration *via* the AKT pathway. *Oncology Reports*, *34*(1), 103–110. https://doi.org/10.3892/or.2015.3979

Yamamoto, Y., & Gaynor, R. B. (2001). Therapeutic potential of inhibition of the NF-κB pathway in the treatment of inflammation and cancer. *The Journal of Clinical Investigation*, *107*(2), 135–142. https://doi.org/10.1172/JCI11914

Yamamoto, K., Murphy, G., & Troeberg, L. (2015). Extracellular regulation of metalloproteinases. *Matrix Biology*, *44*, 255–263. https://doi.org/10.1016/j.matbio.2015.02.007

Yang, Z., Xu, B., Wu, S., et al (2022). Exosomal microRNA-551b-3p from bone marrow-derived mesenchymal stromal cells inhibits breast cancer progression regulating TRIM31/Akt signaling. *Human Cell, 35*(6), 1797–1812. https://doi.org/10.1007/s13577-022-00753-x

Yuan, G., Qian, L., Song, L., et al (2008). Heregulin-β promotes matrix metalloproteinase-7 expression *via* HER2-mediated AP-1 activation in MCF-7 cells. *Molecular and Cellular Biochemistry, 318*, 73–79. https://doi.org/10.1007/s11010-008-9858-6

Yu, H., Pardoll, D., & Jove, R. (2009). STATs in cancer inflammation and immunity: A leading role for STAT3. *Nature Reviews Cancer, 9*(11), 798–809. https://doi.org/10.1038/nrc2734

Yu, Q., & Stamenkovic, I. (2000). Cell surface-localized matrix metalloproteinase-9 proteolytically activates TGF-β and promotes tumor invasion and angiogenesis. *Genes & Development, 14*(2), 163–176.

Zhang, M., Liu, J., Li, M., et al (2018a). Insulin like growth factor 1/insulin like growth factor 1 receptor signaling protects against cell apoptosis through the PI3K/AKT pathway in glioblastoma cells. *Experimental and Therapeutic Medicine, 16*(2), 1477–1482. https://doi.org/10.3892/etm.2018.6336

Zhang, W., Xu, J., Fang, H., et al (2018b). Endothelial cells promote triple-negative breast cancer cell metastasis *via* PAI-1 and CCL5 signaling. *The FASEB Journal, 32*(1), 276–288. https://doi.org/10.1096/fj.201700237RR

Zhang, M., Zhang, Y. Y., Chen, Y., et al (2021). TGF-β signaling and resistance to cancer therapy. *Frontiers in Cell and Developmental Biology, 9*, 786728. https://doi.org/10.3389/fcell.2021.786728

Zhang, H., Zhang, X., Wu, X., et al (2012). Interference of Frizzled 1 (FZD1) reverses multidrug resistance in breast cancer cells through the Wnt/β-catenin pathway. *Cancer Letters, 323*(1), 106–113. https://doi.org/10.1016/j.canlet.2012.03.039

19 Application of Metalloproteases in Cancer Stem Cell Therapy

Uju Dorathy Iliemene and Paul T. Olonishuwa
Science & Technology Faculty, Bingham University, Karu, Nigeria

Mubarak Labaran Liman
Nuhu Bamalli Polytechnic Zaria, Zaria, Nigeria

19.1 INTRODUCTION

Cancer poses a significant global health burden, and conventional treatment modalities often fall short of achieving complete remission. Stem cell therapy offers a unique approach by leveraging stem cells' regenerative potential and ability to selectively target tumors. Proteases, present in plants, animals, and microbes, are ubiquitous entities in nature. The multifaceted actions of proteases have far-reaching consequences, influencing diverse biological procedures encompassing DNA division and transcription, cell ontogenesis and adaptation, tissue development and remodeling, unfolded protein responses and heat shock, ovulation, angiogenesis, wound healing, neurogenesis, fertilization, hemostasis, stem cell mobilization, blood coagulation, autophagy, inflammation, immunity, senescence, apoptosis, and necrosis. The fundamental roles played by proteases in cellular behavior, as well as in the survival and demise of all organisms, are underscored by the fact that dysregulation of proteolytic systems underlies various pathological conditions such as cardiovascular and inflammatory diseases, cancer, and neurodegenerative disorders. Consequently, proteases have garnered significant observation in the medicinal drug industry as prospective reference points for drug development, as well as diagnostic and prognostic biomarkers (Turk, 2006). Due to the chemical process of catalysis, proteases can be categorized into six different groups: metalloproteases, aspartic, cysteine, serine, glutamic, and threonine proteases. In mammals, glutamic proteases are yet to be found (Carlos and Judith, 2008).

Metalloproteases, which play pivotal roles in cell migration, invasion, and tissue remodeling, have emerged as potential candidates for enhancing the efficacy of cancer alleviants using stem cells. Metalloproteases, the most extensive group of proteases in humans, stand out based on the existence of a catalytic metal ion situated in their active site. Among this family of enzymes, you can find MMPs, disintegrin, and metalloproteases (ADAMs), as well as ADAMs with a thrombospondin motif (ADAMTSs) (Apte, 2020). In the human system, scientists have characterized 24 matrix metalloproteinase (MMPs) (Table 19.1), which can be exudate or bound in the membrane. The primary role of MMPs is associated with the extracellular matrix (ECM) remodeling. These enzymes perform a crucial function in breaking down different structural parts of the ECM, like native collagens, fibronectin, elastin, laminin, and proteoglycans. Additionally, they are instrumental in activating pro-MMPs through proteolytic processes.

DOI: 10.1201/9781003394693-22

TABLE 19.1
List of the 24 Identified Human Matrix Metalloproteases (MMPs)

MMP	Alternative Name	MMP	Alternative Name
1	Collagenase-1 Interstitial collagenase	16	Type-3 membrane MMP
2	Gelatinase A 72-kDa type IV collagenase	17	Type-4 membrane MMP
3	Stromelysin-1 Transin-1	18	Collagenase-4
7	Matrilysin Pump-1	19	None MMP-18 human ortholog of Xenopus RASI-1
8	Collagenase-2 Neutrophil collagenase	20	Enamelysin
9	Gelatinase B 92-kDa type IV collagenase	21	None xMMP human ortholog of Xenopus
10	Stromelysin-2	23	Cysteine array MMP Femalysin MMP-22
11	Stromelysin-3	24	Type-5 membrane MMP (MT5-MMP)
12	Macrophage metalloelastase	25	Membrane type-6 MMP(MT5-MMP) Leukolysin
13	Collagenase-3	26	Matrilysin-2/endometase
14	Type-1 membrane MMP	27	None
15	Type-2 membrane MMP	28	Epilysin

19.2 CANCER STEM CELL

Cancer stem cells (CSCs), also known as tumor-initiating cells, are a minute subpopulation of cells inside tumors that are chiefly available in a state of dormancy and have characteristics reminiscent of normal stem cells of adults. They dwell in distinct microenvironments characterized as niches that maintain the inactive state of cancer stem cells or stimulate their rapid growth and differentiation through the intercellular crosstalk and physical components. Niches also influence CSC metastatic growth capability, as well as play a part in protecting them over the immune system (Oshimori et al., 2021). The CSCs differ from the stem cells of normal adults by their capability to self-rejuvenate yet differentiate through biological groups of progenies to reach mature, functionally specialized cell types (i.e., asymmetric division). Those mature cells form the majority of the cancer tumor mass, which is originated and sustained by CSCs, allowing rapid development of tumors (Batlle and Clevers, 2017).

Some proteases including metalloproteases play essential roles within the CSC niches. They degrade the ECM niche and prejudice many CSC characteristics, including their invasion, niche retention, self-rejuvenation, therapy resistance, immune circumlocution, and stemness (Figure 19.1). They either inhibit (–) or support (+) CSC-related processes. They take part in the suppression of tumors and the downregulation of cancer. The inability of tumors to create their own blood supply limits their growth. Therefore, the importance of anti-angiogenic therapeutic agents is evident as it prevents the start of fresh blood vessel and/or tumor-associated vasculature normalization. Penetration of therapeutic agents can be facilitated by returning the tumor-associated vasculature to its normal state. Angiogenesis inhibitors used in cancer treatment have recorded evidential revitalization in the general subsistence and elongated progression-free survival (Yu et al., 2016).

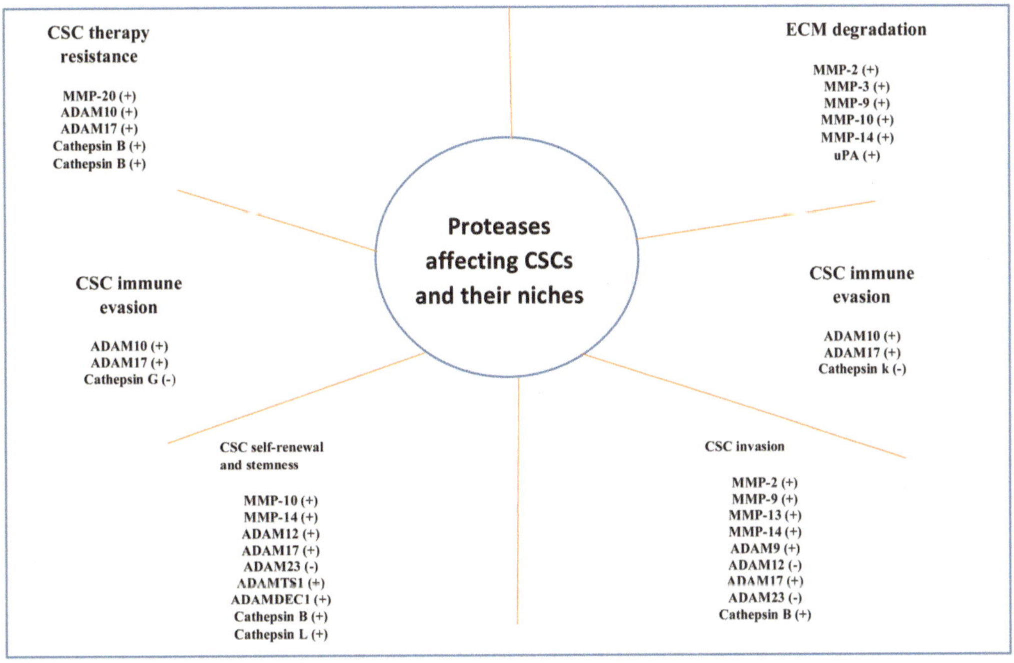

FIGURE 19.1 The effects of proteases on cancer stem cell functions and their niches.

19.3 METALLOPROTEASES: MECHANISMS OF ACTION

Tumor-hosted angiogenesis involving inhibiting enzymes has been identified as a potential medium in anti-cancer therapy. Broad-spectrum and reasonably specific MMP inhibitors have been acknowledged for their anti-angiogenic properties. Most substances that block MMPs have a chemical group that binds to a metal ion in the MMPs and makes it easier for another molecule to attach to it. Most MMP inhibitors incorporate a hydroxamic acid group that binds to the active site's Zn^{2+} ion, enabling covalent alteration by a nucleophile within the active site. Furthermore, the entry of MMP to the nucleus could be explained due to the nuclear localization signal content in their sequences. Conversely, it might be transported into the nucleus via RNA-binding cargo proteins (Frolova et al., 2020). On the other hand, it could be brought into the nucleus through the RNA-binding cargo proteins (Frolova et al., 2020). Later studies uncovered a process in which deprotonation adjacent to the sulfone group took place in the methylene, initiating the opening of the thiirane ring and resulting in the creation of a stable Zn^{2+}-thiolate complex. Thiirane is a three-membered ring structure that is highly reactive and contains a sulfur atom. Its characteristic properties differ from that of oxiranes, alkyl, and aryl sulfides. Thiirane through episulfidation, can be synthesized from alkenes. The use of alkyl lithium dialkyl magnesium halide in ring-opening reactions of thiirane dioxide demonstrates a selective and powerful mechanism-based inhibition strategy targeting the gelatinase subclass within the zinc-dependent MMP family. In other words, thiol-containing structures associated with the opening of thiirane 1 exhibit significant inhibitory activity against MMPs. This occurs because thiirane 1 functions as a thiol in a concealed manner, becoming selectively unmasked within the active sites of gelatinases. Metalloproteases typically employ a nucleophilic attack involving a water molecule as part of the amide bond hydrolysis process. The addition of water to the amide carbonyl leads to the formation of a tetrahedral transition state (Fields, 2019). It plays a role in the therapeutic benefits of stem cells in cancer treatment by involving various mechanisms, which are discussed next.

19.3.1 Extracellular Matrix Remodeling and Cell Invasion

In both regular tissue maintenance and cancer, the actions of adult stem cells are significantly shaped by the microenvironment. Referred to as the stem cell niche, this microenvironment is where neighboring cells and the surrounding ECM offer specific signals and factors. These signals and factors serve to sustain the integrity of stem cells while regulating their self-renewal and proliferation to prevent uncontrolled growth (Scadden, 2006). MMPs, thanks to their capability to cleave, break down, and reorganize ECM molecules, possess the potential to influence a variety of stem cell niches.

However, the role of MMPs in regulating stem cells goes beyond simple ECM degradation and is more complex. Recent studies have revealed the significance of MMP-3, also referred to as stromelysin-1, in terms of maintaining mature epithelial stem cells within the mammary gland. More precisely, MMP-3 has been identified as binding to and inactivating the non-canonical Wnt ligand Wnt5b, thereby functioning as a modulator of Wnt signaling.

Through its interaction with Wnt5b, MMP-3 can shift the equilibrium of signaling pathways, promoting the canonical pathway and resulting in enhanced signaling via β-catenin, ultimately leading to the expansion of the mammary stem cell population (Figure 19.2). Interestingly, MMP-3 appears to have the potential to induce these effects even when its proteolytic activity is not involved. The overexpression of inactive mutants of MMP-3, or solely the hemopexin domain, was enough to trigger hyperplastic growth in the mammary gland (Kessenbrock et al., 2013). Another MMP closely related to MMP-3 is MMP-10, also known as stromelysin-2. In the context of Kras-driven lung cancer, MMP-10 has been associated with lung tumor formation by promoting the proliferation of bronchioalveolar stem cells.

Although the direct substrates or interaction partners of MMP-10 in this process have not been identified, research has shown that MMP-10 plays a direct causal role. The absence of MMP-10 in mice led to a substantial reduction in both the quantity and size of Kras-induced lung tumors (Regala et al., 2011). These investigations offer convincing instances of how modifications in the microenvironment, such as the heightened expression or absence of particular MMPs, can modify signaling pathways linked to the expansion of stem cells, consequently elevating the susceptibility

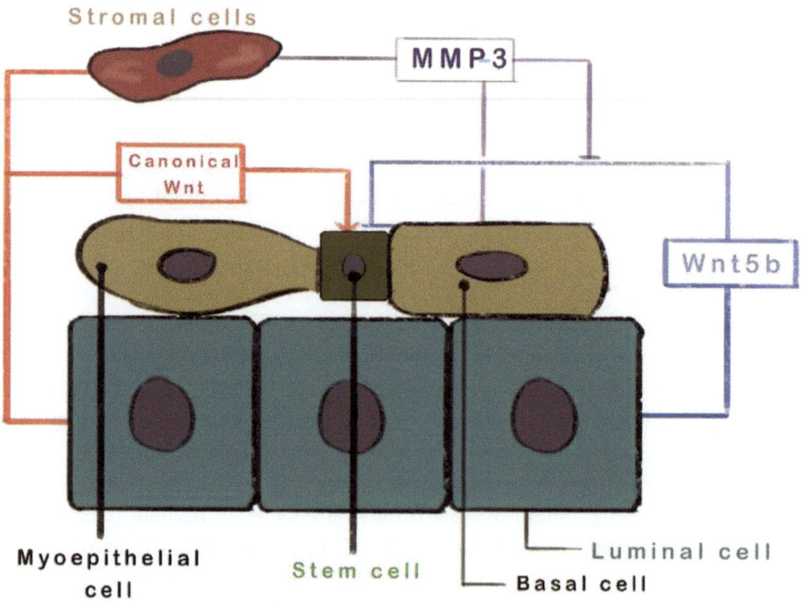

FIGURE 19.2 This mechanism elucidates the specific interaction between MMP-3 and the non-canonical Wnt ligand Wnt5b, which acts as an inhibitor of canonical Wnt signaling.

to abnormal tissue growth. It highlights the intricate interplay between MMPs, ECM remodeling, and stem cell regulation in both normal and pathological conditions.

The successful spread of cancer cells, known as metastasis, relies not only on a supportive local environment for primary tumor growth but also on the establishment of a specialized niche at the metastatic site (Psaila and Lyden, 2009). This specialized environment encompasses diverse elements of the ECM and the enzymes tasked with modifying it, including MMPs. Additionally, different cell types, including bone marrow-derived cells (BMDCs), endothelial cells, and fibroblasts, contribute to the formation of this niche. For instance, prior to the arrival of tumor cells, certain types of cancer cells, like Lewis lung carcinoma (LLC) cells, increase the production of fibronectin at potential metastatic sites. These elevated fibronectin levels promote the clustering of BMDCs through the $\alpha4\beta1$ integrin receptor (Kaplan et al., 2005). These recruited BMDCs express MMP-9, which aids in breakdown of the basement membrane, facilitating the arrival of tumor cells to the metastatic site. As a result, the early stages of metastasis involve the alteration of the ECM, facilitating the establishment and proliferation of circulating tumor cells in distant organs.

Besides fibronectin, a range of other constituents within the ECM also play a role in shaping the metastatic niche. For example, in the early stages, breast cancer cells often exhibit the presence of tenascin C, a protein typically associated with stem cell niches (Oskarsson et al., 2011). Later, stromal cells also produce tenascin C to support the survival and growth of lung metastases. Another constituent of the ECM, known as periostin, known for its role in bone and tooth development, plays a significant role in promoting metastatic colonization. Periostin attracts Wnt ligands, leading to an elevation in Wnt signaling within cancer stem cells, which promotes their proliferation (Malanchi et al., 2011).

Fibroblasts induce the expression of periostin via transforming growth factor-β (TGFβ) in response to infiltrating tumor cells. Periostin expression is crucial for the maintenance of cancer stem cells. It's noteworthy that endothelial tip cells, located at the forefront of newly formed blood vessels, release both periostin and TGFβ, promoting angiogenesis and facilitating the initial growth of micrometastases (Ghajar et al., 2013). Additionally, in the case of LLC, the conditioned medium contains a large chondroitin sulfate proteoglycan called versican, which is upregulated by various human tumors. Versican triggers the activation of macrophages via Toll-like receptor 2 (TLR2), resulting in the synthesis of interleukin-6 (IL-6) and tumor necrosis factor (TNF). This establishes an inflammatory microenvironment that supports metastatic growth.

In gliomas, osteopontin fosters stem-cell-like characteristics and resistance to radiation therapy via CD44 signaling (Pietras et al., 2014). Systematic approaches to identify responses to the ECM have revealed additional ECM molecules that promote metastasis (Reticker-Flynn et al., 2012). Together, these investigations underscore the regulatory function of the ECM in metastasis and propose that focusing on the ECM niche could present a hopeful therapeutic approach in the treatment of cancer. Metalloproteases have the capacity to break down elements of the ECM, thereby assisting tumor cell invasion and enhancing the penetration of therapeutic substances. This remodeling process enables stem cells to reach and target the tumor microenvironment effectively.

19.3.2 Angiogenesis Regulation of Metalloproteases

Tumor blood vessels play multiple roles, supplying oxygen and nutrients, eliminating waste products, and facilitating tumor cell metastasis and immune cell infiltration. These vessels exhibit structural and functional abnormalities compared to non-malignant tissues, contributing to cancer progression by inducing reduced blood flow, resulting in hypoxia and an acidic pH within the tumor microenvironment (Fukumura et al., 2018). The presence of a hypoxic microenvironment due to compromised tumor blood perfusion can enhance tumor cell invasion while hindering the immune cells' anti-cancer capabilities. These immune cells are tasked with generating chemokines, cytokines, proteases, and microvesicles. Inflammatory chemokines and vascular endothelial growth factor (VEGF) serve as both key pro-angiogenic factors and immune modulators, leading to heightened angiogenesis and immune suppression (Albini et al., 2018).

In recent years, numerous forms of tumor angiogenesis have been observed, including vasculogenic mimicry (VM) and sprouting angiogenesis (Lugano et al., 2020). The process of sprouting angiogenesis involves upregulating various hydrolases, such as MMPs and tissue plasminogen activators located within the vascular basement membrane. This results in the degradation and redesigning of the basement membrane and the ECM (Viallard and Larrivée, 2017). Cancer-associated fibroblasts (CAFs) release MMPs, which break the basement membrane of endothelial cells and release stored VEGF, initiating angiogenesis. In cases like pancreatic neuroendocrine tumors, increased MMP-9 secretion leads to the release of VEGF from the matrix, causing a shift from quiet blood vessels to active angiogenesis (Bergers et al., 2000). When MMP-2 activity is inhibited in lung cancer cells, its interaction with integrin-AVB3 is reduced. As a result, it suppresses the VEGF expression by affecting the downstream PI3K/AKT signaling pathway, resulting in a reduction in angiogenesis (Chetty et al., 2009).

VM is a relatively recent model through which invasive tumors create new blood vessels, thereby establishing a blood supply to support tumor growth (Luo et al., 2020). Studies have shown a strong connection between the initial hypoxic conditions in solid tumors and VM. Additionally, hypoxia plays a pivotal role in modulating the expression and function of MMPs (Ling et al., 2017). It has been shown that hypoxia-inducible factor-1a (HIF-1a) directly regulates the expression of MMP-14, MMP-9, and MMP-2 (Li et al., 2021). In the context of tumor cell invasion, the targeted degradation of ECM components by proteases is crucial for crossing tissue barriers. The primary enzymes responsible for ECM degradation are MMPs, which can directly degrade ECM components or indirectly facilitate the process by releasing factors that stimulate growth and receptors on cell surfaces (Marozzi et al., 2021). MMPs engaged in ECM degradation can be classified into two primary categories: MMPs anchored to the cell membrane and MMPs that are soluble (Figure 19.3).

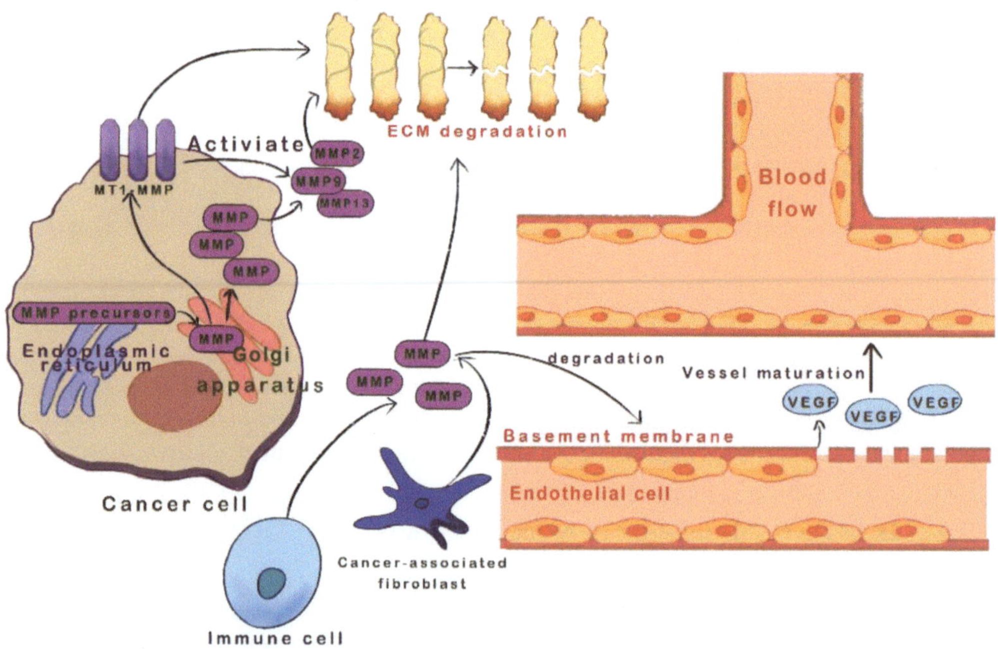

FIGURE 19.3 Metalloproteinases are essential for breaking down the extracellular matrix (ECM) and promoting angiogenesis. They start as inactive zymogens in the endoplasmic reticulum, then move to the Golgi apparatus, where they become either membrane-type 1 matrix metalloproteinases (MT1-MMP) or other soluble MMPs. These enzymes are produced by immune cells and cancer-associated fibroblasts (CAFs). The degradation of ECM primarily relies on the activation of soluble MMPs like MMP-2, MMP-9 (gelatinases), and MMP-13 (collagenase) by MT1-MMP. These MMPs break down the basement membrane of vascular endothelial cells, releasing bound vascular endothelial growth factor (VEGF) from the ECM, which initiates vessel maturation. (From Wang et al., 2022.)

19.3.3 Immune Response Regulation of MMPs

MMPs have been shown to initiate or terminate immunological activities within tumor tissue by modifying proteins found outside the cell that can either be soluble or bound to the cell membrane (Khokha et al., 2013). Recently, there has been a significant increase in enthusiasm for MMPs and their role in controlling the immune response against tumors, specifically concerning immune checkpoints. Endogenous MMPs have the ability to regulate the proteolytic cleavage of fibroblast-expressed programmed cell death protein 1 ligand (PD-L1) (Dezutter-Dambuyant et al., 2016). MMPs -9 and -13, which did not possess the PD-1 binding domain, were found to have a role in cleaving PD-L1 and PD-L2. As a consequence, apoptotic signals on phytohemagglutinin-activated T cells were reversed. On the contrary, the decrease in PD-L1 ligand expression on breast CAFs due to MMP activity resulted in a decrease in the external elimination of exhausted malfunctioning memory CD8+ T cells via apoptosis. This allowed these T cells to exert their inhibitory functions (Dezutter-Dambuyant et al., 2016). Studies demonstrated that recombinant MMP-13 and -7 have the ability to break down purified, unaltered PD-L1, with no detection of fragments with low molecular weight. This suggests the breakdown may be taking place in the PD-1 binding domain. Furthermore, when squamous cell carcinoma cells from humans, known as OSC-20, were treated with a specific MMP-13 inhibitor (CL82198), the expression of PD-L1 was restored. This indicates that the release or slicing of PD-L1 by MMP-13 might have a part in providing protection against invasion and metastasis (Hira-Miyazawa et al., 2018). It has also been demonstrated in a mouse model of syngeneic pancreatic cancer that administering an anti-PD-L1 antibody led to a reduction in the levels of both MMP-2 and MMP-9 mRNA and protein in pancreatic cancer and liver metastases (Figure 19.4) (Zhao et al., 2017). In another study, subcutaneous immunization tests using murine melanoma B16F10-Nex2 cells with SOCS1 protein knockdown indicated a decline in MMP-2, MMP-9, and PD-L1. This resulted in an anti-melanoma immune response, activating CD8+ T cells with effector functions against the tumor (Berzaghi et al., 2017).

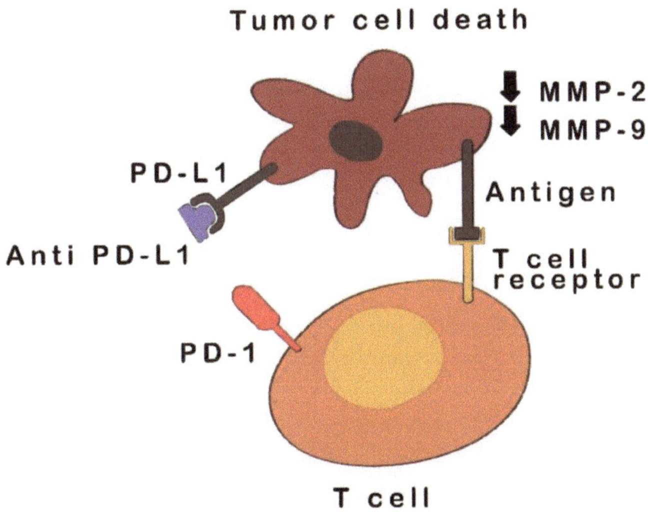

FIGURE 19.4 Blocking PD-L1 reduces MMP-2 and MMP-9 in pancreatic and liver tumors, stopping tumor growth and metastasis. (From Roy et al., 2019.)

19.4 INTERACTION BETWEEN METALLOPROTEASES AND STEM CELLS

19.4.1 Mesenchymal Stem Cells in Cancer Therapy

Mesenchymal stem cells (MSCs) include a population of adult stem cells that intuitively occur in the human body. The initial discovery was made by Friedenstein and colleagues within the supportive matrix of bone marrow (Friedenstein et al., 1968). Despite their significance, there is still limited understanding regarding the comprehensive characteristics and specific *in vivo* MSC localization. Apart from bone marrow, MSCs have been found in numerous other tissues of fetal and adult, including the adipose tissue, heart, skeletal muscles, amniotic fluid, synovial tissue, cord blood, circulating blood, pancreas, and placenta. It has been proposed that MSCs are present in essentially all organs that contain connective tissue (Kalervo, 2005).

Various MSCs are derived from adult human tissues and possess the capacity for differentiation and self-renewal to different mesenchymal lineages, including chondrocytes, osteocytes, and adipocytes. Compared to other sources like neural stem cells, the extraction of MSCs is generally less invasive and does not raise significant ethical concerns (Corsten and Shah, 2008). Furthermore, MSCs can transform into diverse tissue types within the same lineage or across germ lines (Anderson et al., 2001).

The initial crucial step in the homing of MSCs to tumors involves the process of mobilizing them from the bone marrow and other organs. During various injury conditions, naturally occurring MSCs have been noticed to move from the bone marrow and other tissues into the bloodstream, such as normoxia, hypoxia, and inflammation (Hong et al., 2009). The exact mechanism responsible for MSC movement through endothelial cells and their arrival in specific tissues remains incompletely understood. However, it is well-established that one of the inherent functions of MSCs is their capacity to move toward injured tissues and contribute to the tissue repair processes.

Cell receptor transactivation is a process where one receptor is activated by another and is thought to contribute to the process of cell migration. It is a crucial process in cell movement and programmed cell death, but its dysregulation can contribute to pathological conditions including cancer (Javan et al., 2019). MMPs, such as MMP-2 and MMP-9, play a role in this process. MMPs are a group of enzymes with proteolytic activity, responsible for splitting protein precursors such as growth factors, adhesion molecules, cytokines, and their respective receptors. MSCs traversing the endothelial barrier of the bone marrow have been reported to be facilitated by MMP-1 and the inhibitor of metalloproteinase-3 within the tissue (De Becker et al., 2007). Separate research showed that elevated MMP-2 levels were responsible for the migration of MSCs obtained from umbilical cord blood into the damaged tissue mediated by the C1q complement protein (Qiu et al., 2012). In various microenvironments, MSCs demonstrate diverse functions through the secretion of multiple mediators (Figure 19.5). While engaging in communication with tumor cells, MSCs secrete several signaling molecules that play roles in three key aspects of tumor-cell behavior: migration to specific locations, proliferation, and stem cell characteristics.

19.4.2 Anti-angiogenic and MSCs Cancer Therapy

Angiogenesis is a crucial process for tumor growth and progression (Jain et al., 2007). One approach to restricting tumor growth and metastasis is the use of MSCs as carriers of anti-angiogenic agents. These modified MSCs have shown an affinity for cancerous tissue and can deliver anti-angiogenic compounds without significant side effects (Ghaedi et al., 2011). However, continued systemic administration of anti-angiogenic agents can lead to toxicity and inadequate blood supply, which leads to a reduced administration of chemotherapy drugs to the tumor cells (Toi et al., 1994).

Endostatin, an inherent angiogenesis inhibitor, has been generally employed in anti-angiogenic therapy for varying cancers (Zheng et al., 2012). The MSCs derived from the human placenta were

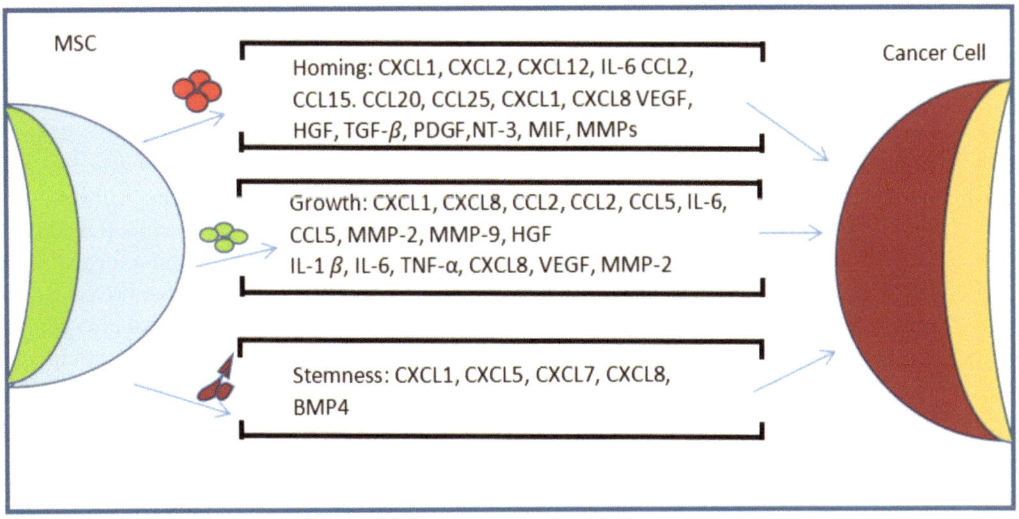

FIGURE 19.5 MSCs secrete various cytokines and mediators during their interaction with tumor cells, contributing to the regulation of tumor cell homing, growth, and stemness at three critical interfaces. (From Javan et al., 2019.)

genetically modified to express endostatin introduced into the cells via adenoviral transduction and subsequently given to immunodeficient mice. The MSCs engineered with human endostatin successfully homed to the tumor tissue and significantly reduced tumor size without notable systemic toxicity. The positive outcomes of the treatment were linked to enhanced tumor cell apoptosis, inhibition of neovascularization, and consequent inhibition of the proliferation of tumor cells and expansion (Zheng et al., 2012).

Furthermore, the administration of anti-angiogenic agents resulted in the normalization of the abnormal structure and impaired function of the blood vessels in the phase II clinical study, further resulting in a significant reduction in brain swelling linked to the tumor (Batchelor et al., 2007). This normalization of blood vessels was associated with decreased vessel diameter and permeability, along with increased coverage of small vasculature by mural cells (Hormigo et al., 2007). MSCs have shown the ability to specifically target tumor vasculature after being treated with intratumoral administration, making them promising candidates for precise transport of anti-angiogenic substances, particularly in cancers with a well-developed vascular network (Bexell et al., 2009).

However, it is important to carefully evaluate the properties of MSCs as they may exhibit certain undesirable characteristics. Safety is a major concern that needs to be thoroughly assessed before the use of MSCs in clinical settings for humans. Despite several clinical reports and ongoing trials, limited information is available regarding the extended-term safety of MSCs in human subjects (Javan et al., 2019). Additionally, the possible fibrosis-promoting ability of MSCs, as observed in liver fibrosis induced by the immunogenicity of human bone marrow mesenchymal stem cells (BM-MSCs) presents another constraint to their therapeutic application.

The variability of MSCs as a cell source for clinical applications poses another challenge. MSCs obtained from different donors may exhibit heterogeneity due to factors such as age, sex, genetics, and epigenetics (McLeod and Mauck, 2017). Furthermore, when utilizing MSCs in cancer-related applications, with the goal of suppressing angiogenesis, there is a concern that MSCs themselves may promote angiogenesis, which is contrary to the desired therapeutic effect (Krueger et al., 2019). Despite these obstacles, researchers are actively exploring strategies to enhance the effectiveness of MSC-based cancer therapy specifically in suppressing angiogenesis.

19.4.3 Immune Modulation via MSCs Cancer Therapy

Interleukins comprising a class of certain immune response-regulating cytokines have been proposed to exhibit anti-tumor properties by exerting directly inducing tumor cell death or enhancing the immune system's response against tumor cells (Okada and Pollack, 2004). In this context, researchers have explored the utilization of MSCs to deliver interleukins. MSCs modified to produce interleukins have been utilized to enhance immune monitoring against cancer, this activation results in the stimulation of cytotoxic lymphocytes and natural killer (NK) cells (Okada and Pollack, 2004). Additionally, MSCs expressing IL-12 have shown the ability to inhibit metastasis to lymph nodes, and various organs, and also promote In animal models of melanoma, breast cancer, and hepatoma, tumor cell apoptosis (Chen et al., 2008). IL-12-expressing MSCs have also demonstrated anti-tumor effects in mice with cervical tumors (Seo et al., 2011) and renal cell carcinomas (Gao et al., 2010). Moreover, MSCs obtained from umbilical cord blood and genetically modified to express IL-21 have exhibited the treatment's effectiveness in mice implanted with xenografts of ovarian cancer (Hu et al., 2011).

Interferon-beta (IFN-β) has demonstrated the treatment exhibits Inhibitory cell growth and inducer of apoptosis in tumor cells (Chawla-Sarkar et al., 2001). However, systemic administration of IFN-β has been restricted due to associated harmful effects. To overcome this limitation, genetic modifications have been made to human MSCs to enable them to produce IFN-β, allowing targeted delivery for the treatment of metastatic melanoma, breast cancer models, lung metastasis, and gliomas (Nakamizo et al., 2005). In mice with ovarian xenografts, intraperitoneal injection of MSCs expressing IFN-β resulted in complete tumor eradication in 70% of the treated mice (Dembinski et al., 2013). These findings validate the capacity of MSCs to serve as precise carriers for generating IFN-β within tumor tissues. Moreover, the intravenous administration MSCs that produce IFN-β led to a notable decrease in the weight of prostate tumors and improved comparing animal survival to control groups (Wang et al., 2012). The anti-cancer effects of IFN-alpha (IFN-α), which exhibit several explorations into regulatory effects have also been undertaken. IFN-α can also be used as an adjunct therapeutic approach aimed at eradicating cases featuring micrometastatic deposits using a heightened likelihood of systemic recurrence after attempting alternative treatment strategies (Lens, 2006). MSCs were engineered to express IFN-α to efficiently target tumor tissues in a mouse model of plasmacytoma. The delivery of the drug was subcutaneous and IFN-α-expressing MSCs significantly restricted tumor growth and increased mouse survival by triggering programmed cell death in cancerous cells and reducing tumor blood vessel growth (Sartoris et al., 2010). Moreover, the introduction of MSCs expressing IFN-α through systemic administration led to a decrease in the proliferation of cancer cells and significantly increased survival within a laboratory mouse model of metastatic melanoma by promoting the induction of apoptosis in tumor cells and impeding the formation of tumors present in blood vessels (Lens, 2006).

Targeting proteases has been a promising therapeutic strategy in anti-cancer therapies to combat the invasive behavior of cancer cells. While early clinical outcomes were not as successful as anticipated, innovative approaches have arisen that utilize proteases not as direct targets for treatment, but rather as triggers for therapeutic interventions. These advancements allow for the precise enabling and direct administration of anti-cancer treatments to the tumor location, where proteases facilitate the initiation or liberation of them present within the tumor microenvironment, including MMPs, cathepsins, or urokinase plasminogen activator (uPA).

19.5 CONCLUSION

Cancer is a debilitating disease globally and the search for effective treatment regimens that will lead to total remission is an ongoing stride in the biomedical field. Stem cell therapy offers a unique approach by leveraging the reformative latency of stem cells and their ability to selectively target tumors by preventing the starting of a fresh blood vessel and the normalization of tumor-associated vasculature and enhancing anti-angiogenic effects. Metalloproteases, which play pivotal roles in cell migration, invasion, and tissue remodeling, have emerged as potential candidates for enhancing the effectiveness of stem-cell-based cancer management. They can degrade the ECM niche and stall many cancer stem cell properties, which include invasion, withholding in the niche, self-restoration, therapy resistance, immune circumlocution, and stemness, thereby promoting tumor suppression and cancer downregulation. MSCs have shown demonstrable promise in cancer therapeutic strategies. Although early clinical outcomes were not as successful as anticipated, innovative approaches have arisen that utilize some proteases, not as direct targets for treatment, but rather as triggers for therapeutic interventions. These advancements allow for precise anti-cancer therapeutics discharge directly to the tumor location, where their activation or release is facilitated by proteases present within the tumor microenvironment.

ACKNOWLEDGMENT

The authors wish to thank all the staff of the library section of Bingham University, Nigeria, and Nuhu Bamali Polytechnic Zaria, Nigeria, for providing us with a conducive environment to write this chapter.

DECLARATIONS

Conflict of interest: The authors declare no conflicts of interest.

Consent for publishing: All authors have approved the final version of the manuscript and given consent for publication.

Funding: Not Applicable.

REFERENCES

Albini, A., Bruno, A., Noonan, D.M., Mortara, L., 2018. Contribution to tumor angiogenesis from innate immune cells within the tumor microenvironment: Implications for immunotherapy. *Front Immunol.* 9:527.

Anderson, D.J., Gage, F.H., Weissman, I.L., 2001. Can stem cells cross lineage boundaries? *Nat Med.* 7:393–6.

Apte, S.S., 2020. ADAMTS proteins: Concepts, challenges, and prospects. *Methods Mol Biol.* 2043:1–12.

Batchelor, T.T., Sorensen, A.G., di Tomaso, E., Zhang, W.T., Duda, D.G., Cohen, K.S., et al., 2007. AZD2171, a pan-VEGF receptor tyrosine kinase inhibitor, normalizes tumor vasculature and alleviates edema in glioblastoma patients. *Cancer Cell.* 11:83–95.

Batlle, E., Clevers, H., 2017. Cancer stem cells revisited. *Nat Med.* 23(10):1124–34. doi:10.1038/nm.4409

Bergers, G., Brekken, R., McMahon, G., Vu, T.H., Itoh, T., Tamaki, K., et al., 2000. Matrix metalloproteinase-9 triggers the angiogenic switch during carcinogenesis. *Nat Cell Biol.* 2:737–44.

Berzaghi, R., Maia, V.S., Pereira, F.V., Melo, F.M., Guedes, M.S., Origassa, C.S., et al., 2017. SOCS1 favors the epithelial-mesenchymal transition in melanoma, promotes tumor progression and prevents antitumor immunity by PD-L1 expression. *Sci Rep.* 7:40585.

Bexell, D., Gunnarsson, S., Tormin, A., Darabi, A., Gisselsson, D., Roybon, L., et al., 2009. Bone marrow multipotent mesenchymal stroma cells act as pericyte-like migratory vehicles in experimental gliomas. *Mol Ther.* 17:183–90.

Carlos, L., Judith, S.B., 2008. Proteases: Multifunctional enzymes in life and disease. *J Biol Chem.* 283(45):30433–7.

Chawla-Sarkar, M., Leaman, D.W., Borden, E.C., 2001. Preferential induction of apoptosis by interferon (IFN)-β compared with IFN-α2. *Clinical Cancer Res.* 7:1821–31.

Chen, X., Lin, X., Zhao, J., Shi, W., Zhang, H., Wang, Y., et al., 2008. A tumor selective biotherapy with prolonged impact on established metastases based on cytokine gene-engineered MSCs. *Mol Ther.* 16:749–56.

Chetty, C., Lakka, S.S., Bhoopathi, P., Rao, J.S., 2009. MMP-2 alters VEGF expression via aVb3 integrin-mediated PI3K/AKT signaling in A549 lung cancer cells. *Int J Cancer.* 127:1081–95.

Corsten, M.F., Shah, K., 2008. Therapeutic stem-cells for cancer treatment: Hopes and hurdles in tactical warfare. *Lancet Oncol.* 9:376–84.

De Becker, A., Van Hummelen, P., Bakkus, M., Broek, I.V., De Wever, J., De Waele, M., et al., 2007. Migration of culture-expanded human mesenchymal stem cells through bone marrow endothelium is regulated by matrix metalloproteinase-2 and tissue inhibitor of metalloproteinase-3. *Haematologica.* 92:440–9.

Dembinski, J.L., Wilson, S.M., Spaeth, E.L., Studeny, M., Zompetta, C., Samudio, I., et al., 2013. Tumor stroma engraftment of gene-modified mesenchymal stem cells as anti-tumor therapy against ovarian cancer. *Cytotherapy.* 15:20–32.

Dezutter-Dambuyant, C., Durand, I., Alberti, L., Bendriss-Vermare, N., Valladeau-Guilemond, J., Duc, A., et al., 2016. A novel regulation of PD-1 ligands on mesenchymal stromal cells through MMP-mediated proteolytic cleavage. *Oncoimmunology.* 5:e1091146.

Fields, G.B., 2019. Mechanisms of action of novel drugs targeting angiogenesis-promoting matrix metalloproteinases. *Front Immunol.* 10:1278. doi: 10.3389/fimmu.2019.01278

Friedenstein, A.J., Petrakova, K.V., Kurolesova, A.I., Frolova, G.P., 1968. Heterotopic of bone marrow. Analysis of precursor cells for osteogenic and hematopoietic tissues. *Transplantation.* 6:230–47.

Frolova, A.S., Petushkova, A.I., Makarov, V.A., Soond, S.M., Zamyatnin Jr, A.A., 2020 Unravelling the network of nuclear matrix metalloproteinases for targeted drug design. *Biology (Basel).* 9:480.

Fukumura, D., Kloepper, J., Amoozgar, Z., Duda, D.G., Jain, R.K., 2018. Enhancing cancer immunotherapy using antiangiogenics: Opportunities and challenges. *Nat Rev Clin Oncol.* 15:325–40.

Gao, P., Ding, Q., Wu, Z., Jiang, H., Fang, Z., 2010. Therapeutic potential of human mesenchymal stem cells producing IL-12 in a mouse xenograft model of renal cell carcinoma. *Cancer Lett.* 290:157–66.

Ghaedi, M., Soleimani, M., Taghvaie, N.M., Sheikhfatollahi, M., Azadmanesh, K., Lotfi, A.S., et al., 2011. Mesenchymal stem cells as vehicles for targeted delivery of anti-angiogenic protein to solid tumors. *J Gene Med.* 13:171–80.

Ghajar, C.M. et al., 2013. The perivascular niche regulates breast tumour dormancy. *Nature Cell Biol.* 15:807–17.

Hira-Miyazawa, M., Nakamura, H., Hirai, M., Kobayashi, Y., Kitahara, H., Bou-Gharios, G., 2018. Regulation of programmed-death ligand in the human head and neck squamous cell carcinoma microenvironment is mediated through matrix metalloproteinase mediated proteolytic cleavage. *Int J Oncol.* 52:379–88.

Hong, H.S., Lee, J., Lee, E., Kwon, Y.S., Lee, E., Ahn, W., et al., 2009. A new role of substance P as an injury-inducible messenger for mobilization of CD29+ stromal-like cells. *Nat Med.* 15:425–35.

Hormigo, A., Gutin, P.H., Rafii, S., 2007. Tracking normalization of brain tumor vasculature by magnetic imaging and proangiogenic biomarkers. *Cancer Cell.* 11:6–8.

Hu, W., Wang, J., Dou, J., He, X., Zhao, F., Jiang, C., et al., 2011. Augmenting therapy of ovarian cancer efficacy by secreting IL-21 human umbilical cord blood stem cells in nude mice. *Cell Transpl.* 20:669–80.

Jain, R.K., Di, T.E., Duda, D.G., Loeffler, J.S., Sorensen, A.G, Batchelor, T.T., 2007. Angiogenesis in brain tumours. *Nat Rev Neurosci.* 8:610–22.

Javan, M.R., Khosrojerdi, A. Moazzeni, S.M., 2019. New insights into implementation of mesenchymal stem cells in cancer therapy: Prospects for anti-angiogenesis treatment. *Front Oncol.* 9:840.

Kalervo, V.H., 2005. Mesenchymal stem cells. *Ann Med.* 37:469–79.

Kaplan, R.N., et al., 2005. VEGFR1-positive haematopoietic bone marrow progenitors initiate the pre-metastatic niche. *Nature.* 438:820–7.

Kessenbrock, K., Dijkgraaf, G.J., Lawson, D.A., Littlepage, L.E., Shahi, P., Pieper, U., et al., 2013. A role for matrix metalloproteinases in regulating mammary stem cell function via the Wnt signaling pathway. *Cell Stem Cell.* 13:300–13.

Khokha, R., Murthy, A., Weiss, A., 2013. Metalloproteinases and their natural inhibitors in inflammation and immunity. *Nat Rev Immunol.* 13:649–65.

Krueger, T.E., Thorek, D.L., Meeker, A.K., Isaacs, J.T., Brennen, W.N., 2019. Tumor infiltrating mesenchymal stem cells: Drivers of the immunosuppressive tumor microenvironment in prostate cancer? *Prostate.* 79:320–30.

Lens, M., 2006. Cutaneous melanoma: Interferon alpha adjuvant therapy for patients at high risk for recurrent disease. *Dermatol Ther.* 19:9–18.

Li, H., Huang, H., Cui, Y., Li, W., Zhang, S., Chen, Y., 2021. Study on the mechanism of capillary leakage caused by hypoxia-inducible factor-1a through inducing high expression of matrix metalloproteinase-9. *J Oncol.* 2021:1–12.

Ling, B., Watt, K., Banerjee, S., Newsted, D., Truesdell, P., Adams, J., et al., 2017. A novel immunotherapy targeting MMP-14 limits hypoxia, immune suppression and metastasis in triple-negative breast cancer models. *Oncotarget.* 8:58372–85.

Lugano, R., Ramachandran, M., Dimberg, A., 2020. Tumor angiogenesis: Causes, consequences, challenges and opportunities. *Cell Mol Life Sci.* 77:1745–70.

Luo, Q., Wang, J., Zhao, W., Peng, Z., Liu, X., Li, B., et al., 2020. Vasculogenic mimicry in carcinogenesis and clinical applications. *J Hematol Oncol.* 13:19.

Malanchi, I. et al., 2011. Interactions between cancer stem cells and their niche govern metastatic colonization. *Nature.* 481:85–89.

Marozzi, M., Parnigoni, A., Negri, A., Viola, M., Vigetti, D., Passi, A., et al., 2021. Inflammation, extracellular matrix remodeling, and proteostasis in tumor microenvironment. *IJMS.* 22:8102.

McLeod, C., Mauck, R., 2017. On the origin and impact of mesenchymal stem cell heterogeneity: New insights and emerging tools for single cell analysis. *Eur Cell Mater.* 34:217–31.

Nakamizo, A., Marini, F., Amano, T., Khan, A., Studeny, M., Gumin, J., et al., 2005. Human bone marrow–derived mesenchymal stem cells in the treatment of gliomas. *Cancer Res.* 65:3307–18.

Okada, H., Pollack, I.F., 2004. Cytokine gene therapy for malignant glioma. *Expert Opin Biol Ther.* 4:1609–20.

Oshimori, N, Guo, Y, Taniguchi, S., 2021. An emerging role for cellular crosstalk in the cancer stem cell niche microenvironment. *J Pathol.* 254(4):384–94. doi:10.1002/path.5655

Oskarsson, T. et al., 2011. Breast cancer cells produce tenascin C as a metastatic niche component to colonize the lungs. *Nature Med.* 17:867–74.

Pietras, A. et al., 2014. Osteopontin-CD44 signaling in the glioma perivascular niche enhances cancer stem cell phenotypes and promotes aggressive tumor growth. *Cell Stem Cell.* 14:357–69.

Psaila, B., Lyden, D., 2009. The metastatic niche: Adapting the foreign soil. *Nat Rev Cancer.* 9:285–93.

Qiu, Y., Marquez Curtis, L.A., Janowska-Wieczorek, A., 2012. Mesenchymal stromal cells derived from umbilical cord blood migrate in response to complement C1q. *Cytotherapy.* 14:285–95.

Regala, R.P., Justilien, V., Walsh, M.P., Weems, C., Khoor, A., Murray, N.R., et al., 2011. Matrix metalloproteinase-10 promotes Kras-mediated bronchio-alveolar stem cell expansion and lung cancer formation. *PLOS ONE.* 6:e26439.

Reticker-Flynn, N.E. et al., 2012. A combinatorial extracellular matrix platform identifies cell extracellular matrix interactions that correlate with metastasis. *Nature Commun.* 3:1122.

Roy, R., Golnaz, M., Andrej, J., Marsha, A.M., 2019. Metalloproteinases and their roles in human cancer. *Anat Rec.* 6:1557–72

Sartoris, S., Mazzocco, M., Tinelli, M., Martini, M., Mosna, F., Lisi, V., et al., 2010. Efficacy assessment of interferon-alpha–engineered mesenchymal stromal cells in a mouse plasmacytoma model. *Stem Cells Dev.* 20:709–19.

Scadden, D.T., 2006. The stem-cell niche as an entity of action. *Nature.* 441:1075–9.

Seo, S., Kim, K., Park, S., Suh, Y., Kim, S., Jeun, S., et al., 2011. The effects of mesenchymal stem cells injected via different routes on modified IL-12-mediated antitumor activity. *Gene Ther.* 18:488–95.

Toi, M., Hoshina, S., Takayanagi, T., Tominaga, T., 1994. Association of vascular endothelial growth factor expression with tumor angiogenesis and with early relapse in primary breast cancer. *Cancer Sci.* 85:1045–9.

Turk, B., 2006. Targeting proteases: successes, failures and future prospects. *Nat Rev Drug Discov.* 5:785–99.

Viallard, C., Larrivée, B., 2017. Tumor angiogenesis and vascular normalization: Alternative therapeutic targets. *Angiogenesis.* 20:409–26.

Wang, G., Zhan, Y., Hu, H., Wang, Y., Fu, B., 2012. Mesenchymal stem cells modified to express interferon-β inhibit the growth of prostate cancer in a mouse model. *J Int Med Res.* 40:317–27.

Wang, Q., Wang, K., Tan, X., Li, Z., Wang, H., 2022. Immunomodulatory role of metalloproteases in cancers: Current progress and future trends. *Front Immunol.* 13:1064033.

Yu, J., Zhang, Y., Leung, L.H., Liu, L., Yang, F., Yao, X., 2016. Efficacy and safety of angiogenesis inhibitors in advanced gastric cancer: A systematic review and meta-analysis. *J Hematol Oncol.* 9:111. doi: 10.1186/s13045-016-0340-8

Zhao, L., Li, C., Liu, F., Zhao, Y., Liu, J., Hua, Y., 2017. A blockade of PD-L1 produced antitumor and anti-metastatic effects in an orthotopic mouse pancreatic cancer model via the PI3K/Akt/mTOR signaling pathway. *Onco Targets Ther.* 10:2115–26.

Zheng, L., Zhang, D., Chen, X., Yang, L., Wei, Y., Zhao, X., 2012. Antitumor activities of human placenta-derived mesenchymal stem cells expressing endostatin on ovarian cancer. *PLOS ONE.* 7:e39119.

20 Regulatory Mechanisms and Clinical Potential of Epigenetic Targeting in Glioma

Yetirajam Rajesh, Suvendu Nandi, Priya Ghosh, and Mahitosh Mandal
School of Medical Science and Technology, Indian Institute of Technology Kharagpur, Kharagpur, India

20.1 INTRODUCTION

Modern medicine faces a significant obstacle in the form of glioblastoma (GBM), which is widely acknowledged as the most severe and invasive type of glioma. The location of the tumor, its heterogeneity, its resistance to radiation and chemotherapy, and its propensity for infiltration and recurrence are some of the factors that make GBM treatment complex. Two main mechanisms, epithelial-mesenchymal transition (EMT) and extracellular matrix (ECM) remodeling, drive the pathophysiology of GBM. GBM can develop when the ECM undergoes changes that promote tumor growth and invasion. However, EMT is associated with GBM's infiltrative potential because it enables epithelial cells to become more like mesenchymal cells, which in turn increases their invasiveness, resistance to cell death, and production of extracellular matrix components [1].

Finding new therapeutic targets for GBM is critical due to these complications and epigenetic modulators have emerged as potential solutions. Several biological processes rely on epigenetic regulation, which is essential for cell differentiation, development, and response to environmental signals. This type of regulation involves alteration in gene expression without alterations to the underlying gene sequence. When it comes to cancer, abnormal epigenetic alterations can promote tumorigenesis by dysregulating tumor suppressor proteins and oncogenes [2].

Enzymes that alter histones and DNA are known as epigenetic modulators, and they can control the transcription of tumor suppressor and oncogene proteins. The methylation and acetylation of these histones, among other modifications, have a profound effect on gene expression. An intriguing avenue for therapeutic intervention could be the discovery that some GBM tumors have mutations in histone-modifying enzymes [3]. Moreover, important signaling pathways that influence the proliferation of GBM cells and stem-like cells are regulated by epigenetic regulators. Among these are the pathways known as the Notch, Hedgehog (Hh), and Wingless (WNT). An important part of cell–cell communication, the Notch signaling cascade influences cell growth, differentiation, and programmed cell death. Embryogenesis and tissue regeneration rely on the Hh pathway to control cell growth and differentiation, while organogenesis, neural patterning, cell migration, polarity, and cell fate decisions are all regulated by the WNT pathway [4]. There is strong evidence that these pathways are dysregulated and contribute to GBM pathogenesis. The fact that they interact with epigenetic regulators further strengthens the case for developing epigenetic therapies specifically for GBM.

The complex relationship between genetic and epigenetic factors in GBM pathogenesis poses both obstacles and possibilities. Since this chapter offers a more in-depth comprehension of the fundamental epigenetic mechanisms and the role that they play in the development of glioma, we

DOI: 10.1201/9781003394693-23

can pave the way for the creation of innovative therapeutic approaches. These hold great promise for enhancing GBM patient outcomes and prognoses, providing a glimmer of hope amidst this overwhelming disease.

20.2 EPIGENETIC REGULATION IN GLIOBLASTOMA (GBM)

Epigenetic modifications have been significant in GBM development and progression. Epigenetic inheritance is primarily mediated through methylation of DNA, modifications in histone protein, chromatin restructuring, and non-coding RNAs. The methylation status of the O-6-methylguanine-DNA methyltransferase gene (MGMT) defines the GBM into categories of MGMT that repair the DNA damage by temozolomide (TMZ) and tumor containing methylated MGMT. DNA methylation and histone modification in association with chaperone molecules are involved in chromatin remodeling contributing plasticity to the genetic machinery. The World Health Organization (WHO) classification also further defines subgroups of glioma in accordance with genetic and epigenetic parameters. The central nervous system (CNS) tumors have also been further categorized by DNA methylation [5]. The epigenetic regulations of different signaling pathways play a pivotal role in the advancement of GBM.

20.2.1 Hh Signaling and Its Epigenetic Control in GBM

Hh signaling plays a crucial role in various cancers including GBM. This pathway is largely inactivated or weakly activated in adult organisms. It can be switched on if required, for instance in repairing wounds. Indian hedgehog (IHh), Sonic hedgehog (SHh), and desert hedgehog (DHh) are three proteins that belong to Hh family of signaling molecules. These Hh proteins function as ligands for membrane-spanning receptors including patched 1 (PTCH1) and patched 2 (PTCH2). Multiple co-receptors including GAS1, CDO, and BOC facilitate PTCH binding to Hh ligands. The activity of smoothened (SMO) is blocked by PTCH when there is no Hh ligand attached to it. This prevents the Hh signaling pathway from being turned on. But when PTCH binds with Hh ligand, it cannot inhibit SMO protein, and the Hh signaling pathway is activated [6]. This pathway in mammalian cells relies heavily on primary cilia, which are microtubule-based organelles emanating from the cell membrane. When Hh ligand is not present, PTCH stays at the bottom of primary cilia and blocks SMO's activity by preventing it from translocating within primary cilia. This causes GliFL, a full-length glioma-associated oncogene, to be split into GliR, a Gli repressor, after being phosphorylated by several protein kinases including protein kinase A (PKA), casein kinase 1 (CK1), and glycogen synthase kinase 3 (GSK3) [7]. GliR attaches to the promoter of Hh target genes and turns them off. The interaction between the Hh ligand and the PTCH receptor triggers the Hh signaling pathway. The complex of Hh-PTCH is engulfed by the cell and destroyed in lysosomes [8]. SMO is no longer inhibited and the Hh signal is transmitted downstream through a cytoplasmic protein complex consisting of GliFL, suppressor of fused (SUFU), and kinesin protein (KIF7) [9]. This is a canonical signaling pathway where Hh controls the Gli transcription factors. In contrast, the non-canonical signaling pathway occurs through Gli-independent mechanisms. The process of tumorigenesis is mostly mediated by the canonical Hh pathway. Dysregulation of this signaling pathway plays a crucial role in the growth and differentiation of tumor cells. Hh pathway elements, such as PTCH1, SUFU, and SMO are often mutated in medulloblastoma, a type of brain tumor that frequently affects children. In glioma, SHh—one of the Hh ligands—is often elevated in tumor cells as well as in the adjacent parenchymal cells [10]. Moreover, this Hh pathway, which regulates neural stem cell division throughout embryonic brain formation, is also necessary for the development and advancement of glioma by maintaining cancer stem cells (CSCs) [11]. Brain tumor stem cells release DHh ligands to trigger the DHh-PTCH2 signaling cascade, which enhances vascular permeability and angiogenesis and

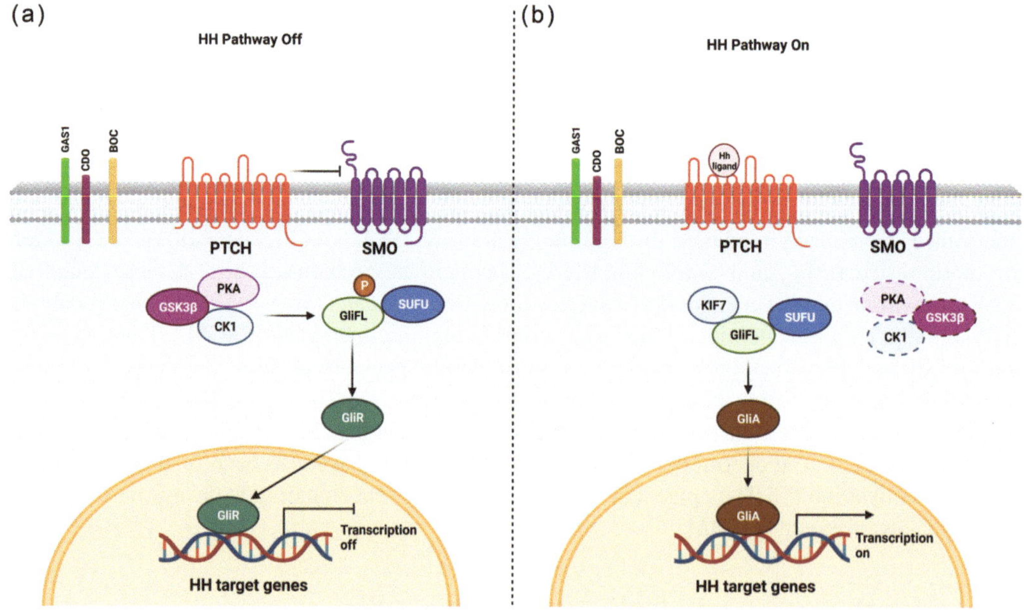

FIGURE 20.1 Visual depiction of the intracellular signaling involving SHh pathway. (a) PTCH prevents SMO from enabling Gli to move to the nucleus and activate genes that cause cell growth. (b) When Hh ligand is present, PTCH no longer inhibits SMO, which leads Gli to enter in the nucleus and turn on genes that enhance the malignancy of tumor cells. (Created using Biorender software.)

supports glioblastoma development [12]. The expression of Gli, a downstream mediator of the Hh pathway, is elevated in glioblastoma, which is connected with the advancement of tumors [13]. The bromodomain and extra-terminal (BET) motif and p300/CBP-associated factor (PCAF) are the most important epigenetic modulators of Hh signaling pathway. A well-characterized member of the BET protein family is BRD4, which regulates the expression of Gli1 and Gli2 by directly attaching to their promoter region [14]. PCAF binds to the promoter of Hh target genes by interacting with Gli1 after the activation of Hh pathway. This interaction caused H3K9 acetylation of the promoter of Hh target genes, which upregulates the Hh signaling pathway [15]. Figure 20.1 illustrates the intracellular signaling of SHh pathway, emphasizing the key stages in the signaling cascade.

20.2.2 WNT Signaling and Its Epigenetic Control in GBM

WNT signaling is a key factor that regulates the formation of the brain and spinal cord. Dysregulation of this signaling pathway is involved in causing abnormal growth and development of brain tumors. Through a network of signal transduction pathways, WNT signaling is able to transfer signals via membrane-bound receptors from outside into the cell cytoplasm and, ultimately, into the nucleus. The canonical WNT pathway, non-canonical WNT/calcium pathway, and non-canonical planar cell polarity pathway are three distinct WNT pathways that have been identified. The canonical signaling cascade is mainly responsible for neural stem cell division, self-renewal, and cell fate determination [16]. When WNT ligands bind to the cysteine-rich regions of membrane-spanning receptors including frizzled (FZD) and low-density lipoprotein receptor-related protein (LRP), both the canonical pathway and non-canonical pathways become activated. The canonical WNT pathway is turned on when a complex containing adenomatous polyposis coli (APC), glycogen synthase kinase 3β (GSK-3β), CK1, axis inhibition proteins 1 and/or 2 (Axin/2) is disassembled, which allows β-catenin to be stable and also helps in its accumulation in cytoplasm. Consequently, β-catenin undergoes translocation into the nucleus from cell cytoplasm where it combines with T cell factor/lymphocyte enhancer factor (TCF/LEF) to increase the expression of several genes such as c-jun, cyclin D1, and c-myc. A recent study demonstrated that FoxM1, a transcription factor, enhances the movement and stability of β-catenin in the nucleus of GBM cells by binding with free β-catenin. It implies that FoxM1 can trigger canonical WNT signaling without a ligand [17]. Mutation of the signaling components of this pathway leads to aberrant activation of this signaling pathway and helps in tumor formation. The participation of the non-canonical (β-catenin independent) WNT pathway in the formation and progression of cancer is fuzzy.

Genes that are epigenetically silenced and may act as tumor suppressors in GBM are DK1, WIf1, and SFRP1, which are known to impede canonical WNT signaling. The promoter regions of DK1, WIf1, and SFRP1 contain DNA methylation and histone modification that epigenetically silence these tumor suppressors. Additionally, it has been proved that administering trichostatin A (TSA), a histone deacetylase inhibitor to T98 cells, results in the restoration of all three genes' expression and reduces cell division as well as increases sensitivity to apoptosis [18]. Other research revealed that in different GBM cell lines (such as LN18, A172, U87, and U183), SFRP1-5 is hypermethylated, and the 5-Aza treatment resulted in demethylation. After the removal of methyl groups from SFRP gene promoters, the level of corresponding proteins will increase, which interferes with the canonical WNT signaling pathway and reduces tumor growth [19]. Figure 20.2 presents a visual depiction of the canonical WNT pathway's activation, illustrating the key components and interactions involved in this critical signaling cascade.

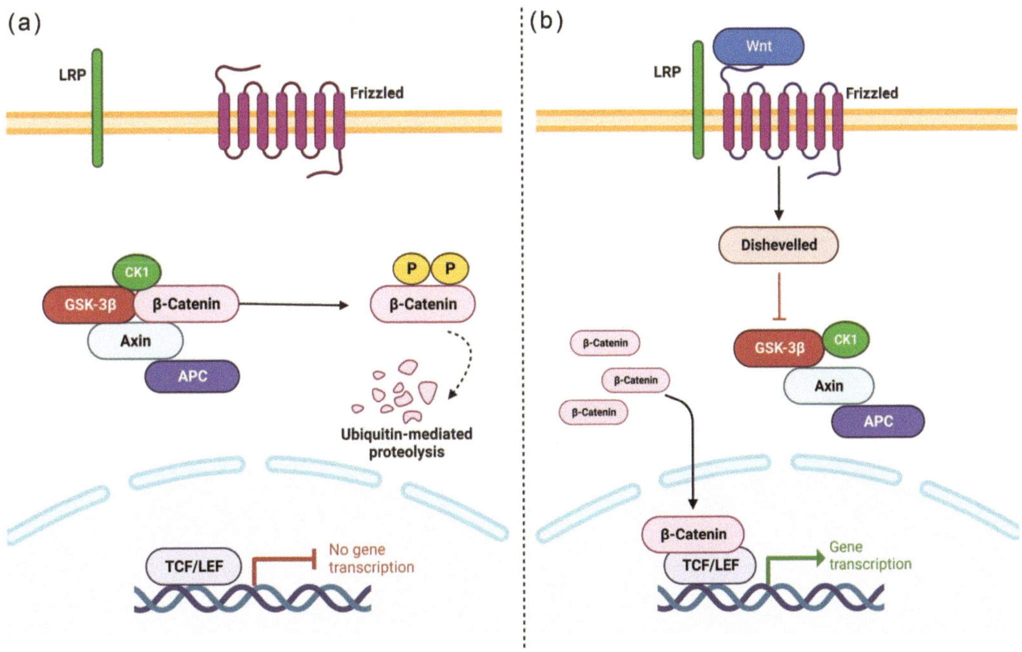

FIGURE 20.2 Synopsis of the canonical WNT signaling pathway's activation. (a) Inactivation of the WNT signaling pathway when its ligand is not present. β-catenin is phosphorylated by Dishevelled, Axin, APC, and GSK-3β complex, and ultimately causes the proteasomal breakdown of β-catenin. (b) On target cells, WNT ligands bind themselves to FZD receptors. Nuclear entry to β-catenin is made possible by FZD's co-receptor LRP5/6, which prevents β-catenin from being phosphorylated by GSK-3β. WNT downstream proteins including cyclin-D1, c-myc, and c-jun are expressed when β-catenin binds to the transcription factor TCF/LEF. (Created using Biorender software.)

20.2.3 NOTCH Signaling and Its Epigenetic Control in GBM

Notch signaling controls a number of biological processes such as cell proliferation, cell fate determination, programmed cell death, and maintenance of stem cells. It is an evolutionarily conserved mechanism among multicellular organisms. As shown in Figure 20.3, NOTCH signaling pathway is activated when the receptors and ligands on adjacent cells bind to each other. Notch serves as a cytoplasmic receptor and there are four types of receptors Notch 1–4. These receptors can bind with five types of ligands called Jagged (Jag) 1–2, Delta-like 1, Delta-like 3, and Delta-like 4. Notch receptor and its ligands can interact in two different ways: cis, when both of them are present on the same cell, or trans, when they are on adjacent cells. Notch target genes are expressed when the intracellular domain of notch (NICD), is freed from the cellular membrane following γ-secretase cleavage, moves into the nucleus, and binds itself to the transcription factor. Dysregulated Notch signaling has been found to be associated with various solid tumors such as brain tumors. It has been reported that increased Notch 1 expression in glioblastoma may be caused by hypoxic conditions in the tumor microenvironment [20]. Methylation of DNA is one of the most frequent epigenetic changes that has been linked to cancer initiation and progression. In contrast to a normal healthy brain, there are low levels of methylation on CpG islands within HEY1 promoter in glioblastoma specimen, which led to its overexpression [21]. Upon the addition of sodium butyrate, an inhibitor of histone deacetylase in human glioma xenograft cell lines, HEY1 expression decreases, and DNA methyltransferase1 level increases [21]. Some human cancers have both hypomethylation and hypermethylation of Notch and several genes related to Notch signaling such as DLL1, DTX1, NOTCH2, Jag1, and HDAC1 are epigenetically silenced in different types of cancer. However, this has not been specifically demonstrated in glioblastoma [22].

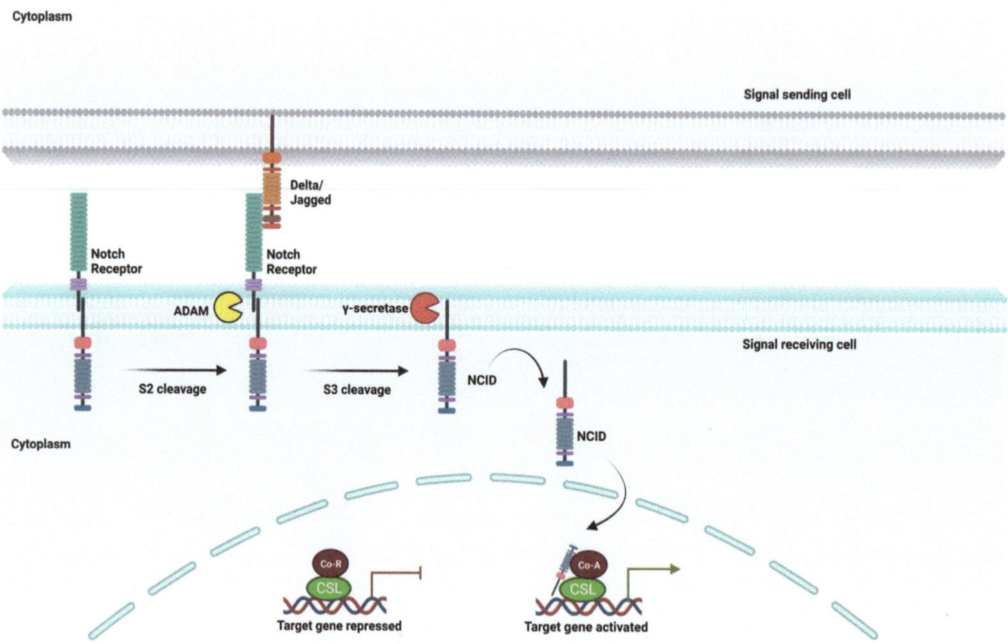

FIGURE 20.3 Diagram illustrating the Notch signaling pathway. When the extracellular domain of signal receiving cell's Notch is bound by signal-sending cell's delta-like or jagged receptor, notch signaling is initiated. ADAM carries out the extracellular domain's proteolytic cleavage, and then γ-secretase cleaves it, releasing the intracellular domain, which then enters into the nucleus to function as a transcription factor. (Created using Biorender software.)

20.3 FUNDAMENTAL EPIGENETIC MECHANISMS INVOLVED IN GLIOMA

20.3.1 DNA Methylation

The end product of DNA methylation is 5-methylcytosine, initiated by DNA methyltransferases (DNMTs), and ends with adding methyl groups to cytosine molecules, increasing the suppression of transcription [23]. Metabolic processing occurs in approximately 70% of CpG dinucleotides in the human genome. A high concentration of the DNA bases guanine and cytosine defines these CpG islands. To regulate gene expression, DNA methylation prevents transcription factors from binding to specific regions called promoters. Additionally, methylated CpG sites attract methyl-CpG-binding proteins and transcriptional repressors, leading to transcriptional repression through the alteration of chromatin structure. When oxidized to 5-hydroxymethylcytosine, 5-methylcytosines are less effective at binding to DNA. In CpGs, the methyl groups transfer from S-adenosyl-methionine to cytosine through a family of DNMTs. DNMT3A and DNMT3B are the most prominent *de novo* DNMTs in humans. Neurons express DNMT1 and DNMT2 methylates aspartic acid transfer RNA [24].

20.3.2 Histone Modifications

When it comes to DNA replication, transcription, and repair, histone modifications are paramount among epigenetic features. Glycosylation, SUMOylation, acetylation, phosphorylation, biotinylation, ADP-ribosylation, and methylation are among the many kinds of modifications [25]. To activate chromatin or transcription, enzymes like histone acetyltransferase add an acetyl group to lysine residues, a process known as histone acetylation. Histone deacetylases, on the other hand, are responsible for histone deacetylation, which in turn deactivates chromatin and represses transcription. These changes occur in gene expression but do not impact the DNA sequence. These alterations are critical for comprehending cancer progression and the efficacy of treatment.

When histone-modifying enzymes and chromatin modifiers do not work right, it changes the structure of chromatin and throws off gene expression patterns, which is a key step in the development of gliomas. Mutations in the gene encoding for these enzymes and modifiers, including ATRX, ARID1A, SMARCA4, and others, cause these modifications [26].

20.3.3 Chromatin Remodeling

Epigenetic regulation of gliomas is heavily reliant on chromatin remodeling, which affects glioma progression and development. A number of mechanisms dynamically regulate chromatin organization and compaction. For example, stable heterochromatin regulates the silencing of transposable elements, genome integrity, ATP-dependent chromatin remodeling complexes, and accessibility of transcription factors to DNA. The coactivators or corepressors aid in the activation or repression of their transcriptional activity [27].

Oncogene reactivation can occur as a result of this process, which includes changes to chromatin insulator elements (IEs), which in turn cause abnormal chromatin remodeling through the formation of DNA loops. Isocitrate dehydrogenase (IDH) 1 and 2 gene mutations cause IE dysfunction and genome-wide DNA hypermethylation; these mutations are common in gliomas [28]. Glioma stem cells (GSCs) rely on chromatin remodeling factors like BRG1 (Brahma-related gene 1) and HELLS (Helicase, Lymphoid Specific) to stay in their stem-like condition. BRG1 is involved in the drug resistance of GSCs and controls genes essential for glycolysis. GSCs express HELLS at a higher rate than either non-malignant brain cells or their differentiated tumor offspring [29, 30].

20.3.4 Post-Translational Histone Modifications

The acetylation, ubiquitylation/SUMOylation (lysine residues), methylation (lysine, arginine/histidine residues), and phosphorylation (serine, threonine/tyrosine residues) are the post-translational changes affecting transcription, DNA replication, and DNA repair [31]. The process relies on histone lysine

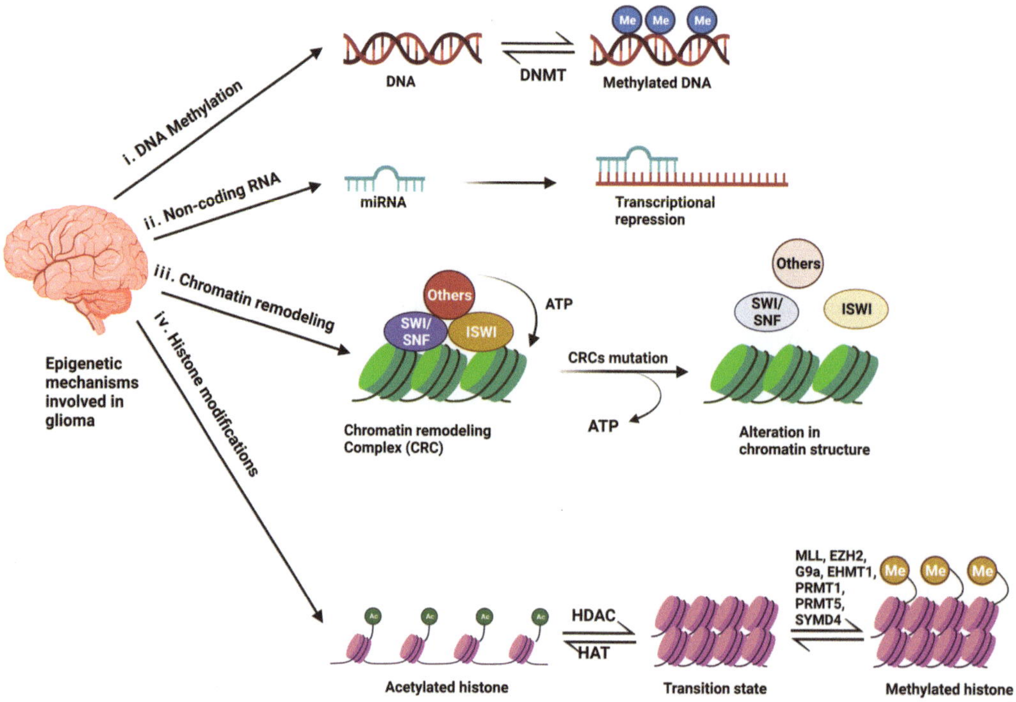

FIGURE 20.4 Snapshot of epigenetic mechanisms that are highly associated with the formation of glioma. (i) DNA methylation at CpG sites by DMNT leads to glioma pathogenesis. (ii) Various non-coding RNAs are involved in multiple aspects of glioma pathogenesis and progression by binding with target genes. (iii) The chromatin remodeling complex uses ATP breakdown to power the alteration of the chromatin structure. If mutation occurs to the important proteins in this complex, it affects the expression of genes that control cell growth and division, which can eventually lead to the growth of glioma. (iv) Acetylation, methylation, and phosphorylation occur at various sites of histones, which affects the interaction between DNA and histone proteins, altering chromatin structure and gene expression, which is associated with glioma pathogenesis. (Created using Biorender software.)

acetyltransferases (HATs), which such as CREB-binding protein/CBP and E1A binding protein P300, lysine acetyltransferase-2A, 2B, 6, 7, and 8 [32]. By facilitating the transfer of an acetyl group from acetyl-CoA to ε-N-acetyl lysine, they catalyze the acetylation of histone proteins.

20.3.5 Non-coding RNAs

Eukaryotic genomes rely on non-coding RNAs (ncRNAs) that are actively in the epigenetic control of glioma [33]. The interactions between the secondary structures of these ncRNAs and proteins or genomic DNA allow them to mediate epigenetic regulation [34]. Some of the most important ncRNAs in glioma are microRNAs (miRNAs), long non-coding RNAs (lncRNAs), and circular RNAs (circRNAs). LncRNAs have an impact on major glioma cell processes like carcinogenesis, progression, and enhanced treatment resistance.

20.4 EPIGENETIC MODIFICATIONS CORRELATING TO GLIOMA DEVELOPMENT

Studies highlight that glioma pathogenesis have distinct epigenetic features regulating glioma progression, molecular classification, such as alterations in IDH1/IDH2, H3F3A/HIST1H3B [35]

TABLE 20.1
Epigenetic Mechanisms Regulating Glioma Pathogenesis

Epigenetic Modifications	Genes/Proteins	Pathway	References
DNA Methylation	p16/INK4a, p14ARF, PTPRD, P73, NDRG2, SLC22A18	Cell proliferation, apoptosis, invasion, and migration	[44–46]
microRNA			[47–52]
miR-873	GLI1, Bcl2	Downregulated	
miR-7	EGFR, PI3K/ATK, Raf/MEK/ERK		
miR-491-5p	EGFR, CDK6 and Bcl-xL		
miR-491-3p	IGFBP2 and CDK6		
miR-93	integrin-beta8, IL-8	Upregulated	
miR-125b-2	mitochondrial pathway of apoptosis		
miR-296-3p	EAG1		
miR-451	SMAD		
Enzymes/Inhibitors			[53–55]
Histone deacetylases	HDAC2, HDAC9, HDAC1, HDAC3,	Gene repression	
Histone methyl transferases	MLL, G9a, EHMT1, SMYD4, EZH2, PRMT5, PRMT1		

(See Table 20.1). Low-grade gliomas in young adults are usually characterized by IDH1/IDH2 point mutations [36]. Oligodendrogliomas are comprised of IDH-mutant (IDHmt) with codeletion of chromosomal arms 1p/19q associated with telomerase reverse transcriptase regulatory motif change of gene [37]. The pediatric midline high-grade glioma involves histone mutation H3K27M and the hemispheric high-grade glioma involves H3G34R/V mutation [38]. Distinctive DNA methylation features drive the epigenetic mutations in gliomas exhibiting distinctive age-associated distribution patterns and location [39]. DNA is methylated when cytosine residues at CpG sites undergo a chemical modification at their fifth carbon position, which is the most common epigenetic change in cancer leading to 5-methylcytosine formation [40]. During replication, DNMT1 methylates hemimethylated DNA for protecting DNA methylation patterns. On the other hand, unmethylated DNA is the only target of DNMT3A, DNMT3B, and DNMT3L for *de novo* methylation [41]. 2-Hydroxyglutarate (2-HG) assists in DNA hypermethylation via TET2 inhibition, and advances the glioma CpG island methylator phenotype (G-CIMP) through IDH mutations [42]. When it comes to epigenetic suppression in gliomas, the MGMT gene is very important, especially in GBM. It is well-known that MGMT methylation is a significant prognostic indicator for TMZ therapy efficacy evaluation in these cases [43].

20.5 EPIGENETIC THERAPY FOR GLIOMA

Our knowledge of the epigenetic shifts that occur during the progression of glioma has been expanded thanks to recent technological developments such as deep sequencing. As a result of this, epigenetic inhibitors have emerged as a potentially useful alternative treatment option. Epigenetic regulators in GBM interact with essential pathways that drive cell and stem-like cell proliferation, most notably the Notch, Hh, and WNT pathways. Recent research has shed light on these interactions and provided evidence to support the hypothesis that targeting these pathways with epigenetic modulators could be an effective therapeutic strategy. Histone deacetylase inhibitors have demonstrated promising results in the treatment of GBM, particularly in terms of their ability to affect CSCs in GBM. These CSCs, which are distinguished by their resistance to conventional treatment protocols for cancer, play a significant part in the aggressive relapse of GBM. Inhibitors of DNMT, which include the chemical compound known as decitabine (DAC), constitute another important category of epigenetic modulators. Antitumor activity is demonstrated by DAC against gliomas that have mutations in IDH1. Clinical trials are currently being conducted to investigate whether DAC is effective in treating IDHmt gliomas [56].

BET inhibitors constitute yet another important class of compounds. BET proteins play a crucial role in regulating the transcription of oncogenes like the MYC and NUT fusion oncoproteins [57]. Enhancer of zeste homolog 2 (EZH2) inhibitors are another type of epigenetic drug that could be used to treat cancer [58].

Furthermore, for glioma treatment, a multimodal treatment strategy must be researched, approved, and implemented. The clinical status of epigenetic drugs has been listed in Table 20.2.

TABLE 20.2
Some Epigenetic Drugs for Glioma

Drug	Structure	Clinical Trial Phase	Target	Reference
Vorinostat	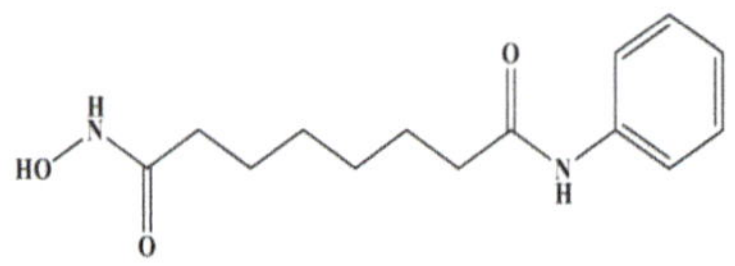	Phase II	Recurrent GBM, refractory solid tumors	[59]
Panobinostat	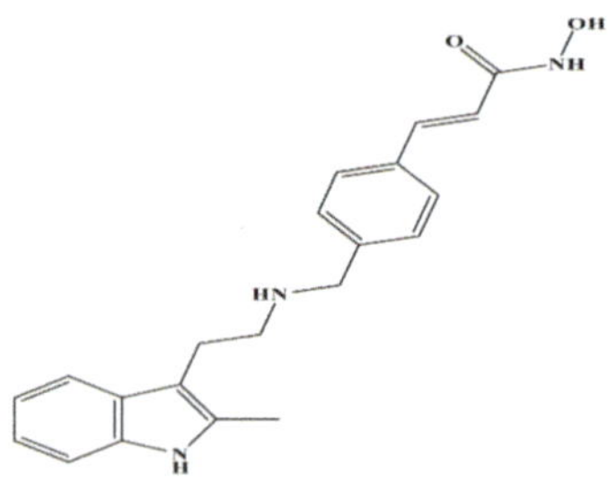	Phase II	Inhibits HDAC, recurrent glioblastoma, anaplastic glioma, mutant diffusive midline glioma	[60]
Romidepsin	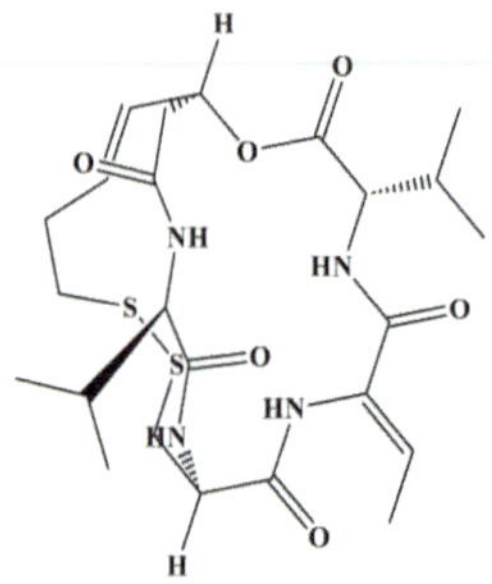	Phase 1	HDAC inhibitor, recurrent GBM	[56]
Valproic acid	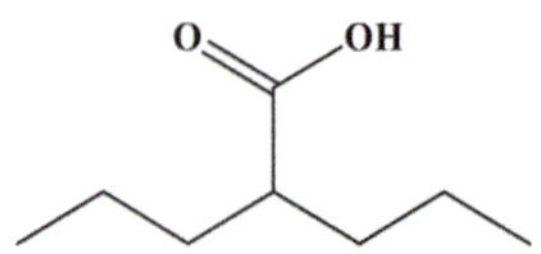	Phase 1	Refractory solid or CNS tumors	[59]
Guadecitabine	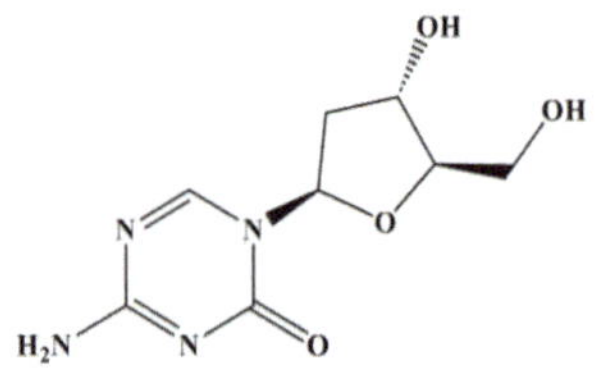	Phase III	DNA methyltransferase (DNMT), used in glioblastoma (GBM) and anaplastic astrocytoma	[61]

TABLE 20.2 *(Continued)*
Some Epigenetic Drugs for Glioma

Drug	Structure	Clinical Trial Phase	Target	Reference
Tazemetostat	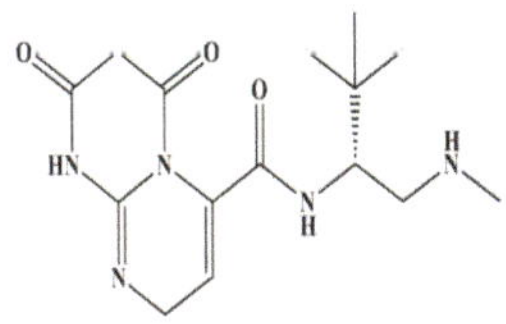	Phase I	Zeste homolog 2 inhibitor (EZH2i), diffusive midline glioma	[58]
Mocetinostat	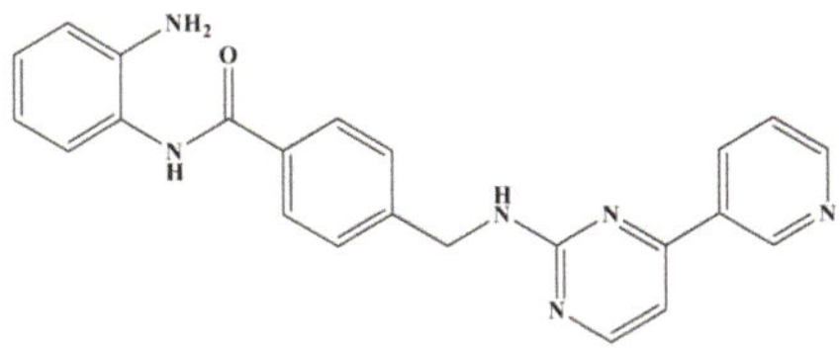	Preclinical phase	Cell-cycle arrest, autophagic cell death, and senescence, glioblastoma	[62]
MC1746	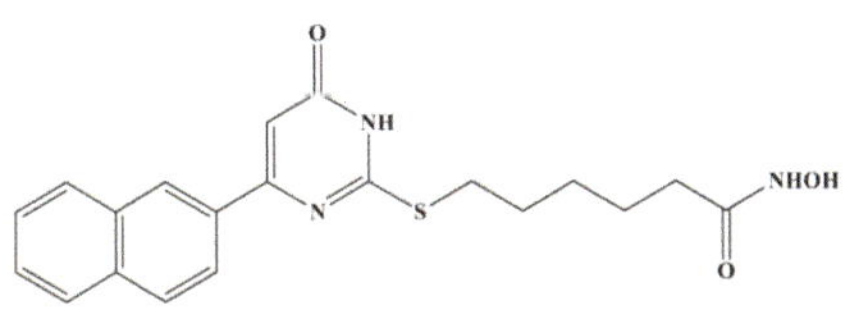	Preclinical phase	HDAC inhibition, glioblastoma	[63]
MC2129	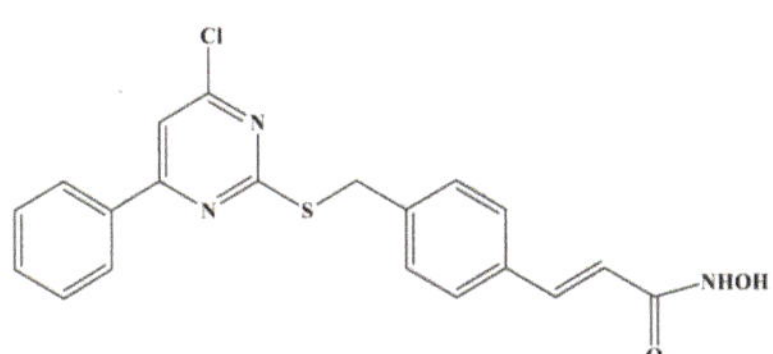	Preclinical phase	HDAC inhibition, glioblastoma	[63]
5-Azacytidine	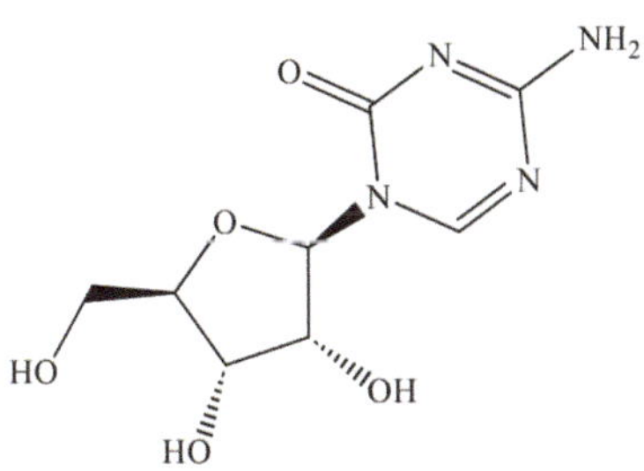	Preclinical phase	DNMT1 inhibitor, used for IDH-mutant glioma	[64]
Domatinostat	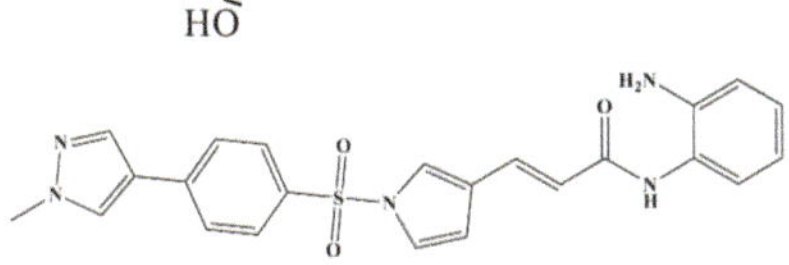	Preclinical studies	HDAC inhibitors, glioblastoma	[65]

20.6 CONCLUSION

The median survival of glioma patients has not significantly improved despite therapeutic developments because of developing radio/chemo-resistance and tissue infiltration, leading to implausible surgical removal. Epigenetic modifications are emerging as potential therapeutic targets since they can be considered clinical biomarkers for glioma. Molecules driving epigenetic changes modulate the plasticity of the tumor environment and facilitate modifications in the epigenomic environment, restoring the susceptibility of tumor cells to conventional therapy. Glioma development is largely dependent on multiple epigenetic alterations. Some examples of these changes include new patterns of DNA methylation, new ways that histones are processed after translation, and generalized changes to the structure of chromatin. These things are very important for making progress in both diagnostic tools and treatments. Epigenetic enzymes and microRNAs driving mutations and deletions of genomic regions have been demonstrated as specific therapeutic targets. Addressing epigenetic changes in glioma treatment is achieved through the use of compounds that target mutant IDH, BET motif proteins, histone deacetylases, DNA methylation, and EZH2.

ACKNOWLEDGMENTS

This work was supported by the Ministry of Human Resource Development (MHRD), Govt. of India; Science and Engineering Research Board (SERB), DST, Govt. of India (JCB/2019/000008); and the Indian Council of Medical Research (ICMR) (Sanction Letter No: 5/13/53/2020-NCD-III). We would like to acknowledge the University Grants Commission, New Delhi, Ministry of Human Resource Development, Govt. of India, and the Indian Institute of Technology, Kharagpur, for providing fellowship and necessary infrastructural support.

DECLARATIONS

Conflict of interest: The authors declare no conflicts of interest.

Consent for publishing: All authors have approved the final version of the manuscript and given consent for publication.

Funding: Not Applicable.

REFERENCES

1. Rajesh Y, Biswas A, Mandal M (2017) Glioma progression through the prism of heat shock protein mediated extracellular matrix remodeling and epithelial to mesenchymal transition. Exp Cell Res 359:299–311.
2. Clarke J, Penas C, Pastori C, et al (2013) Epigenetic pathways and glioblastoma treatment. Epigenetics 8:785–795. https://doi.org/10.4161/epi.25440
3. Allen BK, Stathias V, Maloof ME, et al (2015) Epigenetic pathways and glioblastoma treatment: Insights from signaling cascades. J Cell Biochem 116:351–363. https://doi.org/10.1002/jcb.24990
4. Takebe N, Miele L, Harris PJ, et al (2015) Targeting Notch, Hedgehog, and Wnt pathways in cancer stem cells: clinical update. Nat Rev Clin Oncol 12:445–464. https://doi.org/10.1038/nrclinonc.2015.61
5. Romani M, Pistillo MP, Banelli B (2018) Epigenetic targeting of glioblastoma. Front Oncol 8:448. https://doi.org/10.3389/fonc.2018.00448
6. Marigo V, Davey RA, Zuo Y, et al (1996) Biochemical evidence that Patched is the Hedgehog receptor. Nature 384:176–179. https://doi.org/10.1038/384176a0
7. Price MA, Kalderon D (2002) Proteolysis of the Hedgehog signaling effector Cubitus interruptus requires phosphorylation by Glycogen Synthase Kinase 3 and Casein Kinase 1. Cell 108:823–835. https://doi.org/10.1016/S0092-8674(02)00664-5
8. Mastronardi FG, Dimitroulakos J, Kamel-Reid S, Manoukian AS (2000) Co-localization of patched and activated sonic hedgehog to lysosomes in neurons. Neuroreport 11:581–585. https://doi.org/10.1097/00001756-200002280-00030
9. Rubin LL, de Sauvage FJ (2006) Targeting the Hedgehog pathway in cancer. Nat Rev Drug Discov 5:1026–1033. https://doi.org/10.1038/nrd2086
10. Ehtesham M, Sarangi A, Valadez JG, et al (2007) Ligand-dependent activation of the hedgehog pathway in glioma progenitor cells. Oncogene 26:5752–5761. https://doi.org/10.1038/sj.onc.1210359
11. Bar EE, Chaudhry A, Lin A, et al (2007) Cyclopamine-mediated hedgehog pathway inhibition depletes stem-like cancer cells in glioblastoma. Stem Cells 25:2524–2533. https://doi.org/10.1634/stemcells.2007-0166
12. Azzi S, Treps L, Leclair HM, et al (2015) Desert hedgehog/patch2 axis contributes to vascular permeability and angiogenesis in glioblastoma. Front Pharmacol 6:281. https://doi.org/10.3389/fphar.2015.00281

13. Zhou A, Lin K, Zhang S, et al (2017) Gli1-induced deubiquitinase USP48 aids glioblastoma tumorigenesis by stabilizing Gli1. EMBO Rep 18:1318–1330. https://doi.org/10.15252/embr.201643124
14. Robbins DJ, Fei DL, Riobo NA (2012) The hedgehog signal transduction network. Sci Signal 5: re6. https://doi.org/10.1126/scisignal.2002906
15. Malatesta M, Steinhauer C, Mohammad F, et al (2013) Histone acetyltransferase PCAF is required for Hedgehog–Gli-dependent transcription and cancer cell proliferation. Cancer Res 73:6323–6333. https://doi.org/10.1158/0008-5472.CAN-12-4660
16. Zhang L, Yang X, Yang S, Zhang J (2011) The Wnt/β-catenin signaling pathway in the adult neurogenesis. Eur J Neurosci 33:1–8. https://doi.org/10.1111/j.1460-9568.2010.7483.x
17. Zhang N, Wei P, Gong A, et al (2011) FoxM1 promotes β-catenin nuclear localization and controls Wnt target-gene expression and glioma tumorigenesis. Cancer Cell 20:427–442. https://doi.org/10.1016/j.ccr.2011.08.016
18. Klaus A, Birchmeier W (2008) Wnt signalling and its impact on development and cancer. Nat Rev Cancer 8:387–398. https://doi.org/10.1038/nrc2389
19. Schiefer L, Visweswaran M, Perumal V, et al (2014) Epigenetic regulation of the secreted frizzled-related protein family in human glioblastoma multiforme. Cancer Gene Ther 21:297–303. https://doi.org/10.1038/cgt.2014.30
20. Fan X, Mikolaenko I, Elhassan I, et al (2004) ***Notch1*** and ***Notch2*** have opposite effects on embryonal brain tumor growth. Cancer Res 64:7787–7793. https://doi.org/10.1158/0008-5472.CAN-04-1446
21. Tsung AJ, Guda MR, Asuthkar S, et al (2017) Methylation regulates HEY1 expression in glioblastoma. Oncotarget 8:44398–44409. https://doi.org/10.18632/oncotarget.17897
22. Aktaş S, Zadeoğlulari Z, Erçetin P, Olgun N (2010) The effect of differentiating and apoptotic agents on notch signalling pathway in hepatoblastoma. Hepatogastroenterology 57:891–898.
23. Yin Y, Morgunova E, Jolma A, et al (2017) Impact of cytosine methylation on DNA binding specificities of human transcription factors. Science 356:eaaj2239. https://doi.org/10.1126/science.aaj2239
24. Lin C-C, Chen Y-P, Yang W-Z, et al (2020) Structural insights into CpG-specific DNA methylation by human DNA methyltransferase 3B. Nucleic Acids Res 48:3949–3961. https://doi.org/10.1093/nar/gkaa111
25. Zhao Z, Shilatifard A (2019) Epigenetic modifications of histones in cancer. Genome Biol 20:245. https://doi.org/10.1186/s13059-019-1870-5
26. Ciechomska IA, Jayaprakash C, Maleszewska M, Kaminska B (2020) Histone modifying enzymes and chromatin modifiers in glioma pathobiology and therapy responses. Adv Exp Med Biol 1202: 259–279.
27. Phillips RE, Soshnev AA, Allis CD (2020) Epigenomic reprogramming as a driver of malignant glioma. Cancer Cell 38:647–660. https://doi.org/10.1016/j.ccell.2020.08.008
28. Turcan S, Makarov V, Taranda J, et al (2018) Mutant-IDH1-dependent chromatin state reprogramming, reversibility, and persistence. Nat Genet 50:62–72. https://doi.org/10.1038/s41588-017-0001-z
29. Ganguly D, Sims M, Cai C, et al (2018) Chromatin remodeling factor BRG1 regulates stemness and chemosensitivity of glioma initiating cells. Stem Cells 36:1804–1815. https://doi.org/10.1002/stem.2909
30. Zhang G, Dong Z, Prager BC, et al (2019) Chromatin remodeler HELLS maintains glioma stem cells through E2F3 and MYC. JCI Insight 4:e126140. https://doi.org/10.1172/jci.insight.126140
31. García-Giménez JL, Romá-Mateo C, Pallardó F V. (2019) Oxidative post-translational modifications in histones. BioFactors 45:641–650. https://doi.org/10.1002/biof.1532
32. Lan R, Wang Q (2020) Deciphering structure, function and mechanism of lysine acetyltransferase HBO1 in protein acetylation, transcription regulation, DNA replication and its oncogenic properties in cancer. Cell Mol Life Sci 77:637–649. https://doi.org/10.1007/s00018-019-03296-x
33. Wei J-W, Huang K, Yang C, Kang C-S (2017) Non-coding RNAs as regulators in epigenetics. Oncol Rep 37:3–9. https://doi.org/10.3892/or.2016.5236
34. Chen J, Xue Y (2016) Emerging roles of non-coding RNAs in epigenetic regulation. Sci China Life Sci 59:227–235. https://doi.org/10.1007/s11427-016-5010-0
35. Masui K, Onizuka H, Cavenee WK, et al (2019) Metabolic reprogramming in the pathogenesis of glioma: Update. Neuropathology 39:3–13. https://doi.org/10.1111/neup.12535
36. Yan H, Parsons DW, Jin G, et al (2009) *IDH1* and *IDH2* mutations in gliomas. N Eng J Med 360: 765–773. https://doi.org/10.1056/NEJMoa0808710
37. van den Bent MJ, Chang SM (2018) Grade II and III oligodendroglioma and astrocytoma. Neurol Clin 36:467–484. https://doi.org/10.1016/j.ncl.2018.04.005
38. Wesseling P, Capper D (2018) WHO 2016 Classification of gliomas. Neuropathol Appl Neurobiol 44:139–150. https://doi.org/10.1111/nan.12432

39. Ferris SP, Hofmann JW, Solomon DA, Perry A (2017) Characterization of gliomas: From morphology to molecules. Virchows Archiv 471:257–269. https://doi.org/10.1007/s00428-017-2181-4
40. Michalak EM, Burr ML, Bannister AJ, Dawson MA (2019) The roles of DNA, RNA and histone methylation in ageing and cancer. Nat Rev Mol Cell Biol 20:573–589. https://doi.org/10.1038/s41580-019-0143-1
41. DaRosa PA, Harrison JS, Zelter A, et al (2018) A bifunctional role for the UHRF1 UBL domain in the control of hemi-methylated DNA-Dependent histone ubiquitylation. Mol Cell 72:753–765.e6. https://doi.org/10.1016/j.molcel.2018.09.029
42. Jin G, Reitman ZJ, Duncan CG, et al (2013) Disruption of wild-type IDH1 suppresses D-2-hydroxyglutarate production in IDH1-mutated gliomas. Cancer Res 73:496–501. https://doi.org/10.1158/0008-5472.CAN-12-2852
43. Kundu M, Das S, Nandi S, et al (2023) Magnolol and temozolomide exhibit a synergistic anti-glioma activity through MGMT inhibition. Biochim Biophys Acta Mol Basis Dis 1869:166782. https://doi.org/10.1016/j.bbadis.2023.166782
44. Lee SH, Kim MS, Kwon HC, et al (2000) Growth inhibitory effect on glioma cells of adenovirus-mediated p16/INK4a gene transfer *in vitro* and *in vivo*. Int J Mol Med 6:559–63.
45. Watanabe T, Katayama Y, Yoshino A, et al (2007) Aberrant hypermethylation of $p14^{ARF}$ and O^6-methylguanine-DNA methyltransferase genes in astrocytoma progression. Brain Pathol 17:5–10. https://doi.org/10.1111/j.1750-3639.2006.00030.x
46. Veeriah S, Brennan C, Meng S, et al (2009) The tyrosine phosphatase PTPRD is a tumor suppressor that is frequently inactivated and mutated in glioblastoma and other human cancers. Proc Nat Acad Sci 106:9435–9440. https://doi.org/10.1073/pnas.0900571106
47. CHEN X, ZHANG Y, SHI Y, et al (2015) miR-873 acts as a novel sensitizer of glioma cells to cisplatin by targeting Bcl-2. Int J Oncol 47:1603–1611. https://doi.org/10.3892/ijo.2015.3143
48. Zhang J-S, Zhao Y, Lv Y, et al (2017) miR-873 suppresses H9C2 cardiomyocyte proliferation by targeting GLI1. Gene 626:426–432. https://doi.org/10.1016/j.gene.2017.05.062
49. Liu X, Li G, Su Z, et al (2013) Poly(amido amine) is an ideal carrier of miR-7 for enhancing gene silencing effects on the EGFR pathway in U251 glioma cells. Oncol Rep 29:1387–1394. https://doi.org/10.3892/or.2013.2283
50. Li X, Liu Y, Granberg KJ, et al (2015) Two mature products of MIR-491 coordinate to suppress key cancer hallmarks in glioblastoma. Oncogene 34:1619–1628. https://doi.org/10.1038/onc.2014.98
51. Fang L, Deng Z, Shatseva T, et al (2011) MicroRNA miR-93 promotes tumor growth and angiogenesis by targeting integrin-β8. Oncogene 30:806–821. https://doi.org/10.1038/onc.2010.465
52. Bai Y, Zhang Q-G, Wang X-H (2014) Downregulation of TES by hypermethylation in glioblastoma reduces cell apoptosis and predicts poor clinical outcome. Eur J Med Res 19:66. https://doi.org/10.1186/s40001-014-0066-4
53. Lucio-Eterovic AK, Cortez MA, Valera ET, et al (2008) Differential expression of 12 histone deacetylase (HDAC) genes in astrocytomas and normal brain tissue: Class II and IV are hypoexpressed in glioblastomas. BMC Cancer 8:243. https://doi.org/10.1186/1471-2407-8-243
54. Chang Y, Zhang X, Horton JR, et al (2009) Structural basis for G9a-like protein lysine methyltransferase inhibition by BIX-01294. Nat Struct Mol Biol 16:312–317. https://doi.org/10.1038/nsmb.1560
55. Cheung HC, Yatsenko SA, Kadapakkam M, et al (2012) Constitutional tandem duplication of 9q34 that truncates *EHMT1* in a child with ganglioglioma. Pediatr Blood Cancer 58:801–805. https://doi.org/10.1002/pbc.23219
56. Iwamoto FM, Lamborn KR, Kuhn JG, et al (2011) A phase I/II trial of the histone deacetylase inhibitor romidepsin for adults with recurrent malignant glioma: North American Brain Tumor Consortium Study 03-03. Neuro Oncol 13:509–516. https://doi.org/10.1093/neuonc/nor017
57. Shi J, Vakoc CR (2014) The mechanisms behind the therapeutic activity of BET bromodomain inhibition. Mol Cell 54:728–736. https://doi.org/10.1016/j.molcel.2014.05.016
58. Jing L, Qian Z, Gao Q, et al (2023) Diffuse midline glioma treated with epigenetic agent-based immunotherapy. Signal Transduct Target Ther 8:23. https://doi.org/10.1038/s41392-022-01274-7
59. Tarasenko N, Chekroun-Setti H, Nudelman A, Rephaeli A (2018) Comparison of the anticancer properties of a novel valproic acid prodrug to leading histone deacetylase inhibitors. J Cell Biochem 119:3417–3428. https://doi.org/10.1002/jcb.26512
60. Neth BJ, Balakrishnan SN, Carabenciov ID, et al (2022) Panobinostat in adults with H3 K27M-mutant diffuse midline glioma: A single-center experience. J Neurooncol 157:91–100. https://doi.org/10.1007/s11060-022-03950-8

61. McCarl L, Kumar R, Nigam S, et al (2019) IMMU-16. Guadecitabine (SGI-110) immunosensitizes murine C57BL/6-syngeneic glioma and DIPG MODELS. Neuro Oncol 21:ii96. https://doi.org/10.1093/neuonc/noz036.137
62. Khathayer F (2022) 85P Mocetinostat (MGCD0103) or MG0103 is an isotype-selective histone deacetylase (HDAC) inhibitor induce apoptosis and suppress tumor in glioblastoma cell lines C6 and T98G. Ann Oncol 33:S1409. https://doi.org/10.1016/j.annonc.2022.09.086
63. Was H, Krol SK, Rotili D, et al (2019) Histone deacetylase inhibitors exert anti-tumor effects on human adherent and stem-like glioma cells. Clin Epigenetics 11:11. https://doi.org/10.1186/s13148-018-0598-5
64. Miller JJ, Cahill DP, Arrillaga-Romany I (2022) Enhancing demethylation-induced differentiation in IDH-mutant glioma. Neuro Oncol 24:724–725. https://doi.org/10.1093/neuonc/noac056
65. Nakagawa-Saito Y, Saitoh S, Mitobe Y, et al (2022) HDAC class I inhibitor domatinostat preferentially targets glioma stem cells over their differentiated progeny. Int J Mol Sci 23:8084. https://doi.org/10.3390/ijms23158084

21 Associations of Proteases and Long Non-Coding RNAs in Cancer

Gaurav Kumar Bhagat
University of Kalyani,
Kalyani-Nadia, India

Dipanjan Guha and Angshuman Bagchi
Bioinformatics Infrastructure Facility Center
University of Kalyani,
Kalyani-Nadia, India

21.1 INTRODUCTION

Proteases are widely distributed enzymes that play an important role in maintaining cellular homeostasis by controlling gene expression, migration, differentiation, immune defense, cell death, etc. Malignant tumor formation and development usually occur due to dysregulation in the activities of proteases. Therefore, many proteases are being studied for potential therapeutic use to suppress the growth of a variety of cancer cells. Proteases can be divided into six clusters: threonine, serine, cysteine, aspartate proteases, metalloproteases and glutamic acid revolving around the core of amino acid in active site and method of breakage of the peptide bond (Berg et al., 2002; Eatemadi et al., 2017).

The tumor-promoting properties of different proteases have increased interest in them as molecular targets for different types of cancer (Besingi & Clark, 2015). Understanding the structural characteristics of proteases helps in determining the mechanism of action of anti-proteases that prevent cancer development and metastasis. There are two major mechanisms by which protease - anti-protease interactions occur: (i) irreversible trapping interactions that begin by breaking the inhibitor's inner peptide bond and permanently change its conformation and (ii) reversible tight-binding reactions that stick to the protease's active site directly (Farady & Craik, 2010).

The human genome codes for less than 2% of protein-coding genes; about 98% of RNAs don't have any protein-coding functions. Among this 98% of RNA, long non-coding RNAs (lncRNAs) are a group of RNA that play an important role in regulating cancer development. The lengths of these RNAs are greater than 200 nucleotides. Transcriptional and post-transcriptional gene regulations are mediated by lncRNAs. Thus, lncRNAs play a critical role in controlling various cellular functions like cell cycle and growth, apoptosis, tumor formation and cancer. LncRNAs are known to regulate the induction of several cancer types, which include gastrointestinal, lung, gynecological, prostate, head and neck cancers, etc.

DOI: 10.1201/9781003394693-24

Associations of proteases and lncRNAs in cancer are vital for advancing our understanding of cancer biology. The interplay between proteases and lncRNAs in cancer progression, metastasis, and therapeutic resistance has been discussed in this chapter.

21.2 ROLE OF PROTEASE IN CANCER

The specific processes through which the hemostatic system may promote the spread of cancer are yet unknown. Two well-known coagulant factors that are necessary for tumor invasion are tissue factor (TF) and thrombin. They can be produced in the tumor microenvironment and mediate a variety of cell-signaling processes by stimulating protease-activated receptors (PARs). The compounds like thrombin, TF, and PARs (primarily PAR-2 and PAR-1) play critical roles in tumor cell infringement and metastasis by promoting angiogenesis, tumor cell movement and synergy with host vascular cells. The impact of TF and thrombin and their activator role in the activation of PARs in the initiation of carcinoma is well documented. Treatment alternatives targeting PARs have also been suggested for cancer patients.

Dr. Trousseau and his colleagues were the first to identify the link between blood coagulation, specifically venous thrombosis, and the onset of cancer (Trousseau, 1865). A notable case of cancer-related thromboembolism is Trousseau's syndrome. The synergy between cancer cells, endothelial cells (ECs), extracellular matrix proteins and blood cells have a direct impact on the invasiveness and metastasis of tumors (Menter et al., 2014). A complex network of enzymes, cytokines, receptors and bioactive lipids is necessary for effective intracellular regulation of proteases and their effectors/inhibitors.

PARs, along with PAR-2 and PAR-1, and mRNA expression in primary tumor cells seem to be an important determinant in the invasiveness of primary tumors (Sedda et al., 2014).

21.2.1 Protease-Activated Receptors (PARs)

PARs, a G-protein-coupled receptor (GPCR) receptor, has four isoforms PARs; e.g. PAR-1, PAR-2, PAR-3, and PAR-4, which are encoded by specific genes (Ramachandran et al., 2012). Non-coding RNA upstream of F2R/PAR-1 (ncRuPAR: an RNA gene, is affiliated with the lncRNA class), is a recently deciphered lncRNA that is involved in the upregulation of PAR-1 during embryonic growth. ncRuPAR was found to be highly down regulated in gastric cancer tissues in comparison to the normal cells. ncRuPAR expression was found to be markedly reduced in gastric cancer tissues (Liu et al., 2014a).

21.2.2 PARs and Proteases: Cooperation in Cancer Progression

TF, a membrane glycoprotein preserved on sub-endothelial cells, initiates blood coagulation. Due to endothelial failure, TF is exposed to coagulation agents in the bloodstream. The most important cancer cell procoagulant, TF, also affects the growth of tumors. Stem cells, myofibroblasts, ECs and macrophages have all been discovered to express TF on their surfaces, as well as the tumor vasculature and the tumor microenvironment (Riewald et al., 2002; Fan *et al.*, 2005; Hobbs *et al.*, 2007).

The lncRNA, ncRuPAR inhibits the development of gastric cancer, and its underlying method involves PAR-1 inhibition (Liu et al., 2014a). Vascular endothelial growth factor (VEGF) messenger RNA (mRNA) and PAR-1 levels were found to be higher in gastric cancer tissues in comparison to nearby normal tissues. Also, the ncRuPAR expression level was found to be highly correlated with distant metastasis, lymph node metastasis, tumor size, tumor invasion depth and tumor–nodes–metastasis (TNM) stage (Liu et al., 2014a).

21.3 PROTEASE AND lncRNA ROLE IN VARIOUS CANCER TYPES

21.3.1 Angiogenesis

The growth and development of tumors depend heavily on angiogenesis, often known as the formation of new blood vessels. The lncRNA ncRuPAR inhibits the development of gastric cancer, and its underlying method involves PAR-1 inhibition. Vascular endothelial growth factor (VEGF) messenger RNA (mRNA) and PAR-1 levels were found to be higher in gastric cancer tissues in comparison to nearby normal tissues (Liu et al., 2014a). When thrombin-mediated PAR cleavage takes place in vessel-wall cells, blood cells and cancer, a number of proangiogenic genes, including VEGF and its receptor (VEGFR), MMP-2, TF, MAP, basic fibroblast growth factor (bFGF), angiopoetin-2 (Ang-2) and PI3 kinases, are activated.

21.3.2 Melanoma

Recent studies have suggested PAR-2's dual function in melanoma, which was previously unknown. Thrombin, a procoagulant produced in melanoma cells, acts as a primary activator of PAR-1 and its synthesis is TF-dependent (Mueller et al., 1992). Many lncRNAs are also involved in melanoma malignancy. The dysregulation of several lncRNAs like MALAT1, H19, CCAT1 and PVT1 results in spreading melanoma. Some of these lncRNAs are exclusively expressed in melanoma cell lines. Therefore, they can act as biomarkers for melanoma malignancy; e.g. the expression of SNHG5, PVT1 and SPRY4-IT1 proteins is different in normal individuals than in patients suffering from melanoma. LncRNAs regulate critical pathways in melanoma carcinogenesis like NF-kB and P13K/Akt signaling pathways (Safa et al., 2020).

21.3.3 Breast Cancer

The evidence for PARs, especially PAR-1 and PAR-2, being effective cell invasion mediators in epithelial malignancies is growing (Morris et al., 2006). Breast cancer is a widely examined instance of PAR-1 and PAR-2 function, and breast cancer cells express both of them (Su et al., 2009; Jiang et al., 2004).

Inhibition of PAR activation reduced cell invasion in highly metastatic MDA-435 breast carcinoma cells (Even-Ram et al., 2001). The lncRNA HOTAIR interacts with the transcription factor EZH2, preventing its binding to the E-cadherin promoter. This leads to reduced E-cadherin expression and increased expression of the protease uPA, promoting tumor invasion and metastasis (Gupta et al., 2010). The fact that PAR mediates the spread of breast cancer has been demonstrated by *in vivo* studies. For breast xenografts to develop and invade, PAR-1 expression is required when thrombin interacts with ErbB and EGFR or when fibroblast-derived MMP-1 mediates the Ca^{2+} pathway (Booden et al., 2004).

A schema (Figure 21.1) is shown to provide a quick outline about the interaction of lncRNAs/miRNAs with different proteases. These interactions are important for the initiation or to control the propagation of cancer.

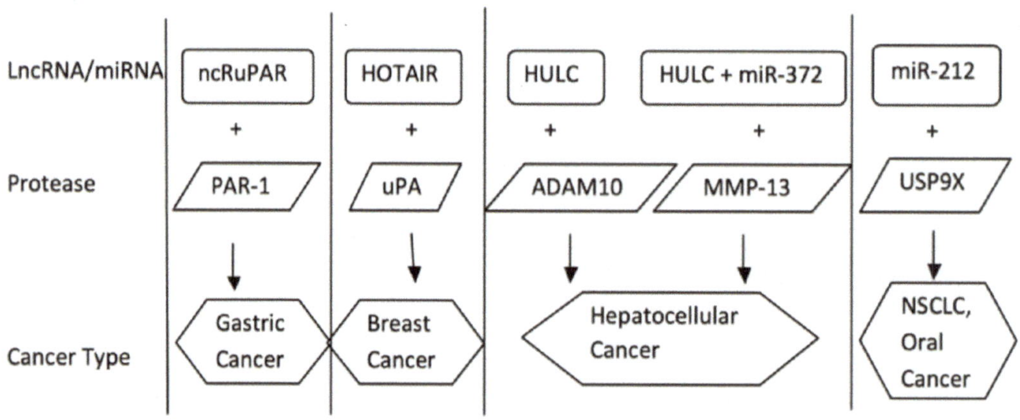

FIGURE 21.1 Schema indicating the interaction of lncRNAs/miRNAs with different proteases to spread/reduce cancer.

21.4 CROSSTALK BETWEEN NON-CODING RNA (ncRNA) AND PROTEASE IN CANCER INDUCTION OR REGULATION

21.4.1 Regulation of Protease Expression through LncRNA

LncRNAs have been found to regulate the expression of proteases in cancer. LncRNAs can directly interact with the genomic regions of protease genes or indirectly modulate their expression through interactions with transcription factors or epigenetic modifiers. By doing so, lncRNAs can exert control over the levels and activity of proteases, influencing cancer-related processes.

One mechanism by which lncRNAs regulate protease expression is through the recruitment of chromatin modifiers to the promoter regions of protease genes. LncRNAs can act as scaffolds, guiding chromatin-modifying complexes to specific genomic loci. For example, in gastric cancer, the lncRNA HOTTIP interacts with the WDR5/MLL complex and promotes histone H3 lysine 4 trimethylation at the promoter region of the MMP-9 gene, leading to enhanced expression of MMP-9 and enhanced tumor invasion.

LncRNAs have the ability to divert a transcription factor from its destination; i.e. the promoters of target genes. This can delay the binding of transcription factors and consequently obstruct their ability to activate or repress the expression of protease genes. This competitive binding can result in altered expression levels of proteases and subsequent impact on cancer progression. An example of this mechanism is seen in breast cancer, where the lncRNA HOTAIR interacts with the transcription factor EZH2, preventing its binding to the E-cadherin promoter. This leads to reduced E-cadherin expression and increased expression of the protease uPA, promoting tumor invasion and metastasis (Gupta et al., 2010).

In summary, lncRNAs can regulate protease expression in cancer through various mechanisms, including chromatin remodeling, interaction with transcription factors, and miRNA sponging. By modulating the expression levels and activity of proteases, lncRNAs contribute to the complex regulatory networks underlying cancer progression. Understanding the intricate interplay between lncRNAs and proteases may provide valuable insights into cancer biology and effectively help in the development of innovative therapeutic strategies targeting these regulatory pathways.

21.4.2 Regulation of LncRNA Expression through Proteases

While the primary role of proteases is protein degradation, emerging evidence suggests that proteases can also regulate the expression of lncRNAs. Proteases can indirectly influence lncRNA expression through their involvement in signaling pathways or by modifying the activity of transcription factors and other regulatory proteins.

Proteases can participate in signaling cascades that ultimately affect lncRNA expression. Certain proteases can activate specific signaling pathways, like MAPK or NF-κB pathways, which can modulate lncRNA expression. Activation of these pathways by proteases leads to the recruitment of transcription factors and the subsequent regulation of target gene expression, including lncRNAs. Proteases are also found to directly influence the activity of transcription factors or other regulatory proteins that control lncRNA expression. Proteases can cleave or degrade transcription factors, altering their stability, DNA-binding capacity, or protein-protein interactions. This proteolytic regulation can result in changes in the expression of lncRNAs that are under the control of these transcription factors. For example, in hepatocellular carcinoma, the protease ADAM10 cleaves and activates Notch1, which in turn promotes the expression of the lncRNA HULC (Giovannini et al., 2021; Hu et al., 2010).The regulation of lncRNA expression by proteases is still an emerging area of research. However, the existing evidence suggests that proteases can play a role in modulating lncRNA expression through various indirect mechanisms involving signaling pathways, transcription factors, RNA-binding proteins, and RNA degradation machinery and thereby can regulate cancer formation or growth.

Understanding the link between proteases and lncRNAs has the potential to provide insights into the intricate regulatory networks underlying cancer and other diseases. It also highlights the relationship between different molecular processes and the complexity of gene expression regulation. Future research will likely unveil additional details about the crosstalk between proteases and lncRNAs, potentially leading to new therapeutic strategies targeting these regulatory interactions.

21.4.3 Role of MicroRNAs (miRNAs)

MicroRNAs (miRNAs) have important functions in several biological processes, which include differentiation, proliferation, cell cycling, apoptosis and migration. LncRNAs can regulate protease expression through post-transcriptional mechanisms, such as miRNA sponging (Fontemaggi, 2023; Liu et al., 2014b). LncRNAs containing complementary sequences to specific miRNAs can compete with the target mRNAs for miRNA binding, acting as molecular sponges and preventing miRNA-mediated degradation or repression. This can indirectly result in increased expression of proteases targeted by the miRNAs. For instance, in hepatocellular carcinoma, the lncRNA HULC acts as a sponge for miR-372, relieving its suppression of the protease MMP-13 and promoting tumor invasion and metastasis (Cao et al., 2019).

Recent evidence suggests that the ubiquitin-specific protease 22 (USP22), is a cancer stem cell marker that appears to have a major role in epithelial malignancies and other solid tumors. Inhibition of cellular proliferation, an increase of cells in the G1 phase, a decrease in apoptosis, and down-regulation of major vault protein (MVP) expression were the effects of miRNA's knockdown of USP22 protein expression. Additionally, USP22 controls the transcription of the MVP drug-resistant gene as a key transcriptional factor. Targeting USP22 may provide new cancer treatment options.

Numerous malignancies, including non-small cell lung cancer (NSCLC) and oral carcinoma, were shown to have elevated levels of the miR-212 gene, which is situated at chromosome 17p13.3. Due to their ability to specifically deconjugate ubiquitin from ubiquitylated substrates, ubiquitin-specific proteases (USPs), the biggest category of deubiquitinases (DUBs), have a significant function in the ubiquitin system. USP9X overexpression has reportedly been linked to tumor growth, apoptosis, chemoresistance, and metastasis. Recent research has shown that miR-212 overexpression dramatically reduced NSCLC cells' ability to invade and migrate by targeting USP9X.

21.5 CONCLUSION AND FUTURE PERSPECTIVES

This chapter investigates the critical roles played by proteases and lncRNAs in the realm of cancer biology. The promising therapeutic potential of targeting proteases in cancer treatment is well-established, with protease inhibitors emerging as promising candidates, some of which are currently undergoing clinical development. Nonetheless, the intricate nature of protease networks and the possibility of unintended side effects necessitate ongoing research to refine these therapeutic approaches. On the other hand, lncRNAs exhibit multifaceted functions encompassing gene regulation, epigenetic modifications, and influence over the tumor microenvironment and are briefly shown in Figure 21.2. The chapter also explores the potential of lncRNAs as diagnostic and prognostic biomarkers, as well as therapeutic targets across various cancer types. The need for in-depth characterizations of specific proteases and lncRNAs, along with their intricate interactions, coupled with multi-omics integration, is underscored. This collective knowledge can be utilized for the development of precise and targeted therapies, heralding a new era in personalized cancer treatment. This chapter summarizes the pivotal roles played by proteases and lncRNAs in the context of cancer and highlights the potential for targeted interventions to reshape the landscape of cancer therapy, offering renewed hope to patients worldwide.

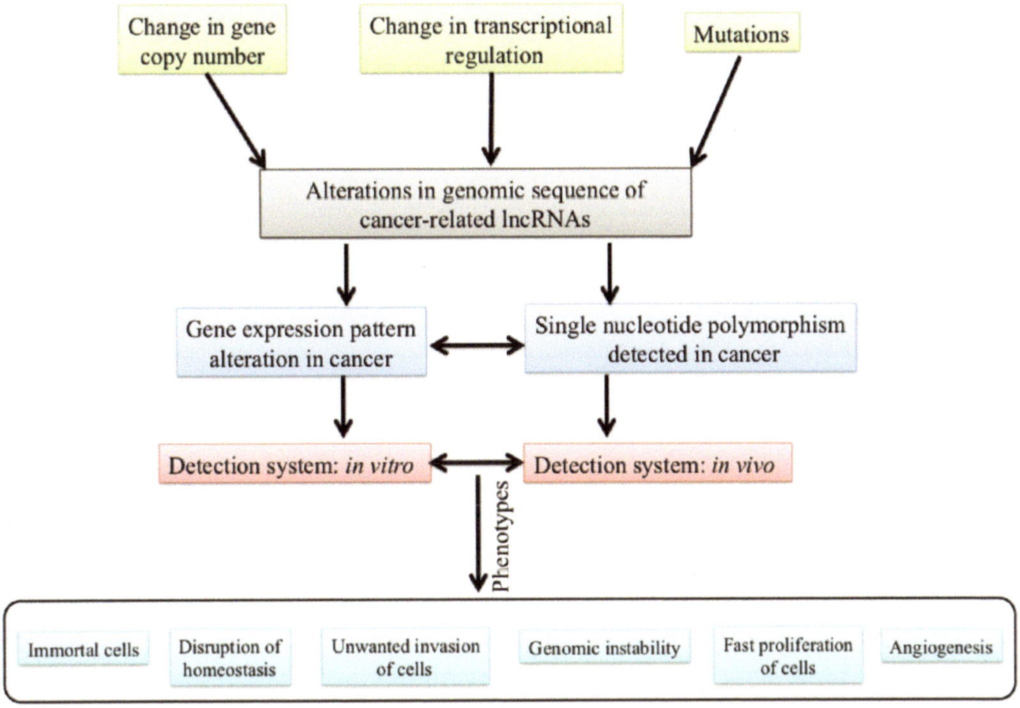

FIGURE 21.2 Changes in the genetic architecture of tumor cells are responsible for the initial findings to show the association of lncRNAs with cancer. Studies were done in both *in vitro* and *in vivo* systems to find out the direct link between lncRNAs and cancer propagation.

ACKNOWLEDGMENTS

The authors acknowledge the financial and infrastructural support from the Department of Biotechnology (DBT), the Government of India, and the University of Kalyani.

DECLARATIONS

Conflict of interest: The authors declare no conflicts of interest.

Consent for publishing: All authors have approved the final version of the manuscript and given consent for publication. We declare that no portions including the figures are taken from a published article.

Funding: The financial support was received from the DBT funded Bioinformatics Infrastructure Facility Center (Sanction no.: BT/PR40162/BTIS/137/48/2022 and Sanction no.: BT/PR40192/BTIS/137/69/2023) sanctioned to Prof. Angshuman Bagchi. GKB receives fellowship from DBT, BINC

REFERENCES

Berg J. M., Tymoczko J. L., Stryer L. (2002). Catalytic strategies. Biochemistry, 5th ed. New York: W H Freeman.

Besingi R. N., Clark P. L. (2015). Extracellular protease digestion to evaluate membrane protein cell surface localization. Nature Protocols, 10(12):2074–2080.

Booden, M. A., Ekert, L., Der, C. J., Trejo, J. (2004). Persistent signaling by dysregulated thrombin receptor trafficking promotes breast carcinoma cell invasion. Molecular Cell. Biology, 24(5), 1990–1999.

Cao Q., Guo Z., Yan Y., Wu J., Song C. (2019). Exosomal long noncoding RNAs in aging and age-related diseases. IUBMB Life, 71(12):1846–1856. https://doi.org/10.1002/iub.2141.

Eatemadi A., Aiyelabegan H. T., Negahdari B., et al. (2017). Role of protease and protease inhibitors in cancer pathogenesis and treatment. Biomedicine & Pharmacotherapy, 86:221–231.

Even-Ram, S. C., Maoz, M., Pokroy, E., Reich, R., Katz, B. Z., Gutwein, P., et al. (2001). Tumor cell invasion is promoted by activation of protease activated receptor-1 in cooperation with the alpha vbeta 5 integrin. Journal of Biological Chemistry, 276(14):10952–10962.

Fan, L., Yotov, W. V., Zhu, T., Esmailzadeh, L., Joyal, J. S., Sennlaub, F., et al. (2005). Tissue factor enhances protease-activated receptor-2-mediated factor VIIa cell proliferative properties. Journal of Thrombosis and Haemostasis, 3(5):1056–1063.

Farady C. J., Craik C. S. (2010). Mechanisms of macromolecular protease inhibitors. Chembiochem, 11(17):2341–2346.

Fontemaggi G. (2023). Non-coding RNA regulatory networks in post-transcriptional regulation of VEGFA in cancer. IUBMB Life, 5(1):30–39. https://doi.org/10.1002/iub.2620.

Giovannini C., Fornari F., Piscaglia F., Gramantieri L. (2021). Notch signaling regulation in HCC: From hepatitis virus to non-coding RNAs. Cell, 10(3):521. https://doi.org/10.3390/cells10030521.

Gupta R. A., Shah N., Wang K. C., Kim J., Horlings H. M., Wong D. J., et al. (2010). Long non-coding RNA HOTAIR reprograms chromatin state to promote cancer metastasis. Nature, 464(7291):1071–1076. https://doi.org/10.1038/nature08975

Hobbs, J. E., Zakarija, A., Cundiff, D. L., et al. (2007). Alternatively spliced human tissue factor promotes tumor growth and angiogenesis in a pancreatic cancer tumor model. Thrombosis Research, 120(Suppl 2):S13–S21.

Hu X., Zhang X., Qiu S., Yu D., Lin S. (2010). Salidroside induces cell-cycle arrest and apoptosis in human breast cancer cells. Biochemical and Biophysical Research Communications, 398(1):62–7. https://doi.org/10.1016/j.bbrc.2010.06.033.

Jiang, X., Bailly, M. A., Panetti, T. S., Cappello, M., Konigsberg, W. H., Bromberg, M. E. (2004). Formation of tissue factor- factor VIIa-factor Xa complex promotes cellular signaling and migration of human breast cancer cells. Journal of Thrombosis and Haemostasis, 2(1):93–101.

Liu L., et al. (2014a). ncRuPAR inhibits gastric cancer progression by down-regulating protease activated receptor-1. Tumor Biology, 35(8):7821–7829. https://doi.org/10.1007/s13277-014-2042-6.

Liu X. H, Sun M., Nie F. Q., Ge Y. B., Zhang E. B., Yin D. D., Kong R., Xia R., Lu K. H., Li J. H., De W., Wang K. M., Wang Z. X. (2014b). Lnc RNA HOTAIR functions as a competing endogenous RNA to regulate HER2 expression by sponging miR-331-3p in gastric cancer. Molecular Cancer, 13:92. https://doi.org/10.1186/1476-4598-13-92.

Menter, D. G., Tucker, S. C., Kopetz, S., Sood, A. K., Crissman, J. D., Honn, K. V. (2014). Platelets and cancer: A casual or causal relationship: Revisited. Cancer and Metastasis Reviews, 33(1):231–269.

Morris, D. R., Ding, Y., Ricks, T. K., Gullapalli, A., Wolfe, B. L., Trejo, J. (2006). Protease-activated receptor-2 is essential for factor VIIa and Xa-induced signaling, migration, and invasion of breast cancer cells. Cancer Research, 66(1):307–314.

Mueller, B. M., Reisfeld, R. A., Edgington, T. S., Ruf, W. (1992). Expression of tissue factor by melanoma cells promotes efficient hematogenous metastasis. Proceedings of the National Academy of Sciences of the United State of America, 89(24):11832–11836.

Ramachandran, R., Noorbakhsh, F., Defea, K., Hollenberg, M. D. (2012). Targeting proteinase-activated receptors: therapeutic potential and challenges. Nature Reviews Drug Discovery, 11(1):69–86.

Riewald, M., Petrovan, R. J., Donner, A., Mueller, B. M., Ruf, W. (2002). Activation of endothelial cell protease activated receptor-1 by the protein C pathway. Science, 296(5574):1880–1888.

Safa A, Gholipour M, Dinger ME, Taheri M, Ghafouri-Fard S (2020). The critical roles of lncRNAs in the pathogenesis of melanoma. ExpMolPathol, 117:104558.

Sedda, S., Marafini, I., Caruso, R., Pallone, F., Monteleone, G. (2014). Proteinase activated-receptors-associated signaling in the control of gastric cancer. World Journal of Gastroenterology, 20(34): 11977–11984.

Su, S., Li, Y., Luo, Y., Sheng, Y., Su, Y., Padia, R. N., et al. (2009). Proteinase-activated receptor 2 expression in breast cancer and its role in breast cancer cell migration. Oncogene, 28(34):3047–3057.

Trousseau, A. (1865). Phlegmasiadolens. Clinique Medicale de l'Hotel-Dieu de Paris, 3, 490–515.

22 Stem Cell-Mediated Interventions in Cancers Caused by Proteases

Farheen Showket, Sabra Parveen, Neetu Badesra, Mohmmad Saleem Dar, Aubaidah Akhtar, Md Mehedi Hossain, Mir Owais Ayaz, and Mohd Jamal Dar
Academy of Scientific & Innovative Research,
Ghaziabad, India
and
CSIR-Indian Institute of Integrative Medicine,
Jammu, India

22.1 INTRODUCTION

Proteases, also known as proteinases or peptidases, are a diverse group of enzymes that act as protein scissors and catalyze the cleavage of peptide linkages to bring about the hydrolysis of proteins. Proteases mediate specific modifications of proteins and selectively carry out vital biological processes such as blood clotting, activating proenzymes, lysis of fibrin clots, and processing and transporting secretory proteins across bio-membranes [1]. Proteases are involved in myriad physiological and pathological activities by tightly regulating signaling pathways. Abnormal protease functioning becomes a ground for pathological alterations ranging from inflammatory diseases to neurological disorders and cancer [2]. The diversity of proteases owing to both their structure and function is enormous. Thus, based on their mechanisms of action and the nature of their active site, proteases are categorized into six classes: Matrix metalloproteinase (MMPs), cysteine proteases, serine proteases, aspartic proteases, threonine proteases, and glutamic proteases. The activity of these proteases is dependent upon different factors. Some are active at different but optimal pH conditions, while some acid proteases mostly have an active site that contains a catalytic aspartic or glutamic acid. Pepsin, for instance, is effective at pH 2. Alkaline or basic proteases act well at levels ranging from pH 8 to 11. At pH 8, trypsin and chymotrypsin, for example, perform to their fullest. Neutral proteases are usually produced by mast cells and trigger complement and kinin activation [3].

22.2 PROTEASE-MEDIATED TUMORIGENESIS

There is a significant compilation of research on the precise mechanism of proteases in cancer development that reveals that proteases can directly or indirectly impact cancer cell properties like cell migration, which is a crucial step in invasion and metastasis [4]. One of the primary functions of proteases observed in tumor invasion is the degradation of the extracellular matrix (ECM) which is a complex protein network that provides structural support to tissues [5]. MMPs, a family of zinc-binding proteinases, cathepsins, and urokinase-type plasminogen activators (uPAs), can break down

DOI: 10.1201/9781003394693-25

ECM proteins, including the basement membrane, promoting detachment of cancer cells from the primary tumor and their invasion to nearby tissues as well as lymphatic and blood vessels, thereby promoting metastasis [5]. MMPs modulate the balance between pro- and antiangiogenic factors and thereby mediate angiogenic response [6]. Thus, by modifying the extracellular environment and promoting cell motility, proteases facilitate the movement of cancer cells from the primary tumor site to distant organs [7].

22.3 STEM CELLS AND THEIR TYPES

A renowned German biologist Ernst Haeckel coined the idea of the "stem cell" in 1868 to refer to the characteristics of a fertilized egg that produces all of the organism's cells. The origin of stem cell therapy began in 1888 after two German zoologists named Theodor Heinrich Boveri and Valentin Haecker defined a stem cell. Their goal was to identify the unique population of cells in an embryo that had the ability to differentiate into more specialized cells. Studies conducted in 1902 by bone marrow researchers Franz Ernst Christian Neumann and Alexander Maximov showed the existence of shared progenitor cells that give rise to adult blood cells, a process known as hematopoiesis [8]. Stem cells have the capacity for unlimited self-renewal, in the context of the lifespan of the organism, while retaining the potential for asymmetric divisions in a way that one daughter cell is itself a stem cell while the other is committed to undergo a terminal differentiation. Stem cells also have the capacity to clonally regenerate all of the different kinds of cells that make up the adult tissue from which they originated. It is challenging to meet this requirement. It was proposed that all self-regenerating progenitor cells that have the broadest developmental potential and are accessible within a specific tissue at a specific time are considered to be stem cells [9]. However, later, the definition of a stem cell became more stringent and it was established that the attainment of stemness is a critical interplay of different cell events like self-renewal, differentiation, apoptosis, and migration, which work in a perfect balance in the living compartments or microenvironment, popularly called a "niche," which is constituted by stromal support cells, blood vessels, neural inputs, and ECM proteins that nurture stem cell activity [10].

22.4 SPECTRUM OF POTENCY SEEN ACROSS STEM CELLS

Developmental potency is an indispensable feature of all stem cells that makes stem cells developmentally flexible to give rise to cells of different lineages. However, the degree of plasticity recedes, and the flexibility of developmental fates narrow, beginning hierarchically from totipotent cells (Figure 22.1), which represent the epitome of stem cell characteristics and can give rise to all embryonic as well as extraembryonic tissues. A fertilized egg from the stage of a zygote to blastomeres represents a totipotent cell [11]. The division of a zygote results in the formation of cells that are more lineage-restricted than totipotent cells and are called pluripotent cells from which nearly 200 different cell types can arise [11]. The latter are called multipotent cells, for instance lymphoid and myeloid stem cells. Last in developmental flexibility are unipotent cells, with their potential to give rise to only one cell type [12]. The organs of the human body such as the liver, skeletal muscle, olfactory mucosa, and skin are abodes of resident stem cells called adult stem cells, which are recruited by biological systems under biological cases such as architectural damage to a tissue, injury, organ degeneration, etc [13]. A mini-library of factors that are necessary and sufficient for pluripotency was worked out by Yamanaka which are utilized in inducing pluripotency in cells [14]. This marks a milestone in stem cell biology.

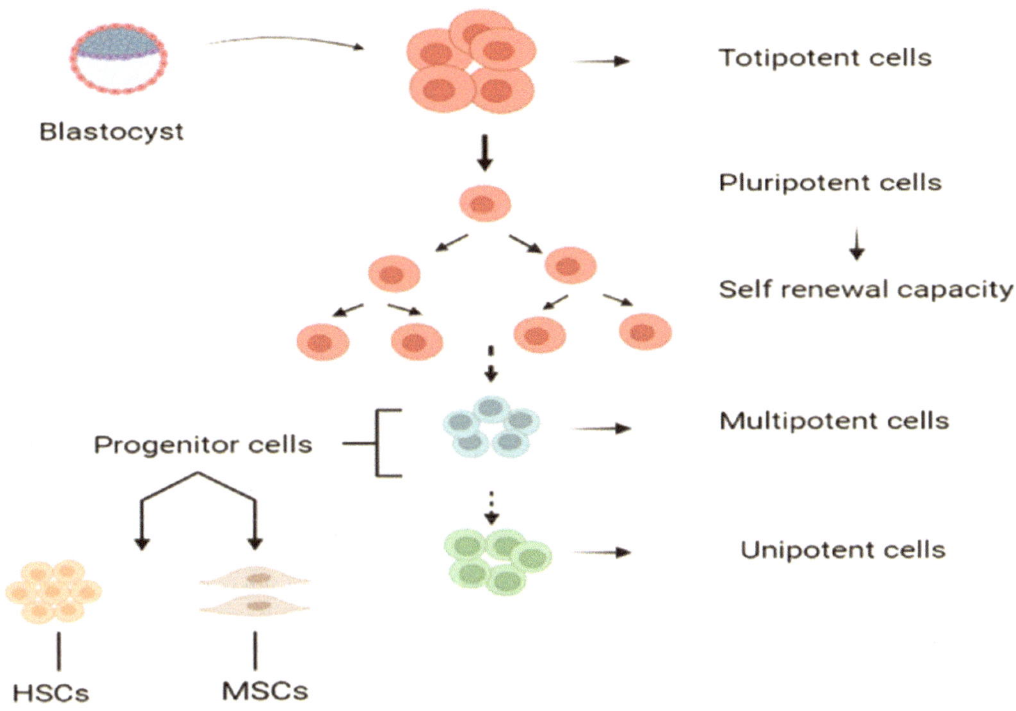

FIGURE 22.1 Hierarchy of stem cells is briefly described, based on the potency. Stem cells are categorized into totipotent cells (having the ability to differentiate into any cell form), pluripotent cells (lower potency than totipotent cells), multipotent stem cells (can give rise to multiple cell types, e.g., HSCs and MSCs), and unipotent stem cells, which can differentiate into only one cell type.

22.5 CANCER STEM CELLS

Cancer cells, occurring in a variety of systematic arrangements and operationally different cell states, display a high degree of heterogeneity within a tumor mass. The contemporary cancer stem cell (CSC) hypothesis places CSCs at the highest rung in the tumor developmental hierarchy and at the "base" in the process of cancer advancement. CSCs, often referred as tumor-initiating cells, are a tiny fraction of the cells in regions of malignancy that resemble normal adult stem cells [15]. The unique self-renewal capacity while also differentiating via progeny production into mature, substantially differentiated cell types is a crucial feature shared by regular SCs and CSCs. The majority of tumor masses consist of cells that are initiated and sustained by CSCs, leading toward robust tumor development. CSCs, resemble typical SCs in a way that they are usually found in a state of dormancy. Others features like CSCs can experience an epithelial-to-mesenchymal-like transition (EMT) in the tumor microenvironment, allowing them to migrate as well as invade and metastasize into the surrounding parenchyma via a variety of proteases [16]. Despite their shared "stemness," the niche for typical SCs is distinct from the microenvironment that CSCs live in. Different habitats have been recognized for various cancer types, for instance, a hypoxic, and invasive habitat characterizes glioblastoma. CSCs in the perivascular niche have also been linked to various diseases such as like brain tumors and malignancies. Intercellular interaction as well as physical microenvironment components either preserve the dormant state of CSCs or drive their proliferation and differentiation, thereby influencing their metastatic potential, protecting CSCs from the immune system. The microenvironment in the niche may potentially guard CSCs from therapeutic targeting. The stemness of a CSC is a result of a nexus of signaling pathways that eventually effect the expression of genes that have been studied to be imperative for stem cell properties. Some of the

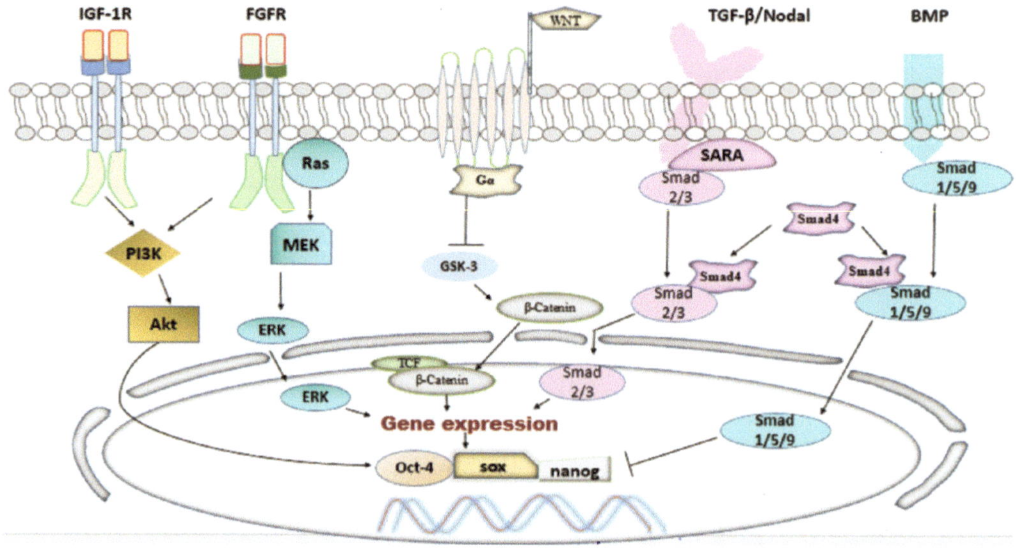

FIGURE 22.2 Signaling pathways regulating pluripotency, self-renewal, and differentiation of cancer stem cells. A network of IGF-1R signaling, FGF signaling, Wnt signaling, TGF-β, and BMP signaling regulate gene expression of oct-4, sox, and nanog via their downstream effectors: AKT, ERK, β-catenin, smad 2/3, and smad 1/5/9, respectively, and thereby control as well as maintain various attributes of cancer stem cells such as potency, lineage commitment, and perpetual self-renewal.

well-documented pathways involved in CSC proliferation and lineage commitment include IGF-1R, FGFR, Wnt/β-Catenin, and BMP4 (Figure 22.2).

22.6 PROTEASES INVOLVED IN CANCER DEVELOPMENT AND PROGRESSION

Protease activity needs to be strictly regulated since it causes the irreversible cleavage of a polypeptide chain. As a result, organisms have evolved a variety of regulatory mechanisms for proteolysis, including temporal and spatial regulation, zymogen activation, protease degradation, and protease inhibition through the use of macromolecular inhibitors [17]. MMPs have a predominant role in ECM disintegration, which qualifies MMPs as one of the initial protease targets to be given a fair deal of consideration in the fight against cancer. Thus, a number of MMP inhibitors were created that may be broadly categorized under two main groups: hydroxamates and non-hydroxamates. Many of the MMP inhibitors, including the hydroxamates batimastat (BB94; British Biotech), marimastat (BB2516; British Biotech), and prinomastat (AG-3340; Aguron) have been investigated in progressed clinical trials. However, these trials were unsuccessful altogether due to adverse effects and possibly minimal clinical efficacy [18]. One of the trendiest topics in protease research at the moment is the investigation of the activity of proteases *in vivo*. Proteases, as previously indicated, were initially studied as important enzymes rebuilding the ECM, which is now widely acknowledged as an effective biochemically and functionally dynamic SC niche component that bears a direct influence on CSC regulation as well (Figure 22.3). Increased expression of several proteases aids breast CSCs in the ECM breakdown and promotes cancer cell invasion. Furthermore, some ECM components can be partially broken down by proteases to produce bioactive peptide fragments, or "matrikines," which can control cellular functions [19]. Proteases ADAM10 and ADAM17 release molecules from the cell surface to influence different signaling pathways. Two members of the ADAM protein family, ADAM10 and ADAM17, have been investigated for their potential use in CSC treatment [20]. Zinc-dependent protease ADAM10 splits a wide range of significant substrates, including EGFR/HER ligands, ICOS-L, TACI, Notch, PD-L1, and MIC-A, as well as UL-16

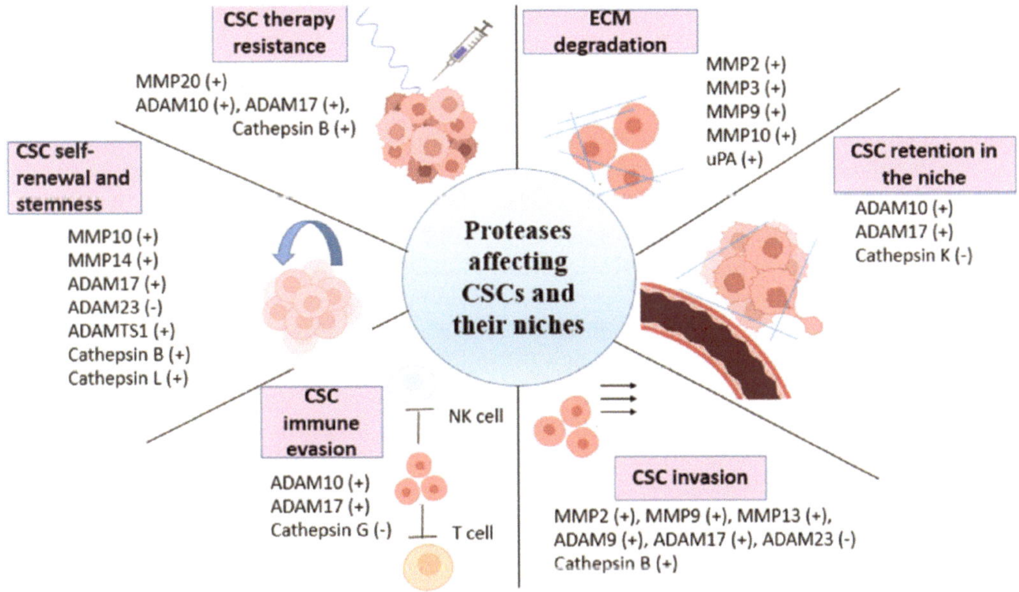

FIGURE 22.3 Proteases affect various cancer stem cell (CSC) properties. Proteases remodel the microenvironment of cancer stem cells and pave a way toward cancer development and progression. Various proteases exhibit oncological roles (promote [+] or inhibit [–] the process) by affecting CSC properties such as extracellular matrix degradation, retention of CSC in the niche, CSC invasion to other parts, immune evasion of CSCs, self-renewal, and resistance of CSC toward therapies.

binding proteins, which are the "stress-related molecules." ADAM17 has demonstrated anticancer efficacy in a number of preclinical systems [21]. Numerous receptors found on the cell surface stimulate stemness, augmenting signals and thereby sustaining the CSCs. These receptors include integrins, discoidin domain receptors, and CD44, which also allows CSCs to connect with the ECM [22]. Atypical papain-clan cysteine cathepsin K, initially described as an osteoclast-produced collagenolytic protease, is overexpressed in several malignancies, most notably glioblastoma. Cathepsin K can hydrolyze a variety of additional fibrillar proteins and proteoglycans as well as different kinds of native collagen molecules [15]. Barbara and the group *in vitro* demonstration of cathepsin K's capacity to cleave and lend SDF-1α non-functional provided legitimacy to a proposed idea that migration of CSCs out of their niches is mediated by cathepsins [23]. It is mediated by associations between SDF-1α, which is delivered by bone-forming cells, and CXCR4, which is expressed by hematopoietic stem cells. Furthermore, since the suppression of ADAM10 and ADAM17 was reported to be accompanied with CSC migration and differentiation, Hiratsuka et al. hypothesized a potential function for both proteases in the maintenance of CSCs in the tumorigenic niche. CSCs have the ability to migrate and undergo EMT, which can start the process of metastasis. Through accelerated angiogenesis, or the creation of new vascular structures through vasculogenic mimicry, CSCs can aid in the growth of tumors. The induction of transmembrane metalloproteases MMP14 and MMP9 is necessary for vascularogenic mimicry since they facilitate the proteolytic cascade involved in melanoma CSC penetration. CSCs line the vasculogenic mimicking channels in breast cancer and work in coordination with them to develop in the perivascular niche [24]. According to Li et al., CSC invasion is boosted when MMPs involved in fibronectin breakdown are activated, creating pathways that aid in CSC migration. It was demonstrated that a number of metalloproteases, including ADAM9 and ADAM17 in addition to MMP2, MMP9, and MMP13, promoted the invasiveness of CSCs [25]. Additionally, stromal cell proteases directly encourage CSC invasion. Ye et al., discovered that tumor-associated macrophages and microglia accelerate CSC invasiveness by releasing and activating the TGF-β1 signaling pathway, which in turn results in an increase in MMP9, whereas CSC-derived MMP2 is activated by microglia-derived MMP14 [26]. Alapati et al. looked into the functions of cathepsin B and urokinase-type plasminogen activator receptor (uPAR) in GSCs and found that parallel inhibition of both proteins prevented integrin–nucleus communication via protein kinase C and interfered with the nucleus–cytoskeleton interaction [2]. MMP14 is one of the most expressed proteases in breast CSCs, particularly in hypoxic environments that cause it to relocalize to the cell surface [27]. This enzyme facilitates the process by which static CSCs become invading CSCs, which is assumed to be the process responsible for CSC metastasis. It was additionally demonstrated that MMP9, MMP14, ADAM12, and ADAM17 control CSC proliferation in various malignancies. As mentioned earlier, proteases may function as suppressors of tumor under certain circumstances. For instance, in lung cancer, ADAM23 inhibits integrin αvβ3 activity, which reduces CSC movement and metastasis. MMP14 induces CSC characteristics and advances the EMT in a cell line of oral squamous cell cancer. In line with these findings, inhibition of MMP14 had a significant impact on CSC characteristics in breast cancer SCs. In an encouraging approach, the Deshaie's laboratory developed and produced PROTAC-1, an active molecule that recruits the ubiquitin-proteasome system to sustain protein homeostasis within a cell, in 2001. This process degrades methionine aminopeptidase-2 (MetAP-2) [28]. The three covalently linked parts make up PROTAC-1: a domain with ovalicin (a MetAP-2 inhibitor), which is identified by the second component, Skp1-Cullin-F-box, and a component called the linker that connects these two domains [29]. Afterward, scientists created several PROTACs to eradicate the proteins that promote sickness, including aryl hydrocarbon receptor (AHR), androgen receptor (AR), estrogen receptor (ER), and FK506 binding protein (FKBP12). These PROTACs are peptide-based, which results in their limited stability and low cell permeability, owing to the reduced lipophilicity (conferred by backbone constituted by peptides) or favorability to pass biomembranes. [30]. Targeted cancer treatments try to tamper with the oncogenic cellular processes of cancer-associated macromolecules (such as oncoproteins) in cancerous cells. However, scientists are currently looking for a more precise and effective way to target cancer-related oncoproteins because of low therapeutic effectiveness of other treatments, drug resistance, and off-target effects [29].

22.7 STEM CELL-BASED THERAPEUTIC INTERVENTIONS MEDIATED BY TARGETING PROTEASES

A promise that revolutionized the therapeutic world and questioned the traditional and age-old approach of combating cancer came in the form of CSC therapy. The therapy gained momentum back in 1957 when a leukemic case was treated successfully with the aid of a bone marrow transplant. This has been a promising approach ever since. The therapy is based on the mechanism of the rebuilding and renewal capacity of the stem cell and relies on the memory of these cells to rebuild the biological system precisely when stimulated either by the intraneous or extraneous signals. Owing to the advancement in our understanding about the regenerative potential of stem cells and an enormous compilation of the successful results, stem cell therapy offers a bright beam of hope in most aggravated medical conditions. The first case report, published in 1939, detailed the transplantation of bone marrow for the treatment of aplastic anemia. Two decades later, in 1958, French oncologist George Mathe used bone marrow transplantation for the treatment of six nuclear researchers who had unintentionally been exposed to radioactive materials—this was the first stem cell transplantation [31], the first clinical attempt to treat aplastic anemia, using aspirated bone marrow. This attempt did not have a successful outcome. The major reason for failure was that the bone marrow had a distinct origin (animal) and was injected into a human. But later in 1958, French oncologist and immunologist Dr. Georges Mathe performed bone marrow transplantation, but in this study he was successfully able to elucidate the host vs graft rejection in his patients. Although he failed in his transplantation procedure, which led to the deaths of his six patients, the experimental observations from this study paved the way for successful follow-up marrow transplantations.

Dr. E. Donnell Thomas performed the first bone marrow transplant in the United States in 1969, but the effectiveness of the allogeneic treatment remained exclusive [32]. The first successful allogeneic transplantation for aplastic anemia and acute myeloid leukemia was documented in a 16-year-old girl in 1972, which was the year of the discovery of cyclosporine (the immune-suppressing medication) [31]. Friendenstein and colleagues conducted a series of studies on bone marrow discharges from the 1960s to the 1970s that revealed the link between osteogenic differentiation and a tiny subset of bone marrow cells. The aforementioned cells were later shown to be distinct from the hematopoietic population and to grow fast in tissue culture tubes as adherent cells. Another significant finding made by Friendenstein's group was the discovery of how bone marrow cells, when seeded as suspension culture, could develop into a colony-forming unit, followed by differentiation into osteoblasts, adipocytes, and chondrocytes, implying how these cells result in the ability to proliferate and differentiate into various types of cells. Stem cells allow researchers to look into the systems that control embryonic development, cellular differentiation, and organ maintenance. Stem cells have enormous potential for the creation of innovative cell-based therapeutics due to their proliferative and differentiation abilities. CSCs are known to be impacted by proteases, which along with tumors and surrounding tissue exhibit disturbed activity of proteolytic enzymes. As a result of the complex interplay between cell-bound and released substances from various SC niche cell types, as well as the niche ECM, resident SCs are either kept in a quiescent state or are stimulated to proliferate and differentiate. Targeting proteases in anticancer therapy has historically been seen as a promising therapeutic approach to combat the invading nature of cancerous cells. The choice of proteases as drug targets appears to be an effective alternative and has lately received affirmations and nods in the area of selective cancer therapy. Thus, utilizing proteases to target or take advantage of CSC habitats may help eradicate quiescent CSCs. For stem cell-based therapies to be used in more clinical settings, a deeper comprehension of basic stem cell functions will enhance stem cell-based regenerative medicine and anticancer techniques. Moreover, utilizing the role of proteases in remodeling the microenvironment of CSC has a tremendous potential in stem cell therapeutic approaches.

22.8 SOME OBSTACLES AND SOME SOLUTIONS

Two of the biggest roadblocks that stand in the way of turning promise into practice are stem cell sourcing and safety. Finding autologous, non-immunogenic, easily obtainable stem cells with the fewest ethical issues and a comparatively low degree of *in vitro* complication without necessitating immortalization is the primary goal of sourcing. Autologous stem cells for transplantation are advantageous mostly because of their immunologic connectivity, which significantly impacts the longevity of transplanted cells. The major challenge with non-autologous stem cell therapy is that they may be easily annihilated by unfavorable immunological conditions before exerting their antitumor effects in the search for microscopic metastases [33]. The utilization of adult stem cell sources would be supported by ethical arguments about the procurement of embryonic stem cells. The utility of embyonic stem cells as autologous vectors is also prohibited because embryos do not survive stem-cell extraction. Lastly, a practical obstacle to ensuring the security of embryonic stem cells in the near future is their tendency to develop teratomas upon transplantation. When it comes to neuronal stem cell treatment, it is not possible to extract neural stem cells from adult human brains, despite the hopeful nature of neural stem cell experiments. An increasing number of reports describe the use of adult somatic cells to reverse differentiation and produce pluripotent stem cells. Because of their diversity and adaptability, stem cells generated from bone marrow may be a useful substitute source of stem cells. Apart from the probable transdifferentiation of mesenchymal stem cells and endothelium, mesenchymal stem cells have the capacity to target tumors originating from different sources. Both cell types appear to be desirable sources of autologous stem cells for tumor targeting since they are relatively easy to extract from patient peripheral blood and bone marrow. Adult stem cells that have not been immortalized do not carry the same hazards and do not risk patients [34]. With an in-depth knowledge of stem cell biology, migration, and fates, research and medicine can harness the immense potential of stem cells in tumor treatment without risking the creation of new tumors, and clinical implementation can bypass conventional approaches like chemotherapy. One challenge is that integrated stem cells release hazardous substances, which can be a concern after the particular malignant process has been eradicated. The capacity to selectively turn off and regulate gene expression can be a solution to that problem.

22.9 CONCLUSION

The structural integrity of proteins because of the peptide linkage is remarkable; proteins in food, despite being boiled, never break down into amino acids. The most stable biological polymers, the proteins, are disintegrated into separate amino acids within hours in our digestive system. The breakdown of a peptide bond, caused by attack on the carbonyl group, is known as proteolysis. Hydrolysis can be brought about chemically or enzymatically—the diverse group of enzymes that catalyze peptide bond cleavage are called proteases. Under healthy physiological settings, proteases display a wide range of cellular functions, including differentiation, gene expression, and cell death. Any perturbance in the regulation of proteases becomes a basis of pathologies including that of cancer. Proteases are closely linked to the advancement of cancer at both primary and metastatic sites due to their capacity to break down extracellular matrices and proteins. One of the hallmarks of malignancy, the invasion and migration of cancer cells through the basal lamina, depends on this digestion of the ECM, which is mediated by various classes of proteases. Several intracellular signaling pathways essential to the preservation of CSCs are impacted by proteases of various types. The idea of a degradome (the entire set of proteases expressed by a tissue or organism) is a recent highlight. There was some uncertainty around how proteases can be utilized to target CSCs, which are considered as propagates of metastasis. From protease inhibitors such as marimastat to modulation therapies targeting proteases in the ECM surrounding the CSC niche, an extensive number of diagnostic and therapeutic strategies against cancer are focused at proteases. The targeted treatment approaches interfere with the oncogenic cellular processes by blocking the functioning of cancer-associated macromolecules (such as oncoproteins) in cancerous cells. Targeted treatments have made significant progress against cancer over the last few decades, however, there are numerous lacunae awaited to be filled in the wide arena of interventions in CSCs by targeting proteases.

DECLARATIONS

Conflict of interest: The authors declare no conflicts of interest.

Consent for publishing: All authors have approved the final version of the manuscript and given consent for publication.

Funding: Not Applicable.

REFERENCES

1. Craik, C.S., M.J. Page, and E.L. Madison, *Proteases as therapeutics*. Biochem J, 2011. **435**(1): p. 1–16.
2. Lopez-Otin, C. and J.S. Bond, *Proteases: multifunctional enzymes in life and disease*. J Biol Chem, 2008. **283**(45): p. 30433–7.
3. Hofer, F., et al., *Catalytic Site pK(a) Values of Aspartic, Cysteine, and Serine Proteases: Constant pH MD Simulations*. J Chem Inf Model, 2020. **60**(6): p. 3030–42.
4. van Zijl, F., G. Krupitza, and W. Mikulits, *Initial steps of metastasis: cell invasion and endothelial transmigration*. Mutat Res, 2011. **728**(1–2): p. 23–34.
5. Lu, P., et al., *Extracellular matrix degradation and remodeling in development and disease*. Cold Spring Harb Perspect Biol, 2011. **3**(12): p. a005058.
6. Quintero-Fabian, S., et al., *Role of Matrix Metalloproteinases in Angiogenesis and Cancer*. Front Oncol, 2019. **9**: p. 1370.

7. Gonzalez, H., C. Hagerling, and Z. Werb, *Roles of the immune system in cancer: from tumor initiation to metastatic progression*. Genes Dev, 2018. **32**(19–20): p. 1267–84.
8. Hoang, D.M., et al., *Stem cell-based therapy for human diseases*. Signal Transduct Target Ther, 2022. **7**(1): p. 272.
9. Morrison, S.J., N.M. Shah, and D.J. Anderson, *Regulatory mechanisms in stem cell biology*. Cell, 1997. **88**(3): p. 287–98.
10. Kumar, A., J.K. Placone, and A.J. Engler, *Understanding the extracellular forces that determine cell fate and maintenance*. Development, 2017. **144**(23): p. 4261–70.
11. Mitalipov, S. and D. Wolf, *Totipotency, pluripotency and nuclear reprogramming*. Adv Biochem Eng Biotechnol, 2009. **114**: p. 185–99.
12. Jaenisch, R. and R. Young, *Stem cells, the molecular circuitry of pluripotency and nuclear reprogramming*. Cell, 2008. **132**(4): p. 567–82.
13. Schmidt, M., et al., *Adult stem cells at work: regenerating skeletal muscle*. Cell Mol Life Sci, 2019. **76**(13): p. 2559–70.
14. Cai, Y., Dai, X., Zhang, Q, et al., *Gene expression of OCT4, SOX2, KLF4 and MYC (OSKM) induced pluripotent stem cells: identification for potential mechanisms*. Diagn Pathol, 2015. 10: p. 35. doi: 10.1186/s13000-015-0263-7
15. Habic, A., et al., *Proteases Regulate Cancer Stem Cell Properties and Remodel Their Microenvironment*. J Histochem Cytochem, 2021. **69**(12): p. 775–94.
16. Huang, T., et al., *Stem cell programs in cancer initiation, progression, and therapy resistance*. Theranostics, 2020. **10**(19): p. 8721–43.
17. Farady, C.J. and C.S. Craik, *Mechanisms of macromolecular protease inhibitors*. Chembiochem, 2010. **11**(17): p. 2341–6.
18. Turk, B., *Targeting proteases: successes, failures and future prospects*. Nat Rev Drug Discov, 2006. **5**(9): p. 785–99.
19. Bonnans, C., J. Chou, and Z. Werb, *Remodelling the extracellular matrix in development and disease*. Nat Rev Mol Cell Biol, 2014. **15**(12): p. 786–801.
20. Saha, N., et al., *ADAM proteases: Emerging role and targeting of the non-catalytic domains*. Cancer Lett, 2019. **467**: p. 50–57.
21. Smith, T.M., Jr., A. Tharakan, and R.K. Martin, *Targeting ADAM10 in Cancer and Autoimmunity*. Front Immunol, 2020. **11**: p. 499.
22. Huang, J., et al., *Extracellular matrix and its therapeutic potential for cancer treatment*. Signal Transduct Target Ther, 2021. **6**(1): p. 153.
23. Breznik, B., et al., *Cysteine cathepsins B, X and K expression in peri-arteriolar glioblastoma stem cell niches*. J Mol Histol, 2018. **49**(5): p. 481–97.
24. Hiratsuka, S., et al., *MMP9 induction by vascular endothelial growth factor receptor-1 is involved in lung-specific metastasis*. Cancer Cell, 2002. **2**(4): p. 289–300.
25. Xiao, L.J., et al., *ADAM17 targets MMP-2 and MMP-9 via EGFR-MEK-ERK pathway activation to promote prostate cancer cell invasion*. Int J Oncol, 2012. **40**(5): p. 1714–24.
26. Ye, X.Z., et al., *Tumor-associated microglia/macrophages enhance the invasion of glioma stem-like cells via TGF-beta1 signaling pathway*. J Immunol, 2012. **189**(1): p. 444–53.
27. Hillebrand, L.E., et al., *MMP14 empowers tumor-initiating breast cancer cells under hypoxic nutrient-depleted conditions*. FASEB J, 2019. **33**(3): p. 4124–40.
28. Deshaies R, et al., Protacs: *Chimeric molecules that target proteins to the Skp1–Cullin–F box complex for ubiquitination and degradation*. Proceedings of the National Academy of Sciences, 2001. 98(15): p. 8554–8559.
29. Li, R., et al., *Proteolysis-Targeting Chimeras (PROTACs) in Cancer Therapy: Present and Future*. Molecules, 2022. **27**(24): p. 8828.
30. Cecchini, C., et al., *From Conception to Development: Investigating PROTACs Features for Improved Cell Permeability and Successful Protein Degradation*. Front Chem, 2021. **9**: p. 672267.
31. Yin, J.Q., J. Zhu, and J.A. Ankrum, *Manufacturing of primed mesenchymal stromal cells for therapy*. Nat Biomed Eng, 2019. **3**(2): p. 90–104.
32. Zhou, T., et al., *Challenges and advances in clinical applications of mesenchymal stromal cells*. J Hematol Oncol, 2021. **14**(1): p. 24.
33. Zhang, C.L., et al., *Stem cells in cancer therapy: opportunities and challenges*. Oncotarget, 2017. **8**(43): p. 75756–75766.
34. Romito, A. and G. Cobellis, *Pluripotent Stem Cells: Current Understanding and Future Directions*. Stem Cells Int, 2016. **2016**: p. 9451492.

23 Stem Cell Therapy in Bone Marrow Cancer

Arjun K. R., Girish Kanavi K., Sudipta Choudhury, Mythreyi R., and Kanthesh M. Basalingappa
School of Life Sciences, JSS Academy of Higher Education & Research, Mysuru, India

23.1 INTRODUCTION

Transcriptome profiling analyses have discovered several properties that are thought to be shared by all stem cells. The fundamental traits of "stemness" are: (i) Janus kinase signal transducers and transcription activators stimulate the TGF-β and Notch signaling pathways; (ii) integrins help cells detect growth cues and interact with the extracellular matrix; (iii) its role in cell cycle, whether stalled in G1 or continuation; (iv) a high stress resistance due to increased DNA repair, detoxifier mechanisms, ubiquitination and protein folding; (v) DNA methylases, histone deacetylases, and DNA helicases all act on modified chromatin; and (vi) translation is controlled by vasa type RNA helicases (Ramalho-Santos *et al.*, 2002). The hematopoietic stem cell (HSC) is the most extensively researched adult stem cell. The success of bone marrow (BM) transplants may be due in great part to HSCs, which have been widely used in clinical settings for more than four decades. HSCs are very scarce and difficult to extract in larger quantities from their *in vivo* environment, like many other adult stem cells. In BM, mesenchymal stem cells (MSCs) are found in low numbers (Short *et al.*, 2003). HSCs can exit this niche, travel through the bloodstream, and then return. Their proximity to endothelial cells may make it easier for HSCs to move from the BM into the bloodstream (Wright *et al.*, 2001). The periosteal capillary network and the nutritive arteries of the bone, which connect to the sinuses via the Haversian canals, provide arterial blood to the venous sinuses in the BM, which supplies both the bone and the BM (Watt, 2002). Multiple myeloma (MM) is defined by a 10% or higher increase in monoclonal plasma cells in BM. It also refers to the clonal development of plasma cells in the BM that results in the generation of paraproteins. In elderly adults and parents, it happens more frequently. The clinical presentation is typically characterized by a multitude of non-specific symptoms, including fatigue, appetite loss, and weight loss. Thrombocytopenia, anemia, leukopenia, sleepiness, fever, chills, thirst, excessive urine loss, pathological fractures, bone pain, and hypercalcemia causing stomach discomfort are some common presenting symptoms (Figure 23.1).

A premalignant condition termed monoclonal gammopathy of unknown significance (MGUS) nearly always precedes MM. About 2% of people over the age of 50 have MGUS, which advances to overt MM at a rate of 1% every year. At any stage of plasma cell dyscrasia, secondary translocations, occasionally including an Ig gene, might take place. Fifteen percent of MM tumors had activating mutations in either NRAS or KRAS. In some malignancies, mutations that occur throughout the tumor's development lead to constant activation of the nuclear factor B (NFB) pathway (Korde *et al.*, 2011). A peculiar combination of entire chromosome gains and losses, non-random chromosomal translocations that lead to deregulation of genes at breakpoints, and point mutations are the hallmarks of MM. Additionally, epigenetic modifications like gene methylation (Gonzalez-Paz *et al.*, 2007) and a cluster of tiny, localized lesions have recently been found by aCGH in the genome of MM cells (Carrasco *et al.*, 2006). Various oncogenic processes facilitate the biological shift

 DOI: 10.1201/9781003394693-26

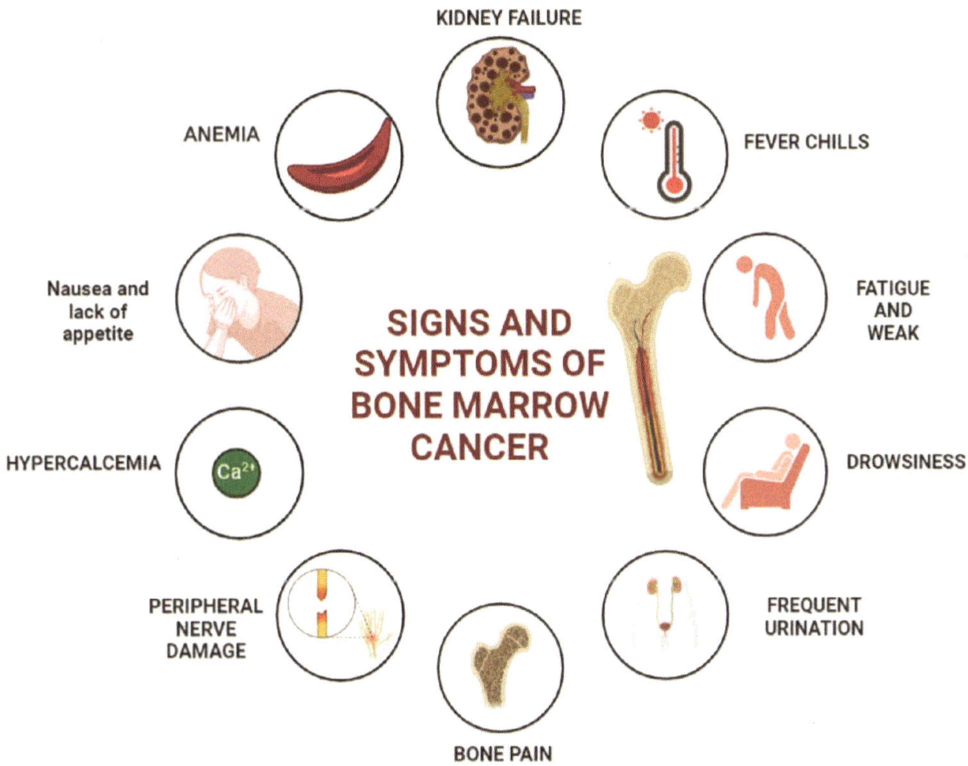

FIGURE 23.1 Signs and symptoms of bone marrow cancer.

from smoldering multiple myeloma (SMM) to MM and from normal plasma cells to MGUS. Early occurrences of MGUS and MM have been linked to a dysregulated cyclin D gene (Bergsagel *et al.*, 2005, Hideshima *et al.*, 2007). Secondary translocations can develop during any stage of plasma cell dyscrasia, occasionally involving an Ig gene.

23.2 PROTEASES: CANCER BIOMARKERS

A biomarker is a physiologically relevant molecule specific to a disease that might serve as a signal of an existing or impending disease state. It can be measured quantitatively in a laboratory. When there is different expression of particular molecules in cancer cells, a link between the diseased state and the altered level can be established. Alternately, certain chemicals can be employed as biomarkers to detect tumorigenesis because they are only present in tumors and are distinct from their equivalent normal tissue (Jemal *et al.*, 2005).

Cysteine cathepsins (CCs), a family of papain-related lysosomal proteases, are frequently elevated in a variety of human malignancies (Turk, 2006). Normal cells generally have CCs in their lysosomes. Still, as cancer spreads, these enzymes shift to the surface of the cell, where they can be secreted into the extracellular environment through a variety of different methods to encourage tumor invasion (Turk & Gučnar, 2003). Though their primary role in endo-lysosomal compartment protein breakdown and recycling is becoming recognized, CCs also make important proteolytic contributions to a variety of immunological activities. They attract and differentiate progenitor cells inside the HSC niche via extracellular proteolytic activity. CCs, by exposing major histocompatibility complex class II antigens, and cathepsins S and L enhance antigen-induced adaptive immunity. In contrast, T-cell motility is regulated by cathepsin X. CCs modulate toll-like receptor signaling and cytokine release to excite innate immune cells and affect their functional differentiation. Natural killer cells and cytotoxic T lymphocytes demonstrate the cytoplasmic expression of cathepsins H and C, which have a role in transforming pro-granzymes into forms that are active on proteases. The adaptive immune response's homeostasis is maintained by cathepsins B and L's cytoplasmic activities, which regulate T and B lymphocyte cell death. The function of CCs in immune cells is intimately related to their expression pattern, activity, and location. Moreover, the contact that produces immune cell energy between cancer and immune cells is mediated by CCs and their endogenous inhibitors (Perišić Nanut *et al.*, 2021). Serine proteases, known as human kallikreins (hKs), are extensively expressed in a variety of tissues and are involved in many regular physiological processes. Cancer and other illnesses are linked to dysregulated enzyme expression. They may play a part in the invasion and metastasis of cancer. They are therefore promising diagnostic indicators for ovarian, breast, and prostate cancers, among others (Paliouras *et al.*, 2007). Although HSCs, the essential elements of BM transplantation, have been widely utilized in therapeutic contexts in the past, stem cells from other organs may one day be used in therapy (Reya *et al.*, 2001). For the majority of cancers, the target cell of changing mutations is unclear, although there is strong evidence that certain kinds of leukemia are caused by mutations that build up in HSCs (Bonnet & Dick, 1997). To put it simply, the machinery required for self-renewing cell divisions, which is typically expressed in stem cells, may be scavenged by newly formed cancer cells (Reya *et al.*, 2001).

23.3 STEM CELLS IN ADULT BM

The BM produces new blood cells, which makes up the core and epiphysis of bones. It is well-known as a primary hematopoietic organ (Zhao *et al.*, 2012). However, there are other immune cell subsets with significant and distinctive properties that are abundant in BM. It is commonly recognized that the BM is the site of B-lymphocyte development and proliferation. Most of the long-lived plasma cells that may produce antibodies specific to an antigen are found in the BM. Consequently, the humoral immune system is stimulated by BM. Both the initial immune response and the memory response can occur in BM, which can function as secondary lymphoid tissue. In the BM microenvironment, the immune system is controlled by soluble substances like cytokines as well as by cell-to-cell interactions. Therefore, as an immune regulatory organ, BM may have a significant effect on systemic immunity as well as the effectiveness of immunotherapy and vaccinations as treatments (Zhao *et al.*, 2012).

MSCs refer to a class of adult progenitor cells that can differentiate into various mesenchymal lineages. Although numerous tissues can be used to separate them, BM has been the most popular. Bone, cartilage, muscle, and marrow stroma are just a few of the mesenchymal tissues that may benefit from MSC repair and regeneration. Moreover, the cells produce cytokines and beneficial development substances that may aid in the healing of other tissues. Additionally, there is proof that they can differentiate into non-mesenchymal lineages, although this study will not take that into account (Chu *et al.*, 2020).

The BM contains the majority of adult MSCs, which make up 0.01% to 0.1% of all BM cells. These multipotent progenitor cells are non-hematopoietic and can differentiate into a variety of tissues of mesenchymal origin (Caplan, 1991). Several hematological and genetic illnesses are now treated of choice with stem cell transplantation from BM, blood, or cord blood (Thomas *et al.*, 2006). Allogeneic BM transplantation has been shown in clinical studies to result in long-lasting remission and ultimately cure (Tyndall & Gratwohl, 1997).

The precursors of skeletal tissue parts such as bone, cartilage, the stroma that supports hematopoiesis, and adipocytes are found in the BM. Therefore, they serve as a crucial example of postnatal non-hematopoietic stem cells and a convenient source for future therapeutic applications (Bergfeld & DeClerk, 2010). Following the differentiation of early osteogenic cells, the production of the first bone, and the vascular invasion of bone rudiments, a complicated chain of events occurs that establishes the primitive marrow stroma during development (Bianco & Riminucci, 1998). Even in the mature marrow, there is still this close connection between the stromal cells and the marrow's vascularity. In the postnatal skeleton, the vascular beds of the bone and BM are shared to a large degree (Ascenzi, 1976).

23.4 IMPLICATIONS OF CANCER TREATMENT

Surgery is the most common kind of cancer treatment, although it has a very poor success rate due to the disease's rapid development and the presence of both local and distant metastases at the time of diagnosis. A multidisciplinary approach to treating cancer should include adjuvant chemotherapy and radiation. Resistance to chemotherapy and radiation, however, may lead to treatment failure or possibly a return of the cancer. Cancer stem cells' resilience, cell-cycle dysregulation, the repair response to radiation-induced DNA damage, and the epithelial-mesenchymal transition (EMT) are some of the factors that contribute to radioresistance in cancer (Liu *et al.*, 2021).

23.4.1 Effects of Radiation on BM

Radiation therapy effectively kills cancer cells and shrinks tumors, and it is still an extensively used treatment for malignant malignancies. Reactive oxygen species (ROS), which are produced as a result of irradiation, cause DNA damage. Better illness outcomes and a reduction in radiation therapy's unfavorable side effects have been achieved through improvements in treatment regimens (Chandra *et al.*, 2019). Human BM skeletal stem cells (hSSCs) are resistant to high-dose radiation, while low-dose irradiation (2.5 Gy) has been shown to have no observable effects on them (Preciado *et al.*, 2018).

23.4.2 Effects of Hematopoietic Growth Factor and Corticosteroids

Hemopoietic growth factor therapy has been well tolerated, both the granulocyte-macrophage factor and the granulocyte factor have the potential to produce low-grade fevers, bone discomfort, and capillary leak syndrome (Morstyn *et al.*, 1988). Most of the time, these symptoms are manageable with minimal effort. Both substances can be injected subcutaneously, resulting in 24-hour-long effective plasma concentrations (Cebon *et al.*, 1988). The positive effects of adrenal androgens on bone growth are likewise diminished by corticosteroids' suppression of their synthesis. Prednisone dosages greater than 20 mg per day reduce gonadotropin-releasing hormone production, which in turn reduces luteinizing hormone synthesis and results in a secondary hypogonadism condition. Anxiety, fear, agitation, indifference, hypomania, lethargy, distractibility, insomnia, irritability, insomnia, mood lability, tearfulness, restlessness, and pressured speech are among the most prevalent mental symptoms (Kershner & Wang-Cheng, 1989). Blood pressure rises have been linked to corticosteroids. The oxidative phosphorylation of skeletal muscle, protein synthesis, muscle membrane excitability, and glucose metabolism are all affected by corticosteroids. Type IIb muscle fiber atrophy is thought to be the source of the muscular weakening linked to corticosteroid usage (Polsonetti *et al.*, 2002).

23.4.3 Effects of Chemotherapy on BM

The majority of chemotherapeutic drugs have a dose-dependent myelosuppressive effect. The substances that are most harmful to BM include alkylating medications, pyrimidine analogs, anthracyclines, anthraquinones, nitrosoureas, methotrexate, hydroxyurea, and mitomycin C. An early issue during cancer treatment is acute myelosuppression, which follows chemotherapy (Zein *et al.*, 1986). The intricate network of signaling pathways that control lineage commitment, differentiation, and balance within the marrow cavity is disrupted by long-term cancer treatment, which results in the dysregulation of osteogenic/adipogenic balance inside the marrow of age-related osteoporotic patients (Georgiou *et al.*, 2010). Doxorubicin and etoposide treatment of MSCs *in vitro* led to improved differentiation potential toward the adipogenic lineage instead of osteogenic differentiation (Buttiglieri *et al.*, 2011, Liu *et al.*, 2021). A decline in trabecular bone volume and an increase in marrow adiposity occur throughout treatment in a rat model of methotrexate and 5-FU chemotherapy. Furthermore, the differentiation of BMSCs toward the adipogenic lineage at the cost of the osteogenic lineage has been shown in the *ex vivo* culture of methotrexate-treated rats (Georgiou *et al.*, 2012).

23.5 CONCLUSION AND FUTURE PERSPECTIVES

Since it has been established that clonal plasma cells and the BM microenvironment interact actively to sustain the pathology of MM, the transition from an inactive MGUS stage to an active MM stage is facilitated by several autocrine or paracrine developmental mechanisms. We now have a preclinical reason available for evaluating possible therapy techniques to improve the target of the BM milieu in addition to the MM cell clone, preventing MM from spreading based on our understanding of the communication between BMSC and MM cells, and vice versa. There have been some recent developments in our understanding of the cellular and molecular mechanisms by which chemotherapy and IR continue to affect the BM. If we have a better understanding of the underlying mechanisms, we will be able to develop novel medicines to avoid the long-term BM toxicity caused by chemotherapy and IR. It will also boost cancer patients' compliance with upcoming consolidation therapy and hasten the restoration of hematological function following autologous and allogeneic BM transplantation.

One of the biggest obstacles is still finding HSCs for all of their potential uses. One of the key objectives of research is still the efficient proliferation of HSCs in culture. Future advancements in gene therapy, proteomics, and genomics might broaden the scope of possible therapeutic applications. Also, pluripotency and self-renewal are the two major properties of HSCs that support a futuristic perspective for the advancement of stem cell therapies. To increase the potential and safety of therapy utilizing HSCs, current and future research will focus on identifying HSC regulators and developing strategies for manipulating stem cells *in vitro* and *in vivo*, such as genome editing. New technology is enabling the development of more targeted and adjuvant therapies that reduce toxicity to normal cells and decrease the likelihood of cancer cell resistance. Multiple efforts are underway to develop a range of combination therapies that include vaccines, immune checkpoint inhibitors, and mutation-specific drugs. These discoveries are paving the way for personalized cancer treatments that significantly improve patient quality of life and outcomes. This knowledge is being refined by new findings from continuing stem cell research, which also opens the door to continued therapy advancements.

DECLARATIONS

Conflict of interest: The authors declare no conflicts of interest.

Consent for publishing: All authors have approved the final version of the manuscript and given consent for publication.

Funding: Not Applicable.

REFERENCES

Ascenzi, A. (1976). Physiological relationship and pathological interferences between bone tissue and marrow. *The Biochemistry and Physiology of Bone*, *4*, 403–445.

Bergfeld, S. A., & DeClerck, Y. A. (2010). Bone marrow-derived mesenchymal stem cells and the tumor microenvironment. *Cancer and Metastasis Reviews*, *29*, 249–261.

Bergsagel, P. L., Kuehl, W. M., & Zhan, et al. (2005). Cyclin D dysregulation: an early and unifying pathogenic event in multiple myeloma. *Blood*, *106*(1), 296–303.

Bianco, P., & Riminucci, M. (1998). The bone marrow stroma *in vivo*: ontogeny, structure, cellular composition and changes in disease. In *Marrow stromal cell culture* (pp. 11–24). Cambridge University Press, Cambridge.

Bonnet, D., & Dick, J. E. (1997). Human acute myeloid leukemia is organized as a hierarchy that originates from a primitive hematopoietic cell. *Nature Medicine*, *3*(7), 730–737.

Buttiglieri, S., Ruella, M., Risso, A., et al. (2011). The aging effect of chemotherapy on cultured human mesenchymal stem cells. *Experimental Hematology*, *39*(12), 1171–1181.

Caplan, A. I. (1991). Mesenchymal stem cells. *Journal of Orthopaedic Research*, *9*(5), 641–650.

Carrasco, D. R., Tonon, G., Huang, Y., et al. (2006). High-resolution genomic profiles define distinct clinico-pathogenetic subgroups of multiple myeloma patients. *Cancer Cell*, 9(4), 313–325.

Cebon, J., Dempsey, P., Fox, R., et al. (1988). Pharmacokinetics of human granulocyte-macrophage colony-stimulating factor using a sensitive immunoassay. *Blood*, *72*(4), 1340–1347.

Chandra, A., Park, S. S., & Pignolo, R. J. (2019). Potential role of senescence in radiation-induced damage of the aged skeleton. *Bone*, *120*, 423–431.

Chu, D.-T., Nguyen, T. T., Tien, N. L., et al. (2020). Recent progress of stem cell therapy in cancer treatment: Molecular mechanisms and Potential Applications. *Cells*, *9*(3), 1–19.

Georgiou, K. R., Foster, B. K., & Xian, C. J. (2010). Damage and recovery of the bone marrow microenvironment induced by cancer chemotherapy-potential regulatory role of chemokine CXCL12/receptor CXCR4 signaling. *Current Molecular Medicine*, *10*(5), 440–453.

Georgiou, K. R., King, T. J., Scherer, M. A., et al. (2012). Attenuated Wnt/β-catenin signaling mediates methotrexate chemotherapy-induced bone loss and marrow adiposity in rats. *Bone*, *50*(6), 1223–1233.

Gonzalez-Paz, N., Chng, W. J., McClure, R. F., et al. (2007). Tumor suppressor p16 methylation in multiple myeloma: Biological and clinical implications. *Blood*, *109*(3), 1228–1232.

Hideshima, T., Mitsiades, C., Tonon, G., et al. (2007). Understanding multiple myeloma pathogenesis in the bone marrow to identify new therapeutic targets. *Nature Reviews Cancer*, *7*(8), 585–598.

Jemal, A., Murray, T., Ward, E., et al. (2005). Cancer statistics. *A Cancer Journal for Clinicians*, *55*(1), 10–30.

Kershner, P., & Wang-Cheng, R. (1989). Psychiatric side effects of steroid therapy. *Psychosomatics*, *30*(2), 135–139.

Korde, N., Kristinsson, S. Y., & Landgren, O. (2011). Monoclonal gammopathy of undetermined significance (MGUS) and smoldering multiple myeloma (SMM): novel biological insights and development of early treatment strategies. *Blood, The Journal of the American Society of Hematology*, *117*(21), 5573–5581.

Liu, Y. P., Zheng, C. C., Huang, Y. N., et al. (2021). Molecular mechanisms of chemo-and radiotherapy resistance and the potential implications for cancer treatment. *MedComm*, *2*(3), 315–340.

Morstyn, G. L. L. N. J. M., Keech, J., Sheridan, W., et al. (1988). Effect of granulocyte colony stimulating factor on neutropenia induced by cytotoxic chemotherapy. *The Lancet*, *331*(8587), 667–672.

Paliouras, M., Borgono, C., & Diamandis, E. P. (2007). Human tissue kallikreins: the cancer biomarker family. *Cancer Letters*, *249*(1), 61–79.

Perišić Nanut, M., Pečar Fonović, U., Jakoš, T., & Kos, J. (2021). The role of cysteine peptidases in hematopoietic stem cell differentiation and modulation of immune system function. *Frontiers in Immunology*, *12*, 680279.

Polsonetti, B. W., Joy, S. D., & Laos, L. F. (2002). Steroid-induced myopathy in the ICU. *Annals of Pharmacotherapy*, *36*(11), 1741–1744.

Preciado, S., Muntión, S., & Rico, A., et al. (2018). Mesenchymal stromal cell irradiation interferes with the adipogenic/osteogenic differentiation balance and improves their hematopoietic-supporting ability. *Biology of Blood and Marrow Transplantation*, *24*(3), 443–451.

Ramalho-Santos, M., Yoon, S., Matsuzaki, Y., et al. (2002). "Stemness": transcriptional profiling of embryonic and adult stem cells. *Science*, *298*(5593), 597–600.

Reya, T., Morrison, S. J., Clarke, M. F., & Weissman, I. L. (2001). Stem cells, cancer, and cancer stem cells. *Nature*, *414*(6859), 105–111.

Short, B., Brouard, N., Occhiodoro-Scott, T., et al. (2003). Mesenchymal stem cells. *Archives of Medical Research*, *34*(6), 565–571.

Thomas, T., Corcoran, L. M., Gugasyan, R., et al. (2006). Monocytic leukemia zinc finger protein is essential for the development of long-term reconstituting hematopoietic stem cells. *Genes & Development*, *20*(9), 1175–1186.

Turk, B. (2006). Targeting proteases: successes, failures and future prospects. *Nature Reviews Drug Discovery*, *5*(9), 785–799.

Turk, D., & Gunčar, G. (2003). Lysosomal cysteine proteases (cathepsins): promising drug targets. *Acta Crystallographica Section D: Biological Crystallography*, *59*(2), 203–213.

Tyndall, A., & Gratwohl, A. (1997). Blood and marrow stem cell transplants in autoimmune disease. A consensus report written on behalf of the European League against Rheumatism (EULAR) and the European Group for Blood and Marrow Transplantation (EBMT). *British Journal of Rheumatology*, *36*(3), 390–392.

Watt, F. M. (2002). Role of integrins in regulating epidermal adhesion, growth and differentiation. *The EMBO Journal*, *21*(15), 3919–3926.

Wright, D. E., Wagers, A. J., Gulati, A. P., et al. (2001). Physiological migration of hematopoietic stem and progenitor cells. *Science*, *294*(5548), 1933–1936.

Zein, T. A., Friedberg, N., & Kim, H. (1986). Bone marrow suppression after intravesical mitomycin C treatment. *The Journal of Urology*, *136*(2), 459–460.

Zhao, E., Xu, H., Wang, L., et al. (2012). Bone marrow and the control of immunity. *Cellular & Molecular Immunology*, *9*, 11–19.

Section IV

Nanotechnology-Based Therapies

24 Protease-Targeting Nanoparticles for Cancer Therapy and Imaging

Sandra Petrovic and Andreja Leskovac
Vinca Institute of Nuclear Sciences, National Institute of the Republic of Serbia, University of Belgrade, Belgrade, Serbia

24.1 INTRODUCTION

Despite significant progress in oncology, cancer remains one of the leading causes of death worldwide. Although chemotherapy is one of the most common treatment options for cancer patients, it often leads to systemic toxicity-related side effects due to its specific biodistribution and activity, causing toxicity in healthy tissues [1]. Consequently, attempts to create alternative chemotherapy treatments have prioritized the development of nanoscale agents that exhibit enhanced specificity toward cancer cells while improving therapeutic efficacy through passive and active targeting.

In recent years, theranostic nanomedicine has emerged as an up-and-coming solution by combining diagnostic and therapeutic capabilities within a single platform. By enhancing tumor selectivity and optimizing chemotherapy effectiveness, theranostics enable the diagnosis and monitoring of therapeutic responses [1]. Specifically designed nanoparticles (NPs) could selectively bind to cancer cells by targeting various enzymes, making them a promising strategy for enhancing cancer theranostics [2]. NPs offer unique advantages such as controllable size, shape, and biocompatibility, which facilitate the manipulation of their chemical properties and biological functions, thereby enhancing their effectiveness in enzyme targeting [3]. Moreover, the surfaces of NPs can be modified, enabling additional customization and functionalization [4].

Proteases are a group of enzymes essential for many physiological processes, including cell-cycle regulation, cell growth and differentiation, immune response, and programmed cell death [5]. However, dysregulated protease activity can promote the initiation and progression of cancer, which presents an opportunity for targeted intervention using NPs. Matrix metalloproteinases (MMPs), urokinase-type plasminogen activators (uPAs), cathepsins, and caspases have been identified as the most promising targets for cancer diagnosis and therapy [6]. By blocking or inhibiting their activity, it may be possible to impede tumor invasion, metastasis, and angiogenesis and enhance the effectiveness of other cancer treatments [7].

This chapter briefly discusses recent progress in cancer imaging and drug delivery by utilizing protease-responsive nanomaterials. Additionally, different theranostic approaches that combine imaging and drug delivery to target cancer-expressed proteases are described, emphasizing their great promise in advancing cancer treatment.

24.2 PROTEASE-RESPONSIVE NANOMATERIALS FOR CANCER IMAGING

Proteases have a crucial function in tumor invasion, angiogenesis, and metastasis, which makes them a promising target for imaging probes in the early detection and treatment of cancer [5].

DOI: 10.1201/9781003394693-28

Conventional diagnostic imaging techniques, such as optical/fluorescence imaging, magnetic resonance imaging (MRI), computed tomography (CT), positron emission tomography (PET), and single-photon emission computed tomography (SPECT), provide imaging of affected organs, enabling the identification of tumor location and size. However, although these methods are valuable for diagnosis, they have certain limitations [7].

In recent years, progress in the field of proteases has led to the development of technologies capable of detecting proteases with high sensitivity in different settings. Fluorescence imaging has become an efficient method for investigating biomolecules, signal pathways, and biological reactions due to fluorescent probes' high resolution and good biocompatibility. In parallel, there has been significant progress in developing fluorescent dyes, including organic dyes such as small molecule dyes, NPs with aggregation-induced emission (AIE), semiconductor polymer nanoparticles (SPNs), as well as inorganic materials like semiconductor quantum dots (QDs), metal nanoclusters, rare earth-doped NPs, and nanodiamonds [8, 9]. Fluorescence-based biosensors using fluorescent materials, including organic dye and fluorescent NPs, have also been widely used to measure extracellular proteases [10].

Fluorescence resonance energy transfer (FRET) is a prominent analytical technique utilized in proteolytic detection. It operates on the principle of distance-dependent transfer of fluorescent energy between a donor molecule and an acceptor molecule [11]. For example, near-infrared (NIR) FRET-based probes conjugated to gold nanoparticles (AuNPs), which act as a fluorescence quencher, have demonstrated efficacy in detecting proteases and have the potential for early cancer diagnosis [12]. AuNPs are commonly used for fluorescence imaging due to their favorable biocompatibility and effective quenching capabilities [13].

Protease-sensitive nanomaterials suitable for optical imaging modalities include also QDs, which could have a wide range of applications for *in vivo* imaging studies [14]. QDs are fluorescent nanocrystals known for their outstanding features in biological imaging, including adjustable spectral range, strong resistance to photobleaching, and easy surface modification [14]. For instance, Pham-Nguyen and coworkers described a tumor-sensing system that utilized AuNPs and QDs connected by an MMP-cleavable linker enabling the monitoring of tumor-specific enzymes *in vivo* [15].

In addition to the enhanced fluorescence intensity, ratiometric fluorescent probes are reported to detect biological systems accurately. These types of probes rely on changes in enzyme activity, where alterations in fluorescence intensity at two different wavelengths occur proportionally. This ratio-based approach allows for precise monitoring and measurement of enzyme activity within the biological system [8].

Promising results are also emerging through the combination of microfluidics, leading to rapid and simultaneous detection of activity patterns for different proteases [14].

Surface-enhanced Raman spectroscopy (SERS) could also be an efficient biosensing method for protease detection. For example, Choi and coworkers introduced an Ag-coated hollow polypyrrole (hPPy) nanohorn platform with a specific peptide sequence, which exhibited SERS activity [16]. Likewise, electrochemical methods have also been proposed for protease detection [17].

24.3 PROTEASE-RESPONSIVE NANOMATERIALS FOR CANCER THERAPY

Nanomaterials play a crucial role in nanomedicine and have been extensively developed as carriers to deliver a wide range of bioactive substances, including hydrophilic or hydrophobic drugs, peptides, nucleic acids, and proteins [18, 19]. Given the significant involvement of proteases in cancer pathology, a variety of protease-responsive nanomaterials have been developed, offering the distribution of bioactive agents within the body, enhancing diagnostic sensitivity, and promoting synergistic effects between diagnosis and treatment [20].

As comprehensively discussed by Rudzińska and coworkers, the therapeutic application of protease inhibitors faces challenges due to the diverse types of proteases involved in different stages of tumor development. Therefore, developing suitable protease inhibitors requires designing and testing small molecule inhibitors specific to particular diseases and tissues [21]. By loading protease inhibitors into nanocarriers, their bioavailability can be enhanced, circulation time can be prolonged, and specific accumulation in tumor tissues can be achieved [21].

Thus far, various enzyme-specific substrate peptides have been utilized as carriers to load chemotherapeutic drugs, radiotherapy sensitizers, photosensitizers, or photothermal agents. These peptides can be attached to the agents through physical adsorption or chemical bonding to create nanoplatforms that effectively tackle the challenging issues associated with tumor treatment [8].

Moreover, multifunctional protease-responsive nanomaterials or systems that simultaneously diagnose and treat cancer have been developed. These nanomaterials could enable real-time monitoring of the treatment response, visualization of protease activity, and assessment of therapeutic efficacy. By combining therapy and diagnostics, theranostic strategies offer the potential for personalized and precise approaches to cancer treatment.

24.3.1 Matrix Metalloproteinase–Responsive Systems

The strategy of developing and functionalizing smart nanoscale drug-delivery systems (DDS) using MMPs is reported to have significant potential in cancer treatment [20]. Mesoporous silica nanoparticles (MSNs) show great promise as new-generation nano delivery carriers. In a recent study, a nanocarrier was developed by capping MSNs with AuNP-biotin conjugate, which included an MMP-2-cleavable linker [22]. This nanocarrier was designed to respond to changes in pH and provide controlled delivery of doxorubicin (DOX), demonstrating excellent tumor destruction capabilities with minimal side effects [22]. Many other chemotherapeutics, including paclitaxel, gemcitabine, coumarin, trichosanthin, methotrexate, pirfenidone, pirarubicin, cabazitaxel, and SN38, were used for the development of promising nanoDDS for cancer treatment [20, 23, 24].

Photodynamic therapy (PDT) and photothermal therapy (PTT) represent non-invasive treatments with high safety and selectivity. PDT employs photosensitizers to produce NIR light-induced cytotoxic reactive oxygen species to induce apoptosis in cancer cells [8]. However, most photosensitizers are incapable of targeting cancer and penetrating tissue. In that sense, it was proposed that MMP-responsive probes can improve the tissue penetration of photosensitizers. For example, Liang and coworkers designed nanoprobes consisted of ferritin nanocage (PFRT)-encapsulated zinc phthalocyanine photosensitizers (ZnPc) and dendritic MSNs loaded oxygen-supplied hemoglobin (Hb)/matrix remodeling reagent (iTGFb). These two parts of nanoprobes were linked by MMP-2 substrate peptide, forming the core-satellite nanoframeworks (T-PFRT). When nanoframeworks were cleaved by MMP-2, the released Hb and iTGFb led to the normalization of the tumor microenvironment, which further improved deep-tissue penetration of PFRT [25]. Considering PTT, it was demonstrated that gold nanorods modified with MMP-9-sensitive peptide sequence displayed enhanced accumulation in tumors and improved treatment efficacy [26].

The use of versatile nanocomposites may enable efficient diagnosis, targeted therapy, and assessment of prognosis for cancer. Numerous MMP-2-responsive nanomaterials have been developed, offering promising opportunities [20]. For instance, the DOX-gelatin/epigallocatechin gallate/AuNPs nanoconjugates, which inhibit prostate cancer by utilizing both fluorescence imaging and DOX, have been developed [27]. In addition, pH-responsive gold nanoclusters for NIR fluorescence imaging and chemo-photodynamic therapy of MMP-2- positive lung cancer are considered effective [28]. Furthermore, dual-modal photoacoustic/NIR fluorescence imaging-enhanced PTT/PDT systems were designed for lung cancer diagnosis and therapy [29]. Likewise, Shi and coworkers developed a magnetic theranostic platform by constructing Gd^{3+}-doped CuS magnetic SPNs (T-MAN). They further modified the SPNs by covalently attaching a cRGD-targeted peptide, NIR fluorophore (Cy5.5), and a quencher QSY21-labeled MMP-2 cleavable peptide substrate for gastric cancer treatment [30]. Additionally, there are many other theranostic nanosystems that utilize optical principles, such as AIE and FRET [20].

Numerous MMP-9-stimulated nanoDDS have been developed for cancer theranostics. For example, the construction of nanocarriers containing PEG-derived polymeric nanogels as MMP-9-cleavable motifs has played a crucial role in the delivery of chemotherapeutics [31]. Furthermore, MMP-9-sensitive curcumin/p53 DNA co-delivery nanoDDS were designed to reverse cisplatin resistance in ovarian cancer [32]. Recently, a novel MMP-9-responsive nanoplatform (PMP@USPIO/DOX, PMPSD) with T2–T1 conversion characteristics were designed to enable tumor imaging and synergistic chemo-photothermal therapy [33]. In addition, Wang and coworkers successfully created multifunctional metal-organic frameworks that were customized with a bone-targeting peptide and a cell-penetrating peptide, both conjugated with MMPs substrate peptide. These frameworks were specifically designed for MRI-guided chemo-photothermal therapy of bone tumors [34]. Chen and coworkers also developed a drug-delivery strategy using supramolecular AIE nanodots for image-guided cancer therapy. The nanodots were composed of alpha-cyclodextrin modified with gemcitabine and TPR, along with a combination of cell-penetrating peptide (R8) and zwitterionic stealth sequence (EK6), connected by the MMP-2 peptide substrate. Upon MMP-2 degradation, the EK6 shielding sequence was removed, enhancing the intracellular internalization of R8 and triggering the release of gemcitabine in response to the intracellular reduction microenvironment. The nanodots demonstrated improved fluorescence signaling and antitumor activity compared to the control group, validating the effectiveness of MMP-2 targeting in tumors [35]. Various multipurpose polymeric nanoplatforms have been utilized to achieve accurate theranostics of cancer. These nanoplatforms possess dual sensitivity to MMP-2/MMP-9 and intelligently load and present different cytotoxic therapeutics such as DOX, paclitaxel, camptothecin, immune modulators, etc. [20].

24.3.2 Cathepsin-responsive Systems

Cathepsins B, L, and S have been widely used as responders in delivering imaging contrast agents and/or therapeutics.

Numerous cathepsin B-responsive nanoDDS have been designed to promote cancer-specific drug release. For example, a nanocomplex was created by combining QDs and MSNs loaded with the DOX and linked to a cell-penetrating peptide. When the nanocomplex encountered high levels of cathepsin B at the tumor site, a specific peptide was cleaved, which triggered the DOX release. The increased drug concentration at the tumor site improved antitumor activity, even in cells resistant to DOX and expressing high levels of cathepsin B [36]. Another study involved the development of a fluorescent probe consisting of two cyanine moieties, CyA and CyB, linked by a GFLG peptide activated by cathepsin B based on a FRET. Once cleaved by cathepsin B, CyA and CyB separated, leading to a strong fluorescence signal that enabled the visualization of tumor progression and the monitoring of treatment effectiveness. Moreover, the probe also possessed NIR fluorescence imaging and PDT properties, making it a versatile tool for both imaging and therapeutic purposes [37]. Recently, novel, target-specific theranostic NPs that can be selectively cleaved by cathepsin B to release the prodrug monomethyl auristatin E in glioblastoma were developed. It was found that reported theranostic NPs accumulated in glioblastoma caused cancer cell death and had synergistic effects with radiation therapy [38]. The outstanding attainments of cathepsin B-responsive nanoplatforms in the targeted therapy and synergistic theranostics of cancer are comprehensively discussed by Li and coworkers [39].

Considering cathepsin L, Wang and coworkers developed a cathepsin L-responsive acetylated azidomannosamine to facilitate tumor-specific labeling *in vivo* that further improved DOX aggregation in tumor lesions and enabled ablation of colorectal cancer and metastatic breast cancer in mice [40]. In a recent study, RNA interference was utilized to knock down the expression of cathepsin L in the T-47D breast cancer cell line. This process resulted in a deceleration of breast cancer cell progression by increasing the rate of apoptosis [41].

Cathepsin S has also been integrated into nanocomplexes for protease-sensitive cancer imaging or targeted therapy. For instance, Stefin A-conjugated nanosized liposomes were demonstrated to bind to cathepsins S and L *in vivo* in a mouse breast cancer model, demonstrating the potential of this approach for cancer diagnosis and treatment [42].

24.3.3 Urokinase-responsive Systems

The uPA and the urokinase-type plasminogen activator receptor (uPAR) are actively participating in cancer progression and metastasis, making them promising targets for developing anticancer therapies. Various applications of uPAR as a target in different therapeutic approaches have been proposed. These include nanoplatforms carrying therapeutic agents, platforms for PTT and PDT, oncolytic virotherapy, gene therapy technologies, monoclonal antibody therapy, and tumor immunotherapy [43]. Among the different uPAR-targeted therapeutic strategies, uPAR-targeted nanoplatforms show significant potential for transitioning from laboratory findings to clinical applications. These nanoplatforms, modified with uPA/ATF/AE105/AE147/PAI-2/U11, have the ability to specifically deliver therapeutic agents to tumor cells without causing substantial harm to healthy tissue; this offers the possibility of overcoming the limitations of conventional chemotherapy or PTT/PDT [43]. In addition, by utilizing a combination of bioengineered drugs along with nanobins, it is possible to target cancer cells accurately, facilitate the delivery of cytotoxic drugs directly to the cancer cells, and enhance the cytotoxic effects [44].

24.3.4 Caspase-responsive Systems

Cancer therapies, such as radiotherapy, chemotherapy, and immunotherapy, commonly induce apoptosis by activating caspase-3. However, while some investigations found a correlation between the expression levels of caspase-3 and the prognosis of the cancer outcome, other reports indicate that caspase-3 levels in cancer samples are remarkably low or undetectable [14]. These discrepancies suggest the necessity for additional studies to fully understand the role of caspases in cancer and their potential as disease biomarkers. In a recent comprehensive investigation, Zhou and coworkers unveiled the precise molecular process through which caspase-3 promotes resistance to chemoradiotherapy and facilitates tumor metastasis in colorectal cancer. Additionally, it is anticipated that inhibiting caspase-3 could significantly enhance colorectal cancer treatment when combined with a cytotoxic treatment regimen [45].

Considering therapy affecting caspase-3, it was found that RGD-coated gold nanostars are involved in PTT-induced apoptosis in HepG2 cells through a mechanism that depended on caspase activation [46]. Also, Xu and coworkers developed multifunctional NPs consisting of a superparamagnetic iron oxide core coated with PEG/PEI/polysorbate 80. These NPs were designed to encapsulate the DOX and demonstrated efficient drug delivery specifically for gliomas. The therapeutic mechanism of the DOX@Ps 80-SPIONs involved the induction of apoptosis through the caspase-3 pathway [47].

24.4 CONCLUSIONS AND PROSPECTS

The interdisciplinary approach involving molecular biology, biomedicine, nanotechnology, and polymer chemistry has greatly enhanced protease research aiming to create technologies that can accurately measure protease activity in real time. These technologies have already demonstrated promising results in preclinical studies for cancer diagnosis, disease staging, therapy selection, and evaluation of treatment response. Theranostics of cancer refers to developing and utilizing multifunctional protease-responsive nanomaterials or systems that combine therapeutic and diagnostic capabilities, specifically targeting proteases involved in cancer progression. By combining therapy and diagnostics, theranostics of cancer holds great promise in personalized medicine. While considerable advancements have been made, several obstacles impede the widespread use of enzyme-responsive nanosystems for drug delivery. These challenges encompass the intricate nature of tumor-associated proteases, the overlap and partial compatibility of substrates with enzymes, the complex design of specific nanocomplexes, and potential concerns regarding the toxicity and biocompatibility of nanoformulations [20]. Critical areas for improvement include the refinement of enzyme structures to better align with specific diseases, meticulous design and screening of substrate sequences, and the delicate integration of these components into a nanocomposite with minimal toxicity and optimal biocompatibility. These efforts are crucial to enhance the theranostic efficiency of protease-responsive nanosystems for drug delivery and should enable their successful translation into clinical applications.

ACKNOWLEDGMENTS

The authors acknowledge the support provided by the Ministry of Science, Technological Development and Innovation of the Republic of Serbia (No. 451-03-66/2024-03/200017).

DECLARATIONS

Conflict of interest: The authors declare no conflicts of interest.

Consent for publishing: All authors have approved the final version of the manuscript and given consent for publication.

REFERENCES

1. Yildiz, T., Gu, R., Zauscher, S., et al. (2018). Doxorubicin-loaded protease-activated near-infrared fluorescent polymeric nanoparticles for imaging and therapy of cancer. *International Journal of Nanomedicine*, 13, 6961–6986.
2. Laban, B., Ralevic, U., Petrovic, S., et al. (2020). Green synthesis and characterization of nontoxic L-methionine capped silver and gold nanoparticles. *Journal of Inorganic Biochemistry*, 204, 110958.
3. Bondzic, A.M., Leskovac, A.R., Petrovic, S., et al. (2019). Conjugates of gold nanoparticles and antitumor gold(III) complexes as a tool for their AFM and SERS detection in biological tissue. *International Journal of Molecular Sciences*, 20, 6306.
4. Stamenovic, U., Davidovic, S., Petrovic, S., et al. (2021). Antimicrobial and biological effects of polyaniline/polyvinylpyrrolidone nanocomposites loaded with silver nanospheres/triangles. *New Journal of Chemistry*, 45, 12711–12720.
5. Eatemadi, A., Aiyelabegan, H.T., Negahdari, B., et al. (2017). Role of protease and protease inhibitors in cancer pathogenesis and treatment. *Biomedicine & Pharmacotherapy*, 86, 221–231.
6. Anderson, C.F., Cui, H. (2017). Protease-sensitive nanomaterials for cancer therapeutics and imaging. *Industrial & Engineering Chemistry Research*, 56, 5761–5777.
7. Mumtaz, T., Qindeel, M., Asim.ur.Rehman, et al. (2020). Exploiting proteases for cancer theranostic through molecular imaging and drug delivery. *International Journal of Pharmaceutics*, 587, 119712.
8. Wang, Q., Cui, H., Gan, N., et al. (2022). Recent advances in matrix metalloproteinases-responsive nanoprobes for cancer diagnosis and therapy. *Reviews in Analytical Chemistry*, 41, 198–216.
9. Yang, R.Q., Lou, K.L., Wang, P.Y., et al. (2021). Surgical navigation for malignancies guided by near-infrared-II fluorescence imaging. *Small Methods*, 5, e2001066.
10. Choi, J.H. (2023). Proteolytic biosensors with functional nanomaterials: current approaches and future challenges. *Biosensors (Basel)*, 13, 171.
11. Selvin, P.R. (2000). The renaissance of fluorescence resonance energy transfer. *Nature Structural & Molecular Biology*, 7, 730–734.
12. Lee, S., Cha, E.J., Park, K., et al. (2008). A near-infrared-fluorescence-quenched gold-nanoparticle imaging probe for *in vivo* drug screening and protease activity determination. *Angewandte Chemie International Edition*, 47, 2804–2807.
13. Vujacic, A., Vodnik, V., Joksic, G., et al. (2011). Particle size and concentration dependent cytotoxicity of citrate capped gold nanoparticles. *Digest Journal of Nanomaterials & Biostructures (DJNB)*, 6, 1367–1376.
14. Vizovisek, M., Ristanovic, D., Menghini, S., et al. (2021). The tumor proteolytic landscape: a challenging frontier in cancer diagnosis and therapy. *International Journal of Molecular Sciences*, 22, 2514.

15. Pham-Nguyen, O.V., Shin, J., Park, Y., et al. (2022). Fluorescence-shadowing nanoparticle clusters for real-time monitoring of tumor progression. *Biomacromolecules*, 23, 3130–3141.
16. Choi, J.H., El-Said, W.A., Choi, J.-W. (2020). Highly sensitive surface-enhanced Raman spectroscopy (SERS) platform using core/double shell (Ag/polymer/Ag) nanohorn for proteolytic biosensor. *Applied Surface Science*, 506, 144669.
17. Xia, N., Sun, Z., Ding, F., et al. (2021). Protease biosensor by conversion of a homogeneous assay into a surface-tethered electrochemical analysis based on streptavidin-biotin interactions. *ACS Sensors*, 6, 1166–1173.
18. Suljovrujic, E., Krstic, M., Rogic Miladinovic, Z., et al. (2023). Optimization of thermoresponsive hydrogels based on oligomers with lower critical solution temperature (LCST) far below/above physiological temperatures for biomedical applications. *Reactive and Functional Polymers*, 189, 105612.
19. Cveticanin, J., Joksic, G., Leskovac, A., et al. (2010). Using carbon nanotubes to induce micronuclei and double strand breaks of the DNA in human cells. *Nanotechnology*, 21, 015102.
20. Li, Y., Zhang, C., Li, G., et al. (2021). Protease-triggered bioresponsive drug delivery for the targeted theranostics of malignancy. *Acta Pharmaceutica Sinica B*, 11, 2220–2242.
21. Rudzińska, M., Daglioglu, C., Savvateeva, L.V., et al. (2021). Current status and perspectives of protease inhibitors and their combination with nanosized drug delivery systems for targeted cancer therapy. *Drug Design, Development and Therapy*, 15, 9–20.
22. Eskandari, P., Bigdeli, B., Porgham Daryasari, M., et al. (2019). Gold-capped mesoporous silica nanoparticles as an excellent enzyme-responsive nanocarrier for controlled doxorubicin delivery. *Journal of Drug Targeting*, 27, 1084–1093.
23. Iaccarino, G., Profeta, M., Vecchione, R., et al. (2019). Matrix metalloproteinase-cleavable nanocapsules for tumor-activated drug release. *Acta Biomaterialia*, 89, 265–278.
24. Duan, H., Liu, C., Hou, Y., et al. (2022). Sequential delivery of quercetin and paclitaxel for the fibrotic tumor microenvironment remodeling and chemotherapy potentiation via a dual-targeting hybrid micelle-in-liposome system. *ACS Applied Materials & Interfaces*, 14, 10102–10116.
25. Liang, T., Zhang, B., Xing, Z., et al. (2021). Adapting and remolding: orchestrating tumor microenvironment normalization with photodynamic therapy by size transformable nanoframeworks. *Angewandte Chemie International Edition*, 60, 11464–11473.
26. Wu, L., Lin, B., Yang, H., et al. (2019). Enzyme-responsive multifunctional peptide coating of gold nanorods improves tumor targeting and photothermal therapy efficacy. *Acta Biomaterialia*, 86, 363–372.
27. Tsai, L.C., Hsieh, H.Y., Lu, K.Y., et al. (2016). EGCG/gelatin-doxorubicin gold nanoparticles enhance therapeutic efficacy of doxorubicin for prostate cancer treatment. *Nanomedicine (Lond)*, 11, 9–30.
28. Xia, F., Hou, W., Zhang, C., et al. (2018). pH-responsive gold nanoclusters-based nanoprobes for lung cancer targeted near-infrared fluorescence imaging and chemo-photodynamic therapy. *Acta Biomaterialia*, 68, 308–319.
29. Xia, F., Niu, J., Hong, Y., et al. (2019). Matrix metallopeptidase 2 targeted delivery of gold nanostars decorated with IR-780 iodide for dual-modal imaging and enhanced photothermal/photodynamic therapy. *Acta Biomaterialia*, 89, 289–299.
30. Shi, H., Sun, Y., Yan, R., et al. (2019). Magnetic semiconductor Gd-doping CuS nanoparticles as activatable nanoprobes for bimodal imaging and targeted photothermal therapy of gastric tumors. *Nano Letters*, 19, 937–947.
31. Gordon, M.R., Zhao, B., Anson, F., et al. (2018). Matrix metalloproteinase-9-responsive nanogels for proximal surface conversion and activated cellular uptake. *Biomacromolecules*, 19, 860–871.
32. Guo, X., Fang, Z., Zhang, M., et al. (2020). A co-delivery system of curcumin and p53 for enhancing the sensitivity of drug-resistant ovarian cancer cells to cisplatin. *Molecules*, 25, 2621.
33. Chen, A., Lu, H., Cao, R., et al. (2022). A novel MMP-responsive nanoplatform with transformable magnetic resonance property for quantitative tumor bioimaging and synergetic chemo-photothermal therapy. *Nano Today*, 45, 101524.
34. Wang, Y., Cui, J., Chen, J., et al. (2022). Novel bone tumor cell targeting nanosystem for chemo-photothermal therapy of malignant bone tumors. *Chemical Engineering Journal*, 446, 136905.
35. Chen, X., Gao, H., Deng, Y., et al. (2020). Supramolecular aggregation-induced emission nanodots with programmed tumor microenvironment responsiveness for image-guided orthotopic pancreatic cancer therapy. *ACS Nano*, 14, 5121–5134.
36. Li, J., Liu, F., Shao, Q., et al. (2014). Enzyme-responsive cell-penetrating peptide conjugated mesoporous silica quantum dot nanocarriers for controlled release of nucleus-targeted drug molecules and real-time intracellular fluorescence imaging of tumor cells. *Advanced Healthcare Materials*, 3, 1230–1239.

37. Chen, X., Lee, D., Yu, S., et al. (2017). *In vivo* near-infrared imaging and phototherapy of tumors using a cathepsin B-activated fluorescent probe. *Biomaterials*, 122, 130–140.
38. Huang, C.H., Chang, E., Zheng, L., et al. (2023). Tumor protease-activated theranostic nanoparticles for MRI-guided glioblastoma therapy. *Theranostics*, 13, 1745–1758.
39. Li, Y., Mei, T., Han, S., et al. (2020). Cathepsin B-responsive nanodrug delivery systems for precise diagnosis and targeted therapy of malignant tumors. *Chinese Chemical Letters*, 31, 3027–3040.
40. Wang, H., Wang, R., Cai, K., et al. (2017). Selective *in vivo* metabolic cell-labeling-mediated cancer targeting. *Nature Chemical Biology*, 13, 415–424.
41. Wang, Z., Xiang, Z., Zhu, T., et al. (2020). Cathepsin L interacts with CDK2-AP1 as a potential predictor of prognosis in patients with breast cancer. *Oncology Letters*, 19, 167–176.
42. Bratovš, A., Kramer, L., Mikhaylov, G., et al. (2019). Stefin a-functionalized liposomes as a system for cathepsins S and L-targeted drug delivery. *Biochimie*, 166, 94–102.
43. Zhai, B.T., Tian, H., Sun, J., et al. (2022). Urokinase-type plasminogen activator receptor (uPAR) as a therapeutic target in cancer. *Journal of Translational Medicine*, 20, 135.
44. Masucci, M.T., Minopoli, M., Di Carluccio, G., et al. (2022). Therapeutic strategies targeting urokinase and its receptor in cancer. *Cancers (Basel)*, 14, 498.
45. Zhou, M., Liu, X., Li, Z., et al. (2018). Caspase-3 regulates the migration, invasion and metastasis of colon cancer cells. *International Journal of Cancer*, 143, 921–930.
46. Li, Y., Hu, P., Wang, X., et al. (2021). Integrin alpha(v)beta(3)-targeted polydopamine-coated gold nanostars for photothermal ablation therapy of hepatocellular carcinoma. *Regenerative Biomaterials*, 8, rbab046.
47. Xu, H.-L., Mao, K.-L., Huang, Y.-P., et al. (2016). Glioma-targeted superparamagnetic iron oxide nanoparticles as drug-carrying vehicles for theranostic effects. *Nanoscale*, 8, 14222–14236.

25 Nanomedicine's Role in Proteolytic Landscape of Breast Cancer

Sobia Noreen and Mehwish Ishtiaq
Institute of Chemistry, University of Sargodha, Sargodha, Pakistan

Sara Hasan
Institute of Chemistry, University of Sargodha, Sargodha, Pakistan
and
The University of Lahore, Sargodha Campus, Sargodha, Pakistan

Shazia Akram Ghumman
College of Pharmacy, University of Sargodha, Sargodha, Pakistan

Samina Aslam
The Women University Multan, Multan, Pakistan

25.1 INTRODUCTION

Proteases, a class of enzymes recognized for their role in proteolysis, the pivotal process of breaking down peptide bonds, have earned prominence in the realm of biologically induced events. Abnormal proteolysis, increased protease expression, malfunctioning protease signaling or disrupted protease inhibitor equilibrium are frequent hitches during the development and progression of diseases (Turk, Turk, & Turk, 2012). Matrix metalloproteases (MMP2) and cathepsins are two of the protease systems implicated in extracellular proteolysis that are modulated by the interaction between stromal and tumor cells. When triggered on the surface of tumor cells, these enzymes are concocted as inactive precursors, steering malignant cells to rupture the basement membranes. Dysregulated proteases from diverse protease families play a crucial role in the genesis of cancer tumors and their subsequent metastasis (Pandya, Orgaz, & Sanz-Moreno, 2017). These factors include extensive extracellular membrane (ECM) remodeling, epithelial-to-mesenchymal transition (EMT), evasion and manipulation of the immune system, resistance to apoptosis signals, and metastasis progression (Mitschke, Burk, & Reinheckel, 2019; Mohan, Das, & Sagi, 2020). Furthermore, within the convoluted web of cellular signaling, proteases wield a substantial influence over pathways of mitogen-activated protein kinase (MAPK), Akt and tumor necrosis factor (TNF), amongst others, exerting a significant impact on the onset and spread of cancer (López-Otín & Hunter, 2010). Moreover, within the tumor microenvironment, a diverse array of proteases from multiple families harboring various proteolytic functions can be found (Figure 25.1).

DOI: 10.1201/9781003394693-29

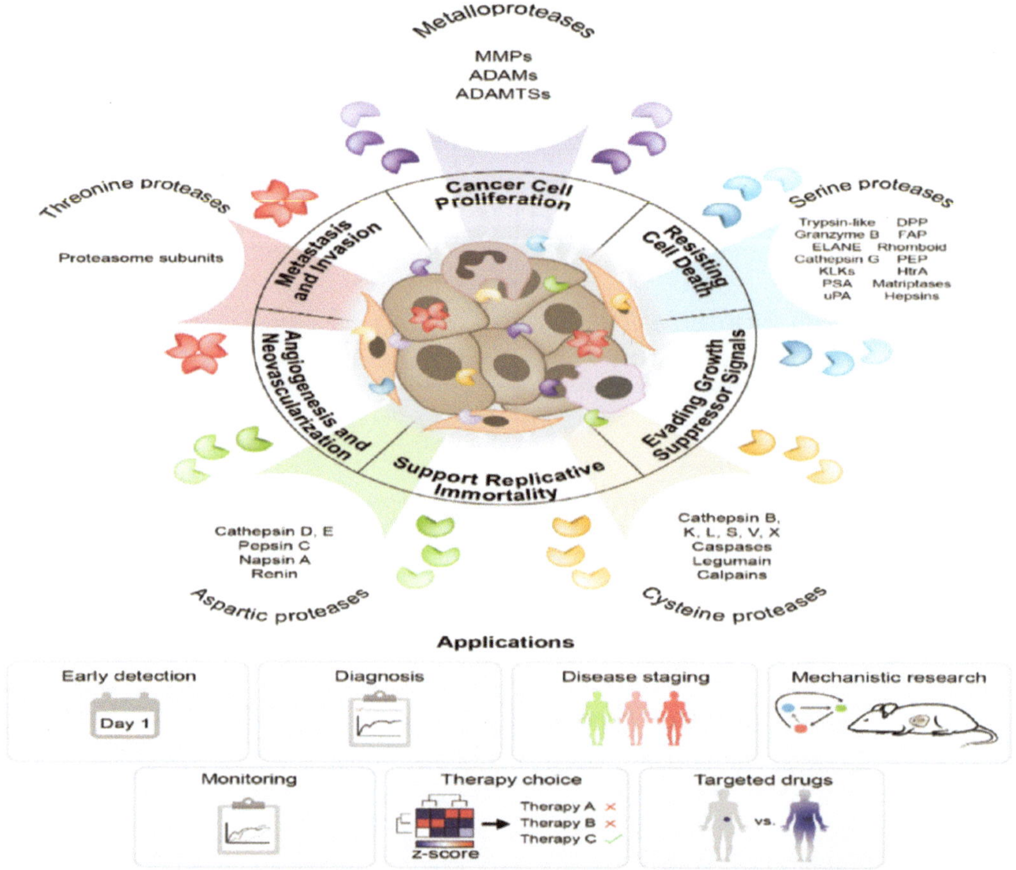

FIGURE 25.1 Mapping the protease landscape: Diverse protease families in the tumor microenvironment drive cancer hallmarks and apprise diagnostic and therapeutic routes. (Reprinted under a Creative Commons Attribution 4.0 International License from Vizovisek, Ristanovic, Menghini, Christiansen, & Schuerle, 2021.)

25.2 ROLE OF PROTEASE IN BREAST CANCER

Proteases assume a crucial role in fostering breast tumor growth and metastasis, profoundly influencing both primary and secondary sites. The complex route of tumors from their primary origin to secondary sites is closely intertwined with the actions of proteases. These enzymes, prominently cathepsins and matrix metalloproteinases, found abundantly in cancerous tissues as well as nearby tissues and organs, catalyze processes crucial for breast cancer (BC) progression. Moreover, the transformation of a normal cell into a cancerous one involves mechanisms such as EMT, where cells disrupt intercellular connections and gain enhanced mobility, facilitating invasion, metastasis and cancer dissemination. In parallel, cancer cells stimulate neovascularization, ensuring a constant supply of essential nutrients and oxygen for their unchecked proliferation by prompting the formation of new blood vessels originating from proliferating endothelial cells. This multifaceted interplay of proteases, EMT, and neovascularization shapes the complex landscape of BC progression (Eatemadi, Aiyelabegan, Negahdari et al., 2017).

25.3 CHALLENGES IN TARGETING PROTEOLYTIC PROCESSES

Challenges in targeting proteolytic processes in BC are multifaceted. BC is not only the most frequently diagnosed cancer but also the second-leading cause of cancer-related deaths in women globally (Siegel, Miller, & Jemal, 2015). Despite notable advancements in fields like immunotherapy and gene therapy, the foremost challenge in BC treatment remains the lack of effective therapies for advanced and metastatic tumors. The complexity deepens when delving into the realm of targeting proteolytic processes. BC is inherently heterogeneous, characterized by a dynamic and ever-evolving proteolytic landscape, which serves as a formidable barrier to the development of precise and effective therapeutic strategies. Achieving specificity in targeting particular proteases relevant to BC, which spares others with essential functions, presents a substantial hurdle. Moreover, the constant emergence of resistance mechanisms adds another intricate layer of complexity to treatment approaches. Ensuring the precise delivery of protease inhibitors within the breast, with its unique microenvironment, is essential but fraught with challenges (Britt, Cuzick, & Phillips, 2020; Masoud & Pagès, 2017; Vizovisek, Ristanovic, Menghini et al., 2021). Overcoming substantial obstacles, including limited tumor selectivity, rapid drug clearance and degradation, as well as the potential for severe toxicity and adverse effects inherent to cancer drug delivery, remains a critical factor for the success of evolving therapeutic approaches in BC (Shi, Kantoff, Wooster, & Farokhzad, 2017).

25.4 PROTEASE TARGETING NANOTECHNOLOGY: PRECISION IN BREAST CANCER THERAPEUTICS

Nanomedicine is a promising approach to overcome challenges related to proteolytic targeting. It is an innovative strategy that has the potential to revolutionize healthcare. Researchers are currently exploring the potential of various types of nanomedicine to improve BC treatment. Nanomedicine-based strategies aim to selectively activate therapeutic agents within the tumor microenvironment, responding to distinct chemical and physical stimuli that set the tumor apart from surrounding healthy tissues. These stimuli, which include variations in pH level and redox potential, offer unique signatures within the tumor. Nanotechnology utilizes these specific stimuli to create a precision delivery system. Additionally, the tumor protease microenvironment has emerged as a focal point of recent attention, holding potential as a resource for advancing cancer nanotherapeutics, nanodiagnostics, and nanotheranostics (Vizovisek, Ristanovic, Menghini et al., 2021). Moreover, the concept of designing nanoparticles (NPs) that become activated in response to elevated concentrations of specific proteases holds promise. Recent advancements have been achieved in systems involving inorganic NPs, liposomes, and polymer nanoparticles in addition to many other NPs that exhibit responsiveness to proteases, offering advantageous outcomes (Smith & Gambhir, 2017). Also, characterizing protease functioning and substrates in the tumor protease microenvironment can help identify potential targets for enzymes and as a result, lead to the formation of efficacious nanomedicine.

Protease-based nanomedicines are particularly interested in the regulation of protumorigenic activity of protease and nano-prodrug approaches, which use NPs as a storehouse to reduce chemotherapeutic effects in tissues that are healthy and focus apoptotic activity among tumor cells (Zhang, Eden, & Chen, 2012). Among the extensively researched drug-delivery systems are liposomes and polymer-based platforms, nanoshells, carbon-60 fullerenes, gold- and iron-based nanoparticles, and an array of other nanodelivery systems encompassing diverse material and physiochemical properties (see Figure 25.2). These systems have been tailored for use in the treatment of BC by protease targeting, aiming to achieve precise drug targeting and enhanced therapeutic outcomes (Bousquet, Ibarboure, Drummond et al., 2008; Hirsch, Stafford, Bankson et al., 2003; Yong, Qian, Roy et al., 2007).

25.4.1 Carbon Nanoparticles

Nanomaterials present captivating physical, chemical, and biological attributes that stem from their intricate structure and composition. These materials hold immense potential across a

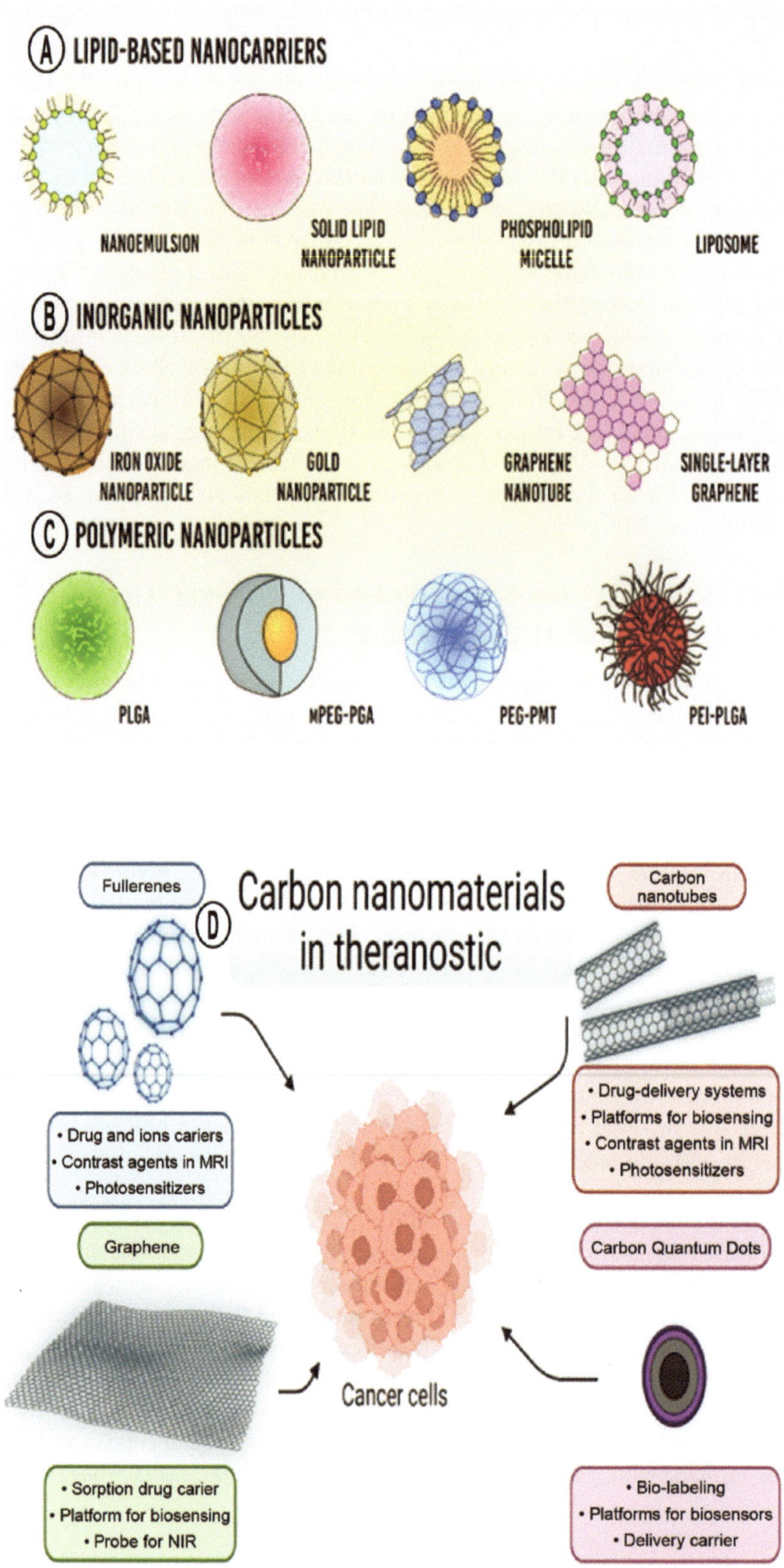

FIGURE 25.2 Nanocarriers for cancer therapeutic delivery: A–Lipid-based, B–inorganic, C–polymeric, D– carbon-based nanomaterials. (Reprinted under a Creative Commons Attribution 4.0 International License Kośсik, Jankowski, & Jagusiak, 2021; Lôbo, Paiva, Silva et al., 2021.)

spectrum of biomedical applications, including the pioneering development of BC treatment routes. Carbon nanomaterials, particularly carbon quantum dots, have captured the attention of the scientific community due to their distinctive physical properties, capacity for chemical functionalization, adaptability, and inherent biocompatibility (Chen, Dougherty, Zhu, & Hong, 2015; Goenka, Sant, & Sant, 2014). Consequently, cancer treatment often necessitates prolonged delivery, posing challenges to minimize off-target effects on healthy tissues. Recent developments have proposed targeted drug delivery using carbon quantum dots (C-Qt). For instance, a novel formulation involving doxorubicin-loaded graphene Qt has shown promise. Su et al., conducted a study with C-Qt, where stable fluorescence was observed at the tumor site for up to 10 days after injections, providing traceability and minimal side effects compared to free drugs. Moreover, C-Qts have been explored for targeted drug delivery in combination with targeted molecules (Su, Guo, Yuan et al., 2020).

Furthermore, C-Qts have found utility in Forster resonance energy transfer (FRET)-based sensors for detecting protease activity. These assays involve tethering a Qt to a protease-specific peptide sequence linked to a fluorescent dye, fluorescent protein, quencher dye or gold nanoparticles. Protease cleavage disrupts FRET, leading to increased C-Qt fluorescence emission. For instance, a highly sensitive biosensor for MMP2 detection was developed using a self-enhanced nanocomposite and a specific peptide. The biosensor employed (nanocomposite-graphene quantum dots) NG-Qts/ Ru@SiO_2 and had a low detection limit of 6.5 pg/mL, showing promise for detecting cancer biomarkers (see Figure 25.3) (Fan, Wang, Liu et al., 2022). C-Qts do possess the potential to improve the diagnostic and therapeutic modalities with protease as targets in BC; however, further research is required to establish their clinical efficiency.

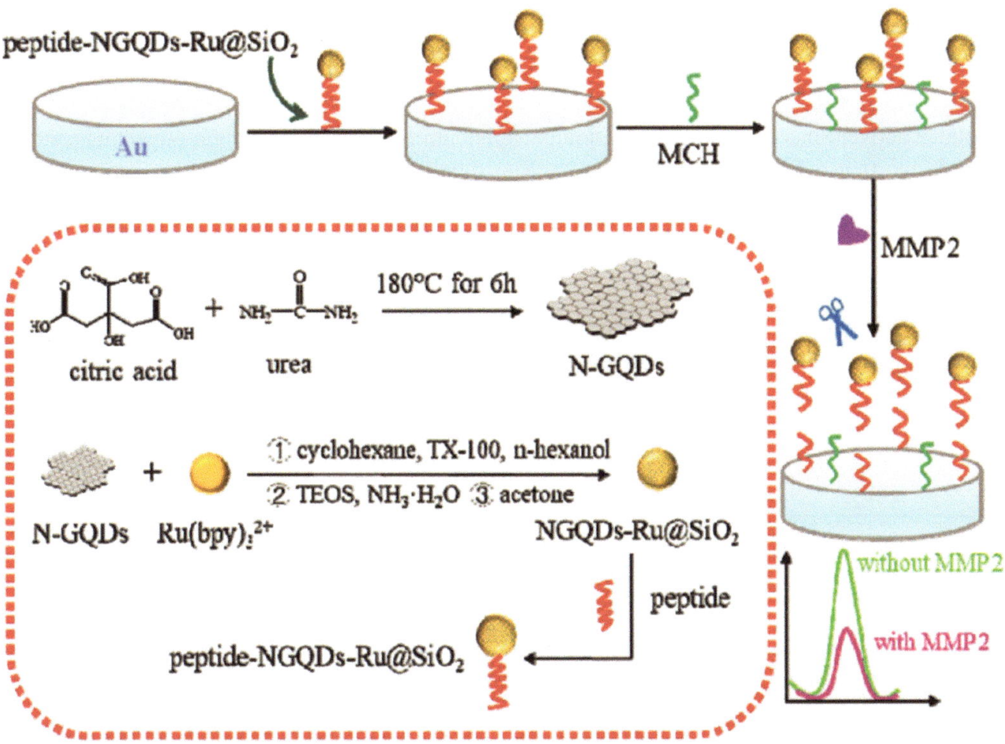

FIGURE 25.3 Electrochemiluminescence biosensors detect MMP2 biomarkers with NG-Qts/Ru@SiO_2 luminophore. (Reprinted with permission from Fan, Wang, Liu et al., 2022.)

25.4.2 Inorganic Nanoparticles

Irrespective of their composition, whether inorganic materials, metals or semiconductors, NPs possess distinctive optical and magnetic properties that hold potential for cancer imaging. Furthermore, a subset of NPs from this category has garnered attention due to their ability to transport molecular medicines, engage in active targeting facilitated by external forces (e.g., magnetic fields), and generate anti-cancer reactive oxygen species (ROS). Consequently, they have been investigated as potential treatments for BC (O'Connell, VandenHeuvel, Kamat, Raghavan, & Godin, 2022). This group often encompasses NPs that can be functionalized to enhance their effectiveness in cancer therapy. For instance, in a specific study, doxorubicin was encapsulated within mesoporous silica nanoparticles and subsequently coated with gelatin to prevent drug leakage. Elevated activity of MMP2 led to the release of the drug payload from the gelatin layer once it accumulated within tumor sites, demonstrating a promising approach to BC (Zou, He, He et al., 2015).

Likewise, in another study, siRNA against protease-activated receptor-1, a gene crucial for tumor metastasis, was conjugated to gold (Au) nanorods and delivered to highly metastatic BC cells. This approach efficiently downregulated protease-activated receptor-1 and reduced the metastatic ability of cancer cells (Zhang, Meng, Ji et al., 2011). Likewise, in another study, a DNA aptamer-based contrast agent was developed to specifically target human MMP9 within the tumor microenvironment, displaying exceptional affinity and sensitivity (Figure 25.4). The agent used plasmonic gold nanospheres modified with MMP9 aptamer, allowing them to aggregate in the presence of human MMP9. This aggregation enhanced optical absorption in the first near infrared window, enabling the detection of these aggregates via ultrasound-guided photoacoustic imaging within the microenvironment. The method showed successful and selective detection of human MMP9 both *in vitro* and *in vivo*, offering potential applications in BC diagnosis (Kim, Anthony, Kubelick, & Emelianov, 2022). This underscores the versatility and potential of inorganic NPs in protease targeting for BC treatment. They offer precise drug delivery mechanisms and opportunities for enhanced therapeutic outcomes.

25.4.3 Liposomes

A notable delivery mechanism with oncological applications is liposomes. Their advantages include a size range that facilitates accumulation within tumors and a surface chemistry well-suited for accommodating various functionalities (Ramishetti & Huang, 2012). A common approach involves

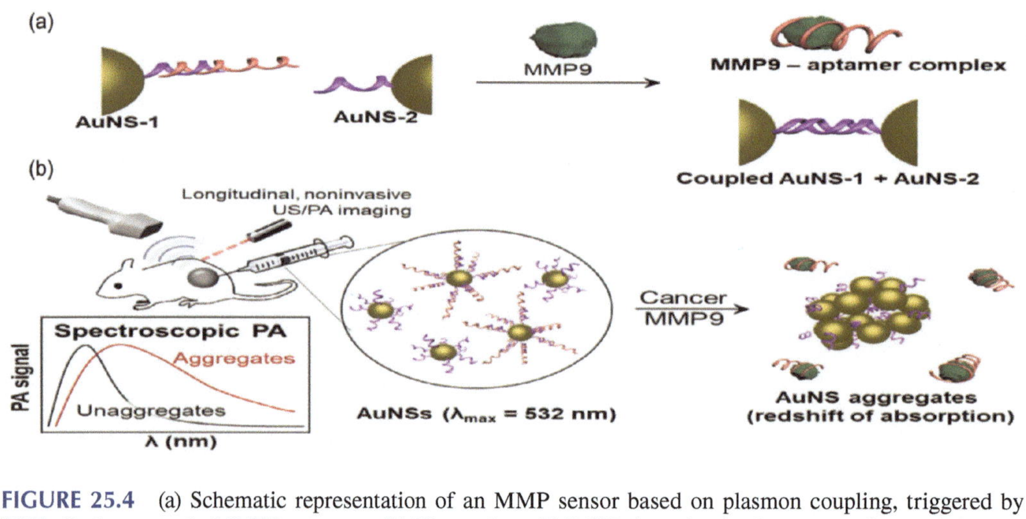

FIGURE 25.4 (a) Schematic representation of an MMP sensor based on plasmon coupling, triggered by DNA displacement in MMP9 presence. (b) Illustration of MMP9 detection in the tumor microenvironment using ultrasound-guided spectroscopic photoacoustic imaging. (Reprinted from Kim, Anthony, Kubelick et al., 2022 under a Creative Common License.)

encapsulating therapeutic agents, such as cytotoxic chemicals, within liposomes equipped with protease-sensitive ligands. The clinical success of liposomal-based chemotherapeutic drugs, exemplified by formulations like Doxil liposomal preparation, has led to their adoption in BC treatment (Khan, 2010). To enhance the effectiveness of cancer genetic treatment, protease-responsive liposomes have been employed, aiming to improve cancer cell transduction (Wan, Han, Fan et al., 2013). Liposomes serve as a valuable drug delivery vehicle for chemotherapeutic agents, enhancing their pharmacological characteristics by influencing drug distribution and the body's response to the drug (Khan, Rezler, Lauer-Fields, & Fields, 2008).

Liu et al., (2022) engineered a multifunctional drug-delivery system using liposomes modified with the EMC peptides. The system effectively encapsulated doxorubicin and tariquidar, designed for targeted drug release within the tumor microenvironment. The system's efficacy as a potential BC therapeutic was attributed to its combination of epidermal growth factor receptor (EGFR) targeting, responsiveness to MMP2 and cathepsin, controlled drug release, and the ability to counteract drug efflux. This approach held great promise for addressing MMP and cathepsin release triple negative BC (Liu, Zhao, Gao et al., 2022). In another research study, specialized thermosensitive liposomes were devised for delivering the MMP inhibitor, marimastat, into the tumor microenvironment. These systems proved highly effective, causing a substantial 50% and 43% reduction in the activity of MMP2 and MMP9 along with a 30% and 43% decrease in the expression of MMP2 and MM9, respectively (Lyu, Xiao, Yin, Yang, & He, 2019). These studies advocate the promising role of liposomal drug-delivery systems in protease targeting within the complex landscape of BC. The multifunctional abilities of liposomes, such as precise targeting, responsiveness to MMP and cathepsin, and controlled drug release, offer a powerful toolset for addressing the challenges of protease-driven metastasis and invasion in BC.

25.4.4 Polymer-Based Nanoparticles

Polymer designs can incorporate responsiveness to disease-associated proteases while also encompassing environmentally friendly and recyclable polymer-based NPs that have undergone extensive study as drug-delivery systems over the past decade (Hu, Zhang, & Liu, 2012). These polymer-based systems can be classified into two categories: natural and synthetic. Natural polymers, including albumin, chitosan, and heparin, have been employed for drug administration and the transport of molecules such as DNA, proteins, and oligonucleotides (Gradishar, Tjulandin, Davidson et al., 2005). Similarly, N-(2-hydroxypropyl)-methacrylamide copolymer, poly-1-glutamic acid, polystyrene maleic anhydride copolymer, and polyethylene glycol are typical examples of synthetic polymers (Li, 2002). Among these, Poly (ethylene glycol) (PEG) is one of the most popular polymers in nanoscale curative architectures for cancer. PEG's biodegradability has contributed to its preference, although the emergence of anti-PEG antibodies among patients has raised concerns, potentially prompting the exploration of alternative polymers in lieu of PEG (Zhang, Sun, Liu, & Jiang, 2016).

Additionally, in a recent study, novel polymeric NPs responsive to MMP were designed and comprehensively characterized, demonstrating exceptional chemotherapy efficacy with minimal side effects in BC models (Pan, Li, Wang et al., 2018). Another approach involved the development of a MMP2/MMP9 sensitive linker and its integration into a nanoscale. This system, comprising biodegradable polymers exhibited enhanced cytotoxicity against BC cells and reduced systemic toxicities (Li, Guo, Duan et al., 2016). Researchers also explored the use of upregulated lysosomal protease cathepsin-B as a target for BC therapy by polymer NPs. They designed a polymer–drug nanocomplex, exhibiting promising results in terms of targeted therapy and reduced systematic toxicity (Wang, Zhang, Zhang et al., 2018). In another study, gelatin NPs were designed for enhanced BC treatment with dual responsiveness potential to photothermal and MMP2 stimuli (Figure 25.5). In both *in vitro* and *in vivo* experiments, the NPs demonstrated potent therapeutic efficacy, suggesting a promising approach for synergistic chemo-photothermal therapy in BC treatment (Chen, Zou, Zhang et al., 2021). Several studies advocate the potential of polymer NPs to precisely deliver drugs with protease as targets in BS, with reduced side effects, however, clinical studies are still lacking.

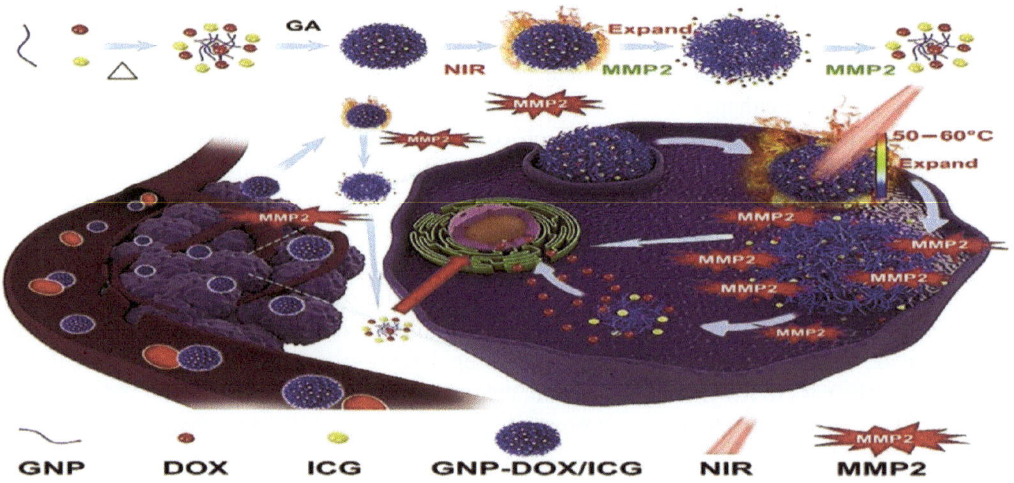

FIGURE 25.5 Photothermal/MMP2 dual responsive polymer (gelatin) nanoparticles for synergistic chemo-photothermal therapy in breast cancer treatment. (Reprinted from Chen, Zou, Zhang et al., 2021 under a Creative Common License.)

25.5 CONCLUSION AND FUTURE HORIZONS

In the domain of BC, proteases have emerged as central figures exerting substantial influence throughout the spectrum of the disease, spanning from benign to malignant states. These enzymatic moieties uphold a finely tuned equilibrium, counteracted by endogenous protease inhibitors that target a diverse array of proteases both formidable challenges and promising prospects for innovative approaches. NPs have ascended as potent tools with the BC paradigm. Operating at cellular and molecular strata, they possess the distinctive properties of biological cues woven into the microenvironment of BC. NPs deftly orchestrate precise drug delivery, thereby ensuring the safe transit of therapeutic cargo to tumor sites while minimizing collateral damage to healthy tissues. Within the ever-evolving landscape of oncology, nanomedicine continuously unveils novel stratagems to propel BC treatments forward. The elucidation of BC's protease panorama has unlocked avenues for pioneering diagnostic and therapeutic modalities. This knowledge begets the development of nanosystems engineered to respond to protease activity, giving rise to protease-centric solutions tailored exclusively for BC. The integration of nano-toolkits designed to utilize protease responsiveness harbors immense potential for the domain of BC nanomedicines. Moreover, to advance BC nanomedicine targeting proteases, it is imperative to establish physiologically relevant preclinical models and steadfastly adhere to stringent characterization protocols during large-scale, reproducible NP manufacturing, all while upholding the tenets of economical and suitable manufacturing practices.

While extensive research and exploration have occurred, the full potential of protease-targeting nanomedicine in BC treatment remains largely untapped. Yet, protease-driven nanomedicine stands poised on the cusp of transformative change, while researchers increasingly embrace emerging concepts in protease-related therapeutics. It is evident that protease-centric nanomedicines will play a central and pivotal role in the future landscape of BC therapeutics.

REFERENCES

Bousquet, A., Ibarboure, E., Drummond, C., Labrugère, C., Papon, E., & Rodriguez-Hernandez, J. (2008). Design of stimuli-responsive surfaces prepared by surface segregation of polypeptide-b-polystyrene diblock copolymers. *International Journal of Biological Macromolecules, 41*(4), 1053–1056.

Britt, K. L., Cuzick, J., & Phillips, K.-A. (2020). Key steps for effective breast cancer prevention. *Nature Reviews Cancer, 20*(8), 417–436.

Chen, D., Dougherty, C. A., Zhu, K., & Hong, H. (2015). Theranostic applications of carbon nanomaterials in cancer: focus on imaging and cargo delivery. *Journal of Controlled Release*, *210*, 230–245.

Chen, X., Zou, J., Zhang, K., Zhu, J., Zhang, Y., Zhu, Z., ... Piao, J.-G. (2021). Photothermal/matrix metalloproteinase-2 dual-responsive gelatin nanoparticles for breast cancer treatment. *Acta Pharmaceutica Sinica B*, *11*(1), 271–282.

Eatemadi, A., Aiyelabegan, H. T., Negahdari, B., Mazlomi, M. A., Daraee, H., Daraee, N., ... Sadroddiny, E. (2017). Role of protease and protease inhibitors in cancer pathogenesis and treatment. *Biomedicine & Pharmacotherapy*, *86*, 221–231.

Fan, X., Wang, S., Liu, H., Li, Z., Sun, Q., Wang, Y., & Fan, X. (2022). A sensitive electrochemiluminescence biosensor for assay of cancer biomarker (MMP-2) based on NGQDs-Ru@ SiO2 luminophore. *Talanta*, *236*, 122830.

Goenka, S., Sant, V., & Sant, S. (2014). Graphene-based nanomaterials for drug delivery and tissue engineering. *Journal of Controlled Release*, *173*, 75–88.

Gradishar, W. J., Tjulandin, S., Davidson, N., Shaw, H., Desai, N., Bhar, P., ... O'Shaughnessy, J. (2005). Phase III trial of nanoparticle albumin-bound paclitaxel compared with polyethylated castor oil–based paclitaxel in women with breast cancer. *Journal of Clinical Oncology*, *23*(31), 7794–7803.

Hirsch, L. R., Stafford, R. J., Bankson, J., Sershen, S. R., Rivera, B., Price, R., ... West, J. L. (2003). Nanoshell-mediated near-infrared thermal therapy of tumors under magnetic resonance guidance. *Proceedings of the National Academy of Sciences*, *100*(23), 13549–13554.

Hu, J., Zhang, G., & Liu, S. (2012). Enzyme-responsive polymeric assemblies, nanoparticles and hydrogels. *Chemical Society Reviews*, *41*(18), 5933–5949.

Khan, D. R. (2010). The use of nanocarriers for drug delivery in cancer therapy. *J Cancer Sci Ther*, *2*(3), 58–62.

Khan, D. R., Rezler, E. M., Lauer-Fields, J., & Fields, G. B. (2008). Effects of drug hydrophobicity on liposomal stability. *Chemical Biology & Drug Design*, *71*(1), 3–7.

Kim, J., Anthony, M. Y., Kubelick, K. P., & Emelianov, S. Y. (2022). Gold nanoparticles conjugated with DNA aptamer for photoacoustic detection of human matrix metalloproteinase-9. *Photoacoustics*, *25*, 100307.

Kościk, I., Jankowski, D., & Jagusiak, A. (2021). Carbon nanomaterials for theranostic use. *C*, *8*(1), 3.

Li, C. (2002). Poly (L-glutamic acid)–anticancer drug conjugates. *Advanced Drug Delivery Reviews*, *54*(5), 695–713.

Li, N., Guo, C., Duan, Z., Yu, L., Luo, K., Lu, J., & Gu, Z. (2016). A stimuli-responsive Janus peptide dendron–drug conjugate as a safe and nanoscale drug delivery vehicle for breast cancer therapy. *Journal of Materials Chemistry B*, *4*(21), 3760–3769.

Liu, C., Zhao, Z., Gao, R., Zhang, X., Sun, Y., Wu, J., ... Chen, C. (2022). Matrix metalloproteinase-2-responsive surface-changeable liposomes decorated by multifunctional peptides to overcome the drug resistance of triple-negative breast cancer through enhanced targeting and penetrability. *ACS Biomaterials Science & Engineering*, *8*(7), 2979–2994.

Lôbo, G. C., Paiva, K. L., Silva, A. L. G., Simões, M. M., Radicchi, M. A., & Báo, S. N. (2021). Nanocarriers used in drug delivery to enhance immune system in cancer therapy. *Pharmaceutics*, *13*(8), 1167.

López-Otín, C., & Hunter, T. (2010). The regulatory crosstalk between kinases and proteases in cancer. *Nature Reviews Cancer*, *10*(4), 278–292.

Lyu, Y., Xiao, Q., Yin, L., Yang, L., & He, W. (2019). Potent delivery of an MMP inhibitor to the tumor microenvironment with thermosensitive liposomes for the suppression of metastasis and angiogenesis. *Signal Transduction and Targeted Therapy*, *4*(1), 26.

Masoud, V., & Pagès, G. (2017). Targeted therapies in breast cancer: new challenges to fight against resistance. *World Journal of Clinical Oncology*, *8*(2), 120.

Mitschke, J., Burk, U. C., & Reinheckel, T. (2019). The role of proteases in epithelial-to-mesenchymal cell transitions in cancer. *Cancer and Metastasis Reviews*, *38*(3), 431–444.

Mohan, V., Das, A., & Sagi, I. (2020). Emerging roles of ECM remodeling processes in cancer. *Seminars in Cancer Biology*, *62*, 192–200.

O'Connell, C., VandenHeuvel, S., Kamat, A., Raghavan, S., & Godin, B. (2022). The proteolytic landscape of ovarian cancer: applications in nanomedicine. *International Journal of Molecular Sciences*, *23*(17), 9981.

Pan, J., Li, P.-J., Wang, Y., Chang, L., Wan, D., & Wang, H. (2018). Active targeted drug delivery of MMP-2 sensitive polymeric nanoparticles. *Chemical Communications*, *54*(79), 11092–11095.

Pandya, P., Orgaz, J. L., & Sanz-Moreno, V. (2017). Modes of invasion during tumour dissemination. *Molecular Oncology*, *11*(1), 5–27.

Ramishetti, S., & Huang, L. (2012). Intelligent design of multifunctional lipid-coated nanoparticle platforms for cancer therapy. *Therapeutic Delivery*, *3*(12), 1429–1445.

Siegel, R. L., Miller, K. D., & Jemal, A. (2015). Cancer statistics, 2015. *CA: A Cancer Journal for Clinicians, 65*(1).

Shi, J., Kantoff, P. W., Wooster, R., & Farokhzad, O. C. (2017). Cancer nanomedicine: progress, challenges and opportunities. *Nature Reviews Cancer, 17*(1), 20–37.

Smith, B. R., & Gambhir, S. S. (2017). Nanomaterials for *in vivo* imaging. *Chemical Reviews, 117*(3), 901–986.

Su, W., Guo, R., Yuan, F., Li, Y., Li, X., Zhang, Y., … Fan, L. (2020). Red-emissive carbon quantum dots for nuclear drug delivery in cancer stem cells. *The Journal of Physical Chemistry Letters, 11*(4), 1357–1363.

Turk, B., Turk, D., & Turk, V. (2012). Protease signalling: the cutting edge. *The EMBO Journal, 31*(7), 1630–1643.

Vizovisek, M., Ristanovic, D., Menghini, S., Christiansen, M. G., & Schuerle, S. (2021). The tumor proteolytic landscape: a challenging frontier in cancer diagnosis and therapy. *International Journal of Molecular Sciences, 22*(5), 2514.

Wan, Y., Han, J., Fan, G., Zhang, Z., Gong, T., & Sun, X. (2013). Enzyme-responsive liposomes modified adenoviral vectors for enhanced tumor cell transduction and reduced immunogenicity. *Biomaterials, 34*(12), 3020–3030.

Wang, Z., Zhang, R. X., Zhang, T., He, C., He, R., Ju, X., & Wu, X. Y. (2018). In situ proapoptotic peptide-generating rapeseed protein-based nanocomplexes synergize chemotherapy for cathepsin-B overexpressing breast cancer. *ACS Applied Materials & Interfaces, 10*(48), 41056–41069.

Yong, K.-T., Qian, J., Roy, I., Lee, H. H., Bergey, E. J., Tramposch, K. M., … Prasad, P. N. (2007). Quantum rod bioconjugates as targeted probes for confocal and two-photon fluorescence imaging of cancer cells. *Nano Letters, 7*(3), 761–765.

Zhang, P., Sun, F., Liu, S., & Jiang, S. (2016). Anti-PEG antibodies in the clinic: current issues and beyond PEGylation. *Journal of Controlled Release, 244*, 184–193.

Zhang, W., Meng, J., Ji, Y., Li, X., Kong, H., Wu, X., & Xu, H. (2011). Inhibiting metastasis of breast cancer cells *in vitro* using gold nanorod-siRNA delivery system. *Nanoscale, 3*(9), 3923–3932.

Zhang, X.-X., Eden, H. S., & Chen, X. (2012). Peptides in cancer nanomedicine: drug carriers, targeting ligands and protease substrates. *Journal of Controlled Release, 159*(1), 2–13.

Zou, Z., He, X., He, D., Wang, K., Qing, Z., Yang, X., … Cai, L. (2015). Programmed packaging of mesoporous silica nanocarriers for matrix metalloprotease 2-triggered tumor targeting and release. *Biomaterials, 58*, 35–45.

26 Proteasome Inhibitors and Nano-Delivery Systems for Therapy of Glioblastoma Multiforme

Mekala Janaki Ramaiah, Janumala Mary Edna, Sahiti Chamarthy, Katari Sukeerthi, and Nallapaty Srilakshmi
Koneru Lakshmaiah Education Foundation, Vaddeswaram, India

Prasanna Srinivasan Ramalingam and Sivakumar Arumugam
Protein Engineering Lab, School of Biosciences and Technology
Vellore Institute of Technology, Vellore, India

26.1 INTRODUCTION

Gliomas are frequent, aggressive brain cancers and constitute 60% of central nervous system (CNS) tumors. High-grade gliomas are characterized by poor prognosis and difficulty in treatment (Koshy, M., et al., 2012). The surgical resection is followed by oral administration of temozolomide (TMZ), followed by conjunction with radiotherapy (Stupp, R., et al., 2005). Surgical resection aims to remove as much of the tumor as possible while preserving neurological function. After surgery, radiation therapy is usually administered to target the remaining tumor cells. Chemotherapy, often with the drug temozolomide, is used in combination with radiation to treat glioblastoma multiforme (GBM). However, traditional treatment poses several limitations, such as invasive surgery, tumor recurrence, resistance to radiation therapy, chemotherapy challenges, proteasome inhibitors, etc.

26.2 PROTEASOME INHIBITORS

The proteasome is a type of multisubunit protein degradation complex machinery that plays a key role in maintaining protein homeostasis through the degradation of regulatory proteins that participate in various cellular processes (Figure 26.1). The US Food and Drug Administration (FDA) has approved three proteasome inhibitors, bortezomib (BTZ), carfilzomib, and ixazomib for treating multiple myeloma; another inhibitor, marizomib, is under intense clinical trials (Figure 26.2). The proteolytic core of the proteasome has three active sites, β5, β2, and β1. All these clinical inhibitors were widely in use as research tools (e.g., epoxomicin, MG-132) to inhibit multiple active sites.

DOI: 10.1201/9781003394693-30

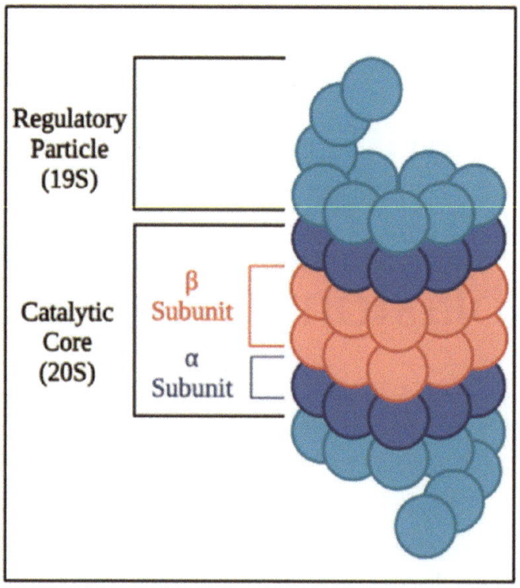

FIGURE 26.1 The structure of 26S proteasome. The proteasome consists of two main components: The 19S regulatory particle (cap) and the 20S core particle (protease core).

Proteasome-mediated degradation is initiated by ubiquitination of targeted proteins and occurs by the involvement of three enzymes: ubiquitin-activating enzyme El, ubiquitin-conjugating enzymes E2, and Ubiquitin ligase E3. These ubiquitinated proteins are cleaved by catalytic subunits composed of three β subunits β_1 (i.e., $PSMB_6$ gene), β_2 ($PSMB_7$ gene), and β_5 ($PSMB_5$ gene), respectively. (Figure 26.3)

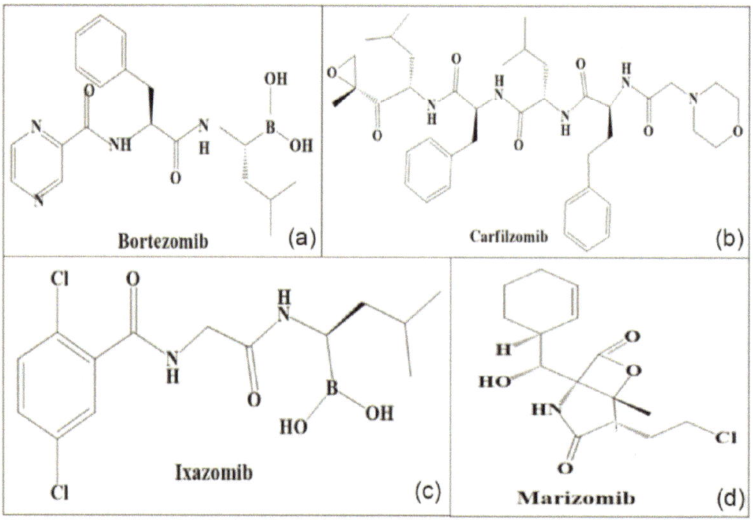

FIGURE 26.2 Chemical structures of (a) bortezomib, (b) carfilzomib, (c) ixazomib, and (d) marizomib.

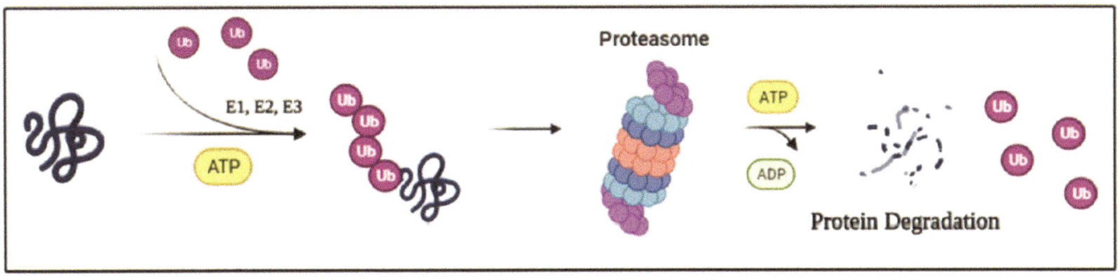

FIGURE 26.3 Mechanism of proteasome inhibitor. Proteasome-mediated degradation is initiated by ubiquitination of targeted proteins and occurs by the involvement of three enzymes El, E2, and E3.

26.3 BORTEZOMIB-BASED NANOFORMULATIONS

BTZ is the first proteasome inhibitor approved by the FDA. It binds to the amino-acid residues of the 26S proteasome, thereby causing the death of cells in mantle cell lymphoma, multiple myeloma, hematological cancers, and solid tumors. The drawbacks of BTZ include poor permeability and low bioavailability. Therefore, in recent years, different BTZ-based drug-delivery systems have been evaluated (Liu, J., et al., 2021). It has been previously shown that the delivery of PIs, such as BTZ, in CD38-targeted cross-linked-chitosan nanoparticles (NPs), has exhibited the toxicity profile of BTZ *in vivo* with low therapeutic efficacy is most likely due to the endothelial cell barrier. Interestingly, NP delivery systems were shown to be capable of targeting large doses of small molecules in the target area while sparing healthy tissues, overcoming the limitations of traditional therapies.

26.4 NANO-DELIVERY VEHICLES

Utilizing NP-based drug delivery has several advantages improved stability, biocompatibility, permeability, and specific targeting The evolution of hybrid NPs also enabled NP-centered drug-delivery systems to target drug resistance associated with cancer. Cancer drug resistance mechanisms encompass elevated expression of drug efflux transporters, impaired apoptotic pathways, and oxygen-deprived environments. NPs designed to target these mechanisms hold the potential to effectively enhance the reversal of multidrug resistance, thereby addressing a significant challenge in cancer therapy.

A variety of NP types are organic NPs, inorganic NPs, and hybrid NPs. Organic NPs consist of liposome-based NPs, polymer-based NPs, and dendrimers. Among polymer-based NPs, polymeric NPs and polymeric micelles are commonly used options. Inorganic NPs encompass gold NPs (Au NPs), carbon nanotubes, silica NPs, magnetic NPs, and quantum dots. Hybrid NPs include lipid-polymer hybrid NPs, organic-inorganic hybrid NPs, and NPs coated with cell membranes (Yao, Y., et al., 2020) (Figure 26.4).

26.5 POLYMERIC NPs

Polymeric NPs produced from synthetic NPs such as poly gamma glutamic acid (PγGA), polylactic acid (PLA), and Poly (lactic-co-glycolic acid) (PLGA) (Begines, B., et al., 2019), which are of 10–1000 nanometers in size. These NPs have specific advantages such as biocompatibility, specific targeting, encapsulation of molecules, drugs, recombinant genes, and controlled drug release avoidance of frequent doses. These NPs induce targeted delivery of proteasome inhibitors to GBM cells in high concentrations without affecting normal cells also known as the enhanced permeation and retention (EPR) effect (Peer, D., et al., 2020). Active targeting entails altering NPs to actively search for distinct molecular targets on afflicted cells, resulting in accurate and discriminating drug delivery. In contrast, passive targeting accumulates naturally due to their size and EPR effect.

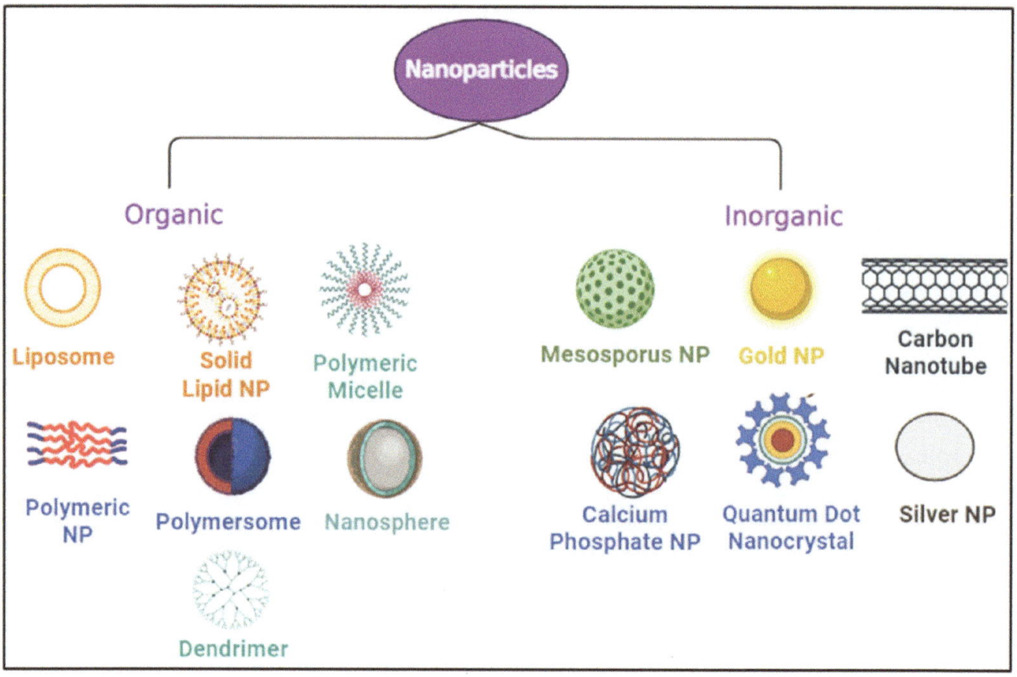

FIGURE 26.4 Types of nanoparticles. Nanoparticles are divided into organic and inorganic nanoparticles. The organic nanoparticles include liposomes, polymeric nanoparticles, dendrimer, solid nanoparticles, polymeric micelles, and nanosphere, while inorganic nanoparticles include mesoporous nanoparticles, gold nanoparticles, calcium phosphate nanoparticles, silver nanoparticles, quantum dots, nanocrystals, etc.

The histone deacetylase (HDAC) inhibitor vorinostat and proteasome inhibitor BTZ combination has exhibited strong scientific and clinical potential but is not warranted for clinical trials. HDAC inhibitor in combination with tyrosine kinase inhibitor and TMZ or antivascular endothelial growth factor (VEGF) antibody and mammalian target of Rapamycin (mTOR) inhibitor failed due to poor penetration to BBB, low solubility, etc. Thus, to overcome these challenges employment of nano-delivery systems (Friday, B. B., et al., 2011).

These NP systems prevent drug degradation, enhance blood–brain barrier (BBB) and blood–brain tumor barrier (BBTB) penetration, and limit solubility. The advantage of polymeric NPs is the easy adjustment of desired size; small enough to pass through the tight junctions of BBB (Bregoli, L., et al., 2016). The other benefits of the NP system are first, it eliminates the risk of becoming degraded or metabolized by enzymes in the bloodstream and, second, it increases the solubility of low-soluble brain-targeted drugs (Bawarski, W. E., et al., 2008). Integrating proteasome inhibitors, and polymeric NPs in the therapy for GBM encompasses a diverse strategy that enhances drug availability in GBM cells. The proteasome inhibitors inhibit the function of proteasomes leading to the accumulation of misfolded and damaged proteins leading to cell-cycle arrest and apoptosis of GBM cells.

The positive effect of polymeric NPs was also affirmed when they observed that the efficacy of BTZ NPs was higher than the free BTZ and that the slow response of free BTZ is almost the same as the sustained release of the drug. A research group showed that out of temozolomide-loaded polymeric nanoparticles (T-PNPs), solid lipid nanoparticles (T-SLNs), and nanostructured lipid carriers (T-NLCs), T-NLCs exhibited most antitumor activity and inhibition against U87 GBM cells compared to the other two.

Dextran (Dex) is an exopolysaccharide (EPS) produced by lactic acid bacteria (LAB) or their enzymes in the presence of sucrose. It consists of linear chain of D-glucose molecules connected by α-(1→6) bonds, and it may contain branches of D-glucose molecules linked by α-(1→4), α-(1→3), or α-(1→2) bonds. Dextran can vary in molecular weight, with some forms having low (<40 kDa) and others having high molecular weight (>40 kDa). Dextran binds with NPs using ester/amide, thereby enhancing stability and prolonged release. A study by a research group revealed that injected RVG peptide liganded multitargeting micelle (RVG-Dex-PTX) penetrated BBB and showed higher anticancer activity compared to the drug alone. Another such dextran-based NP carrier was developed by Li, D., et al. (2020). Polysaccharide-di-drugs conjugate Dex-g-(DOX+BTZ) by the one-pot Schiff base reaction. Doxorubicin (DOX) and BTZ were released at the target organ simultaneously when stimulated by acidic conditions.

26.6 COMBINATION OF BTZ AND HDAC INHIBITORS IN CANCER THERAPY

Unfolded or misfolded proteins are marked by ubiquitin for breakdown through proteasome and aggresome pathways. BTZ, a proteasome inhibitor, causes the buildup of ubiquitin protein aggregates, which are transported to lysosomes for degradation via the aggresome pathway. Aggresome formation relies on microtubule-based movement by dynein motor proteins, with HDAC6 aiding in the interaction of protein complexes. Conversely, deacetylase inhibitor (DACi) compounds inhibiting HDAC6 prevent this interaction. The joint use of proteasome inhibitors and DACi compounds amplifies cellular stress, leading to apoptosis (Figure 26.5).

26.7 MAGNETIC NPs

Studies have indicated that the induction of hyperthermia using NPs, magnetic fluid hypothermia (MFH), in combination with anticancer drugs is an attractive method due to its potential for enhanced anticancer effects. Recent studies have shown that cells treated with MFH are more sensitive to the proteasome inhibitor BTZ compared to cells treated with hot water hyperthermia (HWH) under similar temperature conditions due to microtubule damage, increased aggregation of proteins, and accumulation in the cytoplasm. Magnetic NPs can be produced using various methods such as microemulsion, co-precipitation, chemical precipitation, thermal decomposition, etc. These NPs

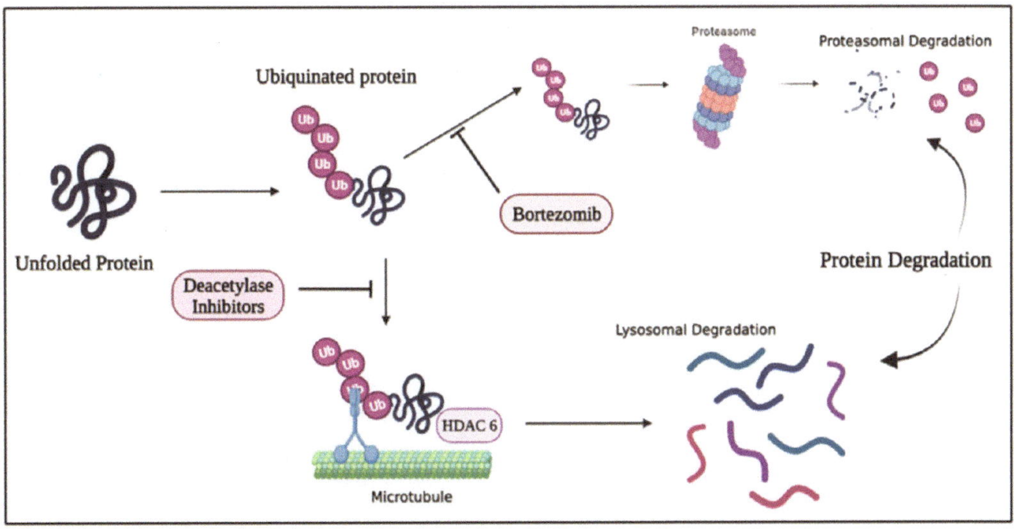

FIGURE 26.5 Inhibitions of the proteasome and aggresome pathways by bortezomib (BTZ) and deacetylase inhibitors (DACi). BTZ mediates the transport of aggregated proteins to lysosomes. Further, histone deacetylase (HDAC) 6 inhibitor helps as a better therapy in a combination of BTZ and HDAC 6 inhibitors.

in GBM have proved to be effective due to their properties, such as the contrast against magnetic resonance imaging (MRI) for visualization of tumors from which accurate size and shape can be determined and imaging can guide the NP to the target organ with the help of an external magnetic field, leading to decreased damage to healthy cells; the heat generated due to superparamagnetic properties can be harnessed and used to damage the sensitive GBM cells; easy penetration of BBB, real-time tracking of the NP in the bloodstream; and a decreased systemic exposure (Gubin, S. P., 2009).

These NPs are coated with a biocompatible material (McBain, S. C., et. al., 2008) and introduced into the bloodstream. This enhances drug delivery or imaging, reducing exposure to healthy tissues. NPs predominantly accumulated in tumor-associated immune cells, not tumor cells. They used 100 nm fluorescent polymeric NPs, and 20 nm magnetic NPs in mouse tumor models. Initial NP distribution depends on vascularization and permeability, but later stages were influenced by cellular uptake. Host phagocytes took up most NPs (>90%), and high tumor-associated macrophages (TAM) counts correlated with increased NP accumulation, highlighting the EPR effect.

In an orthotopic U87MG tumor model, the use of superparamagnetic iron oxide (SPIO) and paclitaxel (PTX) co-loaded PLGA NPs in magnetic targeting treatment led to a substantial improvement in median survival time when compared to the group treated with passive targeting (Ganipineni, L. P., et al., 2018).

Graphene oxide (GO) nanosheets exhibit a unique property of being widespread throughout the tumor. The BTZ molecules were impregnated on the sheets for sustained release. The combined GO-BTZ complex resulted in an enhanced cytotoxic effect as compared to monotherapy of both GO and BTZ (Sharp et al., 2023). When hyperthermia treatment was given, GBM cells overcome DNA repair and MGMT gene resistance toward TMZ. TMZ resistance was decreased, and cells died, but no additional effect on the already sensitive cells was shown. Shirvalilou et al., in 2018 used a multipart NP magnetic GO-based nanocarrier (NGO/SPION/PLGA) for drug delivery that passed very efficiently through BBB. PLGA aids with sustained circulation and NGO's aggregation ability improved due to its hydrophobic surface (Shirvalilou et al., 2018).

26.8 LIPOSOMAL NANO-DELIVERY

BTZ is the first medicine designed to hinder the activity of the 26S proteasome and is used to treat multiple myeloma. The liposomal delivery was developed as an improved method for packaging BTZ using liposomes, to enhance its encapsulation capacity. The ratio of drug to lipid and the pH of the solution affects the liposomal delivery. The combination tested is BTZ with small unilamellar vesicle (SUV) liposomes to make the new form. They used a difference in pH between the inside and outside of the liposomes to help store more BTZ, achieving a high efficiency of 48%, especially when the ratio was one part BTZ to five parts lipid (Butu, A., et al., 2015).

26.9 GOLD (AU) NANOPARTICLES

Au NPs have been employed for the treatment of various cancers such as adenocarcinoma, and prostate cancer. The most important challenge faced is multiple drug resistance (MDR), a clinical observation spanning several decades identified as a phenomenon where human tumors that develop resistance to one therapy also tend to exhibit resistance to multiple other drugs, even when these drugs differ significantly in both structure and how they work (Baguley, B. C. 2010). In adenocarcinoma, the increased toxicity of BTZ at low concentrations with pegylated Au NPs (PEGAuNPs) is due to BTZ's binding to NP surfaces and its cellular uptake, as shown by laser scanning confocal microscopy (LSCM). Dialysis experiments also confirm delayed BTZ release when combined with PEGAuNPs in phosphate buffer saline (PBS).

26.10 PEGYLATED PLGA NP

The most common mutation, ΔF508-CFTR, is known for its sensitivity to temperature and impaired trafficking, resulting in reduced chloride transport and heightened immune response. This mutation causes misfolding, ubiquitination, and premature degradation of ΔF508-CFTR through the proteasome-mediated degradation pathway. Recent studies have shown that targeted inhibition of this pathway using the FDA-approved drug PS-341 (also called Velcade or bortezomib) effectively reduces inflammation in cystic fibrosis (CF) cells. PS-341 is a potent, stable, reversible, and selective inhibitor of chymotryptic threonine protease activity. However, concerns arise regarding the use of the proteasome as a therapeutic target, as it may affect proteostasis and related processes. To address this, a NP-based strategy was developed, involving biodegradable NPs loaded with the drug PLGA-PEGPS-341 for controlled, and sustained drug delivery. *In vitro* release kinetics of the drug from these NPs were evaluated using a proteasomal activity assay over 1–7 days, showing gradual drug release from day 2 to day 7, with maximum inhibition on day 7.

To assess *in vivo* release kinetics and distribution, fluorescently labeled drug-loaded NPs were administered intranasally to C57BL6 mice. Live animal optical imaging demonstrated efficient particle delivery to the mouse lungs 24 hours after treatment, followed by gradual degradation and release over 1–11 days. The efficacy of drug release was measured in CF mice (Cftr-/-) lungs by quantifying changes in proteasomal activity (approximately two fold decrease) and its ability to alleviate *Pseudomonas aeruginosa* LPS (Pa-LPS)-induced inflammation. These findings suggest the potential to alleviate CF lung disease in a murine model by mitigating these inflammatory processes.

There are two different types of BTZ inclusion in PEG-coated Au NPs. One is by association of polymer chains with weak H bonds or dipole charge interactions. The other is adsorption on the surface of Au NPs (Figure 26.6). Emerging organic–inorganic hybrid NPs deliver drugs effectively

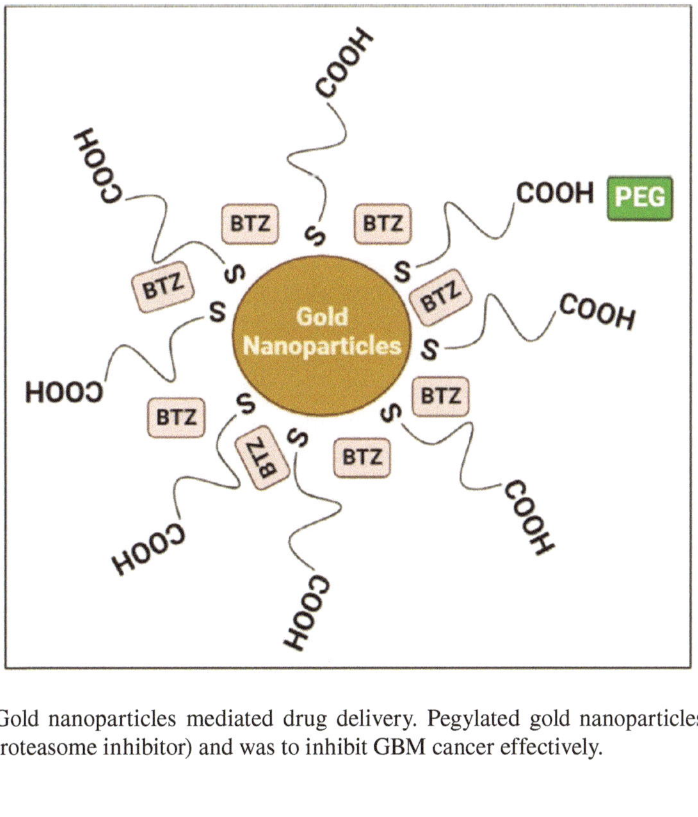

FIGURE 26.6 Gold nanoparticles mediated drug delivery. Pegylated gold nanoparticles was conjugated with BTZ drug (proteasome inhibitor) and was to inhibit GBM cancer effectively.

in cancer cells. Chitosan/gum Arabic with entrapped Au NPs was employed to carry the anticancer drug BTZ. The internalization of the drug was enhanced by using Au NP and BTZ encapsulated in a chitosan/gum Arabic matrix. Due to the positive nature of the nanosystem, the cell growth was reduced, enhancing the permeation and retention was enhanced in cancer cells (Coelho, S. C., et al., 2014).

26.11 STRATEGY SPECIFIC TARGETING

Specific drug-delivery strategies aim for precise targeting to enhance efficacy while minimizing side effects. These methods include ligand-receptor interactions, antibody-drug conjugates, NPs, pH-responsive systems, targeted liposomes, magnetic guidance, gene therapy, smart delivery systems, tissue-specific activation, and combination therapies. These approaches improve drug precision and therapeutic outcomes, with selection depending on the disease and drug involved (Zhao, Z., et al., 2020).

26.11.1 Transferrin Surface-Modified PLGA Nanoparticles

PLGA NPs have distinguishing features including biocompatibility, stability, high encapsulation of many agents and sustained drug release, and improved biodistribution by passive or active targeting to cancer cells (Danhier, F., et al., 2012). The PLGA NPs were formulated with a poloxamer, and further surface modified with transferrin for tumor targeting. The NPs were characterized as polymer carriers of BTZ, and the cellular uptake and growth inhibitory effects were evaluated in pancreatic cells. Cellular internalization of NPs was observed in normal as well as cancer cells, but with higher uptake by cancer cells. The sustained release of the loaded BTZ from PLGA NPs showed cytotoxic effects against pancreatic normal and cancer cells. Noteworthy differential cytotoxicity was attained by transferrin surface-modified PLGA NPs since significant cell growth inhibition by delivered BTZ was only observed in cancer cells. These findings demonstrate that the ligand transferrin enhanced the targeted delivery of BTZ-loaded PLGA NPs to pancreatic cancer cells. These *in vitro* results highlight the transferrin surface-modified PLGA NPs as a promising system for targeted delivery of anticancer drugs.

26.11.2 LDL Receptor

The low-density lipoprotein receptor (LDLR)-related protein (LRP) is a versatile endocytic receptor with diverse cellular and physiological roles. Prior research has indicated that both tyrosine-based and di-leucine motifs located within LRP's cytoplasmic tail are responsible for facilitating its rapid endocytosis (Lillis, A. P., et al., 2008). Nevertheless, there is limited understanding regarding how LRP is targeted for degradation. Through the examination of both the full-length endogenous LRP and a truncated mini-receptor form, it has been observed that the presence of proteasomal inhibitors like MG132 and lactacystin extends LRP's cellular lifespan.

These proteasomal inhibitors also notably elevate the quantity of LRP present at the cell surface. This suggests that the degradation pathway for LRP is obstructed at a stage where the recycling of the receptor to the cell surface still takes place. An investigation, employing immunoelectron microscopy analyses, revealed that proteasomal inhibitor usage led to a reduction in the LRP mini-receptor within the multivesicular bodies' limiting membranes and internal vesicles. These compartments are associated with the degradation process of the receptor (Melman, L., et al., 2002).

26.12 CONCLUSION

Nano-delivery systems help reduce off-target effects by concentrating therapeutic agents within the tumor. This minimizes systemic side effects and preserves healthy brain function. Tailoring proteasome inhibitor-based therapies to individual patient's profile and tumor characteristics is an emerging trend. This personalized approach maximizes treatment efficacy and minimizes adverse effects. Additionally, some nano-delivery systems include imaging agents that enable real-time monitoring of treatment responses and GBM progression, helping clinicians adjust therapies as needed. These innovative approaches have the potential to enhance treatment precision, minimize side effects, and improve the overall quality of life for GBM patients. Continued research and clinical validation are critical for realizing their full therapeutic potential.

DECLARATIONS

Conflict of interest: The authors declare no conflicts of interest.

Consent for publishing: All authors have approved the final version of the manuscript and given consent for publication.

Funding: Not Applicable.

REFERENCES

Baguley, B. C. (2010). Multiple drug resistance mechanisms in cancer. *Molecular Biotechnology*, *46*, 308–316.

Bawarski, W. E., Chidlowsky, E., Bharali, D. J., et al (2008). Emerging nanopharmaceuticals. *Nanomedicine: Nanotechnology, Biology and Medicine*, *4*(4), 273–282.

Begines, B., Alcudia, A., Aguilera-Velazquez, R., et al (2019). Design of highly stabilized nanocomposite inks based on biodegradable polymer-matrix and gold nanoparticles for Inkjet Printing. *Scientific Reports*, *9*(1), 16097.

Bregoli, L., Movia, D., Gavigan-Imedio, J. D., et al (2016). Nanomedicine applied to translational oncology: a future perspective on cancer treatment. *Nanomedicine: Nanotechnology, Biology and Medicine*, *12*(1), 81–103.

Butu, A., Rodino, S., Golea, D., et al (2015). Liposomal nanodelivery system for proteasome inhibitor anticancer drug bortezomib. *Farmacia*, *63*(2), 224–229.

Coelho, S. C., Rocha, S., Pereira, M. C., et al (2014). Enhancing proteasome-inhibitor effect by functionalized gold nanoparticles. *Journal of Biomedical Nanotechnology*, *10*(4), 717–723.

Danhier, F., Ansorena, E., Silva, J. M., et al. (2012). PLGA-based nanoparticles: an overview of biomedical applications. *Journal of Controlled Release*, *161*(2), 505–522.

Friday, B. B., Anderson, S. K., Buckner, J., et al (2011). Phase II trial of vorinostat in combination with bortezomib in recurrent glioblastoma: a north central cancer treatment group study. *Neuro-Oncology*, *14*(2), 215–221.

Ganipineni, L. P., Ucakar, B., Joudiou, N., et al (2018). Magnetic targeting of paclitaxel-loaded poly (lactic-co-glycolic acid)-based nanoparticles for the treatment of glioblastoma. *International Journal of Nanomedicine*, 4509–4521.

Gubin, S. P. (Ed.). (2009). *Magnetic nanoparticles*. John Wiley & Sons.

Koshy, M., Villano, J. L., Dolecek, T. A., et al (2012). Improved survival time trends for glioblastoma using the SEER 17 population-based registries. *Journal of Neuro-Oncology*, 107, 207–212.

Li, D., Su, T., Ma, L., Yin, F., Xu, W., Ding, J., & Li, Z. (2020). Dual-acidity-labile polysaccharide-di-drugs conjugate for targeted cancer chemotherapy. *European Journal of Medicinal Chemistry*, *199*, 112367.

Lillis, A. P., Van Duyn, L. B., Murphy-Ullrich, J. E., et al (2008). LDL receptor-related protein 1: unique tissue-specific functions revealed by selective gene knockout studies. *Physiological Reviews*, *88*(3), 887–918.

Liu, J., Zhao, R., Jiang, X., et al (2021). Progress on the application of bortezomib and bortezomib-based nanoformulations. *Biomolecules*, *12*(1), 51.

McBain, S. C., Yiu, H. H., & Dobson, J. (2008). Magnetic nanoparticles for gene and drug delivery. *International Journal of Nanomedicine*, *3*(2), 169–180.

Melman, L., Geuze, H. J., Li, Y., et al. (2002). Proteasome regulates the delivery of LDL receptor-related protein into the degradation pathway. *Molecular Biology of the Cell*, *13*(9), 3325–3335.

Peer, D., Karp, J. M., Hong, S., et al (2020). Nanocarriers as an emerging platform for cancer therapy. *Nano-Enabled Medical Applications*, 61–91.

Sharp, P. S., Stylianou, M., Arellano, L. M., et al (2023). Graphene oxide nanoscale platform enhances the anti-cancer properties of bortezomib in glioblastoma models. *Advanced Healthcare Materials*, *12*(3), 2201968.

Shirvalilou, S., Khoei, S., Khoee, S., et al (2018). Development of a magnetic nano-graphene oxide carrier for improved glioma-targeted drug delivery and imaging: *in vitro* and *in vivo* evaluations. *Chemico-Biological Interactions*, *295*, 97–108.

Stupp, R., Mason, W. P., Van Den Bent, M. J., et al (2005). Radiotherapy plus concomitant and adjuvant temozolomide for glioblastoma. *New England Journal of Medicine*, *352*(10), 987–996.

Yao, Y., Zhou, Y., Liu, L., et al. (2020). Nanoparticle-based drug delivery in cancer therapy and its role in overcoming drug resistance. *Frontiers in Molecular Biosciences*, *7*, 193.

Zhao, Z., Ukidve, A., Kim, J., et al (2020). Targeting strategies for tissue-specific drug delivery. *Cell*, *181*(1), 151–167.

27 Emerging Proteases for Molecular Imaging and Therapy in Cancer

Presenjit
Radiological Nuclear and Imaging Sciences (RNAIS), Institute of Nuclear Medicine and Allied Sciences, DRDO Delhi, India

Shubhra Chaturvedi and A.K. Mishra
Radiological Nuclear and Imaging Sciences (RNAIS), Institute of Nuclear Medicine and Allied Sciences, (INMAS), DRDO Delhi, India

Vishakha Chaudhary
Radiological Nuclear and Imaging Sciences (RNAIS), Institute of Nuclear Medicine and Allied Sciences, (INMAS), DRDO Delhi, India
University of Delhi, New Delhi, India

V.P. Singh
Institute of Science, Banaras Hindu University, Varanasi, India

Kaman Singh
Bhimrao Ambedkar University, Lucknow, India

27.1 INTRODUCTION

Cancer is among the leading diseases and impacts all facets of the population. The analysis of human mummies and fossilized hominids indicates the prevalence of cancer even in ancient times and, thus, cancer has been recognized throughout human history. In modern times, according to data collected by health agencies like the World Health Organization (WHO), cancer is the second most prevalent cause of mortality globally and can account for one in six fatalities. Cancer is a disease marked by biochemical changes. Imaging, omics analysis, and genome sequencing make it feasible to analyze numerous cancer forms with an extent of specificity and precision previously not conceivable (Nguyen, 2020). Cancer treatment strategies can be based on chemical compounds/drugs (chemotherapy), radiation, or invasive protocols (surgery). Chemotherapeutics can be employed at any stage of cancer, while radiotherapy and surgery are the two main treatment approaches for locoregional and non-metastatic tumors. Severe repercussions typically limit conventional chemotherapy treatments as they cannot distinguish between tumorous and healthy tissues. Targeted therapies incorporating monoclonal antibodies or inhibitors made from small molecules to target particular signaling pathways responsible for blood vessel development,

DOI: 10.1201/9781003394693-31

growth, and survival that are generally out of balance in tumor cells, can be used to lessen these adverse effects (Joo et al., 2013). New functional and molecular imaging biomarkers are being researched to more accurately assess therapy's effects and improve therapy's efficacy. Signature biomarkers like overexpressed proteins, particularly receptors and enzymes, metabolite alteration, tumor microenvironment, and immune response, have been the focus for designing therapeutics. Proteases are widely distributed enzymes that break down the peptide linkage between amino-acid residues in proteins. Proteases, which comprise about 2% of all proteins in healthy cells, play various biochemical regulatory roles, including transcription of genes, cellular division, and cell death. Proteases are crucial for biological activities in normal cells, but for healthy cells, there must be an appropriate equilibrium between them and their antiproteases; otherwise, any disruption of this equilibrium results in disorders, including cancer. Cancer cells change signaling pathways to promote invasion, leading to dysregulated expression patterns of proteases even in adjacent normal cells. Proteases are involved in tumor development, growth, metastasis, and invasion into another site.

27.2 ROLE IN THE PROGRESSION OF CANCER

Proteases are essential for executing crucial biological processes in healthy cells. They control a wide range of functions in the lifecycle of the cell. According to the latest research, proteases have a significant role in tumor initiation and advancement (López-Otín & Matrisian, 2007).

Various proteases in the tumor boost the availability of nourishment and oxygen, which is crucial for the continued growth and metastasis of tumors. Responses from the immune system, inflammatory cell recruitment, tumor invasion, vascular development, dissemination, apoptosis, epithelial to mesenchymal transition (EMT), the mobilization of regular cells from their cellular spaces to aid in metastasis, along with a response to therapy such as chemotherapy, have all been associated with the initial stages protease elevation in several tumors. Some of the critical roles of proteases in cancer are discussed next (Figure 27.1).

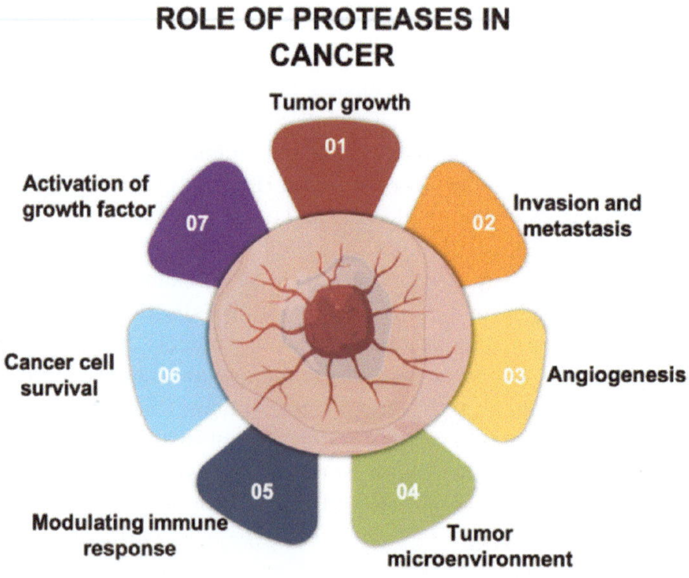

FIGURE 27.1 Some of the functioning of proteases in the cancer development.

27.3 DIFFERENT CLASSES OF PROTEASES

Proteases fall under different categories based on the amino acids at the catalytic site. Major protease categories are: cysteine proteases, serine proteases, aspartic proteases, threonine proteases, and metalloproteases (Rakashanda et al., 2012).

27.3.1 Cysteine Proteases

The members of this protease class are papain, calpain, and caspase. This group's constituents, called cathepsins, are essential to tumor biology because they promote angiogenesis, invasion, metastasis, tumor growth, immune evasion, and drug resistance (Olson & Joyce, 2015). Calpains are intracellular signaling cysteine proteases that are essential for many physiological and pathological processes. Specifically, the widely distributed calpains 1 and 2 are of great importance. Their important roles in apoptosis, migration, and cell transformation have been highlighted by research. According to Leloup and Wells (2011), calpains are seen as potential targets for anticancer therapy. On the other hand, scientists are investigating the possibilities of papain-like cysteine proteinases, which are found in many different types of life forms and have a wide range of enzymatic functions. These proteases are of relevance in the realm of cancer therapy since they are involved in tissue remodeling, virulence, and regulatory pathways (Ozhelvaci & Steczkiewicz, 2023).

27.3.2 Serine Proteases

Serine proteases comprise a group of enzymes with a serine residue in the catalytic site. Serine proteases promote angiogenesis by activating plasminogen, leading to extracellular matrix (ECM) breakdown and the growth of blood vessels that are newly formed. Some serine proteases directly influence cancer cell proliferation and survival, while others modulate the immune response in the tumor microenvironment. Serine proteases have tumor-suppressive functions, such as HtrA1, inducing apoptosis. Furthermore, their altered expression serves as potential biomarkers for cancer diagnosis and prognosis. Understanding the specific functions and mechanisms of serine proteases in cancer is crucial for developing targeted therapies and diagnostic tools to combat this complex disease (Chien, Ota, et al., 2009).

27.3.3 Aspartic Proteases

Aspartic proteases constitute a class of enzymes characterized by two aspartic acid residues at their active site. Notable examples of aspartic proteases include cathepsin D, cathepsin E, BACE1, and BACE2. In cancer biology, particularly cathepsin D and E, aspartic proteases play intricate roles. Cathepsin D, associated with tumor angiogenesis, promotes the formation of new blood vessels, supporting tumor growth. Cathepsin E, studied in gastric cancer, is linked to cancer progression and is involved in tumor invasion, angiogenesis, progression, and apoptosis resistance. The specific functions of these proteases are based on cancer type and the tumor microenvironment, potentially contributing to therapy resistance. Ongoing research aims to unravel their complex mechanisms, offering potential targets for novel cancer therapies. Researchers are developing specific inhibitors and investigating their use in drug-delivery systems to enhance targeted cancer treatments. However, these approaches are still in the experimental stage, requiring further research and clinical validation for their effectiveness and safety. As diagnostic tools, aspartic proteases like cathepsin D and cathepsin E serve as potential biomarkers for cancer diagnosis and prognosis due to their elevated levels in specific tissues and fluids. Scientists are developing assays to measure these proteases, providing insights into disease progression. Additionally, ongoing research explores the use of aspartic proteases in imaging techniques, showing promise for non-invasive cancer detection. Nevertheless, these applications are still under development, necessitating further validation for clinical use (Liaudet-Coopman et al., 2006).

27.3.4 Threonine Proteases

Proteolytic enzymes with a threonine amino acid at the active site, such as TSP50, have structural similarities to serine proteases. While threonine proteases have been extensively studied in various biological processes, their direct involvement in tumor development is gaining attention. Current research suggests potential roles of certain threonine proteases in tumorigenesis, influencing tumor growth and invasion through the regulation of signaling pathways or protein degradation events, although these roles are not as well-characterized as other proteases (Seemüller et al., 1995). TSP50, linked to cell proliferation and tumorigenicity, holds promise in targeted cancer therapy. Strategies like inhibition, gene silencing, combination approaches, and biomarker development are explored for therapeutic potential. In diagnostics, threonine proteases, including TSP50, offer insights into biomarker discovery, disease monitoring, targeted imaging, personalized medicine, and companion diagnostics, enhancing precision and effectiveness in therapeutic interventions (Li et al., 2012).

27.3.5 Matrix Metalloproteases (MMPs)

Matrix metalloproteinases (MMPs) form a class of zinc-dependent extracellular proteases, including collagenases, gelatinases, stromelysins, matrilysins, membrane-type MMPs (MT-MMPs), and others. In the context of cancer, both tumors and stromal cells excessively express MMPs, contributing to matrix breakdown and tumor progression. Upregulation of different MMPs in connective tissues at the onset of tumor growth aggravates the penetration of epithelial tumor cells. Specific MMPs, such as MMP-2, play crucial roles in neo-blood vessel formation, invasion, and dissemination, while MMP-9 exhibits increased proteolytic activity near breast cancer and the vascular system. MMP-13 and MMP-14 are associated with various cancers, including breast, neck, melanoma, lung, and colorectal, and MMP-7 connects directly with the epithelial component of carcinomas (Zhang et al., 2008; Adley et al., 2009). In cancer diagnostics, imaging techniques targeting MMPs offer more precise and effective strategies for diagnosis and management.

27.4 PROTEASE INHIBITORS IN CANCER TREATMENT

Proteases play a crucial role in tumor progression and are considered potential therapeutic targets for cancer treatment due to their involvement in various signaling networks. Developing medications targeting emerging protease targets has been challenging yet fascinating, holding promise for ameliorating malignancies (Horiguchi et al., 2007; Karikari et al., 2007; Choi et al., 2012). Proteasome inhibitors (PSIs), studied for more than a decade, have shown significant success in treating hematological carcinomas (López-Otín & Matrisian, 2007). Various strategies, including designing small molecule inhibitors, constructing generic forms of proteases, and gene therapy, are explored for targeting proteases. However, the complexity of multiple proteases involved at different stages poses challenges in designing universal inhibiting agents. Bortezomib, a protease inhibitor, initially used for relapsed multiple myeloma and mantle cell lymphoma, faced challenges due to low efficacy and economic constraints. In prostate cancer (PC), lentiviral-mediated neprilysin gene transfer showed promise in preventing cancer progression. Another approach involves blocking protease inhibitors to reduce the effectiveness of protective proteases, known as an indirect strategy. In pancreatic cancer, inhibiting caspase inhibitors triggered and amplified the cell death process. A fascinating strategy involves incorporating a biomarker for tumor detection or prognosis by measuring the quantity of protective proteases, offering reliable clinical prognostic predictions. In cases without protective proteases, alternative therapeutic strategies may be needed (Horiguchi et al., 2007; Karikari et al., 2007; Choi et al., 2012; López-Otín & Matrisian, 2007).

27.4.1 Aspartic Proteases Inhibition

To catalyze the dissociation of peptide substrate, aspartic proteases utilize two highly preserved aspartic acid residues in their site of activity. On the substrate, two aspartate residues begin their mechanism of action by regulating a (H_2O) molecule at the intermediate stage (Figure 27.2). These inhibitors formed a strong link within the aspartic protease active site, causing their (OH) group to make a hydrogen bond with any of the aspartic residues, imitating the state of intermediary of the reaction and preventing the protease from releasing the result. The hexapeptide pepstatin is one potent aspartic protease inhibitor containing the uncommon amino acid statin. Due to its unique structure, it can reach the aspartic protease active sites and attach to block them. The protein data bank (PDB) database contains the crystal structures of several aspartic protease inhibitors, including renin, chymosin, pepstatin, γ-secretase, β-secretase, endothelin converting enzyme (ECE), and napsin A (Diment et al., 1989). The progression of cancer has been chiefly linked to CatD and CatE. While CatE expression has been shown to contribute to increased tumor aggressiveness in patients with ovarian or endometrial malignancies, CatD expression is associated with breast cancer. Moreover, CatE can initiate precursor protein activation inside specific cell prelysosomal compartments. Given that human epithelial breast cancer cells express and release CatD at increased levels, it is considered a sign of an unfavorable prognosis in breast cancer. Its presence triggers several processes, including metastasis, fibroblast expansion, angiogenesis, and the proliferation of cancer cells (Ohri et al., 2008; Masson et al., 2011). Procathepsin D (pCD), secreted by malignant cells, can operate as a growth-promoting agent for stromal and cancerous cells. This stimulation strengthens their propensity to spread to other tissues and invade them. Interestingly, napsin A is involved in producing surfactant protein B in lung cancer. Furthermore, it prevents HEK-293 cell tumor formation, and this action is independent of its enzymatic activity (Ueno et al., 2008; Vetvicka & Fusek, 2012).

27.4.2 Cysteine Proteases Inhibition

The catalytic moiety shared by cysteine and thiol proteases is an electron-rich cysteine thiol in a catalytic trio or duo. The catalytic pathway of cysteine proteases comprises the breakdown of peptide linkage, which is accomplished by reducing the proton of a thiol group found in the active

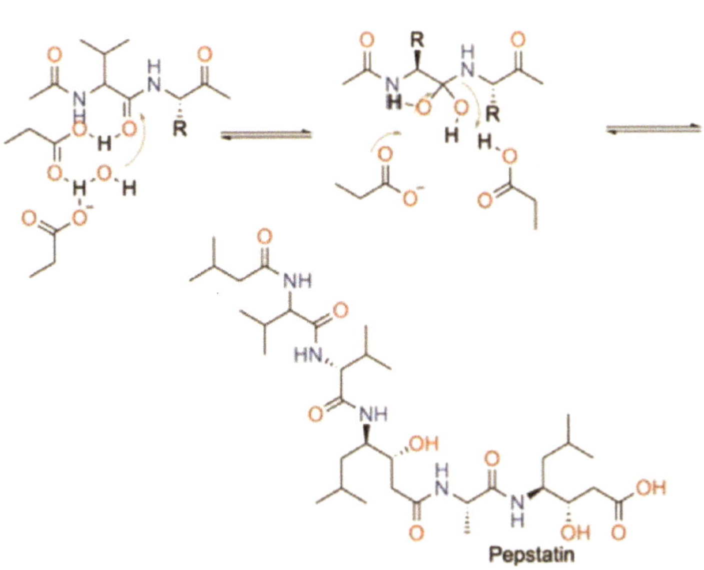

FIGURE 27.2 An aspartic protease method of action as an inhibitor.

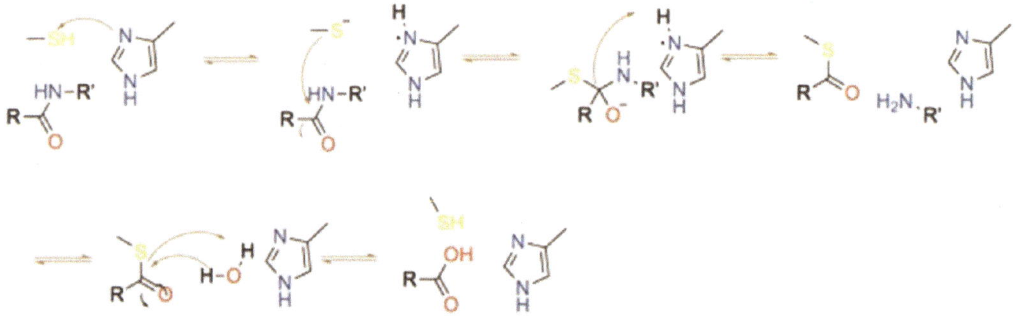

FIGURE 27.3 Cysteine protease mode of function during the inhibition process.

region of the enzyme with a nearby residue bearing an essential side chain, typically a histidine. Deprotonated cysteine could begin a nucleophilic assault on its substrate when a thiol group undergoes deprotonation (Figure 27.3). The ability of cysteine protease inhibitors to attach themselves tightly to the active region of the enzyme, where a ketone group is present, enables the inhibitors to prevent the enzyme from performing its activity (DeClerck et al., 2004). Several malignancies have an overexpression of a class of lysosomal enzymes called cysteine cathepsins. When cancer spreads, these cathepsins are released into the surrounding tissue or migrate to the surface of tumor cells, where they become essential for promoting tumor invasion via many pathways. First, they cause the ECM and basement membrane to degrade directly, allowing tumor cells to move away from the main tumor mass. Second, cathepsins start a proteolytic cascade on the cell surface that releases additional enzymes such as urokinase plasminogen activator (uPA) and MMPs, which in turn promote invasion. Finally, disruption of adherens junctions can promote the migration and invasion of cancer cells by cleaving the cell adhesion protein E-cadherin on the cell surface (Joyce et al., 2004). Elevated cathepsins B and L levels are commonly observed in various tumor types. Furthermore, the significant presence of cathepsin K in osteoclasts has highlighted this enzyme as a critical focus for developing new drugs to treat bone tumors. Specifically, cathepsins B and L are linked to tumor invasion and considered crucial metastasis agents (Mason & Joyce, 2011).

27.4.3 MMP Inhibition

MMPs belong to the protease family, requiring calcium and zinc for their enzymatic activity. The Zn^{2+} atom in the active site of MMP is essential for catalysis. This Zn^{2+} is in the zinc-binding region and attached to three histidine residues. An extremely maintained glutamate acid that is a part of the site of action may be found in this zinc-binding motif. For MMPs, a variety of methods of action are being proposed, one of which is the Manzetti mechanism. According to a report (Figure 27.4), the histidine in the zinc-binding domain may be able to shift the ionic zinc from its original state in order to attach to the substrate, resulting in the development of a temporary condition that causes the substrate's peptide linkage to be broken (Bode et al., 1993; Baumann, 1994). MMPs may process a wide range of bioactive compounds and degrade different proteins found in the ECM. Their activities involve the cleavage of cell surface receptors, the release of apoptotic ligands, and the inactivation of chemokines. Proliferation, migration, differentiation, angiogenesis, apoptosis, and tissue remodeling associated with many physiological or pathological states, including metastasis, and cancer, are all dependent on MMPs. In particular, MMP-2 and MMP-9 are essential to the metastatic process (Van Lint & Libert, 2007).

27.4.4 Serine Proteases Inhibition

Serine is an electron-rich species in the catalytic triad of serine proteases, composed of the three exceptionally stable residues: His, Ser, and Asp. To coordinate the assault of the peptide bond, it

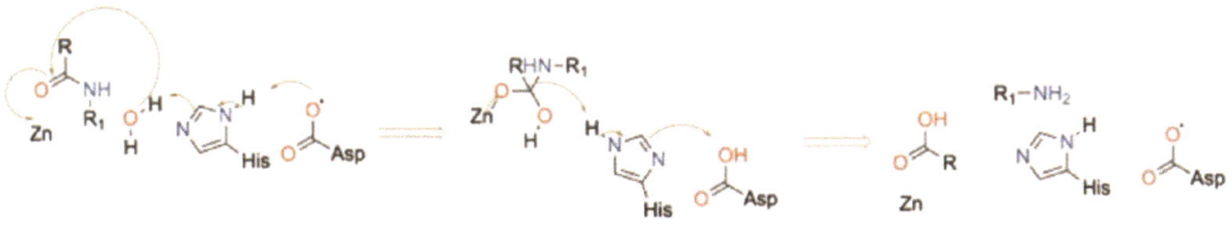

FIGURE 27.4 Manzetti method approach of matrix metalloproteinases activity.

adheres to the carbon of (C=O) of the peptide linkage, and histidine receives the hydrogen bond generated by the serine-OH unit. The C-terminal of the peptide is then expelled after Asp and His form a hydrogen bond (Figure 27.5). Serpins, which resemble proteins in structure, function as inhibitors of serine proteases and prevent serine protease. They cause a significant conformational shift in serine proteases upon binding, interrupting the active site (Frederick Woessner, 1999). The most prominent family of proteases in humans, serine proteases are essential to many physiological and developmental processes in the biological system. Protease-related signaling pathways are strictly regulated, and any alteration in proteolytic activity results in the ECM's breakdown, which is a major contributor to cancer development. Cancer-related proteases are intimately linked to proteases such as pepsin, matriptase-2, TMPRSS4 (Type II transmembrane serine proteases), and secretory serine proteases such as urokinase, plasminogen activator (uPA), kallikreins, and HtrA. They suppress apoptotic pathways in many forms of cancer, upregulating adhesion molecules such as the integrin family, activating MMPs, upsetting the uPA plasminogen system, and starting intracellular signaling cascades. Cell invasion, metastasis, and proliferation are the results of these processes. However, serine protease activity is delicately balanced by serpins, which are naturally occurring inhibitors of serine proteases and aid in maintaining homeostasis. Through its ability to stop cancer cells from invasively spreading, this control eventually controls the formation of tumors (Poddar et al., 2017).

27.4.5 Threonine Proteases Inhibition

The binding region residues of threonine proteases share catalytic characteristics with cysteine and serine proteases. Threonine proteases function similarly by creating a covalent intermediate with their substrate, followed by hydrolysis with water to release the free enzyme (Figure 27.6). Bortezomib, a threonine protease inhibitor, contains pyrazinoic acid,

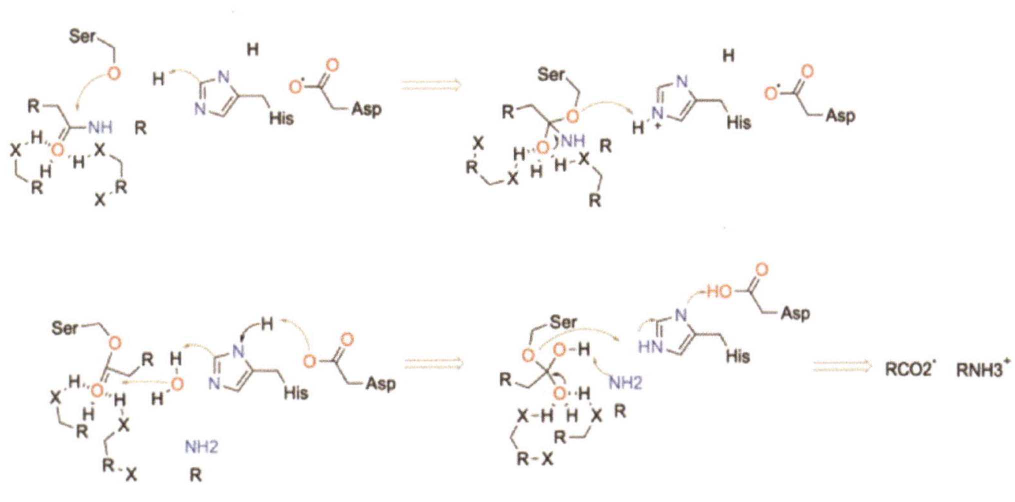

FIGURE 27.5 Mode of action of serine proteases.

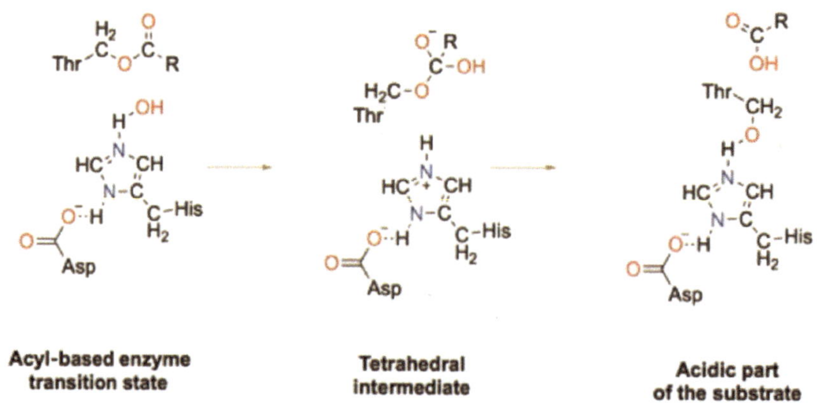

FIGURE 27.6 Threonine protease mode of action.

phenylalanine, and leucine, with a boronic acid replacing a carboxylic acid. The covalent connection between bortezomib's boron atom and the active region of threonine proteases forms a tetrahedral boronate, inhibiting the enzyme due to structural and functional similarities with threonine protease substrates (Brannigan et al., 1995; Hedstrom, 2002). An active site is present at the N-terminal threonine residue of the proteasome, a unique type of threonine protease found in the MB1 subunits of its b-rings. Pharmacophores attached to short peptides have the potential to target this region. PSI selectively blocks the 20S proteasome's ability to behave like chymotrypsin. According to recent research, PSI can stop the growth of osteolytic lesions and lessen tumor accumulation in mice injected with murine 5TGM1 plasmacytoma cells. Dipeptide boronic acid (BA) analogs are reversible, robust, and selective PSIs compared to peptide aldehyde agents. By suppressing the proteasome's chymotryptic activity, BA drugs reduce threonine protease activity and lessen the degradation of cell cycle regulating proteins. Treatment with BA causes apoptosis and inhibits the development of cancer cells (Adams et al., 1999; Meng et al., 1999; Voorhees et al., 2003).

27.5 EMERGING PROTEASES FOR MOLECULAR IMAGING AND THERAPEUTIC APPLICATIONS FOR CANCER

An imaging approach called molecular imaging has been applied to characterize and monitor individual biological molecules as well as pharmacological and cell function processes in distressed cells. Molecular imaging has revolutionized the field of oncology by providing non-invasive and real-time insights into the underlying molecular and cellular processes driving cancer development and progression. Among the various molecular targets explored for cancer imaging, proteases have garnered significant attention due to their dynamic and context-dependent roles in tumorigenesis. The distinct classes of emerging proteases involved in tumor growth are caspases, cathepsins, the high-temperature requirement A (HTRA) family, MT-MMPs, and proteasomes.

27.5.1 Caspases

Apoptosis, or planned cell death, is mediated by an assortment of cysteine proteases called caspases. Dysregulation of cell death functioning is a hallmark of cancer, where the balance between cell survival and death is disrupted, contributing to tumorigenesis and cancer progression. Caspases are classified into initiating caspases, such as (caspase-8 and caspase-9) that become operational in response to apoptotic stimuli and effector caspases (e.g., caspase-3, caspase-7) that carry out the cell death process by cleaving specific substrates within the cell. In cancer, aberrant activation or inhibition of caspases can lead to the evasion of apoptosis, allowing cancerous cells to survive and resist therapeutic interventions. Understanding the contribution of caspases in cancer biology is essential for developing targeted therapies to restore apoptotic signaling and promote cancer cell death (Olsson & Zhivotovsky, 2011).

27.5.1.1 Imaging Approaches for Monitoring Caspase Activity in Tumors

Imaging approaches that can visualize and monitor caspase activity in real time provide valuable tools for understanding apoptotic processes in cancer. Several imaging strategies have been employed to assess caspase activity *in vivo*. One of the most commonly used approaches is the design of activatable imaging probes, also known as "smart probes." These probes are typically based on fluorescence or bioluminescence technologies and are quenched in their inactive state. Upon caspase activation during apoptosis, the probes are cleaved, leading to an increase in fluorescence or luminescence signal, enabling the detection of caspase activity within tumors. In addition to optical imaging, positron emission tomography (PET) and magnetic resonance imaging (MRI) have been explored as non-invasive modalities for imaging caspase activity, providing high-resolution imaging capabilities and potential clinical translation (Shi, 2002; Beroske et al., 2021).

27.5.1.2 Caspase-Based Cancer Therapy

A characteristic of cancer is the deregulation of apoptosis, and several kinds of methods have been investigated for utilizing caspases in cancer treatment. To selectively cause apoptosis in cancer cells, one strategy employs caspase-targeted treatments. For instance, pro-caspase gene therapy has been studied, in which pro-caspases are injected into cancer cells and later triggered to cause apoptosis. Additionally, it has been investigated to find out whether selective small-molecule caspase activators or inhibitors can prevent or enhance apoptosis in cancer cells to reduce adverse effects. These methods show commitment, but for successful clinical translation, issues with sensitivity, transportation, and resistance mechanisms must be resolved (Fulda & Debatin, 2013; Chen et al., 2021).

27.5.2 Cathepsins

Cathepsins are a genus of lysosomal proteases whose presence is essential to several biological activities, such as tissue remodeling and protein breakdown. The secreted cathepsins break down cell–cell attached molecules and the components of outside cells, which aids in the cancer cell's assault and proliferation, invasion, and metastasis of tumor cells. In cancer, cathepsins have emerged as key players in cancer pathogenesis due to their involvement in tumor invasion, metastasis, angiogenesis, and apoptosis resistance. Cathepsins are categorized according to their structure, catalytic processes, and substrate. Serine (cathepsin A and G), aspartic (cathepsin D and E), and lysosomal cysteine proteases (cathepsin B, C, F, H, K, L1, L2/V, O, S, W, X/Z) are the three primary cathepsin categories based on the human genome sequence. Cathepsin B (CTB), one of the 11 proteases in the cathepsin family, has drawn the most attention since it is elevated in a number of malignancies and has been linked to the growth and spread of cancer. As a result, CTB is now more widely acknowledged as a promising target in cancer treatment and tumor imaging. Most cancers have all been linked to overexpression of cathepsin B. Elevated cathepsin development and functionality correlate with tumor aggressiveness and unfavorable outcomes for patients. Cathepsin B contributes to tumor invasion by degradation of the ECM, which promotes cancer cell invasion and migration into surrounding tissues (Figure 27.7). Additionally, cathepsin B promotes angiogenesis by activating pro-angiogenic factors and evades apoptosis, enabling cancer cells to survive and proliferate. Given their multifaceted functions in cancer biology, cathepsins have become attractive targets for cancer research and therapeutic development (Roshy et al., 2003; Jedeszko & Sloane, 2004; Aggarwal & Sloane, 2014).

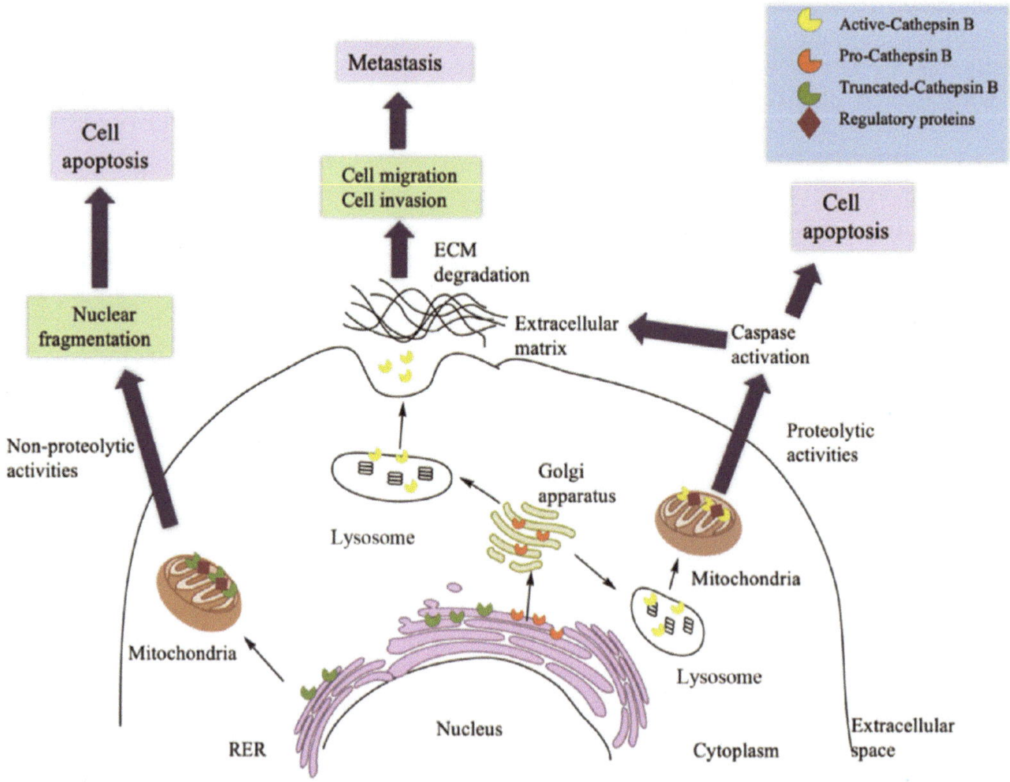

FIGURE 27.7 Role of cathepsin-B in metastasis and cell apoptosis.

27.5.2.1 Cathepsin-Based Imaging Probes for Cancer Visualization

Cathepsin-based imaging probes are designed to bind to active cathepsins in tumor cells selectively and are conjugated with imaging agents, such as fluorescent dyes, radionuclides for PET or single-photon emission computed tomography (SPECT), or contrast agents for MRI. Cathepsin B belongs to the papain family of cysteine proteases and is typically found in the lysosome. However, it can translocate and function as it breaks down ECM constituents. By exploiting the enhanced cathepsin B activity in cancer cells, imaging probes enable the non-invasive detection of tumors and metastases. Cathepsin-B targeted molecular imaging is promising for early cancer detection, accurate tumor staging, and treatment response monitoring. Additionally, these probes can provide valuable information on tumor heterogeneity and identify aggressive cancer subtypes, paving the way for personalized treatment strategies. Covalent bonds are formed between active cysteine cathepsins and the qABP probe BMV109, which has a Cy5 fluorescence for tumor identification. It has shown *in vivo* tumor labelling in mouse breast and colon cancer models when given intravenously and topically applied labelling of human breast, lung, liver, pancreas, and colon cancer tissues (Verdoes et al., 2013; Löser & Pietzsch, 2015).

27.5.2.2 Cathepsin-Targeted Therapeutics in Cancer Treatment

The dysregulation of cathepsins has spurred the development of cathepsin-targeted therapeutics for potential anticancer applications. Small molecule inhibitors selectively block cathepsin activity, disrupting processes like ECM degradation, angiogenesis, and apoptosis resistance, inhibiting tumor advancement and metastasis. Cathepsin-activated pro-drugs, cleaved by cathepsins within tumor cells, offer targeted drug delivery, enhancing therapeutic efficacy while minimizing adverse effects on healthy cells. Cathepsin-targeted immunotherapies, employing antibodies or immune checkpoint inhibitors, aim to boost the immune response against cathepsin-overexpressing cancer

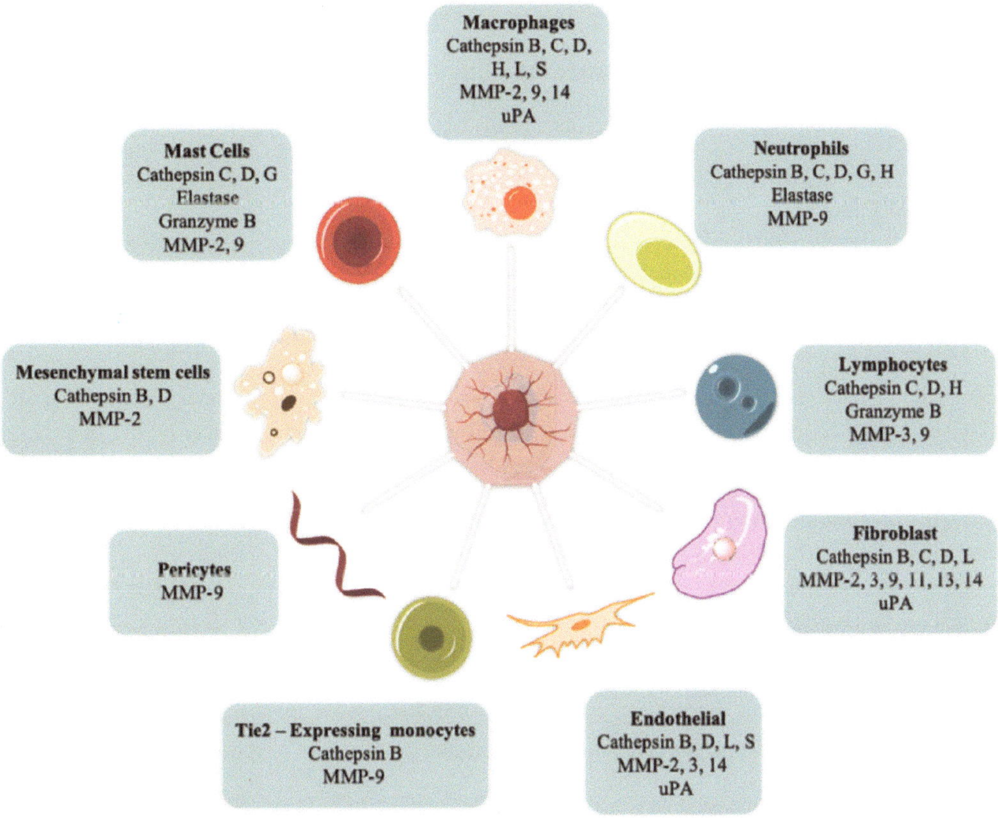

FIGURE 27.8 Proteolytic contribution from various stromal cells.

cells. Combining Taxol with cathepsin inhibitors significantly improves the efficacy of chemotherapy against primary and metastatic tumors, as cathepsin B protection against cell death generated by chemotherapeutic drugs is overcome. Cathepsin B is necessary for a cellular procedure known as lysosomal membrane permeabilization (LMP). Apoptosis and necrosis of cells are caused by LMP, which can be produced by several substances, including tumor necrosis factor (TNF) and chemotherapeutic drugs. TNF is unable to produce LMP in the presence of cathepsin B-deficient hepatocytes. WEHI-S murine fibrosarcoma cells can be protected against apoptosis brought on by TNF or TNF-related apoptosis-inducing ligand (TRAIL) by pharmacological and genetic suppression of cathepsin B. According to these results, cathepsin B is necessary for TNF-induced LMP. The formation of multinucleated cells, an early sign of apoptosis brought on by microtubule-disturbing drugs, can also be avoided in non-small lung cancer cells by limiting cathepsin B activity (Shree et al., 2011; Petersen et al., 2013; Pogorzelska et al., 2018). The involvement of Various stromal cell's contribution in proteolysis is depicted in Figure 27.8

27.5.3 The HTRA Proteases

The HTRA set of proteases is a group of serine proteases that have a variety of roles in physiological functions, which comprise tissue homeostasis, response to stress, and protein quality assurance (Table 27.1). Three serine proteases exist in the HtrA family: HtrA1, HtrA2, HtrA3. Emerging outcomes suggest that HTRA proteases are also involved in cancer progression. HTRA proteases can promote tumorigenesis by influencing critical cell functioning, such as the division of cells, death, and EMT. Dysregulation of HTRA proteases has been linked with cancer aggressiveness,

TABLE 27.1
Significance of HTRA Proteases in Cancer Biology

HTRA Member	Role in Cancer	Reference
HtrA1	To stop cancerous cells from migrating, HTRA degrades XIAP and activates caspase activity	Skorko-Glonek et al., 2013
HtrA2	c-Myc and JunB are increased by WTP1 degradation, which encourages apoptosis	Chien et al. 2009; Chien, 2010
HtrA3	Engaging with TCP1α chaperonin to perform functions of a chaperone	Hartkamp et al., 2010
	XIAP dissociation to support drug-induced cellular death	Wenta et al., 2018

metastasis, and resistance to therapy in several cancer types. The intricate participation of HTRA proteases in cancer biology presents them as intriguing targets for cancer research and therapeutic interventions (Foghsgaard et al., 2001).

27.5.3.1 Molecular Imaging Strategies to Assess HTRA Activity in Tumors

Molecular imaging approaches have been developed to assess HTRA protease activity in tumors, providing valuable insights into their functional roles in cancer. Various imaging modalities, including PET, MRI, and optical imaging, have been explored to visualize and quantify HTRA activity non-invasively. Smart imaging probes, designed to be cleaved by HTRA proteases, allow the specific detection of HTRA activity within tumors. These imaging strategies offer the potential to identify tumors with elevated HTRA protease activity, correlate their presence with disease progression, and stratify patients for personalized treatment approaches based on the level of HTRA protease activity.

27.5.3.2 Therapeutic Implications of Targeting HTRA Proteases in Cancer

The dysregulation of HTRA proteases in cancer opens up promising avenues for therapeutic targeting. Modulating the activity of HTRA proteases can control tumor growth, inhibit metastasis, and sensitize cancer cells to therapy (Wenta et al., 2019). Small molecule inhibitors and peptide-based compounds targeting HTRA proteases are under investigation as potential anticancer agents. Moreover, developing therapeutics targeting HTRA proteases may improve treatment outcomes by modulating tumor microenvironment components, such as ECM remodeling and immune responses. HTRA protease inhibitors may also sensitize cancer cells to existing therapies, rendering them more susceptible to conventional treatments (Chien et al., 2009).

27.5.4 MT-MMPs

MT-MMPs are a subgroup crucial in cancer progression and metastasis due to their ability to localize on cell membranes (Wenta et al., 2019). Unlike other MMPs, MT-MMPs act in the pericellular environment, breaking down the ECM and remodeling the tumor microenvironment. This facilitates tumor invasion, migration, and angiogenesis, making MT-MMPs key players in cancer pathogenesis. Recognized as potential therapeutic targets, they hold promise for cancer diagnosis, prognosis, and therapy. Targeting MT-MMPs through imaging offers significant potential to advance cancer management and improve patient outcomes by understanding their roles in metastasis (Zucker et al., 2003). MT-MMPs serve as an impactful protease play in tumor progression and metastasis, representing key regulators of cancer progression. As membrane-anchored proteases, they localize to the cell surface and facilitate the pericellular breakdown of ECM. This activity is critical for cancer cells to invade surrounding tissues and infiltrate blood vessels and lymphatics, leading to metastatic dissemination to distant organs. MT-MMPs also contribute to tumor angiogenesis by processing and activating pro-angiogenic factors, promoting the production of newly developed blood vessels that sustain tumor growth. Furthermore, they make it easier for cancer cells to spread throughout the body and leave blood arteries to colonize secondary areas, completing the metastatic cascade. The dysregulation of MT-MMP expression and activity is associated with aggressive tumor behavior and poor prognosis in various cancer types. Understanding the precise roles of MT-MMPs in tumor invasion and metastasis provides valuable insights into cancer pathogenesis and presents potential opportunities for targeted therapies aimed at impeding metastatic spread and improving patient outcomes (Zucker et al., 2003).

27.5.4.1 MT-MMP-Targeted Imaging Probes for Cancer Detection

MT-MMP targeted imaging probes have emerged as promising tools for cancer detection and diagnosis. Research studies utilizing mutant mice as a model have shown that the development of tumors was slowed by inhibiting MMPs. Broad-spectrum hydroxamate-based inhibitor, Batimastat that had been tested on humans, was the first inhibitor. Nevertheless, Batimastat did not exhibit an effective response. Individuals with glioblastoma multiforme who received marimastat with temozolomide had a higher rate of survival. A study by Min et al. (2015) developed a radiolabeled MT1-MMP-targeted peptide probe, which demonstrated high specificity and sensitivity for detecting MT1-MMP expression in breast cancer xenograft models using PET imaging. One of the most precise and suitable ways to measure the efficiency of proteases involves implementing optical imaging, particularly with fluorescence and activatable substrates. Initial identification of cancer alongside various pathological disorders will benefit considerably from the accurate detection of MMP-9 enzyme. MT-MMP-targeted imaging probes hold great potential for early cancer detection, monitoring treatment response, and guiding therapeutic interventions, ultimately improving patient outcomes (Rasmussen & McCann, 1997; Krüger et al., 2001).

27.5.4.2 Therapeutic Strategies Exploiting MT-MMPs in Cancer Treatment

Given the crucial role of MT-MMPs in tumor progression, therapeutic strategies that exploit these proteases have been explored as potential anticancer interventions. One method indulges the advancement of small molecule inhibitors that specifically target MT-MMP activity. These inhibitors aim to suppress the proteolytic activity of MT-MMPs, inhibiting tumor invasion and metastasis. Another strategy involves the use of antibody-based therapies targeting MT-MMPs on the surface of cancer cells, leading to immune-mediated destruction of cancer cells expressing MT-MMPs. Additionally, nanotechnology-based approaches have been explored to deliver therapeutic agents specifically to tumor cells with elevated MT-MMP expression. The selective targeting of MT-MMPs in cancer cells may enhance therapeutic efficacy while minimizing off-target effects on normal tissues. These therapeutic strategies hold promise for the development of novel and effective treatments to combat cancer by exploiting the pivotal role of MT-MMPs in tumor invasion and metastasis (Hotary et al., 2000; Gialeli et al., 2011).

27.5.5 Urokinase-type Plasminogen Activator or Urokinase

Urokinase (uPA) is a crucial serine protease that activates plasminogen, playing a significant role in the metastatic cascade. Discovered in urine in 1947, uPA comprises 411 amino-acid residues. Various proteinases activate pro-uPA by breaking the K158-I159 link. The fully active two-chain uPA converts plasminogen into plasmin, promoting protein breakdown and fibrinolysis. uPA and its receptor (uPAR) are involved in angiogenesis, cell adherence, migration, invasion, survival, tumor growth, and metastasis. Elevated uPA and uPAR levels are associated with poor clinical outcomes. The uPA–uPAR interaction activates plasmin, facilitating tumor invasion and spread. uPAR, triggered by uPA, controls cancer cell activities via interaction with cell membrane receptors. These findings propose uPA and uPAR as potential diagnostic, prognostic markers, and targets for anticancer treatments (Koshikawa et al., 2000; Choong & Nadesapillai, 2003).

27.5.5.1 Urokinase-Targeted Therapy in Cancer Treatment

The urokinase (uPA) receptor system, involved in extracellular proteolysis, is influenced by interactions between stromal and tumor cells. A serine protease found in various biological contexts, uPA plays a pivotal role in activating plasminogen, leading to a cascade of proteolytic reactions associated with cancer development and vascular disorders (MacFarlane & Pilling, 1947).

Plasmin is capable of causing uPA to become active. By doing this, a cycle is established, allowing plasmin and uPA to activate one another. Alternately, cathepsin B may stimulate it. Alongside plasmin, cathepsin B and MMPs, particularly MMP-2, are additional immediate or subsequent targets of uPA. Through a process that depends on responses from stromal fibroblasts, uPA is implicated in the activation of MMP-2 (Skrzydlewska et al., 2005; Tang & Han, 2013). Cancer cells rupture the inner membranes as a result of the activation of stromal MMP-2 and uPA, which are first produced as inert substances and then stimulated on the outermost layer of tumor cells (MacFarlane & Pilling, 1947). Furthermore, the enzymes promote the growth of arterial blood vessels to supply the expanding tumor. This discovery emphasizes the intricacy of proteolytic relationships in the tumor microenvironment, which rely on contributions from a variety of stromal cells in addition to the proteases supplied by tumor cells. Conventional anticancer therapies tackle a genetically stable cell population, which means they might not address the resistance brought on by the administration of regular chemotherapeutic agents. The stromal contribution to metastases, migration, and the development of arteries are their targets (MacFarlane & Pilling, 1947).

27.5.6 A Disintegrin and Metalloproteinase with Thrombospondin Motifs (ADAMTSs)

The clan of proteases recognized as ADAMTSs are involved in a number of biological events such as tissue formation, inflammation, and remodeling of ECM. They are part of a family of 19 human enzymes that share certain structural similarities with other MMPs (He et al., 2007). ADAMTS proteases have been examined in the context of cancer for their possible involvement in both cancer treatment and imaging. These proteases influence the development of cancer through the disintegration of the ECM, the control of angiogenic factors, and metastatic cell migration. Their diverse expression in various cancer types suggests both oncogenic and tumor-suppressive functions, highlighting their vital role in cancer biology and potential as therapeutic targets or diagnostic markers.

27.5.6.1 ADAMTSs in Cancer Therapy

The domain organization of ADAMTSs is complicated, and it underwent autoproteolytic fragmentation to produce N- and C-terminal segments that ought to partially include thrombospondin motifs. However, the evidence is not conclusive, some studies have demonstrated that ADAMTS-1 suppresses angiogenesis, slows the growth of tumors, and prevents the creation of metastases. Additional research has revealed that ADAMTS-1's proteolytic state affects its ability to fight tumors because its dissociated fragments have antimetastatic characteristics. The non-angiogenic peptides released by thrombospondins and ADAMTS-1's potential to cloister vascular endothelial growth factor could also have an active part in the protease's propensity to suppress the expansion of endothelial cells in cancer. The secretory protease ADAMTS-8, which has antiangiogenic capabilities, is markedly lowered in cancers of various origins, and this suppression is frequently accompanied by promoted hypermethylation. In the same way, ADAMTS-9 is a putative tumor-suppressor factor since it is minimized or deleted via processes of epigenetic in esophageal malignant tumors. A massive genetic investigation that suggested ADAMTS-15 and ADAMTS-18 belong to the genes that are typically altered in a variety of breast and colorectal malignancies provided surprising support for the necessity of ADAMTSs as tumor-suppressor enzymes. Additionally, several carcinomas typically exhibit ADAMTS-18 genetically disabled phenotypes (Dunn et al., 2004; Dunn et al., 2006., Sjoblom et al., 2006; Lo et al., 2007).

27.6 EMPHASIS ON MOLECULAR IMAGING TECHNIQUES FOR PROTEASE DETECTION

Many imaging techniques have been developed that can detect a relatively significant number of malignancies and reduced levels of cancerous biological indicators, which are connected to them on a molecular level. These methods are now more important than ever for identifying cancer biomarkers as well as identifying individuals who will benefit from a specific treatment and gauging the effectiveness of that treatment. The molecular imaging approaches such as optical imaging, MRI, SPECT, and PET have all been implemented for protease imaging investigation.

27.6.1 Optical Imaging

Optical imaging is more affordable and accessible than various types of imaging like MRI and PET. Fluorescence and bioluminescence imaging are the two optical imaging methods that are most frequently utilized. Using contrast agents, optical imaging is a molecular imaging technique for non-invasively recognizing and analyzing a variety of biological events occurring within living systems. To achieve this, fluorescent moieties must gather at a certain target spot. The fundamental idea is based on the capacity of tissues and cells to cause quantifiable changes in the light coming from the probe, especially in the visible and near infrared (NIR) spectrum. To accomplish preliminary cancer identification, assessment, choice of individualized molecular-based treatments, and therapeutic response, a wide range of molecular probes have been researched and developed. The methods of imaging have become more important for identifying biological pathways indulged in cancer progression and for assessing response to specific therapies. Synthesized activatable probes are commonly used for functional imaging of specific protease activity, employing protease-specific peptide linkers connecting chromophores. These probes exhibit minimal fluorescence due to self-quenching or Forster resonance energy transfer (FRET). Specifically designed for MMPs, which play a crucial role in tumor invasion, spread, and neovascularization, these probes offer on/off fluorescent signals for precise imaging. For instance, a groundbreaking probe for MMP-2 visualization introduced a peptide coupler, dissociated only by MMP-2. Using a dendrimer-based fluorescent precursor known as the proteolytic beacon, MMP-7 behavior was precisely imaged with selective sensitivity. NIR and FRET pairings were developed to enhance protease beacon sensitivity, allowing detection of tumors as small as 0.11 cm. Another innovative probe for MMP-9 imaging utilized a cell-penetrating peptide degraded by MMP-9 or -2, enabling non-receptor-mediated delivery of imaging or therapeutic moieties into cells. These activatable probes offer insights into disease progression and individual response to clinical therapeutics. (Jiang et al., 2004; Yamamoto et al., 2012). Activity-based probes (ABPs) utilizing small compounds for specific target enzymes are crucial for evaluating enzyme quantity, expression, and timing during tumor formation. These probes, particularly targeting cysteine cathepsins, offer insights into cellular processes and disease indicators, including breast adenocarcinoma (Greenbaum et al., 2002; Saghatelian & Cravatt, 2005). An near-infrared fluorescence (NIRF) probe (CY-CDF-PGC) for cathepsin D (Cath-D) in lysosomes, designed for low fluorescence during body transit and intense fluorescence at tumor sites, demonstrated innovative imaging (Saghatelian et al., 2004). Cath-E-specific probes, using fluorogenic peptide precursors, exhibit clear and excellent imaging of Cath-E activity, demonstrating quenching behavior until encountering tumors with high Cath-E levels (Tang & Han, 2013). Optical imaging proves beneficial for caspase imaging, overcoming the challenge of caspase inhibitors blocking cathepsins, with active site-specific probes being selective within the caspase family (Cruz-Monserrate et al., 2012). The uPA system, effective in optical imaging, holds significant therapeutic potential for image-influenced treatment. Targeted uPAR ligands during imaging-guided surgery offer promise in distinguishing malignant areas and reducing postoperative cancer recurrence (Edgington et al., 2009).

27.6.2 PET and SPECT

A crucial molecular imaging technique for quantitative evaluation of a radiolabeled tracer moiety within a biological system is PET. In contrast, SPECT similarly relies on the radio labelled principle but uses the longer-lasting gamma rays. Radioactivity can be administered by injecting, ingesting, or inhaling very small amounts of radio-labeled tracer. Numerous matrix metalloproteinase inhibitors (MMPIs) have been designed with the intent of reducing the proliferation of primary and secondary tumors as well as preventing the spread of cancer. MMPs have been a prominent target for the development of numerous anticancer medicines. The emergence of molecules that are better able to recognize individual MMPs with characteristic active sites and the composition of an activity-based probe for *in vivo* catalysis of MMPs are both made possible by the incorporation of radio-labeled tracer and MMPIs (Persson & Kjaer, 2013). The development of [^{18}F] SAV03M, a pro-drug and carboxylic acid-based MMP-2 inhibitor, has the potential for higher uptake by tumor tissues or organs as compared to normal tissues. HO-[^{123}I]I- and HO-[^{125}I]I-CGS 27023A, as well as other novel radio-iodinated metalloprotease inhibitors created by modification of CGS 27023A, have been documented to be employed for *in vivo* imaging by SPECT. Due to their strong affinity for MMPs as well as their fast clearance from plasma, these iodinated substrates have emerged as important tools for MMP imaging (Brown, 1999). Additionally, *in vivo* imaging involves the utilization of endogenous tissue inhibitors metalloprotease (TIMPs) as a probe for MMP inhibition. It was possible to create a substrate for SPECT imaging by incorporating TIMP-2 with diethylene triamine penta-acetic acid (DTPA) and labelling it with ^{111}In (^{111}In-DTPA-N-TIMP-2). The membrane type 1 (MTI)-MMP, which is abundantly expressed in a variety of malignancies and is important in controlling cell mobility and invasion, displayed a high affinity for ^{111}In-DTPA-N-TIMP-2 and was therefore shown to be useful for *in vivo* imaging of MMP's catalytic action (Furumoto et al., 2003).

SPECT imaging has an enormous amount of therapeutic value considering it can find invasive malignancies and small micrometastatic lesions. The affordability of gamma cameras and the simplicity of radiochemistry, particularly when incorporating ^{99m}Tc, all contribute to the high allure of SPECT imaging of uPAR. Since PET imaging has a high degree of sensitivity and provides data that is genuinely quantitative, it might be regarded as the ideal imaging technique for uPAR. Additionally, the PET field has been the site of the majority of imaging research. More sophisticated investigations on the interaction of uPAR and uPA using radiograph crystallography and surface plasmon resonance have facilitated the establishment of a series of short peptides relevant to non-invasive *in vivo* molecular imaging of uPAR by PET (Seiki, 1999; Edgington et al., 2009). A monomer form of AE120, designated AE105, was conjugated with the metal chelator DOTA at the N-terminus and then labelled with ^{64}Cu in the initial investigation on uPAR-PET imaging (Ploug, 2012; Li et al., 2008). Even though uPAR imaging by PET was shown to be quite successful, ligands for uPAR imaging by SPECT were explored as well. Even though uPA and PAI-1 are thought to be predictive biomarkers, there is still a lack of non-invasive methods to evaluate uPA activity. In response, a SPECT-based *in vivo* uPA imaging probe labelled with (^{111}In) was developed. As [^{111}In] MICA-401 (Figure 27.9), DOTA was radiolabeled with ^{111}In after being conjugated to an uPA inhibitor. As early as approximately 18 hours after the injection, good tumor visualization was made possible using SPECT imaging, and the rim of the tumor showed pronounced radio-tracer uptake. Ninety-five hours after the injection, the ideal tumor contrast was seen. There was also a discernible absorption in the lymph and skin nodes (Ploug et al., 2001).

The most common disease among men is prostate cancer. UPAR has been regarded as a good imaging target for prostate cancer. Researchers have therefore thoroughly studied various radiolabeled-dependent uPAR targeted probes using peptides, antibodies, proteins, and nanoparticles, as non-invasive uPAR mapping in prostate cancer with the potential to distinguish indolent tumors from invasive and aggressive types is anticipated to be a significant diagnostic and prognostic imaging biomarker (see Table 27.2) (Boonstra et al., 2011; Yang et al., 2011; Vangestel et al., 2016).

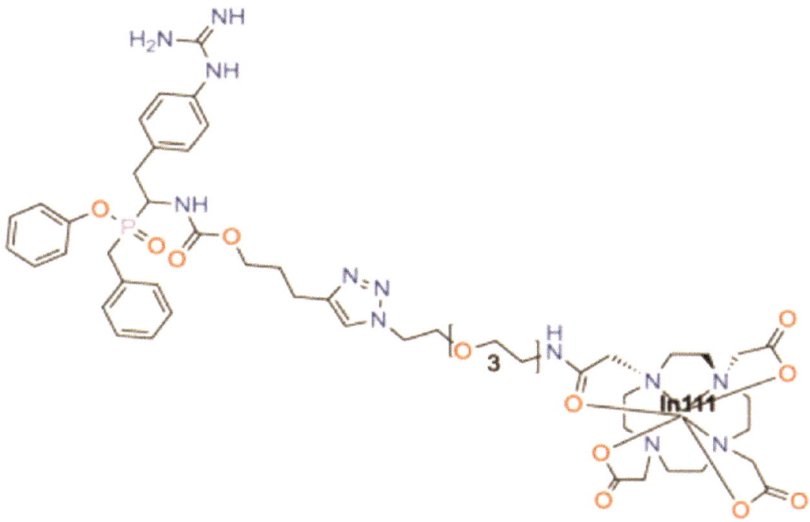

FIGURE 27.9 Chemical structure of [^{111}In] MICA-401. (From Ploug et al., 2001.)

It was discovered that the liver showed a surprisingly high uptake of the first tracer, ^{64}Cu-DOTA-AE105. Two more tracers, ^{64}Cu-CB-TE2A-AE105 and ^{64}Cu-CB-TE2A-PA-AE105, have been evaluated to address this issue. It was discovered that ^{64}Cu-CB-TE2A-PA-AE105 demonstrated a decent tumor–liver ratio through *in vitro* and *in vivo* testing. To lessen the uptake of the biomarker by the non-invaded tissues, ^{68}Ga was employed in place of Cu (Edgington et al., 2009; Yang et al., 2011). The first gallium-68 labelled radiotracers used to image uPAR using a PET scan were ^{68}Ga-DOTA-AE105-NH2, (Figure 27.10) and ^{68}Ga-NOGDA-AE105-NH2, (Figure 27.11) (Edgington et al., 2009).

The uPAR PET ligand ^{18}F-AIF-NOTA-AE105 (Figure 27.12) that has been ^{18}F-labeled was used in the outcome trials. In mice with PC-3 tumors, this ligand was administered intravenously, and PET scans were carried out 0.5, 1.0, and 2.0 hours later. The reconstructed PET images made it simple to recognize tumor lesions, and the ROI evaluation showed that the tumor had a significant uptake (Skovgaard et al., 2017).

27.6.3 MRI

MRI stands out as the widely adopted diagnostic imaging technique for cancer. Utilized globally in medical facilities, it relies on contrast agents, with gadolinium (Gd) being the most commonly used. Gadolinium, a paramagnetic ion with seven unpaired electrons, enhances relaxation and lowers T1

TABLE 27.2
Various PET Imaging Probes Based on the Peptide AE105

^{18}F	^{68}Cu	^{64}Cu
F-AIF-NOTA-AE105	Ga-DOTA-AE105	Cu-DOTA-AE105
F-FB-AE105	Ga-NODAGA-AE105	Cu-CB-TE2A-AE105
F-Click-AE105	Ga-NOTA-AE105	Cu-NOTA-AE105

Source: Yang et al., 2011.

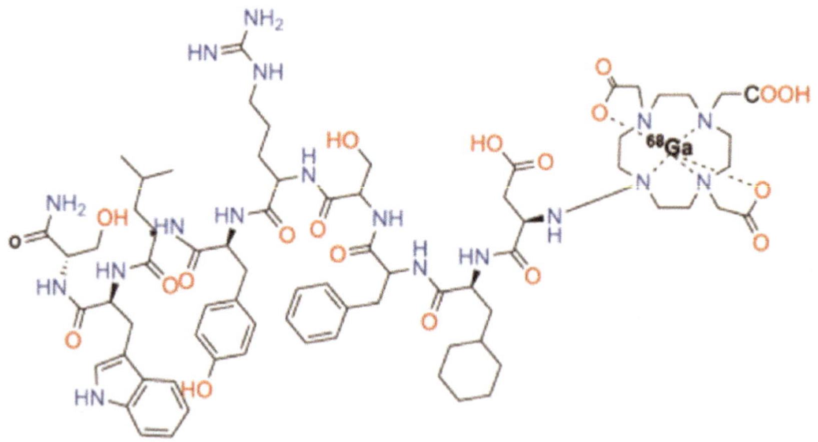

FIGURE 27.10 Chemical structure of ^{68}Ga-DOTA-AE105-NH2.

under magnetic fields. Contrast agents like Gd-DTPA and Gd-DOTA efficiently penetrate tumors through diffusion and undergo renal clearance. Recent advancements involve molecular substrates targeting specific receptors in tumor tissues, enhancing flexibility to allow water entry into gadolinium, ultimately boosting signal intensity for improved cancer cell identification.

According to studies, a substrate was created whose primary function was to impede water access. However, when beta-galactosidase cleaved it, the substrate became unlocked. Recent research has suggested the use of agents that, in an acidic microenvironment, exhibit reduced relaxation as a result of the acidification of the tumor microenvironment (Louie et al., 2000). These studies also proposed that MRI contrast agents can be break through the proteases of the outside domain of a cell, allowing for the assessment and quantification of the excessively expressed proteases linked to either intrinsic or disseminated cancers. They created a novel contrast agent that, when broken down by MMP-7,

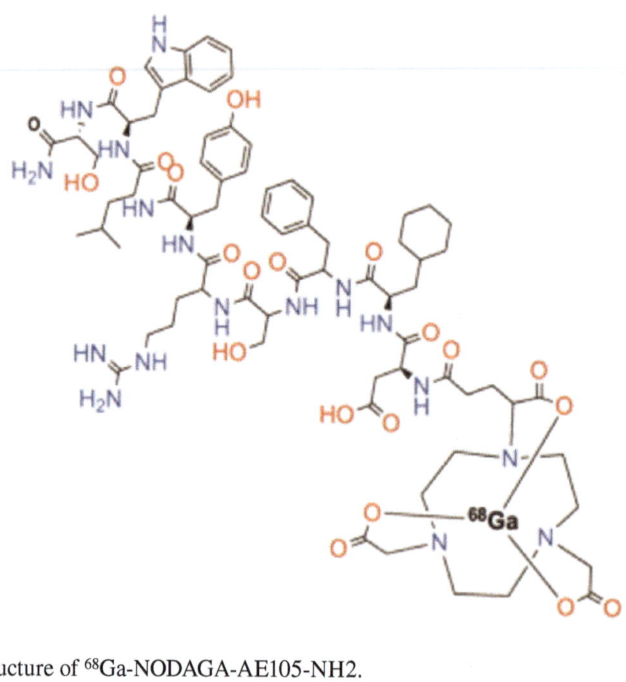

FIGURE 27.11 Structure of ^{68}Ga-NODAGA-AE105-NH2.

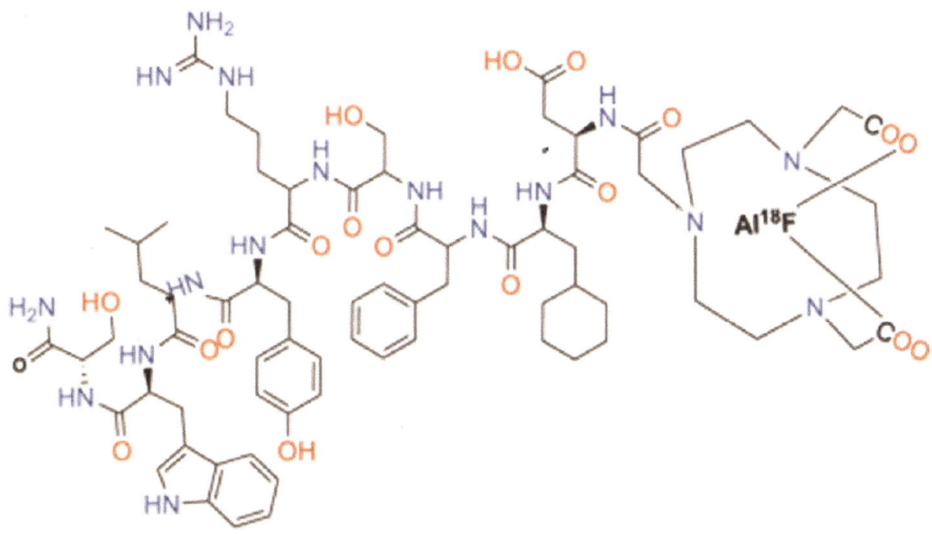

FIGURE 27.12 Chemical structure of ^{18}F-AlF-NOTA-AE105. (From Skovgaard et al., 2017.)

shows lower solubility. By converting solubility from hydrophilic to hydrophobic, these cutting-edge protease-regulated contrast agents (PCAs) evaluate the *in vivo* protease activity by MRI. One of the innovative agents for detecting MMP-7 activity in biological systems is PCA-7. In tumor-bearing animals treated with MMPIs, the PCA-7 switch revealed MMP-7 activity minimization. According to a study, PCA-2 is a parallel agent that may be utilized to identify MMP-2 activity in xenograft tumors that are MMP-2 positive using MRI. While estimated activity was decreased in tumor xenografts that had received treatment, the interaction of PCA-2 with MMP-2 increases MRI contrast (Frullano et al., 2006). Cath-B and Cath-L are the two cathepsins that are most closely associated with the degradation of proteins in lysosomes. The non-invasive imaging techniques for *in vivo* protease activity evaluation have undergone much research. *In vivo* cathepsin (Cath-B and Cath-L) activity was first detected using MRI, and the first MRI probe, poly-l-glutamate (PLG), which is extensively expressed in tumors, was produced. The mechanism behind such imaging is chemical exchange saturation transfer (CEST) between the protons of bulk water and the nitrogen-linked protons of the amine moiety of PLG. Proton transfer from PLG's amide bonds is extremely slow under typical physiological circumstances. The findings of the study demonstrated that Cath-B is favorably overexpressed in brain tumors and that Cath-B levels are necessary for PLG-regulated CEST contrast. Thus, the PLG-based MRI probe evaluates the *in vivo* changes in CEST contrast and is effective in quantifying the expression of cathepsin in various tumor types (Blasi & Sidenius, 2010). Phenomenal anatomical resolution yet exceptionally low sensitivity has both been observed by MRI (Haris et al., 2014). Greater resolution aids in the accurate detection of invasive malignancy. Due to restricted sensitivity, a larger concentration of ligand is typically required in MRI to provide a strong signal. This presents a problem since toxicity could develop at such high doses, in contrast to SPECT and PET where less ligand is needed. Paramagnetic iron oxide nanoparticles are widely used in uPAR-targeted ligands for MRI. mATF-Cy5.5-IO and hATF-Cy5.5-IO, two uPAR targeting MRI ligands, were used in the few research (Yang et al., 2009; Abdalla et al., 2011; James & Gambhir, 2012).

27.7 CONCLUSION AND FUTURE PERSPECTIVES

This work underscores the intricate role of proteases in cancer. Our understanding of this complex topic has evolved as we delve into the functional aspects of proteolytic systems in cancer. Tumor-associated proteases have shifted from being seen as generic pro-metastatic enzymes to being recognized as crucial proteins in the early stages of cancer, revealing multiple functions as antitumor agents. Various proteases are found in tumors and surrounding tissues, meeting the high demand for oxygen and nutrients in the tumor area. Identifying these emerging proteases in the early stages of tumor formation may aid in optimizing cancer treatment. Several studies have demonstrated the potential significance of molecular imaging in identifying different proteases. This non-invasive imaging approach allows *in vivo* evaluation, quantification, and visualization of specific proteases, enhancing image contrast with target-specific molecular probes. Imaging proteases offers advantages, including not requiring deep penetration into intracellular environments, as proteases primarily emerge from extracellular compartments. Furthermore, the biological activity and catalytic function of proteases enhance signal detection. Protease imaging uses fluorescence, bioluminescence, SPECT, PET, MRI, and other imaging modalities. The use of MMP imaging is still being investigated despite the fact that a variety of MMP-targeting probes are available. Bioluminescence may not be appropriate for *in vitro* biological cell research, considering its excellent sensitivity for tracking MMP activity *in vivo.* The widely used imaging technique, MRI, offers good anatomical resolution but low sensitivity, requiring larger probe doses that could be hazardous. In MRI and SPECT imaging, contrast agents like gadolinium and technetium are frequently employed. Radioactive imaging techniques are designed to concentrate at specific action sites, using molecular probes focusing on enzyme activity rather than enzyme levels. This chapter provides valuable insights into emerging proteases and the array of imaging methods employed in studying cancer proteases, contributing to advancements in cancer research and diagnostics.

DECLARATIONS

Conflict of interest: The authors declare no conflicts of interest.

Consent for publishing: All authors have approved the final version of the manuscript and given consent for publication.

Funding: Not Applicable.

REFERENCES

Abdalla, M. O., Karna, P., Sajja, H. K., Mao, H., Yates, C., Turner, T., et al., 2011. Enhanced noscapine delivery using uPAR-targeted optical-MR imaging trackable nanoparticles for prostate cancer therapy. J. Control. Release 149, 314–322.

Adams, J., Palombella, V. J., Sausville, E. A., Johnson, J., Destree, A., Lazarus, D. D., Maas, J., Pien, C. S., Prakash, S., & Elliott, P. J., 1999. Proteasome inhibitors: a novel class of potent and effective antitumor agents. Cancer Res. 59(11), 2615–2622.

Adley, B. P., Gleason, K. J., Yang, X. J., & Stack, M. S., 2009. Expression of membrane type 1 matrix metalloproteinase (MMP-14) in epithelial ovarian cancer: high level expression in clear cell carcinoma. Gynecol. Oncol. 112(2), 319–324.

Aggarwal, N., & Sloane, B. F., 2014. Cathepsin B: multiple roles in cancer. ProteomicsClin. Appl. 8(5-6), 427–437.

Baumann, U., 1994. Crystal structure of the 50 kDa metallo protease from Serratia marcescens. J. Mol. Biol. 242, 244–251.

Beroske, L., Van den Wyngaert, T., Stroobants, S., Van der Veken, P., & Elvas, F., 2021. Molecular imaging of apoptosis: the case of caspase-3 radiotracers. Int. J. Mol. Sci. 22(8), 3948.

Blasi, F., & Sidenius, N., 2010. The urokinase receptor: focused cell surface proteolysis, cell adhesion and signaling. FEBS Lett. 584, 1923–1930.

Bode, W., Gomis-Rüth, F.-X., & Stöckler, W., 1993. Astacins, serralysins, snake venom and matrix Metalloproteinases exhibit identical zinc-binding environments (HEXXHXXGXXH and Met-turn) and topologies and should be grouped into a common family, the 'Metzincins. FEBS Lett. 331, 134–140.

Boonstra, C. M., Verspaget, W. H., Ganesh, F. S., Kubben, J. G. M., Vahrmeijer, A., van de Velde, J. H. C., et al., 2011. Clinical applications of the urokinase receptor (uPAR) for cancer patients. Curr. Pharm. Des. 17, 1890–1910.

Brannigan, J. A., Dodson, G., Duggleby, H. J., Moody, P. C., Smith, J. L., Tomchick, D. R., & Murzin, A. G., 1995. A protein catalytic framework with an N-terminal nucleophile is capable of self- activation. Nature 378, 416–419.

Brown, P. D., 1999. Clinical studies with matrix metalloproteinase inhibitors. APMIS 107, 174–180.

Chen, M., Chen, Z., Castillo, J. B., Cui, L., Zhou, K., Shen, B., … Rao, J., 2021. [18 F]-C-SNAT4: an improved caspase-3-sensitive nanoaggregation PET tracer for imaging of tumor responses to chemo-and immunotherapies. Eur. J. Nucl. Med. Mol. Imaging 48, 3386–3399.

Chien, J., 2010. WT1 as a substrate of HtrA2: a potential pathway for therapeutic targeting by HtrA proteases. Future Oncol. 6(8), 1233–1235.

Chien, J., Campioni, M., Shridhar, V., & Baldi, A., 2009. HtrA serine proteases as potential therapeutic targets in cancer. Current Cancer Drug Targets 9(4), 451–468.

Chien, J., Ota, T., Aletti, G., Shridhar, R., Boccellino, M., Quagliuolo, L., … Shridhar, V., 2009. Serine protease HtrA1 associates with microtubules and inhibits cell migration. Mol. Cell. Biol. 29(15), 4177–4187.

Choi, K. Y., Swierczewska, M., Lee, S., & Chen, X., 2012. Protease-activated drug development. Theranostics 2, 156–179.

Choong, P. F., & Nadesapillai, A. P., 2003. Urokinase plasminogen activator system: a multifunctional role in tumor progression and metastasis. Clin. Orthop. Relat. Res. 415, S46–S58.

Cruz-Monserrate, Z., Abd-Elgaliel, W. R., Grote, T., Deng, D., Ji, B., Arumugam, T., et al., 2012. Detection of pancreatic cancer tumors and precursor lesions by cathepsin E activity in mousemodels. Gut 61 (9), 1315–1322.

DeClerck, Y. A., Mercurio, A. M., Stack, M. S., Chapman, H. A., Zutter, M. M., Muschel, R. J., Raz, A., Matrisian, L. M., Sloane, B. F., Noel, A., Hendrix, M. J., Coussens, L., & Padarathsingh, M., 2004. Proteases, extracellular matrix, and cancer: a workshop of the path B study section. Am. J. Pathol. 164, 1131–1139.

Diment, S., Martin, K., & Stahl, P., 1989. Cleavage of parathyroid hormone in macrophage endosomes illustrates a novel pathway for intracellular processing of proteins. J. Biol. Chem. 264, 13403–13406.

Dunn, J. R. et al. 2006. Expression of ADAMTS-8, a secreted protease with antiangiogenic properties, is downregulated in brain tumors. Br. J. Cancer 94, 1186–1193.

Dunn, J. R. et al. 2004. METH-2 silencing and promoter hypermethylation in NSCLC. Br. J. Cancer 91, 1149–1154.

Edgington, L. E., Berger, A. B., Blum, G., Albrow, V. E., Paulick, M., Lineberry, N., et al., 2009. Noninvasive optical imaging of apoptosis by caspase-targeted activity-based probes. Nat. Med. 15(8), 967.

Foghsgaard, L., Wissing, D., Mauch, D., Lademann, U., Bastholm, L., Boes, M., … Jäättelä, M., 2001. Cathepsin B acts as a dominant execution protease in tumor cell apoptosis induced by tumor necrosis factor. J. Cell Biol. 153(5), 999–1010.

Frederick Woessner, J. Jr, 1999. Matrix metalloproteinase inhibition: from the Jurassic to the third millennium. Ann. N. Y. Acad. Sci. 878, 388–403.

Frullano, L., Tejerina, B., & Meade, T. J., 2006. Synthesis and characterization of a doxorubicin-Gd (III) contrast agent conjugate: a new approach toward prodrug-procontrast complexes. Inorg. Chem. 45, 8489–8491.

Fulda, S., & Debatin, K. M., 2013. Caspase activation in cancer therapy. In Madame Curie Bioscience Database [Internet]. Landes Bioscience.

Furumoto, S., Takashima, K., Kubota, K., Ido, T., Iwata, R., & Fukuda, H., 2003. Tumor detection using 18F-labeled matrix metalloproteinase-2 inhibitor. Nucl. Med. Biol. 30(2), 119–125.

Gialeli, C., Theocharis, A. D., & Karamanos, N. K., 2011. Roles of matrix metalloproteinases in cancer progression and their pharmacological targeting. FEBS J. 278(1), 16–27.

Greenbaum, D., Baruch, A., Hayrapetian, L., Darula, Z., Burlingame, A., Medzihradszky, K. F., et al., 2002. Chemical approaches for functionally probing the proteome. Mol. Cell. Proteomics 1, 60–68.

Haris, M., Singh, A., Mohammed, I., Ittyerah, R., Nath, K., Nanga, R. P. R., et al., 2014. *In vivo* magnetic resonance imaging of tumor protease activity. Sci. Rep. 4, 6081.

Hartkamp, J., Carpenter, B., & Roberts, S. G., 2010. The Wilms' tumor suppressor protein WT1 is processed by the serine protease HtrA2/Omi. Molecular Cell 37(2), 159–171.

He, Y., Liu, X. D., Chen, Z. Y., Zhu, J., Xiong, Y., Li, K., Dong, J. H., & Li, X., 2007. Interaction be- tween cancer cells and stromal fibroblasts is required for activation of the uPAR-uPA-MMP-2 cascade in pancreatic cancer metastasis. Clin. Cancer Res. 13, 3115–3124.

Hedstrom, L., 2002. Serine protease mechanism and specificity. Chem. Rev. 102, 4501–4524.

Horiguchi, A., Zheng, R., Goodman, O. B. Jr., Shen, R., Guan, H., Hersh, L. B., & Nanus, D. M., 2007. Lentiviral vector neutral endopeptidase gene transfer suppresses prostate cancer tumor growth. Cancer Gene Ther. 14, 583–589.

Hotary, K., Allen, E., Punturieri, A., Yana, I., & Weiss, S. J., 2000. Regulation of cell invasion and morphogenesis in a three-dimensional type I collagen matrix by membrane-type matrix metalloproteinases 1, 2, and 3. J. Cell Biol. 149(6), 1309–1323.

James, M. L., & Gambhir, S. S., 2012. A molecular imaging primer: modalities, imaging agents, and applications. Physiol. Rev. 92, 897–965.

Jedeszko, C., & Sloane, B. F., 2004. Cysteine cathepsins in human cancer. Biol. Chem. 385, 1017–1027.

Jiang, T., Olson, E. S., Nguyen, Q. T., Roy, M., Jennings, P. A., & Tsien, R. Y., 2004. Tumor imaging by means of proteolytic activation of cell-penetrating peptides. Proc. Natl. Acad. Sci. 101, 17867–17872.

Joo, W. D., Visintin, I., & Mor, G., 2013. Targeted cancer therapy–are the days of systemic chemotherapy numbered. Maturitas 76(4), 308–314.

Joyce, J. A., Baruch, A., Chehade, K., Morse, N. M., Giraudo, E., Tsai, F. Y., Greenbaum, D. C., Hager, J. H., Bogyo, M., & Hanahan, D., 2004. Cathepsin cysteine proteases are effectors of invasive growth and angiogenesis during multistage tumorigenesis. Cancer Cell 5, 443–453.

Karikari, C. A., Roy, I., Tryggestad, E., Feldmann, G., Pinilla, C., Welsh, K., Reed, J. C., Armour, E. P., Wong, J., Herman, J., Rakheja, D., & Maitra, A., 2007. Targeting the apoptotic machinery in pancreatic cancers using small-molecule antagonists of the X-linked inhibitor of apoptosis protein. Mol. Cancer Ther. 6, 957–966.

Koshikawa, N., Giannelli, G., Cirulli, V., Miyazaki, K., & Quaranta, V., 2000. Role of cell surface metalloprotease MT1-MMP in epithelial cell migration over laminin-5. J. Cell Biol. 148(3), 615–624.

Krüger, A., Soeltl, R., Sopov, I., Kopitz, C., Arlt, M., Magdolen, V., … Schmitt, M., 2001. Hydroxamate-type matrix metalloproteinase inhibitor batimastat promotes liver metastasis. Cancer Res. 61(4), 1272–1275.

Leloup, L., & Wells, A., 2011. Calpains as potential anti-cancer targets. Expert Opin. Ther. Targets 15(3), 309–323.

Liaudet-Coopman, E., Beaujouin, M., Derocq, D., Garcia, M., Glondu-Lassis, M., Laurent-Matha, V., … Vignon, F., 2006. Cathepsin D: newly discovered functions of a long-standing aspartic protease in cancer and apoptosis. Cancer Lett. 237(2), 167–179.

Li, Y. Y., Bao, Y. L., Song, Z. B., Sun, L. G., Wu, P., Zhang, Y., … Li, Y. X., 2012. The threonine protease activity of testes-specific protease 50 (TSP50) is essential for its function in cell proliferation. PLoS One 7(5), e35030.

Li, Z.-B., Niu, G., Wang, H., He, L., Yang, L., Ploug, M., et al., 2008. Imaging of urokinase-type plasminogen activator receptor expression using a 64Cu-labeled linear peptide antagonist by microPET. Clin. Cancer Res. 14, 4758–4766.

Lo, P. H. et al. 2007. Identification of a tumor suppressive critical region mapping to 3p14.2 in esophageal squamous cell carcinoma and studies of a candidate tumor suppressor gene, ADAMTS9. Oncogene 26, 148–157.

López-Otín, C., & Bond, J. S., 2008. Proteases: multifunctional enzymes in life and disease. J. Biol. Chem. 283(45), 30433–30437.

López-Otín, C., & Matrisian, L. M., 2007. Emerging roles of proteases in tumor suppression. Nat. Rev. Cancer 7(10), 800–808.

Löser, R., & Pietzsch, J., 2015. Cysteine cathepsins: their role in tumor progression and recent trends in the development of imaging probes. Front. Chem. 3, 37.

Louie, A. Y., Huber, M. M., Ahrens, E. T., Rothbacher, U., Moats, R., Jacobs, R. E., et al., 2000. *In vivo* visualization of gene expression using magnetic resonance imaging. Nat. Biotechnol. 18(3), 321–325.

MacFarlane, R., & Pilling, J., 1947. Fibrinolytic activity of normal urine. Nature 159(4049), 779.

Mason, S. D., & Joyce, J. A., 2011. Proteolytic networks in cancer. Trends Cell Biol. 21, 228–237.

Masson, O., Prébois, C., Derocq, D., Meulle, A., Dray, C., Daviaud, D., Quilliot, D., Valet, P., Muller, C., & Liaudet-Coopman, E., 2011. Cathepsin-D, a key protease in breast cancer, is up-regulated in obese mouse and human adipose tissue, and controls adipogenesis. PLoS One 6, e16452.

Meng, L., Mohan, R., Kwok, B. H., Elofsson, M., Sin, N., & Crews, C. M., 1999. Epoxomicin, a potent and selective proteasome inhibitor, exhibits *in vivo* antiinflammatory activity. Proc. Natl. Acad. Sci. 96(18), 10403–10408.

Min, K., Ji, B., Zhao, M., Ji, T., Chen, B., Fang, X., & Ma, Q., 2015. Development of a radiolabeled peptide-based probe targeting MT1-MMP for breast cancer detection. PLoS One 10(10), e0139471.

Nguyen, O. T., 2020. The global challenge of cancer. Nat. Cancer 1(1), 1–2.

Ohri, S. S., Vashishta, A., Proctor, M., Fusek, M., & Vetvicka, V., 2008. The propeptide of cathepsin D increases proliferation, invasion and metastasis of breast cancer cells. Int. J. Oncol. 32, 491–498.

Olson, O. C., & Joyce, J. A., 2015. Cysteine cathepsin proteases: regulators of cancer progression and therapeutic response. Nat. Rev. Cancer *15*(12), 712–729.

Olsson, M., & Zhivotovsky, B., 2011. Caspases and cancer. Cell Death Differ. 18(9), 1441–1449.
Ozhelvaci, F., & Steczkiewicz, K., 2023. Identification and classification of papain-like cysteine proteinases. J. Biol. Chem., 104801.
Persson, M., & Kjaer, A., 2013. Urokinase-type plasminogen activator receptor (uPAR) as a promising new imaging target: potential clinical applications. Clin. Physiol. Funct. Imaging 33, 329–337.
Philipps-Wiemann, P., 2018. Proteases—general aspects. In Nunes, C. S., Kumar, V. (eds.), Enzymes in Human and Animal Nutrition (pp. 257–266). Academic Press.
Ploug, M., 2012. Structure-driven design of radionuclide tracers for non-invasive imaging of uPAR: the tale of a synthetic peptide antagonist. Theranostics 3, 467–476.
Ploug, M., Østergaard, S., Gårdsvoll, H., Kovalski, K., Holst-Hansen, C., Holm, A., et al., 2001. Peptide-derived antagonists of the urokinase receptor. Affinity maturation by combinatorial chemistry, identification of functional epitopes, and inhibitory effect on cancer cell intravasation. Biochemistry 40, 12157–12168.
Poddar, N. K., Maurya, S. K., & Saxena, V., 2017. Role of serine proteases and inhibitors in cancer. In Chakraborti, S., Dhalla, N. (eds.), Proteases in Physiology and Pathology (pp. 257–287). Springer.
Pogorzelska, A., Żołnowska, B., & Bartoszewski, R., 2018. Cysteine cathepsins as a prospective target for anticancer therapies—current progress and prospects. Biochimie 151, 85–106.
Rakashanda, S., Rana, F., Rafiq, S., Masood, A., & Amin, S., 2012. Role of proteases in cancer: a review. Biotechnol. Mol. Biol. Rev. 7(4), 90–101.
Rasmussen, H. S., & McCann, P. P., 1997. Matrix metalloproteinase inhibition as a novel anticancer strategy: a review with special focus on batimastat and marimastat. Pharmacol. Therap. 75(1), 69–75.
Roshy, S., Sloane, B. F., & Moin, K., 2003. Pericellular cathepsin B and malignant progression. Cancer Metastasis Rev. 22, 271–286.
Saghatelian, A., & Cravatt, B. F., 2005. Assignment of protein function in the postgenomic era. Nat. Chem. Biol. 1, 130.
Saghatelian, A., Jessani, N., Joseph, A., Humphrey, M., & Cravatt, B. F., 2004. Activity-based probes for the proteomic profiling of metalloproteases. Proc. Natl. Acad. Sci. 101, 10000–10005.
Seemüller, E., Lupas, A., Stock, D., Löwe, J., Huber, R., & Baumeister, W., 1995. Proteasome from Thermoplasmaacidophilum: a threonine protease. Science 268(5210), 579–582.
Seiki, M., 1999. Membrane-type matrix metalloproteinases. Apmis 107, 137–143.
Shi, Y., 2002. Mechanisms of caspase activation and inhibition during apoptosis. Mol. Cell 9(3), 459–470.
Shree, T., Olson, O. C., Elie, B. T., Kester, J. C., Garfall, A. L., Simpson, K., … Joyce, J. A., 2011. Macrophages and cathepsin proteases blunt chemotherapeutic response in breast cancer. Genes Dev. 25(23), 2465–2479.
Sjoblom, T. et al. 2006. The consensus coding sequences of human breast and colorectal cancers. Science 314, 268–274.
Skorko-Glonek, J., Zurawa-Janicka, D., Koper, T., Jarzab, M., Figaj, D., Glaza, P., & Lipinska, B., 2013. HtrA protease family as therapeutic targets. Curr. Pharm. Des. 19(6), 977–1009.
Skovgaard, D., Persson, M., & Kjaer, A., 2017. Imaging of prostate cancer using urokinase-type plasminogen activator receptor PET. PET Clinics 12, 243–255.
Skrzydlewska, E., Sulkowska, M., Koda, M., & Sulkowski, S., 2005. Proteolytic-antiproteolytic balance and its regulation in carcinogenesis. World J. Gastroenterol. 11, 1251–1266.
Tang, L., & Han, X., 2013. The urokinase plasminogen activator system in breast cancer invasion and metastasis. Biomed. Pharmacother. 67, 179–182.
Ueno, T., Elmberger, G., Weaver, T. E., Toi, M., & Linder, S., 2008. The aspartic protease napsin A suppresses tumor growth independent of its catalytic activity. Lab. Investig. 88, 256.
Van Lint, P., & Libert, C., 2007. Chemokine and cytokine processing by matrix metalloproteinases and its effect on leukocyte migration and inflammation. J. Leukoc. Biol. 82, 1375–1381.
Vangestel, C., Thomae, D., Van Soom, J., Ides, J. L., Wyffels, P., Pauwels, S., et al., 2016. Preclinical evaluation of [111In] MICA-401, an activity-based probe for SPECT imaging of *in vivo* uPA activity. Contrast Media Mol. Imaging 11, 448–458.
Verdoes, M., Bender, K. O., Segal, E., van der Linden, W. A., Syed, S., Withana, N. P., et al. 2013. Improved quenched fluorescent probe for imaging of cysteine cathepsin activity. J. Am. Chem. Soc. 135:14726–30.
Vetvicka, V., & Fusek, M., 2012. Procathepsin D as a tumor marker, anti-cancer drug or screening agent. Anti-Cancer Agents Med. Chem. 12, 172–175.
Voorhees, P. M., Dees, E. C., O'Neil, B., & Orlowski, R. Z., 2003. The proteasome as a target for cancer therapy. Clin. Cancer Res. 9(17), 6316–6325.

Wenta, T., Rychlowski, M., Jurewicz, E., Jarzab, M., Zurawa-Janicka, D., Filipek, A., & Lipinska, B., 2019. The HtrA3 protease promotes drug-induced death of lung cancer cells by cleavage of the X-linked inhibitor of apoptosis protein (XIAP). FEBS J. 286(22), 4579–4596.

Wenta, T., Zurawa-Janicka, D., Rychlowski, M., Jarzab, M., Glaza, P., Lipinska, A., … Lipinska, B., 2018. HtrA3 is a cellular partner of cytoskeleton proteins and TCP1α chaperonin. J. Proteom. 177, 88–111.

Yamamoto, N., Bryce, N. S., Metzler-Nolte, N., & Hambley, T. W., 2012. Effects of enzymatic activation on the distribution of fluorescently tagged MMP-2 cleavable peptides in cancer cells andspheroids. Bioconjugate Chem. 23, 1110–1118.

Yang, Y., Adelstein, S. J., & Kassis, A. I., 2011. General approach to identifying potential targets for cancer imaging by integrated bioinformatics analysis of publicly available genomic profiles. Mol. Imaging 10, 123–134.

Yang, Y., Hong, H., Zhang, Y., & Cai, W., 2009. Molecular imaging of proteases in cancer. Cancer Growth Metastasis 2, 13–27.

Zhang, B., Cao, X., Liu, Y., Cao, W., Zhang, F., Zhang, S., … Hao, X., 2008. Tumor-derived matrix metalloproteinase-13 (MMP-13) correlates with poor prognosis of invasive breast cancer. BMC Cancer 8(1), 1–10.

Zucker, S., Pei, D., Cao, J., & Lopez-Otin, C., 2003. Membrane type-matrix metalloproteinases (MT-MMP). Curr. Topics in Dev. Bio. 54,1–74. https://doi.org/10.1016/s0070-2153(03)54004-2.

28 Novel Nano Technologies Associated with Car-T Cell Therapy in Breast Cancer

Avipsa Hazra and Kanthesh M. Basalingappa
School of Life Sciences, JSS Academy of Higher Education & Research, Mysuru, India

Ragini Khanna
School of Life Sciences, JSS Academy of Higher Education & Research, Mysuru, India

Parthiban R.
Malabar Cancer Centre, Thalassery, India

Gopenath T. S.
JSS Academy of Higher Education & Research, Mysuru, India

28.1 INTRODUCTION

Breast cancer was the most occurring cancer in 2020, surpassing lung cancer, with an estimated 2.3 million new cases, which accounts for about 11.7% of all the cancer cases (Mehrotra & Yadav, 2022). According to data published by Globocan in 2020, breast cancer was responsible for 13.5% of all the cases of cancer and 10.6% of all mortalities (Mehrotra & Yadav, 2022). The causes and risks of breast cancer include a variety of factors like changes in lifestyle, inherent or genetic risk factors, reproductive risk factors, and iatrogenic risk factors (Zhu et al., 2023). Breast cancer can be lobular cancer, ductal cancer, a rare form of breast cancer called Paget disease of the breast, phyllodes tumour, or angiosarcoma. The breast may look normal from the outside, thus regular self-check-ups as well as screening tests become important. In most cases, the tumour is a hard mass with irregular edges, without any pain. But it may be soft, round and tender or in some cases even painful (Koo et al., 2017). Other symptoms may involve dimpling of the breast, pain in the breast or nipple region, nipple retraction, discharge from the nipple other than breast milk, reddening and drying of the nipple, etc. (Koo et al., 2017).

28.2 IMMUNOTHERAPY AND THE ADVENT OF CAR-T CELL THERAPY IN BREAST CANCER

Multiple methods have been developed and have been used to diagnose, detect, and treat breast cancer. The methods usually circle around surgery, hormones, and chemical injections. A few such conventional methods are depicted in Figure 28.1.

DOI: 10.1201/9781003394693-32

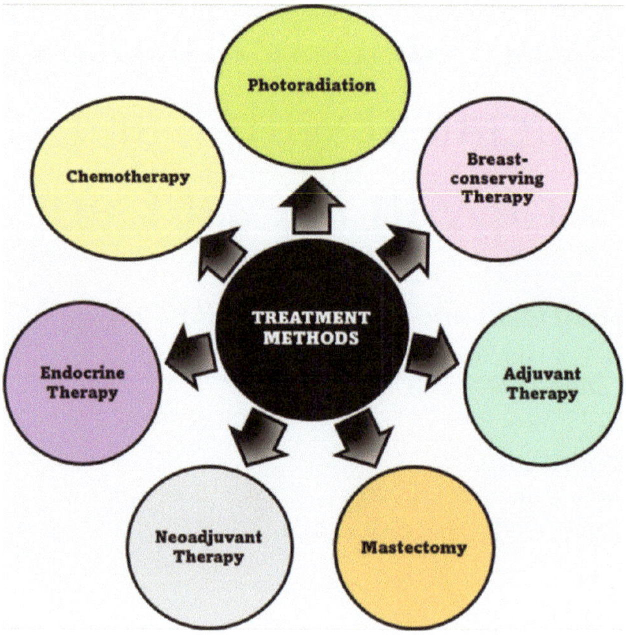

FIGURE 28.1 Conventional treatment methodologies.

Currently, immunotherapy seems to be the most effective form of therapy for any form of cancer. These include immune checkpoint inhibitors, T-cell transfer therapy, monoclonal antibodies, treatment vaccine, immune system modulators, etc. (Oiseth & Aziz, 2017). Immunotherapy mainly exploits the T cell capacity to identify cancer cells and restrict tumour growth as well as develop immunologic memory for faster response to fight against cancer and, thus, has the potential to be used as part of the treatment of breast cancer (Henriques et al., 2021). For this, a form of immunotherapy has been employed that uses engineered T cells, i.e. chimeric antigen receptor (CAR)-T cell therapy (Figure 28.2). In 1989, the Eshhar's group at the Weizmann Institute of Science in Israel developed the first chimeric receptor (Miliotou & Papadopoulou, 2018). CARs are receptors that have been engineered to direct the lymphocytes, mostly T lymphocytes, to recognise and target cancer cells expressing the specific target antigen on its surface (Sterner & Sterner, 2021).

The structure of CARs includes four constituents: (1) a target antigen-binding domain; (2) a hinge region; (3) a transmembrane domain; and (4) single or multiple intracellular signalling domains. The antigen-binding domain provides the target antigen specificity via its scFvs region, which binds to the specific antigen. The hinge region is the structural region that exists extracellularly that helps to hold out the binding units from the transmembrane domain. The transmembrane domain helps to harbour the CAR to the membrane of the T cell, and according to some studies, it also plays a role in the CAR T-cell function, expression level, stability as well as signalling. CAR-T cells are used as a form of therapy via a process called adoptive T-cell transfer (ACT). It involves stable and safe genetic modification of the patient's autologous T cell such that they express a CAR specific for the particular tumour antigen. This genetic modification may be brought about by viral-based gene transfer methods, DNA-based transposons, CRISPR/Cas9 technology, or direct transfer of *in vitro* transcribed-mRNA by electroporation, nanoparticles (NPs), and liposomes. The different important breast cancer target antigens for CAR-T cell therapy are discussed in Table 28.1 (Yang et al. 2022).

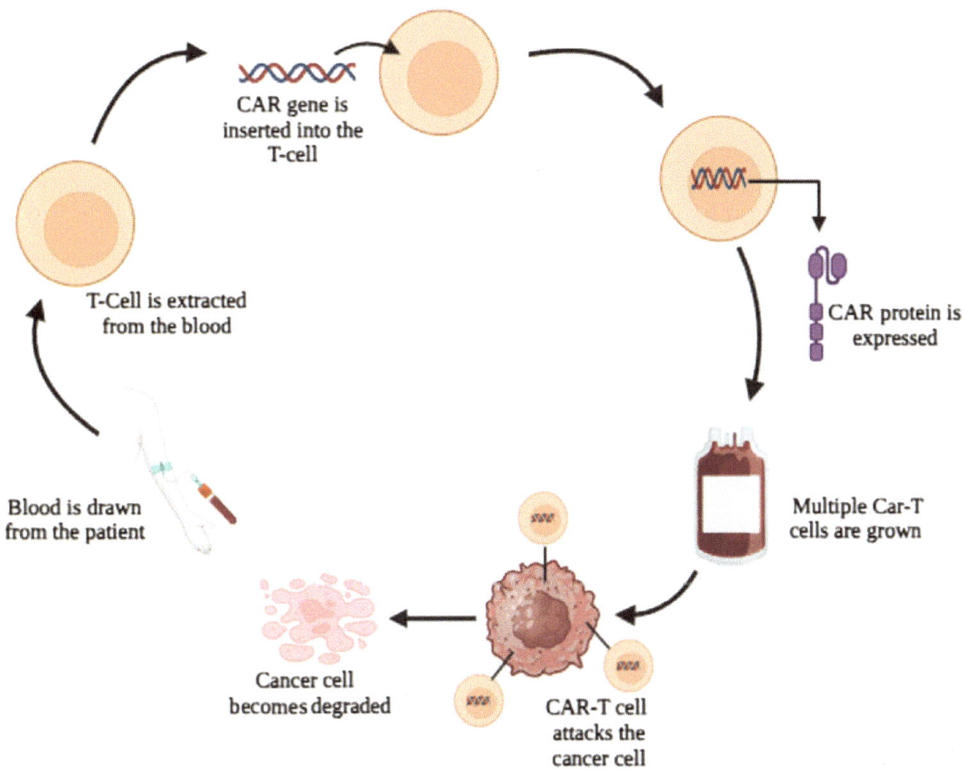

FIGURE 28.2 Process of CAR-T cells production in a laboratory.

TABLE 28.1
Antigens Presented on CAR-T Cells and Their Functions in Cancer Cells

Antigens Presented	Function
Receptor Tyrosine Kinase (RTK) family	Activate pathways that help in tumour development
HER 2 or ERBB 2	Increase expression of EMT genes
Epithelial growth factor receptor (EGFR)	Expressed in triple-negative breast cancer
Hepatocyte growth factor receptor (HGFR)	Its silencing leads to decreased cell proliferation and migration
Receptor tyrosine kinase-like orphan receptor (ROR) 1	Have cytolytic and cytokine secreting activity
AXL	Leads to cell survival, provides metastatic signals, and causes chemoresistance
Mucin (MUC) 1	Secretes mucus to act as chemical barriers against pathogens
Inverted cytokine receptor (ICR)	Regulates interleukins and triggers T-Cell Receptor signals to enable expansion of CAR-T cell at tumour site
Mesothelin (MSLN)	Activates intracellular signalling pathways for development and progression of cancer
CD70	Acts as a target for therapy
CD133	Acts as a biomarker of malignant precursors in case of solid tumours
CD44 containing variant exon v6	Cell surface glycoprotein for proliferation, motility, and survival of tumour
Epithelial cell adhesion molecule (EpCAM)	Blocking prevents tumour growth and metastasis
Chondroitin sulphate proteoglycan (CSPG) 4	Manages the tumour microenvironment to improve T-cell activities and acts as a target for limiting tumour progression

(Continued)

TABLE 28.1 *(Continued)*
Antigens Presented on CAR-T Cells and Their Functions in Cancer Cells

Antigens Presented	Function
Intercellular adhesion molecule (ICAM) 1	Molecular adhesion receptor that regulates cell adhesion, migration, and signal transduction
Tumour endothelial marker (TEM) 8	Associated with increased risk of tumour relapse
Trophoblast cell surface protein (TROP) 8	Oncogene associated with breast cancer for cell differentiation, spreading, invasion, and metastasis
Folate receptor (FR) alpha	Associated with the manufacture of peptides, methylation chemicals, and nucleotide bases that are involved in the emergence of cancer
Disialoganglioside (GD 2)	Promotes interaction of ECM with the tumour
Natural Killer group 2 member D (NKG2D)	Controls the generation of cytokines to regulate cytotoxicity and survival
Carcinoembryonic antigen (CEA)	Serum indicator in case of metastatic breast cancer

28.3 NANOTECHNOLOGY AND ITS COMBINATION WITH CAR-T CELL THERAPY IN BREAST CANCER

Nanotechnology can help in increasing drug concentration in target organs and reducing distribution in healthy tissues, extending drug retention at the site of tumours through improved permeability along with retention effects resembling conditions *in vivo* and other benefits, as shown in Figure 28.3 (Zeng et al., 2023).

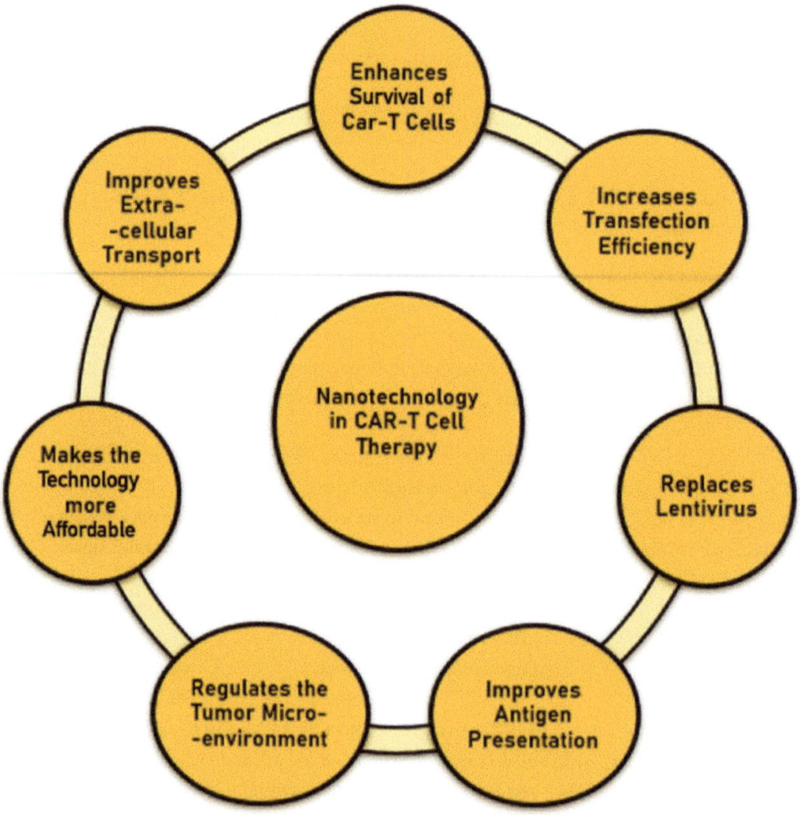

FIGURE 28.3 Advantages of combining nanotechnology with CAR-T cells.

The T-cell activation phase, which involves incubating cells with a viral vector encoding the CAR sequence, is the most important step in CAR-T cell construction. This is typically done using a retrovirus, which is usually an HIV-like lentivirus that penetrates the T-cells then implants its RNA, resulting in a T-cell that expresses the intended CAR receptor. However, because we are unsure of its possible consequences on humans, this stable insertional mutagenesis represents one of the riskiest treatments. The NPs are dependent on DNA mechanisms to successfully implant CAR genes into the nucleus T-cells, causing those cells to recognise leukaemia cells in particular (Smith, 2017). To prevent the capacity of any tumour to evade the human defence mechanism, the programmed death-ligand 1 expressed on the tumour cells should interact with its receptors on the T cells and prevent their effectors from showing their activity (Zou, 2005). By administering nanovaccines to lymph nodes that drain tumours, a similar improved toxic CD8+ T-cell reaction along with suppression of tumour growth have been seen (Jeanbart, 2014). It has also been shown that gold nanoparticles covered with cationic antigen polypeptide enhance *in vivo* antigen-specific CD8+ response (Zhang et al., 2015). Bispecific aptamers aid T lymphocytes in recognising and binding to tumour cells to combat the off-target effect brought on by tumour-associated antigens (Yang et al., 2020). Researchers frequently collaborate CAR-T treatment with the administration of cytokines and supplement it with antibodies to overcome the challenge of the hostile microenvironment. CAR-T cells, cytokines, and antibodies become dispersed, preventing their accumulation in tumour areas, which solves one of the issues (Beckman et al., 2007) (Chames et al., 2009). Another cutting-edge technique was created to successfully eliminate cancerous cells *in situ* by programming a large number of circulating T cells. Anti-CD3e f(ab′)2 segments are combined with the biodegradable poly-(beta-amino ester) (PBAE) complex is built inside the NPs with the help of nuclear localisation signals (NLS) and microtubule-associated sequences (MTAS), which promotes gene transfer in the T cells' nucleus. These easily produced and stored stable polymer NPs offer a convenient, affordable, and broadly accessible approach for CAR-T treatment (Smith et al., 2017). Utilising lipid nanoparticles can result in greater payloads and lower toxicities (Khawar et al., 2023). Polymer-NP hydrogels were developed by Grosskopf in 2022 to distribute CAR-T cells and stimulating cytokines simultaneously. This architecture promotes T-cell retention in tumour microenvironment (TME) over an extended period of time and continuous exposure, improving therapy (Khawar et al., 2023). In contrast, nanozymes exhibit regulation of immunosuppression in TME. In order to establish proof of increased treatment efficacy, (Zhu et al., 2021) modified a combinatorial technique by mixing nanozymes with CAR-T cells (Khawar et al., 2023). By employing nanozymes, they developed a unique nanotechnology strategy to enhance CAR effectiveness by disrupting TME. As a result, there was an increase in blood perfusion and CAR-T cell penetration into the TME. Nanozymes were used in combination in an efficient way to reverse the immunosuppression of the tumour environment and encourage CAR-T cell activation by Zhao and team (Zhao et al., 2021). They created a photothermal and catalytic HA@Cu2xS-PEG (PHCN) nanozyme that targets tumours (Mi et al., 2022). The tumour extracellular matrix (ECM) is harmed by the conversion of light energy into local heat following exposure to a near-infrared (NIR) laser. Additionally, by increasing the production levels of important cytokines including interferon, tumour necrosis factor along with tumour-specific antigens, the reactive oxygen species produced by the nanocatalysed tumour therapy activated the matching CAR T cells at the tumour site (Mi et al., 2022). CAR-macrophages have been created for intrinsic nanocomplex-mediated invasion into TME. It has been proposed that using the same *in vivo* method for CAR-T cells to enter TME will lead to the development of a more successful tumour effective method (Khawar et al., 2023). In addition to evoking strong anti-tumour activity and biosafety via immunological feedback, Liu and colleagues created interleukin (IL)-12 nanostimulant-engineered CAR-T cell (INS-CAR-T) hybrids that also enabled variable treatment effects (Liu et al., 2022). The INS-CAR T biohybrids exhibited increased expression of anti-tumour factors, effective recruitment, deep penetration, and selective proliferation (Liu et al., 2022). CAR-T cells enter the tumour site and cause the initial death. A positive anti-tumour cycle from CAR-T cells to T cells could be formed by secreting pro-inflammatory substances at a suitable moment to activate immune cells and endogenous T cells. CAR-T cells that give peptide antigen to cancer cells may be able to recognise and destroy cancer cells that have lost the CAR antigen (Liu et al., 2022). According to a different approach proposed by several studies, CAR-T cells that are designed to create extracellular vesicles that contain RN7SL1 by dispensing the pattern recognition receptor agonists can successfully promote anti-tumour immunity (Liu et al., 2022). T-cell homing NPs have the ability to rewire T cells *in vivo* without taking them out of the subject's body. Following administration, the NPs transmit T cells-only CAR encoding DNA. When it came to halting the progression of leukaemia in a pre-clinical mouse model, *in vivo* reprogrammed CAR-T cells were just as effective as *ex vivo*-produced conventional CAR-T cells. But there is a possibility of accidental gene transfer into non-target cells when CAR-delivery NPs are administered *in vivo* and, thus, further research is necessary before it can be applied to breast cancer (Schmidts & Maus, 2018). It has been observed to enhance and promote the conventional CAR-T therapy against solid tumours by combining photothermal therapy and adoptive transfer of CAR-T cells. The NIR dye indocyanine green (ICG) is encased in poly(lactic-co-glycolic acid) NPs. ICG acts as a photothermal agent that is liberated into solid tumours after being exposed to NIR light irradiation. The tumour's compact structure is disturbed by mild heat, which also lowers interstitial fluid pressure (IFP), enhances blood perfusion, and liberates tumour-specific antigens that may strongly excite CAR-T cells. About one-third of the animals receiving treatment had no tumour cells found after about 20 days, and tumour growth had been severely suppressed. Other researchers created CAR-T cells that were coupled to indocyanine green nanoparticles (INPs) via the biorthogonal process. The ECM was ablated, tumour tissues became loose, tumour vasculature grew, blood perfusion increased, and modest photothermal intervention boosted blood flow. INPs created specifically for CAR-T biohybrids accumulated and entered the tumour heavily in this way, remodelling the TME and resuming the immune reaction, which increased the effectiveness of CAR-T immunotherapy. To accomplish a synergistic photothermal and immunological treatment, CAR-T cells directed against the antigen chondroitin sulphate proteoglycan-4 (CSPG4) were intravenously administered. In this process, mild tumour heating can partially or completely destroy tumour cells or the ECM in addition to directly killing them. This reduces the density of solid tumours and high IFP, dilates tumour blood vessels, and increases CAR.CSPG4+ T infiltration and accumulation within the tumour (Li et al., 2021). In photothermal therapy, anti-HER2 antibody linked silica-gold nanoshells are particularly targeted against human breast carcinoma cells (*in vitro*) and breast carcinoma cells that are HER2+. This can be extended to CAR-T cell therapy as well (Zeng et al., 2023). Nie et al. (2019) linked magnetic nanoclusters (NCs) and the PD-1 antibody (aP) together to produce NC-Ap by using a benzoic-imine bond, which is sensitive to pH and also inverse electron demand Diels-Alder cycloaddition (Nie et al., 2019). Due to the expression of PD-1 on effector T cells, the created NC-aP binds to these cells. Through magnetisation, magnetic resonance imaging (MRI) directed T cells and aP to supplement solid tumours. The aP is hydrolysed after being freed from the benzoic-imine bond because of the acidic tumour microenvironment. As a result, with a few adverse effects, the adoptively transferred T cells and aP suppress the growth of solid tumours. Adenosine is present in many tumours, such as lung, prostate, uterine, and breast cancers (Nawaz et al., 2020). Adenosine is one of the immunosuppressive chemicals that blocks the immunological activity of CD4+ and CD8+ T lymphocytes. The A2a adenosine receptor (A2aR) was expressed on the surface of activated T cells and caused adenosine to build up extracellularly, suppressing T-cell multiplication and blocking IFN-γ production. Therefore, SCH-58261 (SCH), a small chemical molecule that inhibits A2aR, can be efficiently transported to CAR-T cells in tumours utilising nanotechnology. According to their report, Wang et al. combined CAR-T therapy with SCH-loaded cross-linked multilamellar liposomes (cMLV) that greatly slowed the tumour expansion and increased the survival rates of the groups that received the treatment, the rate at which T cells infiltrated the tumour, and the IFN-γ expression level *in vivo*. This technique improves CAR T-cell effectiveness in solid tumours by restoring tumour-resident T-cell hypofunction (Mi et al., 2022).

Immunosuppressive molecules like CTLA-4 and PD-L1 prevent immune cell activation and allow tumour cells to evade detection. Inhibitors that target these molecules and CAR-T treatment were employed to reset the suppressive solid tumour microenvironment. Nanoparticulate RNA vaccines are cutting-edge treatments that give cancer patients liposomal antigen-encoding RNA (RNA-LPX) such that T cells are activated. To attain the controlled multiplication of CAR-T cell development based on RNA-LPX dose, recently, researchers integrated CAR-T with the nanoparticulate RNA vaccination. The mechanism entails intravenous delivery of the antigen to the antigen-presenting cells that are present in the spleen, lymph nodes, and bone marrow, followed by the start of a type-I IFN-driven immune-stimulatory programme that is dependent on the toll-like receptor (Mi et al., 2022). Additionally, Chan et al. successfully overcame the issue of restricted antigen presentation, encouraged CAR-T cell proliferation *in vivo*, and improved the efficacy of solid tumour therapy using the customised nanoemulsion (Clec9A-TNE) vaccine. To deliver CAR-T cells intra-tumourally, one research team created an injectable hydrogel-encapsulated porous immune-microchip system (i-G/MC). The thin immune-MCs (i-MCs), that were coupled with HEMOXCell (Hemo; an oxygen carrier)-loaded alginate, were created from IL-15-loaded alginate microspheres for the injectable i-G/MC system. The alginate self-assembles into a gel layer. As a result of the i-MCs' high porosity and connectivity, CAR-T cell trafficking is made possible. A marine extracellular haemoglobin called Hemo has a significant ability to store oxygen and can bind up to 156 oxygen molecules per molecule. On injecting the i-G/MC into the solid tumour, the hydrogel layer immediately broke down, Hemo provided oxygen to the TME and CAR-T cells, and HIF-1 expression was reduced. The findings showed that the immunological niche enhances hypoxic TEM and encourages CAR-T cell penetration and survival in solid tumours (Mi et al., 2022). Targeted nanocarriers can be used to supply a combination of immune-modulating substances that have shown to be able to kill pro-tumour cell populations while also activating anti-tumour effector cells. To switch the tumour microenvironment of solid tumours from suppressive to stimulatory, repeated infusion was done of lipid NPs that were overlaid with the tumour-targeting peptide iRGD and packed with a mixture of a PI3K inhibitor to block immune-suppressive tumour cells and a -GalCer agonist of therapeutic T cells. This therapy produced a 2-week therapeutic window that allowed CAR-T cells specific to the tumour to target the lesion, grow rapidly, and cause tumour shrinkage. Outside of this therapeutic window, CAR-T cell administration demonstrated no curative results. The widespread adoption of this therapy is constrained by the systemic autoimmune toxicity and high cost.

Lipid NPs containing a strong chemical cocktail have been developed that can simultaneously stimulate essential anti-tumour immune cells and block suppressor cells in the tumour microenvironment. This was done by utilising the *in vivo* mouse 4T1 syngeneic breast cancer model. It was found that injecting CAR-T cells that are tumour-specific during this time period caused disease eradication in half of the animals that were treated and increased the viability of the other animals by more than two-fold. Infused T cells were able to efficiently penetrate tumour lesions, lead to efficient expansion, and eventually remove malignancies. To determine whether liposomes can open a therapeutic window for tumour-specific T cells that have been externally conditioned to be successful, a synergistic combination of nanoparticle conditioning and adoptive T-cell treatment has been used. To achieve this, *in vivo* studies using the 4T1 breast cancer model were conducted. Liposome infusions were this time administered either (i) before the infusion of CAR-T cells, nine doses of PI-3065/7DW8-5 liposomes, or (ii) following the infusion of CAR-T cells, nine doses of PI-3065/7DW8-5 liposomes. Serial imaging of mice revealed that concurrent injection of PI-3065/7DW8-5 liposomes and ROR1-28z CAR-T cells only slightly (yet significantly) improved the ability of the cells to grow in established 4T1 tumours (3.4-fold by the sixth day after infusion). In contrast, on injecting T cells into mice that did not receive preconditioning, accumulation of T cells infused after 3 weeks of PI-3065/7DW8-5 liposome preconditioning was observed in breast tumours at a 22-fold higher level, by the sixth day after infusion. Liposome preconditioning complied with CAR-T-cell therapy eliminated 4T1 tumours in four out of the eight mice that were treated, and the other four exhibited significant tumour reduction with double (average 42-day gain) the viability as compared to the conventional CAR-T-cell therapy. This regimen was effective and generally supported. Although complete clearance from PI-3065/7DW8-5 liposomes supplied before and after T-cell therapy was not achieved, their use extended life by 17 days in comparison to T-cell injections alone. CAR-T cells were given, then nine doses (150 g each) of PI-3065 were given to see if this may increase the tumour-eliminating activity of the adoptively transferred CAR-T cells in this model. Findings show that the remodelling of the tumour microenvironment caused by NPs is secure and can significantly improve the therapeutic effectiveness of CAR-T cells. The best results from this adjuvant nanotherapy are obtained when it is repeated several weeks before T-cell infusion. This gives the delivered immune modifying medicines time to finish impacting the tumour microenvironment (Zhang et al., 2018). Before reinfusion, nano backpacks can bind to CAR-T cells in the backpack approach to CAR-T therapy to assist their return to the tumour site and to support the direct and gradual release of the carrying immune booster or other substances in the TME to increase the effectiveness of therapy.

To build a "backpack" system, researchers linked T cells to NPs with several functions, including targeting, tracing, controlling, and regulating. The activation of cytokines from the backpack via "autocrine" mechanisms improve the therapeutic effectiveness of CAR-T. Recently, Cai's team created CAR-T cells and IL-12-loaded HSA-based NPs for the treatment of solid tumours. The effectiveness of CAR-T cells in solid tumours was greatly increased by IL-12 nanopacks, according to the results. Finally, immunomodulators that are first loaded onto nanopacks and then added to the surface of CAR-T cells can significantly enhance the therapeutic effect on solid tumours. Additionally, designing backpack NPs with drug delivery technology that are coupled to CAR-T cell surfaces is a suitable technique for combination therapy since it increases the accumulation of CAR-T cells in the tumour microenvironment. Numerous implanted biomaterials and nanobackpacks that support CAR-T treatment also incorporate immunopotentiators. The adjuvant effect of the immunopotentiator can boost the immunogenicity of the antigen when associated with it, restore low immune function to normal, and hasten the induction of immune response tumour (Li et al., 2021). Additionally, NPs can help CAR-T cell treatment by delivering antigens to lymphoid regions and encouraging endogenous APCs to present antigens. To target lymph nodes, Leyuan Ma et al. created an albumin-bound phospholipid polymer by joining a CAR small molecule or peptide ligand to the lipid tail. As a result, CAR ligand was successfully transferred to the lymph nodes by albumin and incorporated into the APC membrane using the lipid tail to change macrophages and DCs, and following endogenous cell activation subsequently boosted CAR-T cell multiplication and increased anti-tumour effects (Huang et al., 2020). The amphiphile CAR-T ligands (amph-ligands), a type of peptide vaccination, have also been reported by the Leyuan Ma team to have totally eradicated the tumours that cause melanoma, breast cancer, and glioblastoma in mice (Zhang et al., 2022). Fang et al. created lipid NPs that carry a combination of medications that block suppressor cells in the solid TME and instantly boost T lymphocytes.

A TGF-inhibitor-encapsulated PEGylated liposome targeting T lymphocytes was created by Irvine and colleagues. Through CD90 (an internalising receptor) or CD45 (a non-internalising receptor), they contrasted two cellular strategies. Both techniques greatly increased the effectiveness of T-cell functioning and caused a substantial tumour decline in a mouse model, according to their *ex vivo* and *in vivo* effectiveness demonstrations (Nawaz et al., 2020). For successful target specific cancer cell suppression, a final hypothesis is proposed that includes the synergistic co-delivery of a hybrid polymeric nanoparticle (HPNPs) adsorbed to the modified CAR T-cells. A monoclonal antibody that is specific to CAR T-cells could be used to bind the HPNPs. In this case, Cancer stem cells (CSC) could be precisely targeted by loading a conditionally replicative adenoviral vector (CRAdV) onto the lipid NPs. By expressing a diphtheria toxin gene that was previously inserted into it, the CRAdV may thus be able to block CSC. Triple-negative breast cancer is a more aggressive kind of tumour with few treatment choices, hence this detection and inhibition strategies have been proposed with regard to triple-negative breast cancer (Nayak et al., 2023). There is a lot of hope for CAR-T cells because numerous studies have demonstrated the potential of nanotechnology in conjunction with CAR-T therapy for solid tumours. About 40 CAR-T targets are now being tested in clinical trials for solid tumours, many outnumbering haematological tumours. The majority of CAR-T studies for solid tumours had modest response rates, ranging from 0% to 25%. (Mi et al., 2022). Different modes of nanotechnology are used for enhancing the activity of CAR-T cells for developing effective medications to cancer cells. Research is still ongoing in the field and current discoveries aim to combine NPs with CRISPR/Cas9 genome editing for targeting CAR-T-cell for better effective treatment (Lou et al., 2021).

28.4 CONCLUSION AND FUTURE PERSPECTIVES

The study of cancer immunotherapy has entered a new phase with the advancement of nanotechnology, particularly with regard to nanomaterials. Novel nanomedicines against breast cancer cannot be commercialised because innovative research concepts and original research concepts are not being produced and used in collaboration with the pharmaceutical industry within academia. A top focus should be the transfer of these advanced tools across the bench to the bedside. Innovating complex and advanced technologies which would be costly to develop, impractical for scaling up, or unsteady in their final form is a difficult task. There are considerable difficulties regarding technique and formulation development with reference to the ramping up of the operations for complex nanoparticulate carriers. Next-generation CAR-T cell treatments will concentrate on targeting multiple antigens, CAR modulation via intrinsic biosensing, reducing effectors of solid TME, and finally on-target toxicities. These methods should also generate fresh concepts for customised immunotherapy. Nanomaterials have a special place in targeted administration for cancer immunotherapy, but their capacity to target is constrained by their controllability. Therefore, the establishment of monitoring and characterisation methodologies as well as the ongoing updating of clinical data are necessary for the application of nanomaterials. The availability of the given medicine at tumour locations is restricted by the selectivity and efficiency of nanocarriers, directly limiting the clinical use of nanomaterials. The biosafety of NPs must be further taken into account as a medication carrier. It is necessary to conduct in-depth research on the pharmacokinetic behaviour of nanomaterials for animal models as well as the assessment of immune system responses by diverse components. However, to maximise the advantages of early detection, diagnostic tests must be exceedingly accurate to prevent misdiagnosis and overtreatment. A relatively more controllable cancer environment for patients and physicians will result from the rapid advancement of nanotechnology in other aspects like cancer evaluations, chemotherapies, and radiotherapies as well. Even though cancer has a dynamic aspect that makes it difficult to treat, innovation is still moving forward, and the convergence of several technologies shows promise.

DECLARATIONS

Conflict of interest: The authors declare no conflicts of interest.

Consent for publishing: All authors have approved the final version of the manuscript and given consent for publication.

Funding: Not Applicable.

REFERENCES

Beckman RA, Weiner LM, & Davis HM. (2007). Antibody Constructs in Cancer Therapy - Protein Engineering Strategies to Improve Exposure in Solid Tumors. Cancer. 109(2):170–179.

Chames P, Van Regenmortel M, Weiss E, et al. (2009). Therapeutic Antibodies: Successes, Limitations and Hopes for the Future. Br J Pharmacol. 157(2):220–233.

Henriques B, Mendes F, & Martins D. (2021). Immunotherapy in Breast Cancer: When, How, and What Challenges? Biomedicines. 9(11):1687.

Huang M, Deng J, Gao L, et al. (2020). Innovative Strategies to Advance CAR T Cell Therapy for Solid Tumors. Am J Cancer Res. 10(7), 1979–1992.

Jeanbart L. (2014). Enhancing Efficacy of Anticancer Vaccines by Targeted Delivery to Tumor-Draining Lymph Nodes. Cancer Immunol Res. 2:436–447.

Khawar MB, Afzal A, Abbasi, MH, et al. (2023). Nano-Immunoengineering of CAR-T Cell Therapy Against Tumor Microenvironment: The Way Forward in Combating Cancer. OpenNano. 10:100124.

Koo MM, von Wagner C, Abel GA, et al. (2017). Typical and Atypical Presenting Symptoms of Breast Cancer and Their Associations with Diagnostic Intervals: Evidence from a National Audit of Cancer Diagnosis. Cancer Epidemiol. 48:140–146.

Li Y, Miao W, He D, et al. (2021). Recent Progress on Immunotherapy for Breast Cancer: Tumor Microenvironment, Nanotechnology and More. Front Bioeng Biotechnol. 9:680315.

Liu Z, Zhou Z, Dang Q, et al. (2022). Immunosuppression in Tumor Immune Microenvironment and Its Optimization from CAR-T Cell Therapy. Theranostics. 12(14), 6273–6290.

Lou J, Heater A, & Zheng G. (2021). Improving the Delivery of Drugs and Nucleic Acids to T Cells Using Nanotechnology. Small Struct. 2:2100026.

Mehrotra R, & Yadav K. (2022). Breast Cancer in India: Present Scenario and the Challenges Ahead. World J Clin Oncol. 13(3):209–218.

Mi J, Ye Q, & Min Y. (2022). Advances in Nanotechnology Development to Overcome Current Roadblocks in CAR-T Therapy for Solid Tumors. Front Immun. 13:849759.

Miliotou AN, & Papadopoulou LC. (2018). Car T-Cell Therapy: A New Era in Cancer Immunotherapy. Curr Pharm Biotechnol. 19(1):5–18.

Nawaz W, Xu S, Li Y, et al. (2020). Nanotechnology and Immunoengineering: How Nanotechnology Can Boost Car-T Therapy. Acta Biomater. 109:21–36.

Nayak V, Patra S, Singh KR, et al. (2023). Advancement in Precision Diagnosis and Therapeutic for Triple-Negative Breast Cancer: Harnessing Diagnostic Potential of CRISPR-Cas & Engineered Car T-Cells Mediated Therapeutics. Environ Res. 235:116573.

Nie W, Wei W, Zuo L, et al. (2019). Magnetic Nanoclusters Armed With Responsive PD-1 Antibody Synergistically Improved Adoptive T-Cell Therapy for Solid Tumors. ACS Nano. 13(2):1469–1478.

Oiseth SJ, & Aziz MS. (2017). Cancer Immunotherapy: a Brief Review of the History, Possibilities, and Challenges Ahead. J Cancer Metastasis Treat. 3:250–261.

Schmidts A, & Maus MV. (2018). Making Car T Cells a Solid Option for Solid Tumors. Front Immunol. 9:2593.

Smith TT. (2017). In Situ Programming of Leukaemia-Specific T Cells Using Synthetic DNA Nanocarriers. Nat Nanotechnol. 12:813.

Smith TT, Stephan SB, Moffett HF, et al. (2017). In Situ Programming of Leukaemia-Specific T Cells Using Synthetic DNA Nanocarriers. Nat Nanotechnol. 12(8):813–820.

Sterner RC, & Sterner RM. (2021). CAR-T Cell Therapy: Current Limitations and Potential Strategies. Blood Cancer J. 11:69.

Yang Y, Sun X, Xu J, et al. (2020). Circular Bispecific Aptamer-Mediated Artificial Intercellular Recognition for Targeted T Cell Immunotherapy. ACS Nano. 14(8):9562–9571.

Yang, Y. H., Liu, J. W., Lu, C., & Wei, J. F. (2022). CAR-T Cell Therapy for Breast Cancer: From Basic Research to Clinical Application. *International Journal of Biological Sciences*, *18*(6), 2609–2626.

Zeng Q, Liu Z, Niu T, et al. (2023). Application of Nanotechnology in Car-T-Cell Immunotherapy. Chinese Chem Lett. 34(3):107747.

Zhang F, Stephan SB, Ene CI, et al. (2018). Nanoparticles That Reshape the Tumor Milieu Create a Therapeutic Window for Effective T-Cell Therapy in Solid Malignancies. Cancer Res. 78(13):3718–3730.

Zhang P, Chiu YC, Tostanoski LH, et al. (2015). Polyelectrolyte Multilayers Assembled Entirely from Immune Signals on Gold Nanoparticle Templates Promote Antigen-Specific T Cell Response. ACS Nano. 9:6465–6477.

Zhang Z, Wang T, Wang X, et al. (2022). Improving the Ability of Car-T Cells to Hit Solid Tumors: Challenges and Strategies. Pharmacol Res. 175:106036.

Zhao Y, et al. (2021) Nanozyme-Initiated In Situ Cascade Reactions for Self-Amplified Biocatalytic Immunotherapy. *Advanced Materials (Deerfield Beach, Fla.)* 33(3):e2006363. doi:10.1002/adma.202006363

Zhu LP, Liu J, Zhou GY, et al. (2021). Remodeling of Tumor Microenvironment by Tumor-Targeting Nanozymes Enhances Immune Activation of CAR T Cells for Combination Therapy. Small. 17:2102624.

Zou W. (2005). Immunosuppressive Networks in the Tumour Environment and Their Therapeutic Relevance. Nat Rev Cancer. 5:263.

Section V

Bioinformatics and System Biology

29 System Biology Approach of Protease Cathepsin K in Glioblastoma Multiforme

Prasanna Srinivasan Ramalingam and Sivakumar Arumugam
Protein Engineering Lab, School of Biosciences and Technology
Vellore Institute of Technology, Vellore, India

Purushothaman Balakrishnan
TanBio R. and D. Solution, Thanjavur, India

Janaki Ramaiah Mekala
Koneru Lakshmaiah Education Foundation,
Green Fields, Vaddeswaram, Guntur, India

29.1 CATHEPSIN: TYPES, ROLE, AND FUNCTION

Cathepsins are lysosomal cysteine proteases that are involved in normal physiological conditions and play a vital role in the antigen processing system, homeostasis maintenance, and are involved in energy metabolism and the degradation of unnecessary proteins through lysosomal degradation (Yadati T et al., 2020). In pathologic conditions, cathepsin levels are observed to be elevated and its dysregulation is involved in various pathologic conditions such as cancer, neurological disorders, and metabolic syndrome. The overexpression and upregulation of cathepsins in pathologic conditions identified them as potential prognostic and diagnostic markers and therapeutic targets in various diseases including cancer (Vidak et al., 2019). Cathepsins are generally classified into three main types, namely, aspartic proteases, serine proteases, and cysteine cathepsins with expression levels of 4%, 31%, and 25%, respectively. Cathepsins D and E belong to the aspartic proteases, cathepsins A and G belong to the serine proteases, and cathepsins B, C, F, H, K, L, O, S, V, X, and W belong to cysteine cathepsins (Mijanović et al., 2022) (Table 29.1).

29.2 SPECIAL NOTE ON THE ROLE, FUNCTION, AND INHIBITORS OF CATHEPSIN K

Cysteine cathepsin K (CTSK) is a cysteine protease that degrades collagen and remodels the extracellular matrix (ECM) and also plays a crucial role in bone resorption due to its overexpression in osteoclasts. Several studies indicate that the overexpression of CTSK is significantly associated with various cancers such as bone cancer, renal cancer, liver cancer, breast cancer, lung cancer, colon cancer, and glioblastoma (Mijanović et al., 2022). CTSK overexpression also regulates various cancer signaling pathways such as the Wnt/β-catenin signaling pathway, PI3K-mTOR signaling pathway, MAPK signaling pathway, and RANK/RANKL signaling pathway, respectively, and thus

DOI: 10.1201/9781003394693-34

TABLE 29.1
Cathepsin Types, Enzyme Number and Their Role in Various Cancers

Cathepsin	Type	Enzyme Commission Number	Cancer	Mechanism	Reference
Cathepsin A	Serine protease	3.4.16.5	Lung adenocarcinoma	Overexpressed in A549 cells and its knockdown reduces the proliferation and invasion	Hu B et al., 2020
Cathepsin B	Cysteine protease	3.4.22.1	Prostate carcinoma	Overexpressed in PC3M cells and its inhibition by CA-074 (Cathepin B inhibitor) reduced the Matrigel invasion	Szpaderska AM & Frankfater A., 2001
Cathepsin C	Cysteine protease	3.4.14.1	Breast cancer	Promotes breast-to-lung metastasis by the formation of neutrophil extracellular traps (NETs) and by regulating neutrophils recruitment	
Cathepsin F	Cysteine protease	3.4.22.41	Gastric cancer	Promotes tumor differentiation, invasion, and metastasis, and its downregulation inhibits apoptosis and promotes proliferation	Ji C et al., 2018
Cathepsin G	Serine protease	3.4.21.20	Acute myeloid leukemia	Promotes proliferation and metastasis	Khan M et al., 2018
Cathepsin K	Cysteine protease	3.4.22.38	Glioblastoma	Precents homing of malignant and therapy-resistant glioma stem-like cells in niches	Hira VV et al., 2017
Cathepsin L	Cysteine protease	3.4.22.15	Ovarian cancer	Overexpressed in OV-90 and SKOV3 cells, and their downregulation inhibits the proliferation, invasion and metastasis	Zhang L et al., 2015
Cathepsin V	Cysteine protease	3.4.22.43	Colorectal cancer	Promotes tumor metastasis and invasion, and also acts as poor prognosis marker	Yang L et al., 2022

promotes proliferation and metastasis (Dai et al., 2020). Cathepsin K is a 215 amino acid length cysteine protease that contains signaling peptide (1–16 AA), pro-domain (17–99 AA), and a mature domain (100–215 AA) in its structure as shown in Figure 29.1.

29.2.1 Role of Cathepsin K in Various Cancers

Cathepsin K plays a crucial role in the progression, proliferation, invasion, and metastasis of various cancers via the regulation of various molecular targets (Qian et al., 2022). In breast cancer, CTSK promotes proliferation by cyclin C1 overexpression, promotes metastasis by ECM degradation and remodeling, and also promotes the matrix metalloproteinase (MMP) expression levels. In bone cancer, CTSK was reported to degrade the collagen matrix and bone loss, and even its overexpression promotes bone metastasis. Additionally, the CTSK overexpression is associated with bone cancer proliferation and it potentially acts as a prognostic marker (van der Heijden et al., 2014). In lung adenocarcinoma, CTSK overexpression promotes proliferation and invasion, and it is found to be highly expressed in tumor-associated macrophages (TAMs). The elevated CTSK binds with the

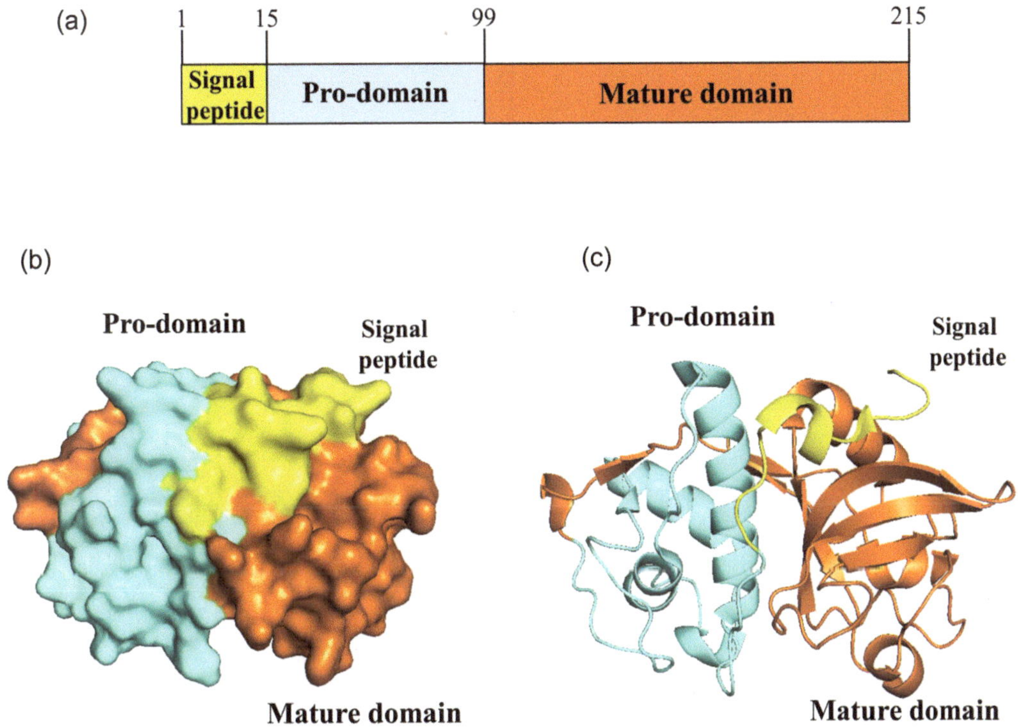

FIGURE 29.1 Domain organization (a), surface view (b), and cartoon view (c) of cathepsin K (215 AA). The signaling peptide (1–16 AA), pro-domain (17–99 AA), and mature domain (100–215 AA) are shown in yellow, magenta, and orange colors, respectively.

mTORC1 complex and regulates the mTOR pathway. In colorectal cancer, CTSK promotes metastasis and is associated with poor prognosis, and also been reported to regulate the NF-κB pathway and could elevate the expression interleukin (IL) 10 and 17, thereby mediating the M2 polarization of TAMs (Li et al., 2019). In renal cancer, CTSK is associated with tuberous sclerosis complex (TSC) mutations (especially TSC2) and transcription factor binding to IGHM enhancer 3 (TFE3) rearrangements, ultimately playing a role as a prognostic marker. In renal Caki cells, CTSK inhibitors halt the mTOR phosphorylation and reduce bone resorption (Palsgrove et al., 2018). In prostate cancer, the CTSK is highly expressed and associated with proliferation and metastasis. In both *in vitro* and *in vivo,* the inhibition of CTSK by a potent CTSK inhibitor reduces metastatic prostate cancer-induced bone resorption (Liang et al., 2019). In ovarian cancer, CTSK overexpression is associated with proliferation and epithelial–mesenchymal transition (EMT), while its inhibition reverses this effect. Also, the N-myc downstream regulatory gene 1 (NDRG1) inhibition downregulated the expression of CTSK in ovarian cancer cells (Tingting et al., 2020). In melanoma proliferation, CTSK was reported to promote tumor proliferation, invasion, migration, and metastasis, and also plays a vital role in the regulation and degradation of collagen and matrix proteins (Petricevic et al., 2017). In gastric cancer, CTSK promotes cancer metastasis via acting as a downstream effector of coronin 3 protein, and the positive correlation of CTSK and MMP-9 expression levels was observed indicating their role in the progression of gastric cancer (Digklia & Wagner, 2016).

29.2.2 Role of Cathepsin K in Glioblastoma

Glioblastoma (GBM) is a grade IV aggressive brain cancer with poor prognosis and survival rates in patients. GBM recurrence is very common even after surgery and chemotherapy, due to the

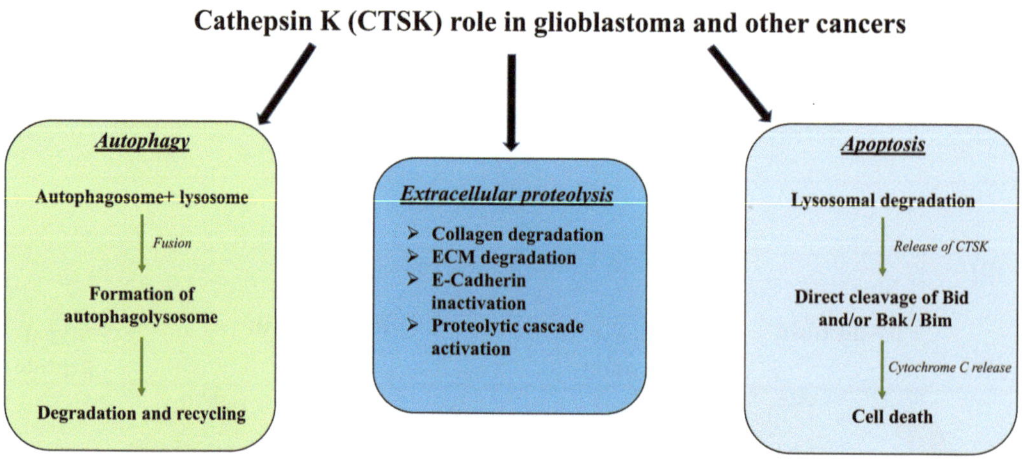

FIGURE 29.2 Various roles of cathepsin K in glioblastoma and other cancers.

therapeutic resistance of glioblastoma stem cells (GSCs) in the tumor microenvironment (TME) (Louis et al., 2016). Several reports indicated the presence and expression of lysosomal cysteine cathepsins including cathepsin K, cathepsin B, and cathepsin X that are associated with the tumor-cell infiltration and expression of TAMs and play a role in the maintenance of TME (Paw et al., 2015). In a study, the localization pattern of cathepsin K along with cathepsin X was evaluated in grade IV GBM patients, and their results showed that these cathepsins are overexpressed (both mRNA and protein levels) in GBM. Also, they are localized in the CD133 and CD68-rich regions, and observed in TME stromal cells, indicating their role in tumor vasculature and maintenance of GSCs. In another study, CTSK was reported to cleave and inactivate the function of stromal cell-derived factor 1 α (SDF-1 α) and thus prevents the invasion of GBM cells and plays a crucial role in the CXCR4/7-positive therapy-resistant glioma stem-like cells (Hira et al., 2017). Therapeutic targeting of the CTSK is associated with the reduction of MHC I-mediated immune evasion in GSCs. Cathepsin K plays a crucial role in various molecular mechanisms in autophagy, apoptosis, and extracellular proteolysis in GBM as well as in other cancer types (Paw et al., 2015). In apoptosis, it plays a major role in the regulation of pro-apoptotic and anti-apoptotic proteins, thus mediating cellular death; through autophagy, it is involved in the autophagolysosome formation and degradation of unnecessary proteins; and in extracellular proteolysis, it is involved in collagen and ECM degradation, E-cadherin inactivation, and proteolytic cascade activation. A detailed overview of cathepsin K's role in GBM and other cancers is shown in Figure 29.2.

29.2.3 Cathepsin K Inhibitors

Inhibition of cathepsin K reduces the proliferation, invasion, and metastasis of several cancer types including GBM and, thus, CTSK has emerged as a potential cancer target in recent years. Several cathepsin K inhibitors such as K11777, odanacatib (MK-0822), L-873724, Gü2602, Gü1303, L-006235, balicatib (AAE581), calpeptin, relacatib, ONO-5334, MK-0674, and 2-cyanopyrimidine have been successfully discovered and developed. The 2D structures of these inhibitors are depicted in Figure 29.3. The structure-activity relationship (SAR) activity of these compounds plays a major role in the inhibitory potential and cell penetration potential of these compounds and contributes to the high affinity and selectivity of these compounds toward cathepsin K. Several *in vitro* and *in vivo* studies reported that cathepsin K promotes various cancers including GBM and the inhibition of cathepsin K by potent inhibitor inhibits proliferation, invasion, differentiation, and metastasis. Cathepsin K inhibitors are reported to increase the expression levels of TNF-related apoptosis-inducing ligand (TRAIL), upregulation of pro-apoptotic proteins such as Bam and Bax, and induce intracellular reactive oxygen species levels. Although several inhibitors with versatile modes of

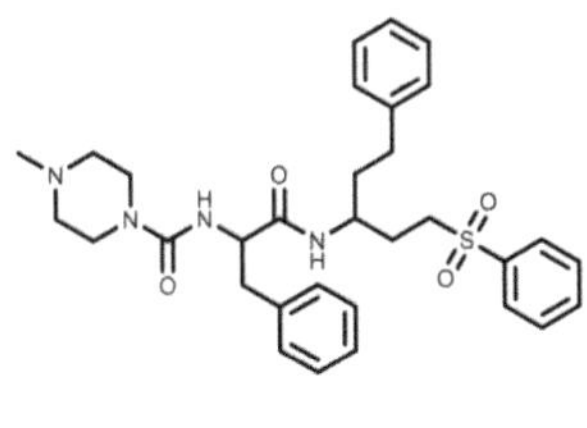

K11777

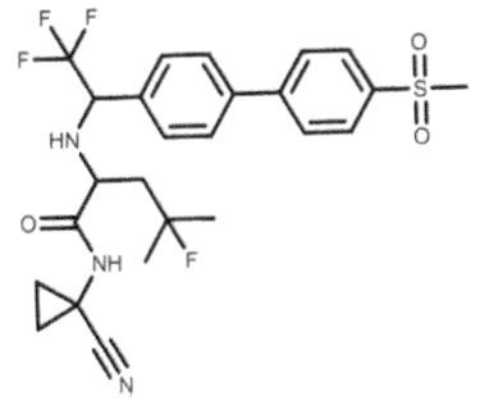

Odanacatib

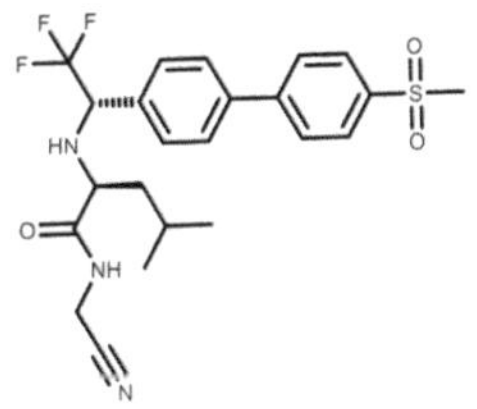

L-873724

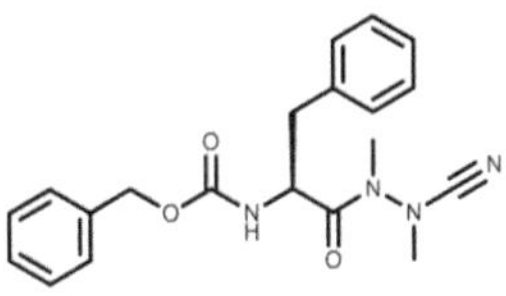

Gü1303

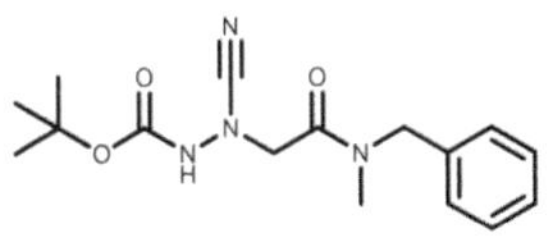

Gü2602

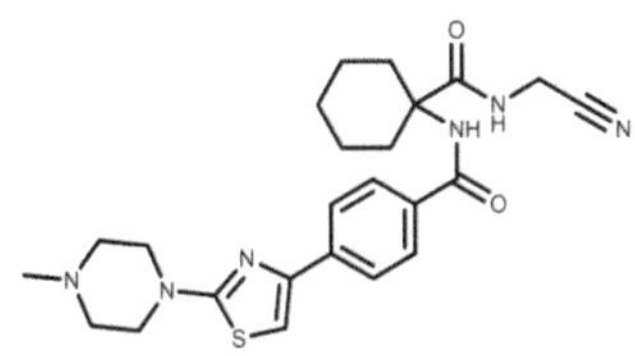

L-006235

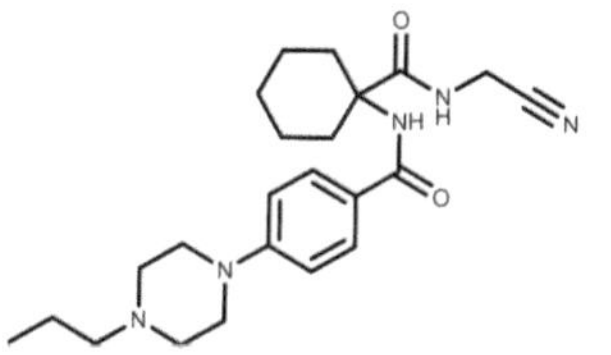

Balicatib

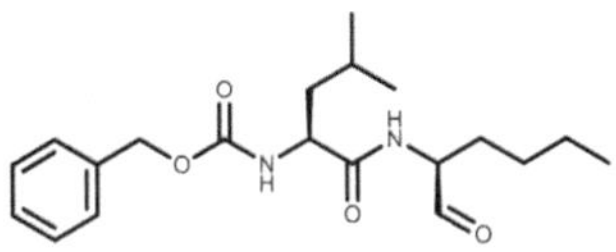

Calpeptin

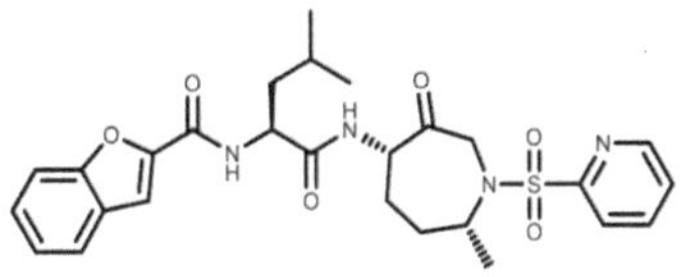

Relacatib

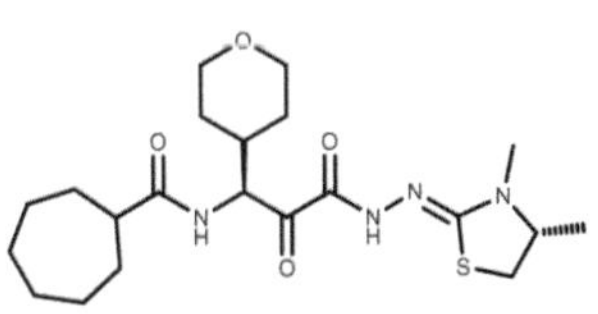

ONO-5334

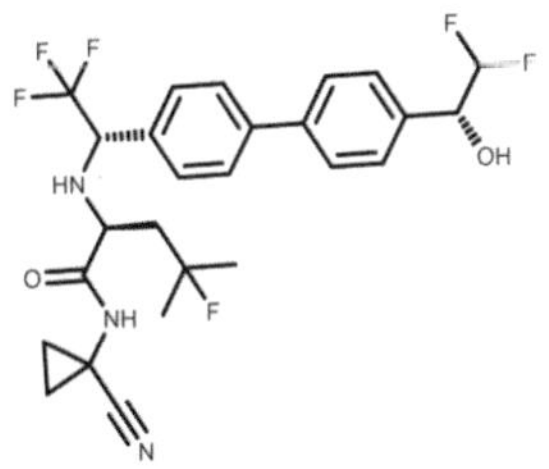

MK-0674

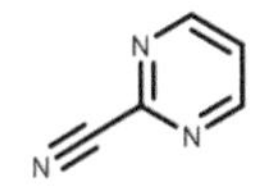

2-Cyanopyrimidine

FIGURE 29.3 2D structures of cathepsin K inhibitors.

action have been discovered, there is no US Food and Drug Administration (FDA)-approved drug available for the treatment of cathepsin K overexpressed cancers. However, the usage of computational systems biology approaches and integrative omics approaches will aid in drug design, development and discovery, target validation, therapeutic development, and precision decision-making.

29.3 BIOINFORMATICS AND OMICS APPROACHES TOWARD CATHEPSIN K

A gene expression analysis experiment of nearly 311 lysosomal proteases was carried out using The Cancer Genome Atlas (TCGA) dataset and found that the cathepsin K was upregulated and

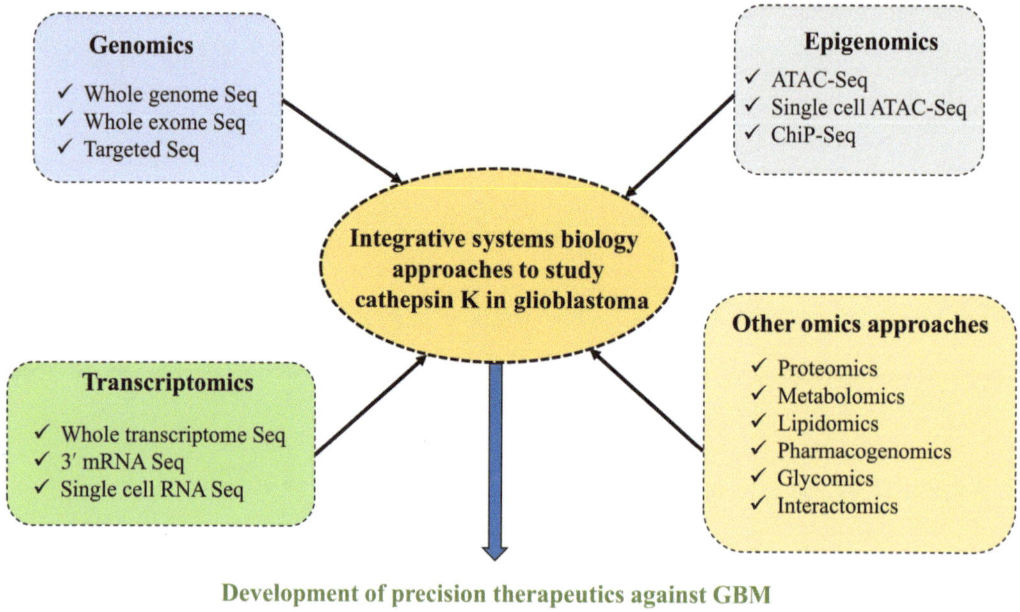

FIGURE 29.4 Multi-omics and system biology approaches against cathepsin K in glioblastoma.

overexpressed in GBM than in normal glioma cells. Additionally, the IHC and protein expression studies revealed the overexpression of CTSK. In another study, through *in situ* hybridization, researchers found that the CTSK has been selectively overexpressed in osteoclasts than in other tissue and showed higher expression than the other tested cathepsins such as cathepsins S, B, and L. The integrative omics approach is a versatile tool to predict the key player in the pathologies, identify the crucial and significant disease target, and develop cancer prevention and therapeutics (Ramalingam et al., 2023). Omics approaches such as genomics, proteomics, transcriptomics, epigenomics, metabolomics, lipidomics, glycomics, pharmacogenomics, and interactomics are widely used in cancer research (Reel et al., 2021). Whole-genome sequencing, whole-exome sequencing, and targeted sequencing are commonly used in genomics; whole-transcriptome sequencing, 3′ mRNA sequencing, and single-cell whole RNA sequencing are commonly used in transcriptomics; ATAC-sequencing, ChiP-sequencing, and single-cell ATAC-sequencing are commonly used in epigenomics, while peptide/protein sequencing is commonly used in proteomics. The complete overview of the integrative systems biology approach to studying the cathepsin K in GBM is shown in Figure 29.4.

29.4 OMICS METHODS FOR DATA INTEGRATION AND DECISION-MAKING

Several omics methods were used for the precision decision on cancer prevention and treatment, and some of the methods are as follows:

Statistical-based: iCluster+ is a widely used statistical-based tool to profile genomic, epigenomic, and transcriptomic data, and the Bayesian integrative clustering approach is also used to classify the subtypes and range of tumors. Pathway representation and analysis by direct reference on graphical models (PARADIGM), Bayesian consensus clustering, and integrative Bayesian analysis of genomics data (iBAG) are also commonly used statistical-based methods (Sathyanarayanan et al., 2020).

Multivariate-based: Joint non-negative matrix factorization is the commonly used method that creates common and individual submatrices by disintegrating multiple transcriptomics data. Also, joint and individual variation explained (JIVE) and MoCluster are widely used due to their robustness and precision in multivariate analysis.

Similarity-based: Similarity network fusion is the most commonly used method that constructs the network based on available patient data and could potentially fuse with spectrum data. Also, the regularized multiple kernel learning locality preserving projections (rMKL-LPP), and neighborhood based multi-omics clustering (NEMO) are often used to construct similarity-based network data.

Network-based: The iOmicsPASS tool predicts molecular-level interactions as well as subnetworks and employs the nearest shrunken centroid algorithm to provide precision decisions at the phenotypic level (Koh et al., 2019).

Fusion-based: Pattern fusion analysis compiles individual data into global data with more precision and less bias and is, thus, widely used in cancer omics approaches.

Correlation-based: Canonical correlation analysis is commonly used to analyze the post-translational modifications such as methylation and glycosylation of gene transcripts, and could provide an efficient correlation between the groups.

29.5 CONCLUSION AND FUTURE PERSPECTIVES

Cathepsins are lysosomal cysteine proteases that are overexpressed and associated with various cancers such as lung, breast, colon, and GBM. In this chapter, we have extensively compiled the data regarding the types, families, roles, and functions of various cathepsins in different cancers with special concern on cathepsin K (CTSK) in GBM. Also, commonly available CTSK along with its physiochemical data and mechanism of action have been discussed. Additionally, the integrative systems biology approaches including various omics approaches and their significance in the precision therapeutic development for CTSK-associated GBM have been discussed. Finally, we conclude that by employing these systems biology approaches, we could predict the precision response in the CTSK-associated GBM and develop novel therapeutics. Furthermore, it will also help the novel and more potent inhibitor against CTSK through drug discovery and development approaches.

ACKNOWLEDGMENTS

The authors would like to thank KL University, Andhra Pradesh, India, and Vellore Institute of Technology (VIT), Vellore, India, for providing the necessary facilities to carry out this work.

DECLARATIONS

Conflict of interest: The authors declare no conflicts of interest.

Consent for publishing: All authors have approved the final version of the manuscript and given consent for publication.

Funding: Not Applicable.

REFERENCES

Dai R, Wu Z, Chu HY, et al. (2020) Cathepsin K: The action in and beyond bone. Front Cell Dev Biol, 8, 433.

Digklia A, Wagner AD (2016) Advanced gastric cancer: Current treatment landscape and future perspectives. World J Gastroenterol, 22(8), 2403–2414.

Hira VV, Verbovšek U, Breznik B, et al. (2017) Cathepsin K cleavage of SDF-1α inhibits its chemotactic activity towards glioblastoma stem-like cells. Biochim Biophys Acta Mol Cell Res, 1864(3), 594–603.

Hu B, Zhu X, Lu J (2020) Cathepsin A knockdown decreases the proliferation and invasion of A549 lung adenocarcinoma cells. Mol Med Rep, 21(6), 2553–2559.

Ji C, Zhao Y, Kou YW, et al. (2018) Cathepsin F knockdown induces proliferation and inhibits apoptosis in gastric cancer cells. Oncol Res, 26(1), 83–93.

Khan M, Carmona S, Sukhumalchandra P, et al. (2018) Cathepsin G is expressed by acute lymphoblastic leukemia and is a potential immunotherapeutic target. Front Immunol, 8, 1975.
Koh HWL, Fermin D, Vogel C, et al. (2019) iOmicsPASS: network-based integration of multiomics data for predictive subnetwork discovery. NPJ Syst Biol Appl, 5, 22.
Li L, Zhu Z, Zhao Y, et al. (2019) FN1, SPARC, and SERPINE1 are highly expressed and significantly related to a poor prognosis of gastric adenocarcinoma revealed by microarray and bioinformatics. Sci Rep, 9(1), 7827.
Liang W, Wang F, Chen Q, et al. (2019) Targeting cathepsin K diminishes prostate cancer establishment and growth in murine bone. J Cancer Res Clin Oncol, 145(8), 1999–2012.
Louis DN, Perry A, Reifenberger G, et al. (2016) The 2016 world health organization classification of tumors of the central nervous system: A summary. Acta Neuropathol, 131(6), 803–820.
Mijanović O, Jakovleva A, Branković A, et al. (2022) Cathepsin K in pathological conditions and new therapeutic and diagnostic perspectives. Int J Mol Sci, 23(22), 13762.
Palsgrove DN, Li Y, Pratilas CA, et al. (2018) Eosinophilic Solid and Cystic (ESC) renal cell carcinomas harbor TSC mutations: Molecular analysis supports an expanding clinicopathologic spectrum. Am J Surg Pathol, 42(9), 1166–1181.
Paw I, Carpenter RC, Watabe K, et al. (2015) Mechanisms regulating glioma invasion. Cancer Lett, 362(1), 1–7.
Petricevic SJ, Pavlovic A, Capkun V, et al. (2017) Cathepsin K expression in melanoma is associated with metastases. Histol Histopathol, 32(7), 711–716.
Qian D, He L, Zhang Q, et al (2022) Cathepsin K: A versatile potential biomarker and therapeutic target for various cancers. Curr Oncol, 29(8), 5963–5987.
Ramalingam PS, Priyadharshini A, Emerson IA, et al. (2023) Potential biomarkers uncovered by bioinformatics analysis in sotorasib resistant-pancreatic ductal adenocarcinoma. Front Med (Lausanne), 10, 1107128.
Reel PS, Reel S, Pearson E, et al. (2021) Using machine learning approaches for multi-omics data analysis: A review. Biotechnol Adv, 49, 107739.
Sathyanarayanan A, Gupta R, Thompson EW, et al. (2020) A comparative study of multi-omics integration tools for cancer driver gene identification and tumour subtyping. Brief Bioinform, 21(6),1920–1936.
Szpaderska AM, Frankfater A. (2001) An intracellular form of cathepsin B contributes to invasiveness in cancer. Cancer Res, 61(8), 3493–500.
Tingting Z, Xiaojing L, Xiaoyan T, et al. (2020) The antisense long noncoding RNA AGAP2-AS1 regulates cell proliferation and metastasis in epithelial ovarian cancer. J Cancer, 11(18), 5318–5328.
van der Heijden L, Dijkstra PD, van de Sande MA, et al. (2014) The clinical approach toward giant cell tumor of bone. Oncologist, 19(5), 550–561.
Vidak E, Javoršek U, Vizovišek M, et al (2019) Cysteine cathepsins and their extracellular roles: Shaping the microenvironment. Cells, 8(3), 264.
Yadati, T., Houben, T., Bitorina, A., & Shiri-Sverdlov, R. (2020). The ins and outs of cathepsins: physiological function and role in disease management. Cells, 9(7), 1679.
Yang L, Zeng Q, Deng Y, et al. (2022) Glycosylated cathepsin V serves as a prognostic marker in lung cancer. Front Oncol, 12, 876245.
Zhang L, Wei L, Shen G, et al. (2015) Cathepsin L is involved in proliferation and invasion of ovarian cancer cells. Mol Med Rep, 11(1), 468–474.

30 Omics Technologies for Matrix Metalloproteinases

MMPs as Cancer Biomarkers

Sherlin Rosita Arockiaraj, Jay Shree Mathivanan, Shivani Jambulingam, and Thomas Jebastin
Bishop Heber College (Autonomous), Tiruchirappalli, India

Karnati Balaji and Mahima Bhardwaj
Vignan's Foundation for Science, Technology and Research, Vadlamudi, Guntur, India

Pushparaj Annadurai
C.P.R. Environmental Education Center, Alwarpet, Chennai, India

Abhinav Parashar
PDPIAS, CHARUSAT, Changa, India

Daniel A. Gideon
St. Joseph's University, Bengaluru, India

30.1 INTRODUCTION TO OMICS TECHNOLOGIES AND THEIR USES IN BIOMARKER DISCOVERY

One of the most significant studies or technologies employed in modern day biological research is "omics." Proteomics, metabolomics, transcriptomics, genomics, interactomics, immunomics, redoxomics, and lipidomics are a few of the types of omics technologies. The study of an organism's entire gene and protein sequence, as well as its function and features, is referred to as genomics. The biological function of the gene product is revealed through transcriptomics, proteomics, and metabolomics. Identifying all kinds of non-targeted gene products (transcripts, proteins, and metabolites) present in a given organism is the primary objective of omics technologies (Leng et al., 2019). Multiomics can unearth previously unidentified incidental illness mediators and biological insights that single omics investigations cannot (Leiserson et al., 2015). The discovery of several potential biomarkers has been made possible over the past few decades by the rapid development of high-throughput omics technology (Yan et al., 2023).

30.1.1 Omics in Cancer Studies

The history of cancer genomics can be broadly divided into two categories, before next-generation sequencing (NGS), or pre-NGS, and after NGS, or post-NGS. Alongside the advancement of technology to examine the genomic, transcriptomic, proteomic, and metabolomics landscapes of breast

DOI: 10.1201/9781003394693-35

cancer, our knowledge of its natural history has changed. While each of these omics dimensions was taken into account separately in first-generation "omics analyses," it is becoming obvious that more comprehensive, integrative methods are needed to properly comprehend complex biological systems (Parsons & Francavilla, 2020).

30.1.2 Genomics

Genomics techniques such as shotgun whole-genome sequencing (WGS) and next-generation sequencing (NGS) have become widely popular (Seehausen et al., 2014). Using genomics, the methylome, mutations of critical genes, as well as the sequence of individuals' genomes can be determined. Genomics studies provide crucial details about mutations, structural changes, epigenetic problems, and genetic abnormalities in individuals. Thereby, it aids in disease risk prediction and precision oncology, effectively promoting a better understanding of the contributions of genetic, ethnic, sex-specific, as well as environmental factors involved in diseases such as cancer and metabolic syndrome.

30.1.3 Transcriptomics

Transcriptomics technologies are used to study an organism's transcriptome, i.e., the sum of all of its RNA transcripts. While mRNA serves as a transient intermediary molecule in the information network, non-coding RNAs perform additional diverse functions. A transcriptome captures a snapshot in time of the total transcripts present in a cell (Lowe et al., 2017). The expression of an organism's genes in various tissues, environments, or differing time periods may be measured to learn more about the biology of the organism and mechanism of gene expression regulation. Understanding human illness has been greatly aided by the study of how gene expression varies in various species using transcriptomic analysis. Comprehensive gene expression analysis enables the identification of large-scale coordinated patterns that more focused tests are unable to detect (Dong et al., 2013).

30.1.4 Proteomics

A genome's whole collection of expressed proteins is referred to as its "proteome." A genome is mostly unchanging, but when genes are activated or inactivated in response to environmental cues, the expression levels of proteins inside a given cell can undergo significant changes. Examination of the differentially expressed genes (DEGs) and protein expression levels within a diseased cell or tissue may provide a more thorough picture of the illness, as proteins are directly engaged in both normal and disease-associated biochemical activities. Some researchers refer to all the proteins generated by a particular cell in a single time frame as the "functional proteome," which reflects the dynamic character of the proteome (Cho, 2007). A comprehensive categorization of protein–protein interfaces and protein interactomes is a useful tool for modeling protein complexes and comprehending the fundamentals of molecular recognition, as these interactions are essential for signal transmission, gene regulation, and metabolism. It is cumbersome to use biochemical methods such as yeast two-hybrid assay to detect protein–protein interactions *in vitro*. Moreover, it is difficult to perform western blotting studies for each factor/protein involved in a defined signal transduction pathway that can be influential in cancer studies. Hence, advancements in proteomics such as whole-cell mass spectrometry aids in both qualitative as well as quantitative detection of expression levels of proteins (He et al., 2015; Cuypers et al., 2022). For example, the expression of ADAM12 in the urine of breast cancer patients aids as a biomarker for predicting the stage of cancer. Another protein, ADAM17, has been detected in high-grade breast carcinoma tissue, making it a candidate marker for assessing the seriousness of breast cancer (Coleman, 2017). Multiomics approaches go beyond routinely employed diagnostic methods such as computed tomography and magnetic resonance imaging; medical practitioners are now able to devise strategies to curb disease progression and also identify novel biomarkers of disease and tailor-make drugs to realize the goal of personalized medicine.

30.1.5 Metabolomics

The study of cells via profiling all or a significant portion of their metabolites is known as metabolomics. Metabolomics involves an elaborate process—handling data/information, processing raw analytical data, handling standards and ontology related to metabolomics, performing statistical analysis and data mining, integrating data, and modeling metabolic networks mathematically within a systems biology framework. The main metabolomics methodologies are examined in light of these particular bioinformatics requirements, as are the state-of-the-art analytical tools for data creation (Shulaev, 2006). Metabolite profiling may also be used as a technique in systems biology, where metabolite snapshots are utilized to examine biological dynamics using mathematical models (Idle & Gonzalez, 2007).

30.2 MATRIX METALLOPROTEINASE NETWORKS IN CANCER

30.2.1 Matrix Metalloproteinases in Cancer

The extracellular matrix (ECM) preserves the three-dimensional structure of tissues by anchoring cells together. Matrix metalloproteinases (MMPs) were initially believed to assist in metastasis. Further research has revealed that MMPs are essential for cell survival, motility, differentiation, and proliferation and are not solely involved in ECM remodeling alone. However, it is now understood that they have various biological functions in different stages of carcinogenesis, in regulating the growth of both primary and secondary tumors. The ECM components that serve as the physical impediments to cell migration must be locally broken down for a tumor cell to spread from the tumor's primary location to other organs. There are six fundamental changes that lead to cancer progression, including self-support in growth signals, insensitivity to growth inhibitory signals, evasion of apoptosis, infinite replication, sustained angiogenesis, invasion, and metastasis. Recent studies have shown that MMPs play a role in multiple stages of carcinogenesis, not just invasion and metastasis (Visse & Nagase, 2003). MMPs are the primary protease enzymes in charge of ECM remodeling in several diseases (Gialeli et al., 2011). A few differentially expressed MMPs are employed as tissue-specific cancer prognostic biomarkers. Several research reports have revealed that suppression of these indicators by a number of inhibitors is beneficial in cancer treatment trials (Cathcart et al., 2015). As one of the most widely investigated MMPs, MMP-9 plays vital roles in many biological processes. To control ECM remodeling, MMP-9 can cleave a wide variety of ECM and cell-surface proteins. Numerous studies have linked MMP-9 to cancer pathophysiology, revealing its involvement in diverse cancer-related processes such as invasion, metastasis, and angiogenesis.

30.2.2 MMPs Types and Their Roles in Cancer

MMPs are a family of zinc-dependent ECM-altering endopeptidases that have the ability to degrade all components of the ECM (Yadav et al., 2014). MMPs play a major role in the tumor growth and metastasis, which includes cell–cell and cell–ECM interactions, and angiogenesis (Gialeli et al., 2011). MMPs are classified into collagenases (MMP-1, MMP-8, MMP-13) (Mallya et al., 1990; Lovejoy et al., 1999), gelatinases (MMP-2 and MMP-9) (Fridman et al., 1992; Lovejoy et al., 1999), stromelysins (MMP-3, MMP-10, and MMP-11) (Kaur et al., 2002), matrilysins (MMP-7 and MMP-26) (Pavlovsky et al., 1999), membrane-type MMPs (MMP-14, MMP-15, MMP-16, MMP-17, MMP-24, and MMP-25) and others (MMP-19, MMP-21, MMP-23A, MMP-27, and MMP-28) (Visse & Nagase, 2003; Sbardella et al., 2012; Fischer & Riedl, 2021) based on structural features and substrates.

Colitis and colitis-associated cancer (CAC) are characterized by elevated levels of most MMPs, including MMP-2, MMP-3, MMP-7, MMP-9, MMP-10, MMP-12, and MMP-13. As colitis is the ultimate cause of CAC, lowering the levels of these MMPs may therefore successfully stop the onset of inflammation and CAC (He et al., 2023). Higher expression of a particular MMP can be reported and regarded with its relativity to cancers and is closely associated with the progression of cancer and poor prognosis (Ibrahim & Ngom, 2015). The regulation of various biological pathways or processes and jointly found to be involved with gene or protein clusters (Anwar et al., 2014) and hence, analyzing complete networks of possibly implicated signaling molecules would be one of the potentially most effective ways to examine these complicated interactions (Xu et al., 2016). Even though the factors that cause carcinogenesis vary depending on cancer type, the processes that result in metastasis are generally the same for all types of solid tumors and mainly rely on MMPs (Niland et al., 2021). The emergence of secondary malignancies at a distant site, or metastasis, continues to be the main contributor to cancer death. There are 23 endopeptidases that make up the MMP family of enzymes that degrade the tissue's ECM and basement membranes' protein components. They can be divided into five classes based on their substrate specificity and domain structure, this includes membrane-type, gelatinases, collagenases, stromelysins, and matrilysins (Decock et al., 2011). Among them, collagenases break down collagen types 1, 2, and 3 (MMP-1, -8, -13, -14, and -18), gelatinases break down type 4 of collagen and gelatin (MMP-2 and -9), stromelysins will hydrolyze a wide range of ECM components, such as different kinds of collagen, elastin, proteoglycans, and glycoproteins (MMP-3, -7, -10, -11, -26, and -27), elastases will debase elastin (MMP-12), and membrane type MMPs have a range of activities that include the activation of MMP-2 (MMP-14, -15, -16, -17, -24, and -25) (Carey et al., 2021).

30.2.3 Protein–Protein Interactions

It is significantly challenging to study tumor metastasis, as tumors from a primary site can spread anywhere in the body to one or more secondary sites (Jemal et al., 2011). Breast cancer is the second most prevalent disease in women globally and the most commonly diagnosed malignancy (Cai et al., 2017) and hence, some focus on this specific cancer type is given in this chapter. Protein–protein interactions (PPIs) play important roles in a number of intracellular and intercellular metabolic processes. Two different proteins in a PPI are most likely to share similar functions and therefore, selective inhibition of particular MMPs is necessary for enhancing the therapeutic benefit of cancer treatment (Radisky et al., 2017). The constructed PPI networks are topologically analyzed for assessing the functional significance of selected genes of breast cancer using bioinformatics pathways database such as KEGG. Biophysical methods for protein separation and characterization have advanced significantly in the past ten years, with notable advancements being made in liquid chromatography (LC), isotope-coded affinity tag (ICAT) labeling, two-dimensional polyacrylamide gel electrophoresis (2D-GE), and a few other methods (Aslam et al., 2016). Each breast cancer type, for example, can have unique molecular features and etiological factors. In personalized oncomedicine, there is a need to identify the level of expression of genes of interest and to quantify the protein levels.

The combined utility of genomics, epigenomics, transcriptomics, and metabolomics poses an avenue to enhance cancer diagnostics. Other related fields such as phenomics/cellomics, connectomics, secretomics, matrisomics, exosomics, angiomics, chaperomics, epichapereomics, phosphoproteomics, ubiquitinomics, metalloproteomics, terminomics, degradomics, metadegradomics, adhesomics, stressomics, microbiomics, immunomics, salivomics, materiomics as well as other biomics (Neagu et al., 2021) are exciting avenues for exploration of proteomic alterations in cancers, allowing for patient-specific profiling of disease etiology, stage and treatment prognosis.

30.3 MMPs AS BIOMARKERS OF CANCERS

The American National Cancer Institute (NCI) defines a biomarker as a biological molecule that can be discovered in tissues, body fluids, or blood and serves as a marker indicating the presence of either a healthy/unhealthy condition or disease. In the healthy individuals, such biomarkers are either absent altogether, or are expressed at higher concentrations. Biomarkers are biomolecules—they can be mutations in gene sequences, higher expression of mRNA of particular genes, higher expression of small RNA (miRNA) or long non-coding RNA molecules, increased expression of protein/enzyme molecules, or the metabolic products of catalysis by enzymes. The identification of biomarkers is now frequently carried out using a "discovery" approach, utilizing methods like high-throughput sequencing, gene expression arrays, and mass spectrometry to quickly identify individual or groups of biomarkers that differ between cohorts (Yadav et al., 2014).

In the extremely rare case of male breast cancer, serum protein alterations were recently studied in male breast cancer patients using 2D-GE and MALDI-TOF MS. On comparison with healthy controls, the differentially expressed serum proteins such as CDC7, mitochondrial aldehyde dehydrogenase, and dimethyladenosine transferase-1 were explored. Also, alterations in (a) lipid metabolism and transport by APOA1 and APOE, (b) apoptosis and immune response—CD5, CD5L, clusterin, C-C motif chemokine-14 (CCL14), (c) transcription factor levels—e.g. STAT3 and SSX3, (d) invasion and apoptosis related factors, e.g. alpha-2-HS-glycoprotein, (e) estrogen synthetic factors—e.g. aromatase (CYP19A1), as well as MMP-27 (Zografos et al., 2019) were studied. Identification of novel differentially expressed proteins and genes can provide valuable information and pave the way for pharmacological control of transcription/translation as well as enzymatic/structural activity of these proteins. For example, MMP-7 is a key biomarker of several cancers as well as inflammation-related diseases; higher MMP-7 expression was found to support cancer proliferation, differentiation, metastasis and invasion (Liao et al., 2021). MMP-9 was found to be both a pharmacological target as well as crucial biomarker in melanoma (Napoli et al., 2020). When MMPs, particularly MMP-1, were examined, it revealed that triple-negative breast cancer tissues had higher expression of MMP-1 than did estrogen receptor and human epidermal growth factor 2 receptor in breast cancer tissues (Wang et al., 2019). Matrilysins, namely, MMP-7 and MMP-26, are members of the MMPs group that have the potential for use as markers for breast cancer (Piskór et al., 2021). Through the degradation of the ECM, MMP-11 facilitates the invasion and metastasis of cancer. Targeting of hub genes associated with MMP-11 may improve clinical care and research for breast cancer patients (Kim et al., 2021). Tumor growth and prognosis are significantly influenced by membrane-type MMP-14. Previous research indicates that MMP-14 is highly expressed in several forms of cancer and promotes tumor invasion and metastasis by changing the ECM. The immunohistochemistry results of a study demonstrated that breast cancer tissues displayed higher levels of MMP-14 than the surrounding normal breast tissues (Di et al., 2018). Colorectal cancer (CRC), also called colon or rectal cancer, is a significant medical issue throughout the world with a higher mortality rate. Both environmental and genetic factors are involved in the etiology and pathophysiology of CRC (Shi et al., 2023). In CRC patients, there was a strong correlation seen between the serum levels of several biomarkers and clinicopathologic characteristics. Interestingly, a combination biomarker model comprising CEA, TIMP-1, and MMP-7 may be able to accurately and precisely diagnose CRC (Huang et al., 2021).

Cervical cancer is the main cause of cancer-related mortality in developing countries and is one of the most prevalent gynecological carcinomas in the world. A major factor in the development of cervical cancer is the ongoing research of high-risk human papillomavirus (HPV) infection (Doghish et al., 2023). Increased MMP-9 levels have been linked to a poor outcome in cervical cancer (Padežnik et al., 2023). Through ECM disintegration and angiogenesis, the expression of MMP-2, MMP-9, and MMP-14, tissue inhibitors of metalloprotease 1 and 2 (TIMP-1, TIMP-2), and vascular endothelial growth factor A (VEGF-A) contributes to tumor invasion and metastasis. MMP-9 expression is associated with a favorable prognosis in cervical cancer patients (Azevedo Martins et al., 2020). P53 activity regulated MMP-1, and as MMP transcription is involved in the breakdown of ECM during tumor invasion, its dysregulation would encourage the spread of tumors (Kurnia et al., 2022). A few cancer phenotypes that are more aggressive and invasive are frequently linked to elevated MMP levels. MMPs are biomarkers that can offer important information about the tumor microenvironment, assisting in the early identification and tracking (and staging) of the cancer. Certain MMP profiles in cancer patients' blood, tissues, or other body fluids can be measured, providing physicians and researchers with a non-invasive means to gauge the severity of the disease and decide on the course of treatment. Several types of tumors have various MMPs that play role as biomarkers (Huang et al., 2021). In case of brain tumors, MMPs are formed in the brain, and compounds like amyloid and pro-inflammatory cytokines may boost this synthesis (Lee et al., 2012). MMP-9 levels considerably differ between tumor types and stages, showing a positive correlation with aggressiveness. Elevated levels of MMP-9 in brain tumor tissues indicate poor prognosis. MMP-2 and MMP-9 have been found in both latent and active forms in the cerebrospinal fluid of individuals with brain tumors. Patients' CSF and urine both reflected the overexpression of MMP-2 and -9 in the source tumor tissue (Miller et al., 2004). Additionally, to MMP-9 being expressed differently in each molecular subgroup, triple-negative and HER2-positive breast tumors have also been shown to overexpress MMP-9. The considerable correlation between MMP-9 overexpression and a higher risk of metastasis and cancer recurrence provides compelling evidence that higher MMP-9 expression could help segregate subsets of aggressive breast cancer into clinically meaningful subtypes (Lowe et al., 2017). Several collagenases, including MMP-1, MMP-8, and MMP-13, can initially cleave denatured collagens before being degraded by MMP-9. In addition, type IV collagen, the primary constituent of the basement membrane, is degraded by MMP-9. It has a variety of roles in the spreading process, including tumor invasion, tumor-induced angiogenesis, and tumor microenvironment immunomodulation (Shulaev, 2006). MMP-13 may be a useful prognostic biomarker for invasive breast cancer, as shown by immunohistochemical studies. The relationship observed between the expression of Her2/neu and TIMP-1 and tumor-derived MMP-13 suggests that MMP-13 could be helpful in foretelling breast cancer development and severity (Idle & Gonzalez, 2007). MMP-2 and MMP-9 have been studied as potential prognostic biomarkers of CRC. MMP-1, -2, -7, -13, and TIMP-1, -2, -3, and -4 are studied through immunohistochemical staining of primary colon or rectal cancer. Reduced survival of colon cancer patients has been linked to high expression of MMP-2 in the stroma and malignant epithelium. Strong TIMP-3 epithelial and stromal cytoplasmic staining was linked to a longer survival in rectal cancer patients (Dobra et al., 2023). Table 30.1 shows the chromosomal location, molecular weight, tissue of expression, cellular location, as well as function of MMPs and their role as biomarkers (as reported in literature) in disease, particularly cancer. Table 30.2 shows the relevance of key types of MMPs as biomarkers which are monitored in some cancers.

TABLE 30.1
Overview of MMP Types, Tissue Expression, Location and Function in Humans

Human MMP Type/Chromosome Location	M.wt.	Tissue(s) Expressed	Location	Function/Biomarker in Disease
MMP-1 (Collagenase-1) Chr. 11	54 kDa	Almost all tissues, but highest in GI tract, liver and gallbladder, kidney and urinary bladder, female reproductive tissues, and bone marrow and lymphoid tissues Uniprot id: P03956	ECM, secreted	• Cleavage of type I, II, and III collagens as well as type VII and X collagens • Biomarker in bladder, colon, breast, and brain cancers
MMP-2 (Collagenase–4) or Gelatinase-A Chr. 16q13-q21	72 kDa	Gall bladder, urinary bladder, female and male reproductive tissues, smooth muscles, adipose tissue, pancreas and skin Secreted by cardiomyocytes, fibroblasts, and myofibroblasts	*Isoform 1*: ECM, secreted and nucleus *Isoform 2*: mitochondria and cytoplasm	• Type IV collagenase / Gelatinase-A • Biomarker in breast, bladder, and cervical cancers
MMP-3 (Stromelysin-1) Chr. 11q22.2	54 kDa	Salivary gland, adipose tissue, skin, skeletal muscle, breast, and endometrium Uniprot id: P08254	ECM, secreted, cytoplasm, and nucleus	• Degrade fibronectin, laminin, gelatins of type I, III, IV, and V • Collagens III, IV, X, and IX, and cartilage proteoglycans • Biomarker in breast, colon, bladder, and colorectal cancers
MMP-7 (Matrilysin-1) Chr. 11q22.2	30 kDa	Gallbladder, salivary gland, urinary bladder, kidney, and breast Uniprot id: P09237	ECM, secreted	• Degrades casein, gelatins of types I, III, IV, and V, and fibronectin. Activates procollagenase • Prognostic biomarker in the early stage of several tumors and prevents the development of many cancers
MMP-8 (Collagenase-2) Chr. 11q22.2	53 kDa	Bone marrow and spleen Uniprot id: P22894	Cytoplasmic granule, ECM, secreted	• Can degrade fibrillar type I, II, and III collagens • Biomarker in gastric, colorectal, hepatocellular, and lung cancers along with TIMP-1
MMP-9 (Gelatinase-B) Chr. 20q13.2	78 kDa	Bone marrow, spleen, lymph node, appendix, tonsil, lung/ thymus/adipose tissue. Uniprot id: P14780	ECM, secreted	• Plays an essential role in local proteolysis of the extracellular matrix and in leukocyte migration • Biomarker in primary human breast cancer, squamous cell cancer, esophageal cancer, ovarian cancer, bladder cancer, gastric cancer, and in most cancers
MMP-10 (Stromelysin-2) Chr. 11q22.2	54 kDa	Endometrium, appendix, stomach, and urinary bladder. Uniprot id: P09238	ECM, secreted	• Degrade fibronectin, gelatins of type I, III, IV, and V • Weakly collagens III, IV, and V. Activates procollagenase • Biomarker in bladder, head and neck, early breast cancer, oral squamous cell carcinoma, cervical tumors, non-small-cell lung cancer, and hepatocellular carcinoma

(Continued)

TABLE 30.1 *(Continued)*
Overview of MMP Types, Tissue Expression, Location and Function in Humans

Human MMP Type/Chromosome Location	M.wt.	Tissue(s) Expressed	Location	Function/Biomarker in Disease
MMP-11 (Stromelysin-3) Chr. 22q11.23	55 kDa	Endometrium, cervix, placenta, ovary, smooth muscle, heart muscle. Uniprot id: P24347	ECM, secreted	• May play an important role in the progression of epithelial malignancies • Biomarker in metastatic breast cancer, pancreatic cancer, colon, endometrial, oral cancer, and bladder cancer
MMP-12 (Mφ elastase) Chr. 11q22.2	54 kDa	Appendix, tonsil, urinary bladder, rectum, and colon. Uniprot id: P39900	ECM, secreted	• Involved in tissue injury and remodeling; significant elastolytic activity • Accept large and small amino acids at the P1' site • Biomarker in hepatocellular carcinoma, oral cancer, ovarian cancer; lung, gastric, and colorectal cancers
MMP-13 (Collagenase-3) Chr. 11q22.2	54 kDa	Urinary bladder, lung, pituitary gland, bone marrow, and vagina. Uniprot id: P45452	ECM, secreted	• Degradation of extracellular matrix proteins including fibrillar collagen, fibronectin, TNC, and ACAN • Cleaves triple helical collagens, including type I, type II, and type III collagen • Role in wound healing, tissue remodeling, cartilage degradation, bone development, bone mineralization and ossification • Biomarker in invasive breast cancer, esophageal squamous cell carcinoma, breastbone metastasis, and prostate cancer
MMP-14 (MT1-MMP) Chr. 14q11.2	66 kDa	Cervix, endometrium, adipose tissue, gallbladder, skin, placenta, and ovary/vagina Uniprot id: P50281	Melanosome, cytoplasm, and membrane	• Degrade the collagen • Activates progelatinase A • involved in actin cytoskeleton reorganization by cleaving PTK7 • Gastric cancer, vestibular schwannomas, non-small-cell lung cancer, ovarian cancer, digestive system carcinomas, glioma, pleural mesotheliomas, etc.
MMP-15 (MT2-MMP) Chr. 16q21	76 kDa	Liver, thyroid gland, heart muscle, duodenum, testis, small intestine/stomach Uniprot id: P51511	Membrane	• Degrades various components of the extracellular matrix • Activates progelatinase A • Biomarker in salivary gland adenoid cystic carcinoma; also in bladder, pancreatic, papillary thyroid, and colorectal cancer etc.

TABLE 30.1 *(Continued)*
Overview of MMP Types, Tissue Expression, Location and Function in Humans

Human MMP Type/Chromosome Location	M.wt.	Tissue(s) Expressed	Location	Function/Biomarker in Disease
MMP-16 (MT3-MMP) Chr. 8q21.3	70 kDa	Cerebral cortex, amygdaia, hyphothalamus, hippocampal formation, and gallbladder Uniprot id: P51512	Cell membrane, ECM, and secreted	• Degrades the extracellular matrix, such as collagen type III and fibronectin • Activates progelatinase A • Involve matrix remodeling of blood vessels • Biomarker in serous ovarian cancer, hepatic carcinoma, non-small-cell lung cancer, kidney clear cell renal carcinoma, osteosarcoma, and bladder cancers
MMP-17 (MT4-MMP) Chr. 12q24.33	67 kDa	Hippocampal formation, cerebral cortex/basal ganglia, amygdala, hypothalamus, and ovary/cervix Uniprot id: Q9ULZ9	Cell membrane, ECM, and secreted	• Degrades fibrin • Involved in the activation of membrane-bound precursors of growth factors or inflammatory mediators • Biomarker in vestibular schwannomas, mesothelioma, hepatocellular carcinoma, and gastric cancer
MMP-18 (Collagenase-4)	92 KDa	Detected in mammary gland, lung, small intestine, colon, and almost all other tissues, except for brain, skeletal muscle, kidney, liver, or peripheral blood leucocytes Uniprot id: **O13065**	ECM, secreted	• Physiological roles in several normal processes like wound healing, apoptosis, bone elongation, embryo development, uterine involution, angiogenesis, etc. • Biomarker in cutaneous squamous cell carcinoma, hepatocellular, bladder, and esophageal cancers
MMP-19 (RASI-1) Chr. 12q13.2	57 kDa	Adipose tissue, spleen, lung, gallbladder, and endometrium Uniprot id: Q99542	ECM and secreted	• Degrades aggrecan and cartilage oligomeric matrix protein (comp), during development, hemostasis and pathological conditions (arthritic disease) • Biomarkers in non-small-cell lung cancer, pancreatic cancer (pancreatic ductal adenocarcinoma), colon cancer etc.
MMP-20 (Enamelysin) CHr. 11q22.2	54 kDa	Testis and tonsil Uniprot id: O60882	ECM and secreted	• Degrades amelogenin, the major protein component of the enamel matrix and two of the macromolecules characterizing the cartilage extracellular matrix, aggrecan, and the cartilage oligomeric matrix protein (COMP) • Biomarker in mesothelioma, breast, colon, and lung cancers
MMP-21 Chr. 10q26.2	65 kDa	Epididymis, ovary, lymph node, skin, and retina Uniprot id: Q8N119	Secreted	• Plays an essential role in the generation of left-right asymmetry during embryogenesis • Biomarker in esophageal squamous cell carcinoma, hepatocellular, gastric cancer, etc.

(Continued)

TABLE 30.1 *(Continued)*
Overview of MMP Types, Tissue Expression, Location and Function in Humans

Human MMP Type/Chromosome Location	M.wt.	Tissue(s) Expressed	Location	Function/Biomarker in Disease
MMP-23 Chr. 1p36.33	44 kDa	Ovary, cervix, endometrium, lung, and urinary bladder/prostate Uniprot id: O75900	Endoplasmic reticulum membrane	• Protease regulating the surface expression of some potassium channels by retaining them in the endoplasmic reticulum • Biomarker in non-small-cell lung cancer, malignant melanoma, breast cancer, and gastric cancer
MMP-24 (MT5-MMP) Chr. 20q11.2	73 kDa	Cerebellum, cerebral cortex, kidney, lung, and pancreas Uniprot id: Q9Y5R2	Cell membrane, Golgi apparatus, trans-network membrane, ECM, and secreted	• Cleavage of N-cadherin (CDH2) and acts as a regulator of neuro-immune interactions and neural stem cell quiescence • Involved in cell–cell interactions between nociceptive neurites and mast cells • Key regulator of neural stem cells • Biomarker in human breast cancer, hepatocellular carcinoma, brain tumors, and non-small-cell lung cancer, etc.
MMP-25 (MT6-MMP) Chr. 16q13.3	63 kDa	Spleen, bone marrow, lung, appendix, and adipose tissue Uniprot id: Q9NPA2	Cell membrane, ECM, and secreted	• Activates progelatinase A • Biomarker in advanced gastric cancer, salivary gland adenoid cystic carcinoma, endometrial cancer, etc.
MMP-26 (Matrilysin-2) Chr. 11p15.4	30 kDa	Endometrium, prostate, hypothalamus, thymus, and bone marrow Uniprot id: Q9NRE1	ECM and secreted	• Hydrolyzes collagen type IV, fibronectin, fibrinogen, beta-casein, type I gelatin and alpha-1 proteinase inhibitor. • Activate progelatinase B. • Biomarker in breast cancer, prostate cancer, gliomas, and endometrial cancer.
MMP-27 Chr. 11q22.2	59 kDa	Breast, bone marrow, skin, adipose tissue, salivary gland, and tongue Uniprot id: Q9H306	Endoplasmic reticulum membrane	• Degrade protein components of the extracellular matrix such as fibronectin, laminin, gelatins, and/or collagens • Biomarker in oral squamous cell carcinoma, hepatocellular carcinoma, colorectal cancer, bladder cancer, and breast cancer
MMP-28 (Epilysin) Chr. 17q12	59 kDa	Skin, testis, colon, urinary bladder, lung/basal ganglia, and amygdale Uniprot id: Q9H239	ECM and secreted	• Degrade casein • Play a role in tissues homeostasis and repair • Biomarker in bladder, pancreatic, gastric, and hepatocellular cancers

TABLE 30.2
Different Types of Cancer with Reported MMPs as Biomarkers with the Location and Method of Analysis

Cancer Type	MMPs	Location	Method of Analysis
Brain	MMP-2	Tissue	IHC, ELISA
	MMP-9	Tissue	IHC, ELISA
	MMP-2, MMP-9	Tissue, cerebrospinal fluid	IHC, ELISA
Breast	MMP1, TIMP-1	Tissue, serum, nipple aspirates	Gene analysis, ELISA, IHC
	MMP-9	Tissue, serum, urine	IHC, gelatin zymography
	MMP-13	Tissue	IHC
Colorectal	MMP-2	Tissue	IHC
	MMP-9	Tissue	IHC
	MMP-2, MMP-9	Plasma	ELISA
	MMP-7	Serum	ELISA
	MMP-1	Tissue	ELISA
Prostate	MMP-2, MMP-9	Tissue, plasma	ELISA, IHC
	MMP-2	Tissue	IHC
	MMP-9	Urine	Gelatin zymography

30.4 MMPs IN CANCER

30.4.1 Significance of MMPs Pathways in Cancer

The exploration of MMP pathways through gene ontology (GO) analysis unveils intricate networks crucial for understanding their multifaceted roles in cancer progression. Generally, GO results provide information about the functional characteristics of a set of genes or proteins. These results are usually organized as molecular function (MF), biological process (BP), and cellular component (CC). The various biological processes that MMPs coordinate are particularly illuminated by GO BP keywords, which connect these proteases to important pathways that are involved in ECM remodeling, cell adhesion, angiogenesis, inflammatory responses, wound healing, and cell migration. The MF describes the specific role of genes in molecular level whereas the CC focuses on the active locations of the genes within the cell. Genes are shortlisted from the NCBI gene and used for both GO and pathway enrichment analysis (Carbon et al., 2009). Those MMPs include MMP-1, MMP-2, MMP-3, MMP-7, MMP-8, MMP-9, MMP-10, MMP-11, MMP-12, MMP-13, MMP-14, MMP-15, MMP-16, MMP-17, MMP-19, MMP-20, MMP-21, MMP-23A, MMP-23B, MMP-25, MMP-26, MMP-27, and MMP-28. All these are then used as an input for GO analysis. The p-value in gene GO data analysis is a measure of the statistical significance of the enrichment of specific gene sets within a given set of genes. The significance threshold is chosen to be 0.05, which is often referred to as alpha level threshold. The p-values lesser than the selected threshold are considered to be statistically significant. Here, the analysis with these genes reveals that all the three GO results such as MF, BP, and CC was statistically significant. All these results were also represented as chord plots (see Figures 30.1–30.3). From this, we could also infer the specific BPs that correlate

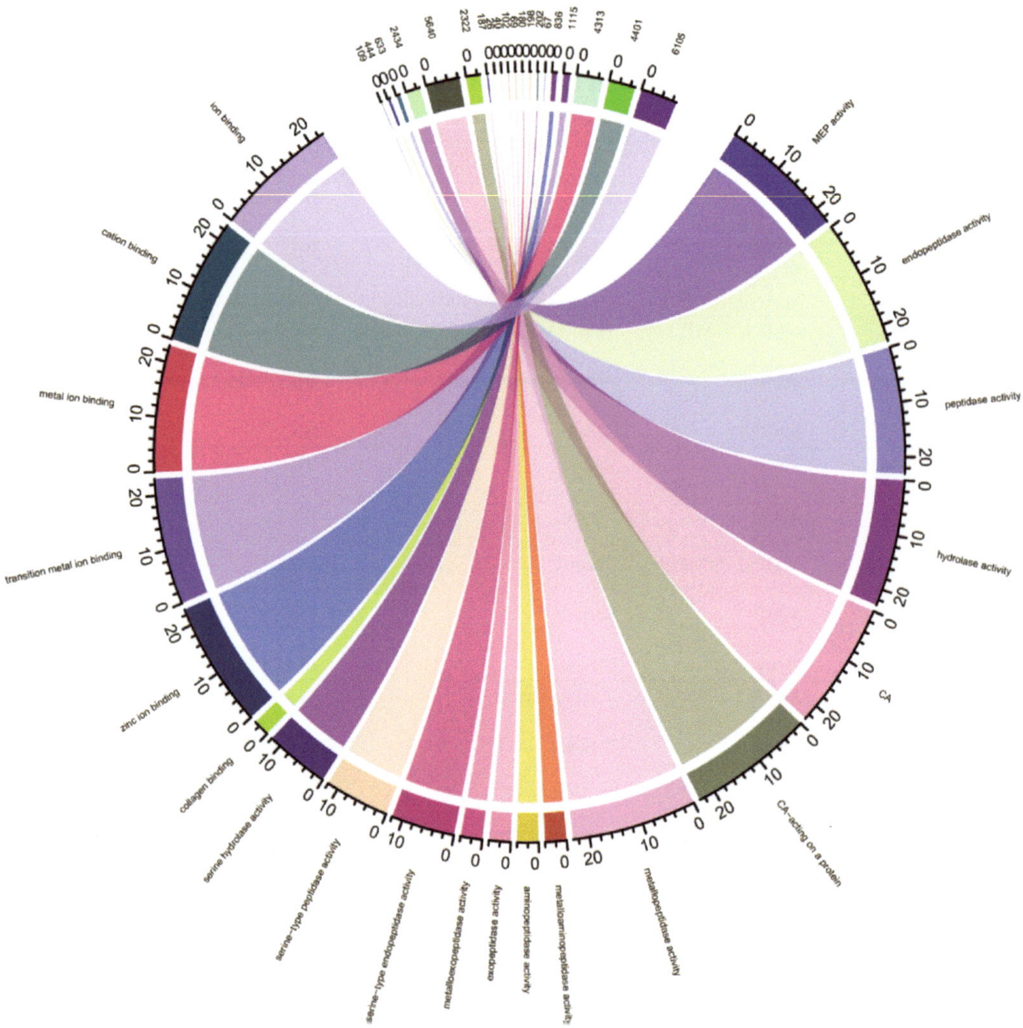

FIGURE 30.1 Chord plot indicating molecular function of matrix metalloproteinases (MMPs). The functional roles of MMPs are indicated by the chord plot, showing their multiple roles such as metallopeptidase/metalloendopeptidase activity, hydrolase activity, aminopeptidase activity, serine hydrolase activity, collagen binding, and Zn^{2+}-binding roles, etc. These gene ontology classifications aid in identification of the functions of MMPs.

with the cancer progression, including collagen catabolic process, ECM disassembly, ECM organization, response to amyloid-beta, response to peptide, endochondral ossification, and proteolysis (Table 30.1). Table 30.3 shows the gene ontology (GO) process involved and the gene expression P-values for those genes in cancer progression.

Similarly, the pathway enrichment analysis, carried out using SRplot on these MMPs reveals certain pathways that are directly associated with cancer progression. Pathways such as tumor necrosis factor (TNF) signaling were involved in amelioration of cancers such as breast and colorectal. In this context, the interleukin (IL)-17-signaling pathway was also involved in breast cancer, and estrogen-signaling pathways have been playing a role in both breast and cervical cancer. The enrichment analysis and the MMPs associated with each signaling pathway are represented in Figures 30.4 and 30.5.

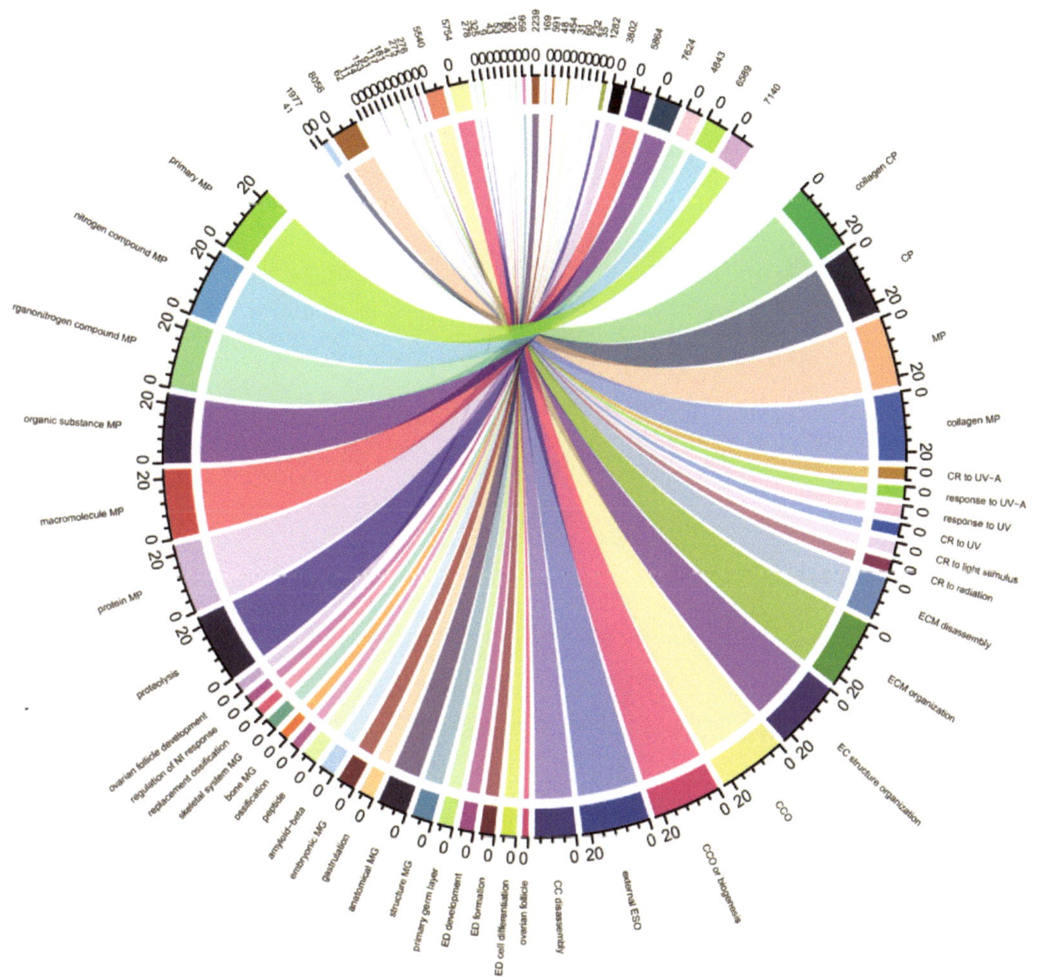

FIGURE 30.2 Chord plot indicating the biological processes that matrix metalloproteinases (MMPs) are associated with. The diverse types of functions as per gene ontology (GO), as indicated by the image, reveal the major and minor functions of MMPs. The key GO functions are collagen binding and metabolism, extracellular matrix (ECM) organization, ECM disassembly, and other minor functions, such as gastrulation, ovarian follicle development, primary germ layer development, etc.

30.4.2 PPI

PPI interactions play a huge role in understanding the functional features of the proteins and coding genes responsible for them. MMPs, as well as a wide range of other regulatory proteins, are all involved in these interactions. Comprehending the distinct PPI networks linked to MMPs is crucial to deciphering the intricate mechanisms that underlie their activity and control. There are about 23 MMPs considered for the evaluation of PPI networks, which was first considered with the finding of the GO datasets retrieved from the Gene Ontology database (Ashburner et al., 2000; Aleksander et al., 2023). This escalated to finding the GO BP, GO MF, and GO CC. The protein functional class for MMPs is considered for the understanding of the functional enrichment of the genes selected. This deciphers the choice of the metalloproteinase class in the PANTHER database, which is provided as the tool for GO enrichment analysis in the Gene Ontology database (Thomas et al., 2022). The data is evaluated for the construction of a network

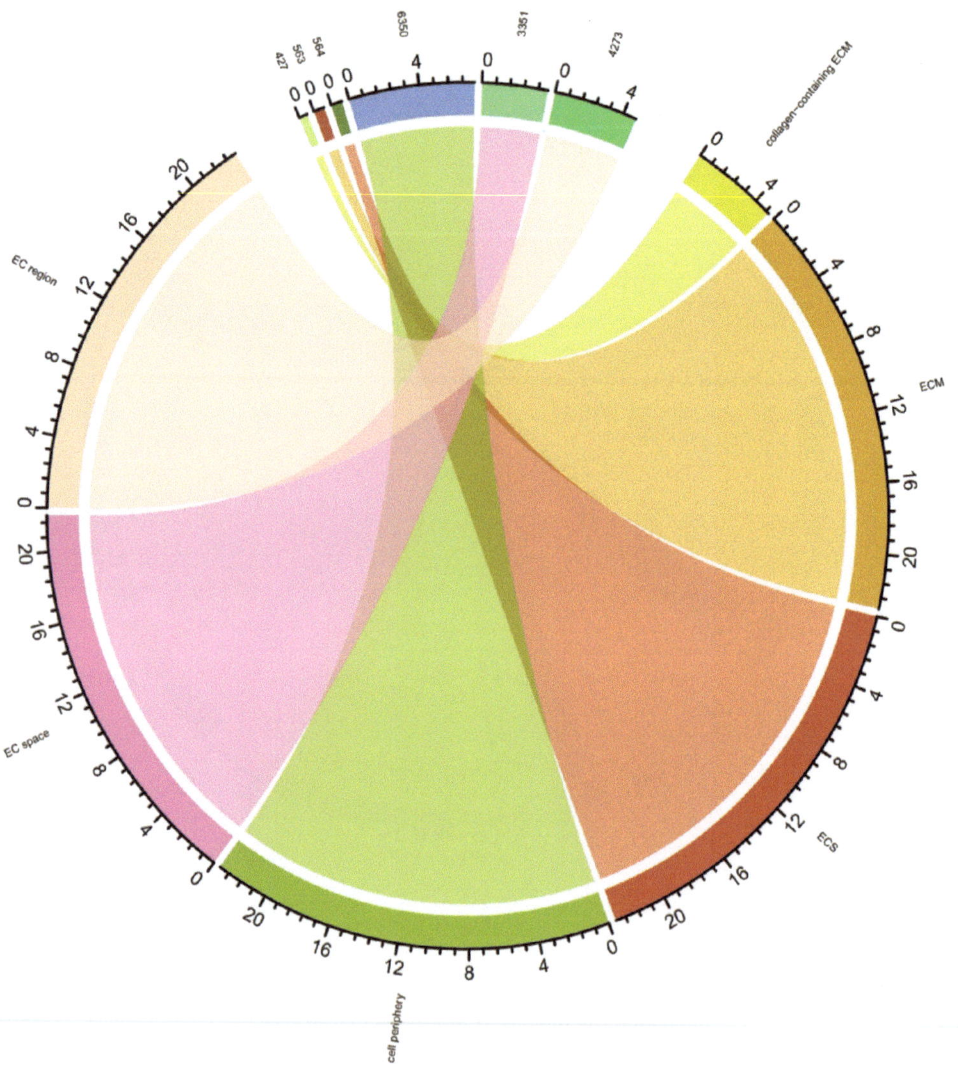

FIGURE 30.3 Chord plot indicating cellular location of matrix metalloproteinases (MMPs). As this image indicates, MMPs are predominantly associated with the extracellular matrix (EC region, EC space, ECS, ECM) and with cellular membranes (cell periphery).

TABLE 30.3
Gene Ontology Results of Biological Process Specifying the Certain Processes in Cancer Progression

Gene Ontology Biological Process	Raw p-value
Collagen catabolic process	2.28E-60
Extracellular matrix disassembly	6.31E-31
Extracellular matrix organization	2.03E-43
Response to amyloid-beta	3.01E-09
Response to peptide	1.29E-04
Endochondral ossification	7.10E-06
Proteolysis	2.26E-28

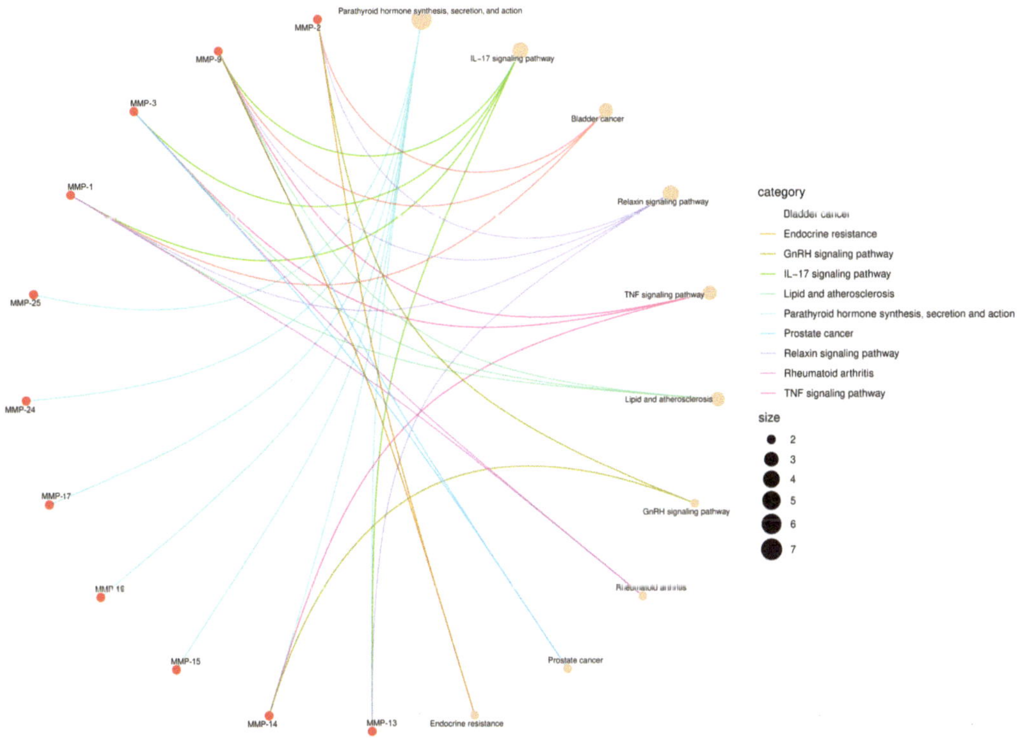

FIGURE 30.4 Pathway enrichment showing the link between different matrix metalloproteinases (MMPs) and the signaling pathways. (Generated using SRplot.)

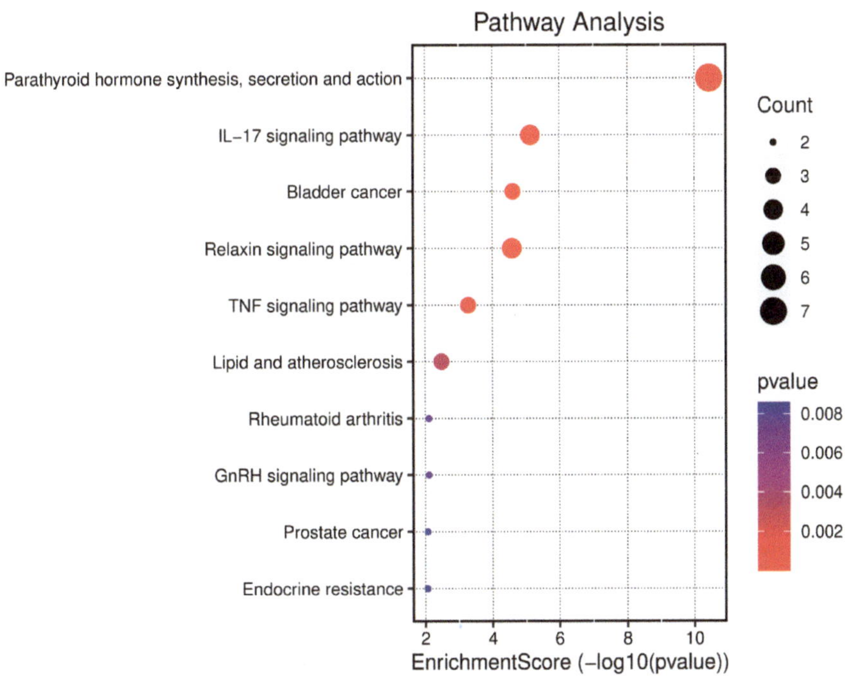

FIGURE 30.5 Pathway enrichment score represented in dot plot. The plot indicates the top pathways that are connected to the functioning of matrix metalloproteinases (MMPs). The dots are colored based on the enrichment score (p-value) and the size of the dot indicates the count. The greatest enrichment scores were obtained for parathyroid hormone synthesis, secretion, and action, interleukin-17 signaling, bladder cancer, relaxin signaling, and the tumor necrosis factor signaling pathway. (Generated using SRplot.)

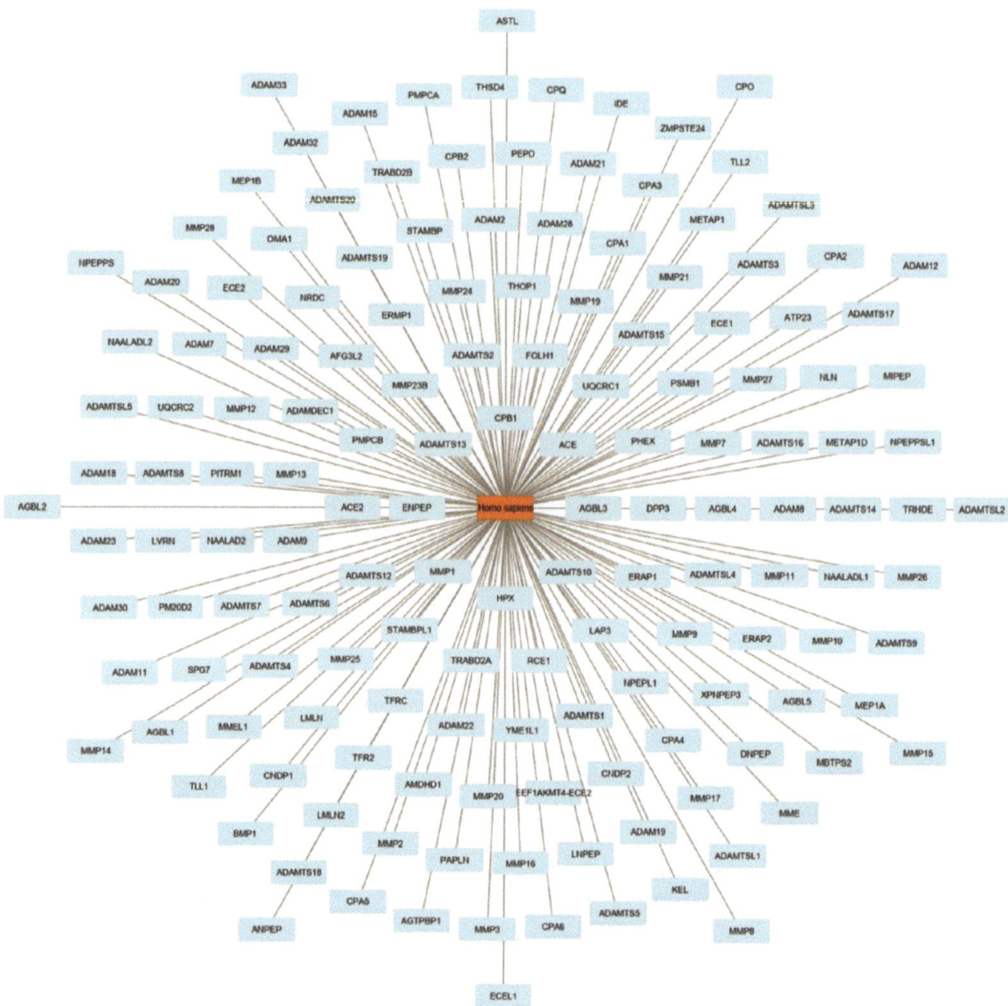

FIGURE 30.6 All matrix metalloproteinase (MMP) association with metalloproteinase activity in *Homo sapiens* based on gene ontology data. This comprehensive list of human proteins with MMP activity is indicated and these individual proteins can be studied in different diseases to see whether a selected protein can serve as a unique biomarker of the disease.

using the Cytoscape software (Shannon et al., 2003). The selection of the protein functional class, especially metalloproteinase, has been linked with their association with all of the classifications of MMPs, such as collagenases, stromelysins, matrilysins, and gelatinases. These MMPs have been associated with significant pathways that correlate with tumorigenesis, angiogenesis, and apoptosis. The GO data have revealed 155 genes, including the MMPs associated with metalloproteinase. The constructed network is provided in Figure 30.6. The identified genes are evaluated for their functional relationships using the GeneMANIA database. GeneMANIA is a public database that allows identification of functionally related genes based on a variety of gene association networks. The interaction networks in Figure 30.7 between genes are shown as edges and the graphs show the associations between the genes. It makes use of a range of data sources, including co-expression information, genetic relationships, PPI, and protein domain similarity (Warde-Farley et al., 2010).

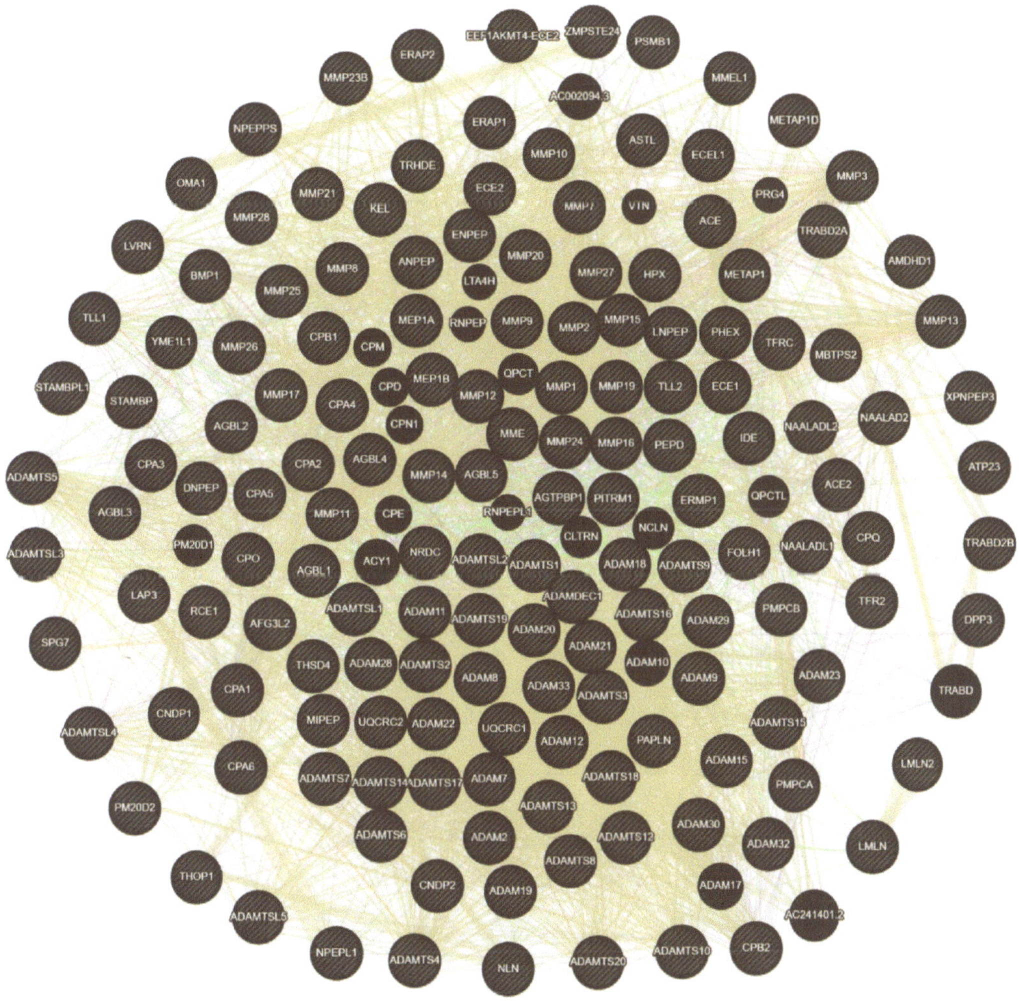

FIGURE 30.7 Network image depicting the functional annotation of 155 genes associated with matrix metalloproteinases (MMPs). Using GeneMANIA, the predicted functional relationships of interacting genes also reveal the protein–protein interactions between them. The genes from Figure 30.6 were studied for their interactions based on gene ontology enrichment analysis. Both known and predicted protein–protein interactions reveal the functional relationship between interacting genes or protein partners.

30.5 CONCLUSION

Omics technologies such as transcriptomics, proteomics, metabolomics, and genomes have completely changed the landscape of cancer research. While other omics techniques show the biological roles of gene products, genomics offers insights into the complete genetic composition and function of an organism. MMPs are one of these that have become significant tissue-specific cancer markers. Because they are involved in ECM remodeling, MMPs are essential for the growth of cancer, and aggressive cancer phenotypes are linked to higher MMP levels. Critical information for diagnostic and therapy choices is provided by the detection of distinct MMP profiles in different forms of cancer. Moreover, the amalgamation of omics methodologies in conjunction with comprehensive investigations on MMPs has significant possibilities for augmenting our comprehension of cancer biology and formulating more accurate treatment methods. We are getting closer to realizing the promise of precision medicine as we continue to understand the complexity of MMPs and their roles in different forms of cancer.

DECLARATIONS

Conflict of interest: The authors declare no conflicts of interest.

Consent for publishing: All authors have approved the final version of the manuscript and given consent for publication.

Funding: Not Applicable.

REFERENCES

Aleksander, S. A., et al. (2023). "The gene ontology knowledgebase in 2023." *Genetics*, *224*(1), iyad031.

Anwar, S., Dikhit, M. R., Singh, K. P., et al. (2014). Interaction between Nbp35 and Cfd1 proteins of cytosolic Fe-S cluster assembly reveals a stable complex formation in *Entamoeba* histolytica. *PLoS One*, *9*(10), e108971.

Ashburner, M., Ball, C. A., Blake, J. A., et al. (2000). Gene ontology: Tool for the unification of biology. *Nature Genetics*, *25*(1), 25–29.

Aslam, B., Basit, M., Nisar, M. A., et al. (2016). Proteomics: Technologies and their applications. *Journal of Chromatographic Science*, *55*(2), 1–15.

Azevedo Martins, J. M., Rabelo-Santos, S. H., do Amaral Westin, M. C., et al. (2020). Tumoral and stromal expression of MMP-2, MMP-9, MMP-14, TIMP-1, TIMP-2, and VEGF-A in cervical cancer patient survival: A competing risk analysis. *BMC Cancer*, *20*(1), 1–11.

Cai, Y. D., Zhang, Q., Zhang, Y. H., et al. (2017). Identification of genes associated with breast cancer metastasis to bone on a protein–protein interaction network with a shortest path algorithm. *Journal of Proteome Research*, *16*(2), 1027–1038.

Carbon, S., et al. (2009). "AmiGO: Online access to ontology and annotation data." *Bioinformatics*, *25*(2), 288–289.

Carey, P., et al. (2021). "Metalloproteinases in ovarian cancer." *International Journal of Molecular Sciences*, *22*(7), 3403.

Cathcart, J., Pulkoski-Gross, A., & Cao, J. (2015). Targeting matrix metalloproteinases in cancer: Bringing new life to old ideas. *Genes & Diseases*, *2*(1), 26–34.

Cho, W. C. (2007). Proteomics technologies and challenges. *Genomics, Proteomics & Bioinformatics*, *5*(2), 77–85.

Coleman, W. B. (2017). Next-generation breast cancer omics. *The American Journal of Pathology*, *187*(10), 2130–2132.

Cuypers, E., Claes, B. S., Biemans, R., et al. (2022). "On the spot" digital pathology of breast cancer based on single-cell mass spectrometry imaging. *Analytical Chemistry*, *94*(16), 6180–6190.

Decock, J., et al. (2011). "Matrix metalloproteinases: Protective roles in cancer." *Journal of Cellular and Molecular Medicine*, *15*(6), 1254–1265.

Di, D., Chen, L., Guo, Y., Wang, L., et al. (2018). Association of BCSC-1 and MMP-14 with human breast cancer. *Oncology Letters*, *15*(4), 5020–5026.

Dobra, G., Gyukity-Sebestyén, E., Bukva, M., et al. (2023). MMP-9 as prognostic marker for brain tumours: A comparative study on serum-derived small extracellular vesicles. *Cancers*, *15*(3), 712.

Doghish, A. S., Ali, M. A., Elyan, S. S., et al. (2023). miRNAs role in cervical cancer pathogenesis and targeted therapy: Signaling pathways interplay. *Pathology-Research and Practice*, *224*, 154386.

Dong, H., Lei, J., Ding, L., et al. (2013). MicroRNA: Function, detection, and bioanalysis. *Chemical Reviews*, *113*(8), 6207–6233.

Fischer, T., & Riedl, R. (2021). Challenges with matrix metalloproteinase inhibition and future drug discovery avenues. *Expert Opinion on Drug Discovery*, *16*(1), 75–88.

Fridman, R., Fuerst, T. R., Bird, R. E., et al. (1992). Domain structure of human 72-kDa gelatinase/type IV collagenase. Characterization of proteolytic activity and identification of the tissue inhibitor of metalloproteinase-2 (TIMP-2) binding regions. *Journal of Biological Chemistry*, *267*(22), 15398–15405.

Gialeli, C., Theocharis, A. D., & Karamanos, N. K. (2011). Roles of matrix metalloproteinases in cancer progression and their pharmacological targeting. *The FEBS Journal*, *278*(1), 16–27.

He, L., Kang, Q., Chan, K. I., et al. (2023). The immunomodulatory role of matrix metalloproteinases in colitis-associated cancer. *Frontiers in Immunology*, *13*, 1093990.

He, M., Guo, S., & Li, Z. (2015). In situ characterizing membrane lipid phenotype of breast cancer cells using mass spectrometry profiling. *Scientific Reports*, *5*(1), 11298.

Huang, X., Lan, Y., Li, E., et al. (2021). Diagnostic values of MMP-7, MMP-9, MMP-11, TIMP-1, TIMP-2, CEA, and CA19-9 in patients with colorectal cancer. *The Journal of International Medical Research*, *49*(5), 3000605211012570.

Ibrahim, Z. M., & Ngom, A. (2015). The relative vertex clustering value-a new criterion for the fast discovery of functional modules in protein interaction networks. *BMC Bioinformatics*, *16*, 1–14.

Idle, J. R., & Gonzalez, F. J. (2007). Metabolomics. *Cell Metabolism*, *6*(5), 348–351.

Jemal, A., Bray, F., Center, M. M., et al. (2011). Global cancer statistics. *CA: A Cancer Journal for Clinicians*, *61*(2), 69–90.

Kaur, K., Zhu, K., Whittemore, M. S., et al. (2002). Identification of the active site of gelatinase B as the structural element sufficient for converting a protein to a metalloprotease. *Biochemistry*, *41*(15), 4789–4797.

Kim, H. S., Kim, M. G., Min, K. W., et al. (2021). High MMP-11 expression associated with low CD8+ T cells decreases the survival rate in patients with breast cancer. *PLoS One*, *16*(5), e0252052.

Kurnia, I., Rauf, S., Hatta, M., et al. (2022). Molecular Patho-mechanisms of cervical cancer (MMP1). *Annals of Medicine and Surgery*, *77*, 103415.

Lee, Y. W., Cho, H. J., Lee, W. H., & Sonntag, W. E. (2012). Whole brain radiation-induced cognitive impairment: Pathophysiological mechanisms and therapeutic targets. *Biomolecules & Therapeutics*, *20*(4), 357.

Leiserson, M. D., Vandin, F., Wu, H. T., et al. (2015). Pan-cancer network analysis identifies combinations of rare somatic mutations across pathways and protein complexes. *Nature Genetics*, *47*(2), 106–114.

Leng, A., Jing, J., Nicholas, S., et al. (2019). Catastrophic health expenditure of cancer patients at the end-of-life: A retrospective observational study in China. *BMC Palliative Care*, *18*, 1–10.

Liao, H. Y., Da, C. M., Liao, B., & Zhang, H. H. (2021). Roles of matrix metalloproteinase-7 (MMP-7) in cancer. *Clinical Biochemistry*, *92*, 9–18.

Lovejoy, B., Welch, A. R., Carr, S., Luong, C., Broka, C., Hendricks, R., ... Browner, M. F. (1999). Crystal structures of MMP-1 and-13 reveal the structural basis for selectivity of collagenase inhibitors. *Nature Structural Biology*, *6*(3), 217–221.

Lowe, R., Shirley, N., Bleackley, M., et al. (2017). Transcriptomics technologies. *PLoS Computational Biology*, *13*(5), e1005457.

Mallya, S. K., Mookhtiar, K. A., Gao, Y., Brew, K., Dioszegi, M., Birkedal-Hansen, H., & Van Wart, H. E. (1990). Characterization of 58-kilodalton human neutrophil collagenase: Comparison with human fibroblast collagenase. *Biochemistry*, *29*(47), 10628–10634.

Miller, W., Makova, K. D., Nekrutenko, A., & Hardison, R. C. (2004). Comparative genomics. *Annual Review of Genomics and Human Genetics*, *5*, 15–56.

Napoli, S., Scuderi, C., Gattuso, G., et al. (2020). Functional roles of matrix metalloproteinases and their inhibitors in melanoma. *Cells*, *9*(5), 1151.

Neagu, A. N., Whitham, D., Buonanno, E., et al. (2021). Proteomics and its applications in breast cancer. *American Journal of Cancer Research*, *11*(9), 4006.

Niland, S., Riscanevo, A. X., & Eble, J. A. (2021). "Matrix metalloproteinases shape the tumor microenvironment in cancer progression. *International Journal of Molecular Sciences*, *23*(1), 146.

Padežnik, T., Oleksy, A., Cokan, A., Takač, I., & Sobočan, M. (2023). Changes in the extracellular matrix in endometrial and cervical cancer: A systematic review. *International Journal of Molecular Sciences*, *24*(6), 5463.

Parsons, J., & Francavilla, C. (2020). Omics approaches to explore the breast cancer landscape. *Frontiers in Cell and Developmental Biology*, *7*, 395.

Pavlovsky, A. G., Williams, M. G., Ye, Q. Z., et al. (1999). X-ray structure of human stromelysin catalytic domain complexed with nonpeptide inhibitors: Implications for inhibitor selectivity. *Protein Science*, *8*(7), 1455–1462.

Piskór, B. M., Przylipiak, A., Dąbrowska, E., et al. (2021). Plasma concentrations of matrilysins MMP-7 and MMP-26 as diagnostic biomarkers in breast cancer. *Journal of Clinical Medicine*, *10*(7), 1436.

Radisky, E. S., Raeeszadeh-Sarmazdeh, M., & Radisky, D. C. (2017). Therapeutic potential of matrix metalloproteinase inhibition in breast cancer. *Journal of Cellular Biochemistry*, *118*(11), 3531–3548.

Sbardella, D., Fasciglione, G. F., Gioia, M., et al. (2012). Human matrix metalloproteinases: An ubiquitarian class of enzymes involved in several pathological processes. *Molecular Aspects of Medicine*, *33*(2), 119–208.

Seehausen, O., Butlin, R. K., Keller, I., et al. (2014). Genomics and the origin of species. *Nature Reviews Genetics*, *15*(3), 176–192.

Shannon, P., et al. (2003). "Cytoscape: A software environment for integrated models of biomolecular interaction networks." *Genome Research*, *13*(11), 2498–2504.

Shi, X., Yu, J., Lu, C., Luo, Q., Xu, C., Li, J., & Wang, W. (2023). Screening of the shared pathogenic genes of ulcerative colitis and colorectal cancer by integrated bioinformatics analysis. *Journal of Cellular and Molecular Medicine*. https://doi.org/10.1111/jcmm.17878

Shulaev, V. (2006). Metabolomics technology and bioinformatics. *Briefings in Bioinformatics*, *7*(2), 128–139.

Thomas, P. D., et al. (2022). "PANTHER: Making genome-scale phylogenetics accessible to all." *Protein Science*, *31*(1), 8–22.

Visse, R., & Nagase, H. (2003). Matrix metalloproteinases and tissue inhibitors of metalloproteinases: Structure, function, and biochemistry. *Circulation Research*, *92*(8), 827–839.

Wang, Q. M., Lv, L., Tang, Y., et al. (2019). MMP-1 is overexpressed in triple-negative breast cancer tissues and the knockdown of MMP-1 expression inhibits tumor cell malignant behaviors *in vitro*. *Oncology Letters*, *17*(2), 1732–1740.

Warde-Farley, D., et al. (2010). "The geneMANIA prediction server: Biological network integration for gene prioritization and predicting gene function." *Nucleic Acids Research* 38(suppl_2), W214–W220.

Xu, J., Yao, Y., Ren, S., Wang, G., & Jin, H. (2016). Matrix metalloproteinase expression and molecular interaction network analysis in gastric cancer. *Oncology Letters*, *12*(4), 2403–2408.

Yadav, L., Puri, N., Rastogi, V., Satpute, P., Ahmad, R., & Kaur, G. (2014). Matrix metalloproteinases and cancer-roles in threat and therapy. *Asian Pacific Journal of Cancer Prevention*, *15*(3), 1085–1091.

Yan, T., Cai, Y., Wei, Y., et al. (2023). Expression of matrix metalloproteinases and their association with clinical characteristics of solid tumors. *Gene*, *850*, 146927.

Zografos, E., Anagnostopoulos, A. K., Papadopoulou, A., et al. (2019). Serum proteomic signatures of male breast cancer. *Cancer Genomics & Proteomics*, *16*(2), 129–137.

31 Designing Novel Serine Protease Inhibitors against Potential Biomarkers Specific to Lung Adenocarcinoma through Systems Biology and Structural Bioinformatics Approaches

Anamika Mishra and Soundharya H.
School of Bio Sciences and Technology, Vellore Institute of Technology, Vellore, Tamil Nadu, India

Aniket Naha
Medical Biotechnology and Computational Drug Designing Laboratory, Pushpagiri Research Centre, Pushpagiri Medical Society, Tiruvalla, Kerala, India

Soumitra Nath
Gurucharan College, Silchar, Assam, India

Nibu Varghese
Mar Athanasios College for Advanced Studies (MACFAST), Tiruvalla, Kerala, India

Soumya R. S.
Pushpagiri Research Centre, Pushpagiri Medical Society, Tiruvalla, Kerala, India

31.1 INTRODUCTION

31.1.1 Lung Adenocarcinoma (LUAD)

Lung adenocarcinoma (LUAD) refers to malignant tumour in the epithelial cells of lungs, which is the leading cause of global cancer-related death, claiming 1,796,144 lives in 2020, making up nearly 18% of all fatalities. More than 2 million cases were diagnosed worldwide with non-small-cell lung cancer (NSCLC) comprising more than 81% in the United States with the average age at diagnosis being 71 and older (Myers & Wallen, 2022). Morphologically, the tumour displays an epithelial

DOI: 10.1201/9781003394693-36

neoplasm with glandular differentiation or occasionally mucin production. It often develops as a peripheral mass with pleural puckering and thickening, as well as central fibrosis (Mullangi & Lekkala, 2020). Though smoking contributes as a major risk factor associated to this cancer, non-smokers account for the majority of instances. However, in women, a decline of 4% was observed, which attributes to reduced smoking rates, improved treatments, and early detection. The severity and incidence rate of LUAD depends on the exposure rate of carcinogens emitted from smoking, radiation, genetic mutations and environmental contaminants (Myers & Wallen, 2022).

LUAD has the potential to spread to nearby sites such as the pleura, diaphragm, pericardium layer and bronchi. In advanced stages, it may invade the mediastinum, major blood vessels, trachea, oesophagus, vertebral column and adjacent lung lobes. The malignancy can also extend to nearby lymph nodes, progressing from peri bronchial nodes to mediastinal or subcarinal nodes and ultimately, lymph nodes in the adjacent lung. Distant metastasis involves the spread to adjacent lung areas, pleural nodules, pericardial or pleural malignant effusion, or distant organs like the brain, bones or liver (Myers & Wallen, 2022). During the early stages, LUAD may not present specific symptoms, with detection often occurring through CT scans and chest X-rays. Bronchoscopy and lung tissue biopsy confirm the presence of malignant tumour cells. General symptoms include a persistent cough with mucus or blood, difficulty breathing, chest pain, wheezing, fatigue, loss of appetite, swelling in the neck veins and recurring lung infections (Myers & Wallen, 2022).

31.1.2 Subtypes of LUAD

According to the World Health Organization (WHO) classification of lung cancer, adenocarcinoma of the lung is further categorized into various subtypes, namely, acinar, lepidic, papillary, micropapillary, invasive mucinous, minimally invasive, solid, colloid, fetal and enteric adenocarcinoma, which is briefly illustrated in Figure 31.1 (Nicholson et al., 2022). Each subtype is categorized

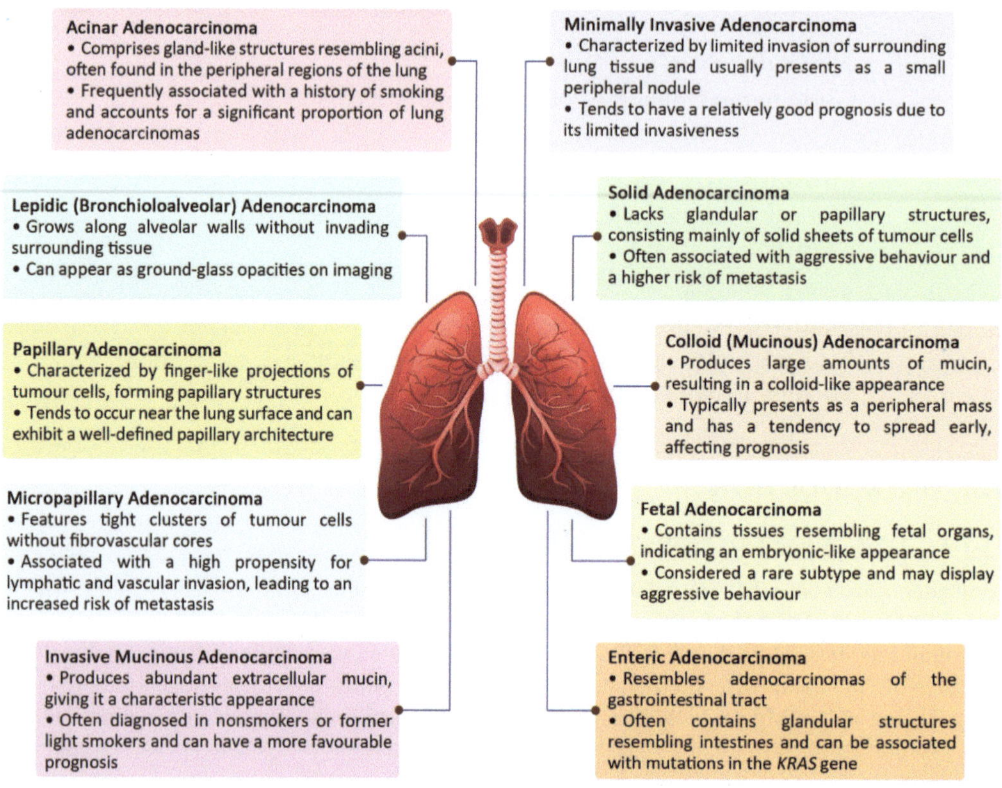

FIGURE 31.1 Types and subtypes of lung adenocarcinoma (LUAD).

according to one main growth pattern and morphologic characteristics, depending on the size of the tumour, its invasiveness, the histologic site of invasion and its cellular arrangement pattern. The most prevalent of these categories is lepidic tumour growth, which includes the development of tumours surrounding pre-existing alveolar structures with increased alveolar septa with significant prognostic features. Acinar adenocarcinoma, on the other hand, is usually found with tumour cells invading the glandular structures of a fibroelastic stroma, with poor prognostic characteristics (Zheng, 2016). Papillary adenocarcinoma is characterized by a papillary growth pattern with finger-like extensions of cancer cells in the lining of fibrovascular centres. Micropapillary adenocarcinoma, on the other hand, results from the formation of cellular tufts without the involvement of the fibrovascular core. One of the aggressive subtypes of LUAD with a poor prognosis in the lungs is micropapillary adenocarcinoma. Solid adenocarcinoma is another subtype of LUAD that often develops as layers of cancer cells without any discernible pattern of growth. It is made up of cells that are tightly packed and lacks glandular or papillary like cell growth. The lepidic, acinar, papillary, micropapillary and solid adenocarcinoma together comprise invasive adenocarcinoma, while the rest of the subtypes like fetal, colloid and enteric adenocarcinoma are rare variants and are categorized as invasive mucinous adenocarcinoma (Inamura, 2017).

31.1.3 Genomic Alterations of LUAD

LUAD is subjected to numerous mutations which trigger the invasion of malignant tumours. There are several somatic genomic alterations that drive LUAD. According to the comprehensive molecular profiling report of The Cancer Genome Atlas (TCGA), 18 significant genomic mutations were reported. Amongst all, mutations in *KRAS, STK11, EGFR, NF1, TP53, KEAP1* and *BRAF* significantly enrich tumour incidence (The Cancer Genome Atlas Research Network, 2014). The copy number alterations like gain of chromosome 5p also influence LUAD. The amplification of telomerase reverse transcriptase (*TERT*) is responsible for the 5p15 alteration, which results in enabling replicative immortality to the cancer cells (Alidousty et al., 2021). Similarly, amplification of the *MYC, EGFR, MET, KRAS, MDM2* and *ERBB2* genes and deletions in *LRP1B, PTPRD*, and *CDKN2A* are some recurrent alterations reported in LUAD. Furthermore, the majority of genomic alterations (~75%) were found to promote the RTK/RAF/RAS signalling pathway while genomic alterations of *ALK, ROS1* and *RET* have shown high incidence of LUAD in younger patients (Luo et al., 2018).

31.2 THERAGNOSTIC BIOMARKERS PROPOSED FOR LUAD

Given the complexity of the tumour, it is crucial to recognize important biomarkers in order to facilitate accurate cancer diagnosis, prognosis and the development of tailored treatments. There are several types of tumour biomarkers such as genetic mutations or alterations, epigenetic modifications, proteomic and metabolic changes, level of miRNAs, expression level of tumour-related proteins, circulating immune cells, tumour cells and so on. Among these, protein biomarkers are the most promising indicators for diagnosis (Zamay et al., 2017). *ALK, EGFR* and *KRAS* are the three specific genes that are typically used to characterize LUAD. Since epidermal growth factor receptor (EGFR) regulates cellular homeostasis, proliferation and growth, *EGFR* mutations have a substantial impact on *EGFR* signalling and in turn accelerate the growth and proliferation of cancer cells. The mutated *EGFR* causes *EGFR* signalling to impact the aerobic glycolysis, thus facilitating the increased uptake of glucose by cancer cells. This mutation increases the ability of the malignant cells to proliferate further and activates the antiapoptotic signalling pathway (Calvayrac et al., 2017).

Anaplastic lymphoma kinase (*ALK*) rearrangement involving the fusion of *ALK* with *EML4* and helicase enzyme *DEAD/DEAH* helicase 11 (*DDX11*) serves as prominent biomarkers for the diagnosis of LUAD. Recently, an important factor in the development of cancer was then reported as the elevated expression of genes such *CENPA*, *CDC20* and *CDK1*. The role of *CDC20* and *CDK1* in cell proliferation and cell division regulation were associated with the increasing growth of tumour cells, but the role of centromere protein A (*CENPA*) is not clearly understood in the development of

LUAD. The mutual overexpression of these three genes has been utilized as a network of diagnostic biomarkers in LUAD (Liu et al., 2018). Several circulating circular RNAs (circRNAs) have been identified to induce tumourigenesis and thus play a significant role as diagnostic biomarkers. In contrast to healthy people, patients with LUAD were shown to have higher rates of three differentially expressed serum and serum exosomal circRNAs, hsa_circ_0001492, hsa_circ_0001439 and hsa_circ_0000896 (Kang et al., 2022). Feng in 2018 reported that the exosomal miRNAs in the serum showed notable upregulation in LUAD patients. Ex-miR21-5p, Ex-miR126-3p and Ex-miR140-5p expression levels were higher in LUAD patients than in healthy controls, indicating that these three ex-miRNAs may be more accurate, sensitive and specific diagnostic molecular markers of early adenocarcinoma (Feng et al., 2018). *HER2* is an important tumour biomarker belonging to the *ERBB* receptor tyrosine kinase family which is responsible for encoding *HER2* protein. *HER2* plays a significant role in promoting cell proliferation by downstream activation of the *PI3K-AKT* and *MEK-ERK* signalling pathways. Mutations in *HER2* have become crucial targets for developing novel therapeutic approaches, one such being dacomitinib.

Liu *et al.*, have reported apolipoprotein E (*APOE*) as a potential biomarker of LUAD that actively involves in angiogenesis, metastasis, DNA synthesis and cell proliferation. The overexpression of *APOE* serves as an efficient early diagnostic biomarker and a therapeutic biomarker to validate and monitor the effect of treatment (Liu et al., 2014). Apart from these, *ALKBH2* and *CYP4B1* genes have also been investigated for their effective prognostic and therapeutic target potential. Elevated *ALKBH2* expression were found to be highly linked with decreased overall survival (OS) and progression-free survival (PPS) while dysregulated *ALKBH2* expression affected the ability of immune cells to invade the LUAD malignancy (Wu et al., 2021).

31.3 PROTEASE INHIBITORS AND THEIR CLASSES

Protease inhibitors are found to be a prominent therapeutic agent because of their ability to intervene in the processing of the growth of cancer, its invasion and metastasis. Protease inhibitors can be any chemical that includes proteins, peptides or even small molecules but majorly classified as reversible, irreversible and engineered. The reversible inhibitors (bortezomib, nirmatrelvir) interact with the target proteins via a non-covalent interaction, while the inhibition can be reversed once the inhibitor is removed from the system. The irreversible inhibitors covalently link to their targets and the inhibition cannot be reversible or deactivated. The engineered inhibitors have high specificity and selectivity to bind to the active site or allosteric sites of proteases and hinders the spread of LUAD upon confining them to a particular position (Song et al., 2021).

There are various classes of protease inhibitors that have displayed high clinical effectiveness like tyrosine kinase inhibitors (erlotinib, afatinib, osimertinib) which specifically target the EGFR, and it is widely employed for curing advanced NSCLC that includes LUAD. Similarly, ALK inhibitors like alectinib, brigantinib (Alunbrig) and crizotinib that are administered against NSCLC with ALK gene rearrangements (Trezza et al., 2020).

31.4 DESIGNING PROTEASE INHIBITORS AGAINST LUAD

The protease inhibitors designed for LAUD involve targeting proteases that play a crucial role in the development, progression, metastasis and invasiveness of cancer. The first step involves thorough research to identify the proteases that are overexpressed in LAUD. The proteases that are highly involved in LAUD are matrix metalloproteinases (MMPs), cathepsins and urokinase plasminogen activator (uPA), that could be potential targets for inhibition. Thereafter, rational drug designing is carried out upon designing inhibitors with high specificity and affinity from the available 3D structures of target proteases. Computer-aided molecular modelling and docking studies are employed to predict the intermolecular interactions between inhibitors and proteases' active sites. The third step involves high-throughput screening, which is performed to screen large compound libraries for potential lead inhibitors that show promising activity against the target proteases. This approach can identify novel scaffolds and chemical entities with inhibitory potential. Following this, the peptide-based inhibitors are investigated, these are designed to mimic the natural substrates of the target proteases. These peptides could be modified to improve stability, cell permeability and specificity.

Recently, combination therapies were widely used that had shown an increased therapeutic spectrum for treating diseases (Naha et al., 2021; Naha & Ramaiah, 2023). Similarly, protease inhibitors in combination with other targeted therapies or chemotherapy may lead to synergistic effects, enhancing the overall efficacy of treatment. Thereafter, strategies for selective delivery of protease inhibitors to target sites needs to be optimized for minimizing potential side effects on normal tissues and maximizing the therapeutic effect. Following rigorous *in vitro* optimization and validation, pre-clinical and clinical trials are finally conducted to evaluate the safety and efficacy of the inhibitors in patients with LUAD (Eatemadi et al., 2017).

31.5 MECHANISM OF ACTION OF PROTEASE INHIBITORS

There are several steps involved in the mechanism of protease inhibitors for combating LAUD, which includes the inhibition of tumour invasion and metastasis. Proteases such MMPs and serine proteases are involved in degrading the extracellular matrix, allowing cancer cells to invade surrounding tissues and spread to distant sites. Protease inhibitors can block the activity of these enzymes, thereby preventing cancer cells from breaking down the extracellular matrix and inhibiting their ability to invade surrounding tissues and metastasize to other organs. Thereafter, the induction of apoptosis occurs where proteases like caspases are key players in the programmed cell death processes. Dysregulation of apoptosis is a common characteristic of cancer cells. Protease inhibitors can restore apoptosis in cancer cells by inhibiting antiapoptotic proteins or promoting the activation of proapoptotic proteins.

Furthermore, the immune response modulates some of the proteases involved in immune response modulation by processing cytokines and chemokines. Protease inhibitors can affect the tumour microenvironment by influencing immune-cell recruitment and activity, potentially enhancing the antitumour immune response. Targeting the oncogenic pathways is the next step in this process, where certain proteases can activate or process signalling molecules involved in oncogenic pathways. Protease inhibitors can disrupt these pathways by preventing the cleavage of signalling molecules or their receptors, leading to the inhibition of cancer cell growth and survival. Finally, medication is the last step which can be achieved through personalized medicine, targeted therapy, genetic mutations and alterations in proteases. Protease inhibitors can be tailored to target specific mutated proteases, offering a personalized approach to treatment based on the molecular characteristics of the tumour (Manasanch & Orlowski, 2017). Table 31.1 and Figure 31.2 highlight the examples of protease inhibitors alongside their mechanisms of action.

TABLE 31.1
Different Types of Protease Inhibitors and Their Mechanism of Action

Protease Inhibitor	Mechanism of Action	Example
Erlotinib (Tarceva)	EGFR tyrosine kinase inhibitor, blocks signalling	EGFR is overexpressed in LAUD
Afatinib (Gilotrif)	Irreversible EGFR/HER2 inhibitor, disrupts signalling	EGFR mutations in LAUD
Osimertinib (Tagrisso)	Selective EGFR inhibitor, targets T790M mutation	EGFR T790M mutation in LAUD
Alectinib (Alecensa)	ALK inhibitor, blocks ALK signalling	ALK rearrangements in LAUD
Brigatinib (Alunbrig)	ALK inhibitor targets ALK and ROS1	ALK or ROS1 mutations in LAUD
Crizotinib (Xalkori)	ALK inhibitor, targets ALK and MET	ALK rearrangements or MET amplification
Pembrolizumab (Keytruda)	Immune checkpoint inhibitor, targets PD-1	Enhances immune response in LAUD

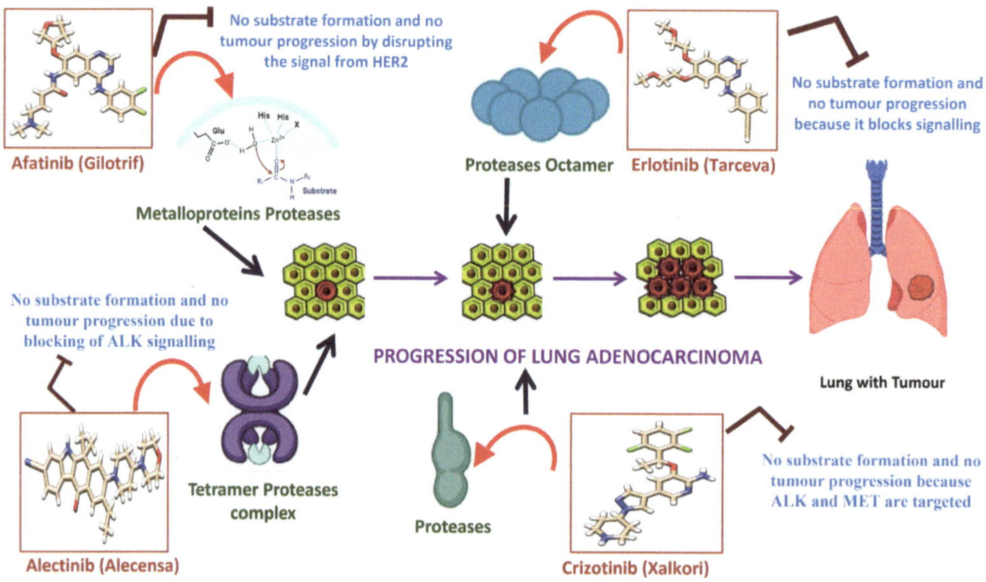

FIGURE 31.2 Schematic representation of the inhibitory action mechanisms of some potent protease inhibitors administered to combat LAUD.

31.6 DATABASE/SERVERS FOR *IN SILICO* CANCER ANALYSIS

There are various databases and servers that amalgamate curated data from various sources (experimental, theoretical or predicted) for studying different stages of cancer progression. The data available on these platforms include genetic mutations, gene expression profiles, copy number alterations and clinical annotations from a multitude of patients diagnosed with LAUD. These resources facilitate in gaining insights into the genetic drivers of LAUD identify biomarkers, understand disease progression and predict patient outcomes. These databases help to make a user-friendly environment, interactive visualizations and powerful analysis tools that empower users to extract meaningful information from complex datasets (Myers & Wallen, 2022).

TABLE 31.2
Different Data Repositories and Servers Used in Lung Cancer Research

Database/ Server	Description	URL
TCGA	Provides comprehensive genomic and clinical data for cancers, including LAUD	https://www.cancer.gov/tcga
GDC	Hosts and provides access to TCGA data for analysis and visualization	https://portal.gdc.cancer.gov/
cBioPortal	Platform to explore genetic alterations, pathways and clinical outcomes	https://www.cbioportal.org/
UCSC Cancer Genomics Browser	Interface for visualizing and analysing cancer genomics data	http://xena.ucsc.edu/welcome-to-ucsc-xena/
ICGC	Offers genomic data for various cancer types, including LAUD	https://dcc.icgc.org/
COSMIC	Provides somatic mutation data for various cancers, including LAUD	https://cancer.sanger.ac.uk/cosmic
Lung Cancer Explorer	Comprehensive analysis of lung cancer data, including LAUD	https://lce.biohpc.swmed.edu/lungcancer/
FIREbrowse	Analysis and visualization of TCGA data, including LAUD	http://firebrowse.org/
OncoLnc	Provides survival analysis based on gene expression levels in various cancers, including LAUD	http://www.oncolnc.org/

The TCGA is one such comprehensive database that provides genomic and clinical data for a wide range of cancers, including LUAD. It offers high-quality data on genetic mutations, gene expression, and other molecular features of cancer. Genomic Data Commons (GDC) is the source that provides information about the hosts and provides access to the TCGA data. Its highly cross referenced and gives access to several other servers and databases including cBioPortal, UCSC Cancer Genomics Browser, Lung Cancer Mutation Consortium (LCMC), Lung Adenocarcinoma Gene Expression Database (LAUDGED) and the International Cancer Genome Consortium (ICGC) alongside many others. Table 31.2 provides detailed information about the uses of various servers and databases specific to LUAD and databases which host information pertaining to LUAD (Ashok & Ramaiah, 2022).

31.7 SYSTEMS BIOLOGY APPROACHES FOR LAUD

A systems biology approach to studying LAUD can enhance understanding the disease as a complex system of interconnected components, including genes, proteins, pathways and interactions (Shen et al., 2021). The descriptive depiction of the systems biology approach is illustrated in Figure 31.3.

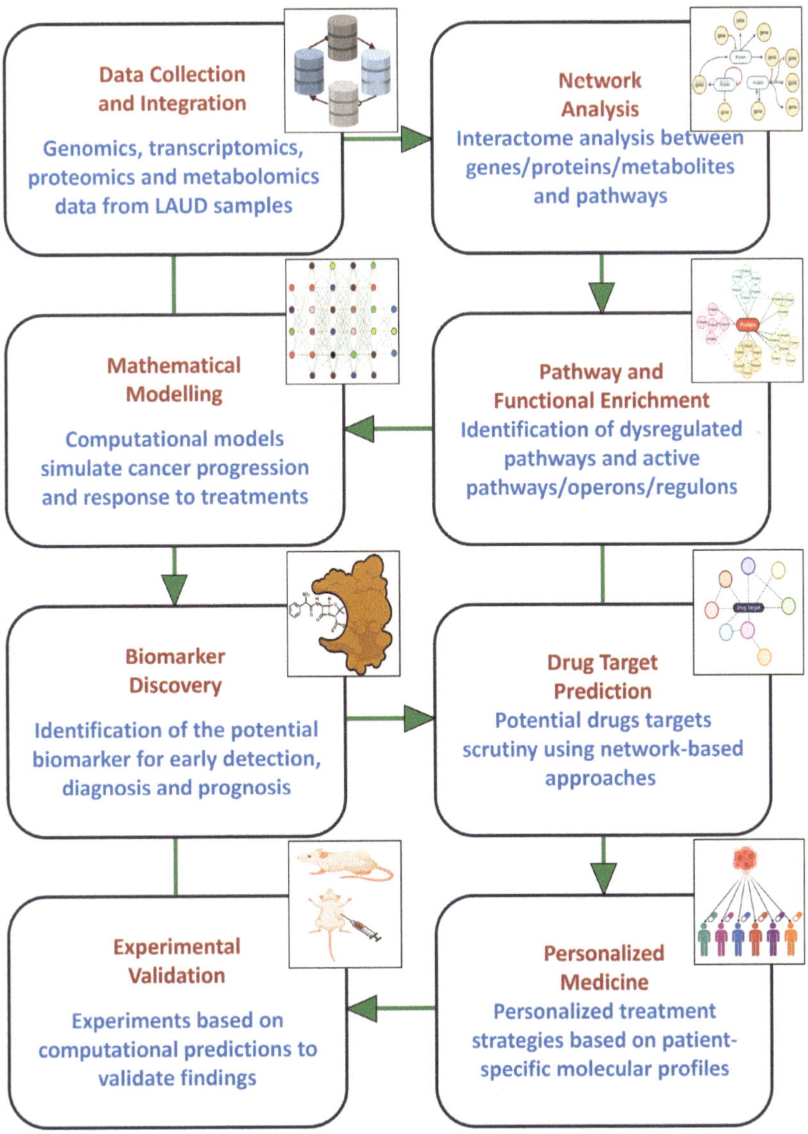

FIGURE 31.3 A streamline methodology highlighting the important aspects of the systems biology approach.

31.7.1 Gene Interaction or Protein–Protein Interaction Network Analysis

Construction of a detailed gene interaction network for LAUD involves integrating omics data, identifying key genes and analysing their interactions within the context of pathways and biological process. Thereafter, transcriptomics, genomics and proteomics data need to be collected, integrated and evaluated in gene or protein interaction prediction servers (STRING, BioGRID, IntAct, GeneMANIA) to build the possible interactome based on both experimental validations and theoretical predictions. Following, data normalization might be required to ensure consistency and removal of technical and biological errors. Differential gene expression can be studied upon extracting data from NCBI-GEO database and analysis using GEO2R, edgeR, DESeq2 or limma packages to disseminate the patterns of upregulation or downregulation. The gene interaction or protein–protein interactome, co-expression patterns and pathway associations can be finally visualized using Cytoscape, which represents genes as nodes and interactions as edges. Elucidation of the functionality of interacting genes/proteins and their metabolic pathways are done upon functional enrichment analysis that depicts the gene ontology (biological process, molecular function and cellular components), domain enrichments (Pfam, InterPro) and pathway enrichments (KEGG, Reactome). Finally, validation and clinical relevance entails experimental validation, benchmarking, and cross-validation, facilitating a comparison of predicted interactions with known interactions from trusted sources such as curated databases (e.g., BioGRID, IntAct) to assess the predictive accuracy of the software (Naha et al., 2020, 2022).

31.7.2 Multiomics Mutational and Epigenomics Network Analysis for LAUD

Epigenetic modifications, such as DNA methylation, histone modification and regulation of non-coding RNAs may lead to activation of oncogenes and inactivation of tumour suppressor genes. Dysregulation of epigenetic modifications can impact signalling pathways in cancerous cells, and these changes can provide valuable biomarkers and potential therapeutic targets for cancer. On the other hand, copy number variations (CNV) can lead to abnormal gene expression and contribute to genome and molecular phenotype heterogeneity. DNA CNV is associated with a higher risk of LAUD development. Hence, the comprehensive analysis of CNV is essential for understanding its role in LAUD progression. Therefore, comprehensive analysis of mutations, epigenetics and transcriptomics using multiomics approaches is crucial for the proper treatment of LAUD.

The network is designed based on data retrieval from RNA-seq, CNV, methylation data and clinical information for LAUD from the TCGA and other databases. The clinical data including gender, age, TNM stage, tumour grade, and OS are retrieved from the GDC portal. The gathered data are processed by identifying identical regions with overlaps, removing regions with fewer than five probes, calculating gene CNV values, removing sites with insufficient detection, filling missing values using the k-Nearest Neighbour (KNN) algorithm and filtering specific methylation sites. Correlation coefficients between gene expression levels needs to be calculated to identify CNV and methylation-related genes, as well as mutated and epigenetically modified genes. To enhance the understanding of genes associated with LAUD and to examine the expression patterns, icluster multiomics clustering analysis and TIMMER analysis are good options. The final validation involves the Kaplan-Meier survival analysis for OS, PPS survival, and disease-specific survival. KMplot utilizes the gene expression profiles to validate the prediction and assess its prognostic significance through the GEPIA database. This comprehensive validation process contributes to the credibility and generalizability of the initial analysis (Zhao et al., 2021).

31.8 STRUCTURAL BIOINFORMATICS APPROACHES FOR LAUD

The structural biology approach plays an important role in the design of novel inhibitors for LUAD. This approach involves studying the three-dimensional structures of proteins involved in LUAD

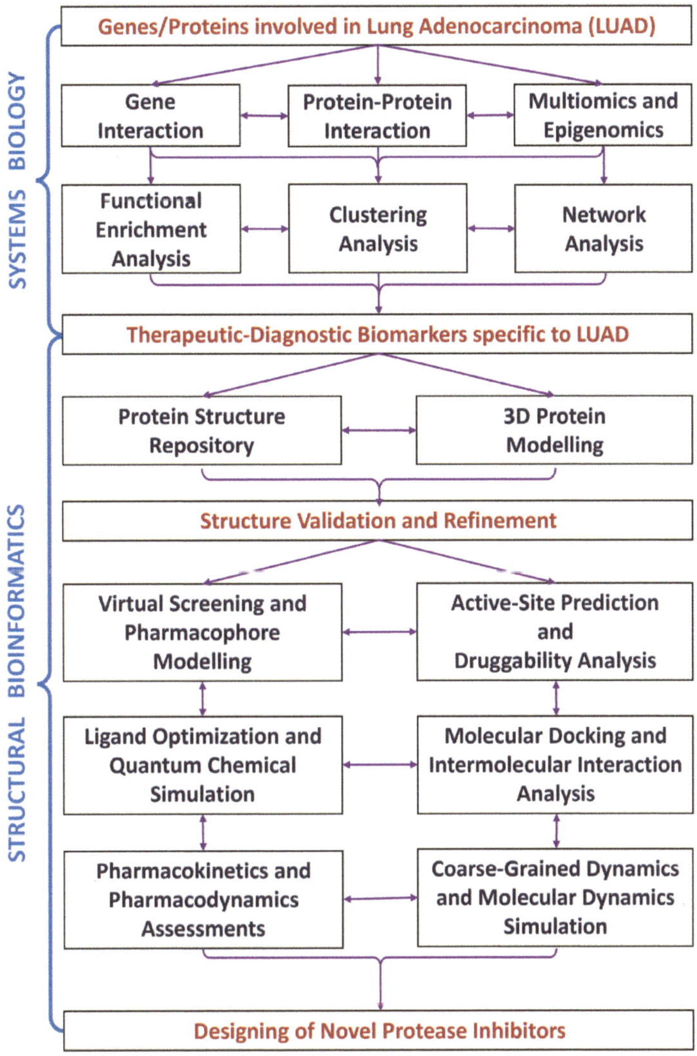

FIGURE 31.4 A conjoint methodology integrating systems biology and structural bioinformatics approaches to design novel lead molecules against theragnostic biomarkers specific to lung adenocarcinoma.

pathogenesis to identify potential binding sites and interactions with inhibitors (Bassetto et al., 2016). A detailed pipeline explaining the interlinking of systems biology and structural bioinformatics pivotal to cancer research is illustrated in Figure 31.4.

31.8.1 Drug Analogues

Lead compound identification initiates with the selection of an existing drug or a known compound with activity against the target protein. This is followed by structural characterization and rational design; wherein binding interactions are analysed to identify specific functional groups crucial for binding. Structural modifications are then applied to the lead compound, resulting in analogues with altered chemical structures while preserving the core pharmacophore (Bora-Singhal et al., 2020). Virtual screening is conducted to assess potential analogues and structure-based optimization is performed to enhance binding affinity, selectivity and drug-like properties, guided by the structural information of the protein-ligand complex. Pharmacophore mapping identifies essential features in the lead compound, guiding the design of analogues that maintain or enhance these features for improved interaction with targets. Scaffold hopping explores different chemical scaffolds while retaining critical pharmacophoric elements to generate diverse structures (Batool et al., 2019).

31.8.2 Pharmacophore Modelling

Pharmacophore modelling is regarded as a crucial component in the design of novel inhibitors for LUAD within structural biology approaches. It involves the identification and representation of key structural and chemical features that are essential for a ligand to bind to its target protein. There are many servers and software that play an important role in modelling. The Pharmer server allows one to search for the database of protein-ligand complexes to generate pharmacophore models based on known interactions. Understanding ligand–receptor interactions involves the analysis of interactions between known ligands and the target protein in LUAD, with the identification of critical interactions contributing to binding and inhibition. The next step is feature identification, which helps in identifying the key chemical and structural features such as hydrogen bond donors/acceptors, aromatic rings, hydrophobic regions and positively/negatively charged groups. The LigandScout server is used in the generation of pharmacophore models that help in knowing about the essential features required for ligand binding. Validation and optimization are carried out using the Discovery Studio server, known for generating and validating pharmacophore models through a user-friendly interface (Wadanambi & Mannapperuma, 2021).

Interaction mapping visualizes the alignment of potential inhibitors with the pharmacophore model to understand how different ligands match the required features. The final step involves iterative optimization, where the pharmacophore model is iteratively refined, and compounds are designed to improve binding affinity, specificity, and other drug-like properties. Pharmacophore modelling, guiding the rational design of novel inhibitors for LUAD, focuses on essential interactions necessary for effective binding. This approach aids in the identification of compounds capable of disrupting specific protein-ligand interactions and provides insights for the development of targeted therapies (Weng et al., 2020).

31.8.3 Lead Optimization and Validation Through Quantum Chemical Calculations and Density Functional Theory Calculations

The effective lead molecule or analogue needs further optimization and validation to decipher its stability and reactivity parameters. The LAUD microenvironment, being acidic, facilitates in immune suppression, tumour progression and, thus, metastasis. Therefore, lead molecules with electropositive sidechains or electrophilic centres would be pivotal as they are capable of inducing electrophilic addition/substitution reactions. The electronegative centres in the drug could help in mobilizing the electrons towards them, thus facilitating in efficient drug binding with high affinity. These features, including the global and local reactivity parameters, could be adjudged through density functional theory (DFT) calculations and quantum chemical simulations. Ligand designing and optimization can be carried out either in open-source servers like AutoGrow4 or standalone platforms like GAMESS, Multiwfn, ORCA or Gaussian-09 quantum chemistry computational software packages. Visualization of their subsequent results informed of spectral curves or maps are done through standalone GaussView, GaussSum, IboView and Avogadro softwares. The intrinsic properties and chemical reactivity of the molecule could be examined through natural bond orbital (NBO), frontier molecular orbitals (FMO), molecular electrostatic potential (MEP) and local descriptors or Fukui functions. NBO analysis substantiates the charge delocalization from bonding (donor) to antibonding (acceptor) orbitals. The bonding–antibonding interactions contributed by the electronegative donor atomic orbitals to electron-deficient acceptor atomic acceptor orbitals impart the molecular stability of the lead (Naha et al., 2022; Usharani et al., 2023).

Global descriptors describe the overall properties and characteristics of the molecule which could be analysed from the energy profiles displayed by FMOs. Highest occupied molecular orbital (HOMO) and the lowest unoccupied molecular orbital (LUMO) explains the overall electron affinity (A) and ionization potential (I) of the molecule. Local descriptors further elucidate the potential sites for chemical reactivity and site selectivity that are prone to electrophilic, nucleophilic or radical attacks. According to Parr et al. (1999), the highest value of the Fukui function determines the reactive centre in the lead molecule which are expressed as f_k^+, f_k^- and f_k^0 for nucleophilic, electrophilic and radial attacks, respectively. The dual descriptor function (Δf) efficiently predicts the reactive sites favouring nucleophilic attack ($\Delta f > 0$), electrophilic attack ($\Delta f < 0$) and radial attack ($\Delta f = 0$) (Parr et al., 1999).

31.8.4 *In Silico* Pharmacokinetics and Pharmacodynamics (PK-PD)

The pharmacokinetic and pharmacodynamic (PK-PD) parameters play an important role in understanding the behavioural patterns of the inhibitors inside the system and the response of the system upon acquiring the inhibitor molecule. As temperature and pressure play an important attribute in the nature and reactivity of the drug molecule, the analysis could be performed under ambient room temperature (298.15K) and atmospheric pressure of 1.00 atm. The pharmacodynamic metrics were understood upon studying the enthalpy (H), entropy (S), heat capacity (C_v), Gibb's free energy (G), polarizability (α) and dipole moment (μ) of the studied drug molecules. The thermodynamics and dipole moment indicate the potency of molecular stacking and formation of highly ordered self-assembly making the molecule conducive for charge carrier diffusion (Naha et al., 2022; Usharani et al., 2023). However, excessively high dipole moment results in vigorous undesirable aggregate formation further causing difficulty in electronic excitation, which ultimately results in poor phase separation at donor-acceptor interphase. The pharmacokinetics parameters include the absorption, distribution, metabolism, excretion, and toxicity (ADMET) assessments which can be predicted from several online servers like admetSAR, SwissADME, pkSCM and ProTox-II servers (Naha et al., 2022).

31.8.5 Molecular Docking and Intermolecular Interaction Analysis

Molecular docking analysis between the protease inhibitor and a therapeutic biomarker specific to LUAD can be performed to understand the binding affinity of the ligand to that of its target. Molecular docking can be performed using several online servers like the PatchDock, ParDock, SwissDock and SeamDock servers and standalone software platforms like AutoDock and AutoDock Vina. Most of these servers/software use semi-empirical free energy scoring function to predict the binding energies of docked poses with high reproducibility. Based on the binding energy (ΔG_{bind}) which is calculated upon the summation of energies from van der Waals, hydrogen bonding, electrostatic, torsional, solvation, rotational and vibrational interactions.

$$\Delta G_{bind} = \Delta G_{vdW} + \Delta G_{hbond} + \Delta G_{elec} + \Delta G_{tor} + \Delta G_{solv} + \Delta G_{rot} + \Delta G_{vib}$$

Upon performing the molecular docking, the protein-ligand complex would be studied for assessing the intermolecular interaction analysis, which provides a detailed outlook of the various interactions established by the ligand with the amino-acid residues of the target. The various interactions include conventional H-bond, C-H bond, van der Waals interactions and several non-canonical interactions including chalcone bonds, alkyl, π-alkyl, π-sulphur, π-cation/anion and π-sigma bonds (Miryala et al., 2021; Naha et al., 2021, 2022).

31.8.6 Coarse-Grained and Molecular Dynamics Simulation (MDS)

Coarse-grained dynamics would thereafter be performed to determine the residue-level fluctuations in Angstrom (Å) scale concerning the most favourable conformation of the protein. CABSflex is a robust online platform that uses a coarse-grained protein model as a simulation engine for the CABS model design. Stability was ascertained from the Monte-Carlo simulation upon determining the root mean square fluctuation (RMSF) trajectories in a nanometre (nm) scale. The trajectories are generated based on default restraints that provide maximum and minimum ranges on a pair of atoms for undergoing their specific dynamic orientations within defined spaces. The fluctuations beyond the assigned ranges were penalized and designated as unstable (Jamroz et al., 2013).

Thereafter, molecular dynamics simulation (MDS) is performed to estimate and evaluate the stability when both the native (unbound state) drug targets or drug targets complexed with the drug or lead molecule (bound state). The analysis was performed mimicking the *in vitro* environment inside the *in silico* setup. The Groningen machine for chemical simulation (GROMACS) suite is a highly efficient, open-source and flexible MDS suite integrating tutorials for simulating simple proteins in water to highly complex simulations incorporating membrane proteins, biphasic systems, protein-ligand complexes and construction of virtual sites. The stability of the protein or protein-ligand complexes are then adjudged under this dynamic condition to evaluate their stability patterns. Stability is determined through the deviations and fluctuations of the protein backbone or Cα atoms. MDS analysis governs the critical understanding of a protein's stability in a simulated cellular environment (aqueous or lipid) under a favourable timescale. The protein topology needs to be built with several force-field mechanics (GROMOS, CHARMM, OPLS) and simple point charge water models. Thereafter, solvation, energy minimization, and equilibration of the system need to be attained prior to the MD-run. MDS needs to be analysed upon studying the various trajectories like root-mean-square deviation (RMSD), root-mean-square fluctuation (RMSF), radius of gyration (Rg), energy parameters, H-bond analysis, solvation parameters like solvent accessible surface area and free energy of solvation profiles. Upon correlating molecular docking and dynamics simulation results, effector protease inhibitors could be designed against suitable therapeutic biomarkers of LUAD for establishing a successful treatment regimen to combat the disease (Lemkul, 2019; Basu et al., 2022; Naha et al., 2022; Naha & Ramaiah, 2022).

31.9 FUTURE PERSPECTIVES OF CANCER THERAPY

Future perspectives in cancer therapy involve leveraging various advancements in understanding metabolic disorders and the role of protease in cancer development, particularly in lung cancer. Structural bioinformatics and systems biology techniques are crucial in analysing biomarkers' impact on cancer onset and progression. Integrating gene expression studies with protein interaction data and regulatory pathway investigations enables the design of precise, personalized cancer treatments. Novel approaches like machine learning (ML) aid in investigating biomarkers' functions and developing tailored therapies. ML and artificial intelligence (AI) have emerged as promising platforms for planning, assessing, and overseeing cancer-related therapies. They enable the design of tailored treatments, including inhibitors specifically targeting tumour-associated mutations and metabolic alterations. ML is instrumental in analysing multiomics data, which encompasses non-invasive methods for predicting therapy effectiveness and prognosis based on novel theragnostic biomarkers (Gao et al., 2023). Efforts are also being made to develop advanced prediction models which can predict the survival rate of patients undergoing specific treatment, deep learning approaches are employed to design such models. One such sophisticated prediction model, the cytokine-based immune checkpoint inhibitor (ICI) response index, was used to predict the OS rate of NSCLC patients receiving ICI therapy (Wei et al., 2023). While ML and AI have demonstrated tremendous potential in many medical applications, including cancer research, it is crucial to recognize that they have their specific limitations. It is still essential to test AI-generated theories in the lab and turn them into effective treatments. Collaboration between AI specialists, biologists and doctors will be crucial for the effective application of cancer treatment options as the area continues to develop.

31.10 CONCLUSION

To address the heterogeneity of LUAD, one prominent way is to target particular biomarkers which are responsible for the development and progression of the disease. This enables a more specialized method of treatment which involves designing novel inhibitors that effectively surpass the effects of these potential biomarkers. These inhibitors can be tailored according to the genetic profile of the affected individual, thereby increasing the efficacy of the treatment. The current study focuses on designing protease inhibitors using multidisciplinary approaches which involve structural bioinformatics and systems biology analysis. These computational methods have shown to offer a brief understanding of the genetic variations and their effects on the structure and functional role of tumour-related proteins and have also enabled the identification of significant markers via omics data analysis. These approaches are significant for their capacity to offer a thorough knowledge of the molecular metabolism and mechanism of the underlying disease, which is necessary for the plausible formulation of efficient and specific inhibitors. It holds the potential to drastically change the manner in which we comprehend LUAD and other malignancies, resulting in better outcomes and targeted therapeutics.

DECLARATIONS

Conflict of interest: The authors declare no conflicts of interest.

Consent for publishing: All authors have approved the final version of the manuscript and given consent for publication.

Funding: Not Applicable.

REFERENCES

Alidousty, C., Duerbaum, N., Wagener-Ryczek, S., Baar, T., Martelotto, L. G., Heydt, C., Siemanowski, J., Holz, B., Binot, E., Fassunke, J., Merkelbach-Bruse, S., Wolf, J., Kron, A., Buettner, R., & Schultheis, A. M. (2021). Prevalence and potential biological role of TERT amplifications in ALK translocated adenocarcinoma of the lung. *Histopathology*, *78*(4), 578–585. https://doi.org/10.1111/his.14256

Ashok, G., & Ramaiah, S. (2022). A critical review of datasets and computational suites for improving cancer theranostics and biomarker discovery. *Medical Oncology*, *39*(12), 206. https://doi.org/10.1007/s12032-022-01815-8

Bassetto, M., Massarotti, A., Coluccia, A., & Brancale, A. (2016). Structural biology in antiviral drug discovery. *Current Opinion in Pharmacology*, *30*, 116–130. https://doi.org/10.1016/j.coph.2016.08.014

Basu, S., Naha, A., Veeraraghavan, B., Ramaiah, S., & Anbarasu, A. (2022). In silico structure evaluation of BAG3 and elucidating its association with bacterial infections through protein–protein and host-pathogen interaction analysis. *Journal of Cellular Biochemistry*, *123*(1), 115–127. https://doi.org/10.1002/jcb.29953

Batool, M., Ahmad, B., & Choi, S. (2019). A structure-based drug discovery paradigm. *International Journal of Molecular Sciences*, *20*(11), 2783. https://doi.org/10.3390/ijms20112783

Bora-Singhal, N., Mohankumar, D., Saha, B., Colin, C. M., Lee, J. Y., Martin, M. W., Zheng, X., Coppola, D., & Chellappan, S. (2020). Novel HDAC11 inhibitors suppress lung adenocarcinoma stem cell self-renewal and overcome drug resistance by suppressing Sox2. *Scientific Reports*, *10*(1), 4722. https://doi.org/10.1038/s41598-020-61295-6

Calvayrac, O., Pradines, A., Pons, E., Mazières, J., & Guibert, N. (2017). Molecular biomarkers for lung adenocarcinoma. *European Respiratory Journal*, *49*(4), 1601734. https://doi.org/10.1183/13993003.01734-2016

Eatemadi, A., Aiyelabegan, H. T., Negahdari, B., Mazlomi, M. A., Daraee, H., Daraee, N., Eatemadi, R., & Sadroddiny, E. (2017). Role of protease and protease inhibitors in cancer pathogenesis and treatment. *Biomedicine & Pharmacotherapy*, *86*, 221–231. https://doi.org/10.1016/j.biopha.2016.12.021

Feng, M., Zhao, J., Wang, L., & Liu, J. (2018). Upregulated expression of serum exosomal microRNAs as diagnostic biomarkers of lung adenocarcinoma. *Annals of Clinical & Laboratory Science*, *48*(6), 712–718.

Gao, Q., Yang, L., Lu, M., Jin, R., Ye, H., & Ma, T. (2023). The artificial intelligence and machine learning in lung cancer immunotherapy. *Journal of Hematology & Oncology*, *16*(1), 55. https://doi.org/10.1186/s13045-023-01456-y

Inamura, K. (2017). Lung cancer: Understanding its molecular pathology and the 2015 WHO classification. *Frontiers in Oncology*, *7*. https://doi.org/10.3389/fonc.2017.00193

Jamroz, M., Kolinski, A., & Kmiecik, S. (2013). CABS-flex: Server for fast simulation of protein structure fluctuations. *Nucleic Acids Research*, *41*(1), 427–431. https://doi.org/10.1093/nar/gkt332

Kang, Y., You, J., Gan, Y., Chen, Q., Huang, C., Chen, F., Xu, X., & Chen, L. (2022). Serum and serum exosomal CircRNAs hsa_circ_0001492, hsa_circ_0001439, and hsa_circ_0000896 as diagnostic biomarkers for lung adenocarcinoma. *Frontiers in Oncology*, *12*. https://doi.org/10.3389/fonc.2022.912246

Lemkul, J. (2019). From proteins to perturbed Hamiltonians: A suite of tutorials for the GROMACS-2018 molecular simulation package [Article v1.0]. *Living Journal of Computational Molecular Science*, *1*(1), 1–53. https://doi.org/10.33011/livecoms.1.1.5068

Liu, W. T., Wang, Y., Zhang, J., Ye, F., Huang, X.-H., Li, B., & He, Q.-Y. (2018). A novel strategy of integrated microarray analysis identifies CENPA, CDK1 and CDC20 as a cluster of diagnostic biomarkers in lung adenocarcinoma. *Cancer Letters*, *425*, 43–53. https://doi.org/10.1016/j.canlet.2018.03.043

Liu, Z., Gao, Y., Hao, F., Lou, X., Zhang, X., Li, Y., Wu, D., Xiao, T., Yang, L., Li, Q., Qiu, X., & Wang, E. (2014). Secretomes are a potential source of molecular targets for cancer therapies and indicate that APOE is a candidate biomarker for lung adenocarcinoma metastasis. *Molecular Biology Reports*, *41*(11), 7507–7523. https://doi.org/10.1007/s11033-014-3641-4

Luo, W., Tian, P., Wang, Y., Xu, H., Chen, L., Tang, C., Shu, Y., Zhang, S., Wang, Z., Zhang, J., Zhang, L., Jiang, L., Liu, L., Che, G., Guo, C., Zhang, H., Wang, J., & Li, W. (2018). Characteristics of genomic alterations of lung adenocarcinoma in young never-smokers. *International Journal of Cancer*, *143*(7), 1696–1705. https://doi.org/10.1002/ijc.31542

Manasanch, E. E., & Orlowski, R. Z. (2017). Proteasome inhibitors in cancer therapy. *Nature Reviews Clinical Oncology*, *14*(7), 417–433. https://doi.org/10.1038/nrclinonc.2016.206

Miryala, S. K., Basu, S., Naha, A., Debroy, R., Ramaiah, S., Anbarasu, A., & Natarajan, S. (2021). Identification of bioactive natural compounds as efficient inhibitors against *Mycobacterium tuberculosis* protein-targets: A molecular docking and molecular dynamics simulation study. *Journal of Molecular Liquids*, *341*, 117340. https://doi.org/10.1016/j.molliq.2021.117340

Mullangi, S., & Lekkala, M. R. (2020). *Adenocarcinoma*.

Myers, D. J., & Wallen, J. M. (2022). *Lung adenocarcinoma*. StatPearls Publishing.

Naha, A., & Ramaiah, S. (2022). Structural chemistry and molecular-level interactome reveals histidine kinase EvgS to subvert both antimicrobial resistance and virulence in *Shigella flexneri* 2a str. 301. *3 Biotech*, *12*(10), 258. https://doi.org/10.1007/s13205-022-03325-w

Naha, A., & Ramaiah, S. (2023). Novel antimicrobial peptide SAAP mutant as a better adjuvant to sulbactam-based treatments against clinical strains of XDR *Acinetobacter baumannii*. *Probiotics and Antimicrobial Proteins*. https://doi.org/10.1007/s12602-023-10067-5

Naha, A., Banerjee, S., Debroy, R., Basu, S., Ashok, G., Priyamvada, P., Kumar, H., Preethi, A. R., Singh, H., Anbarasu, A., & Ramaiah, S. (2022). Network metrics, structural dynamics and density functional theory calculations identified a novel Ursodeoxycholic Acid derivative against therapeutic target Parkin for Parkinson's disease. *Computational and Structural Biotechnology Journal*, *20*, 4271–4287. https://doi.org/10.1016/j.csbj.2022.08.017

Naha, A., Kumar Miryala, S., Debroy, R., Ramaiah, S., & Anbarasu, A. (2020). Elucidating the multi-drug resistance mechanism of *Enterococcus faecalis* V583: A gene interaction network analysis. *Gene*, *748*, 144704. https://doi.org/10.1016/j.gene.2020.144704

Naha, A., Vijayakumar, S., Lal, B., Shankar, B. A., Chandran, S., Ramaiah, S., Veeraraghavan, B., & Anbarasu, A. (2021). Genome sequencing and molecular characterisation of XDR Acinetobacter baumannii reveal complexities in resistance: Novel combination of sulbactam–durlobactam holds promise for therapeutic intervention. *Journal of Cellular Biochemistry*, *122*(12), 1946–1957. https://doi.org/10.1002/jcb.30156

Nicholson, A. G., Tsao, M. S., Beasley, M. B., Borczuk, A. C., Brambilla, E., Cooper, W. A., Dacic, S., Jain, D., Kerr, K. M., Lantuejoul, S., Noguchi, M., Papotti, M., Rekhtman, N., Scagliotti, G., van Schil, P., Sholl, L., Yatabe, Y., Yoshida, A., & Travis, W. D. (2022). The 2021 WHO classification of lung tumors: Impact of advances since 2015. *Journal of Thoracic Oncology*, *17*(3), 362–387. https://doi.org/10.1016/j.jtho.2021.11.003

Parr, R. G., Szentpály, L. V., & Liu, S. (1999). Electrophilicity index. *Journal of the American Chemical Society*, *121*(9), 1922–1924. https://doi.org/10.1021/ja983494x

Shen, Z., Liu, S., Liu, J., Liu, J., & Yao, C. (2021). Weighted Gene co-expression network analysis and treatment strategies of tumor recurrence-associated hub genes in lung adenocarcinoma. *Frontiers in Genetics*, *12*. https://doi.org/10.3389/fgene.2021.756235

Song, R., Qiao, W., He, J., Huang, J., Luo, Y., & Yang, T. (2021). Proteases and their modulators in cancer therapy: Challenges and opportunities. *Journal of Medicinal Chemistry*, *64*(6), 2851–2877. https://doi.org/10.1021/acs.jmedchem.0c01640

The Cancer Genome Atlas Research Network. (2014). Comprehensive molecular profiling of lung adenocarcinoma. *Nature*, *511*(7511), 543–550. https://doi.org/10.1038/nature13385

Trezza, A., Cicaloni, V., Pettini, F., & Spiga, O. (2020). Potential roles of protease inhibitors in anticancer therapy. In *Cancer-Leading Proteases* (pp. 13–49). Elsevier. https://doi.org/10.1016/B978-0-12-818168-3.00002-4

Usharani, N., Naha, A., Anbarasu, A., Ramaiah, S., Kanth, S. V., & Natarajan, S. (2023). Green synthesis and characterization of water soluble nanocarnosine: A prospective drug delivery system. *Applied Materials Today*, *32*, 101812. https://doi.org/10.1016/j.apmt.2023.101812

Wadanambi, P. M., & Mannapperuma, U. (2021). Computational study to discover potent phytochemical inhibitors against drug target, squalene synthase from leishmania donovani. *Heliyon*, *7*(6), e07178. https://doi.org/10.1016/j.heliyon.2021.e07178

Wei, F., Azuma, K., Nakahara, Y., Saito, H., Matsuo, N., Tagami, T., Kouro, T., Igarashi, Y., Tokito, T., Kato, T., Kondo, T., Murakami, S., Usui, R., Himuro, H., Horaguchi, S., Tsuji, K., Murotani, K., Ban, T., Tamura, T., & Sasada, T. (2023). Machine learning for prediction of immunotherapeutic outcome in non-small-cell lung cancer based on circulating cytokine signatures. *Journal for Immuno Therapy of Cancer*, *11*(7), e006788. https://doi.org/10.1136/jitc-2023-006788

Weng, C.-W., Li, J.-H., Tsai, J.-Y., Lin, S.-H., Chang, G.-C., Liu, C.-C., & Chen, J. J. (2020). Pharmacophore-based virtual screening for the identification of the novel Src inhibitor SJG-136 against lung cancer cell growth and motility. *American Journal of Cancer Research*, *10*(6), 1668.

Wu, G., Yan, Y., Cai, Y., Peng, B., Li, J., Huang, J., Xu, Z., & Zhou, J. (2021). ALKBH1-8 and FTO: Potential therapeutic targets and prognostic biomarkers in lung adenocarcinoma pathogenesis. *Frontiers in Cell and Developmental Biology*, *9*. https://doi.org/10.3389/fcell.2021.633927

Zamay, T., Zamay, G., Kolovskaya, O., Zukov, R., Petrova, M., Gargaun, A., Berezovski, M., & Kichkailo, A. (2017). Current and prospective protein biomarkers of lung cancer. *Cancers*, *9*(12), 155. https://doi.org/10.3390/cancers9110155

Zhao, Y., Gao, Y., Xu, X., Zhou, J., & Wang, H. (2021). Multi-omics analysis of genomics, epigenomics and transcriptomics for molecular subtypes and core genes for lung adenocarcinoma. *BMC Cancer*, *21*(1), 257. https://doi.org/10.1186/s12885-021-07888-4

Zheng, M. (2016). Classification and pathology of lung cancer. *Surgical Oncology Clinics of North America*, *25*(3), 447–468. https://doi.org/10.1016/j.soc.2016.02.003

32 System Biology and Bioinformatics Approaches in HTLV-1 Protease Targeting Adult T-Cell Leukemia Disease

Chandrabose Selvaraj
CsrDD Lab, CGHR, Saveetha Medical College, Saveetha Institute of Medical and Technical Sciences, Chennai, India

Sanjeev Kumar Singh
Science Block, Alagappa University, Karaikudi, India

Devadasan Velmurugan
School of Bioengineering, SRM Institute of Science & Technology, Kattankulathur, Chennai, India

32.1 INTRODUCTION

The viral replication mechanism relies heavily on the viral-generating enzyme known as viral protease, which plays a critical part in many viruses' lifecycle replication and maturation (Majerova & Novotny, 2021). These viral proteases are enzymes with the specialized function of cleaving the polypeptide chain at specified sites, and the cleaved portions initiate as functional proteins (Zephyr, Kurt Yilmaz, & Schiffer, 2021). Functionally, most viruses must replicate their genetic material and produce new viral particles (virions) to infect other cells (Konvalinka, Krausslich, & Muller, 2015). The HIV-1 protease, for instance, is essential for the replication of the HIV virus. It cleaves the HIV precursor polyprotein into mature viral proteins (Yost & Marcotrigiano, 2013). Similarly, the HTLV protease (HTLV-PR) is indispensable for the replication and function of the Human T-lymphotropic virus (HTLV). It cleaves the HTLV precursor polyprotein into mature HTLV viral proteins (Rucker, Horn, Meiselbach, & Sticht, 2011). In addition to their role in viral replication, proteases produced by viruses play crucial roles in viral release from cells and immune system evasion. Some viral proteases, for instance, can cleave proteins that are essential for cell adhesion (Gorkhali et al., 2021). Viral protease is responsible for cleaving viral proteins in predetermined places, as shown in the Figure 32.1. In addition to their role in viral replication, proteases produced by viruses play crucial roles in viral release from cells and immune system evasion (Rao, Tanksale, Ghatge, & Deshpande, 1998). Other viral proteases can cleave immune-related host cell proteins, allowing the virus to hide from the defense system of the body (Rahbar Saadat, Hosseiniyan Khatibi, Zununi Vahed, & Ardalan, 2021).

Host cells engage in antigen presentation, in which viral antigens are presented to T cells, a type of white blood cell that plays a crucial role in the adaptive immune response (Samuel, 2001; Selvaraj, Abhirami, Vijayakumar, Alfaiz, & Singh, 2022). The presentation of viral antigens by host cells to T cells is known as antigen presentation (Sorci, Cornet, & Faivre, 2013; Selvaraj et al.,

 DOI: 10.1201/9781003394693-37

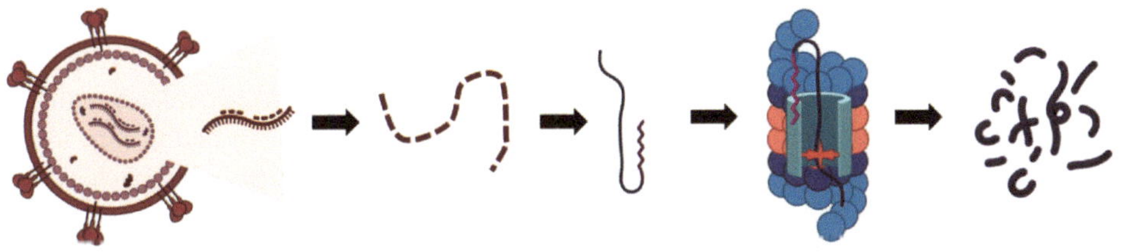

FIGURE 32.1 Mechanism of protease in viral polypeptide cleavage into fragments.

2022). The ability of the immune system to recognize and respond to a virus is hindered when viral proteases hamper antigen presentation. The immune system generates antibodies—proteins that recognize and neutralize foreign invaders—but viral proteases can mutate to evade detection (Selvaraj, Pravin, Alhoqail, Nayarisseri, & Singh, 2022; Selvaraj et al., 2023).

32.2 VIRUS WITH PROTEASE ENZYME

32.2.1 Multiple Virus Proteases

Multiple RNA and DNA viruses have the protease enzyme for its polypeptide-cleaving activity. For example, replication of the hepatitis C virus depends on the multifunctional enzyme NS3 protease (HCV). This serine protease is necessary for viral replication, assembly, and release because it cleaves the HCV polyprotein precursor into mature viral proteins (Raney, Sharma, Moustafa, & Cameron, 2010). The HCV NS3 protease involves multiple processes throughout the HCV lifecycle, including immune evasion. The NS3 protease from HCV is a multifaceted enzyme that serves several purposes (Abdullah, McWhirter, & Suo, 2023).

The folding and activity of the NS3 protease domain are dependent on the NS4A cofactor domain. The HCV NS3 protease cleaves these host cell proteins to facilitate viral replication and immune evasion. The family of flaviviruses, including Zika, dengue, and West Nile, rely on a specific serine protease for replication (Wahaab et al., 2021). It catalyzes the maturation of the polyprotein precursor of flaviviruses, allowing for viral replication, assembly, and release. The complete molecule consists of two domains, NS3 (a protease domain) and NS2B (a cofactor domain). A catalytic triad of amino acids triggers proteolysis by the NS3 protease domain (Shiryaev & Strongin, 2010). Several host cell proteins are cleaved by picornavirus proteases, which the virus uses to evade the host immune system and replicate (Lei & Hilgenfeld, 2017). Antiviral drug development has focused on picornavirus proteases because they are central to viral replication (Sun, Chen, Cheng, & Wang, 2016). Multiple host-cell proteins are cleaved by herpes virus proteases, which aid the virus in evading the host immune system and facilitating viral replication (Liu & Zhou, 2007). Proteases from different herpes viruses are similar because of their high degree of conservation. Herpes virus protease inhibitors that target a specific herpes virus, thus, are challenging to develop. Herpes virus proteases play a role in immune system evasion and other processes throughout their lifecycle. This suggests that using protease inhibitors against herpes viruses may have unintended consequences (Sucharita, Krishnagopal, & van Drunen Littel-van den Hurk, 2023). These polyproteins must be broken down into their component parts for viral replication and assembly. The adenovirus protease cleaves the polyproteins at defined sites in the protein (Readler, Burke, Sharma, Excoffon, & Kolawole, 2021). The main protease (Mpro) and the papain-like protease (PLpro) are the two most abundant in coronaviruses (Smak, Chandrabose, Tvaroska, & Koca, 2021). Mpro cleaves the viral polyproteins at specific sites, releasing individual non-structural proteins required for viral replication and transcription. In addition to its role in SARS-CoV-2 replication, Mpro plays a role in other elements of the virus's life cycle, including immune evasion. SARS-CoV-2 also contains a significant amount of the protease enzymes known as papain-like protease (PLpro) (Selvaraj et al., 2020).

Proteases in coronavirus family like PLpro and the Mpro are essential for the reproduction and maturation of the virus. SARS-papain-like CoV-2's protease (PLpro), like the Mpro, is responsible for digesting viral polyproteins (Selvaraj, Dinesh, Panwar, Boura, & Singh, 2021). An effective method for learning about the genetic variety, evolutionary connections, and practical consequences of distinct viral strains is multiple sequence analysis (MSA). Comparisons are made between many sequences of biological macromolecules, such as DNA, RNA, or amino-acid sequences (Perovic, Glisic, Veljkovic, Paessler, & Veljkovic, 2023). Similarities and homologies between the sequences can be utilized to infer their functional or evolutionary links through the alignment. Since MSAs can be used to locate conserved sections of viral genomes that are crucial for viral replication or pathogenicity, they are frequently applied to this field of study (Pearson, 2013). MSAs can be used to find residues that play key roles in protein–protein interactions and conserve and highlight regions under selective pressure (Scheel & Rice, 2013). Overall, MSAs are potent tools for studying viral proteases and have been utilized to significantly increase our understanding of these enzymes (Solanki, Putatunda, Kumar, Bhatia & Walia, 2021). In Figure 32.2, we see the multiple sequence alignment of different viruses, with the high allocation of blue-colored regions representing the highly conserved regions in different viral proteases (Thompson, Linard, Lecompte, & Poch, 2011). Based on this, we can see that more blue-colored regions are spotted in Figure 32.2, showing the top conserved residues aligned between the protease from various viruses.

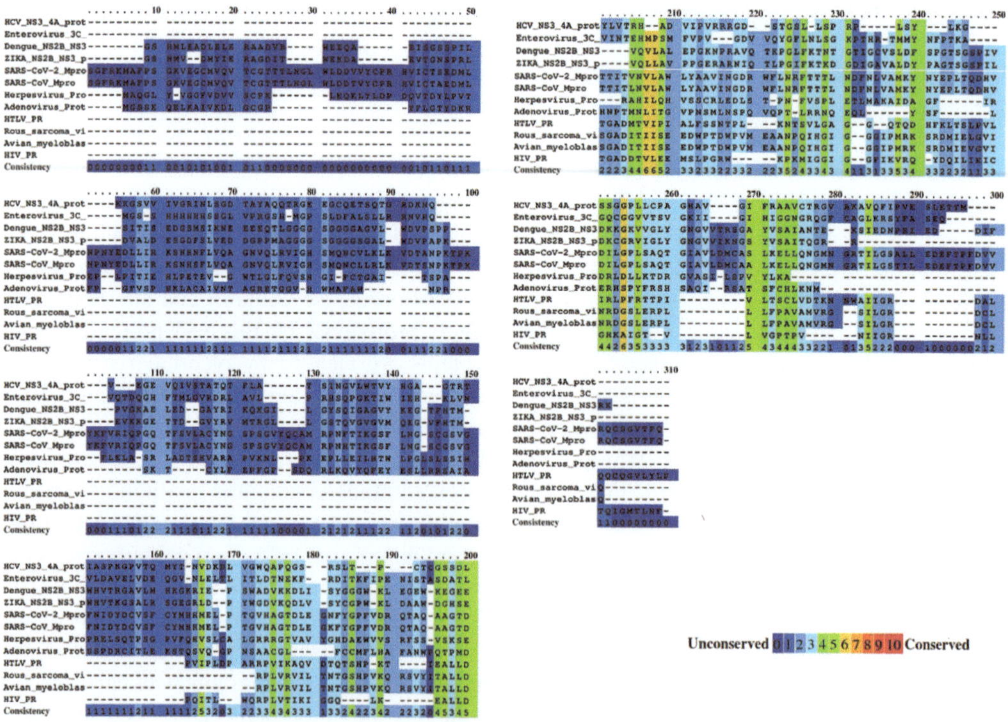

FIGURE 32.2 Sequence analysis of various viral proteases.

FIGURE 32.3 Phylogenetic analysis of various viral proteases.

Phylogenetic analysis is an effective method for investigating virus evolutionary relationships. Here, we can see how various viruses are connected by building phylogenetic trees by comparing their DNA sequences (Karamipour, Mojbafan, & Mazaheri Nezhad Fard, 2023). Here, several types of viral proteases, such as those found in influenza, coronaviruses, hepatitis viruses, HTLV, and HIV, have been analyzed using phylogenetic analysis. Figure 32.3 displays the results of a phylogenetic study based on multiple sequence alignment, which reveals substantial differences among the proteases despite their similar biological roles. These investigations have revealed important information about the origins and evolution of these viruses, as well as their capacity to escape the host immune system and cause disease. Since all of these viruses have a protease enzyme, their sequences are extremely well conserved and show family-specific changes in a phylogenetic tree; within this context, HIV-PR and HTLV-PR display inter-node similarities (Bozek & Lengauer, 2010). Understanding the similarities and differences between viral proteases relies heavily on comparing their structural details. By comparing their structures in Figure 32.4, we can learn more about the evolutionary history, biological roles, and

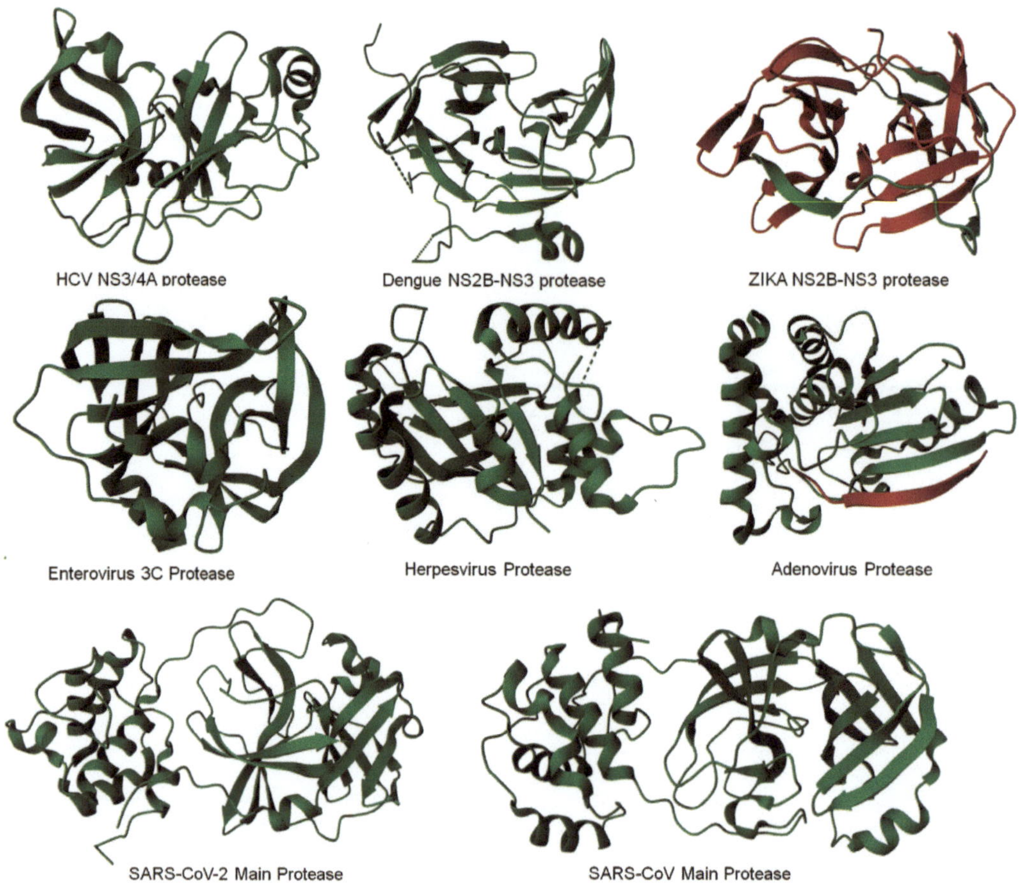

FIGURE 32.4 Various viral protease structures with similar function of polypeptide cleavage.

therapeutic targets of these enzymes. Structurally, the proteases are classified into various proteases, they are aspartyl proteases (Asp), cysteine proteases (Cys), serine proteases (Ser), and threonine proteases (Thr).

32.2.2 Retrovirus Protease

Retrovirus protease is an enzyme present in retroviruses, an RNA virus family. This protease is involved in the maturation of viral particles and plays an important function in the viral life cycle (Okafor et al., 2023). The enzyme retrovirus protease degrades viral polyproteins into individual functional proteins. These polyproteins are created during viral replication and must be processed to become mature viral particles (Selvaraj, Singh, & Singh, 2014b). HIV-PR is also involved in other aspects of HIV replication, such as the maturation of HIV envelope proteins and the release of virus particles from infected cells (Reddy, Singh, Dessalew, Tripathi, & Selvaraj, 2012).

The catalytic triad breaks the peptide connections between amino acids in the HIV polyprotein precursor, allowing the mature viral proteins to be released (Rahimi, Taghdir, & Abasi Joozdani, 2023). There is a relative virus with the same enzyme and a similar structural feature as HIV-1 protease called the HTLV-1 virus. The Human T-cell leukemia virus type 1 protease, also known as the HTLV-1 protease, is an enzyme discovered in the virus (HTLV-1).

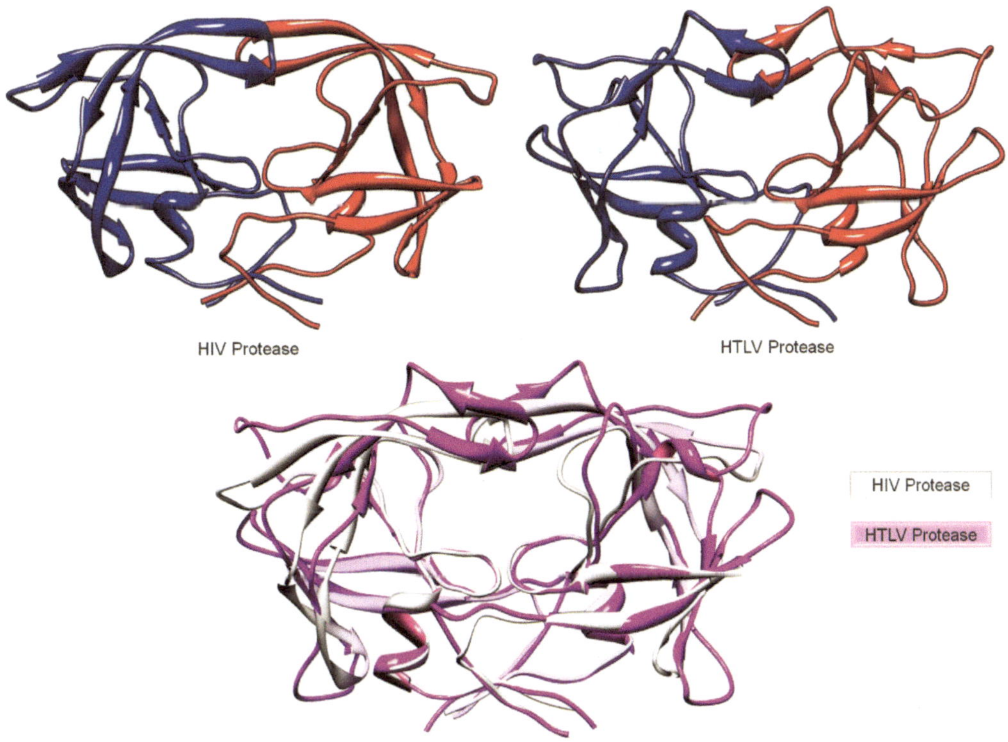

FIGURE 32.5 Structural comparison of HIV-protease and HTLV protease showing structural similarity.

32.2.3 Comparative Analysis of HTLV-PR with HIV-PR

HTLV-1 was the first identified human retrovirus, and it is related to various diseases such as adult T-cell leukemia-lymphoma (ATL) and tropical spastic paraparesis/HTLV-1-associated myelopathy. Acute T-cell lymphoblastic leukemia (ATLL) is a deadly form of T-cell cancer. This HIV and HTLV-PR structure, provided in Figure 32.5, contains a protease enzyme that is required for virus maturation and replication, and that has almost similar function and structure (Selvaraj, Singh, & Singh, 2014a). Even though they share similar structural features, their sequence similarity shows they are at variance in between both the proteases (Figure 32.6). Furthermore, several amino acids shared by both PR enzymes (HIV-1/HTLV-1) fulfil the role of active sites in somewhat different positions; In order: ARG8/10 LEU23/30 ASP29/32/36 ALA28/35 GLY27/49/34/58 as shown in Figure 32.7. Comparable protease macromolecules adopt similar steric structures and form structural

```
CLUSTAL 2.0.12 multiple sequence alignment


2B7F_A|PDBID|CHAIN|SEQUENCE      PVIPLDPARRPVIKAQVDTQTSHPKTIEALLDTGADMTVIPIALFSSNTP 50
2UXZ_A|PDBID|CHAIN|SEQUENCE      PQITL--WQRPLVTIKIGGQLK-----EALLDTGADDTVLEEMSLPG--R 41
                                 * *.*   :**::. ::. * .     ********* **:      :..

2B7F_A|PDBID|CHAIN|SEQUENCE      LKNTSVLGAGGQTQDHFKLTSLPVLIRLPFRTTPIVLTSCLVDTKNNWAI 100
2UXZ_A|PDBID|CHAIN|SEQUENCE      WKPKMIGGIGG----FIKVRQYDQIL-IEICG-HKAIGTVLVGPTP-VNI 84
                                  * . : * **    .:*: .   :: : :     .: : **...    *

2B7F_A|PDBID|CHAIN|SEQUENCE      IGRDALQQCQGVLYLP 116
2UXZ_A|PDBID|CHAIN|SEQUENCE      IGRNLLTQIGCTLNF- 99
                                 ***: * *   .* :
```

FIGURE 32.6 Sequence comparison of HIV-protease and HTLV protease showing sequence variations.

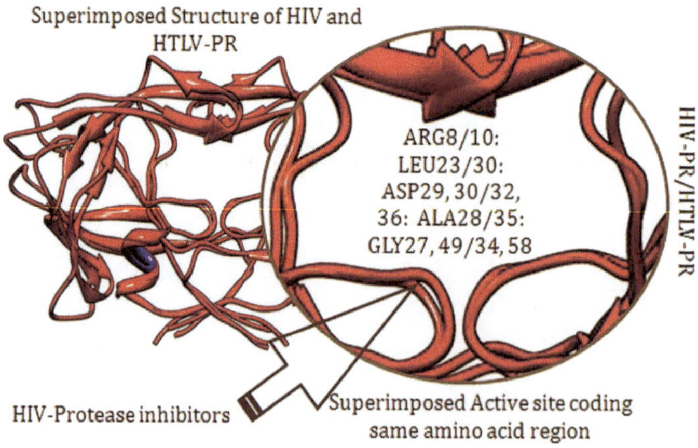

FIGURE 32.7 Active site residue comparison between the HIV-PR and HTLV-PR.

sites (ligand binding sites) in the middle of both chains. These pockets can form strong bonds with man-made or natural compounds like medicines and biomolecules (Selvaraj, Omer, Singh & Singh, 2015). Structural similarity between the ligand-binding domains of HIV-1 PR and HTLV-1 PR is essential for establishing a causal relationship between the two proteins (Petrillo, Dinh, Vogt, & Ma, 2023).

HIV and HTLV share structural similarities in the active site between their two monomeric chains, making them both attractive research targets. The active sites, mechanism of interactions, and binding conformation of protease inhibitors are just some of the similarities between the two retroviral

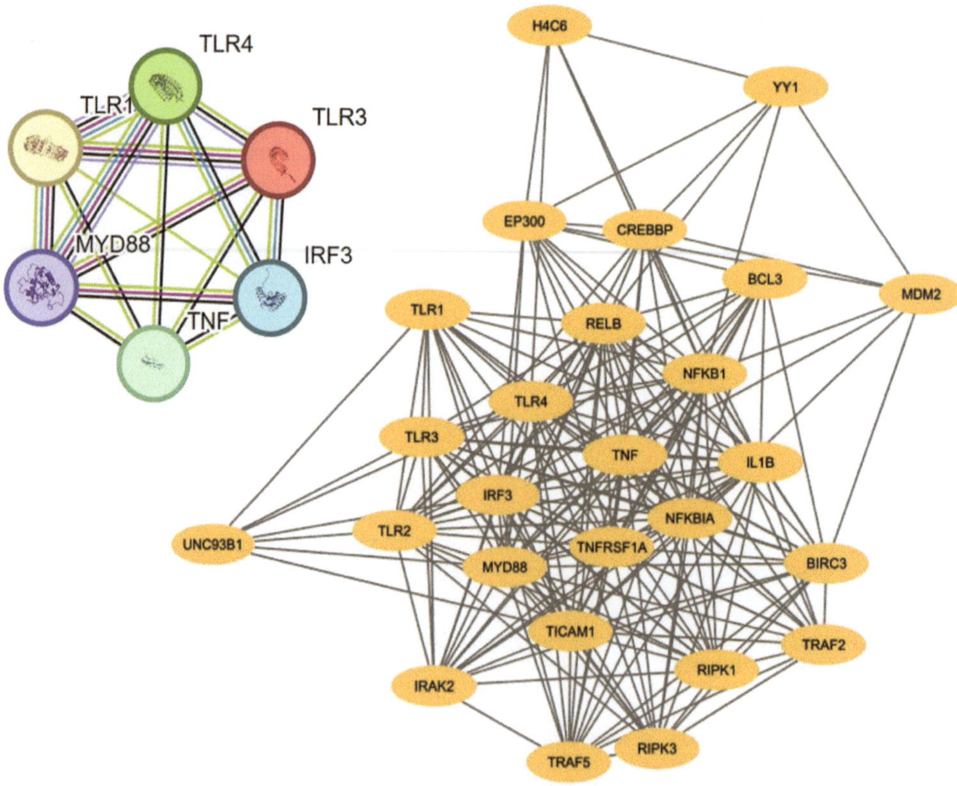

FIGURE 32.8 (a) Protein–protein interactions of HIV-PR in the host–viral protein interactions.

proteases that have been uncovered through comparative investigations (Dakshinamoorthy, Asmita, & Senapati, 2023). The two proteases, HIV and HTLV, share many structural features, including their overall shape and the location of their active sites, but some of the same amino acids fulfil the job of the active site in different positions. As shown in Figure 32.7, the residues ARG8/10, LEU23/30, ASP29/30/32/36, ALA28/35, GLY27/49/34,58 are sharing common amino acids as active sites residues (Selvaraj, Omer, Singh, & Singh, 2015).

32.2.4 System Biological Pathway Analysis of HIV and HTLV-PR

In both HIV-PR and HTLV-PR, the biological pathway of protein–protein interactions is crucial to understanding the mechanism in the host cell interacting patterns. The system biology of pathogenic viruses relies heavily on host–pathogen interactions, particularly with retroviruses (Durmus Tekir & Ulgen, 2013). This concept recognizes the multifaceted nature of retroviral infection beyond the simple virus–host cell contact. There is a complex interplay between the virus, the host cell, and the host immune system. Using systems biology methods, researchers have uncovered several previously unrecognized host characteristics that facilitate retroviral infection. Proteins, RNAs, and lipids are all examples of such factors (Finlay & McFadden, 2006). These models can be used to model the progression of retroviral infections and estimate the effectiveness of potential therapies (Nisole & Saib, 2004). Retroviral protease protein–protein interactions are important to the virus's life cycle, making them prime targets for antiretroviral therapy. Using protease inhibitors or other techniques to disrupt these connections can successfully reduce retroviral replication and manage the infection. The study of retroviral protease and its intricate protein–protein interaction network is a hotspot for researchers in virology and the pharmaceutical industry.

The innate immune system relies heavily on the Toll-like receptor (TLR) family of pattern recognition receptors (PRRs). Among the many types of microbiological patterns that TLRs may identify are those of viral proteins, lipids, and nucleic acids. The activity of HIV proteases can also be

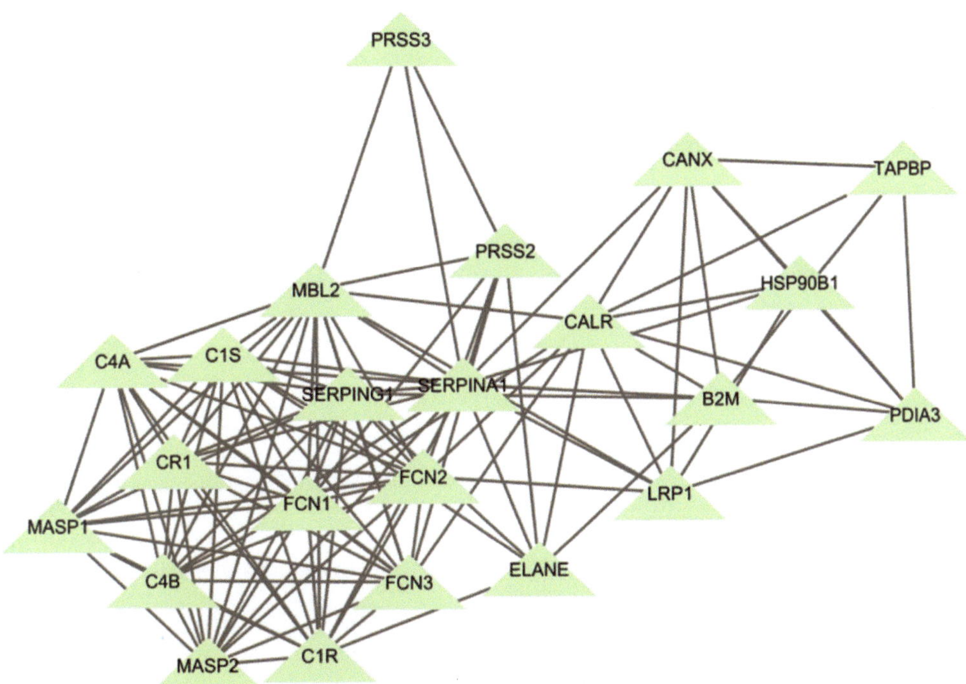

FIGURE 32.8 (b) Protein–protein interactions of HTLV-PR in the host–viral protein interactions.

modulated by direct interaction between TLRs and these proteins (Lester & Li, 2014). Figure 32.8a shows a graphical representation of the relevance of these connections, showing how TLR4 activation of the NF-kB signaling pathway might result in the expression of the HIV protease gene. The HTLV-PR protein can interact with the nuclear factor kappa B (NF-kB) protein, a transcription factor that controls the expression of several genes involved in inflammation and cell proliferation. Adult T-cell leukemia (ATL) growth may be aided by this interaction, leading to NF-kB activation (Martin, Maldonado, Mueller, Zhang & Mansky, 2016). Figure 32.8b shows protein–protein interactions involving the protein MASP2, which has been demonstrated to interact *in vitro* with HTLV-PR. The HTLV-PR shows the main interactions with MBL2 to initiate HTLV-PR activations, in which these enzyme interactions are a significant search for protein–protein blockers (Kalinichenko, Komkov, & Mazurov, 2022). Even though interactions are observed in Figure 32.8b, it is not yet clear what role PRSS2 has in HTLV biology or disease mechanisms.

32.3 FUTURE PERSPECTIVES

More research is needed to completely understand how the atom-level process of atomic transfer in the active region of HTLV-PR responds to cleavage formation and generates new proteins. This understanding could pave the way for the creation of more effective preventive strategies and targeted therapeutics. There is a need for more sensitive and specific assays to detect HTLV-1 infection, as the current approaches have their drawbacks. Better diagnostic tools would allow for earlier diagnosis, which would then allow for earlier intervention and treatment. At the moment, there is no HTLV-1 vaccine available. Scientists are working on vaccines that will stimulate an immune response strong enough to counteract the virus. In the future, artificial intelligence (AI) systems can be educated using massive collections of clinical data and photos to detect indicators of HTLV infection. Using this data, AI-powered diagnostic tools for detecting HTLV in biological samples like blood and tissue can be created. AI can be used to speed up the drug discovery process by pinpointing new therapeutic targets and estimating the efficacy of possible medication candidates. Treatments for HTLV infection may be improved as a result of this discovery.

DECLARATIONS

Conflict of interest: The authors declare no conflicts of interest.

Consent for publishing: All authors have approved the final version of the manuscript and given consent for publication.

Funding: Not Applicable.

REFERENCES

Abdullah, M. A. F., McWhirter, S. M., & Suo, Z. (2023). Modulation of kinase activities in vitro by hepatitis C virus protease NS3/NS4A mediated-cleavage of key immune modulator kinases. *Cells*, *12*(3). doi:10.3390/cells12030406

Bozek, K., & Lengauer, T. (2010). Positive selection of HIV host factors and the evolution of lentivirus genes. *BMC Evol Biol*, *10*, 186. doi:10.1186/1471-2148-10-186

Dakshinamoorthy, A., Asmita, A., & Senapati, S. (2023). Comprehending the structure, dynamics, and mechanism of action of drug-resistant HIV protease. *ACS Omega*, *8*(11), 9748–9763. doi:10.1021/acsomega.2c08279

Durmus Tekir, S. D., & Ulgen, K. O. (2013). Systems biology of pathogen-host interaction: networks of protein-protein interaction within pathogens and pathogen-human interactions in the post-genomic era. *Biotechnol J*, *8*(1), 85–96. doi:10.1002/biot.201200110

Finlay, B. B., & McFadden, G. (2006). Anti-immunology: Evasion of the host immune system by bacterial and viral pathogens. *Cell*, *124*(4), 767–782. doi:10.1016/j.cell.2006.01.034

Gorkhali, R., Koirala, P., Rijal, S., Mainali, A., Baral, A., & Bhattarai, H. K. (2021). Structure and function of major SARS-CoV-2 and SARS-CoV proteins. *Bioinform Biol Insights*, *15*, 11779322211025876. doi:10.1177/11779322211025876

Kalinichenko, S., Komkov, D., & Mazurov, D. (2022). HIV-1 and HTLV-1 transmission modes: Mechanisms and importance for virus spread. *Viruses*, *14*(1). doi:10.3390/v14010152

Karamipour, S., Mojbafan, M., & Mazaheri Nezhad Fard, R. (2023). Comparative analysis of 198 SARS-CoV-2 genomes from Iran and West Asia, February 2020 to December 2021. *Iran J Pathol*, *18*(3), 289–298. doi:10.30699/IJP.2023.557658.2935

Konvalinka, J., Krausslich, H. G., & Muller, B. (2015). Retroviral proteases and their roles in virion maturation. *Virology*, *479–480*, 403–417. doi:10.1016/j.virol.2015.03.021

Lei, J., & Hilgenfeld, R. (2017). RNA-virus proteases counteracting host innate immunity. *FEBS Lett*, *591*(20), 3190–3210. doi:10.1002/1873-3468.12827

Lester, S. N., & Li, K. (2014). Toll-like receptors in antiviral innate immunity. *J Mol Biol*, *426*(6), 1246–1264. doi:10.1016/j.jmb.2013.11.024

Liu, F., & Zhou, Z. H. (2007). Comparative virion structures of human herpesviruses. In A. Arvin, G. Campadelli-Fiume, E. Mocarski, P. S. Moore, B. Roizman, R. Whitley & K. Yamanishi (Eds.), *Human Herpesviruses: Biology, Therapy, and Immunoprophylaxis*. Cambridge: Cambridge University Press.

Majerova, T., & Novotny, P. (2021). Precursors of viral proteases as distinct drug targets. *Viruses*, *13*(10). doi:10.3390/v13101981

Martin, J. L., Maldonado, J. O., Mueller, J. D., Zhang, W., & Mansky, L. M. (2016). Molecular studies of HTLV-1 replication: An update. *Viruses*, *8*(2). doi:10.3390/v8020031

Nisole, S., & Saib, A. (2004). Early steps of retrovirus replicative cycle. *Retrovirology*, *1*, 9. doi:10.1186/1742-4690-1-9

Okafor, S. N., Meyer, A., Gadsden, J., Ahmed, F., Guzman, L., Ahmed, H., … Angsantikul, P. (2023). Drug reprofiling to identify potential HIV-1 protease inhibitors. *Molecules*, *28*(17). doi:10.3390/molecules28176330

Pearson, W. R. (2013). An introduction to sequence similarity ("homology") searching. *Curr Protoc Bioinformatics, Chapter 3*, 3 1 1-3 1 8. doi:10.1002/0471250953.bi0301s42

Perovic, V., Glisic, S., Veljkovic, M., Paessler, S., & Veljkovic, V. (2023). Novel entropy-based phylogenetic algorithm: A new approach for classifying SARS-CoV-2 variants. *Entropy (Basel)*, *25*(10). doi:10.3390/e25101463

Petrillo, N., Dinh, K., Vogt, K. A., & Ma, S. (2023). Catalytic mechanism of human T-cell leukemia virus type 1 protease investigated by combined QM/MM molecular dynamics simulations. *J Chem Inf Model*, *63*(12), 3865–3877. doi:10.1021/acs.jcim.3c00440

Rahbar Saadat, Y., Hosseiniyan Khatibi, S. M., Zununi Vahed, S., & Ardalan, M. (2021). Host serine proteases: A potential targeted therapy for COVID-19 and influenza. *Front Mol Biosci*, *8*, 725528. doi:10.3389/fmolb.2021.725528

Rahimi, M., Taghdir, M., & Abasi Joozdani, F. (2023). Dynamozones are the most obvious sign of the evolution of conformational dynamics in HIV-1 protease. *Sci Rep*, *13*(1), 14179. doi:10.1038/s41598-023-40818-x

Raney, K. D., Sharma, S. D., Moustafa, I. M., & Cameron, C. E. (2010). Hepatitis C virus non-structural protein 3 (HCV NS3): A multifunctional antiviral target. *J Biol Chem*, *285*(30), 22725–22731. doi:10.1074/jbc.R110.125294

Rao, M. B., Tanksale, A. M., Ghatge, M. S., & Deshpande, V. V. (1998). Molecular and biotechnological aspects of microbial proteases. *Microbiol Mol Biol Rev*, *62*(3), 597–635. doi:10.1128/MMBR.62.3.597-635.1998

Readler, J. M., Burke, M. R., Sharma, P., Excoffon, K., & Kolawole, A. O. (2021). Adenovirus co-opts neutrophilic inflammation to enhance transduction of epithelial cells. *Viruses*, *14*(1). doi:10.3390/v14010013

Reddy, K. K., Singh, S. K., Dessalew, N., Tripathi, S. K., & Selvaraj, C. (2012). Pharmacophore modelling and atom-based 3D-QSAR studies on N-methyl pyrimidones as HIV-1 integrase inhibitors. *J Enzyme Inhib Med Chem*, *27*(3), 339–347. doi:10.3109/14756366.2011.590803

Rucker, P., Horn, A. H., Meiselbach, H., & Sticht, H. (2011). A comparative study of HIV-1 and HTLV-I protease structure and dynamics reveals a conserved residue interaction network. *J Mol Model*, *17*(10), 2693–2705. doi:10.1007/s00894-011-0971-1

Samuel, C. E. (2001). Antiviral actions of interferons. *Clin Microbiol Rev*, *14*(4), 778–809,doi:10.1128/CMR.14.4.778-809.2001

Scheel, T. K., & Rice, C. M. (2013). Understanding the hepatitis C virus life cycle paves the way for highly effective therapies. *Nat Med*, *19*(7), 837–849. doi:10.1038/nm.3248

Selvaraj, C., Abhirami, R., Vijayakumar, R., Alfaiz, F. A., & Singh, S. K. (2022). Immunological insights of selectins in human disease mechanism. *Adv Protein Chem Struct Biol*, *129*, 163–188. doi:10.1016/bs.apcsb.2021.11.005

Selvaraj, C., Dinesh, D. C., Panwar, U., Boura, E., & Singh, S. K. (2021). High-throughput screening and quantum mechanics for identifying potent inhibitors against Macl domain of SARS-CoV-2 Nsp3. *IEEE/ACM Trans Comput Biol Bioinform*, *18*(4), 1262–1270. doi:10.1109/TCBB.2020.3037136

Selvaraj, C., Dinesh, D. C., Pedone, E. M., Alothaim, A. S., Vijayakumar, R., Rudhra, O., & Singh, S. K. (2023). SARS-CoV-2 ORF8 dimerization and binding mode analysis with class I MHC: computational approaches to identify COVID-19 inhibitors. *Brief Funct Genomics*, *22*(2), 227–240. doi:10.1093/bfgp/elac046

Selvaraj, C., Omer, A., Singh, P., & Singh, S. K. (2015). Molecular insights of protein contour recognition with ligand pharmacophoric sites through combinatorial library design and MD simulation in validating HTLV-1 PR inhibitors. *Mol Biosyst*, *11*(1), 178–189. doi:10.1039/c4mb00486h

Selvaraj, C., Panwar, U., Dinesh, D. C., Boura, E., Singh, P., Dubey, V. K., & Singh, S. K. (2020). Microsecond MD simulation and multiple-conformation virtual screening to identify potential anti-COVID-19 inhibitors against SARS-CoV-2 main protease. *Front Chem*, *8*, 595273. doi:10.3389/fchem.2020.595273

Selvaraj, C., Pravin, M. A., Alhoqail, W. A., Nayarisseri, A., & Singh, S. K. (2022). Intrinsically disordered proteins in viral pathogenesis and infections. *Adv Protein Chem Struct Biol*, *132*, 221–242. doi:10.1016/bs.apcsb.2022.06.002

Selvaraj, C., Shri, G. R., Vijayakumar, R., Alothaim, A. S., Ramya, S., & Singh, S. K. (2022). Viral hijacking mechanism in humans through protein-protein interactions. *Adv Protein Chem Struct Biol*, *131*, 261–276. doi:10.1016/bs.apcsb.2022.05.001

Selvaraj, C., Singh, P., & Singh, S. K. (2014a). Molecular insights on analogs of HIV PR inhibitors toward HTLV-1 PR through QM/MM interactions and molecular dynamics studies: Comparative structure analysis of wild and mutant HTLV-1 PR. *J Mol Recognit*, *27*(12), 696–706. doi:10.1002/jmr.2395

Selvaraj, C., Singh, P., & Singh, S. K. (2014b). Molecular modeling studies and comparative analysis on structurally similar HTLV and HIV protease using HIV-PR inhibitors. *J Recept Signal Transduct Res*, *34*(5), 361–371. doi:10.3109/10799893.2014.898659

Shiryaev, S. A., & Strongin, A. Y. (2010). Structural and functional parameters of the flaviviral protease: A promising antiviral drug target. *Future Virol*, *5*(5), 593–606. doi:10.2217/fvl.10.39

Smak, P., Chandrabose, S., Tvaroska, I., & Koca, J. (2021). Pan-selectin inhibitors as potential therapeutics for COVID-19 treatment: In silico screening study. *Glycobiology*, *31*(8), 975–987. doi:10.1093/glycob/cwab021

Solanki, P., Putatunda, C., Kumar, A., Bhatia, R., & Walia, A. (2021). Microbial proteases: Ubiquitous enzymes with innumerable uses. *3 Biotech*, *11*(10), 428. doi:10.1007/s13205-021-02928-z

Sorci, G., Cornet, S., & Faivre, B. (2013). Immune evasion, immunopathology and the regulation of the immune system. *Pathogens*, *2*(1), 71–91. doi:10.3390/pathogens2010071

Sucharita, S., Krishnagopal, A., & van Drunen Littel-van den Hurk, S. (2023). Comprehensive analysis of the tegument proteins involved in capsid transport and virion morphogenesis of alpha, beta and gamma herpesviruses. *Viruses*, *15*(10). doi:10.3390/v15102058

Sun, D., Chen, S., Cheng, A., & Wang, M. (2016). Roles of the picornaviral 3C proteinase in the viral life cycle and host cells. *Viruses*, *8*(3), 82. doi:10.3390/v8030082

Thompson, J. D., Linard, B., Lecompte, O., & Poch, O. (2011). A comprehensive benchmark study of multiple sequence alignment methods: Current challenges and future perspectives. *PLoS One*, *6*(3), e18093. doi:10.1371/journal.pone.0018093

Wahaab, A., Mustafa, B. E., Hameed, M., Stevenson, N. J., Anwar, M. N., Liu, K., & Ma, Z. (2021). Potential role of flavivirus NS2B-NS3 proteases in viral pathogenesis and anti-flavivirus drug discovery employing animal cells and models: A review. *Viruses*, *14*(1). doi:10.3390/v14010044

Yost, S. A., & Marcotrigiano, J. (2013). Viral precursor polyproteins: Keys of regulation from replication to maturation. *Curr Opin Virol*, *3*(2), 137–142. doi:10.1016/j.coviro.2013.03.009

Zephyr, J., Kurt Yilmaz, N., & Schiffer, C. A. (2021). Viral proteases: Structure, mechanism and inhibition. *Enzymes*, *50*, 301–333. doi:10.1016/bs.enz.2021.09.004

33 Protease Profiling in Lung Cancer Using Bioinformatics Tools

Gowrav Baradwaj, Avipsa Hazra, Dhanu A. S., and Kanthesh M. Basalingappa
School of Life Sciences, JSS Academy of Higher Education & Research, Mysuru, India

33.1 INTRODUCTION

Cancer in general terms is defined as 'an ailment where certain cells of the body multiply unrestrainedly and escalate to other body parts', by the National Cancer Institute (Brown *et al.*, 2023). It describes cancer as the uncontrolled abnormal growth of cells which leads to the formation of a tumour, a nodule, or a lump in any organ (Hanahan & Weinberg, 2011). Over the years, cancer has become one of the most rapidly increasing fatal diseases, with its risks constantly being on the rise and it is expected to peak even into the 21st century, due to the worsening of environmental and lifestyle conditions, with increase in pollution and lack of physical activity, respectively (Torre *et al.*, 2016). Currently every fourth individual in the population is at the risk of contracting cancer, with about 14 million new cases reported annually, out of which India contributes about 11 lakh of those reports (Hanahan & Weinberg, 2011). As cancer is triggered by the uncontrolled division of cells, depending upon where the mass of cells has started multiplying, cancer can be of multiple different types – depending on the organ of occurrence, stage of cancer, and other factors. There have been about 227 different kinds of cancer that have been recorded (Hassanpour & Dehghani, 2017). Cancer of the blood, brain, and lymph nodes is seen to be more prevalent in the younger population (Thun *et al.*, 2018; Yoo & Shin, 2003). In the male population, the most frequently occurring cancers are those caused in the prostate, the colorectal region, the bronchi, the lung, or in the bladder of the urinary tract. In women, breast, lung, bronchi, uterine, thyroid, and colorectal cancer occur most often. The common cancers between both the genders are colon, rectum, bronchi, and lung (Miller *et al.*, 2016). Out of them all, epidemiological studies show that lung cancer is the highest occurring cancer in the world, contributing about 12.3% of all cancers (Parkin *et al.*, 2001). Lung cancer can be divided into multiple subtypes based on histology, as seen in Figure 33.1. These classifications have eased the diagnosis and treatment of the disease (Travis, 2011; Schabath & Cote, 2019).

Like most cancers, lung cancer also sees the involvement of p53 gene, whose mutation leads to abnormality in apoptosis, angiogenesis, DNA metabolism, and cell division. The p53 mutation causes disruption in the DNA-binding position, which then combines with CDK-P2 and CDC2 to restrict the cancer cells to the G1and G2 phases of the cell cycle ((Taylor & Stark, 2001; Chae *et al.*, 2011). Lung cancer also is characterized by the hypomethylation of the LINE-1 transposable element, that leads to further gene deletion, leading to chromosome instability. This hypomethylation often leads to deranged expression of oncogenes (Hassanpour & Dehghani, 2017). These deranged expressions lead to elimination of the HAT genes, which causes modifications in the

DOI: 10.1201/9781003394693-38

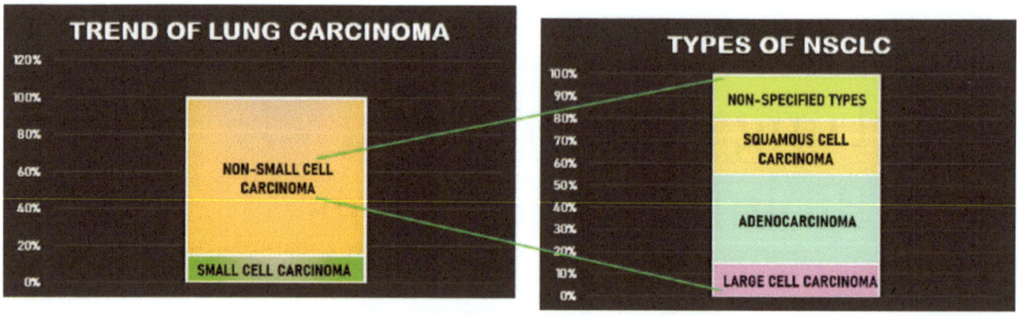

FIGURE 33.1 Incidence rates of various lung cancer types.

histone deacetylases expression for histone acetylation. These all lead to the formation of lung cancer (Noonan *et al.,* 2009). Lung cancer has been responsible for 1.1 million deaths in the population since the 2000s, which is estimated to be about 17.8% of the deaths due to cancer (Lippman & Spitz, 2001). However, over time, the peak that was observed for lung cancer in the 2000s seems to plateaued and is on the verge of decline. This is predominantly because lung cancer is caused due to smoking (Figure 33.2), the control of which can control the disease. Despite the progress, lung cancer remains an undefeated epidemic disease (Alberg *et al.,* 2005).

33.2 PROTEASES AND THEIR ROLES IN LUNG CANCER

Proteases, like numerous other proteins crucial for maintaining healthy physiology, can potentially lose control, initiating and advancing various diseases, including cancer (Lee *et al.,* 2004). In cancer, proteases have complex and dynamic functions due to the change in expression in tumour cells.

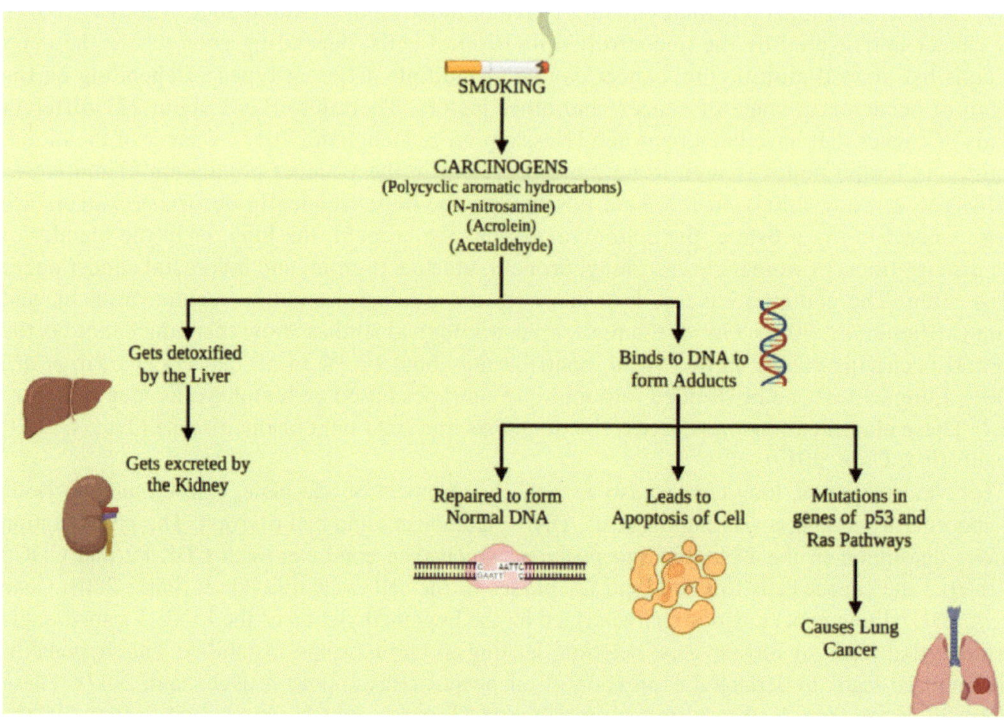

FIGURE 33.2 Carcinogens in smoking as a leading contributor to lung cancer.

Human protease classes are classified into five categories depending on their catalytic mechanism: aspartic, cysteine, metalloproteinases, serine, and threonine. Proteolytic enzyme involvement may differ between cancer types and phases of malignant transformation, and they may contain multiple classes of proteases and multiple proteolytic pathways within a single malignancy. Significant knowledge on the many activities of proteases in cancer has been gathered during the past four decades (Rakashanda *et al.*, 2012). The role of proteolytic enzymes in the development of tumours, angiogenesis, invasion, and metastasis has been extensively substantiated. Various proteolytic enzyme families, such as plasmin, plasminogen activators, serine proteases, kallikreins, and matrix metalloproteinases (MMPs), have been implicated in tumour expansion, metastatic progression, invasion, and the promotion of angiogenesis. Tumour-associated trypsinogen (TAT) is a trypsin-like protease found in numerous malignancies that produces a metastatic phenotype in cancer cells (Koivunen *et al.*, 1990; Vilen *et al.*, 2008). Furthermore, membrane-bound proteins, including G protein-coupled receptors and receptor tyrosine kinases, are known to be shed by the ADAMs to either activate or inhibit their related signalling pathways. As a result, numerous extracellular proteases are involved in cancer progression and metastasis (Park *et al.*, 2020). Proteases have been linked to tumour growth and progression, which is highly dependent on nutrition and oxygen availability, according to research (Eatemadi *et al.*, 2017). Proteases such urokinase cathepsins B, D, and L, and plasminogen activator as well as other metalloproteases, play critical roles in invasion and metastasis. These proteases destroy the interstitial matrix and basement membranes, allowing cancer cells to enter and disseminate locally and to distant locations. Proteases are valuable new prognostic markers since they are directly and causally involved in spread of cancer. Similarly, the urokinase plasminogen activator, a robust and an independent prognostic marker in cancer, plays an important role in detection of lung cancer. It is also critical in the progression of cancer (Duffy, 1996). MMPs like MMP-7 and MMP-9 are found to be abundant in lung malignancies and are much higher in non-small cell lung cancer (NSCLC) than in normal tissue (Safranek *et al.*, 2009). MMP-1 and MMP-3 are both promoters of polymorphisms and are linked to NSCLC susceptibility and increased probability of lymphatic metastasis of these tumours (Fang *et al.*, 2005). Also, it was found that on treating mice with CH1104I (a dual inhibitor of MMP-2 and -9) that lung carcinoma metastasis become primarily restricted, implying that inhibiting MMP-2 and -9 could significantly inhibit tumour invasion and metastasis (Chen *et al.*, 2008). In lung cancer, multiple proteases have shown their influences, depending on the kind of cancer. As in squamous-cell lung carcinoma (SQCLC), the levels of two lysosomal cysteine protease proteins, cathepsin L (CTSL), cathepsin B (CTSB), as well as three cysteine protease inhibitor proteins, cystatin C (CNC), stefin A (SFA), and stefin B (SFB), are much greater, especially the levels of CTSB protein, CTSB activity, and SFA protein in the Stage 1. SQCLCs and lung adenocarcinoma (LACs) displayed not only considerably increased CTSB activity but also significantly stronger inhibitory capacity against endogenous CTSB activity as compared to normal lungs. Leupeptin, a small CTSB inhibitor, was able to protect CTSB from preincubation-induced inhibition in lung cancer and lung parenchyma extracts, indicating that CTSB inhibition by endogenous cystatins is active-site directed and competitive. Ultrafiltration-passed protein from SQCLC extracts inhibited much more activity of pure bovine spleen cell carcinoma (CC) than equivalent protein preparations from normal lungs, showing that CC in the extracts may be complexed with specific cystatins. The key components of the inhibitory potential against CTSB and CTSC in SQCLCs and the lungs are quantitatively different cystatin combinations (Krepela *et al.*, 1998). Recent research has revealed that in both the SQCLC and adenocarcinoma (ADC) subtypes of NSCLC, the expression levels of the serpin family, particularly the gene encoding alpha1-antitrypsin (AAT), are notably lower in tumour tissue compared to the surrounding normal lung tissue. AAT serves as a versatile inhibitor of serine proteases and functions as an acute-phase serum glycoprotein. Notably, AAT acts as an inhibitor of transmembrane serine protease 2 (TMPRSS2) activity, impacting tumorigenic pathways by mitigating acute lung damage through the reduction of inflammation and cell death. Increased TMPRSS2 activity promotes signal transmission between cancer cells and the extracellular environment, influencing a variety of biological responses(Wilson *et al.*, 2005; Jairaman *et al.*, 2015; Lam *et al.*, 2015; Ercetin *et al.*, 2019; Azouz *et al.*, 2021; Schneider *et al.*, 2022). These different proteases involved in aggravating and repairing as well as acting as signals for lung cancer may or may not occur at a single or at different stages.

33.3 BIOINFORMATIC TOOLS IN DETECTION OF PROTEASE ENZYMES

Lung cancer is responsible for more than 85% of all cases worldwide, establishing it as one of the foremost contributors to mortality, of which NSCLC is most frequent, accounting for about 80% of all cases (Chen *et al.*, 2016). Although chemotherapy is essential in the treatment of NSCLC, it is often accompanied by unfavourable side effects (Cersosimo, 2002). Hence the need for development of new techniques with minimal side effects. One such technique involves the detection of the proteases that are involved in the cancer and treating them. These proteases include MMP, serine proteases, cysteine proteases, and inflammatory proteases. Various bioinformatics tools have been developed to detect and analyse these proteases and study their abnormalities that lead to cancer and develop specific treatments with minimal side effects.

33.3.1 Matrix Metalloproteinases

In the context of NSCLC, microarray techniques were utilized to examine protease MMP. In this study, differentially expressed genes (DEG)were analysed using the R programming language on two distinct datasets: one comprising microarray data from erlotinib-resistant NSCLC cells and the other from NSCLC tissue samples. Gene Ontology (GO) and Kyoto Encyclopedia of Genes and Genomes (KEGG) analyses were used to functionally annotate DEGs, and Oncomine, GeneMANIA, R2, Coremine, and FunRich were used to further functionally analyse DEGs. These investigations demonstrated that erlotinib treatment resistance in NSCLC may be conferred by MMP-1 and could potentially have a connection with COPS5. They also demonstrated that erlotinib resistance affected immunological response, the cytokine-mediated signalling system, and inflammatory responses (Zhou *et al.*, 2020). Another study investigated the involvement of MMP-10 in mouse lung cancer stem cells (CSCs). RNAi-mediated gene knockdown was used to generate enriched CSCs from CMT167 and LLC cells in this investigation and the results showed that MMP-10 is overexpressed in human NSCLC and Kras-transformed BASC lung cancer starting cells (Justilien *et al.*, 2012).

33.3.2 Cysteine Proteases

Cysteine proteases are typically involved in growth, immunological response, and apoptosis. These protease expression levels in tumour cells have been examined using microarray-based gene expression analysis (Huisman *et al.*, 2017). In a study, the expression of CST1 and CST2 in colorectal cancer was investigated using immunohistochemical and microarray chip analysis and was compared using GEPIA and Oncomine analysis, and the results were contrasted with other cancer types. These results demonstrated that CST1 and CST2 expression was much higher in lung adenocarcinoma when compared to normal cells using the TIMER database (Xie *et al.*, 2021). Another investigation delved into exploring the involvement of CST1 in lung adenocarcinoma by examining its expression in various cell lines, including BEAS2B and human LUAD cell lines (Yang *et al.*, 2022). These results were later contrasted with information gathered from various databases like UCSC Xeno database, TCGA, and GTEx. The expression analysis of tumour cells and normal cells was compared using GEPIA and were also contrasted with information from the Kaplan-Meier Plotter database (Gan *et al.*, 2020). By comparing the data obtained from TCGA-GTEx and GEPIA, it was found that lung adenocarcinoma cells exhibit higher levels of CST1 expression and the Kaplan-Meier Plotter database analysis showed a correlation between poor prognosis and high CST1 expression as well as overall survival and progression-free survival (Yang *et al.*, 2022).

33.3.3 Cathepsins

Cathepsins belong to the proteolytic enzyme family, which are crucial for several physiological processes but are always kept under check to prevent any uncontrolled proteolytic activity (Zavašnik-Bergant & Turk, 2006). It has been discovered that cathepsin B contributes to the breakdown of collagen and extracellular matrix (ECM), aiding in tumour invasion and metastasis (Tan *et al.*, 2013). On NSCLC, a study was conducted on the immunological characteristics and expression pattern of Cathepsin F (CTSF). The analysis made use of several bioinformatics tools, including UALCAN and the Kaplan-Meier Plotter, GTEx, TIMER, and TCGA. This study showed that CTSF was normally expressed in almost all cell types, but the analysis of CTSF in pan-cancers using TIMER showed a significant decrease in the expression compared to normal cells. Elevated CTSF expression was strongly correlated with improved OS in LUAD patients and with better PFS in patients with lung squamous cell carcinoma (LUSC), suggesting that CTSF may serve as a tumour-suppressor gene in both LUAD and LUSC (Song *et al.*, 2021). Another study examined the effect of CTSA knockdown in LUAD. In this work, CTSA-expressing LUAD cells were compared to the A549 cell line, which had CTSA knocked down using siRNA. TCGA provided the RNA sequencing information for LUAD and normal tissues, and R software was used to compare the expression of CTSA in LUAD and normal tissues. There was a significant overexpression of CTSA in LUAD based on the examination of data from TCGA. The examination of CTSA expression in the knockdown cells showed that the expression of CTSA was significantly lower than in the normal cells which decreased the proliferation of A549 cells (Hu *et al.*, 2020).

33.4 CONCLUSION

Lung cancer is a major global health concern, responsible for a significant number of deaths. While conventional therapies like chemotherapy and radiotherapy have shown effectiveness, they often come with adverse side effects. Prolonged use of these treatments can also lead to the development of resistance in cancer cells. Leveraging bioinformatics tools to identify specific enzymes and molecules that play crucial roles in cancer cell proliferation, invasion, and metastasis holds promise for developing more targeted treatment strategies. By analysing individual patient data through bioinformatics tools, tailored treatment plans can be devised based on their unique conditions. This approach allows for the early intervention and suppression of cancer progression. Moreover, the molecules identified through data analysis can serve as valuable biomarkers for cancer diagnosis. Monitoring the levels of these biomarkers can offer insights into the severity and stage of cancer progression. Numerous bioinformatics tools are currently employed in cancer diagnosis and treatment, however, further research in this field is essential to comprehensively identify all potential proteases involved in cancer, enhancing our understanding of the disease and improving treatment outcomes. In the years ahead, the fight against lung cancer will undergo a profound transformation towards highly individualized and patient-focused approaches. Bioinformatics tools will assume a pivotal role in driving this transformation and will entail the emergence of increasingly sophisticated computational models, powered by artificial intelligence and machine learning, aimed at unravelling the complex molecular landscapes of individual lung cancer cases. The incorporation of real-time monitoring, facilitated by wearable devices and liquid biopsies, will become standard practice. As our understanding of the genetic and proteomic intricacies of lung cancer deepens, we can expect to uncover a wider range of biomarkers, enabling swifter and more accurate diagnoses, thus making efforts to improve global collaborative efforts. As these advancements materialize, the future offers the prospect of more efficient, precisely targeted, and patient-tailored lung cancer treatments, ultimately lessening the global burden of this disease and enhancing the quality of life for individuals grappling with it.

DECLARATIONS

Conflict of interest: The authors declare no conflicts of interest.

Consent for publishing: All authors have approved the final version of the manuscript and given consent for publication.

Funding: Not Applicable.

REFERENCES

Alberg, A. J., Brock, M. V., & Samet, J. M. (2005). Epidemiology of lung cancer: looking to the future. Journal of Clinical Oncology, 23(14), 3175–3185.

Azouz, N. P., Klingler, A. M., Callahan, V., et al. (2021). Alpha 1 antitrypsin is an inhibitor of the SARS-CoV-2-priming protease TMPRSS2. Pathogens & Immunity, 6(1), 55–74.

Brown, J. S., Amend, S. R., Austin, R. H., Gatenby, R. A., Hammarlund, E. U., & Pienta, K. J. (2023). Updating the Definition of Cancer. Molecular Cancer Research: MCR, 21(11), 1142–1147.

Cersosimo, R. J. (2002). Lung cancer: a review. American Journal of Health-System Pharmacy, 59(7), 611–642.

Chae, S. W., Sohn, J. H., Kim, D. H., et al. (2011). Over expressions of cyclin B1, cdc2, p16 and p53 in human breast cancer: the clinicopathologic correlations and prognostic implications. Yonsei Medical Journal, 52(3), 445–453.

Chen, M. H., Cui, S. X., Cheng, Y. N., et al. (2008). Galloyl cyclic-imide derivative CH1104I inhibits tumor invasion through suppressing matrix metalloproteinase activity. Anti-Cancer Drugs, 19(10), 957–965.

Chen, W., Zheng, R., Baade, P. D., et al. (2016). Cancer statistics in China, 2015. CA: A Cancer Journal for Clinicians, 66(2), 115–132.

Duffy, M. J. (1996). Proteases as prognostic markers in cancer. Clinical Cancer Research: An Official Journal of the American Association for Cancer Research, 2(4), 613–618.

Eatemadi, A., Aiyelabegan, H. T., Negahdari, B., et al. (2017). Role of protease and protease inhibitors in cancer pathogenesis and treatment. Biomedicine & Pharmacotherapy, 86, 221–231.

Ercetin, E., Richtmann, S., Martinez Delgado, B., et al. (2019). Clinical significance of SERPINA1 gene and its encoded alpha1-antitrypsin protein in NSCLC. Cancers, 11(9), 1306.

Fang, S., Jin, X., Wang, R., et al. (2005). Polymorphisms in the MMP1 and MMP3 promoter and non-small cell lung carcinoma in North China. Carcinogenesis, 26(2), 481–486.

Gan, J., Meng, Q., & Li, Y. (2020). Systematic analysis of expression profiles and prognostic significance for FAM83 family in non-small-cell lung cancer. Frontiers in Molecular Biosciences, 7, 572406.

Hanahan, D., & Weinberg, R. A. (2011). Hallmarks of cancer: the next generation. Cell, 144(5), 646–674.

Hassanpour, S. H., & Dehghani, M. (2017). Review of cancer from perspective of molecular. Journal of Cancer Research and Practice, 4(4), 127–129.

Hu, B., Zhu, X., & Lu, J. (2020). Cathepsin A knockdown decreases the proliferation and invasion of A549 lung adenocarcinoma cells. Molecular Medicine Reports, 21(6), 2553–2559.

Huisman, S. M., Van Lew, B., Mahfouz, A., et al. (2017). BrainScope: interactive visual exploration of the spatial and temporal human brain transcriptome. Nucleic Acids Research, 45(10), e83.

Jairaman, A., Yamashita, M., Schleimer, R. P., et al. (2015). Store-operated Ca2+ release-activated Ca2+ channels regulate PAR2-activated Ca2+ signaling and cytokine production in airway epithelial cells. The Journal of Immunology, 195(5), 2122–2133.

Justilien, V., Regala, R. P., Tseng, I. C., et al. (2012). Matrix metalloproteinase-10 is required for lung cancer stem cell maintenance, tumor initiation and metastatic potential. PLoS One, 7(4), e35040.

Koivunen, E., Itkonen, O., Halila, H., et al. (1990). Cyst fluid of ovarian cancer patients contains high concentrations of trypsinogen-2. Cancer Research, 50(8), 2375–2378.

Krepela, E., Prochazka, J., Karova, B., et al. (1998). Cysteine proteases and cysteine protease inhibitors in non-small cell lung cancer. Neoplasma, 45(5), 318–331.

Lam, D. K., Dang, D., Flynn, A. N., et al. (2015). TMPRSS2, a novel membrane-anchored mediator in cancer pain. Pain, 156(5), 923.

Lee, M., Fridman, R., & Mobashery, S. (2004). Extracellular proteases as targets for treatment of cancer metastases. Chemical Society Reviews, 33(7), 401–409.

Lippman, S. M., & Spitz, M. R. (2001). Lung cancer chemoprevention: an integrated approach. Journal of Clinical Oncology: Official Journal of the American Society of Clinical Oncology, 19(18 Suppl), 74S–82S.

Miller, K. D., Siegel, R. L., Lin, C. C., et al. (2016). Cancer treatment and survivorship statistics, 2016. CA: A Cancer Journal for Clinicians, 66(4), 271–289.
Noonan, E., Place, R. F., Pookot, D., et al. (2009). miR-449a targets HDAC-1 and induces growth arrest in prostate cancer. Oncogene, 28(14), 1714–1724.
Park, K. C., Dharmasivam, M., & Richardson, D. R. (2020). The role of extracellular proteases in tumor progression and the development of innovative metal ion chelators that inhibit their activity. International Journal of Molecular Sciences, 21(18), 6805.
Parkin, D. M., Bray, F. I., & Devesa, S. S. (2001). Cancer burden in the year 2000. The global picture. European Journal of Cancer, 37, 4–66.
Rakashanda, S., Rana, F., Rafiq, S., et al. (2012). Role of proteases in cancer: a review. Biotechnology and Molecular Biology Reviews, 7(4), 90–101.
Safranek, J., Pesta, M., Holubec, L., et al. (2009). Expression of MMP-7, MMP-9, TIMP-1 and TIMP-2 mRNA in lung tissue of patients with non-small cell lung cancer (NSCLC) and benign pulmonary disease. Anticancer Research, 29(7), 2513–2517.
Schabath, M. B., & Cote, M. L. (2019). Cancer progress and priorities: lung cancer. Cancer Epidemiology, Biomarkers & Prevention, 28(10), 1563–1579.
Schneider, M. A., Richtmann, S., Gründing, A. R., et al. (2022). Transmembrane serine protease 2 is a prognostic factor for lung adenocarcinoma. International Journal of Oncology, 60(4), 1–12.
Song, L., Wang, X., Cheng, W., et al. (2021). Expression signature, prognosis value and immune characteristics of cathepsin F in non-small cell lung cancer identified by bioinformatics assessment. BMC Pulmonary Medicine, 21, 1–17.
Tan, G. J., Peng, Z. K., Lu, J. P., et al. (2013). Cathepsins mediate tumor metastasis. World Journal of Biological Chemistry, 4(4), 91.
Taylor, W. R., & Stark, G. R. (2001). Regulation of the G2/M transition by p53. Oncogene, 20(15), 1803–1815.
Thun, M. J., Linet, M. S., Cerhan, J. R., et al. (2018). Schottenfeld and Fraumeni Cancer Epidemiology and Prevention (4th ed.). Oxford University Press.
Torre, L. A., Siegel, R. L., Ward, E. M., et al. (2016). Global cancer incidence and mortality rates and trends—an update. Cancer Epidemiology, Biomarkers & Prevention, 25(1), 16–27.
Travis, W. D. (2011). Pathology of lung cancer. Clinics in Chest Medicine, 32(4), 669–692.
Vilen, S. T., Nyberg, P., Hukkanen, M., et al. (2008). Intracellular co-localization of trypsin-2 and matrix metalloprotease-9: possible proteolytic cascade of trypsin-2, MMP-9 and enterokinase in carcinoma. Experimental Cell Research, 314(4), 914–926.
Wilson, S., Greer, B., Hooper, J., et al. (2005). The membrane-anchored serine protease, TMPRSS2, activates PAR-2 in prostate cancer cells. Biochemical Journal, 388(3), 967–972.
Xie, Q., Liu, L., Chen, X., et al. (2021). Identification of cysteine protease inhibitor CST2 as a potential biomarker for colorectal cancer. Journal of Cancer, 12(17), 5144.
Yang, J., Luo, G., Li, C., et al. (2022). Cystatin SN promotes epithelial-mesenchymal transition and serves as a prognostic biomarker in lung adenocarcinoma. BMC Cancer, 22(1), 589.
Yoo, K. Y., & Shin, H. R. (2003). Cancer epidemiology and prevention. Epidemiology and Health, 25(1), 1–15.
Zavašnik-Bergant, T., & Turk, B. (2006). Cysteine cathepsins in the immune response. Tissue Antigens, 67(5), 349–355.
Zhou, H., Xiang, Q., Hu, C., et al. (2020). Identification of MMP1 as a potential gene conferring erlotinib resistance in non-small cell lung cancer based on bioinformatics analyses. Hereditas, 157(1), 1–10.

Index

Note: Page numbers in *italics* refer to tables.